Lecture Notes in Artificial Intelligence 13280

Subseries of Lecture Notes in Computer Science

More information about this subseries at https://link.springer.com/bookseries/1244

João Gama · Tianrui Li · Yang Yu ·
Enhong Chen · Yu Zheng · Fei Teng (Eds.)

Advances in Knowledge Discovery and Data Mining

26th Pacific-Asia Conference, PAKDD 2022
Chengdu, China, May 16–19, 2022
Proceedings, Part I

Springer

Editors
João Gama
Laboratory of Artificial Intelligence
and Decision Support
University of Porto
Porto, Portugal

Tianrui Li 🆔
School of Computing and Artificial
Intelligence
Southwest Jiaotong University
Chengdu, China

Yang Yu
National Key Laboratory for Novel
Software Technology
Nanjing University
Nanjing, China

Enhong Chen
School of Computer Science and Technology
University of Science and Technology
of China
Hefei, China

Yu Zheng
JD iCity, JD Technology & JD Intelligent
Cities Research
Beijing, China

Fei Teng
School of Computing and Artificial
Intelligence
Southwest Jiaotong University
Chengdu, China

ISSN 0302-9743　　　　　　　ISSN 1611-3349　(electronic)
Lecture Notes in Artificial Intelligence
ISBN 978-3-031-05932-2　　　ISBN 978-3-031-05933-9　(eBook)
https://doi.org/10.1007/978-3-031-05933-9

LNCS Sublibrary: SL7 – Artificial Intelligence

This Springer imprint is published by the registered company Springer Nature Switzerland AG
The registered company address is: Gewerbestrasse 11, 6330 Cham, Switzerland

General Chairs' Preface

On behalf of the Organizing Committee, it is our great pleasure to welcome you to the 26th Pacific-Asia Conference on Knowledge Discovery and Data Mining (PAKDD2022), held in Chengdu, China, during May 16–19, 2022. Starting in 1997, PAKDD has long established itself as one of the leading international conferences in data mining and knowledge discovery. PAKDD provides an international forum for researchers and industry practitioners to share their new ideas, original research results, and practical development experiences from all Knowledge Discovery and Data Mining (KDD) related areas. In response to the COVID-19 pandemic and the need for social distancing, PAKDD 2022 was held as a hybrid conference for both online and onsite attendees.

Our gratitude goes first and foremost to the researchers, who submitted their work to PAKDD 2022. We would like to deliver our sincere thanks for their efforts in research, as well as in preparing high-quality presentations. We also thank all the collaborators and sponsors for their trust and cooperation. It is our great honor that three eminent keynote speakers joined the conference: Jian Pei (Simon Fraser University, Canada), Bernhard Schölkopf (Max Planck Institute for Intelligent Systems, Germany) and Ji-Rong Wen (Renmin University, China). They were extremely professional and have high reputations in their respective areas. We enjoyed their participation and talks, which made the conference one of the best academic platforms for knowledge discovery and data mining.

We would like to express our sincere gratitude to the contributions of Steering Committee members, Organizing Committee members, Program Committee members and anonymous reviewers, led by Program Committee Co-chairs: João Gama (University of Porto), Tianrui Li (Southwest Jiaotong University), and Yang Yu (Nanjing University). We are also grateful for the hosting organization Southwest Jiaotong University which is continuously providing institutional and financial support to PAKDD 2022. We feel beholden to the PAKDD Steering Committees for their constant guidance and sponsorship of manuscripts.

Finally, our sincere thanks go to all the participants and volunteers. We hope all of you enjoyed PAKDD 2022.

April 2022

Enhong Chen
Yu Zheng

PC Chairs' Preface

It is our great pleasure to present at the 26th Pacific-Asia Conference on Knowledge Discovery and Data Mining (PAKDD 2022) as the Program Committee Chairs. PAKDD is one of the longest established and leading international conferences in the areas of data mining and knowledge discovery. It provides an international forum for researchers and industry practitioners to share their new ideas, original research results, and practical development experiences from all KDD related areas, including data mining, data warehousing, machine learning, artificial intelligence, databases, statistics, knowledge engineering, big data technologies and foundations.

This year PAKDD received 627 submissions, among which 69 submissions were rejected at a preliminarily stage due to the policy violations. There were 320 Program Committee members and 45 Senior Program Committees members involved in reviewing process. Each submission was reviewed by at least three different reviewers. Over 67% of those submissions were reviewed by four or more reviewers. Eventually, 121 submissions were accepted and recommended to be published, resulting in an acceptance rate of 19.30%. Out of these, 29 submissions were about applications, 4 submissions were related to big data technologies, 46 submissions were on data science and 42 submissions were about foundations. We would like to appreciate all PC members and reviewers, who offered a high-quality program with diligence on PAKDD 2022.

The conference program featured keynote speeches from distinguished researchers in the community, most influential paper talks, cutting-edge workshops and comprehensive tutorials.

We wish to sincerely thank all PC members and reviewers for their invaluable efforts in ensuring a timely, fair, and highly effective PAKDD 2022 program.

April 2022

João Gama
Tianrui Li
Yang Yu

Organization Committee

Honorary Co-chairs

Dan Yang   Southwest Jiaotong University, China
Zhi-Hua Zhou   Nanjing University, China

General Co-chairs

Enhong Chen   University of Science and Technology of China, China
Yu Zheng   JD.com, China

Program Committee Co-chairs

Joao Gama   University of Porto, Portugal
Tianrui Li   Southwest Jiaotong University, China
Yang Yu   Nanjing University, China

Workshop Co-chairs

Gill Dobbie   University of Auckland, New Zealand
Can Wang   Griffith University, Australia

Tutorial Co-chairs

Gang Li   Deakin University, Australia
Tanmoy Chakraborty   Indraprastha Institute of Information Technology Delhi, India

Local Arrangement Co-chairs

Yan Yang   Southwest Jiaotong University, China
Chuan Luo   Sichuan University, China
Xin Yang   Southwestern University of Finance and Economics, China

Sponsor Chair

Xiaobo Zhang   Southwest Jiaotong University, China

Publicity Co-chairs

Xiangnan Ren   Group 42, United Arab Emirates
Hao Wang   Zhejiang Lab, China
Junbo Zhang   JD.com, China
Chongshou Li   Southwest Jiaotong University, China

Proceedings Chair

Fei Teng   Southwest Jiaotong University, China

Web and Content Co-chairs

Xiaole Zhao   Southwest Jiaotong University, China
Zhen Jia   Southwest Jiaotong University, China

Registration Chairs

Hongmei Chen   Southwest Jiaotong University, China
Jie Hu   Southwest Jiaotong University, China
Yanyong Huang   Southwestern University of Finance and
                                    Economics, China

Steering Committee

Longbing Cao   University of Technology Sydney, Australia
Ming-Syan Chen   NTU
David Cheung   University of Hong Kong, China
Gill Dobbie   University of Auckland, New Zealand
Joao Gama   University of Porto, Portugal
Zhiguo Gong   University of Macau, China
Tu Bao Ho   Japan Advanced Institute of Science and
                                    Technology, Japan
Joshua Z. Huang   Shenzhen Institutes of Advanced Technology,
                                    Chinese Academy of Sciences, China
Masaru Kitsuregawa   Tokyo University, Japan
Rao Kotagiri   University of Melbourne, Australia
Jae-Gil Lee   Korea Advanced Institute of Science &
                                    Technology, South Korea

Ee-Peng Lim	Singapore Management University, Singapore
Huan Liu	Arizona State University, USA
Hiroshi Motoda	AFOSR/AOARD and Osaka University, Japan
Jian Pei	Simon Fraser University, Canada
Dinh Phung	Monash University, Australia
P. Krishna Reddy	International Institute of Information Technology, Hyderabad, India
Kyuseok Shim	Seoul National University, South Korea
Jaideep Srivastava	University of Minnesota, USA
Thanaruk Theeramunkong	Thammasat University, Thailand
Vincent S. Tseng	NCTU
Takashi Washio	Osaka University, Japan
Geoff Webb	Monash University, Australia
Kyu-Young Whang	Korea Advanced Institute of Science & Technology, South Korea
Graham Williams	Australian National University, Australia
Min-Ling Zhang	Southeast University, China
Chengqi Zhang	University of Technology Sydney, Australia
Ning Zhong	Maebashi Institute of Technology, Japan
Zhi-Hua Zhou	Nanjing University, China

Host Institute

Southwest Jiaotong University

Contents – Part I

Big Data Technologies

Contents – Part II

Contents – Part III

Data Science

Data Science

PGADA: Perturbation-Guided Adversarial Alignment for Few-Shot Learning Under the Support-Query Shift

Siyang Jiang[1,2], Wei Ding[1], Hsi-Wen Chen[1], and Ming-Syan Chen[1(✉)]

[1] Department of Electrical Engineering, National Taiwan University, Taipei, Taiwan
{syjiang,wding,hwchen}@arbor.ee.ntu.edu.tw, mschen@ntu.edu.tw
[2] School of Mathematics and Statistics, Huizhou University, Huizhou, China

Abstract. Few-shot learning methods aim to embed the data to a low-dimensional embedding space and then classify the unseen query data to the seen support set. While these works assume that the support set and the query set lie in the same embedding space, a distribution shift usually occurs between the support set and the query set, i.e., *the Support-Query Shift*, in the real world. Though optimal transportation has shown convincing results in aligning different distributions, we find that the small perturbations in the images would significantly misguide the optimal transportation and thus degrade the model performance. To relieve the misalignment, we first propose a novel adversarial data augmentation method, namely *Perturbation-Guided Adversarial Alignment (PGADA)*, which generates the *hard* examples in a self-supervised manner. In addition, we introduce *Regularized Optimal Transportation* to derive a smooth optimal transportation plan. Extensive experiments on three benchmark datasets manifest that our framework significantly outperforms the eleven state-of-the-art methods on three datasets. Our code is available at https://github.com/772922440/PGADA.

Keywords: Few-shot learning · Adversarial data augmentation · Optimal transportation

1 Introduction

Recently, deep learning models have celebrated success in several computer vision tasks [17, 18, 22], which require large amounts of labeled data. However, collecting sufficient labels to train the model parameters involves considerable human effort, which is unacceptable in practice [12]. Moreover, the labels of training data and testing data are usually disjoint, i.e., the labels in the testing phase are unseen in the training phase. In contrast, few-shot learning aims to learn the model parameter by a handful of training data (support sets) and adapt to the testing data (query sets) [2, 11, 30], effectively. In the testing phase, the model classifies the query set into the support sets according to the distance of their embeddings.

While these approaches extract rich contextual features from the images, the embeddings from the support set and the query set usually exhibit certain distribution shifts, i.e., *the Support-Query Shift* [2]. For example, the images are captured by various devices, e.g., smartphones and single-lens reflex cameras in different environments,

J. Gama et al. (Eds.): PAKDD 2022, LNAI 13280, pp. 3–15, 2022.
https://doi.org/10.1007/978-3-031-05933-9_1

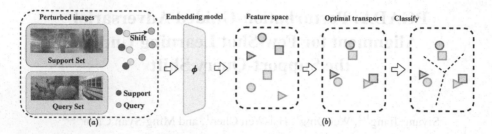

Fig. 1. (a) Illustration of the support-query shift in a 3-way 1-shot classification task, where the support set and the query set are embedded into different distributions. (b) First, we embed the support set and the query set via an embedding model ϕ into a feature space. Next, optimal transportation is employed to align the support set (red) and the query set (green). But, the small perturbations may misguide the transported results, leading to wrong predictions, i.e., classify the green circle to the red triangle. (Color figure online)

e.g., foggy and high-luminance. Since the learned embeddings are located in different spaces, degrading the model performance, *optimal transportation* [8] has been proposed to stage the embeddings from different domains into the same latent space. However, in this paper, we theoretically prove that optimal transport can be easily misguided by small perturbations in the images (as illustrated in Fig. 1).

Meanwhile, several training techniques [13, 36] have been proposed to derive a more robust embedding model against the perturbations. On the one hand, data augmentation methods transform a single image [25] or combine multiple images [33, 34] to create more training samples at pixel level. However, these methods cannot create new information which is not included in the given data [36]. On the other hand, adversarial training methods such as projected gradient descent (PGD) [23], AugGAN [16] are used to find the perturbed images to confuse the model, i.e., predicting an incorrect label, as the additional training samples. However, these methods usually require numerous iterations to generate the adversarial examples by optimizing a predefined adversarial loss, which is computationally intensive. Besides, a clear trade-off has been shown between the accuracy of a classifier and its robustness against adversarial examples [13].

To address the above issues, we propose *Perturbation-Guided Adversarial Alignment (PGADA)* to relieve the negative effect caused by the small perturbations in the support-query shift. PGADA aims to generate the perturbed data as the *hard* examples, i.e., less similar to the original data point in the embedding space but still classified into the same class. Next, the model is trained on these generated data by minimizing the empirical risk to enhance the embedding model's robustness of noise tolerance. We further introduce *smooth optimal transport*, which regularizes the negative entropy of the transportation plan to take more query data points as the anchor nodes, leading to a higher error tolerance of the transportation plan.

The contributions of this work are summarized as follows.

- We formally investigate how the perturbation in the images would affect the results of optimal transportation under *the Support-Query Shift*.
- We propose *Perturbation-Guided Adversarial Alignment (PGADA)* to relieve the misalignment problem from small perturbations via deriving a more robust feature extractor and a smooth transportation plan under distribution shifts.

– Extensive experiments manifest that PGADA outperforms eleven state-of-the-art methods by 10.91% on three public datasets.

2 Preliminary

2.1 Few-shot Learning

Given a labeled support set $\mathcal{S} = \cup_{c \in \mathcal{C}} \mathcal{S}^c$, with $\mathcal{C}$ classes, where each class c has $|\mathcal{S}^c|$ labeled examples, the goal of few-shot learning is to classify the query set $\mathcal{Q} = \cup_{c \in \mathcal{C}} \mathcal{Q}^c$ into these $\mathcal{C}$ classes. Let ϕ denotes the embedding model $\phi(x) \in \mathbb{R}^d$, which encodes the data point x to the d-dimensional feature. ϕ is learned from a labeled training set $\mathcal{D} = \{x_i, y_i\}_{i \in [1, |\mathcal{D}|]}$, where x_i is the data point and y_i is the corresponding label. The embedding model can be learned by empirical risk minimization (ERM),

$$\min_{\phi, \theta} E_{\{x,y\} \sim \mathcal{D}}[L(\theta(\phi(x)), y)],$$

where θ is a trainable parameter to map the embedding $\phi(x_i)$ to the class y_i.

Through the embedding model ϕ, we can encode the data points in support set (i.e., $x_{s,i} \in \mathcal{S}$) and query set (i.e., $x_{q,j} \in \mathcal{Q}$) to the feature $\phi(x_{s,i})$ and $\phi(x_{q,j})$, respectively. These features are used as input to a comparison function M, which measures the distance, e.g., l_2-norm, between two samples. Specifically, we classify the query example $\phi(x_{q,j})$ by averaging the embedding $\phi(x_{s,i}^c)$ of the support set in class $\mathcal{S}^c$, which can be written as follows.

$$\phi^c(x_s) = \frac{1}{|\mathcal{S}^c|} \sum_{x_{s,i} \in \mathcal{S}^c} \phi(x_{s,i}), \quad y_q = \arg\min_c M(\phi^c(x_s), \phi(x_{q,j})).$$

2.2 The Support-Query Shift and Optimal Transportation

The conventional few-shot learning methods assume the support set and the query set lie in the same distribution. A more realistic setting is that the support set $\mathcal{S}$ and the query set $\mathcal{Q}$ follow different distributions, i.e., the support-query shift [2]. While these two sets are sampled from different distributions μ_s and μ_q, the embeddings for the support set $\mathcal{S}$ (i.e., $\phi(x_s)$) and the query set $\mathcal{Q}$ (i.e., $\phi(x_q)$) are likely to lie in different embedding spaces. Thus, it would lead to a wrong classification result via the comparison module $M(\phi(x_s), \phi(x_q))$ [2].

To tackle with the support-query shift, optimal transportation [8] is one of the effective techniques to align different distributions by a transportation plan $\pi(\mu_s, \mu_q)$, which can formally be written as follows.

$$W(\mu_s, \mu_q) = \inf_{\pi \in \Pi(\mu_s, \mu_q)} \int w(x_s, x_q) d\pi(x_s, x_q), \tag{1}$$

where $\Pi(\mu_s, \mu_q)$ is the set of transportation plans (or couplings) and w is the cost function, and W is the overall cost of transporting distribution μ_s to μ_q. In our practice, we select l_2-norm of the embedding vector, i.e., $\|\phi(x_s) - \phi(x_q)\|_2^2$, as our distance function w.

Since there are only finite samples for both the support set $x_{s,i} \in \mathcal{S}$ and the query set $x_{q,j} \in \mathcal{Q}$, the discrete optimal transportation adopts the empirical distributions to estimate the probability mass function $\hat{\mu}_s = \sum \delta_{s,i}$, and $\hat{\mu}_q = \sum \delta_{q,j}$, where $\delta_{s,i}$ and $\delta_{q,j}$ is the Dirac distribution. We obtain

$$\pi^* = \arg\min_{\pi} \sum_{\substack{x_{s,i} \sim \hat{\mu}_s \\ x_{q,j} \sim \hat{\mu}_q}} w(x_{s,i}, x_{q,j}) \pi(x_{s,i}, x_{q,j}) \tag{2}$$

Then, Sinkhorn's algorithm [9] is adopted to solve the optimal transportation plan π^*.

Equipped with the optimal plan π^*, we transport the embeddings of the support set $\phi(x_{s,i})$ to $\hat{z}_{s,i}$ by barycenter mapping [8] to adapt the support set to the query set.

$$\hat{\phi}(x_{s,i}) = \frac{\sum_{x_{q,j} \in \mathcal{Q}} \pi^*(x_{s,i}, x_{q,j}) \phi(x_{q,j})}{\sum_{x_{q,j} \in \mathcal{Q}} \pi^*(x_{s,i}, x_{q,j})}. \tag{3}$$

$\hat{\phi}(x_{s,i})$ denotes the transported embedding of $x_{s,i}$. Therefore, we can correctly measure the distance metric $M(\hat{\phi}(x_{s,i}), \phi(x_{q,j}))$ in a shared embedding space.

3 Methodology

Here, we first investigate the misestimation of optimal transport of perturbed images. Then, we illustrate our framework, namely *Perturbation-Guided Adversarial Alignment (PGADA)*, which relieves the perturbation in the images and derives a more robust embedding model. In addition, a regularized optimal transportation is introduced to align the support set and the query set better, which takes more data points from the query set as anchors to enhance the error tolerance.

3.1 Motivation

We observe that optimal transportation has a challenge, which comes from the quality of the embedding $\phi(x)$, i.e., the perturbation in the image may misguide the transportation plan. For example, clean images' embeddings may give a better transportation plan than those of foggy images. With some derivation, we formally estimate the error of transported embedding $\hat{\phi}(x_{s,i})$ in Eq. (3) as follows,

Theorem 1. *The error of the transported embedding is*

$$E[\|\hat{\phi}(x_{s,i}) - \hat{\phi}_\sigma(x_{s,i})\|_2^2] = \sqrt{d(\sigma_s^2 + \sigma_q^2)},$$

where $\hat{\phi}_\sigma(x_{s,i})$ is the transported embedding from the perturbed distribution $W_\sigma(\mu_s, \mu_q)$. $W_\sigma(\mu_s, \mu_q) := W(\mu_s * \mathcal{N}_{\sigma_s}, \mu_q * \mathcal{N}_{\sigma_q})$ denotes the original support and query set distributions μ_s and μ_q being perturbed with Gaussian noises σ_s and σ_q, and $*$ is the convolution operator.

As the noise level, i.e., σ_s, and σ_q, increases, it is more likely to mislead the transportation plan and alleviate the model's performance. Therefore, it's non-trivial to learn a better embedding model ϕ having a better capability of *noise tolerance* such that $\phi(x_p) \approx \phi(x)$, where x_p is original image x with small perturbation.

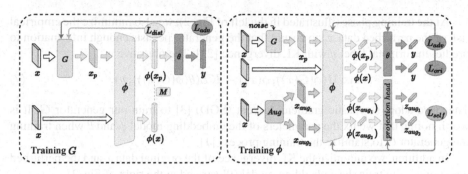

Fig. 2. Illustration of *Perturbation-Guided Adversarial Alignment (PGADA).*

3.2 Perturbation-Guided Adversarial Alignment (PGADA)

According to the Theorem 1, the optimal transport can be easily misguided by considering perturbed images. The goal of PGADA is to generate a set of augmented data to derive a more robust embedding model and relieve the perturbation in images. Recently, MaxUp [13] synthesized augmented data by *minimizing the maximum loss* over the augmented data x_p, which can be formally written as follows.

$$\min_{\phi,\theta} E_{\{x,y\}\sim\mathcal{D}}[\max_{x_p} L(\theta(\phi(x_p)), y)], \qquad (4)$$

and can be easily minimized with stochastic gradient descent (SGD). Specifically, MaxUp samples a batch of augmented data x_p and compute the gradient of the data point which has the highest loss L. Therefore, the model would learn the hardest example over all augmented data x_p.

However, it's hard to collect sufficient labels for each class in few-shot learning. Therefore, instead of maximizing the empirical risk of the labeled data by Eq. (4), we introduce a self-supervised learning-based objective.

$$\min_{\phi} E_{\{x\}\sim\mathcal{D}}[\max_{x_p} M(\phi(x_p), \phi(x))], \qquad (5)$$

where M is the comparison module in few-shot learning, e.g., l_2-norm. By maximizing the distance between $\phi(x_p)$ and $\phi(x)$, PGADA generates perturbed image x_p, which is less similar to the original image x in embedding space, as a *hard* example. To effectively generate the perturbed data, we introduce a semantic-aware perturbation generator to synthesize the augmented data.

$$x_p = G(x), \ s.t. \ \|x_p - x\|_2^2 \le \epsilon, \qquad (6)$$

where G is a model to generate the perturbed image x_p.[1] Besides, we utilize dropout [14] to provide the randomness of our model. Compared to conventional adversarial training techniques [23,31], which usually sample the perturbed images from an i.i.d distribution $x_p \sim P(\cdot|x)$, e.g., Gaussian distribution, our method can encode the semantic of the image x without requiring many samples to achieve convergence [13].

---

[1] We employ a 3-layer convolutional neural network as our G.

At the training phase (illustrated in the left of Fig. 2), we also minimize the empirical risk of the perturbed data x_p to ensure that the generator persists enough information to predict the original label y with KL divergence [22]. The objective is,

$$\max_G M(\phi(G(x)), \phi(x)) - KL(\theta(\phi(G(x))), y). \tag{7}$$

Then, we adopt stochastic gradient descent (SGD) [3] to train our generator G. It is worth noting that we fix the parameters of the embedding model ϕ and θ when training the generator G to stabilize the training process [1].

In addition, we minimize the KL divergence of the original data x and the perturbed data point x_p to train the embedding model (illustrated in the right of Fig. 2).

$$L_{ori} = KL(\theta(\phi(x)), y)), \quad L_{adv} = KL(\theta(\phi(x_p)), y)).$$

To enhance the generalizability of the embeddings, we also leverage the unlabeled data by the auxiliary contrastive self-supervised learning [6]. At each iteration, we sample N images either from training or testing set,[2] and generate 2 augmentations for each images, i.e., each augmented image has 1 positive example and $2N - 1$ corresponding negative examples with the self-supervised loss defined as follows.

$$L_{self} = \frac{1}{2N} \sum_{k=1}^{N} [\ell(2k - 1, 2k) + \ell(2k, 2k - 1)]. \tag{8}$$

ℓ denotes the *NT-Xent* Loss [6], which can be written as,

$$\ell(i, j) = -\log \frac{\exp\left(cos(z_i, z_j)/\tau\right)}{\sum_{k=1}^{2N} \mathbb{1}_{k \neq i} \exp\left(cos(z_i, z_k)/\tau\right)},$$

where z_* is defined as $W\phi(x_*)$, and W is a trainable projection matrix. Summing up, the overall all objective becomes

$$\min_{\phi, \theta} L_{ori} + \lambda_1 L_{adv} + \lambda_2 L_{self}, \tag{9}$$

where λ_1 and λ_2 are the trade-off parameters between each loss. Similarly, the classifier θ is trained by minimizing L_{ori} and L_{adv}. The pseudo code is presented in Algorithm 1.

3.3   Regularized Optimal Transportation

After deriving a robust embedding model, we extend the original transportation (Eq. 2) with negative entropy regularization to align the support set and the query set. Thus, the transport plan is penalized as follows.

$$\pi^* = \arg\min_\pi \sum_{\substack{x_{s,i} \sim \hat{\mu}_s \\ x_{q,j} \sim \hat{\mu}_q}} \beta w(x_{s,i}, x_{q,j}) \pi(x_{s,i}, x_{q,j}) + (1 - \beta)\pi(x_{s,i}, x_{q,j}) \log \pi(x_{s,i}, x_{q,j}),$$

$$\tag{10}$$

---

[2] Note that it is valid to access the images from testing set in few-shot learning, which is named transductive few-shot learning [22].

Algorithm 1. PGADA

Require: Dataset $\mathcal{D}$, comparison module M, learning rate η, trade-off parameters λ_1 and λ_2.
Ensure: Embedding model ϕ
1: Initialize generator G, embedding model ϕ, and classifier θ.
2: **for** $\{x, y\}$ in $\mathcal{D}$ **do**
3:   # fixed ϕ, θ, update G
4:   $x_p = G(x)$, $L_{dist} = -M(\phi(x_p), \phi(x))$, $L_{adv} = KL(\theta(\phi(x_p)), y))$
5:   $G \leftarrow G - \eta \nabla (L_{dist} + L_{adv})$ # Generated less similar data points with perturbations.
6:   # fixed G, update ϕ, θ
7:   $x_p = G(x)$, $L_{ori} = KL(\theta(\phi(x)), y))$, $L_{adv} = KL(\theta(\phi(x_p)), y))$
8:   $\phi \leftarrow \phi - \eta \nabla (L_{ori} + \lambda_1 L_{adv} + \lambda_2 L_{self})$
9:   $\theta \leftarrow \theta - \eta \nabla (L_{ori} + \lambda_1 L_{adv})$ # classifying the generated samples correctly.

---

where β is the weight parameter to determine the smoothness of the transportation plan. In other words, the data points in the support set take more data points in the query set as anchors for alignment. Accordingly, we can better align the support set to the query set because each labeled data point is representative, especially with limited data points and labels.

4 Experiment

We compare PGADA to eleven baselines, including conventional few-shot learning methods and adversarial data augmentation methods on three real datasets.

4.1 Experiment Setup

Datasets. Following [2], we validate our framework on three benchmark datasets in few-shot learning. 1) *CIFAR100* [20] consists of 60,000 images, evenly distributed in 100 classes (64 classes for training, 10 classes for validation, and 25 classes for testing). 2) *miniImageNet* [29] is a subset of ImageNet, with 60,000 images from 100 classes (64 classes for training, 16 classes for validation, and 20 classes for testing). 3) *FEM-NIST* [5] is a dataset with 805,263 handwritten characters in 62 classes (42 classes for training, 10 classes for validation, and 10 classes for testing).

Evaluation. Following [2], the average top-1 accuracy scores with 95% confidence interval from 2000 runs are reported. All experiments are under the 5-way setting. Note that we conduct the tasks of 1-shot and 5-shot with 8-target and 16-target, i.e., 1 or 5 instances per class in the support set and 8 or 16 instances in query set, in CIFAR100, and miniImageNet. While in FEMNIST, we only adopt tasks of 1-shot and 1-target, restricted by the setting of the dataset.

Implementation. We use a 4-layer convolutional network as the embedding model ϕ on CIFAR100 and FEMNIST, and ResNet18 for miniImageNet. As a general framework, we combine PGADA with two classifiers, i.e., ProtoNet and MatchingNet. The

Table 1. Accuracy comparison of the three datasets with two types of baselines.

Dataset	CIFAR100				miniImageNet				FEMNIST
	8-target		16-target		8-target		16-target		1-target
	1-shot	5-shot	1-shot	5-shot	1-shot	5-shot	1-shot	5-shot	1-shot
Few-shot learning									
MatchingNet [30]	$30.71_{\pm0.38}$	$41.15_{\pm0.45}$	$31.00_{\pm0.34}$	$41.83_{\pm0.39}$	$35.26_{\pm0.50}$	$44.75_{\pm0.55}$	$37.20_{\pm0.48}$	$44.22_{\pm0.52}$	$84.25_{\pm0.71}$
ProtoNet [26]	$30.02_{\pm0.40}$	$42.77_{\pm0.47}$	$30.29_{\pm0.33}$	$42.52_{\pm0.41}$	$36.37_{\pm0.50}$	$47.58_{\pm0.57}$	$35.69_{\pm0.45}$	$46.29_{\pm0.53}$	$84.31_{\pm0.73}$
TransPropNet [21]	$34.15_{\pm0.39}$	$47.39_{\pm0.42}$	$34.20_{\pm0.40}$	$44.31_{\pm0.38}$	$24.10_{\pm0.27}$	$27.24_{\pm0.33}$	$25.38_{\pm0.30}$	$28.05_{\pm0.30}$	$86.42_{\pm0.76}$
FTNET [10]	$28.91_{\pm0.37}$	$37.28_{\pm0.40}$	$28.66_{\pm0.31}$	$37.37_{\pm0.33}$	$39.02_{\pm0.46}$	$51.27_{\pm0.45}$	$39.70_{\pm0.40}$	$52.00_{\pm0.37}$	$86.13_{\pm0.71}$
TP [2]	$34.00_{\pm0.46}$	$49.71_{\pm0.47}$	$35.55_{\pm0.41}$	$50.24_{\pm0.39}$	$40.49_{\pm0.54}$	$59.85_{\pm0.49}$	$43.83_{\pm0.51}$	$55.87_{\pm0.42}$	$93.63_{\pm0.63}$
Adversarial data augmentation									
MixUp [34]	$37.82_{\pm0.47}$	$52.57_{\pm0.47}$	$38.52_{\pm0.42}$	$53.33_{\pm0.40}$	$42.98_{\pm0.54}$	$57.22_{\pm0.48}$	$43.64_{\pm0.48}$	$57.33_{\pm0.42}$	$97.22_{\pm0.46}$
CutMix [33]	$39.36_{\pm0.48}$	$54.76_{\pm0.48}$	$40.05_{\pm0.44}$	$55.44_{\pm0.40}$	$35.50_{\pm0.52}$	$45.50_{\pm0.56}$	$35.78_{\pm0.48}$	$44.85_{\pm0.52}$	$96.89_{\pm0.49}$
Autoencoder [24]	$39.05_{\pm0.50}$	$53.24_{\pm0.47}$	$39.82_{\pm0.44}$	$53.88_{\pm0.40}$	$45.36_{\pm0.56}$	$57.69_{\pm0.51}$	$45.65_{\pm0.52}$	$57.39_{\pm0.44}$	$96.53_{\pm0.43}$
AugGAN [16]	$39.54_{\pm0.50}$	$53.05_{\pm0.47}$	$39.50_{\pm0.45}$	$53.42_{\pm0.39}$	$44.65_{\pm0.55}$	$57.55_{\pm0.50}$	$44.91_{\pm0.49}$	$57.10_{\pm0.42}$	$96.42_{\pm0.52}$
MaxEntropy [36]	$38.14_{\pm0.40}$	$51.02_{\pm0.56}$	$38.21_{\pm0.34}$	$51.33_{\pm0.62}$	$48.21_{\pm0.36}$	$57.67_{\pm0.63}$	$48.99_{\pm0.21}$	$59.01_{\pm0.44}$	$97.19_{\pm0.51}$
MaxUp [13]	$34.84_{\pm0.44}$	$47.51_{\pm0.46}$	$35.20_{\pm0.40}$	$47.63_{\pm0.39}$	$37.62_{\pm0.55}$	$48.65_{\pm0.58}$	$38.13_{\pm0.50}$	$49.19_{\pm0.51}$	$96.48_{\pm0.53}$
Ours									
PGADA (ProtoNet)	$42.16_{\pm0.52}$	$\mathbf{56.52}_{\pm0.47}$	$42.73_{\pm0.46}$	$\mathbf{56.83}_{\pm0.40}$	$55.44_{\pm0.61}$	$\mathbf{67.34}_{\pm0.49}$	$55.69_{\pm0.62}$	$\mathbf{66.90}_{\pm0.50}$	$\mathbf{97.97}_{\pm0.40}$
PGADA (MatchingNet)	$\mathbf{42.25}_{\pm0.53}$	$50.98_{\pm0.45}$	$42.60_{\pm0.45}$	$51.80_{\pm0.39}$	$\mathbf{56.15}_{\pm0.61}$	$63.08_{\pm0.49}$	$\mathbf{56.12}_{\pm0.57}$	$63.61_{\pm0.45}$	$97.96_{\pm0.39}$

hyperparameters are selected by grid search with $\eta = 1e - 3$, $b = 128$, $d = 128$, $\lambda_1 = 1$, $\lambda_2 = 1$, $\beta = 0.5$, respectively. Note that we also employ transductive batch normalization [21]. Besides, SGD optimizer [19] is adopted to train all models in 200 epochs with early stopping. All experiments are implemented in a server with a Intel(R) Core(TM) i9-9820X CPU@3.30 GHz, and a GeForce RTX 3090 GPU.

4.2 Quantitative Analysis

Few-Shot Learning. We first compare five state-of-the-art few-shot learning methods, including *i) MatchingNet* [30], *ii) ProtoNet* [26], *iii) TransPropNet* [21], *iv) FTNET* [10], and *v) Transported Prototypes (TP)* [2]. As shown in Table 1, PGADA outperforms the best baseline (TP) by at least 13.12% in CIFAR100, 12.51% in miniImageNet, and 4.65% in FEMNIST, respectively. Note that PGADA achieves the best improvement on miniImageNet, since the image size of miniImageNet is the largest one compared to CIFAR100 and FEMNIST. In addition, PGADA achieves better improvement in the tasks of 1-shot (i.e., 22.46%) than the tasks of 5-shot (14.77%) averagely, which demonstrates the robustness of our method, especially with limited labeled data. Compared with the first four baselines, i.e., ProtoNet, MatchingNet, TransPropNet, and FTNET, our method achieves consistent improvement, 41.48%, 39.78%, 71.61%, and 54.85%, respectively, since these methods do not consider the inherent distribution shift between the support and query set. While TP also utilizes optimal transport to align the support and query set, it still shows relatively weak performance compared to PGADA because the transportation plan of TP is misguided by the small perturbations in the images, as proved in Theorem 1. Since PGADA is a model agnostic adversarial alignment framework, we equip PGADA with different classifiers, e.g., ProtoNet and MatchingNet. Our framework outperforms the baseline by 39.78% and 21.81% in ProtoNet and MatchingNet, respectively, manifesting the generability of PGADA. We observe that ProtoNet outperforms MatchingNet in the 5-shot case as the prototypes reduce the bias by averaging the embedding vectors, which is more robust.

Table 2. The results of ablation studies.

Dataset	CIFAR100				miniImageNet				FEMNIST
	8-target		16-target		8-target		16-target		1-target
	1-shot	5-shot	1-shot	5-shot	1-shot	5-shot	1-shot	5-shot	1-shot
PGADA	$42.16_{\pm0.52}$	$56.52_{\pm0.47}$	$42.73_{\pm0.46}$	$56.83_{\pm0.40}$	$55.44_{\pm0.61}$	$67.34_{\pm0.49}$	$55.69_{\pm0.62}$	$66.90_{\pm0.50}$	$97.98_{\pm0.40}$
Generator									
fixed G	$38.58_{\pm0.48}$	$52.41_{\pm0.47}$	$39.26_{\pm0.43}$	$52.67_{\pm0.39}$	$43.50_{\pm0.55}$	$55.65_{\pm0.50}$	$43.48_{\pm0.51}$	$55.42_{\pm0.43}$	$96.41_{\pm0.52}$
w/o noise	$37.16_{\pm0.47}$	$50.12_{\pm0.46}$	$37.73_{\pm0.41}$	$50.50_{\pm0.38}$	$44.06_{\pm0.56}$	$56.97_{\pm0.48}$	$44.42_{\pm0.49}$	$56.96_{\pm0.42}$	$96.89_{\pm0.48}$
w/o KL	$37.30_{\pm0.47}$	$50.79_{\pm0.46}$	$37.91_{\pm0.42}$	$51.35_{\pm0.39}$	$44.22_{\pm0.54}$	$55.04_{\pm0.49}$	$44.21_{\pm0.49}$	$53.96_{\pm0.41}$	$96.49_{\pm0.48}$
Regularized optimal transport (OT)									
w/o OT	$35.76_{\pm0.41}$	$54.06_{\pm0.45}$	$35.66_{\pm0.35}$	$54.09_{\pm0.38}$	$44.30_{\pm0.52}$	$61.23_{\pm0.53}$	$44.15_{\pm0.46}$	$60.86_{\pm0.48}$	$94.03_{\pm0.48}$
TP [2]	$34.00_{\pm0.46}$	$49.71_{\pm0.47}$	$35.55_{\pm0.41}$	$50.24_{\pm0.39}$	$40.49_{\pm0.54}$	$59.85_{\pm0.49}$	$43.83_{\pm0.51}$	$55.87_{\pm0.42}$	$93.63_{\pm0.63}$
TP w/o OT	$33.07_{\pm0.38}$	$50.99_{\pm0.44}$	$32.96_{\pm0.32}$	$50.71_{\pm0.37}$	$38.07_{\pm0.45}$	$55.31_{\pm0.51}$	$37.94_{\pm0.41}$	$55.11_{\pm0.44}$	$91.84_{\pm0.56}$
Self-supervised learning (SSL)									
w/o SSL	$39.33_{\pm0.50}$	$53.66_{\pm0.47}$	$40.31_{\pm0.44}$	$54.23_{\pm0.40}$	$47.96_{\pm0.57}$	$61.38_{\pm0.49}$	$48.70_{\pm0.52}$	$61.44_{\pm0.43}$	$97.07_{\pm0.48}$

Adversarial Data Augmentation. In addition, as our method is closed to adversarial data augmentation, we also compares six adversarial data augmentation methods, including vi) MixUp [34], vii) CutMix [33], viii) Autoencoder [24], ix) AugGAN [16], x) MaxEntropy [36], and xi) MaxUp [13]. According to Table 1, PGADA outperforms MixUp and CutMix by 9.12% and 8.84% on average. Compared with the Autoencoder and AugGAN, our method improved by 11.77% and 12.41% because they synthesize similar patterns to the original images to alleviate the data sparsity problem, which cannot create new information that is not included in the given data [13]. In contrast, PGADA is able to explore the perturbed data that is most likely to confuse the model, which can be regarded as hard examples. Compared to MaxEntropy, which uses opposite gradients to search the worst data point, we observe that PGADA outperforms MaxEntropy by 6.52% on average since PGADA adds noise to the generator, increasing the uncertainty of the generated data. Then, by adding these data points to the training phase, the model can defense against more unknown perturbations attacking, resulting in higher accuracy. In addition, our method also significantly outperforms MaxUp because the selected worst data point in MaxUp is still near the original images, resulting in less robustness. In addition, PGADA is more suitable for few-shot learning, as it generates the perturbations in a self-supervised manner by comparing the embedding of original images and perturbed images to explore hard examples.

4.3   Ablation Studies

We conduct ablation studies to evaluate the importance of different modules in PGADA. Note that we only present the results of PGADA with ProtoNet as PGADA with MatchingNet shows similar ones.

Effect of Generator. We compare PGADA with three different variants of the generator, including i) fixed G, fixing the parameters of the generator, ii) w/o noise, removing the noise function of the generator, and iii) w/o KL, removing the classification loss of perturbed data. When we fix the parameters of the generator, we can observe that the performance of PGADA drops by 8.46%, 24.31%, and 1.63% in three datasets, respectively. It demonstrates that our trainable generator is able to extract the information from original images x to generate meaningful perturbed images x_p. As we remove the noise, it also shows a consistent decline in performance, which demonstrates that leveraging some randomness in the training process is helpful to explore the perturbation. Lastly, our method without classification loss (i.e., *w/o KL*) still outperforms TP [2]. It demonstrates that the self-supervised-learning-based objective function (Eq. 5) works well to learn the inherent information from the original data, leading to a robust embedding model. However, the classification loss of the perturbed data also boosts the model capability as it regularizes the generator by preserving useful information to predict a correct label, rather than exploration via random walk, which is less efficient.

Effect of Regularized Optimal Transportation. Here, we investigate the effect of regularized optimal transportation (OT), which plays a crucial role during the evaluation phase in few-shot learning under the support-query shift. It shows that OT significantly improves the model's capability since it aligns the distributions of the support and query set. Even though the performance of PGADA drops as we remove OT. PGADA still outperforms the state-of-the-art method TP [2], which also utilizes OT by 4.86%. Also, compared TP to TP w/o OT, we observe that optimal transportation in TP does not perform well as in PGADA. This observation echoes the motivation of this work (Theorem 1), i.e., the small perturbations misguide the optimal transportation plan. Also, this result manifests that a robust embedding model indeed alleviates the support-query shift in few-shot learning.

Effect of Self-supervised Learning. Last, we evaluate the effect of self-supervised learning. Equipped with the contrastive loss, PGADA improves by 5.83%, 12.13%, and 0.94% in three datasets, respectively, since the model leverages the structural information of the unlabeled data in the training phase and thus leaps in model performance. The results also demonstrate that PGADA and self-supervised learning can incorporate together to explore the information from the training set and testing set to improve the generality of the embedding model. Understanding why this combination performs well is interesting for future works.

5 Related Work

Few-Shot Learning. Few-shot learning can be divided into three main categories, including optimization-based method [11], hallucination-based method [15], and metric-based method [30]. On the one hand, the optimization-based methods aim to train a quickly adaptive model and quickly adapt to other tasks by fine-tuning [4, 11]. On the other hand, the hallucination-based methods hallucinate the scarce labeled data by synthesizing representations [15] with Generative Adversarial Networks (GANs) [1]. Our work is most relevant to the metric-based methods where Vinyals et al. [30] and Snell et al. [26] introduce the pairwise (and classwise) metrics to determine the label of the query set according to the support set. Sung et al. [27] model the non-linear relation between class representations and queries by neural networks. Moreover, cross-domain few-shot learning [7] argues that the training and testing sets may not lie in the same distribution. Thus, several techniques such as optimal transport [8], self-supervised learning [22,35] have been introduced to relieve the domain shift. While the domain shift not only occurs between the training set and the testing set but also between the support set and the query set [2]. In addition, the optimal transport plan can be easily misguided by the perturbation in the images and leads to unacceptable performance.

Data Augmentation and Adversarial Training. Data augmentation has been widely used in machine learning with various transformations on a single image [25], e.g., resizing, flipping, rotation, cropping, or multiple images, e.g., MixUp [34] and Cut-Mix [33]. Another line of studies [6,32] works on several self-training schemes by maximizing the pairwise similarity of augmented data. However, the aforementioned methods cannot create new information not included in the given data [28] since they synthesize and train on similar images. In contrast, adversarial training [23,36] has been developed to defend against adversarial attacks, which conducts the attack to generate perturbed examples from clean data that the model misclassifies. Then, the generated examples are used as training data to compensate for the model weaknesses, contributing to model robustness. A similar work, MaxUP [13], generates a set of random augmented data and searches the hardest example that would maximize the classification loss. However, MaxUp requires numerous random augmented data to explore more information, thus leading to ineffectiveness. In contrast, PGADA directly generates hard samples in a self-supervised manner, which is simpler and computationally efficient.

6 Conclusion

In this paper, we propose *Perturbation-Guided Adversarial Alignment (PGADA)* to solve the support-query shift in few-shot learning. Our key idea is to generate perturbed images that are hard to classify and then train on these perturbed data to derive a more robust embedding model and alleviate the misestimation of optimal transportation. In addition, a negative entropy regularization is introduced to obtain a smooth transportation plan. The experiment results manifest that PGADA outperforms eleven baselines by at least 13.12% in CIFAR100, 12.51% in miniImageNet, and 4.65% in FEMNIST, respectively. Future works include applying PGADA to other computer vision tasks and incorporating it with other data augmentation schemes.

Acknowledgement. S. Jiang is supported by the science and technology plan project in Huizhou (No. 2020SD0402030).

References

1. Antoniou, A., Storkey, A., Edwards, H.: Data augmentation generative adversarial networks. In: ICLR (2017)
2. Bennequin, E., Bouvier, V., Tami, M., Toubhans, A., Hudelot, C.: Bridging few-shot learning and adaptation: new challenges of support-query shift. ECML-PKDD (2021)

3. Bottou, L.: Stochastic gradient descent tricks. In: Montavon, G., Orr, G.B., Müller, KR. (eds) Neural Networks: Tricks of the Trade. LNCS, vol. 7700. Springer, Berlin (2012). https://doi. org/10.1007/978-3-642-35289-8_25
4. Boudiaf, M., Masud, Z.I., Rony, J., Dolz, J., Piantanida, P., Ayed, I.B.: Transductive information maximization for few-shot learning. arXiv preprint arXiv:2008.11297 (2020)
5. Caldas, S., et al.: LEAF: a benchmark for federated settings. NeurIPS (2019)
6. Chen, T., Kornblith, S., Norouzi, M., Hinton, G.: A simple framework for contrastive learning of visual representations. In: ICML, pp. 1597–1607. PMLR (2020)
7. Chen, W.Y., Liu, Y.C., Kira, Z., Wang, Y.C.F., Huang, J.B.: A closer look at few-shot classification. arXiv preprint arXiv:1904.04232 (2019)
8. Courty, N., Flamary, R., Tuia, D., Rakotomamonjy, A.: Optimal transport for domain adaptation. IEEE TPAMI (2016)
9. Cuturi, M.: Sinkhorn distances: lightspeed computation of optimal transport. NeurIPS **26**, 2292–2300 (2013)
10. Dhillon, G.S., Chaudhari, P., Ravichandran, A., Soatto, S.: A baseline for few-shot image classification. In: ICLR (2019)
11. Finn, C., Abbeel, P., Levine, S.: Model-agnostic meta-learning for fast adaptation of deep networks. In: ICML (2017)
12. Garcia, V., Bruna, J.: Few-shot learning with graph neural networks. In: ICLR (2017)
13. Gong, C., Ren, T., Ye, M., Liu, Q.: MaxUp: Lightweight adversarial training with data augmentation improves neural network training. In: CVPR, pp. 2474–2483 (2021)
14. Goodfellow, I., et al.: Generative adversarial nets. NeurIPS **27**, 2672–2680 (2014)
15. Hariharan, B., Girshick, R.: Low-shot visual recognition by shrinking and hallucinating features. In: CVPR, pp. 3018–3027 (2017)
16. Huang, S.W., Lin, C.T., Chen, S.P., Wu, Y.Y., Hsu, P.H., Lai, S.H.: AugGAN: cross domain adaptation with GAN-based data augmentation. In: ECCV, pp. 718–731 (2018)
17. Jiang, S., Chen, H.W., Chen, M.S.: Dataflow systolic array implementations of exploring dual-triangular structure in QR decomposition using high-level synthesis. In: ICFPT (2021)
18. Jiang, S., Yao, X., Long, Q., Chen, J., Jiang, H.: Fund investment decision in support vector classification based on information entropy. Rev. Econ. Finance **15**, 57–66 (2019)
19. Kingma, D.P., Ba, J.: Adam: a method for stochastic optimization. In: ICLR (2014)
20. Krizhevsky, A., Hinton, G., et al.: Learning multiple layers of features from tiny images (2009)
21. Liu, Y., et al.: Learning to propagate labels: Transductive propagation network for few-shot learning. In: ICLR (2018)
22. Phoo, C.P., Hariharan, B.: Self-training for few-shot transfer across extreme task differences. In: ICLR (2020)
23. Samangouei, P., Kabkab, M., Chellappa, R.: Defense-GAN: protecting classifiers against adversarial attacks using generative models. In: ICLR (2018)
24. Schonfeld, E., Ebrahimi, S., Sinha, S., Darrell, T., Akata, Z.: Generalized zero-and few-shot learning via aligned variational autoencoders. In: CVPR. pp. 8247–8255 (2019)
25. Simonyan, K., Zisserman, A.: Very deep convolutional networks for large-scale image recognition. In: ICLR (2015)
26. Snell, J., Swersky, K., Zemel, R.S.: Prototypical networks for few-shot learning. In: NeurIPS (2017)
27. Sung, F., Yang, Y., Zhang, L., Xiang, T., Torr, P.H., Hospedales, T.M.: Learning to compare: relation network for few-shot learning. In: CVPR, pp. 1199–1208 (2018)
28. Theagarajan, R., Chen, M., Bhanu, B., Zhang, J.: ShieldNets: Defending against adversarial attacks using probabilistic adversarial robustness. In: CVPR, pp. 6988–6996 (2019)
29. Triantafillou, E., et al.: Meta-dataset: a dataset of datasets for learning to learn from few examples. arXiv preprint arXiv:1903.03096 (2019)

30. Vinyals, O., Blundell, C., Lillicrap, T., Wierstra, D., et al.: Matching networks for one shot learning. In: NeurIPS (2016)
31. Wang, Y.X., Girshick, R., Hebert, M., Hariharan, B.: Low-shot learning from imaginary data. In: CVPR, pp. 7278–7286 (2018)
32. Xie, Q., Luong, M.T., Hovy, E., Le, Q.V.: Self-training with noisy student improves ImageNet classification. In: CVPR, pp. 10687–10698 (2020)
33. Yun, S., Han, D., Oh, S.J., Chun, S., Choe, J., Yoo, Y.: CutMix: regularization strategy to train strong classifiers with localizable features. In: ICCV, pp. 6023–6032 (2019)
34. Zhang, H., Cisse, M., Dauphin, Y.N., Lopez-Paz, D.: mixup: beyond empirical risk minimization. In: ICLR (2017)
35. Zhao, A., et al.: Domain-adaptive few-shot learning. In: WACV, pp. 1390–1399 (2021)
36. Zhao, L., Liu, T., Peng, X., Metaxas, D.: Maximum-entropy adversarial data augmentation for improved generalization and robustness. arXiv preprint arXiv:2010.08001 (2020)

Auxiliary Local Variables for Improving Regularization/Prior Approach in Continual Learning

Linh Ngo Van[✉], Nam Le Hai, Hoang Pham, and Khoat Than

Hanoi University of Science and Technology, No 1, Dai Co Viet Road, Hanoi, Vietnam
{linhnv,khoattq}@soict.hust.edu.vn

Abstract. Regularization/prior approach emerges as one of the major directions in continual learning to help a neural network reduce forgetting the learned knowledge. This approach measures the importance of weights for previous tasks and then imposes a constraint on them in the current task without retraining on past data as well as extending the network architecture. However, regularization/prior-based methods face the problem in which weights can be moved intensively to the parameter region obtaining good performance for the current task but getting bad ones for previous tasks. In this paper, we present a novel solution in order to deal with this problem. Rather than using global variables as in the original methods, we add auxiliary local variables for each task that are considered as adjusting factors to suitably change the global ones to this task. As a result, the global variables can be preserved in a good region for all tasks to reduce the forgetting phenomenon. In particular, by imposing a variational distribution on the auxiliary local variables which are employed as multiplicative noise to the input of layers, we can achieve theoretical properties: Uncorrelated likelihoods, correlated pre-activation, and data-dependent regularization which are missing in the existing methods. These properties bring several benefits as follows: (1) Uncorrelated likelihoods between different data instances lead to reduce the high variance of stochastic gradient variational Bayes; (2) correlated pre-activation helps increase the representation ability for each task; and (3) data-dependent regularization guarantees to preserve the global variables in good region for all tasks. Our extensive experiments show that adding the local variables improves the performances of regularization/prior-based methods with significant magnitudes on several datasets. In particular, it makes several standard baselines approach SOTA results.

Keywords: Continual learning · Regularization/prior-based approach · Variational dropout · Local and global variables

1 Introduction

When working on the sequence of multiple tasks, artificial neural networks (ANN) often suffer from forgetting knowledge acquired from previous tasks if

---

L. N. Van, N. L. Hai and H. Pham—Equal contribution.

© The Author(s), under exclusive license to Springer Nature Switzerland AG 2022
J. Gama et al. (Eds.): PAKDD 2022, LNAI 13280, pp. 16–28, 2022.
https://doi.org/10.1007/978-3-031-05933-9_2

they cannot revisit past data. Meanwhile, humans can not only retain the learned knowledge but also can manipulate this knowledge to adapt to new tasks quickly. Therefore, developing methods that have the same learning ability as humans' has been paid a great deal of attention by researchers. It has become the main topic in continual learning which requires methods to learn in an online fashion from streaming data associated with a series of consecutive tasks.

Regularization/prior approach [6,11,13,16,19,25] emerges as an effective solution to work well on all earlier tasks without storing and retraining on past data. In this approach, methods learn model's parameters while concurrently identify their importance after each task. Based on the importance, a regularization term is added to the objective function to penalize the deviation of parameters when learning the next task. Therefore, "How to estimate the weight importance?" becomes the main question in this approach and several methods with different strategies have been proposed. However, those methods still suffer from reducing performances on previous tasks. In spite of putting a strict constraint on important weights, the optimal solution can be still lain on the region obtaining good performance for the current task but getting bad ones for previous tasks. For instance, if unimportant weights for previous tasks are updated significantly in the current task, they can seriously affect performances on previous tasks. In this paper, we aim to address this issue in order to improve the regularization/prior approach.

Meanwhile, Dropout is a well-known regularization technique for a deterministic neural network (DNN) whose weights are deterministic values. Originally, the dropout rate is fixed and chosen manually, then several mechanisms [7,12,15] are proposed to learn a DNN with an estimated dropout rate based on variational inference. Recently, some studies [5,9,17] investigated the effectiveness of dropout in the context of continual learning and showed that dropout helps reduce the catastrophic forgetting phenomenon. However, there are still several issues that need to be addressed. First, the idea of dropout has not been considered in a Bayesian neural network (BNN) [4] whose weights are represented by a distribution. Therefore, it cannot be applied to the existing prior-based methods. Second, there is a lack of theoretical analyses on why dropout works well in continual learning. Previous investigations are only limited to simple experimental scenarios in DNN. Finally, it is impractical to manually tune dropout rate in continual learning scenarios. Instead, the dropout rate should be separately characterized for each task to adapt well to the sequence of tasks. We emphasize that existing adaptive dropout methods [12,15] are impossible to apply to regularization/prior methods as they only focus on connecting DNN with dropout and BNN. To achieve this goal, they have to consider a limitation of BNN whose prior is fixed, e.g. log uniform distribution [12,18] or the discrete quantised Gaussian prior [7]. Then, dropout rate is also achieved by applying variational inference on the corresponding BNN. Meanwhile, regularization/prior-based methods use the learned model from the previous task as the prior in the current task to preserve the learned knowledge. Consequently, we cannot leverage adaptive dropout for regularization/prior-based methods.

Our contributions are listed as follows: (i) We introduce a novel method which adds auxiliary local variables, namely *ALV*. In ALV, there are two kinds of variables: Local and global variables. The global variables (the original weights of neural networks) are shared across all tasks. Our method can be combined with previous regularization/prior-based methods on the global variables for preserving the acquired knowledge from previous tasks. Meanwhile, the local variables characterize each specific task and adjust the global variables to perform well on corresponding task. Intuitively, adding local variables, which are learned and saved for each specific task, can help the global variables avoid the trap of local optimum that is merely good for this task. (ii) We demonstrate the application of ALV in both BNN and DNN. For the global variables, it is flexible to select the approximate posterior distribution learned from the previous task as the prior distribution for the current task. Meanwhile, the auxiliary local variables are considered as Gaussian multiplicative noise in Dropout when putting a constraint on approximations of their posteriors. We jointly learn both the approximate posteriors of auxiliary local variables for each task and global variables for all tasks. Thus, we provide a mechanism to learn dropout rate for each task in continual learning. (iii) we point out three important properties of ALV that are missing in the existing regularization/prior-based methods for both DNNs and BNNs: *uncorrelated likelihoods* between different data instances which reduce the high variance of stochastic gradient variational Bayes; *correlated pre-activation* which increases the representation ability for each task; *data-dependent regularization* which guarantees to preserve global variables in a good region for all tasks.

We evaluate the practical effectiveness of ALV using three regularization/ prior-based methods including Elastic Weight Consolidation (EWC) [13], Variational Continual Learning (VCL) [19] and Uncertainty-based Continual Learning (UCL) [1]. The experimental results on several benchmark datasets show that ALV can improve the performances of the baselines with significant magnitudes. In particular, ALV can make standard methods approach the state-of-the-art results in several experiments.

In the rest of the paper, the related work and background are briefly summarized in Sect. 2. Section 3 and 4 present our proposal and experiments respectively. The conclusion is drawn in Sect. 5.

2   Related Work and Backgrounds

In this section, we introduce related work and then present some backgrounds.

2.1   Related Work

Recently, a vast number of studies have addressed the problems of continual learning (CL) and they can be divided into three main approaches: Memory-based approach, architecture-based approach and regularization/prior-based approach. In this work, we focus on a regularization/prior-based approach which does not require storing and re-training past data as well as building a dynamic architecture. The main idea of this approach is to impose a constraint on the deviation of model's parameters of the current task from those acquired from previous tasks. The constraint is often constructed based on two main directions:

Adding a regularization term or replacing the prior in the current task by the posterior acquired from the previous task. We will explicitly discuss on some particular methods from the two direction.

In terms of regularization [1,3,11,13,14,23,26], an extra regularization term, which constrains the learning weights of the current task to lie close to the solutions learned from previous tasks, is added to the original loss when training a neural network. While the original loss aims to perform well on the current task, the regularization term is used to reduce the catastrophic forgetting phenomenon. Several strategies are proposed to measure the weight importance that influences the strict level of the constraints in regularization terms. Synaptic Intelligence (SI) [26] estimates the weight importance based on the changes of loss function with respect to each weight. Memory-aware Synapses (MAS) [2] uses the change of outputs instead of the loss function to evaluate the weight importance. Elastic Weight Consolidation (EWC) [13] is derived from Bayesian learning and weight importance is measured by the diagonal Fisher information matrix. Adaptive Group Sparsity based Continual Learning (AGS-CL) [11] focuses on node importance instead of weight importance and uses the average activation of each node to measure.

Regarding prior-based methods [1,6,19,22], online variational inference [8,21] is used to build a constraint between consecutive tasks. In detail, the approximate posterior acquired from previous task is considered as a prior to learn the current task. Variational continual learning (VCL) [19] is the first work that applies it in continual learning. The objective function of VCL consists of two terms: Likelihood term and KL-divergence term. When the prior is the approximate posterior learned from previous tasks, KL-divergence between the current approximate posterior and the prior makes VCL more stable. Uncertainty regularized Continual Learning (UCL) [1] improves VCL by defining node importance and then adding two regularization terms. Based on the node importance, the first term limits the change of weights related to important nodes and the other makes weights more active to learn new tasks.

Dropout is used in continual learning-based methods [5,10,17,20] and it improves noticeably the original methods in performing well on previous tasks. However, dropout's role has not been discussed adequately and the dropout rate is merely selected manually and fixed even when the learning tasks are changed. There is a lack of a mechanism to control this parameter when working on multiple tasks. Moreover, dropout has not been directly considered in BNN. It could be because BNN already has some of dropout's important properties such as uncertainty and regularization. However, we find that the idea of dropout can be appropriate to both DNNs and BNNs in continual learning with multiple tasks. Adding auxiliary variables brings remarkable proprieties: Correlated pre-activation and data-dependent regularization. Meanwhile, variational dropout (VD) [12] provides a mechanism to learn the dropout rate in DNNs. We emphasize that VD is not used for Bayesian neural networks (BNNs) whose weights are random variables. It merely aims to connect DNNs with dropout and BNNs. In detail, a noise $s \sim \mathcal{N}(1, \alpha)$ is multiplied by a deterministic weight

θ_{ij} will result in a random variable $w_{ij} = s\theta_{ij}$ following $\mathcal{N}(\theta_{ij}, \alpha\theta_{ij}^2)$. Notably, this property is only correct if θ is a deterministic value. Therefore, VD is only used for a DNN. Moreover, in order to assure that the objective function (the evidence lower bound) of the corresponding BNN is the same as the objective function of the original DNN with dropout w.r.t parameters θ, VD has to make the KL-term $(KL(q(W)\|p(W)))$ independent from θ. As a result, VD must set the prior distribution $(p(W))$ to a fixed log-uniform distribution. Meanwhile, in prior-based approach for CL, the approximate posterior acquired from previous tasks is considered as the prior distribution to learn the current task. It means that VD is unable to keep the knowledge learned from the previous tasks. In our work, thanks to separating the posteriors of local and global variables, ALV can work on not only DNNs but also BNNs in continual learning.

2.2 Background

Consider a neural network as a probabilistic model $p(\mathbf{y}|\mathbf{x}, \boldsymbol{\theta})$ over output y conditioned on input x and parameters $\boldsymbol{\theta}$ which represent the weight matrices. In the Bayesian approach, $\boldsymbol{\theta}$ are random variables and follow a prior $p(\boldsymbol{\theta})$. In continual learning, data streams arrive and belong to consecutive tasks. Let $(\mathbf{X}_t, \mathbf{Y}_t) = (\mathbf{x}_t^{(i)}, \mathbf{y}_t^{(i)})_{i=1}^{N_t}$ be data of task t where N_t is the number of data instances in this task. We focus on two approaches: Online variational inference (OVI) and regularization. We describe VCL for continual learning.

VCL: leverages OVI to learn tasks continuously. At task t, the true posterior is approximated by a Gaussian variational distribution based on the mean-field approximation: $q_t(\boldsymbol{\theta}) = \prod_j^J \mathcal{N}(\mu_{t,j}, \sigma_{t,j}^2)$ where J is the cardinality of θ, and $\mu_{t,j}, \sigma_{t,j}$ are the mean and standard deviation of θ_j respectively. Inspired by OVI, $q_t(\boldsymbol{\theta})$ is exploited as the prior in the next task $t + 1$. The variational objective of VCL at task t is as follows:

$$\sum_{i=1}^{N_t} E_{q_t(\boldsymbol{\theta})}\left[\log p(\mathbf{y}_t^{(i)}|\boldsymbol{\theta}, \mathbf{x}_t^{(i)})\right] - KL(q_t(\boldsymbol{\theta})\|q_{t-1}(\boldsymbol{\theta})) \tag{1}$$

This objective consists of a Likelihood term and a KL-divergence term. While the former helps the learned model adapt to the current task, the latter prevents it from forgetting the previous tasks. Meanwhile, due to intractability, VCL uses reparameterization trick and Monte-Carlo sampling to calculate the likelihood term: $E_{q_t(\boldsymbol{\theta})}\left[\log p(\mathbf{y}_t^{(i)}|\boldsymbol{\theta}, \mathbf{x}_t^{(i)})\right] \approx \frac{1}{K}\sum_{k=1}^{K}\log p(\mathbf{y}_t^{(i)}|\boldsymbol{\theta}^{(k)}, \mathbf{x}_t^{(i)})$ where $\boldsymbol{\theta}^{(k)} = \boldsymbol{\mu}_t + \boldsymbol{\sigma}_t \odot \boldsymbol{\epsilon}_k$ and $\boldsymbol{\epsilon}_k$ ($k \in \{1, 2, .., K\}$) is sampled from an unit Gaussian, and $\odot$ is the element-wise multiplication.

3   Improving Regularization/Prior-Based Methods with Auxiliary Local Variables in Continual Learning

When training the model for a new task, $\boldsymbol{\theta}$ can be moved intensively to the parameter region leading to good performance for the current task. In case that the constraints on previous tasks are not good and strict enough, the learned local optimum can be outside the region which works well on all tasks. To alleviate this problem, for each task, we introduce auxiliary local variables which help to create a mechanism for adaptation. The goal is to make the global variables $\boldsymbol{\theta}$ capture the characteristics of all tasks while the local variables adapt to the corresponding task. As mentioned above, existing methods based on OVI or regularization usually keep the likelihood term unchanged but modify the KL-term or regularization term to deal with the stability-plasticity dilemma. In contrast, our work directly changes the likelihood term and can be used in a wide range of regularization/prior-based methods.

For brevity, we only present ALV for fully-connected neural networks in the main text. We will describe ALV for a particular layer of these networks and then how it can be applied to VCL. Similar derivation can be applied for EWC and UCL.

3.1 Auxiliary Local Variables in Each Neural Network Layer

Consider a hidden layer l of a fully-connected neural network, let $\mathbf{A}^{(l)}$ be a $M \times D$ matrix of input features, $\boldsymbol{\theta}^{(l)}$ be a $D \times H$ weight matrix where M is the number of data instances in a minibatch, D and H are the number of input and output dimensions respectively. $\boldsymbol{\theta}^{(l)}$ follows a prior distribution $p(\boldsymbol{\theta}^{(l)})$. Denote the pre-activation $\mathbf{B}^{(l)} = \mathbf{A}^{(l)}\boldsymbol{\theta}^{(l)}$. For each task, we add auxiliary variables $s^{(l)}$ to the layer l and use a multiplicative combination: $\mathbf{B}^{(l)} = (\mathbf{A}^{(l)} \odot \mathbf{s}^{(l)})\boldsymbol{\theta}^{(l)}$ where $\odot$ is the element-wise multiplication. $\mathbf{s}^{(l)}$ is a $M \times D$ matrix where $s_{md}^{(l)}$ is a random variable. It is flexible to choose a prior for $s_{md}^{(l)}$. However, log-uniform prior [12] is selected because it often achieves better results than non-informative Gaussian prior in our experiments. In particular, we can reinterpret multiplicative combination for each data instance m as follows:

$$\mathbf{B}_m^{(l)} = (\mathbf{A}_m^{(l)} \odot \mathbf{s}_m^{(l)})\boldsymbol{\theta}^{(l)} - \mathbf{A}_m^{(l)}\tilde{\theta}^{(l)}$$

where $\tilde{\boldsymbol{\theta}}^{(l)} = [diag(\mathbf{s}_m^{(l)})]\boldsymbol{\theta}^{(l)}$ and is considered as the model's parameters for each task. It means that the local variables $\mathbf{s}_m^{(l)}$ can change the global variable $\boldsymbol{\theta}^{(l)}$ to adapt to each data instance in a new task. As a result, it can help the global variable $\boldsymbol{\theta}^{(l)}$ move intensively to the good region for all tasks.

We use variational inference to learn the local variables. Specifically, we put a constraint on the form of the approximate posterior in order to consider the local variable as a Gaussian multiplicative noise. We approximate the true posterior of $\mathbf{s}_m^{(l)}$ ($m \in \{1, .., M\}$ and $\mathbf{s}_m^{(l)} \in \mathbb{R}^{1 \times D}$) by a Gaussian distribution: $q(\mathbf{s}_m^{(l)}) = \prod_{d=1}^{D} \mathcal{N}(1, \alpha_{t,d}^{(l)})$ where the mean is set to 1 and the variance $\alpha_{t,d}^{(l)}$ is learned to capture the change of $\tilde{\boldsymbol{\theta}}^{(l)}$ around $\boldsymbol{\theta}^{(l)}$. In addition, $\alpha_t^{(l)}$ is shared across the inputs of data instances. It is worth noting that although we add a huge number of the auxiliary variables $\mathbf{s}^{(l)}$ (a $M \times D$ matrix), the number of parameters $\alpha_t^{(l)}$ (a D-dimensional vector), which have to be learned, is considerably smaller than the cardinality of the global parameters. We can further reduce $\alpha_t^{(l)}$ to a scalar.

However, using a vector of variational parameters $\boldsymbol{\alpha}_t^{(l)}$ instead of a scalar makes the approximate posterior richer to characterize a new task t.

3.2   ALV for Regularization/Prior-Based Methods

Next, we will present how to apply ALV in continual learning setting. We use VCL as a case study for OVI-based approach.

ALV for VCL: In terms of online variational inference, the approximate posterior of global variables $\boldsymbol{\theta}$ at a task $t-1$ is utilized as the prior in the next task t: $p(\boldsymbol{\theta}) = q_{t-1}(\boldsymbol{\theta})$. We maximize the log likelihood: $\log p(\mathbf{Y}_t|\mathbf{X}_t) = \sum_{i=1}^{N_t} \log p(\mathbf{y}_t^{(i)}|\mathbf{x}_t^{(i)})$. We use mean-field variational inference with variational distributions $q_t(\boldsymbol{\theta}), q_t(\mathbf{s})$ and obtain ELBO:

$$\sum_{i=1}^{N_t} E_{q_t(\boldsymbol{\theta}),q_t(\mathbf{s})} \left[\log p(\mathbf{y}_t^{(i)}|\mathbf{s},\boldsymbol{\theta},\mathbf{x}_t^{(i)}) \right] - KL(q_t(\mathbf{s})\|p(\mathbf{s})) - KL(q_t(\boldsymbol{\theta})\|q_{t-1}(\boldsymbol{\theta}))$$

Note that the $KL(q_t(\boldsymbol{\theta})\|q_{t-1}(\boldsymbol{\theta}))$ term helps the learned model prevent forgetting the previous tasks and we keep this term as in VCL. We use log-uniform prior for $\mathbf{s}$ and the approximation of $KL(q_t(\mathbf{s})\|p(\mathbf{s}))$ as in VD [12]. Moreover, we tune a coefficient to adjust the effect of this term in experiments. For the likelihood term, we propose a variant of local reparameterization trick to calculate this term. In detail, we consider at each layer l: $\mathbf{B}^{(l)} = (\mathbf{A}^{(l)} \odot \mathbf{s}^{(l)})\boldsymbol{\theta}^{(l)}$ where $b_{mh}^{(l)} = \sum_{d=1}^{D}(a_{md}^{(l)} \cdot s_{md}^{(l)})\theta_{dh}^{(l)}$. Reparameterization trick is sequentially adopted for variables $\mathbf{s}^{(l)}$ and $\boldsymbol{\theta}^{(l)}$. Because $q(\mathbf{s}_m^{(l)}) = \prod_{d=1}^{D}\mathcal{N}(1,\alpha_{t,d}^{(l)})$, we can calculate the auxiliary local variables $\mathbf{s}^{(l)}$:

$$s_{md}^{(l)} = 1 + \sqrt{\alpha_{t,d}^{(l)}}\gamma_{md}^{(l)}$$

where $\gamma_{md}^{(l)}$ is sampled from $\mathcal{N}(0,1)$. It is plugged in the pre-activation:

$$b_{mh}^{(l)} = \sum_{d=1}^{D} a_{md}^{(l)}(1 + \sqrt{\alpha_{t,d}^{(l)}}\gamma_{t,md}^{(l)})\theta_{dh}^{(l)}$$

Since, $\theta_{dh}^{(l)} \sim \mathcal{N}(\mu_{t,dh}^{(l)}, (\sigma_{t,dh}^{(l)})^2)$, we can rewrite: $b_{mh}^{(l)} \sim \mathcal{N}(\omega_{mh}^{(l)}, \delta_{mh}^{(l)})$ where

$$\omega_{mh}^{(l)} = \sum_{d=1}^{D} a_{md}^{(l)}(1 + \sqrt{\alpha_{t,d}^{(l)}}\gamma_{md}^{(l)})\mu_{t,dh}^{(l)}$$

$$\delta_{mh}^{(l)} = \sum_{d=1}^{D} \left(a_{md}^{(l)}(1 + \sqrt{\alpha_{t,d}^{(l)}}\gamma_{md}^{(l)}) \right)^2 (\sigma_{t,dh}^{(l)})^2$$

By using reparameterization trick, we have $b_{mh}^{(l)} = \omega_{mh}^{(l)} + \sqrt{\delta_{mh}^{(l)}}\varepsilon_{mh}^{(l)}$ where $\varepsilon_{mh}^{(l)}$ is sampled from $\mathcal{N}(0,1)$. Let $L^{(m)} = E_{q_t(\theta),q_t(s)} \log p(\mathbf{y}_t^{(m)}|\mathbf{s},\theta,\mathbf{x}_t^{(m)})$ be the likelihood of data instance m and is expressed as:

$$L^{(m)} = E_{\gamma_m,\varepsilon_m} \log p(\mathbf{y}_t^{(m)}|\alpha_t,\mu_t,\sigma_t,\gamma_m,\varepsilon_m,\mathbf{x}_t^{(m)})$$

After sampling γ and ε, we can calculate the likelihood term and optimize the objective function with respect to variational parameters α_t, μ_t and σ_t for task t.

3.3 Theoretical Analyses

In this subsection, we theoretically analyze ALV's properties such as uncorrelated likelihoods, correlated pre-activation, data-independent regularization based on adding auxiliary local variables for a particular task t. While the regularization propriety assures that the global variables are lain in good region for all tasks, the other properties help to learn each task well.

Uncorrelated Likelihoods and Correlated Pre-activation. In terms of learning BNNs, most of existing methods, that use usual Monte Carlo gradient estimator or reparamameterization trick, suffer from high variance. The local reparameterization trick [12] is proposed to deal with this problem. However, even though the local reparameterization trick has the property of uncorrelated likelihoods between different data instances to reduce high variance, it does not have correlated pre-activation to improve representation learning for each task. Therefore, we present these two properties to show the advantages of ALV in comparison with the existing methods in learning BNNs.

Uncorrelated Likelihoods Between Data Instances: The likelihood of the mth data instance is: $L^{(m)} = E_{\gamma_m,\varepsilon_m} \log p(\mathbf{y}_t^{(m)}|\alpha_t,\mu_t,\sigma_t,\gamma_m,\varepsilon_m,\mathbf{x}_t^{(m)})$. For two data instances m and m', it is straightforward to see that $\{\gamma_m,\varepsilon_m\}$ and $\{\gamma_{m'},\varepsilon_{m'}\}$ are independent, therefore, $Cov[L^{(m)},L^{(m')}] = 0$. This property helps to reduce the high variance of stochastic gradient variational Bayes [12]. Indeed, the likelihood of all data instances is approximated by minibatch-based Monte Carlo estimator and is written: $L = \frac{N_t}{M}\sum_m^M L^{(m)}$ where M is the size of mini-batch and N_t is the number of all data instances. The variance of L is expressed as follows:

$$\frac{N_t^2}{M^2}\left(\sum_{m=1}^M Var[L^{(m)}] + 2\sum_{m=1}^M \sum_{m'=m+1}^M Cov[L^{(m)},L^{(m')}]\right)$$

We emphasize that because data instances are drawn from empirical distribution, the variance $Var[L^{(m)}]$ and covariance $Cov[L^{(m)},L^{(m')}]$ are computed with respect to the unit Gaussian distributions γ and ε. Since $Cov[L^{(m)},L^{(m')}] = 0$, ALV can achieve a lower variance estimation.

Correlated pre-activation: We prove that, for each data instance m, adding the local variables for each layer l results in correlated pre-activation. Consider an element h, we have $b_{mh} \sim \mathcal{N}(\omega_{mh}, \delta_{mh})$ where ω_{mh}, δ_{mh} are functions over random variable γ_m which is derived from reparameterization trick on the local variables. It means that ALV creates a hierarchical distribution on the pre-activation b_m and can achieve better representation on each layer. We emphasize that the original local reparameterization trick [12] fully factorizes b_m. Therefore, it only obtains the property of uncorrelated likelihoods but does not capture correlated representation.

An Effective Regularization for Continual Learning. Recently, [24] proved that dropout with Bernoulli distribution leads to the regularization term. In this work, we extend the analyses to Gaussian multiplicative noise instead of only Bernoulli dropconnect as in [24]. More specifically, we demonstrate that ALV achieves similar regularization:

$$ Reg_{mult}^{(l)} = \frac{1}{2} \left\langle \boldsymbol{H}_{h^{(l)}(\mathbf{x})}(\bar{L}); \; diag \left(\boldsymbol{\alpha}_t^{(l)} \odot (h^{(l)}(\mathbf{x}))^2 \right) \right\rangle $$

where $\bar{L}$ is the original loss without auxiliary variables, $\boldsymbol{\alpha}_t^{(l)}$ is the deviation of variational distribution $q_t(\boldsymbol{s})$ for task t, $h^{(l)}(\mathbf{x})$ is the lth hidden layer, $\boldsymbol{H}_{h^{(l)}(\mathbf{x})}(\bar{L})$ is the Hessian matrix of $\bar{L}$ w.r.t $h^{(l)}(\mathbf{x})$ and $\langle .; . \rangle$ is the inner product of vectorizations of matrices. Note that the loss is the negative log likelihood in ALV.

Meanwhile, based on this regularization, [17] analyzed that minimizing the second derivative of the loss w.r.t the activation can obtain the flatness of the minima where the model can perform well on all tasks. Similarly, ALV also has this property as discussed in [17], therefore it can guarantee to preserve global variables in good region for all tasks.

4   Experiments

We use three regularization/prior-based methods: EWC, VCL and UCL to evaluate how ALV improves them on five datasets: Split MNIST, Permuted MNIST, Split CIFAR-100, Split CIFAR-10/100 and Split Omniglot. We ignore considering episodic or coreset memory to boost the effectiveness of all methods in our experiments. These datasets are generated from four original ones: MNIST, CIFAR10, CIFAR100 and Omniglot and can be applied to simulate the process of continuous arriving data. In this work, we inherit UCL's experiment setup, which assumes that data arrives in task by task and all data points of a task come at the same time.

Settings: We use again the source codes of EWC and UCL which are released from the original UCL paper[1]. We implement VCL based on the source code of UCL. ALV is injected into several layers of the networks.

---

[1] https://github.com/csm9493/UCL.

Table 1. The effect of ALV on Split MNIST and Permuted MNIST

Method	Split MNIST			Permuted MNIST		
	EWC	VCL	UCL	EWC	VCL	UCL
w/o Dropout	96.23	98.59	99.64	44.63	86.22	95.86
Dropout	97.65	98.42	99.61	91.97	86.05	95.94
ALV	99.79	98.67	99.73	92.22	87.96	96.37

4.1 Effectiveness of ALV on Split and Permuted MNIST Datasets

Table 1 illustrates the average accuracy over all tasks after finishing training the last task on the two datasets. For Split MNIST, consistently, ALV-based methods outperform the respective baselines. In particular, ALV improves EWC significantly from 97.65% to 99.79% average accuracy and can outperform the SOTA method such as UCL. The performances of ALV for VCL and UCL also increase noticeably. However, we found that in VCL and UCL, without Dropout seems to work well on this experiment and is comparable to ALV. This can be explained as the data distribution between tasks may be really similar and there is too much uncertainty in VCL and UCL with Dropout which lack an adaptive mechanism. For Permuted MNIST, after training 10 tasks sequentially, EWC, VCL, and UCL achieve performances at 91.97%, 86.22%, and 95.94% respectively. With ALV, the accuracy of UCL slightly increases by 0.14% while VCL rises from 86.22% to 87.96%. Similarly, EWC with ALV also rises from 91.97%

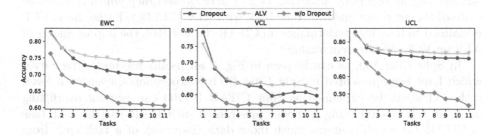

Fig. 1. Experimental results on Split Cifar-10/100.

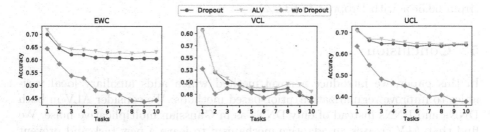

Fig. 2. Experimental results on Split CIFAR-100.

to 92.22% and over double the result of without Dropout. The reason for this significant improvement is that the auxiliary variables increase the uncertainty in EWC, while UCL and VCL themselves already have the property of uncertainty.

Besides that, it is straightforward to see the superiority of auxiliary variables over Dropout in continual learning. Dropout can help to prevent the learned model from being stuck at local optimum that works well for a task in standard cases. Especially, adding local variables to each task achieves even a more flexible adapting mechanism to work well on large number of tasks. Therefore, ALV significantly outperforms Dropout on all the baselines.

4.2   Effectiveness of ALV on Split CIFAR-100 and Split CIFAR-10/100

To reinforce the evidence of the effectiveness of ALV beyond MLP architectures, Split CIFAR-100 and Split CIFAR-10/100 datasets are additionally used in our experiments.

Figure 1 illustrates the effectiveness of ALV in comparison with the baselines on Split Cifar-10/100 dataset. Overall, when applying ALV, all methods have witnessed a remarkable increase in performance compared to their results both with and without Dropout. In addition, the figure also depicts average performances of all the methods after each task. It is clear that, the average accuracy of all methods decrease when evaluating on test sets of previous tasks and the current one, which indicate the forgetting of acquired knowledge. However, ALV can help the baselines mitigate this phenomenon with the support of local variables to achieve better average accuracy through tasks. Specifically, EWC with ALV outperforms all others by achieving 74.23% average accuracy which reaches the state-of-the-art result and overcome UCL with ALV (73.13%). Futhermore, VCL is claimed to have low performance on CIFAR dataset [16], the improvements of ALV on VCL are also remarkable.

In Split Cifar-100, as can be seen in Fig. 2, all results are under the curves which have been presented in Split Cifar-10/100 experiments, even with fewer tasks. This can be explained as Split CIFAR-10/100 can prevent overfitting on Split CIFAR-100 using a model pre-trained on CIFAR-10 (the first task is CIFAR-10) which contains much more data than that of a task split from CIFAR-100. Again, all three methods gain significant improvement after applying ALV. Specifically, ALV on EWC obtains the most noticeable improvement (from 60.56% with Dropout to 63.2%).

5   Conclusion

In this paper, we introduce a novel method which adds auxiliary local variables to improve regularization/prior-based methods. We consider ALV in both DNNs and BNNs instead of only DNNs as in Gaussian multiplicative noise. We find that ALV creates an adaptive mechanism to learn a new task and prevent forgetting the knowledge acquired from previous tasks. In particular, we theoretically demonstrate the advantages of ALV in terms of uncorrelated likelihoods between different data instances, correlated pre-activation, and an effective data-dependent regularization for continual learning. Our extensive experiments show that ALV improves the performances of regularization/prior-based methods with significant magnitudes.

Acknowledgement. This work was funded by Gia Lam Urban Development and Investment Company Limited, Vingroup and supported by Vingroup Innovation Foundation (VINIF) under project code VINIF.2019.DA18

References

1. Ahn, H., Cha, S., Lee, D., Moon, T.: Uncertainty-based continual learning with adaptive regularization. In: Advances in Neural Information Processing Systems, pp. 4392–4402 (2019)
2. Aljundi, R., Babiloni, F., Elhoseiny, M., Rohrbach, M., Tuytelaars, T.: Memory aware synapses: learning what (not) to forget. In: Proceedings of the European Conference on Computer Vision (ECCV), pp. 139–154 (2018)
3. Benzing, F.: Understanding regularisation methods for continual learning. In: Workshop of Advances in Neural Information Processing Systems (2020)
4. Blundell, C., Cornebise, J., Kavukcuoglu, K., Wierstra, D.: Weight uncertainty in neural network. In: International Conference on Machine Learning, pp. 1613–1622. PMLR (2015)
5. De Lange, M., et al.: A continual learning survey: defying forgetting in classification tasks. IEEE Trans. Pattern Anal. Mach. Intell. (2021)
6. Farquhar, S., Gal, Y.: A unifying Bayesian view of continual learning. In: The Bayesian Deep Learning Workshop at Neural Information Processing Systems (2018)
7. Gal, Y., Hron, J., Kendall, A.: Concrete dropout. In: Advances in Neural Information Processing Systems, pp. 3581–3590 (2017)
8. Ghahramani, Z., Attias, H.: Online variational Bayesian learning. In: Slides from talk Presented at NIPS Workshop on Online Learning (2000)
9. Goodfellow, I.J., Mirza, M., Xiao, D., Courville, A., Bengio, Y.: An empirical investigation of catastrophic forgetting in gradient-based neural networks. arXiv preprint arXiv:1312.6211 (2013)
10. Ha, C., Tran, V.D., Van, L.N., Than, K.: Eliminating overfitting of probabilistic topic models on short and noisy text: the role of dropout. Int. J. Approximate Reasoning **112**, 85–104 (2019)
11. Jung, S., Ahn, H., Cha, S., Moon, T.: Continual learning with node-importance based adaptive group sparse regularization. In: Advances in Neural Information Processing Systems (2020)
12. Kingma, D.P., Salimans, T., Welling, M.: Variational dropout and the local reparameterization trick. In: Advances in Neural Information Processing Systems, vol. 28, pp. 2575–2583 (2015)
13. Kirkpatrick, J., et al.: Overcoming catastrophic forgetting in neural networks. Proc. Natl. Acad. Sci. **114**(13), 3521–3526 (2017)
14. Li, Z., Hoiem, D.: Learning without forgetting. IEEE Trans. Pattern Anal. Mach. Intell. **40**(12), 2935–2947 (2017)

15. Liu, Y., Dong, W., Zhang, L., Gong, D., Shi, Q.: Variational Bayesian dropout with a hierarchical prior. In: Proceedings of the IEEE Conference on Computer Vision and Pattern Recognition, pp. 7124–7133 (2019)
16. Loo, N., Swaroop, S., Turner, R.E.: Generalized variational continual learning. In: International Conference on Learning Representation (2021)
17. Mirzadeh, S., Farajtabar, M., Pascanu, R., Ghasemzadeh, H.: Understanding the role of training regimes in continual learning. In: Advances in Neural Information Processing Systems (2020)
18. Molchanov, D., Ashukha, A., Vetrov, D.: Variational dropout sparsifies deep neural networks. In: International Conference on Machine Learning, pp. 2498–2507 (2017)
19. Nguyen, C.V., Li, Y., Bui, T.D., Turner, R.E.: Variational continual learning. In: International Conference on Learning Representation (2018)
20. Nguyen, V.S., Nguyen, D.T., Van, L.N., Than, K.: Infinite dropout for training Bayesian models from data streams. In: IEEE International Conference on Big Data (Big Data), pp. 125–134. IEEE (2019)
21. Sato, M.A.: Online model selection based on the variational Bayes. Neural Comput. **13**(7), 1649–1681 (2001)
22. Swaroop, S., Nguyen, C.V., Bui, T.D., Turner, R.E.: Improving and understanding variational continual learning. In: NeurIPS Continual Learning Workshop (2018)
23. Van Linh, N., Bach, T.X., Than, K.: A graph convolutional topic model for short and noisy text streams. Neurocomputing **468**, 345–359 (2022)
24. Wei, C., Kakade, S.M., Ma, T.: The implicit and explicit regularization effects of dropout. In: Proceedings of the 37th International Conference on Machine Learning. Proceedings of Machine Learning Research, vol. 119, pp. 10181–10192. PMLR (2020)
25. Yin, D., Farajtabar, M., Li, A.: Sola: continual learning with second-order loss approximation. In: Workshop of Advances in Neural Information Processing Systems (2020)
26. Zenke, F., Poole, B., Ganguli, S.: Continual learning through synaptic intelligence. Proc. Mach. Learn. Res. **70**, 3987 (2017)

Emerging Scientific Topic Discovery by Finding Infrequent Synonymous Biterms

Junfeng Wu[1], Guangyan Huang[1(✉)], Roozbeh Zarei[1], Jianxin Li[1],
Guang-Li Huang[1], Hui Zheng[2], Jing He[3], and Chi-Hung Chi[4]

[1] Deakin University, 211 Burwood Highway, Burwood, VIC 3125, Australia
guangyan.huang@deakin.edu.au
[2] Nanjing University of Posts and Telecommunications, Nanjing, China
[3] University of Oxford, Oxford, UK
[4] Nanyang Technological University, Singapore, Singapore

Abstract. With the increasing information load brought by the accelerated growth of research papers, the automatic discovery of a field's emerging scientific topics becomes vital. It enables broad applications, such as optimizing resource allocations for promising research areas, predicting future technology trends, finding knowledge gaps and new concepts, and recommending personalized research directions. However, two challenges - the rareness of emerging-topic publications and the linguistic diversity in the description of emerging topics - hinder existing text analytic methods from effectively identifying the evolving terms in emerging topics. According to our observation, an emerging topic originating from a collaboration of two sub-fields could be represented by a biterm, each term from one sub-field. In this paper, we propose a novel finding Infrequent Synonymous Biterms to discover Emerging Scientific Topics (isBEST) method to overcome the challenges. Our isBEST method reduces linguistic diversity using document-level clustering to find the linguistic variants of each key biterm. The biterms in the same cluster expressing very similar meanings are unified to the most common synonymous biterm. Then, to address the rareness issue, isBEST converts each input document into a vector of coefficients on synonymous biterms and clusters them at the corpus level with cosine similarity. In each document, larger coefficients are assigned to rarer synonymous biterms. The underlying logic is the higher chance of a rarer synonymous biterm to be an emerging topic denoted by the two terms, each from a collaborating sub-field. Experiments on two large scholarly paper datasets demonstrate the accuracy and effectiveness of our isBEST method.

Keywords: Scientific data mining · Imbalance data mining ·
Emerging topic discovery · Text mining · Just-in-time knowledge

© The Author(s), under exclusive license to Springer Nature Switzerland AG 2022
J. Gama et al. (Eds.): PAKDD 2022, LNAI 13280, pp. 29–40, 2022.
https://doi.org/10.1007/978-3-031-05933-9_3

1   Introduction

With the increasingly severe information overload brought by the accelerated production of research papers, it is vital to analyze a field's research trend effectively. A transition is ongoing from "scholarly big data" to "scholarly very-large data" [15] due to the exponential growth of scholarly publications [17], raising the cost to keep pace with the accelerating scientific development. Thus, automatically discovering the field trend or emerging topics become more and more critical for a broad range of applications, such as optimizing resource allocations for promising research areas [10], predicting future scientific topics and technology trends [7], finding knowledge gaps that require new research [6], identifying new concepts in the scientific literature [8], recommending personalized research directions [1] etc.

An emerging topic often originates from the collaboration of multiple super-topics (each denotes a sub-field), enabling the search for the topic's publications in an academic search engine using a search expression like

$$\bigvee_{i=1}^{n} \left(\text{Term}_{i,1} \wedge \text{Term}_{i,2} \right), \tag{1}$$

where for each i, $\text{Term}_{i,1}$ and $\text{Term}_{i,2}$ are two terms (i.e., phrases) that represent two super-topics (each from one collaborating sub-field). This paper aims at maximizing the finding of emerging topics by developing a solver algorithm to find candidates of emerging topics from an input academic database and generate a search expression for each candidate. We evaluate the solver's performance by summing up the scores of every output expression. Each score comes from querying the search engine with the corresponding expression and applying the number of publications per year from search results to a scoring function.

Most existing methods [3,5,9,11] adopt external knowledge (e.g., ontologies) to help automation of emerging topic discovery. For example, in [11], algorithm AUGUR searches CSO ontology for super-topics associative to the terms and finds collaborations between these super-topics. It identifies emerging topics from these collaborations. However, these methods face the following challenges:

(1) The rare emerging-topic papers are often hidden among huge amounts of non-emerging-topic (or developed-topic) publications. Big data analytics often fail to update the knowledge database using the terms of the emerging topics because they have limited capabilities to deal with data imbalance and often take these rare papers as noises.
(2) The linguistic diversity of an emerging topic makes a topic hard to identify. Many emerging topics do not have stable terminologies until they become developed topics. Emerging-topic authors often choose personalized synonyms of the terms, either consciously to suit the unique purpose or unintentionally, causing further trouble for emerging-topic identification.

This paper proposes a novel finding Infrequent Synonymous Biterms to discover Emerging Scientific Topics (isBEST) method to overcome the challenges.

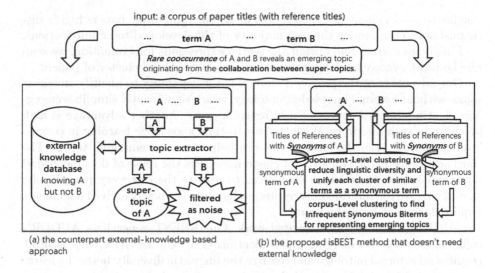

Fig. 1. External-knowledge-based methods versus our isBEST.

We use Fig. 1 as an example to explain the main mechanism of isBEST by comparing it with the state-of-the-art external-knowledge-based approach. The counterpart approach has an obvious limitation, as shown in Fig. 1a, in which a knowledge/ontology database only can identify a super-topic for a known term (Term A) but wrongly filters an unknown term (Term B) as a noise. Our isBEST method, as shown in Fig. 1b adaptively identify super-topic for any term (Terms A and B) using just-in-time knowledge learnt from the reference titles by a document/paper-level clustering to reduce linguistic diversity; this overcomes the counterpart's limitation. Moreover, isBEST conducts a corpus-level clustering by finding infrequent synonymous biterms for representing emerging topics. The three steps of the isBEST method are summarized as below.

First, we borrow the idea of MAG [14] that integrates language similarity and network similarity to resolve the problem of linguistic diversity. As shown in Fig. 1b, a paper's title and its references form a similarity network. Within the network, many synonyms exist. Language similarity implies the word embedding vectors of synonyms are also similar. Therefore, we perform a document-level clustering that groups similar word-embedding vectors to find the linguistic variants of every key biterms and unify each group of linguistic variants to the most common synonymous biterm, which comprises the just-in-time knowledge about the collaboration of two super-topics, one super-topic per term.

Then, we generalize the idea of AUGUR [11] and use synonymous biterms to measure the collaborations of super-topics. The generalization deduces the association of rare synonymous biterms to emerging topics since these rarer biterms often means more innovation in super-topic collaborations. When we conduct a corpus-level clustering, we regard each paper as a vector of coefficients on synonymous biterms, using cosine similarity to compare two papers. We assign higher weights to rarer synonymous biterms to make emerging-topic papers highly distinguishable, even among the vast majority of the developed-topic publications.

Finally, to extract and output the search expressions for evaluation, we scan for the top-k synonymous biterms within each corpus-level cluster of papers.

One advantage of the proposed method is the capability to identify emerging topics without external knowledge/ontology, which significantly simplifies emerging scientific topic discovery of new research areas. Another advantage is high precision in such discovery. With the just-in-time knowledge learning in corpus-level clustering, even the latest terms can help recognize emerging topics. The third advantage is the two levels of clustering allows the reuse of document-level intermediate results and thus enables us to update the discovery periodically; that is, we save the intermediate results so that only the newly arrived documents require document-level clustering.

AUGUR [11] is the most related work. Our isBEST generalizes AUGUR's collaboration-based emerging-topic identification. Unlike AUGUR, isBEST requires no external ontology and resolves the linguistic diversity better by learning and using just-in-time knowledge in document-level clustering.

The contributions of this paper are twofold. (1) We propose isBEST, a new emerging scientific topic discovery method that detects emerging topics with just-in-time knowledge learning. Experimental results show that our isBEST achieves better accuracy than the external-knowledge-based counterpart AUGUR method. (2) We develop a specialized document vectorization strategy to ensure a high-precision comparison of emerging-topic publications. The key technique is to find the rare synonymous biterms of emerging topics and assign larger coefficients to rarer synonymous biterms.

The remainder of this paper is organized as follows. The formal problem statement is in Sect. 2. We present the related works in Sect. 3. In Sect. 4, we detail the proposed method. In Sect. 5, we evaluate the accuracy and effectiveness of our method. Section 6 concludes this paper.

2   Problem Statement

We formalize the research problem as follows to analyze a field's research trend or emerging topics.

Terminology: (1) An **emerging topic** is a topic, which appeared in (or was supported by) only a small number of publications in the past, but recently is supported by an acceleratedly increasing number of publications. (2) A **super-topic** of an emerging topic is one of the broader topics whose collaborations produce the emerging topic; for example, "u-net image segmentation" is an emerging topic, its two super topics can be described by "u-net" and "image segmentation" but the descriptions of super topics have multiple expressions, such as "U-shaped networks", "LCU-Net", "segmentation of muscle images", "ventricle segmentation", etc. (3) A **term** is a scientific/technical word or phrase.

Input: (1) a **scholarly paper database** with information about the title, the publication year, and the references' titles of each paper in the field,

(2) a **target year** to specify which year's emerging topics to find, (3) a **scoring function** returning zero for non-emerging topics and higher positive value for larger emergingness of the topic, the arguments of which are the publication numbers in every single year from several years before the target year to the target year, and (4) a number k indicating how many emerging topic candidates the solver can give for evaluation.

Output: one search expression for each emerging topic candidate, taking the form of Eq. (1) and serving the search of the topic's papers on ScienceDirect.com for evaluation. During the evaluation, the publication numbers in the search result from ScienceDirect.com will be passed to the scoring function to compute the score of this expression.

Constraints: (C1) A solver of this problem cannot access any data or search results from ScienceDirect.com. However, we allow the solver to guess the arguments of the scoring function according to the input database and use emulated evaluations with the function to improve its outputs. (C2) Every emerging topic can only be described by one output expression.

3   Related Work

Emerging research topic discovery is a subfield of Topic Detection and Tracking (TDT). Most existing works adopt external-knowledge-based approaches, such as temporal graphs [5], semantics-based bursty and emerging topic detection [3], the Research Communities Map Builder [9], and a very recent related work, AUGUR [11]. All these methods face the two challenges mentioned in the introduction. They tend to miss the rare but vital information about emerging topics due to the lack of on-demand updates of the external knowledge database.

As a state-of-the-art method, AUGUR [11] finds emerging topics by measuring the collaborations between super-topics. This measure allows efficient narrow-down of the search range because emerging topics happen wherever collaborations accelerate; thus, measuring collaboration helps alleviate the rareness issue. Unfortunately, AUGUR relies on an external ontology to resolve linguistic diversity. The ontology does a poor job due to lacking the latest knowledge on emerging topics. Despite this, its collaboration measure is a great inspiration.

There are also some methods not using any external knowledge, e.g., the scientific network in [12] for the detection of conference-level research communities, and the temporal word embedding approach Leap2Trend in [4]. The study in [12] is unrelated to our research problem. Leap2Trend [4] measures collaborations between super-topics using temporal word embedding, but since it does not find the clusters of papers for intermediate representation of topics, it does not guarantee our research problem's constraint C2. Also, because it detects popular research trends by finding the temporal leaps or jumps in the ranking of two terms' similarities in terms of word2vec embeddings, it is more designed for macro-scale trend detection than for the micro-scale emerging topic discovery aimed by this paper. Our practice has demonstrated Leap2Trend cannot achieve good results; e.g., Leap2Trend's best output expression for the machine learning dataset is a topic represented by a biterm, "monitoring" + "dynamical systems", but the topic already has many publications every year, the number of publications per year from 2017 to 2021 are 171, 252, 336, 429, 643 respectively. This topic is not emerging any more.

There is one pioneer work that affects this subfield very much. Comparison of six evaluation functions for emerging topics is reported in [13]. Its experiments show that among these six functions, SLP_{PI} performs unexpectedly well in emerging topic prediction and beats other scoring functions most of the time. Therefore, the scoring function in our experiments is based on SLP_{PI}.

4   The Proposed Method

An overview of the proposed isBEST method is elaborated in Fig. 1b. We design isBEST as a two-level clustering method to overcome the two tightly coupled challenges, i.e., the linguistic diversity and the rareness of emerging-topic publications. The document-level clustering (illustrated by Fig. 2(a)) deals with linguistic diversity, while the corpus-level clustering (illustrated by Fig. 2(b)) fights against the rareness issue. We adopt the latest KNN-based density peak clustering [2] for both levels. After the two-level clusterings, a final step extracts the output search expressions.

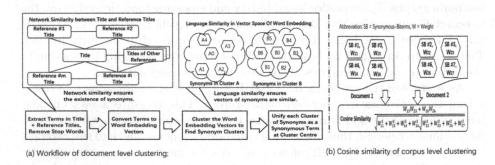

(a) Workflow of document level clustering: (b) Cosine similarity of corpus level clustering

Fig. 2. Two levels of clusterings in our proposed isBEST method.

4.1   Document-Level Clustering

We resolve the linguistic diversity issue by a document-level clustering, which combines language similarity and network similarity by clustering the embedding vectors of terms in each paper's title and reference titles.

Figure 2(a) illustrates the workflow of the document-level clustering. By network similarity, the title is similar to the reference titles, and the reference titles have mutual similarities if belonging to the same super-topic. According to language similarity, the embedding vectors of synonyms are similar, and there are many synonyms in similar texts. Therefore, if we cluster the embedding vectors, we will group the synonyms into clusters. A cluster centre will be the most common synonym in each synonym group, thus, a synonymous-term candidate.

There are still possibilities that a synonymous term in one paper fails to become the cluster centre in another paper. Fortunately, with the weighted average radius of each synonymous term in all of the papers it occurs, where the weights are the numbers of vectors in clusters, we can determine two synonymous terms are mutual synonyms if their cluster centres locate within each other's average radius. We then merge the mutual synonyms and use the most frequent one as their unified form.

At the end of this step, synonymous biterms are combined from cooccurring synonymous terms in each paper and serve as the just-in-time knowledge.

4.2   Corpus-Level Clustering

Because of the rareness, seeking emerging-topic information among a colossal volume of developed-topic data is challenging. External knowledge databases often miss rare but vital information about emerging topics. Therefore, we use the just-in-time knowledge (i.e., synonymous biterms) from document-level clustering instead of the external knowledge databases.

Based on our previous method in [16], which uses synonymous biterms to generalize the super-topic collaboration measure of AUGUR, i.e., using the rareness and accelerating cooccurences of a synonymous biterm to measure the strength of collaborations, we design a specialized strategy of document vectorization for emerging-topic-sensitive comparison, converting each paper to a vector of coefficients on synonymous biterms, assigning more weights to rarer synonymous biterms. More specifically, the weight takes value from the following equation:

$$
w\left(p_1, p_2\right) = \frac{\text{TFIDF}\left(p_1\right)\text{TFIDF}\left(p_2\right)}{2 - \left(0.5 - 0.5\text{CosSim}\left(p_1, p_2\right)\right)\frac{5}{5 + \text{BTF}_y\left(p_1, p_2\right)}}. \tag{2}
$$

Here, variables p_1 and p_2 are the two synonymous terms in the synonymous biterm. They also refer to the word embedding vectors of the terms. The TFIDF in Eq. (2) is computed by first converting the documents to synonymous terms and then calculating the traditional TF-IDF formula. The $\text{BTF}_y\left(p_1, p_2\right)$ in Eq. (2) is the biterm frequency of synonymous biterm p_1, p_2 before target year y. The cosine similarity CosSim is defined as follows.

$$
\text{CosSim}\left(p_1, p_2\right) = \frac{p_1 \cdot p_2}{\sqrt{p_1 \cdot p_1}\sqrt{p_2 \cdot p_2}}. \tag{3}
$$

This vectorization assigns larger coefficients to rarer synonymous biterms because those biterms reveal bigger collaborations than other biterms. Figure 2(b) shows how we use these document vectors to compute the similarity between two documents. Since the rare emerging topic biterms dominate the paper's coefficients, we diminish the similarity between an emerging-topic paper and a developed-topic paper while promoting the similarity between two papers on the same emerging topic.

4.3   The Search Expression Extraction

With corpus-level clustering producing clusters of papers, the proposed method takes the following operations to extract the output expressions.

(1) The method conducts an emulated evaluation using the scoring function with the number of publications per year in the cluster as arguments. It sorts those clusters by scores and keeps the top-k as the final clusters.
(2) For each final cluster, the method collects the synonymous biterms occurring in at least one paper of the cluster, sorts those biterms by frequency, and then in the ascending order of frequency, try to remove unnecessary biterms one by one. It checks whether a biterm is removable by constructing a graph using the papers in this cluster as nodes, linking two papers in the graph if at least one synonymous biterm cooccur in these two papers. A biterm is removable if the graph is connected without it.

5   Experiments

We use two fields of publication data from the 2020-05-29 release of Microsoft Academic Graph [14] as two input databases, which are relevant to machine learning and data mining. There are 417,075 imported papers in the field of machine learning and 418,970 imported papers in data mining. Because of the data range, we choose $y = 2019$ as the target year to find the emerging topics.

We choose these two fields because they are different typical examples. Most emerging works in machine learning cross disciplines, but few in data mining do so. The CSO ontology in the counterpart AUGUR covers only super-topics in the discipline of computer science, making AUGUR suitable for the emerging topic discovery in data mining but not suitable in machine learning. To prove this, we create another counterpart called FoS AUGUR by replacing the CSO ontology in AUGUR with the FoS (Fields of Study) ontology from MAG and comparing both of them with our isBEST. To be clear, we rename the original AUGUR as CSO AUGUR.

We choose the following scoring function based on SLP_{PI} in [13], and use this function to assign scores on how emerging a topic is.

$$\begin{aligned} &\text{score}\left(N_{y-5}, n_{y-4}, n_{y-3}, n_{y-2}, n_{y-1}, n_y\right) \\ &= \text{rareness}\left(N_{y-5} + \sum_{i=1}^{4} n_{y-i}\right) \times \max\left(0, \text{SLP}_{\text{PI}}\left(n_{y-4}, n_{y-3}, \ldots, n_y\right)\right), \end{aligned} \quad (4)$$

where y is the target year to find emerging topics in, N_{y-5} is the total number of publications five years before y, and for $i = 0, \ldots, 5$, n_{y-i} is the number of publications on the topic in the year i years before y. Also,

$$\text{rareness}\left(x\right) = \sqrt{1/(1+x)}, \quad (5)$$

SLP_{PI} is computed as the slope of the sequence below using linear regression:

$$\frac{n_{y-i+1} - n_{y-i} + 0.001}{n_{y-i} + 0.001}, \quad i = 4, 3, 2, 1. \quad (6)$$

Note that in the original SLP$_{PI}$ formula in [13], there is no 0.001 in Eq. (6). We add 0.001 to avoid the division-by-zero errors.

Since the outputs of our research problem are the search expressions on the site ScienceDirect.com, the emergingness scores of the site's query results measure the solver algorithm's performance. The evaluation results are shown in Figs. 3, with k, the number of emerging topic candidates in output, set as 100. Here, the accumulated scores are computed using the following formula.

$$\text{accScore}_i = \sum_{j \le i} \text{score}_j. \tag{7}$$

The variables i and j in Eq. (7) are the emerging-topic candidate indices. Since each solver gives a list of candidates for evaluation, we assign each candidate an index to distinguish them. Similarly, the accumulated positive SLP$_{PI}$ values are computed using the following formula.

$$\text{accPosSLP}_{PI_i} = \sum_{j \le i} \max\left(0, \text{SLP}_{PI_j}\right). \tag{8}$$

Maximizing the accScore metric at $i = 100$ is our aim. The accPosSLP$_{PI}$ metric is just an auxiliary quality metric. We also show results in a slightly larger range of i (i.e., 110 instead of 100) to allow inspecting metrics beyond $i = 100$.

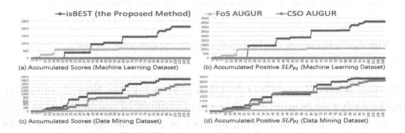

Fig. 3. Comparison of accumulated scores and accumulated positive SLP$_{PI}$.

Figure 3 shows that when the candidate index increases, our proposed method, isBEST, eventually outperforms the counterparts in both datasets, in terms of both accumulated score and accumulated positive SLP$_{PI}$. Figures 3a and 3b show that FoS AUGUR outperforms CSO AUGUR in the machine learning dataset because the FoS ontology covers much more disciplines than CSO. Figures 3c and 3d show that CSO AUGUR outperforms FoS AUGUR in the data mining paper dataset because the CSO ontology is more specialized than FOS. This demonstrates the importance of isBEST: we adopt just-in-time knowledge learning in isBEST to avoid the external knowledge database's weakness.

In addition to numeric evaluations using accumulated scores and accumulated positive SLPPI, Figs. 4 and 5 demonstrate the quality of the emerging topics discovered by our isBest method is better than those discovered by the counterpart

Fig. 4. Comparison of the top 10 output expressions for machine learning.

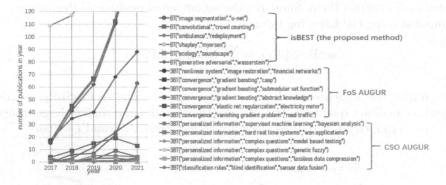

Fig. 5. Comparison of the top 6 output expressions for data mining.

methods (FoS AUGUR and CSO AUGUR). As defined in Sect. 2, emerging topics should satisfy two conditions: (1) a small number of publications in the past; and (2) an acceleratedly increasing number of publications in recent years. So, when we consider the last five years (2017–2021) in the experiments, we define recent years as (2019–2021) and the past years as (2017, 2018). Figure 4 shows the top 10 emerging topics in the field of "machine learning" while Fig. 5 shows the top 6 emerging topics in "data mining". In both figures, "BT", "BTs" and "3BT" stand for biterm, two different biterms and three different biterms respectively. They are used to abbreviate the following search expressions where field (i.e., "machine learning"/"data mining") is used to exclude unrelated results.

- $BT(A, B)$ = search expression: field AND (A AND B).
- $BTs((A, B), (C, D))$ = search expression: field AND ((A AND B) OR (C AND D)).
- $3BT(A, B, C)$ = search expression: field AND ((A AND B) OR (A AND C) OR (B AND C)).

Figure 4 only shows the results of the proposed isBEST method and FoS AUGUR because CSO AUGUR cannot find any emerging topic. Compared with FoS

AUGUR, our isBEST method has discovered high quality emerging topics that are all denoted by BT or BTs, while the FoS AUGUR only found emerging topics denoted by 3BT; obviously, BT and BTs are more precise than 3BT. The top 4 highest quality examples that satisfy the two conditions of the emerging topics well are BT("gan", "Wasserstein"), BTs(("augmentation", "few shot"), ("unsupervised learning", "few-shot")), BTs("challenging", "disentangled representations"), ("variational autoencoders", "disentangled")) and BT("demonstrations", "deep q-network").

Figure 5 also shows our isBEST method has discovered better emerging topics denoted by BT while both FoS AUGUR and CSO AUGUR only found emerging topics represented by 3BT. In the same way, to compare the number of publications changing with the years according to emerging topic's two conditions, the overall trend is that although the numbers of publications to support emerging topics found by FoS AUGUR are increasing quickly in recent years (from 2019–2021), satisfying emerging topic condition 2, the numbers of publications for those topics cannot satisfy the emerging topic condition 1 - an only small number of publications in the past; this means the result topics found by FoS AUGUR are the developed topics that were emerging topics in the past but not any more (i.e., they have small rareness values (Eq. (5)) and thus get bad scores (Eq. (4)). So, after excluding the developed topics, the proposed isBEST method shows better results; two that satisfy the emerging topic conditions well are BT("image segmentation", "u-net") and BT("generative adversarial", "Wasserstein").

6   Conclusions

In this paper, we propose the isBEST method to discover emerging topics with just-in-time knowledge learning. The main mechanism of isBEST lies in using two levels of clustering to overcome two challenges separately. One challenge is the difficulty to find the rare emerging-topic papers in large scholarly paper databases. Another challenge is the complication in defining the topic similarity between papers due to emerging topics' linguistic diversity. Our isBEST overcomes linguistic diversity challenge by combining language similarity and network similarity to find synonymous biterms in the document-level clustering. Also, it solves the rareness challenge in the corpus-level clustering with a vectorization strategy for document comparison that assigns more weights to rarer synonymous biterms. Because those biterms find emerging topics supported by super topics appears in a more accelerated increasing number of papers, this vectorization allows the corpus-level clustering to find emerging topics accurately. Experimental results on two datasets demonstrate the accuracy and effectiveness of the proposed method for emerging topic discovery. Our practice also suggests that just-in-time knowledge learning is superior to using an external knowledge database for more effective emerging topic discovery.

Acknowledgement. This work was partially supported by Australia Research Council (ARC) Discovery Project (DP190100587).

References

1. Alam, M.M., Ismail, M.A.: RTRS: a recommender system for academic researchers. Scientometrics **113**(3), 1325–1348 (2017)
2. Chen, Y., et al.: Fast density peak clustering for large scale data based on KNN. Knowl.-Based Syst. **187**, 104824 (2020)
3. Decker, S.L., Aleman-Meza, B., Cameron, D., Arpinar, I.B.: Detection of bursty and emerging trends towards identification of researchers at the early stage of trends. Ph.D. thesis, University of Georgia Athens (2007)
4. Dridi, A., Gaber, M.M., Azad, R.M.A., Bhogal, J.: Leap2Trend: a temporal word embedding approach for instant detection of emerging scientific trends. IEEE Access **7**, 176414–176428 (2019)
5. Erten, C., Harding, P.J., Kobourov, S.G., Wampler, K., Yee, G.: Exploring the computing literature using temporal graph visualization. In: Visualization and Data Analysis 2004, vol. 5295, pp. 45–56. International Society for Optics and Photonics (2004)
6. Ezzeldin, M., El-Dakhakhni, W.: Metaresearching structural engineering using text mining: trend identifications and knowledge gap discoveries. J. Struct. Eng. **146**(5), 04020061 (2020)
7. Kim, M.: Scientific trend analysis and curation with Korean R&D information. J. Supercomput. **72**(9), 3663–3673 (2016)
8. King, D., Downey, D., Weld, D.S.: High-precision extraction of emerging concepts from scientific literature. In: Proceedings of the 43rd International ACM SIGIR Conference on Research and Development in Information Retrieval, pp. 1549–1552 (2020)
9. Osborne, F., Scavo, G., Motta, E.: A hybrid semantic approach to building dynamic maps of research communities. In: Janowicz, K., Schlobach, S., Lambrix, P., Hyvönen, E. (eds.) EKAW 2014. LNCS (LNAI), vol. 8876, pp. 356–372. Springer, Cham (2014). https://doi.org/10.1007/978-3-319-13704-9_28
10. Prabhakaran, V., Hamilton, W.L., McFarland, D., Jurafsky, D.: Predicting the rise and fall of scientific topics from trends in their rhetorical framing. In: Proceedings of the 54th Annual Meeting of the Association for Computational Linguistics (Volume 1: Long Papers), pp. 1170–1180 (2016)
11. Salatino, A.A., Osborne, F., Motta, E.: Augur: forecasting the emergence of new research topics. In: Proceedings of the 18th ACM/IEEE on Joint Conference on Digital Libraries, pp. 303–312 (2018)
12. Sun, X., Ding, K., Lin, Y.: Mapping the evolution of scientific fields based on cross-field authors. J. Inform. **10**(3), 750–761 (2016)
13. Tseng, Y.H., Lin, Y.I., Lee, Y.Y., Hung, W.C., Lee, C.H.: A comparison of methods for detecting hot topics. Scientometrics **81**(1), 73–90 (2009)
14. Wang, K., Shen, Z., Huang, C., Wu, C.H., Dong, Y., Kanakia, A.: Microsoft academic graph: when experts are not enough. Quant. Sci. Stud. **1**(1), 396–413 (2020)
15. Wu, J., Giles, C.L.: Scholarly very large data: challenges for digital libraries. In: Challenges For Large Scale Networking (LSN) Workshop on Huge Data: A Computing, Networking and Distributed Systems Perspective (2020)
16. Wu, J., Huang, G., Zarei, R.: ETBTRank: ranking biterms in paper titles for emerging topic discovery. In: Long, G., Yu, X., Wang, S. (eds.) AI 2021. LNCS, vol. 13151, pp. 775–784. Springer, Cham (2022). https://doi.org/10.1007/978-3-030-97546-3_63
17. Xia, F., Wang, W., Bekele, T.M., Liu, H.: Big scholarly data: a survey. IEEE Trans. Big Data **3**(1), 18–35 (2017)

Predicting Abnormal Events in Urban Rail Transit Systems with Multivariate Point Process

Xiaoyun Mo[✉], Mingqian Li, and Mo Li

Nanyang Technological University, Singapore, Singapore
{xiaoyun001,mingqian001,limo}@ntu.edu.sg

Abstract. Abnormal events in rail systems, including train service delays and disruptions, are pains of the public transit system that have plagued urban cities for many years. The prediction of when and where an abnormal event may occur, can benefit train service providers for taking early actions to mitigate the impact or to eliminate the faults. Prior works rely on rich sources of sensor or log data that require extensive efforts in sensor deployment, data gathering and preparation. In this article, we aim at predicting abnormal events by leveraging only basic information of historical events (*e.g.*, dates, technical causes) that can be easily obtained from existing open records. We propose a non-trivial method which categorizes event pairs based on their basic information, and then characterizes inter-event influence between event pairs via a multivariate Hawkes process. The proposed method overcomes the major hurdle of data sparsity in abnormal events, and retains its efficacy in capturing the underlying dynamics of event sequences. We conduct experiments with a real-world dataset containing Singapore's 5-year abnormal rail events, and compare with a wide range of baseline methods. The results demonstrate the effectiveness of our method.

Keywords: Abnormal event prediction · Multivariate Hawkes process · Data sparsity

1 Introduction

Mass Rapid Transit (MRT) rail system usually provides the backbone of the public transit system. MRT-related abnormal events including train service delays and disruptions are a crucial problem that has plagued urban cities like Singapore for many years. The occurrence of an abnormal event can impair the journey of thousands to tens of thousands of commuters. The causes of these events vary, but the majority are due to technical faults such as power failures, signal errors, *etc.* On 7th July, 2015, one of the most severe MRT abnormal events in Singapore, which was caused by electrical power trips, crippled two major rail lines in Singapore during evening peak hours and affected up to 413,000 commuters. The operator was fined $5.4 million to take responsibility for this event [22]. Reducing the number of abnormal events, or mitigating their impact on commuters are thus vital tasks for train service providers.

© The Author(s), under exclusive license to Springer Nature Switzerland AG 2022
J. Gama et al. (Eds.): PAKDD 2022, LNAI 13280, pp. 41–53, 2022.
https://doi.org/10.1007/978-3-031-05933-9_4

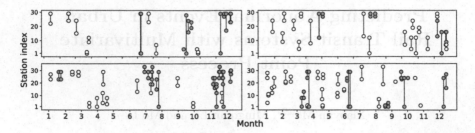

Fig. 1. Distribution of abnormal events from year 2015 (top left) to 2018 (bottom right) on the East-West line of Singapore. Each event is represented by a line segment with two bounding circles, indicating the stretch of abnormal stations. An event is marked in red if it is close to some other event(s) on both temporal (*i.e.,* within 1 week) and spatial (*i.e.,* stretches of stations overlap) scales.

Predictive analysis of MRT abnormal events benefits train service providers and commuters. On one hand, it helps with the predictive maintenance of the MRT system to eliminate hazards proactively, as well as prompt post-event actions to transfer affected commuters. On the other hand, prediction results can enhance public awareness of the operational conditions of the MRT system, and can help them to make decisions about their travel choices. Existing studies related to rail system failures leverage rich sources of data from sensors such as temperature, infrared and strain [8], *etc.,* which are practically hard to execute due to the costly deployment of sensors; as well as data from logs such as maintenance logs, equipment details [11], *etc.,* comprising heterogeneous data sources that require extensive efforts in data gathering, storing and pre-processing.

This paper aims at predicting when (*i.e.,* date) and where (*i.e.,* rail line and stations) will a future abnormal event occur. We leverage historical event data with event attributes including date, the abnormal rail line and stations, as well as the type of technical fault that causes the event. The data are easily accessible from two public channels, namely, official tweets posted by Singapore MRT operators (*i.e.,* SMRT and SBS) and local news feeds (*e.g.,* The Straits Times). We collect data about the abnormal events from January 2015 to December 2019 and perform the study. Figure 1 shows the spatial-temporal distribution of abnormal events on the East-West line, one of the most popular MRT lines in Singapore. The figure suggests certain locality on both temporal and spatial scales when events take place. It is likely that after one event occurs other events of overlapping stretches may follow, and as a result the sequence of events display a clustered dynamic pattern. This paper makes use of such a pattern, *i.e.,* the excitation influence between events, to model event sequence and therefore forecast future events.

Executing this approach, however, entails special challenges due to a major issue of data sparsity, *i.e.,* the number of abnormal events is extremely limited to capture the sophisticated inter-event influences. Specificaly, the influence decays as the interval between two events' timings increases, and events of different technical causes may be distinctive in the pattern of triggering future events. Other factors like whether the two events are on the same rail line, or whether their stretches of abnormal stations overlap, may lead to a difference on the magnitude of influence between them. In addition, it is also necessary to quantify by how much an event occurs innately, *i.e.*, occurs as natural arrivals. We need learn from very limited historical events as the training set in order to derive a unified model to numerically quantify the dependencies.

Contributions. This paper proposes a novel method based on multivariate Hawkes process, PAbEve (**P**redicting **Ab**normal **Eve**nt in MRT system), to address the above challenges in predicting MRT abnormal events. Leveraging the information of historical events, which is lightweight and publicly accessible, PAbEve retains its efficacy in modeling the abnormal event sequence, including the timings and locations, and then utilize it to predict the timings and locations of future events. PAbEve captures non-trivial inter-event influences and its parameters are expressive for those influences. We conduct extensive experiments with a real-world dataset containing Singapore's 5-year MRT abnormal events, and evaluate PAbEve in comparison with a wide range of alternative approaches. The results suggest PAbEve outperforms other methods in overall performance.

2   Related Work

Abnormal Event Prediction. Predicting abnormal events has attracted extensive attention in recent years. We divide existing works into three categories according to the object being studied. The first category studies on time series of instances in equal-length time steps, and treats those of extreme values as abnormal instances, such as key timings of flu seasons [1], congestion in traffic streams [9], and financial crisis in stock price series [4]. The second category of works attempts to construct indicative features or to find precursors of abnormal events, and use them as predictors for future events. For instances, some works conduct predictive analysis on a rich set of sensor, logging (*e.g.*, maintenance logs) and/or contextual data (*e.g.*, weather), to construct meaningful features for the prediction of railway point failures [11], railway service interruptions [8] and medical equipment failures [21]. Some studies focus on mining media articles (*e.g.*, tweets) to find precursors of social events like protests [3,14,23]. The third category of works directly studies event occurrences and utilize the inter-event correlation to predict future events, such as crime, vehicle collision, *etc.* [16]. Generally, the number of abnormal events is limited, resulting in a major challenge of data sparsity. Our work falls into this category. Only a few prior works of abnormal event prediction address the issue of data sparsity [1,15,23]. However, those approaches cannot be applied to our case, as their prediction problems fall into the first/second categories. We are unable to conduct contextual analysis of rich data sources because other relevant information for MRT abnormal events is also limited. Relevant information like the technical cause of event will be instead used as auxiliary covariates in this paper.

Point Process. As a mathematical approach for modeling event sequences, point process has been widely adopted to deal with prediction tasks, such as the prediction of taxi pickup, crime, neuronal activity, *etc.* Generally, a point process is characterized by a conditional intensity function. Classical point processes, such as Hawkes process [7], formulate the conditional intensity function based on a strong assumption on the dynamic pattern of event sequences. In the past decades, many non-trivial models extend these classical models to 3D spatio-temporal or multivariate space [2,12,19,24]. Recent deep learning techniques incubate state-of-the-art point processes, which are usually able to embed long-term memory of historical events and make very few assumptions on the dynamic pattern of event sequences [5,13,17,25]. Some existing works also propose intensity-free models to develop more general point processes using frameworks such as adversarial learning [10,20].

3   Preliminaries

Temporal Point Process. A temporal point process is a random process of event occurrence characterized by a *conditional intensity function,* $\lambda(t|\mathcal{H}_t)$, which is the event rate at time $t \in \mathbb{R}$ conditioned on historical events $\mathcal{H}_t$ before t. For convenience, we omit the notation $\mathcal{H}_t$ in the rest of the paper. The functional form of $\lambda(t)$ is usually designed according to the dynamic pattern of event sequences. For example, Hawkes process is a kind of temporal point process that characterizes the self-exciting dynamic pattern, *i.e.,* the occurrence of an event can raise the event rate in the near future.

Multivariate Hawkes Process. A multivariate Hawkes process can be regarded as a sequence of correlated Hawkes processes of multiple event types. Formally, for a U-dimensional Hawkes process, the conditional intensity function of the u-th event type, $\lambda_u(t)$, $u = 1, ..U$, is defined as

$$\lambda_u(t) = \mu_u + \sum_{i:t_i<t} \alpha_{uu_i} g(t - t_i) \tag{1}$$

where μ_u is the natural arrival rate (*i.e.,* background rate) of the u-th event type, α_{uu_i} is the trigger coefficient between the u-th and u_i-th event types, and $g(\Delta t)$ is the trigger function that usually decays with the increase of Δt.

4   Methodology

4.1   Problem Definition

In the MRT system, suppose there are R rail lines, and U possible technical faults that cause abnormal events. We use r ($r = 1, ..., R$) and u ($u = 1, ..., U$) to denote the indices of rail lines and technical faults, respectively. Each rail line is divided into M equal-size segments so that rail lines are of equal numbers of segments (*i.e.,* "lengths"). We then denote a stretch of stations as $x = [x_-, x_+]$, where x_- and x_+ are the indices of the two bounding segments on a specific rail line, $1 \le x_- \le x_+ \le M$. The list of distinct stretches of the r-th rail line is denoted by X^r, with size $S = |X^r| = \frac{M(M+1)}{2}$ that is identical for all rail lines. We use s to denote the index of a stretch $(s = 1, ..., S)$. Suppose we are given an abnormal event sequence $e_1, ..., e_n$, where $e_i = (t_i, r_i, s_i)$, with $t_i \in \mathbb{Z}$ the time of event in terms of day. The causes of events are denoted by $u_1, ..., u_n$, each of which is the index of technical fault. Given the information of n historical events above, we aim to predict the time, abnormal rail line and stretch of stations of the next event, $e_{n+1} = (t_{n+1}, r_{n+1}, s_{n+1})$.

4.2 Categorization of Event Pairs

We categorize an event pair (e_i, e_j), where $i > j$, hierarchically using the locations and technical faults of both events. The categorization is of three levels. For the first level, we divide event pairs based on their technical faults. For simplicity, we assume that inter-event influence only exists between two events that are caused by the same type of technical fault. According to the official tweets posted by MRT operators, there are 6 main types of technical faults, namely, train fault, track fault, power fault, signal fault, platform fault (mostly the screen door errors), and others. Therefore, for the first level, event pairs are divided into 6 groups of different fault types. For the second level, we distinguish *intra-line* pairs, for which two events occur on the same rail line (*i.e.*, $r_i = r_j$), from *inter-line* pairs, for which two events occur on different rail lines (*i.e.*, $r_i \ne r_j$). The influence between inter-line pair of events is possible as the two rail lines can be run by the same transit operator. For the third level, we further divide event pairs into *overlapping* pairs or *non-overlapping* pairs, according to whether the two events' stretches of abnormal stations overlap (*i.e.*, $X^{r_i}_{s_i} \cap X^{r_j}_{s_j} \ne \emptyset$) or not (*i.e.*, $X^{r_i}_{s_i} \cap X^{r_j}_{s_j} = \emptyset$). Note that the stretches of an inter-line pair can also overlap via interchange stations.

4.3 Multivariate Hawkes Process

We propose a multivariate Hawkes process that can capture the specific inter-event influence of each category of event pairs. We first derive the conditional intensity function, and then present the procedure of prediction.

Conditional Intensity Function. The occurrence rate of abnormal events on the r-th rail line at the s-th stretch of stations in X^r, is specified by a conditional intensity function defined as

$$\lambda(t, r, s) = \sum_{u=1}^{U} \lambda_u(t, r, s) \tag{2}$$

$$\lambda_u(t, r, s) = \mu_{urs} + \sum_{j: t_j < t, u_j = u} \phi_{u_j}(r, r_j, s, s_j) g(t - t_j) \tag{3}$$

where $\lambda_u(t, r, s)$ is a subordinate conditional intensity function of technical fault u, with $u = 1, ..., U$. μ_{urs} represents the natural arrival rate of events of the type indicated by the subscript indices. The second term of $\lambda_u(t, r, s)$ represents the trigger rates that are brought by events before t. According to the 3-level categorization described in Sect. 4.2, we assign each category of event pairs with a distinct trigger coefficient specified by $\phi_u(\cdot)$, and it is defined as

$$\phi_u(r, r', s, s') = \begin{cases} a_u & r \neq r' \text{ and } X_s^r \cap X_{s'}^{r'} = \emptyset \\ b_u & r \neq r' \text{ and } X_s^r \cap X_{s'}^{r'} \neq \emptyset \\ c_u & r = r' \text{ and } X_s^r \cap X_{s'}^{r'} = \emptyset \\ d_u & r = r' \text{ and } X_s^r \cap X_{s'}^{r'} \neq \emptyset \end{cases} \tag{4}$$

in which a_u, b_u, c_u and d_u are the trigger coefficients for the u-th fault type, for inter-line non-overlapping, inter-line overlapping, intra-line non-overlapping and intra-line overlapping event pairs, respectively. The trigger function $g(\Delta t)$ is defined in order to weaken the influence as time elapses. Particularly, it is defined in a non-parametric way, i.e., Δt is discretized as $\Delta t = k\delta t$, for $k = 0, ..., K$. The hyper-parameters K and δt control the span and granularity of time intervals, respectively. Then the trigger function $g(\Delta t)$ is specified by a sequence of scalars $[g_k]_{k=1}^K$. When $\Delta t > K\delta t$, $g(\Delta t)$ equals to zero.

Parameter Learning. The parameters are optimized iteratively using the maximum likelihood estimation. The likelihood of event e_i is defined as

$$L_i = \lambda_{u_i}(t_i, r_i, s_i) \cdot exp\left\{ -\int_{t_{i-1}}^{t_i} \left(\sum_{u=1}^U \sum_{r=1}^R \sum_{s=1}^S \lambda_u(\tau, r, s) \right) d\tau \right\} \tag{5}$$

where the exponential term of Eq. (5) means no event during the time interval (t_{i-1}, t_i) [18]. A lower-bound of the log-likelihood $logL$ of an n-length event sequence is then derived as

$$logL \geq \sum_{i=1}^n \left(p_{ii}log\frac{\mu_{u_i r_i s_i}}{p_{ii}} + \sum_{j:t_j<t_i, u_j=u_i} p_{ij}log\frac{\phi_{u_j}(r_i, r_j, s_i, s_j)g(t_i - t_j)}{p_{ij}} \right)$$
$$- (t_n - t_0)\sum_{u=1}^U \sum_{r=1}^R \sum_{s=1}^S \mu_{urs} - \sum_{r=1}^R \sum_{s=1}^S \sum_{j=1}^n \left(\phi_{u_j}(r, r_j, s, s_j) \int_{t_j}^{t_n} g(\tau - t_j)d\tau \right) \tag{6}$$

based on Jensen's inequality $(log(E[X]) \geq E[log(X)])$. The weights p_{ii} and p_{ij}, $j = 1, ..., i - 1$ are computed following [24].

$$p_{ii} = \frac{\mu_{u_i r_i s_i}}{\lambda_{u_i}(t_i, r_i, s_i)}, \qquad p_{ij} = \frac{\phi_{u_j}(r_i, r_j, s_i, s_j)g(t_i - t_j)}{\lambda_{u_i}(t_i, r_i, s_i)} \tag{7}$$

Specifically, p_{ii} denotes the probability that event e_i arrives naturally, while p_{ij} denotes the probability that it is rather triggered by a previous event e_j.

Given the lower-bound of $logL$, the analytical solutions of the parameters can be obtained by setting the first derivative of the lower-bound with respect to each parameter to zero and then solving the equations. Specifically, the solutions of

---

Algorithm 1: Iterative algorithm for parameter learning

Input: events $[e_i]_{i=1}^n$, faults $[u_i]_{i=1}^n$, randomly initialized weights p_{ii}'s and p_{ij}'s

Output: values of parameters

1 **repeat**
2   Update μ_{urs} by Eq. (8) for $u = 1, ..., U$, $r = 1, ..., R$, and $s = 1, ..., S$;
3   Update a_u by Eq. (9), similarly for b_u, c_u and d_u, for $u = 1, ..., U$;
4   Update g_k by Eq. (10) for $k = 0, ..., K$;
5   **for** $i = 1, ..., n$ *and* $j = 1, ..., i - 1$ **do**
6   Update p_{ii} by Eq. (7);
7   Update p_{ij} by Eq. (7) if $t_i - t_j \leq K\delta t$ and $u_i == u_j$ **else** set $p_{ij} = 0$.

8 **until** p_{ii}'s *and* p_{ij}'s *converge*;

μ_{urs}, a_u (and adapted to b_u, c_u and d_u according to Eq. (4)) and g_k are depicted in Eq. (8), (9) and (10), respectively, where $\mathbb{I}[\cdot]$ is the indicator function.

$$\mu_{urs} = \frac{\sum_{i=1}^n p_{ii}\mathbb{I}[u_i = u, r_i = r, s_i = s]}{t_n - t_0} \tag{8}$$

$$a_u = \frac{\sum_{i=1}^n \sum_{j:t_j < t_i, u_j = u_i} p_{ij}\mathbb{I}[u_j = u, r_i \neq r_j, X_{s_i}^{r_i} \cap X_{s_j}^{r_j} = \emptyset]}{\sum_{r=1}^R \sum_{s=1}^S \sum_{j=1}^n \mathbb{I}[u_j = u, r \neq r_j, X_s^r \cap X_{s_j}^{r_j} = \emptyset] \int_{t_j}^{t_n} g(\tau - t_j)d\tau} \tag{9}$$

$$g_k = \frac{\sum_{i=1}^n \sum_{j:t_j < t_i, u_j = u_i} p_{ij}\mathbb{I}[k\delta t \leq t_i - t_j < (k+1)\delta t]}{\delta t \sum_{r=1}^R \sum_{s=1}^S \sum_{j=1}^n \phi_{u_j}(r, r_j, s, s_j)\mathbb{I}[k\delta t \leq t_n - t_j]} \tag{10}$$

Provided n historical abnormal events, we first initialize the weights p_{ii}'s and p_{ij}'s by random. After that, a loop is used to iteratively optimize the values of all parameters and weights until convergence, *i.e.*, until the values of parameters (or weights p_{ii}'s and p_{ij}'s) do not change substantially in a single iteration. The iterative algorithm is shown in Algorithm 1.

Prediction. Given the conditional intensity functions and historical abnormal events, the probability density function of some $t \in (t_n, +\infty)$, r, s being the time, rail line, and stretch of the next event is given as

$$f(t, r, s) = \lambda(t, r, s) \cdot exp\left\{-\int_{t_n}^t \lambda(\tau)d\tau\right\} \tag{11}$$

We predict the timing of event e_{n+1} by taking its expectation as

$$\hat{t}_{n+1} = E[t|\mathcal{H}_{t_n}] = \frac{\int_{t_n}^{t_n+T} t \cdot \left(\sum_{r=1}^R \sum_{s=1}^S f(t, r, s)\right) dt}{\int_{t_n}^{t_n+T} \left(\sum_{r=1}^R \sum_{s=1}^S f(t, r, s)\right) dt} \tag{12}$$

where T is a sufficiently large time duration (*e.g.*, $T = 150$). After that, the abnormal rail line as well as the stretch of stations are predicted as follows:

$$\hat{r}_{n+1} = \text{argmax}_r f(\hat{t}_{n+1}, r) = \text{argmax}_r \sum_{s=1}^S f(\hat{t}_{n+1}, r, s) \tag{13}$$

Table 1. Distribution of events by fault.

Fault type	Train	Track	Power	Signal	Platform	Others
Num. of events	86	81	19	27	22	19

$$\hat{x}_{n+1} = E[x|\mathcal{H}_{t_n}, \hat{t}_{n+1}, \hat{r}_{n+1}] = \frac{\sum_{s=1}^{S} X_s^{\hat{r}_{n+1}} \cdot f(\hat{t}_{n+1}, \hat{r}_{n+1}, s)}{\sum_{s=1}^{S} f(\hat{t}_{n+1}, \hat{r}_{n+1}, s)} \tag{14}$$

in which $\hat{x}_{n+1} = [\hat{x}_-, \hat{x}_+]$, with $\hat{x}_-$ and $\hat{x}_+$ the indices of the two bounding segments that specify a stretch of stations on the $\hat{r}_{n+1}$-th rail line. All the integrals in the equations above are approximated using summation.

5   Experiments

5.1   Experimental Setup

Dataset. We collect MRT abnormal events from January 2015 to December 2019, from two open sources, *i.e.*, official tweets posted by operators and local news feeds[1]. The provided information includes date, approximate time of the day, rail line, cause, and the stretch of affected stations. We set the causes of a few events with no cause specified as the "others" type of technical fault. After filtering out isolated incidents with system irrelevant causes (*e.g.*, passenger's fall, animal invasion), we finally obtain 254 events, the distribution of which by fault type is shown in Table 1. We sort the events by date and use the first 75% for training and the last 25% for testing. There is no validation set due to the scarcity of observed events. But we provide sensitivity analysis which shows clear trends of the impact of hyper-parameters on the performance.

Baselines. We compare PAbEve with 9 baseline methods, where 5 are for timing prediction only (*i.e.*, NextDay, Auto-regressive, Hawkes parametric, Hawkes nonparametric and NNPP), and the rest 4 for both timing and location prediction (*i.e.*, Poisson loc, MMEL loc, MMEL fault+loc and RMTPP loc).

- NextDay: a naive baseline which uses the next day of the most recent event as the prediction result, *e.g.*, $\hat{t}_{n+1} = t_n + 1$.
- Poisson loc: a homogeneous Poisson process with the conditional intensity function $\lambda(t, r, s) = \frac{\sum_{i=1}^{n} \mathbb{I}[r_i = r, s_i = s]}{t_n - t_0}$ that is constant over time.
- Auto-regressive [6]: which assumes the most recent l inter-event intervals are linearly correlated. We select l as 6.
- Hawkes parametric (Hawkes p) [7]: a temporal Hawkes process with trigger function g defined parametrically as $g(t - t') = e^{-\beta(t-t')}$.

---

[1] An example tweet on 17th February, 2015, is "11:27:37 [EWL] Due to a train fault at Jurong East, there will be no train service from Lakeside to Clementi on the east bound...". The event data is accessible via https://github.com/PAbEve/data.

Table 2. Prediction performance of evaluated approaches.

	NextDay	Auto-reg.	Hawkes p	Hawkes n/p	NNPP
MAE	8.156	7.131	6.634	6.797	74.878
	Poisson loc	MMEL loc	MMEL fault+loc	RMTPP loc	PAbEve
MAE	6.562	6.869	9.588	7.438	**6.156**
Hit rate	0.453	0.503	0.478	0.547	**0.578**
CosSim	0.437	0.485	0.458	0.473	**0.562**

- Hawkes non-parametric (Hawkes n/p): a temporal Hawkes process with trigger function g estimated non-parametrically, i.e., $g = \{g(k\delta t)|k = 0, 1, ...\}$.
- MMEL loc [24]: a multivariate Hawkes process with the u'-th dimensional conditional intensity function $\lambda_{u'}(t)$ given in Eq. (1), where $\{u' = (r, s)|r = 1, ..., R; s = 1, ..., S\}$. Hyper-parameters D is set as 1 and α as 0 for simplicity without losing generality.
- MMEL loc+fault [24]: which uses the same settings as MMEL loc, but with $\{u' = (r, s, v)|r = 1, ..., R; s = 1, ..., S; v = 1, ..., U\}$.
- RMTPP loc [5]: a neural marked temporal point process with its marks being the items in $\{u = (r, s)|r = 1, ..., R; s = 1, ..., S\}$.
- NNPP [17]: a fully neural temporal point process which models the cumulative conditional intensity function and obtain the conditional intensity function via its derivative.

Metrics. We use 3 kinds of metrics to evaluate the performance of compared methods on m test events, including (1) **MAE**, which is the mean absolute error in days between the predicted and ground-truth times, i.e., $\frac{1}{m}\sum_{l=1}^{m}|\hat{t}_{n+l}-t_{n+l}|$; (2) **Hit rate**, which is the proportion of test events where the predicted and ground-truth rail lines are the same, i.e., $\frac{1}{m}\sum_{l=1}^{m}\mathbb{I}[\hat{r}_{n+l} = r_{n+l}]$; (3) **CosSim**, which is the mean cosine similarity between the predicted and ground-truth stretches, i.e., $\frac{1}{m}\sum_{l=1}^{m}\frac{x_{n+l}\cdot\hat{x}_{n+l}}{|x_{n+l}||\hat{x}_{n+l}|}\mathbb{I}[r_{n+l} = \hat{r}_{n+l}]$. For MAE, smaller values are better, while for Hit rate and CosSim, larger values are preferred.

5.2   Experimental Results

We run each evaluated method for 10 rounds, and take the average of 10 rounds for each metric. We set the hyper-parameters as $\delta t = 1$, $K = 25$ and $M = 10$. The results of prediction are summarized in Table 2.

Results of Timing Prediction. In terms of MAE, we may draw the following conclusions. First, parametric methods are neither superior nor inferior to semi-parametric methods according to the experiment results. Semi-parametric method refers to those with a part of the model (e.g., the trigger function) designed in a non-parametric way, including all Hawkes-based evaluated methods except Hawkes p. We see among the top 5 performing methods, there are semi-parametric methods PAbEve with MAE 6.156, Hawkes n/p with MAE 6.797 and MMEL loc with MAE 6.869, and also parametric methods Poisson loc with MAE 6.562 and Hawkes p with MAE 6.634. Second, among the parametric methods, however, only light-weight methods, *i.e.*, Poisson loc, Hawkes p and Autoregressive, can compete with semi-parametric methods. Heavy-weight methods which may not be able to express the sparse data may perform arbitrarily bad, such as NNPP which yield a MAE of 74.878. Particularly, the naive Poisson loc outperforms all the other evaluated methods except PAbEve, which indicates that those non-trivial methods may degrade when trained on insufficient data. Third, among the semi-parametric Hawkes-based methods, we see Hawkes n/p of 6.797 outperforms MMEL loc of 6.869, and MMEL loc outperforms MMEL fault+loc of 9.588. We suspect that increasing the number of inputs (*e.g.*, rail line, technical fault) is probable to worsen the performance, as the number of parameters to learn are increased as well. Overall PAbEve outperforms the others as it properly incorporates all aspects of information via the dedicated design of inter-event influences.

Results of Location Prediction. Location prediction consists of the prediction of abnormal rail line and the stretch of stations. We evaluate 5 methods that can conduct location prediction, and PAbEve outperforms the others. For the prediction of abnormal rail lines, RMTPP loc performs close to PAbEve's, but it uses unadjustable prediction for all events (*i.e.*, standard deviation is zero), and so is Poisson loc. Comparison between MMEL loc and MMEL fault+loc shows increasing the number of inputs may worsen the performance. For the prediction of abnormal stretches, PAbEve again outperforms all others. Among them RMTPP loc predicts trivially using the entire rail line. Both RMTPP loc and Poisson loc predict using the same value for all events.

Model Interpretation. To interpret PAbEve, we visualize the estimated background rates, trigger coefficients, trigger function and weights. The results are shown in Fig. 2. From Fig. 2(a) and (b), we see events of the highest background rates are those on both rail lines caused by train fault or track fault, and the category of inter-line overlapping event pairs caused by "others" fault has the most significant inter-event influences. For the trigger function, as shown in Fig. 2(c), it fluctuates between 0 and 0.67, which is dissimilar to exponential or power law functions and this may simply be resulted from the data sparsity issue. Finally, for each of the 5 Hawkes-based evaluated methods, we investigate the probabilities of an event being natural arrival (represented by the sum of p_{ii}'s) or triggered event (represented by the sum of p_{ij}'s). PAbEve is the one with the largest ratio of background versus trigger (*i.e.*, about 3 to 1).

Results of Sensitivity Tests. We explore the impact of hyper-parameters δt, K and M, on the prediction performance. Each result is averaged over 10 rounds. The results are depicted in Fig. 3. We test the impact of δt by setting K to 25, and changing δt from 1 to 10 with PAbEve. The MAEs shown in Fig. 3(a) depict a clear trend that increasing δt will probably worsen the performance. When $\delta t \leq 2$, PAbEve outperform all other evaluated methods. Similarly, we test the

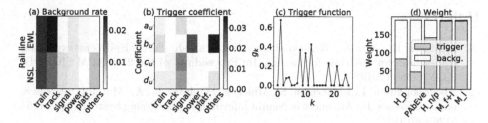

Fig. 2. Visualization of (a) background rates, (b) trigger coefficients, (c) trigger function, and (d) sum of training events' p_{ii}'s ("backg.") and that of p_{ij}'s ("trigger").

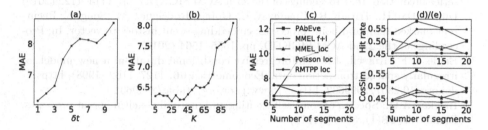

Fig. 3. Sensitivity tests. (a) MAEs under δt from 1 to 10 ($K = 25$); (b) MAEs under K from 5 to 100 ($\delta t = 1$); (c) MAEs, (d) Hit rates and (e) CosSims, of all location-available methods under M from 5 to 20 ($\delta t = 1$, $K = 25$).

impact of K by setting δt to 1, and changing K from 5 to 100 at an interval of 5 with PAbEve. The MAEs depicted in Fig. 3(b) range from 6.156 to 8.056. When $K \leq 65$, PAbEve outperform all other methods with the MAEs oscillating between 6.156 and 6.538. Finally, for the number of rail line segments, M, as shown in Fig. 3(c), there are only mild changes of MAEs for most methods when M changes, except MMEL fault+loc for which the errors increase significantly. For location prediction results shown in Fig. 3(d) and (e), as M increases, there is no specific trend for the hit rate of rail line or the similarity measure of stretch.

6 Conclusion

We present a novel solution to predicting when and where will a future MRT abnormal event occur, based on historical abnormal events. We first categorize event pairs based on basic contextual information, and then design a multivariate Hawkes process to model the sparse sequence of abnormal events. The proposed PAbEve approach retains its efficacy when being trained on extremely limited training events. Experimental results using real-world data from open sources demonstrate the superiority of PAbEve over other alternative solutions.

Acknowledgement. This work is supported by Alibaba Group through Alibaba Innovative Research (AIR) Program and Alibaba-NTU Singapore Joint Research Institute (JRI), and Singapore MOE AcRF Tier 1 RG18/20.

References

1. Adhikari, B., Xu, X., Ramakrishnan, N., Prakash, B.A.: EpiDeep: exploiting embeddings for epidemic forecasting. In: Proceedings of the 25th ACM SIGKDD, pp. 577–586 (2019)
2. Apostolopoulou, I., Linderman, S., Miller, K., Dubrawski, A.: Mutually regressive point processes. In: Advances in Neural Information Processing Systems, pp. 5115–5126 (2019)
3. Deng, S., Rangwala, H., Ning, Y.: Learning dynamic context graphs for predicting social events. In: Proceedings of the 25th ACM SIGKDD, pp. 1007–1016 (2019)
4. Ding, D., Zhang, M., Pan, X., Yang, M., He, X.: Modeling extreme events in time series prediction. In: Proceedings of the 25th ACM SIGKDD, pp. 1114–1122 (2019)
5. Du, N., Dai, H., Trivedi, R., Upadhyay, U., Gomez-Rodriguez, M., Song, L.: Recurrent marked temporal point processes: embedding event history to vector. In: Proceedings of the 22nd ACM SIGKDD, pp. 1555–1564 (2016)
6. Engle, R.F., Russell, J.R.: Autoregressive conditional duration: a new model for irregularly spaced transaction data. Econometrica **66**, 1127–1162 (1998). https://doi.org/10.2307/2999632. https://www.jstor.org/stable/2999632
7. Hawkes, A.G.: Spectra of some self-exciting and mutually exciting point processes. Biometrika **58**(1), 83–90 (1971)
8. Li, H., et al.: Improving rail network velocity: a machine learning approach to predictive maintenance. Transp. Res. Part C: Emerg. Technol. **45**, 17–26 (2014)
9. Li, J., et al.: Predicting path failure in time-evolving graphs. In: Proceedings of the 25th ACM SIGKDD, pp. 1279–1289 (2019)
10. Li, S., Xiao, S., Zhu, S., Du, N., Xie, Y., Song, L.: Learning temporal point processes via reinforcement learning. arXiv preprint arXiv:1811.05016 (2018)
11. Li, Z., Zhang, J., Wu, Q., Gong, Y., Yi, J., Kirsch, C.: Sample adaptive multiple kernel learning for failure prediction of railway points. In: Proceedings of the 25th ACM SIGKDD, pp. 2848–2856 (2019)
12. Marsan, D., Lengline, O.: Extending earthquakes' reach through cascading. Science **319**(5866), 1076–1079 (2008)
13. Mei, H., Eisner, J.M.: The neural Hawkes process: a neurally self-modulating multivariate point process. In: Advances Neural Information Processing Systems, vol. 30, pp. 6754–6764 (2017)
14. Ning, Y., Muthiah, S., Rangwala, H., Ramakrishnan, N.: Modeling precursors for event forecasting via nested multi-instance learning. In: Proceedings of the 22nd ACM SIGKDD, pp. 1095–1104 (2016)
15. Ning, Y., Tao, R., Reddy, C.K., Rangwala, H., Starz, J.C., Ramakrishnan, N.: Staple: spatio-temporal precursor learning for event forecasting. In: Proceedings of the 2018 SIAM International Conference on Data Mining, pp. 99–107 (2018)
16. Okawa, M., Iwata, T., Kurashima, T., Tanaka, Y., Toda, H., Ueda, N.: Deep mixture point processes: spatio-temporal event prediction with rich contextual information. In: Proceedings of the 25th ACM SIGKDD, pp. 373–383 (2019)
17. Omi, T., Ueda, N., Aihara, K.: Fully neural network based model for general temporal point processes. arXiv preprint arXiv:1905.09690 (2019)
18. Rasmussen, J.G.: Lecture notes: temporal point processes and the conditional intensity function. arXiv preprint arXiv:1806.00221 (2018)
19. Reinhart, A., et al.: A review of self-exciting spatio-temporal point processes and their applications. Stat. Sci. **33**(3), 299–318 (2018)

20. Shchur, O., Biloš, M., Günnemann, S.: Intensity-free learning of temporal point processes. arXiv preprint arXiv:1909.12127 (2019)
21. Sipos, R., Fradkin, D., Moerchen, F., Wang, Z.: Log-based predictive maintenance. In: Proceedings of the 20th ACM SIGKDD, pp. 1867–1876 (2014)
22. Tan, C.: SMRT fined record $5.4 million for July 7 breakdown, January 2016. https://www.straitstimes.com/singapore/transport/smrt-fined-record-54-million-for-july-7-breakdown
23. Zhao, L., Sun, Q., Ye, J., Chen, F., Lu, C.T., Ramakrishnan, N.: Multi-task learning for spatio-temporal event forecasting. In: Proceedings of the 21th ACM SIGKDD, pp. 1503–1512 (2015)
24. Zhou, K., Zha, H., Song, L.: Learning triggering kernels for multi-dimensional Hawkes processes. In: International Conference on Machine Learning, pp. 1301–1309 (2013)
25. Zuo, S., Jiang, H., Li, Z., Zhao, T., Zha, H.: Transformer Hawkes process. In: International Conference on Machine Learning, pp. 11692–11702. PMLR (2020)

Are Edge Weights in Summary Graphs Useful? - A Comparative Study

Shinhwan Kang[1], Kyuhan Lee[1], and Kijung Shin[1,2(✉)]

[1] Kim Jaechul Graduate School of AI, KAIST, Daejeon, South Korea
{shinhwan.kang,kyuhan.lee,kijungs}@kaist.ac.kr
[2] School of Electrical Engineering, KAIST, Daejeon, South Korea

Abstract. *Which one is better between two representative graph summarization models with and without edge weights?* From web graphs to online social networks, large graphs are everywhere. Graph summarization, which is an effective graph compression technique, aims to find a compact *summary graph* that accurately represents a given large graph. Two versions of the problem, where one allows edge weights in summary graphs and the other does not, have been studied in parallel without direct comparison between their underlying representation models. In this work, we conduct a systematic comparison by extending three search algorithms to both models and evaluating their outputs on eight datasets in five aspects: (a) reconstruction error, (b) error in node importance, (c) error in node proximity, (d) the size of reconstructed graphs, and (e) compression ratios. Surprisingly, using unweighted summary graphs leads to outputs significantly better in all the aspects than using weighted ones, and this finding is supported theoretically. Notably, we show that a state-of-the-art algorithm can be improved substantially (specifically, 8.2×, 7.8×, and 5.9× in terms of (a), (b), and (c), respectively, when (e) is fixed) based on the observation.

1 Introduction and Related Works

Relationships between objects, such as friendships in online social networks, co-appearance of tags, and hyperlinks between web pages, are universal. They are naturally represented as graphs, whose sizes have grown at a tremendous rate due to advances in web technology. For example, the number of web pages (i.e., nodes in web graphs) increased by about 500× in the past two decades.

Graph compression is a useful technique for efficient utilization of such large graphs. Many techniques have been developed for various purposes, including storage [5, 7, 9, 15, 17, 19, 23, 26, 28], query processing [6, 13, 18, 25], influence analysis [21, 22], pattern mining [16, 27], anomaly/outlier detection [3, 8], privacy preservation [29, 30], visualization [10, 11], and representation learning [12, 32]. We refer to surveys [4, 20] for details of them. Their common goal is to find a compact representation that exactly or approximately describes a given graph.

Among them, we focus on *graph summarization* [2, 13–15, 17, 18, 23, 25, 28]. whose objective is to find a concise *summary graph* G' that accurately describes

J. Gama et al. (Eds.): PAKDD 2022, LNAI 13280, pp. 54–67, 2022.
https://doi.org/10.1007/978-3-031-05933-9_5

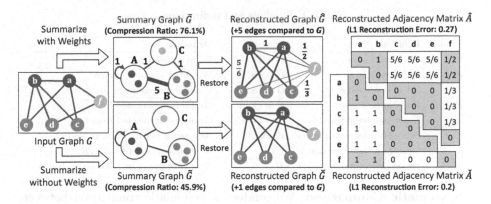

Fig. 1. Examples of weighted and unweighted graph summarization and reconstruction. The summary graph $\tilde{G}$ without edge weights is more concise with smaller reconstruction error and reconstruction size than $\overline{G}$ with edge weights.

a given large graph G, or equivalently, concise G' from which we can restore a graph close to G.[1] Each node in G' is interpreted as a group of nodes in G, and each edge in G' is interpreted as the presence of edges between all pairs of nodes in two groups. Since the output G' is in the form of a graph, other graph compression methods can be applied to G' for further compression [28]. That is, graph summarization can be used as a preprocessing step of other compression methods. Moreover, a wide range of graph algorithms can be approximately executed on G' without full reconstruction (see Appendix A and [25]).

There are two representative graph summarization models: a summary graph with edge weights [2,17,18,25] and one without edge weights [13–15,23,28]. While the latter is typically used with edge corrections for lossless compression, this work focuses on G'. While a number of search algorithms aiming at finding a high-quality summary graph under a given constraint have been developed for each model, there was no systematic comparison between the two models.

Which one is better between the two graph summarization models? Are edge weights in summary graphs useful? For a systematic comparison between the two models, we extend three search algorithms [15,17,18] to both models and evaluate their outputs in eight real-world graphs in five aspects: (a) reconstruction error, (b) error in node importance [24], (c) error in node proximity [31], (d) the number of edges in reconstructed graphs, and (e) compression ratios.

Counterintuitively, we find out that using unweighted summary graphs gives a significantly better trade-off among (a)–(e) than using weighted ones, regardless of search algorithms and datasets (See Fig. 1 for an example). Notably, adapting a state-of-the-art algorithm for the weighted model [17] to the unweighted model leads to 8.2×, 7.8×, and 5.9× improvements in terms of (a), (b), and (c) (when (e) is fixed) and 2.2× improvements in terms of (d) (when (a) is similar).

Our contributions are three-fold:

---

[1] While we use the term "graph summarization" to refer to this specific way of compression, the term has also been used more generally, as surveyed in [20].

Table 1. Symbols and definitions.

Symbol	Definition	Symbol	Definition
V	Set of subnodes	$G = (V, E)$	Input graph
E	Set of subedges	$\overline{G} = (S, P, \omega)$	Weighted summary graph
S	Set of supernodes	$\tilde{G} = (S, P)$	Unweighted summary graph
P	Set of superedges	$\hat{G} = (V, \hat{E}, \hat{\omega})$	Graph reconstructed from $\overline{G}$
ω	Superedge weight function	$\check{G} = (V, \check{E})$	Graph reconstructed from $\tilde{G}$
$\hat{E}, \check{E}$	Sets of reconstructed edges	$A, \hat{A}, \check{A}$	Adjacency matrix of G, $\hat{G}$, and $\check{G}$
$\hat{\omega}$	Subedge weight function	E_{AB}	# of subedges between $A, B \in S$
S_i	Supernode containing $i \in V$	Π_{AB}	# of subnode pairs between $A, B \in S$

- **Systematic Comparison:** We conduct a systematic comparison between two extensively-studied graph summarization models using three search algorithms, eight datasets, and five evaluation metrics.
- **Unexpected Observation:** Our comparison leads to a surprising observation that using unweighted models is significantly better than using weighted ones in all considered aspects. We support this finding theoretically (see Theorem 1).
- **Improvement of the State of the Art:** By exploiting the observation, we can improve a state-of-the-art algorithm [17] substantially in all considered aspects (see Figs. 2, 3 and 4).

Reproducibility: The source code and the datasets are available at [1].

Roadmap: In Sect. 2, we introduce graph summarization models. In Sect. 3, we define problems and present algorithms. In Sect. 4, we provide empirical results. In Sect. 5, we present theoretical results. In Sect. 6, we offer conclusions.

2   Graph Summarization Models

We introduce weighted and unweighted graph summarization models, which are compared throughout this work. See Table 1 for frequently-used symbols.

Input Graph. Consider an undirected graph $G = (V, E)$ with a set of *subnodes* $V = \{1, \cdots, |V|\}$ and a set of *subedges* $E \subseteq \binom{V}{2}$. We use $A \in \mathbb{R}^{|V| \times |V|}$ to denote its *adjacency matrix*. Each entry $A_{ij} = 1$ if $\{i, j\} \in E$ and $A_{ij} = 0$ otherwise.

2.1   Weighted Graph Summarization Model

Definition. A *weighted summary graph* $\overline{G} = (S, P, \omega)$ of $G = (V, E)$ consists of a set of *supernodes* S, a set of *superedges* P, and a *superedge weight function* ω. The set S is a partition of V. That is, supernodes are disjoint sets of subnodes whose union is V. Each superedge $\{A, B\} \in P$ joins two supernodes $A \in S$ and $B \in S$. The function ω takes each superedge $\{A, B\} \in P$ and returns its *weight* ω_{AB}, which is equal to $E_{AB} := |\{\{i, j\} \in E : i \in A, j \in B\}|$, i.e., the number of subedges in G that join subnodes between $A \in S$ and $B \in S$.

Reconstruction. The *reconstructed graph* $\hat{G} = (V, \hat{E}, \hat{\omega})$ obtained from $\overline{G} = (S, P, \omega)$ consists of the set of subnodes V, the set of reconstructed subedges $\hat{E} \subseteq \binom{V}{2}$, and a subedge weight function $\hat{\omega}$. If we let $S_i \in S$ be the supernode containing each subnode $i \in V$ and let Π_{AB} be the number of possible pairs of subnodes between supernodes A and B. That is, $\Pi_{AB} := \binom{|A|}{2}$ if $A = B$ and $\Pi_{AB} := |A| \cdot |B|$ otherwise. The adjacency matrix $\hat{A} \in \mathbb{R}^{|V| \times |V|}$ of $\hat{G}$ is defined as

$$\hat{A}_{ij} = \hat{\omega}_{ij} := \begin{cases} \frac{\omega_{S_i S_j}}{\Pi_{S_i S_j}}, & \text{if } i \neq j \text{ and } \{S_i, S_j\} \in P, \\ 0, & \text{otherwise.} \end{cases} \tag{1}$$

2.2 Unweighted Graph Summarization Model

Definition. An *unweighted summary graph* $\tilde{G} = (S, P)$ of G consists of a set of supernodes S and a set of superedges P. Note that, unlike $\overline{G}$, $\tilde{G}$ does not have the superedge weight function ω.

Reconstruction. The adjacency matrix $\check{A} \in \mathbb{R}^{|V| \times |V|}$ of the graph $\check{G} = (V, \check{E})$ reconstructed from $\tilde{G}$ is defined as

$$\check{A}_{ij} := \begin{cases} 1, & \text{if } i \neq j \text{ and } \{S_i, S_j\} \in P \\ 0, & \text{otherwise.} \end{cases} \tag{2}$$

While $\check{G}$ is typically used with edge corrections for lossless compression [15, 23, 28], this work focuses on $\check{G}$.

3 Problem Formulation and Algorithms

Based on the graph summarization models, we formulate graph summarization as optimization problems. Then, we present six search algorithms for the problems.

3.1 Optimization Problem Formulation

Given a graph G, we aim to minimize the difference between a reconstructed adjacency matrix A' (i.e., $\check{A}$ or $\hat{A}$) and the adjacency matrix A of G. Specifically, we aim to minimize the L_p reconstruction error, i.e.,.

$$RE_p(A, A') := ||A - A'||_p, \tag{3}$$

while constraining the size of the output summary graph G' (i.e., $\overline{G}$ or $\tilde{G}$) to be at most a given constant. The size can be (a) the number of supernodes in G'[2, 18, 25], (b) the number of superedges in G', or (c) the size of G' in bits [17].

Size in Bits of Summary Graphs. The size of a weighted summary graph $\overline{G} = (S, P, \omega)$ in bits is defined as

$$size_{bits}(\overline{G}) := 2|P| \log_2 |S| + |P| \log_2 \omega_{max} + |V| \log_2 |S|, \tag{4}$$

Table 2. The outlines of the considered search algorithms are given in Algorithms 1 and 2, and the details of each algorithm are provided in the table below.

Algorithm	Outline	G'	T	$size()$	$groups()$	$loss()$	$sparsify?()$
k-Grass (W)	Algorithm 1	$\overline{G}$	Infinite	$|S|$	$\{S\}$	Equation (3)	False
k-Grass (U)	Algorithm 1	$\tilde{G}$	Infinite	$|S|$	$\{S\}$	Equation (3)	False
SSumM (W)	Algorithm 1	$\overline{G}$	Finite	Equation (4)	Clusters [6]	Equation (6)	True
SSumM (U)	Algorithm 1	$\tilde{G}$	Finite	Equation (5)	Clusters [6]	Equation (7)	True
MoSSo-Lossy (W)	Algorithm 2	$\overline{G}$	N/A	N/A	Clusters [6]	Equation (6)	N/A
MoSSo-Lossy (U)	Algorithm 2	$\tilde{G}$	N/A	N/A	Clusters [6]	Equation (7)	N/A

Algorithm 1: Batch computation of a summary graph

Input: (1) input graph G, (2) budget k, and (3) # iters: T
Output: summary graph G'
1 initialize G'; $t \leftarrow 1$
2 **while** $size(G') > k$ **and** $t < T$ **do**
3     $C \leftarrow groups()$; $t \leftarrow t + 1$
4   **for each** $C_i \in C$ **do**
5 | merge one or more pairs within C_i to minimize $loss()$
6 **if** $sparsify?()$ **then** sparsify G' until $size(G') \leq k$
7 **return** G'

Algorithm 2: Incremental update of a summary graph

Input: (1) summary graph G' and (2) change in {src, dst}
Output: updated G'
1 $C \leftarrow groups()$;
2 **for each** $u \in \{$src, dst$\}$ **do**
3     $\hat{N}_u \leftarrow$ sample neighbors of u
4   **for each** $w \in \hat{N}_u$ **do**
5 | $P \leftarrow C' \in C$ where $w \in C'$
6 | $v \leftarrow$ draw one in $\hat{N}_u \cap P$
7 | **if** $loss()$ drops **then** move w to S_v
8 **return** G'

---

where ω_{max} is the largest superedge weight in $\overline{G}$, and in our experiments in Sect. 4, $\omega_{max} << |S|$. The three terms on the right side in Eq. (4) correspond to $|P|$ superedges in bits, $|P|$ superedge weights in bits, and the supernode membership of $|V|$ subnodes in bits, respectively. Similarly, the size of an unweighted summary graph $\tilde{G}$ in bits is defined as

$$size_{bits}(\tilde{G}) := 2|P| \log_2 |S| + |V| \log_2 |S|. \tag{5}$$

3.2 Weighted Graph Summarization Algorithms

We introduce three searching algorithms for finding a weighted summary graph $\overline{G} = (S, P, \omega)$ of the input graph G. See Algorithms 1 and 2 for their outlines and Table 2 for details.

k-Grass. K-GRASS [18] first initializes the set S of supernodes so that each subnode forms a singleton supernode. Then, it repeats greedily merging a supernode pair until $|S|$ reaches the target number (i.e., the given constraint). Specifically, in each step, among all supernode pairs, K-GRASS merges a pair whose merger increases Eq. (3) least. During the entire process, K-GRASS creates a superedge between each supernode pair A and B (i.e., $\{A, B\} \in P$) if and only if $E_{AB} > 0$.

SSumM. SSUMM [17] initializes S as in K-GRASS. Then, SSUMM divides S into disjoint groups of supernodes with similar connectivity to find pairs to be merged efficiently. After that, in each group, SSUMM repeats merging a pair of supernodes whose merger decreases Eq. (6) most.

$$size_{bits}(\overline{G}) + \sum_{\{A,B\}\in P} \Pi_{AB} \cdot \mathcal{H}(\frac{E_{AB}}{\Pi_{AB}}) + \sum_{\{A,B\}\notin P} 2E_{AB} \log_2 |V|, \qquad (6)$$

where $\mathcal{H}(\cdot)$ is the entropy function. Equation (6) considers both the size of a summary graph and the reconstruction error. Specifically, the second term is the number of bits for exactly restoring the subedeges between supernodes that are joined by superedges, and the third term is that for the other subedges (see [17] for details). During the process, the superedge between each supernode pair exists only when it decreases Eq. (6). If $size_{bits}(\overline{G})$ (i.e., Eq. (4)) cannot satisfy the given constraint (i.e., the target size) within the given number of iterations, SSUMM sparsifies $\overline{G}$ greedily based on Eq. (3) to satisfy the constraint.

MoSSo-Lossy. MOSSO-LOSSY is a lossy variant of MOSSO [15], which is a lossless graph compression algorithm. While processing subedges incrementally, it updates $\overline{G} = (S, P, \omega)$. Specifically, for each subedge $\{u, v\}$, it samples a fixed number of neighbors of u and v. Then, for each such neighbor w, MOSSO-LOSSY moves w from S_w to the supernode which another sampled subnode with similar connectivity belongs to if this change decreases Eq. (6). As in SSUMM, for each pair of supernodes, a superedge joins them only when it decreases Eq. (6).

3.3 Unweighted Graph Summarization Algorithms

We extend the above algorithms for obtaining an unweighted summary graph $\tilde{G} = (S, P)$ of the input graph G. The differences are highlighted in Table 2 with the outlines in Algorithms 1 and 2.

k-Grass (Unweighted). This variant is different from K-GRASS in that Eq. (2) is used, instead of Eq. (1), when Eq. (3) is computed. Moreover, for each supernode pair, the superedge between them exists only when it decreases Eq. (3).

Table 3. Summary of the eight real-world graphs used in the paper. They are obtained from emails (EE), collaborations (DB), co-purchases (A6), computer networks (SK), online social networks (LJ), and hyperlinks (WS, DP, and WL).

Name	# Nodes	# Edges	Name	# Nodes	# Edges
Email-Enron (EE)	36,692	183,831	DBLP (DB)	317,080	1,049,866
Amazon-0601 (A6)	403,394	2,443,408	WebSmall (WS)	325,557	2,738,969
Skitter (SK)	1,696,415	11,095,298	LiveJournal (LJ)	3,997,962	34,681,189
DBPedia (DP)	18,268,991	126,890,209	WebLarge (WL)	18,483,186	261,787,258

SSumM (Unweighted). Instead of Eq. (6) used in SSumM, this variant uses Eq. (7), whose second term is the number of bits for exactly restoring the subedges between supernodes that are joined by unweighted superedges.

$$size_{bits}(\tilde{G}) + \sum_{\{A,B\} \in P} 2\left(\Pi_{AB} - E_{AB}\right) \log_2 |V| + \sum_{\{A,B\} \notin P} 2E_{AB} \log_2 |V|. \quad (7)$$

MoSSo-Lossy (Unweighted). This variant uses Eq. (7), instead of Eq. (6), which is used in MoSSo-Lossy.

4 Experiments

We review our experiments for comparing weighted and unweighted graph summarization in five aspects. We describe the settings and then present the results.

4.1 Experimental Settings

Machines: We performed our experiments on a desktop with a 3.80GHZ Intel i7-10700K CPU and 64 GB memory.

Datasets: We used the eight datasets summarized in Table 3.

Search Algorithms: We used the six algorithms described in Sect. 3. We implemented them commonly in OpenJDK 12 and set their target size to $\{0.1, 0.2, \cdots, 0.9\}$ of the size in the input graph. We fixed T to 20 in both versions of SSumM. We excluded [15,23,28] from the comparison since they assume extra components (e.g., edge corrections) in addition to a summary graph.

Evaluation Metric: Given the input graph $G = (V, E)$ and a summary graph G' (i.e., $\hat{G}$ or $\tilde{G}$), the compression ratio is defined in bits as $\frac{size_{bits}(G')}{2|E| \log_2 |V|}$.

4.2 Results

Reconstruction Error: The L_1 and L_2 reconstruction error (i.e., $p = \{1, 2\}$ in Eq. (3)) is compared in Fig. 2. Unweighted summary graphs described the input

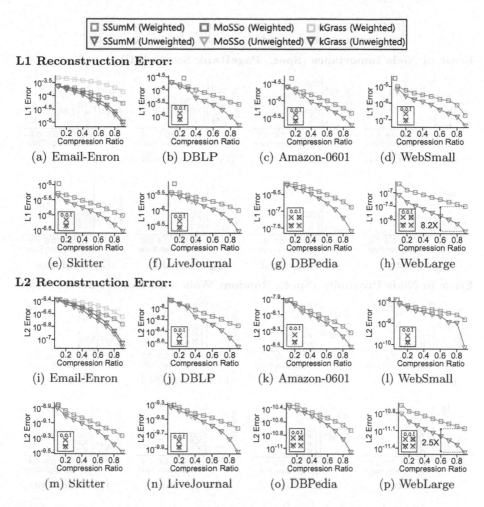

Fig. 2. The reconstruction error is significantly lower in unweighted graph summarization than in weighted summarization. **o.o.t.**: out of time (≥ 48 h).

graph more accurately (specifically, up to 8.2× when comparing SSUMM and its variant) than weighted ones, when compression ratios were the same.

Error in Node Importance: We used PageRank [24] (with the damping factor 0.85) to measure the importance of subnodes. In Fig. 3(a)–(h), we report the sum of absolute difference between PageRank scores obtained from input and summary graphs (see Appendix A for how to compute PageRank scores on a summary graph). When the compression ratios were the same, unweighted summary graphs maintained the node importance more accurately (specifically, up to 7.8× when comparing SSUMM and its variant) than weighted ones.

Error in Node Proximity: We used Random Walk with Restart (RWR) [31] (with the damping factor 0.95) to measure the proximity between subnodes. For

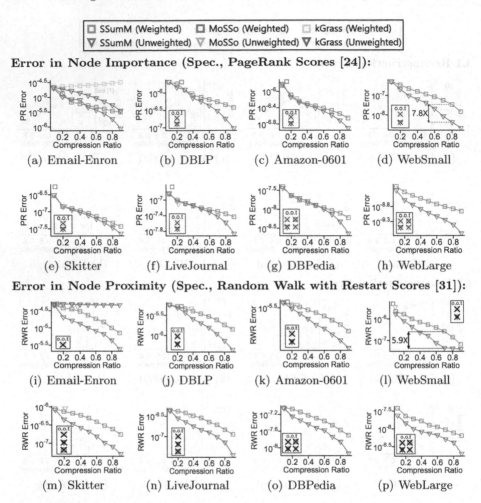

Fig. 3. Importance of nodes and proximity between nodes are preserved more accurately in unweighted graph summarization than in weighted summarization. **o.o.t.:** summarization or RWR computation ran out of time ($\geq$48 h).

each query node, we compute the RWR scores between the query node and the others on input and summary graphs, and we compute the sum of absolute difference (see Appendix A for how to compute the RWR scores on a summary graph). In Fig. 3(i)–(p), we report the difference averaged over 100 randomly-sampled query nodes. Unweighted summary graphs preserved the proximity between nodes more accurately (specifically, up to 5.9× when comparing SSuMM and its variant) than weighted ones, when the compression ratios were the same.

Size of Reconstructed Graphs: As shown in Fig. 4, when L_1 reconstruction errors were similar, graphs reconstructed from unweighted summary graphs had significantly fewer (specifically, up to 2.2× fewer when comparing SSuMM and

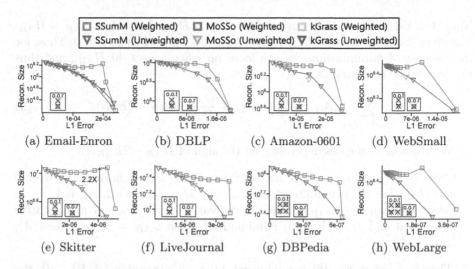

Fig. 4. When reconstruction errors are similar, more concise graphs are reconstructed from unweighted summary graphs than from weighted graphs. **o.o.t.**: out of time (≥48 h). **o.o.r.**: out of range with too many subedges.

its variant) subedges than those reconstructed from weighted ones. When reconstruction errors are similar, fewer reconstructed edges, which lead to faster query processing (see [25] and Appendix A for examples), are preferred.

5   Discussion: Why Can Edge Weights Be Harmful?

As answers to this question, we provide an example in Fig. 1, and we prove in Theorem 1 that at least when the L_1 reconstruction error is the objective, the superedge weight function ω is not useful and even harmful. The theorem, however, is not generalized to other objectives.

Theorem 1. *Consider a graph G and its weighted summary graph $\overline{G} = (S, P, \omega)$. Assume ω is not fixed but variable. When $RE_1(A, A')$ is minimized, for each superedge $\{A, B\} \in P$, the weight $\frac{\omega_{AB}}{\Pi_{AB}}$ of subedges reconstructed from it is either 1 or 0, just as in Eq. (2), where an unweighted summary graph is used.*

Proof. The L_1 reconstruction error can be written as follows:

$$RE_1(A, A') = \sum_{\{A,B\}\in P} \sum_{\{i,j\}\in \Pi_{AB}} |A_{ij} - A'_{ij}| + \sum_{\{A,B\}\notin P} \sum_{\{i,j\}\in \Pi_{AB}} |A_{ij}|,$$

Since the second term on the right side does not depend on ω, we focus on the first term where

$$\sum_{\{i,j\}\in \Pi_{AB}} |A_{ij} - A'_{ij}| = E_{AB}\left|1 - \frac{\omega_{AB}}{\Pi_{AB}}\right| + (\Pi_{AB} - E_{AB})\left|0 - \frac{\omega_{AB}}{\Pi_{AB}}\right|. \qquad (8)$$

Note that Eq. (8) is strictly larger when $\omega_{AB} > \Pi_{AB}$ than when $\omega_{AB} = \Pi_{AB}$. Moreover, Eq. (8) is strictly larger when $\omega_{AB} < 0$ than when $\omega_{AB} = 0$. Thus, for the purpose of minimization, we can focus on when $\omega_{AB} \in [0, \Pi_{AB}]$, and thus Eq. (8) can be rewritten as

$$\sum_{\{i,j\}\in \Pi_{AB}} |A_{ij} - A'_{ij}| = E_{AB} + \frac{\omega_{AB}}{\Pi_{AB}}(\Pi_{AB} - 2E_{AB}). \qquad (9)$$

We consider two cases depending on the sign of $\Pi_{AB} - 2E_{AB}$.

- **Case 1.** $\Pi_{AB} < 2E_{AB}$: Since the derivative w.r.t. ω_{AB} is negative between 0 and Π_{AB}, Eq. (9) is minimized when $\omega_{AB} = \Pi_{AB}$, i.e., when A'_{ij} is 1.
- **Case 2.** $\Pi_{AB} \geq 2E_{AB}$: Since the derivative w.r.t. ω_{AB} is non-negative between 0 and Π_{AB}, Eq. (9) is minimized when $\omega_{AB} = 0$, i.e., when A'_{ij} is 0,

Therefore, when Eq. (8) is minimized, for each superedge $\{A, B\} \in P$, the weight of subedges reconstructed from it (i.e., $\frac{\omega_{AB}}{\Pi_{AB}}$) is either 1 or 0, as in the unweighted model.

6   Conclusion and Future Directions

In this work, we conducted a systematic comparison between two extensively-studied graph summarization models with and without superedge weights. To this end, we extended three search algorithms to both models (Algorithms 1–2 and Table 2) and compared their outputs from eight real-world graphs in five aspects (Figs. 2, 3 and 4). Our empirical comparison revealed a surprising finding that removing superedge weights leads to significant improvements in all five aspects, as in the example in Fig. 1. Then, we developed a theoretical analysis to shed light on this counterintuitive observation (Theorem 1). Noteworthy, we showed in Figs. 2, 3 and 4 that SSuMM [17], a state-of-the-art graph-summarization algorithm, can be improved substantially (specifically, up to 8.2×, 7.8×, and 5.9× in terms of reconstruction error, error in node importance, and error in node proximity, respectively, when the compression ratio was fixed; and 2.2× in terms of the size of reconstructed graphs, when the reconstruction error was similar) based on the observation. As future work, we would like to explore (a) better superedge weighting schemes and (b) combinations of weighted and unweighted superedges.

Reproducibility: The source code and the datasets are available at [1].

Acknowledgements. This work was supported by National Research Foundation of Korea (NRF) grand funded by the Korea government (MSIT) (No. NRF-2020R1C1C1008296) and Institute of Information & Communications Technology Planning & Evaluation (IITP) grant funded by the Korea government (MSIT) (No.2019-0-00075, Artificial Intelligence Graduate School Program (KAIST)).

A   Appendix: Graph Algorithms on Summary Graphs

Given a summary graph G' (i.e., $\overline{G}$ or $\tilde{G}$) and a query node $u \in V$, an approximate set of neighbors of u can be retrieved from G' without reconstructing the entire graph, as described in Algorithm 3. In other words, neighborhood queries can be answered approximately from G'. A wide range of graph algorithms (e.g., DFS, BFS, PageRank, and Dijkstra's) access the input graph only through neighborhood queries, and thus they can be executed approximately on summary graphs without restoring the entire graph. See Algorithm 4 for examples.

Algorithm 3: getNeighbors(G',u)

Input: (1) summary graph G' ($\overline{G}$ or $\tilde{G}$) and (2) query subnode u
Output: approximate neighborhood $\hat{N}_u$ of u with subedge weights

1 $\hat{N}_u \leftarrow \emptyset$
2 **for each** A where $\{A, S_u\} \in P$ **do**
3   **for each** $v \in A$ **do**
4   **if** $v \neq u$ **then**
5   **if** $G' = \overline{G}$ **then**
6   add v to $\hat{N}_u$ with weight $\frac{\omega_{AS_u}}{\Pi_{AS_u}}$
7   **if** $G' = \tilde{G}$ **then**
8   add v to $\hat{N}_u$ with weight 1
9 **return** $\hat{N}_u$

Algorithm 4: PageRank [24] and Random Walk with Restart (RWR) [31] on G'

Input: (1) summary graph G', (2) damping factor d, and
$\qquad\quad$ (3) (only for RWR) query subnode u
Output: score vector $r^{new} \in \mathbb{R}^{|V|}$

1 $V \leftarrow \bigcup_{A \in S} A$
2 $r^{old} \leftarrow \mathbf{0}$; $r^{new} \leftarrow \frac{1}{|V|} \cdot \mathbf{1}$ $\qquad\qquad$ ▷ **0 is the zero vector of size** $|V|$
3 $q \leftarrow \frac{1}{|V|} \cdot \mathbf{1}$ $\qquad\qquad\qquad\qquad$ ▷ **1 is the one vector of size** $|V|$
4 (only for RWR) $q \leftarrow \mathbf{0}$; $q_u \leftarrow 1$
5 **while** $r^{new} \neq r^{old}$ **do**
6      $r^{old} \leftarrow r^{new}$; $r^{new} \leftarrow \mathbf{0}$
7   **for each** $v \in V$ **do**
8          $\hat{N}_v \leftarrow$ **getNeighbors**(G', v)
9          $w_{sum} \leftarrow$ sum of weights in $\hat{N}_v$
10   **for each** neighbor l with weight w in $\hat{N}_v$ **do**
11              $r_l^{new} \leftarrow r_l^{new} + \frac{w}{w_{sum}} r_v^{old}$
12      $r^{new} \leftarrow d \cdot r^{new} + (1 - d \cdot \sum_{v \in V} r_v^{new}) \cdot \mathbf{q}$
13 **return** r^{new}

References

1. Online appendix, source code and datasets (2022). https://github.com/ShinhwanKang/PAKDD22-ComparativeStudy

2. Beg, M.A., Ahmad, M., Zaman, A., Khan, I.: Scalable approximation algorithm for graph summarization. In: Phung, D., Tseng, V.S., Webb, G.I., Ho, B., Ganji, M., Rashidi, L. (eds.) PAKDD 2018. LNCS (LNAI), vol. 10939, pp. 502–514. Springer, Cham (2018). https://doi.org/10.1007/978-3-319-93040-4_40

3. Belth, C., Zheng, X., Vreeken, J., Koutra, D.: What is normal, what is strange, and what is missing in a knowledge graph: unified characterization via inductive summarization. In: WWW (2020)

4. Besta, M., Hoefler, T.: Survey and taxonomy of lossless graph compression and space-efficient graph representations. arXiv preprint arXiv:1806.01799 (2018)

5. Boldi, P., Vigna, S.: The webgraph framework I: compression techniques. In: WWW (2004)

6. Buehrer, G., Chellapilla, K.: A scalable pattern mining approach to web graph compression with communities. In: WSDM (2008)

7. Chierichetti, F., Kumar, R., Lattanzi, S., Mitzenmacher, M., Panconesi, A., Raghavan, P.: On compressing social networks. In: KDD (2009)

8. Davis, M., Liu, W., Miller, P., Redpath, G.: Detecting anomalies in graphs with numeric labels. In: CIKM (2011)

9. Dhulipala, L., Kabiljo, I., Karrer, B., Ottaviano, G., Pupyrev, S., Shalita, A.: Compressing graphs and indexes with recursive graph bisection. In: KDD (2016)

10. Dunne, C., Shneiderman, B.: Motif simplification: improving network visualization readability with fan, connector, and clique glyphs. In: CHI (2013)

11. Dwyer, T., Riche, N.H., Marriott, K., Mears, C.: Edge compression techniques for visualization of dense directed graphs. TVCG 19(12), 2596–2605 (2013)

12. Fahrbach, M., Goranci, G., Peng, R., Sachdeva, S., Wang, C.: Faster graph embeddings via coarsening. In: ICML (2020)

13. Kang, S., Lee, K., Shin, K.: Personalized graph summarization: formulation, scalable algorithms, and applications. In: ICDE (2022)

14. Khan, K.U., Nawaz, W., Lee, Y.-K.: Set-based approximate approach for lossless graph summarization. Computing 97(12), 1185–1207 (2015). https://doi.org/10.1007/s00607-015-0454-9

15. Ko, J., Kook, Y., Shin, K.: Incremental lossless graph summarization. In: KDD (2020)

16. Koutra, D., Kang, U., Vreeken, J., Faloutsos, C.: VOG: summarizing and understanding large graphs. In: SDM (2014)

17. Lee, K., Jo, H., Ko, J., Lim, S., Shin, K.: SSumM: sparse summarization of massive graphs. In: KDD (2020)

18. LeFevre, K., Terzi, E.: Grass: graph structure summarization. In: SDM (2010)

19. Lim, Y., Kang, U., Faloutsos, C.: SlashBurn: graph compression and mining beyond caveman communities. TKDE 26(12), 3077–3089 (2014)

20. Liu, Y., Safavi, T., Dighe, A., Koutra, D.: Graph summarization methods and applications: a survey. CSUR 51(3), 1–34 (2018)

21. Mathioudakis, M., Bonchi, F., Castillo, C., Gionis, A., Ukkonen, A.: Sparsification of influence networks. In: KDD (2011)

22. Mehmood, Y., Barbieri, N., Bonchi, F., Ukkonen, A.: CSI: community-level social influence analysis. In: Blockeel, H., Kersting, K., Nijssen, S., Železný, F. (eds.) ECML PKDD 2013. LNCS (LNAI), vol. 8189, pp. 48–63. Springer, Heidelberg (2013). https://doi.org/10.1007/978-3-642-40991-2_4

23. Navlakha, S., Rastogi, R., Shrivastava, N.: Graph summarization with bounded error. In: SIGMOD (2008)

24. Page, L., Brin, S., Motwani, R., Winograd, T.: The pagerank citation ranking: bringing order to the web. Technical report, Stanford InfoLab (1999)

25. Riondato, M., García-Soriano, D., Bonchi, F.: Graph summarization with quality guarantees. DMKD **31**(2), 314–349 (2017)
26. Rossi, R.A., Zhou, R.: GraphZIP: a clique-based sparse graph compression method. J. Big Data **5**(1), 1–14 (2018)
27. Shah, N., Koutra, D., Zou, T., Gallagher, B., Faloutsos, C.: TimeCrunch: interpretable dynamic graph summarization. In: KDD (2015)
28. Shin, K., Ghoting, A., Kim, M., Raghavan, H.: SWeG: lossless and lossy summarization of web-scale graphs. In: WWW (2019)
29. Shoaran, M., Thomo, A., Weber-Jahnke, J.H.: Zero-knowledge private graph summarization. In: Big Data (2013)
30. Sui, P., Yang, X.: A privacy-preserving compression storage method for large trajectory data in road network. J. Grid Comput. **16**(2), 229–245 (2018)
31. Tong, H., Faloutsos, C., Pan, J.Y.: Random walk with restart: fast solutions and applications. KAIS **14**(3), 327–346 (2008)
32. Zhou, H., Liu, S., Lee, K., Shin, K., Shen, H., Cheng, X.: DPGS: degree-preserving graph summarization. In: SDM (2021)

Mu2ReST: Multi-resolution Recursive Spatio-Temporal Transformer for Long-Term Prediction

Hao Niu[1(✉)], Chuizheng Meng[2], Defu Cao[2], Guillaume Habault[1], Roberto Legaspi[1], Shinya Wada[1], Chihiro Ono[1], and Yan Liu[2]

[1] KDDI Research, Inc., Fujimino, Saitama 356-8502, Japan
{ha-niu,gu-habault,ro-legaspi,sh-wada,ono}@kddi-research.jp
[2] University of Southern California, Los Angeles, CA 90089, USA
{chuizhem,defucao,yanliu.cs}@usc.edu

Abstract. Long-term spatio-temporal prediction (LTSTP) over different resolutions plays a crucial role in planning and dispatching smart city applications, such as smart transportation and smart grid. The Transformer, which has demonstrated superiority in capturing long-term dependencies, was recently studied for spatio-temporal prediction. However, it is difficult to leverage it using both multi-resolution knowledge and spatio-temporal dependencies to aid LTSTP. The challenge typically lies in addressing two issues: (1) efficiently fusing information across multiple resolutions that demands elaborate and complicated modifications to the model, and (2) handling the necessary long-term sequence that makes concurrent space and time attentions too costly to be performed. To address these issues, we proposed a multi-resolution recursive spatio-temporal transformer (Mu2ReST). It implements a novel multi-resolution structure with recursive prediction from coarser to finer resolutions. This proposal reveals that an arduous modification of the model is not the only way to leverage multi-resolution knowledge. It further uses a redesigned lightweight space-time attention implementation to concurrently capture spatial and temporal dependencies. Experiment results using open and commercial urban datasets demonstrate that Mu2ReST outperforms existing methods for multi-resolution LTSTP tasks.

Keywords: Long-term spatio-temporal prediction · Multi-resolution · Recursive prediction · Spatio-temporal transformer

1 Introduction

Many applications that have spatio-temporal data, such as smart transportation and smart grid, demand long-term (e.g., several days ahead or longer) spatio-temporal prediction. Besides, prediction requires to be done on multiple resolutions for effective planning and dispatching purposes. For example, in

---

H. Niu and C. Meng—Equal Contribution.

J. Gama et al. (Eds.): PAKDD 2022, LNAI 13280, pp. 68–80, 2022.
https://doi.org/10.1007/978-3-031-05933-9_6

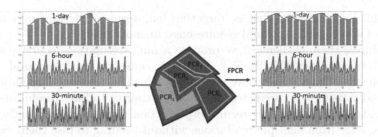

Fig. 1. Example of average household electricity consumption aggregated from different spatial and temporal resolutions (MinMaxScaler is performed on the data totalling 15 days from a Friday). The left and right curves show the average consumption in different temporal resolutions {1-day, 6-h and 30-min} for a postcode region (PCR) and a larger area composed of four postcode regions (FPCR), respectively. The curves show the 30-min resolution has several fluctuations and confounds whatever patterns (e.g., weekday vs. weekend, morning vs. noon vs. evening vs. night, weekly and daily periodicity) are hidden. In contrast, the 6-h/1-day coarser resolution has less/the least fluctuations and with more perceivable patterns emerging.

liberal electricity markets, some retailers require long-term predictions in multiple spatial and temporal resolutions to procure electricity from different markets. Spatial resolutions could be meshes, postal codes, wards and prefectures, while temporal resolutions could be minutes, hours, days, or weeks. Low performance in long-term spatio-temporal prediction (LTSTP) can result to a large gap between actual demand and procurement, which in the end leads to a more costly balance. Improving LTSTP performance is therefore of crucial importance to electricity retailers.

Figure 1 illustrates some spatio-temporal data in multiple resolutions, which generally exhibit different fluctuations and patterns. We posit, for instance, that transferring learned knowledge from coarser to finer resolutions will benefit the prediction task while reducing the impact of data perturbations on the model. Recent advances in spatio-temporal prediction mainly rely on recurrent neural network (RNN), convolutional neural network (CNN) and graph neural network (GNN) based models. These models can be modified with convolution and pooling operations for them to utilize multi-resolution knowledge [15, 17].

Very recent studies have demonstrated the superiority of Transformer-based models in capturing long-term dependencies. These notably include the Informer model [18] that successfully outperforms SOTA on both long-term time series prediction (LTTP) and LTSTP, thanks to its efficient time attention implementation. However, just like in any other Transformer-based models, leveraging multi-resolution knowledge needs elaborate and even complicated modifications in order to effectively utilize information across different resolutions by fusing them. There have been some recent attempts in computer vision tasks [1,6], but to our knowledge, a solution has yet to be proposed for LTSTP. In addition, employing the Transformer to implement concurrent space and time attentions for LTSTP is intractable, that is, targeting long-term predictions over a large number of spatial regions would require huge memory usage.

To address these challenges, we propose a multi-resolution recursive spatio-temporal transformer (Mu2ReST). We first came up with a novel multi-resolution structure with recursive prediction, which handles each resolution one by one and transfers the predicted results from coarser to finer resolutions. Our experiments confirm that this recursive prediction component efficiently leverages knowledge from multiple resolutions without needing to fuse them. Second, compared to existing works, and more importantly, to prevent our model from being costly for LTSTP, we simply incorporated both space and time attentions.

Thus, the contributions of this paper are (1) a multi-resolution structure with a recursive prediction strategy that provides an uncomplicated and effective way of exploiting multi-resolution knowledge, (2) an end-to-end Transformer-based model with a simplified space-time attention implementation that captures spatial and temporal dependencies concurrently, and (3) experiments on both public and proprietary spatio-temporal urban datasets confirming that our model outperforms previous SOTA on average by 8.61% on RMSE and by 9.32% on MAE.

2   Related Works

RNN-/CNN-/GNN-Based Models
Models based on RNN, and its variants such as long short-term memory (LSTM) and gated recurrent unit (GRU), have achieved significant performance in different tasks to capture long-term temporal dependencies [4,12]. For LTSTP problems, combining RNN with CNN/GNN is usually adopted to capture spatio-temporal dependencies simultaneously. For example, LSTNet [8] utilizes both RNN and CNN to (i) discover long-term patterns of time series, and (ii) extract short-term local dependency patterns among variables (i.e., spatial dependencies). GraphWaveNet [15] uses stacked dilated 1D convolution component to seize both short- and long-term temporal dependencies at different granular levels. It also captures spatial dependencies using GNN with pre-defined and self-learned graphs. GMAN [17] adapts an encoder-decoder architecture with multiple spatio-temporal attention blocks to simulate the impact of spatio-temporal factors. In order to capture spatial dependencies at both node and group levels, it particularly applies attention to different node groups using max-pooling.

Transformer-Based Models
Based on a self-attention mechanism [13], Transformer models have recently been used with great success in natural language processing and computer vision [11,14]. However, applying Transformer for LTSTP requires huge memory, which deters reproducing such a success. Some effective methods have then been proposed to lower memory requirement and improve computation speed, such as Sparse Transformer [2], LogSparse Transformer [10], Reformer [7] and Informer [18]. In particular, Reformer uses locality sensitive hashing attention, while Informer introduces ProbSparse self-attention. However, these models do not consider multiple resolutions and generally focus the attention on only one

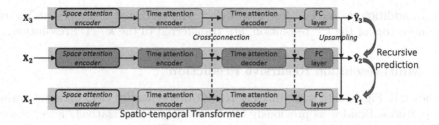

Fig. 2. Architecture of Mu2ReST.

dimension (spatial or temporal). In contrast, spatial-temporal transformer network (STTN) [16] and spacetimeformer (STF) [5] proposed Transformer-based models that have both space and time attentions for spatio-temporal data. However, even though both models can deal with LTSTP, they do not carefully consider the complexity that arises when a large number of variables (spatial regions/nodes) are involved.

3   The Proposed Method

We consider multi-resolution spatio-temporal data as shown in Fig. 1. Assuming there are K temporal resolutions and N regions for all spatial resolutions, $u_{n,k}^t$ is the value of region $n \in \{1, 2, \cdots, N\}$, at temporal resolution $k \in \{1, 2, \cdots, K\}$ for a given time step t. At that time step, we represent the spatio-temporal data with temporal resolution k as $\mathbf{U}_k^t = \left\{ u_{1,k}^t, u_{2,k}^t, \cdots, u_{N,k}^t \right\}$. The N regions can be represented as nodes of a heterogeneous graph. For a specific period $\mathbf{T}$ from time step t, the corresponding multi-resolution spatio-temporal data can be represented as $\mathbb{M}^{\mathbf{T}} = \left\{ \mathbf{M}_1^{\mathbf{T}}, \mathbf{M}_2^{\mathbf{T}}, \ldots, \mathbf{M}_K^{\mathbf{T}} \right\}$, where $\mathbf{M}_k^{\mathbf{T}} = \left\{ \mathbf{U}_k^t, \mathbf{U}_k^{t+1}, \cdots, \mathbf{U}_k^{t+I_k-1} \right\}$ and I_k indicates the time steps of resolution k for period $\mathbf{T}$. For instance, if the target period is 30 days, I_k is equal to 30 for 1-day resolution and 1440 for 30-min resolution. In other words, each $\mathbf{M}_k^{\mathbf{T}}$ has different lengths as they represent different temporal resolutions. The goal for multi-resolution LTSTP is to predict $\mathbf{Y} = \mathbb{M}^{\mathbf{T}^+}$ based on input $\mathbf{X} = \mathbb{M}^{\mathbf{T}^-}$, where $\mathbf{T}^+$ and $\mathbf{T}^-$ are the long-term future and historical periods, respectively, at time step t_0. In addition, we denote $\mathbf{Y}_k = \mathbf{M}_k^{\mathbf{T}^+}$ and $\mathbf{X}_k = \mathbf{M}_k^{\mathbf{T}^-}$ as target and input of the temporal resolution k.

Figure 2 shows the overall architecture of Mu2ReST, which includes two components: (1) a "vertical" multi-resolution structure exploiting multi-resolution knowledge along the time dimension, and (2) a "horizontal" Transformer-based model implementing a simplified space-time attention. Figure 2 illustrates multi-resolution with three temporal resolutions (i.e., $K = 3$). In this example, the higher the k value, the coarser the resolution. Considering the case in Fig. 1, $k = 1, 2$ and 3 correspond to 30-min, 6-h and 1-day resolutions, respectively. We use a dedicated spatio-temporal Transformer for each resolution, and perform the prediction one by one from the coarsest, $k = 3$, to the finest, $k = 1$, resolution. In addition, as further described below, the predicted result $\hat{\mathbf{Y}}_k$ of $\mathbf{Y}_k$, or hidden vectors of the kth resolution, are transferred to the $k - 1^{\text{th}}$ resolution.

3.1   Multi-resolution Recursive Prediction

As shown in Fig. 1, data with different resolutions generally yield different variation patterns. Besides, as previously mentioned, there are relatively fewer noises and fluctuations in the coarser resolution data. With our recursive prediction scheme, the predicted results are passed from coarser to finer resolutions to improve LTSTP. We shall use the instance of Mu2ReST with three temporal resolutions in Fig. 2 to illustrate this idea. We first train the Transformer model for the coarsest resolution based on a loss (e.g., MSE loss) function $\mathcal{L}$:

$$\mathcal{L}_3 = \mathcal{L}\left(\hat{\mathbf{Y}}_3, \mathbf{Y}_3\right). \tag{1}$$

We pass $\mathbf{X}_3$ to the trained model to get the corresponding prediction $\hat{\mathbf{Y}}_3$. From the example in Fig. 1, the third resolution reflects the learned patterns from the 1-day resolution data, such as weekday vs. weekend and weekly periodicity.

As resolutions are linked to each other, by aggregating (e.g., summing or averaging) data with finer resolution k, we can retrieve data with coarser resolution $k + i$ $(i > 0)$. For instance, we can obtain the 1-day resolution data by aggregating every four elements of the 6-h resolution. We apply this constraint as a regularization term that enables to train the model for finer resolutions:

$$\mathcal{L}_2 = \mathcal{L}\left(\hat{\mathbf{Y}}_2, \mathbf{Y}_2\right) + \mathcal{L}\left(\hat{\mathbf{Y}}_{2\to3}, \hat{\mathbf{Y}}_3\right) \text{ and } \mathcal{L}_1 = \mathcal{L}\left(\hat{\mathbf{Y}}_1, \mathbf{Y}_1\right) + \mathcal{L}\left(\hat{\mathbf{Y}}_{1\to2}, \hat{\mathbf{Y}}_2\right), \tag{2}$$

where $\hat{\mathbf{Y}}_{k\to k+1}$ denotes the aggregation of $\hat{\mathbf{Y}}_k$ to $k+1$ resolution. By doing so, we can exploit the learnt patterns from coarser resolutions to help the prediction of finer resolutions. Thus, we can leverage the multi-resolution information without needing to modify the model. The algorithm of the multi-resolution recursive prediction is summarized in Algorithm 1.

In the same way that we downsample (i.e., aggregate) the temporal resolution predictions, we can optionally upsample $\hat{\mathbf{Y}}_{k+1}$ to aid the prediction of $\hat{\mathbf{Y}}_k$. Moreover, we can optionally transfer encoder and/or decoder outputs from coarser to finer resolutions as shown by the dashed lines (cross connections) in Fig. 2.

3.2   Spatio-Temporal Transformer

For LTSTP, it is essential to reduce the memory demand as much as possible to deal with long sequences. Although spatio-temporal transformer has been proposed for spatio-temporal data [5,16], the models are not suitable for LTSTP especially for a large number of variables. STTN [16] alternatively performs space and time attentions, while STF [5] jointly performs them with Performer

---

Algorithm 1: Multi-resolution recursive prediction

Input: Multi-resolution data $(\mathbf{X}_k, \mathbf{Y}_k)$, $k \in \{1, 2, \cdots, K\}$

Ouput: Trained Transformer model f_k for each k

*Train:

1. Initialize parameters

2. Train f_K (coarsest resolution) with $\mathcal{L}_K = \mathcal{L}\left(\hat{\mathbf{Y}}_K, \mathbf{Y}_K\right)$, $k = K$;

3. **while** $k > 1$ **do**

> Perform $\hat{\mathbf{Y}}_k = f_k(\mathbf{X}_k)$, then $k = k - 1$;
>
> Train f_k with $\mathcal{L}_k = \mathcal{L}\left(\hat{\mathbf{Y}}_k, \mathbf{Y}_k\right) + \mathcal{L}\left(\hat{\mathbf{Y}}_{k \to k+1}, \hat{\mathbf{Y}}_{k+1}\right)$

end

*Test/Application: Perform $\hat{\mathbf{Y}}_k = f_k(\mathbf{X}_k)$ with test/real-time $\mathbf{X}_k$ for each k

---

FAVOR+ attention [3]. Consequently, their minimum memory usage requirements are $\mathcal{O}\left(I_k^2 + N^2\right)$ and $\mathcal{O}\left(I_k N\right)$, respectively. To overcome this issue, we designed a lightweight spatio-temporal Transformer for LTSTP (cf. Fig. 2). It applies the space attention only at the head of the encoder part, and incorporates the Informer as base model to acquire both high performance and efficiency.

(1) Space self-attention. Since we use a dedicated spatio-temporal Transformer for each resolution in our model, for simplicity, we omit the subscript k in the following equations. For each input data $\mathbf{U}^t$, we map the value of each region u_n^t onto a D dimension hidden vector $\mathbf{z}_n^t$ using a linear layer L_u: $\mathbf{z}_n^t = L_u(u_n^t)$.

Space self-attention is based on [13]. The corresponding query, key and value vectors are generated using $\mathbf{Z}^t = [\mathbf{z}_1^t, \mathbf{z}_2^t, \cdots, \mathbf{z}_N^t]$ with $\mathbf{Z}^t \in \mathbb{R}^{N \times D}$, as follows:

$$\mathbf{Q}^{t,h} = \mathbf{Z}^t W_{Q_s}^h \in \mathbb{R}^{N \times d}, \mathbf{K}^{t,h} = \mathbf{Z}^t W_{K_s}^h \in \mathbb{R}^{N \times d}, \mathbf{V}^{t,h} = \mathbf{Z}^t W_{V_s}^h \in \mathbb{R}^{N \times d}, \quad (3)$$

where $h \subset \{1, 2, \cdots, H\}$ indicates the index over H attention heads, $d = D/H$. $W_{Q_s}^h \in \mathbb{R}^{D \times d}$, $W_{K_s}^h \in \mathbb{R}^{D \times d}$ and $W_{V_s}^h \in \mathbb{R}^{D \times d}$ are all learnable weight matrices for the space attention. Self-attention in space dimension is then performed as

$$\mathbf{S}^t = \text{Concat}(\mathbf{S}^{t,1}, \mathbf{S}^{t,2}, \cdots, \mathbf{S}^{t,H}) W_{Os} \in \mathbb{R}^{N \times D}, \quad (4)$$

where $\mathbf{S}^{t,h} = \text{Softmax}(\frac{\mathbf{Q}^{t,h}\mathbf{K}^{t,h}}{\sqrt{d}})\mathbf{V}^{t,h}$, and $W_{Os} \in \mathbb{R}^{D \times D}$ is another learnable weight matrix. To further reduce the memory usage when dealing with long sequences, we also apply the ProbSparse self-attention (originally designed for the time dimension in the Informer) to our space attention. Specifically, the ProbSparse self-attention uses a sparse matrix $\overline{\mathbf{Q}}^{t,h}$ instead of $\mathbf{Q}^{t,h}$ to calculate each $\mathbf{S}^{t,h}$. It basically makes it capable to accommodate long sequences (see details in [18]). Note that, for simplicity, we currently do not consider any pre-defined graph structure, and position embedding is not used in the space attention block.

(2) Time self-attention and output module. We select the Informer as our base model to realize these mechanisms. It has proved to handle efficiently long sequence inputs and achieves SOTA performance on both LTTP and LTSTP [18]. The time attention operates like the space attention described above, but performed over the temporal, instead of the spatial, dimension. In addition, like other Transformer-based models, it applies position and time embeddings to help the model incorporate the order and temporal information.

Besides the ProbSparse self-attention, the Informer also includes self-attention distilling and generative-style decoding to further improve its performance. The details of the latter two are summarized as follows. The self-attention distilling consists of applying a convolution operation on the output of a time attention block before inputting it to the next time attention block. Therefore, the length of output of the attention blocks will decrease gradually moving further within the cascade of time attention blocks, which also reduces memory usage. As for the generative-style decoder, a standard Transformer decoder structure plus a fully connected layer is used to realize a long sequence output with only one forward step in order to avoid the time-consuming step-by-step inference.

Lastly, our implementation can achieve a minimum memory usage of $\mathcal{O}\left(I_k \log\right.$ $\left(I_k\right) + N \log\left(N\right))$. This is more efficient than those of STTN ($\mathcal{O}\left(I_k^2 + N^2\right)$) and STF ($\mathcal{O}\left(I_k N\right)$) for long sequence inputs (i.e., large I_k) with a substantial number of spatial regions/nodes (i.e., large N).

4   Experiments

We test Mu2ReST on two different urban datasets: an open dataset, the New York Yellow Taxi Trip Records (NYT)[1] and a commercial one, the Household Electricity Consumption Records (HEC). Their details are presented in Table 1.

Table 1. Dataset description (SR: Spatial Resolution; TR: Temporal Resolution)

	NYT	HEC
Period	2017.01-2019.12	2019.05-2020.07
Coarse SR/regions	11 boroughs	5 FPCRs
Fine SR/regions	67 taxi zones	20 PCRs
TRs	[1-day, 6-h, 30-min]	
Input/Output period	30/10 days	

For NYT, we aggregated the pick-up and drop-off numbers respectively over both taxi zones and boroughs of Manhattan[2], and considered [1-day, 6-h, 30-min] temporal resolutions. For HEC, we use the average household electricity consumption over postcode regions (PCRs) and four postcode regions (FPCRs), and consider the same temporal resolutions. Our task is to perform 10-day prediction (of both pick-up and drop-off for NYT, and household electricity consumption for HEC) using 30-day historical data for each resolution. We generate samples using a sliding window with 1-day steps (the coarsest temporal resolution). In addition, we construct the train, validation, and test datasets with 60%, 20%, and 20%, respectively, of the samples in time order.

---

[1] https://www1.nyc.gov/site/tlc/about/tlc-trip-record-data.page.
[2] The taxi zones and boroughs are according to https://data.cityofnewyork.
  us/Transportation/NYC-Taxi-Zones/d3c5-ddgc and https://www1.nyc.gov/assets/
  doh/downloads/pdf/survey/uhf_map_100604.pdf respectively.

We conducted the experiments on a Linux server with a TITAN V 12GB GPU. In order to handle long-term data, the batch size is set to 2 for all methods. Since we selected Informer as our base model, we kept the default structure of Informer in Mu2ReST and added a space attention block at the head. We kept most of the default parameters of Informer in Mu2ReST. The main modifications are summarized in Table 2. For a fair comparison, we used the same modified parameters on the original Informer, which has no space attention block.

Table 2. Main parameters of Mu2ReST.

Space attention block	1
Time attention block of encoder	2
Time attention block of decoder	1
Batch size	2
Epochs	100
Patience	15
Minimum learning rate	$1e^{-4}$
Dropout	0.05
Loss function	MSELoss

4.1 Results and Analysis

We compared Mu2ReST to eight baseline methods, namely, **GRU** [4], **LSTnet** [8], **GraphWaveNet** [15], **GMAN** [17], **Reformer** [7], **Informer** [18], **STTN** [16], and **STF** [5] (cf. details in Sect. 2). For GraphWaveNet and GMAN, their pre-defined spatial graph are constructed based on traffic connectivity and space proximity for NYT, and on space proximity for HEC. Table 3 shows the LTSTP performance over the different resolutions, in which the upper/lower part shows results for NYT/HEC. Performance is measured in root mean squared error (RMSE) and mean absolute error (MAE). The presence of '-' indicates failure to obtain results due to out-of-memory error, albeit smaller hidden vectors and less hidden layers are used. Further, Mu2ReST in Table 3 performs recursive prediction based only on Eq. 2 without the optional upsampling and cross connections.

We observe that Mu2ReST achieves the best LTSTP performance on both datasets, i.e., outperforms with 10/12 on NYT and 7/12 on HEC, totaling 17/24. While the best baselines (GRU and STF), each succeeds only 2/24. Notably, Mu2ReST outperforms all other baselines for the finest resolution, i.e., the 30-min resolution in fine regions, which is more important in many practical applications. On the other hand, baselines mainly perform best in coarse temporal resolutions. As the data of coarse resolutions have less noise and less drastic fluctuations, it is probably easier to discover some patterns even with the more conventional solutions like GRU. Data with fine resolutions are more complicated and, consequently, more difficult to estimate even with powerful models.

A drop in performance on finer temporal resolutions can be observed in all methods. However, Mu2ReST benefits from the knowledge of coarser resolutions via its recursive implementation. By building up from previously learned coarser patterns, our model achieves better and more robust results compared to all baselines. For instance, a weekly pattern could be unveiled (e.g., weekday vs. weekend) in the 1-day resolution. Knowing this pattern, Mu2ReST targeted daily patterns (e.g., morning vs. noon vs. evening vs. night) in the 6-h resolution. Finally, it paid attention on finding finest patterns in the 30-min resolution.

Table 3. Long-term spatio-temporal prediction (LTSTP) performance for different resolutions with different methods. We repeat each experiment three times and report the mean and standard deviation (results on HEC are magnified 100× for easy reading). Comparison is based on mean values, and then, standard deviations. Best result in each column is emphasized in bold.

Resolution	1 day		6 h		30 min	
Metric	RMSE	MAE	RMSE	MAE	RMSE	MAE
NYT, coarse regions (boroughs)						
GRU	112.79(9.59)	72.94(6.62)	122.20(9.92)	71.34(8.16)	135.71(29.71)	78.18(17.95)
LSTnet	82.21(4.74)	45.87(3.72)	101.12(7.38)	53.84(5.52)	110.04(5.03)	61.35(2.80)
GraphWaveNet	88.36(0.94)	51.15(1.49)	127.48(0.27)	74.53(0.51)	150.66(1.12)	86.43(0.59)
GMAN	83.26(11.80)	46.12(9.62)	239.78(0.12)	145.59(0.71)	-	-
Reformer	112.63(0.84)	76.62(1.89)	119.98(19.79)	71.38(12.55)	132.74(11.16)	78.17(8.20)
Informer	83.93(11.22)	49.57(9.21)	88.18(5.41)	49.47(3.25)	116.81(9.75)	66.99(6.36)
STTN	101.88(0.93)	61.68(2.97)	113.73(9.05)	63.24(4.78)	-	-
STF	**71.06(3.78)**	**44.70(3.64)**	87.34(1.31)	49.85(0.68)	175.54(7.88)	102.38(6.76)
Mu2ReST	79.50(2.36)	46.47(0.97)	**85.49(3.08)**	**46.82(2.33)**	**103.03(6.06)**	**57.32(3.37)**
NYT, fine regions (taxi zones)						
GRU	20.05(1.48)	13.92(1.17)	23.20(1.38)	15.05(0.95)	26.46(3.05)	16.50(1.80)
LSTnet	15.79(1.00)	9.67(0.80)	20.23(1.20)	12.14(1.19)	24.76(0.63)	15.73(0.36)
GraphWaveNet	16.60(0.64)	9.75(0.24)	24.55(0.05)	14.66(0.48)	30.63(0.21)	17.65(0.11)
GMAN	17.87(3.27)	11.49(2.68)	43.21(0.41)	27.58(0.57)	-	-
Reformer	24.01(1.00)	18.24(1.22)	25.71(4.60)	17.48(3.64)	27.77(1.56)	17.52(1.24)
Informer	15.75(2.36)	10.57(2.23)	16.92(1.36)	9.84(0.84)	23.12(1.04)	13.41(0.60)
STTN	21.97(0.30)	15.65(0.41)	23.93(2.30)	15.86(2.24)	-	-
STF	16.76(0.70)	10.88(0.47)	21.15(1.25)	12.32(0.59)	35.90(4.00)	24.47(2.59)
Mu2ReST	**14.27(0.42)**	**8.85(0.45)**	**16.19(1.10)**	**9.25(0.77)**	**21.46(0.30)**	**12.68(0.22)**
HEC, coarse regions (FPCRs)						
GRU	1.99(0.10)	**1.59(0.05)**	2.27(0.10)	1.77(0.08)	2.50(0.02)	1.88(0.00)
LSTnet	2.27(0.06)	1.65(0.03)	2.95(0.05)	2.20(0.03)	2.40(0.22)	**1.84(0.19)**
GraphWaveNet	2.79(1.10)	2.23(0.91)	3.04(0.75)	2.43(0.70)	2.59(0.22)	2.11(0.25)
GMAN	**1.98(0.32)**	1.61(0.30)	3.18(0.11)	2.53(0.02)	3.85(0.50)	3.14(0.45)
Reformer	1.99(0.16)	1.65(0.12)	2.75(0.34)	2.20(0.30)	2.85(0.25)	2.21(0.17)
Informer	2.26(0.08)	1.72(0.04)	2.49(0.06)	1.92(0.05)	2.56(0.12)	1.98(0.14)
STTN	2.07(0.22)	1.60(0.18)	2.35(0.11)	1.81(0.05)	2.50(0.27)	1.94(0.21)
STF	2.87(0.67)	2.28(0.64)	3.03(0.35)	2.30(0.30)	2.77(0.19)	2.13(0.20)
Mu2ReST	2.08(0.12)	1.63(0.06)	**2.08(0.10)**	**1.60(0.06)**	**2.40(0.14)**	1.85(0.09)
HEC, fine regions (PCRs)						
GRU	**2.01(0.07)**	1.60(0.03)	2.33(0.09)	1.81(0.06)	2.57(0.03)	1.93(0.01)
LSTnet	2.31(0.07)	1.68(0.03)	3.02(0.05)	2.24(0.03)	2.49(0.21)	1.91(0.18)
GraphWaveNet	2.72(1.02)	2.20(0.85)	2.98(0.75)	2.36(0.70)	2.61(0.19)	2.11(0.21)
GMAN	2.07(0.37)	1.68(0.35)	3.21(0.10)	2.56(0.03)	4.16(0.93)	3.38(0.79)
Reformer	2.01(0.13)	1.65(0.10)	2.77(0.25)	2.21(0.22)	2.98(0.24)	2.31(0.17)
Informer	2.37(0.13)	1.79(0.08)	2.54(0.04)	1.95(0.03)	2.63(0.12)	2.04(0.13)
STTN	2.06(0.19)	**1.59(0.16)**	2.37(0.11)	1.82(0.05)	2.52(0.24)	1.96(0.19)
STF	2.90(0.68)	2.30(0.63)	3.08(0.33)	2.33(0.29)	2.83(0.15)	2.16(0.17)
Mu2ReST	2.12(0.11)	1.66(0.06)	**2.14(0.08)**	**1.64(0.05)**	**2.49(0.13)**	**1.91(0.08)**

By capturing spatial and temporal dependencies simultaneously, STTN and STF outperform Informer 3/12 on NYT and 12/12 on HEC. The corresponding results show that the spatial dependency is especially effective for data with relatively stable spatial dependencies. For example, the spatial dependencies may not change or only change slightly over the 1-day resolution, but may change drastically over the 30-min resolution. However, their space-time attention implementations limit the ability of LTSTP given a large number of variables. In fact, when performing experiments on the 30-min resolution with the NYT dataset, we could not run STTN, albeit with smaller parameters.

Although Mu2ReST uses a simpler space-time attention implementation in order to account for long sequence inputs and outputs, it outperforms STTN and STF with few exceptions (only some cases of the 1-day resolution prediction). Moreover, Mu2ReST outperforms its base model Informer on all resolutions. On average, Mu2ReST reduces RMSE and MAE by 6.84% and 8.96%, respectively, on NYT dataset, and by 10.38% and 9.67%, respectively, on HEC dataset. These results confirm the effectiveness of introducing multi-resolution based recursive prediction and space attention in Mu2ReST.

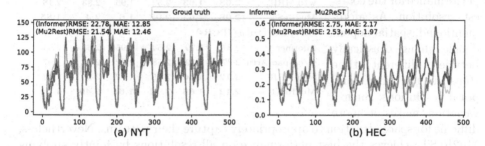

Fig. 3. Examples of predicted results of Informer and Mu2ReST for both datasets on the 30-min resolution and fine spatial resolution.

In addition, Fig. 3 illustrates some predicted results from both datasets. Besides improving RMSE and MAE performance, Mu2ReST matches peak values better than the Informer (which is especially visible for HEC after the 5th predicted day). Such improvements are essential for planning and dispatching. For instance, these long-term refinements could lead to significant cost reduction in electricity procurement in a liberal electricity market scenario.

4.2   Ablation Study

We performed an ablation study to further understand Mu2ReST's performance on both datasets. As Mu2ReST implements two separate components, we can isolate each one in order to define the following specific settings: (1) **w/o SpaceAttn**: adding only recursive prediction structure by deleting the space attention part of Mu2ReST; and (2) **w/o MRR**: adding only space attention to the Informer without using the multi-resolution recursive prediction.

Results are shown in Table 4. Firstly, we can observe that both settings alone cannot beat the Informer in all temporal resolutions. The first setting (w/o SpaceAttn) only improves performance for the finest temporal resolution, which demonstrates the effectiveness of transferring information via the recursive principle. On the other hand, the space attention implementation (w/o MRR setting) only improves the performance for the coarsest resolution. As aforementioned, spatial dependencies change drastically between different time steps in the finer temporal resolutions. Such a variation makes it more

Table 4. Ablation study results on both datasets (due to the space limitation, standard deviations are omitted in this table).

Resolution	1 day		6 h		30 min	
Metric	RMSE	MAE	RMSE	MAE	RMSE	MAE
NYT, coarse regions (boroughs)						
Informer	83.93	49.57	88.18	49.47	116.81	66.99
w/o SpaceAttn	83.93	49.57	93.54	53.68	115.91	67.66
w/o MRR	**79.5**	**46.47**	93.88	53.83	118.33	68.59
Mu2ReST	**79.5**	**46.47**	**85.49**	**46.82**	**103.03**	**57.32**
NYT, fine regions (taxi zones)						
Informer	15.75	10.57	16.92	9.84	23.12	13.41
w/o SpaceAttn	15.75	10.57	17.12	10.44	22.27	13.22
w/o MRR	**14.27**	**8.85**	17.55	10.26	23.17	13.6
Mu2ReST	**14.27**	**8.85**	**16.19**	**9.25**	**21.46**	**12.68**
HEC, coarse regions (FPCRs)						
Informer	2.26	1.72	2.49	1.92	2.56	1.98
w/o SpaceAttn	2.26	1.72	2.51	1.97	2.43	1.92
w/o MRR	**2.08**	**1.63**	2.52	1.96	2.83	2.18
Mu2ReST	**2.08**	**1.63**	**2.08**	**1.6**	**2.4**	**1.85**
HEC, fine regions (PCRs)						
Informer	2.37	1.79	2.54	1.95	2.63	2.04
w/o SpaceAttn	2.37	1.79	2.58	2.02	2.52	1.97
w/o MRR	**2.12**	**1.66**	2.55	1.97	2.9	2.23
Mu2ReST	**2.12**	**1.66**	**2.14**	**1.64**	**2.49**	**1.91**

difficult for space attention to appropriately capture their patterns. Nevertheless, Mu2ReST achieves the best performance on all resolutions by jointly applying recursive prediction and space attention. The former improves prediction performance for finer resolutions after the latter improved it for coarser resolution.

4.3 Further Discussions

(1) Regularization Term. We performed additional experiments to further confirm the effectiveness of the regularization term by modifying it from $\mathcal{L}\left(\hat{\mathbf{Y}}_{k \to k+1}, \ \hat{\mathbf{Y}}_{k+1}\right)$ to $\mathcal{L}\left(\hat{\mathbf{Y}}_{k \to k+1}, \mathbf{Y}_{k+1}\right)$. In Table 5, **RealY** gives the results after this modification. Results from this experiments show that it gives worse performance than the original Mu2ReST for finer resolutions. Indeed, $\mathbf{Y}_{k+1}$ can be obtained from $\mathbf{Y}_k$ and does not contain any additional information compared to $\mathbf{Y}_k$, following the information theory perspective[3].

Nevertheless, $\hat{\mathbf{Y}}_{k+1}$ is the predicted result from the trained model, which reflects the learnt patterns of the data in resolution $k + 1$. Thus, $\mathcal{L}\left(\hat{\mathbf{Y}}_{k \to k+1}, \hat{\mathbf{Y}}_{k+1}\right)$ provides auxiliary information to assist the modeling of resolution k. It is to note that we did not consider recursive prediction along the

---

[3] In fact, aggregation leads to information loss [9].

spatial dimension for simplicity. However, we believe that this scheme should equally be effective along the spatial dimension and it will be addressed in our future work.

(2) Upsampling and Cross Connections. In addition, upsampling and cross connections (XCon) could also be adopted along with recursive prediction, as shown by the solid and dashed lines, respectively, in Fig. 2.

We implemented upsampling based on fully connected layers (FC()) in order to generate $\hat{Y}'_k$ from $\hat{Y}_{k+1}$, and get the final result of $\hat{Y}_k$ using $\hat{Y}_k = FC_2 \left(\text{Concat} \left(\hat{Y}_k + FC_1 \left(\hat{Y}_{k+1} \right) \right) \right)$.

XCon are hidden space connections that transfers the encoder and decoder outputs from coarser to finer

Table 5. Experiment results on different technical concerns. In Mu2ReST, the coarsest resolution are not affected by different recursive prediction implementations, so 1-day resolution results are the same and not shown.

Resolution	6 h		30 min		6 h		30 min	
Metric	RMSE	MAE	RMSE	MAE	RMSE	MAE	RMSE	MAE
	NYT, coarse regions (boroughs)				NYT, fine regions (taxi zones)			
RealY	100.44	59.71	115.85	66.06	18.33	11.12	22.76	13.33
Upsamling	92.27	50.79	106.83	59.74	17.16	9.77	21.92	12.88
XCon	109.18	61.73	123.63	70.9	19.06	10.81	23.05	13.5
Mu2ReST	**85.49**	**46.82**	**103.03**	**57.32**	**16.19**	**9.25**	**21.46**	**12.68**
	HEC, coarse regions (FPCRs)				HEC, fine regions (PCRs)			
RealY	2.38	1.82	2.79	2.18	2.44	1.87	2.85	2.22
Upsamling	2.27	1.74	2.48	1.91	2.34	1.79	2.56	1.97
XCon	2.6	2	2.79	2.13	2.67	2.05	2.87	2.2
Mu2ReST	**2.08**	**1.6**	**2.4**	**1.85**	**2.14**	**1.64**	**2.49**	**1.91**

resolutions. This way, the hidden knowledge acquired in the kth resolution could be concatenated to the one of the resolution $k-1$. However, as currently defined, both upsampling and XCon could not bring additional gain to prediction performance, as shown by the results in Table 5. A more complex implementation of these components shall further leverage the benefit of multi-resolution. However, improving them is beyond the scope of this paper, and will therefore be addressed in the future.

5   Conclusion

We proposed both recursive prediction and space attention in order to leverage the Transformer when performing long-term spatio-temporal prediction (LTSTP). Specifically, we first designed a recursive prediction scheme to study and transfer information from coarser to finer resolutions. This scheme proved to improve the prediction performance for finer resolutions. Afterwards, we applied a simplified spatio-temporal Transformer structure using the Informer (a SOTA Transformer-based model for long-term time series prediction and LTSTP) as the base model to capture both spatial and temporal dependencies. Such an adaptation is especially effective for coarse resolution data with relatively stable spatial dependencies. As a result, these two ideas benefit all resolutions in LTSTP. Furthermore, our recursive prediction proposal is a generic method, which means that it could be applied to any model and not only Transformer.

Besides, the flexibility of our structure allows us to use different models for different resolutions and further improve the performance. In the future, we will evaluate these points as well as further investigate how to improve the attention structure and information transfer methods between resolutions.

Acknowledgements. Chuizheng Meng is partially supported by KDDI Research, Inc. and NSF Research Grant CCF-1837131. Defu Cao is partially supported by KDDI Research, Inc. and the Annenberg Fellowship of the University of Southern California.

References

1. Chen, C.F., Fan, Q., Panda, R.: CrossViT: cross-attention multi-scale vision transformer for image classification. In: ICCV (2021)
2. Child, R., Gray, S., Radford, A., Sutskever, I.: Generating long sequences with sparse transformers. arXiv (2019)
3. Choromanski, K., et al.: Rethinking attention with performers. In: ICLR (2021)
4. Chung, J., Gulcehre, C., Cho, K., Bengio, Y.: Empirical evaluation of gated recurrent neural networks on sequence modeling. In: NeurIPS Workshop (2014)
5. Grigsby, J., Wang, Z., Qi, Y.: Long-range transformers for dynamic spatiotemporal forecasting. arXiv (2021)
6. Ke, J., Wang, Q., Wang, Y., Milanfar, P., Yang, F.: MUSIQ: multi-scale image quality transformer. In: ICCV (2021)
7. Kitaev, N., Kaiser, Ł., Levskaya, A.: Reformer: the efficient transformer. In: ICLR (2020)
8. Lai, G., Chang, W.C., Yang, Y., Liu, H.: Modeling long-and short-term temporal patterns with deep neural networks. In: SIGIR (2018)
9. Lee, B.H., Park, J.: A spectral measure for the information loss of temporal aggregation. J. Stat. Theory Pract. **14**, 1–23 (2020)
10. Li, S., et al.: Enhancing the locality and breaking the memory bottleneck of transformer on time series forecasting. In: NeurIPS (2019)
11. Parmar, N., et al.: Image transformer. In: ICML (2018)
12. Torres, J.F., Hadjout, D., Sebaa, A., Martínez-Álvarez, F., Troncoso, A.: Deep learning for time series forecasting: a survey. Big Data **9**(1), 3–21 (2021)
13. Vaswani, A., et al.: Attention is all you need. In: NeurIPS (2017)
14. Wolf, T., et al.: Transformers: state-of-the-art natural language processing. In: EMNLP (2020)
15. Wu, Z., Pan, S., Long, G., Jiang, J., Zhang, C.: Graph wavenet for deep spatial-temporal graph modeling. In: IJCAI (2019)
16. Xu, M., et al.: Spatial-temporal transformer networks for traffic flow forecasting. arXiv (2020)
17. Zheng, C., Fan, X., Wang, C., Qi, J.: GMAN: a graph multi-attention network for traffic prediction. In: AAAI (2020)
18. Zhou, H., et al.: Informer: beyond efficient transformer for long sequence time-series forecasting. In: AAAI (2021)

LCAN: Light Cross-Attention Network for Collaborative Filtering Recommendation

Lin Liu[1], Wei Zhou[1], Junhao Wen[1(✉)], Yihao Zhang[2], Yu Wang[1], and Hanwen Zhang[1]

[1] School of Bigdata and Software Engineering, Chongqing University, Chongqing, China
{linliu,zhouwei,jhwen,wang_y,hanwenzhang}@cqu.edu.cn
[2] School of Artificial Intelligence, Chongqing University of Technology, Chongqing, China
yhzhang@cqut.edu.cn

Abstract. In recent years, an emerging research work in recommendation systems aimed at exploring users' potential interaction preferences. However, most existing methods can only capture information about the user's purchase (or click) history. To estimate users' potential interaction preferences more accurately, it is necessary to consider auxiliary information when modeling user-item interactions. In this paper, a Light Cross-Attention Network (LCAN) is proposed. LCAN makes full use of existing information in three parts: 1) User-Item interaction graph. The interaction history is an important signal, and the user's interaction preferences can be obtained directly from the interaction history. 2) User-User and Item-Item relationships. The user-user and item-item graphs are additionally constructed based on the similarity between users and items to alleviate data sparseness. 3) Complementarity between graphs. Information between different graphs is interrelated, and a graph-level cross-attention network is used to capture the complementarity between graphs. Extensive experiments have been conducted by comparing state-of-the-art methods, and it shows that our LCAN model can outperform the most advanced recommendation methods.

Keywords: Recommender system · Collaborative filtering · Cross-attention network · Multi-graph convolution

1 Introduction

Personalized recommendations play a central role in many online content-sharing platforms, helping users to discover items of interest. Accurately capturing user preferences is the core element of the recommendation system. As an effective solution, Collaborative Filtering (CF) [15] has attracted wide attention because of its simplicity and efficiency. The basic idea is that users who have purchased

© The Author(s), under exclusive license to Springer Nature Switzerland AG 2022
J. Gama et al. (Eds.): PAKDD 2022, LNAI 13280, pp. 81–92, 2022.
https://doi.org/10.1007/978-3-031-05933-9_7

similar items in the past will also tend to purchase similar items in the future. In essence, user-item interaction can be modeled as a user-item bipartite graph. Therefore, graph neural networks (GNNs), especially graph convolutional networks (GCNs), have become a popular method in current collaborative filtering recommendation systems [5,12,13].

The most common paradigm for CF is to learn latent features (a.k.a. embedding) to represent a user and an item, and perform predictions based on the embedding vectors [3,10]. Matrix factorization (MF) [7] is an early such model, which directly maps the single ID of a user to his embedding. Some follow-on studies [1,15] introduce personal history as the pre-existing feature of a user, and integrate embeddings of historical items to enrich the user's representation. More recent works [4,11,12] organize all historical interactions as a bipartite user-item graph to integrate the multi-hop neighbors into the representations and achieve the state-of-the-art performance. These significant improvements are attributed to the modeling of user-item relationships, which evolved from using only a single ID to personal history, and then a holistic interaction graph.

Despite their effectiveness, two critical flaws limit their performance. First, the previous modeling of user-item relationships is coarsegrained, because they supposed that users purchase items with uniform motivation. However, the motivations behind the user's decision-making are multiple in the real world. For example, some users purchase items because of their high-cost performance, while others purchase items because of their eye-catching appearance. If the user's purchase motivation is not considered, it will lead to sub-optimal recommendations. Second, the traditional methods do not make full use of existing information. They only consider the modeling of user-item relationships and ignore the user-user and item-item relationships. These relationships are also a significant signal. When the data is sparse, it often fails to achieve reasonable results.

In this paper, we propose a novel Light Cross-Attention Network For Collaborative Filtering Recommendation (LCAN) approach, an end-to-end deep model that considers the diversity and heterogeneity of latent preferences in a uniform framework. The model consists of three parts: preference extractor, fusion layer, and prediction layer. First, given an edge of the user-item bipartite graph, the preference extractor is to identify the latent preferences by decomposing the edge into multiple latent spaces. Considering the inherent differences between the user and item graphs, graph-level cross-attention is used to capture the complementarity and heterogeneity between graphs. Second, the fusion layer with preference-level attention automatically recognizes the importance of different preferences and filters out unimportant ones to generate a unified embedding for prediction. Third, the recommendation score is put forward through the prediction layer.

The contributions made in this paper are as follows:

- LCAN is a novel collaborative filtering approach based on graph neural networks, capturing more fine-grained user preferences hidden behind user behavior based on a semantic transformation.

– LCAN models user-user and item-item relationships explicitly: user-user and item-item graphs are constructed. It conducts learning simultaneously on all three graphs and employs cross-attention to capture the complementarity between graphs.
– Extensive experiments on three real-world datasets are conducted to verify the effectiveness of the proposed model.

2 Related Work

2.1 Collaborative Filtering

Given a user-item rating matrix, collaborative filtering methods usually map both users and items into the same low-dimensional latent space [15]. MF is a simple and efficient collaborative filtering method that decomposes the score matrix into the product of two low-rank matrices. PMF [7] optimizes the maximum likelihood function by minimizing the mean square error between the observed value and the predicted value. BiasMF [6] introduces a bias term and a regular term to improve recommendation accuracy and alleviate the overfitting problem. AutoRec [8] uses an autoencoder to decompose the score matrix, and then reconstruct it to predict the score, and obtain competitive results directly on many benchmark data sets. DMF [14] takes the interaction history of the users and items as a feature vector and inputs it to a multi-layer perceptron to learn the latent expression of users and items. GC-MC [1] stacks a graph convolution layer followed by a dense layer to accumulate messages aggregated according to different types of edges as node representations. However, these methods assume that users have a unified purchase motivation and cannot capture more fine-grained user interaction preferences.

2.2 Graph Neural Networks

Graph neural networks [16], especially graph convolutional neural networks (GCN), have gained considerable interest. The main idea of GCN is how to iteratively aggregate the feature information of neighbors and integrate the aggregated information with the current central node representation. LR-GCCF [2] uses the residual method to represent the user-item bipartite graph in the learning process. NGCF [10] iteratively propagates user and item embeddings in the graph to distill high-hop collaborative signals with graph convolutions. BGNN [17] considers the importance of interaction between neighbors and uses a bilinear aggregator to perform element-wise product between neighbors of each node. DGCF [11] and MCCF [12] hold that treating user-item interactions as isolated data instances is not sufficient to capture the diversity of user preferences on adopting an item, may lead to suboptimal representations. NIA-GCN [9] deployed a pairwise neighborhood aggregation layer to capture relationships between pairs of neighbors. LightGCN [4] conducts ablation analyses on GCN and improves the model's performance and scalability by removing the feature

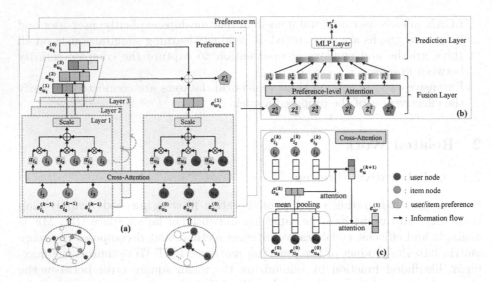

Fig. 1. The overview framework of the proposed model. (a) Preference extractor. The figure shows the preference extraction process of the user part, and the item part is similar to it; (b) Fusion layer and prediction layer; (c) Cross-Attention. This example predicts the rating User u_1 would give to the Item I_4.

transformation and nonlinear activation operation. However, these methods only use the basic user-item bipartite graph. They ignore other rich information provided by users and items, which reduces the performance of recommendations.

3 Approach

There are two parts in the proposed method, the user part and the item part. The general framework is shown in Fig. 1. This section uses the user part as an example to introduce our model. First, the model takes the user-item interaction graph, user-user graph, and item-item graph as input, and then predicts the user's rating of the item. Specifically, features of the purchased items and features of neighbor users are aggregated as the user's embedding. Second, in the process of feature aggregation, the user's potential interaction preferences are considered through the following two modules: (1) The preference extractor with cross-attention recognizes the user's preferences from edges in different graphs. (2) The fusion layer with preference-level attention automatically recognizes the importance of different preferences and recombines them to obtain a unified user embedding. Third, the predicted score is output through the Multilayer Perceptron (MLP) layer.

3.1 Embedding Layer

Following the mainstream recommender models, we describe a User u (an Item i) with an embedding vector $e_u \in \mathbb{R}^d (e_i \in \mathbb{R}^d)$, where d denotes the embedding size. Assuming that the user's interaction is driven by M preferences, in order to distinguish the different preferences of users (items), M preference-specific transformation matrices $\{Q_k\}_1^m$ and $\{R_k\}_1^m$ are designed, respectively, to extract different features that correspond to particular preference. Under the m-th preference, the embedding of the user (item) is as follows:

$$E^m = [e_{u_1}^m, \cdots, e_{u_{u_N}}^m, e_{i_1}^m, \cdots, e_{i_{i_M}}^m], \tag{1}$$

where $e_u^m = e_u Q_m, e_i^m = e_i R_m$, u_N is the number of users, i_M is the number of items, and $m \in \{1, 2, \cdots, M\}$ represents the m-th user interaction preference. There is a similar embedding process for the user-user graph and item-item graph. In the following description, for the sake of simplicity, the preference indicator superscript m is ignored, and $m = 1$ is taken as an example.

Multi-graph Encoding. In addition to the user-item bipartite graph, the relationships between user-user and item-item are also explicitly modeled to alleviate the data sparse problem in collaborative filtering. Generally speaking, the graph is constructed by calculating the pairwise cosine similarity on the rows (columns) of the scoring matrix. In order to avoid the over-smoothing problem caused by graph convolutional networks, only first-order neighbors are aggregated in the user-user graph and item-item graph. At the same time, first-order neighbors are used to generate attention coefficients for each layer in the user-item graph convolutional network.

3.2 Cross-Attention Layer

Existing research treats each graph separately and then combines them, ignoring the importance that different graphs are related to each other. For example, it is possible to infer items that the target user may be interested in based on the features of similar users. As shown in Fig. 1(c), a new graph-level cross-attention mechanism is defined to calculate neighbor attention coefficients. The mechanism is mainly divided into two parts: User-based item attention and Item-based user attention.

User-Based Item Attention. A single mean-pooling is used to summarize the similar users representation: $\bar{G}_u^{(k)} = \frac{1}{N_S} \sum_{n=1}^{N_S} e_{u_n}^{(0)}$. Next, combine the $e_{i_l}^{(k)}$ with $\bar{G}_u^{(k)}$:

$$h_i^{(k)} = \tanh(\mathbf{W}_i^{(k)} e_{i_l}^{(k)}) \otimes \tanh(\mathbf{W}_u^{(k)} \bar{G}_u^{(k)}), \tag{2}$$

where N_s represents the number of similar users of User u, $\tilde{G}_m^u$ represents the matrix filled by K column $\bar{G}_u^{(k)}$. $\otimes$ is used to denote the Element-wise multiplication of two matrices. Specially, $h_i^{(k)}$ represents the probability that User u will purchase Item i due to the m-th preference. Softmax function is used to obtain the weight coefficient $\alpha_{i \rightarrow u}^{(k)}$:

$$\alpha_{i \rightarrow u}^{(k)} = softmax(\mathbf{W}_h^{(k)} h_i^{(k)}), \tag{3}$$

Since the attention probability of each item feature under m-th preference is calculated from the above formula, the new representation of the item is the weighted sum of the item features:

$$e_u^{(k+1)} = \sum_l \alpha_{i \to u}^{(k)} e_{i_l}^{(k)} \tag{4}$$

Item-Based User Attention. User-based item attention can distinguish which items are purchased by the User u due to m-th preference. In the same way, Item-based user attention is used to distinguish which users have the same purchase preference as the target user. Using the new item representation $e_u^{(k+1)}$ combined with the original user features $e_{u_n}^{(0)}$ to screen out users who share the same purchase preference as User u. The process is as follows:

$$h_u^{(k)} = \tanh(\mathbf{W}_u^{(k)} e_{u_n}^{(0)}) \otimes \tanh(\mathbf{W}_u^{(k)} \tilde{e}_u^{(k+1)}), \tag{5}$$

$$\alpha_{u' \to u}^{(0)} = softmax(\mathbf{W}_h^{(k)} h_u^{(k)}), \tag{6}$$

where $\tilde{e}_u^{(k+1)}$ represents the matrix filled by K column $e_u^{(k+1)}$. Since the attention probability of each user feature under m-th preference is calculated from the above formula, the new representation of the user is the weighted sum of the similar users features:

$$e_{u'}^{(1)} = \sum_n \alpha_{u' \to u}^{(0)} e_{u_n}^{(0)} \tag{7}$$

3.3 Preference Extractor

After using the first-hop neighbors, we further stack more cross-attention layers in the user-item bipartite graph to collect influential signals from high-order neighbors. In the aggregation process, feature transformation and nonlinear activation are removed, and only the neighborhood aggregation operation in the standard GCN is retained. The message aggregation function of the k-th layer is defined as:

$$e_u^{(k)} = \frac{1}{\sqrt{\mathcal{N}_u}} \sum_{i \in \mathcal{N}_u} \alpha_{i \to u}^{(k-1)} e_i^{(k-1)} \tag{8}$$

Considering that the norm of embedding after aggregation may be large, similar to LightGCN [4], we scale these embedding by $\frac{1}{\sqrt{\mathcal{N}_u}}$.

After k layers, we sum the embeddings at different layers up as the final embeddings:

$$Z_u = \sum_{K=0}^{K} e_u^{(k)} + e_{u'}^{(1)} \tag{9}$$

Z_u is the final embedding of the User u under the first preference ($m = 1$), plus the preference indicator superscript as Z_u^1. User's embedding set $\{Z_u^1, Z_u^2, \cdots, Z_u^M\}$ is abstained under different interaction preferences.

Remark: The preference extraction of User u concentrates all the attention points because the item part also has a similar process, and the embedded set $\{Z_i^1, Z_i^2, \cdots, Z_i^M\}$ of Item i under different preferences can be obtained.

3.4 Fusion Layer

Obviously, not every preference is equally important. One or two preferences dominate our interactions for most people, which leads us to propose a preference-level attention mechanism to filter out unimportant preferences.

Preference-Level Attention. A new fusion layer is proposed, which integrates $\{Z_u^m, Z_i^m\}, m \in \{1, \cdots, M\}$ into a unified representation, with the formula as follows:

$$\vartheta_u^1, \vartheta_u^2, \cdots, \vartheta_u^M = att(Z_u^1 \infty Z_u^2 \infty \cdots \infty Z_u^M), \tag{10}$$

where att is a two-layer neural network, which plays the role of preference-level attention mechanism. ∞ represents a join operation. $\{\vartheta_u^m\}$ is normalized by the softmax layer, and obtained the importance of m-th preference β_u^m as follows:

$$\beta_u^m = softmax(\vartheta_u^m) \tag{11}$$

The greater of β_u^m, the more dominant user's preference will be. Through these weight factors β_u^m, the final embedding of user u is obtained through weighted sum:

$$Z_u = \sum \beta_u^m Z_u^m \tag{12}$$

In the same way, the final embedding Z_i of Item i is obtained.

3.5 Prediction Layer

Once obtaining the final embeddings of user u and item i from the user and item part separately, connect them in series and pass them through MLP to predict the rating r'_{ui} from u to i as:

$$f_1 = [Z_u \| Z_i],$$
$$f_2 = \sigma(W_2 \cdot f_1 + b_2),$$
$$\cdots \tag{13}$$
$$f_l = \sigma(W_l \cdot f_{l-1} + b_l),$$
$$r'_{ui} = w^T \cdot f_l,$$

where l is the index of a hidden layer.

In this work, the training objective function is formulated as follows:

$$\mathcal{L}_r = \frac{1}{2|\mathcal{O}|} \sum_{(u,i) \in \mathcal{O}} (r'_{ui} - r_{ui})^2 + \lambda \|\theta\|_2^2, \tag{14}$$

where $\mathcal{O}$ is the set of observed ratings, and r_{ui} is the ground truth rating by the User u on the Item i. λ and θ represent the regularization weight and the parameters of the model.

4 Experiments

In this section, experiments are performed on three real-world datasets, Movielens, Amazon and Yelp. First, the experimental setup is introduced, and then it is compared with other methods. Second, an ablation experiment is carried out to verify whether the multi-graph auxiliary information can alleviate the problem of data sparseness and the effectiveness of the cross-attention mechanism. Third, a hyperparameter discussion is carried out.

Datasets. Movielens, Amazon, and Yelp are used in the experiments. MovieLens-100k has been widely used to evaluate recommendations, which contains $100,000$ ratings from 943 users to $1,682$ movies. Amazon is a widely used product recommendation dataset containing $65,170$ ratings from $1,000$ users to $1,000$ items. Yelp is a local business recommendation dataset containing $30,838$ ratings from $1,286$ users to $2,614$ items. For each dataset, randomly select the historical score of 80% as the training set, and treat the remaining as the test set.

Baselines. To evaluate the effectiveness of our model, we compare LCAN with several state-of-the-arts, including matrix factorization methods: PMF [7], BiasMF [6]; auto-encoders based methods: AutoRec [8]. Typically, we use I-AutoRec to represent the item-based setting, which has better performance than the user-based; GCN based methods: GC-MC [1], LightGCN [4], MCCF [12].

Implementation. All the programs are executed on a computer with dual Intel Xeon E5-2678 v3 processors and an RTX 3090 GPU with 24-G memory. LCAN is implemented using PyTorch. We test the learning rate in the range of $\{1e^{-4}, 5e^{-4}, 1e^{-3}, 5e^{-3}\}$, and the coefficient of L2 normalization in $\{1e^{-1}, 1e^{-2}, 1e^{-3}, 1e^{-4}\}$. We have tried preference number M in the range of $\{1, 2, 3, 4\}$, and the embedding dimension d in range $\{8, 16, 32, 64, 128, 256, 512\}$, and batch size in the range of $\{64, 128, 256, 512\}$. The model parameters are initialized with a Gaussian distribution with a mean value of 0 and a standard deviation of 0.1. Adam is used as the optimizer. At the same time, dropout is applied to multi-preference fusion and the drop rate tests in the range of $\{0.1, 0.4, 0.5, 0.6\}$. Two layers of neural parts are explicitly used for the neural network, and ReLU is used as the activation function. Experiments are repeated three times, and the average results are recorded.

Two widely used evaluation protocols: Root Mean Square Error (RMSE) and Absolute Error (MAE), are used as evaluation metrics.

4.1 Performance Comparison

The comparative results are summarized in Table 1. From this table, there are the following observations:

Table 1. Overall comparison. The best performance is highlighted in bold, and the second is underlined.

Models		PMF	BiasMF	I-AutoRec	GC-MC	LightGCN	MCCF	LCAN
Movielens	RMSE	0.9638	0.9257	0.9435	0.9145	0.9092	_0.9070_	**0.8823**
	MAE	0.7559	0.7258	0.7370	0.7165	0.7072	_0.7050_	**0.6815**
Amazon	RMSE	0.9339	0.9028	0.9213	0.8946	0.8898	_0.8876_	**0.8653**
	MAE	0.7113	0.6759	0.7064	0.6619	0.6521	_0.6428_	**0.6274**
Yelp	RMSE	0.3967	0.3902	0.3817	0.3850	_0.3721_	0.3806	**0.3451**
	MAE	0.1571	0.1616	0.1201	0.1354	_0.0997_	0.1029	**0.0925**

– LCAN substantially outperforms baselines in most cases, which verifies the effectiveness of our model. Compared to the strongest baseline, MAE is improved by 3.33%, 2.40%, and 7.22% and RMSE is improved by 2.72%, 2.51%, and 7.26% on the three datasets, respectively. This significant improvement is attributed to the preference extraction capabilities and the auxiliary information of multi-graph.
– The GCN-based model outperforms the CF-based and AutoEncoder-based models on the three datasets. These improvements are attributed to the graph convolutional layer. At the same time, we found that LightGCN and MCCF have similar performance, which further verifies that feature transformation and nonlinear activation reduce the performance of the graph-based collaborative filtering method.
– Our method has achieved better results on Yelp. The reason is that the Yelp dataset is more sparse, and only considering user-item interactions is not enough to obtain reasonable embedding. Adding auxiliary information can significantly improve performance.

4.2 Ablation Experiments

The user-user and item-item relationships are explicitly modeled to alleviate the problem of data sparseness in collaborative filtering. At the same time, a cross-attention network is used to capture the complementarity between different graphs, and preference-level attention is used to filter out unimportant preferences. To evaluate and verify the effectiveness of these components, an ablation study is conducted on the Yelp Dataset.

From Table 2, It can be concluded that all three main components of our proposed model, the Multi-Graph Encoding, the Cross-Attention, and the Preference-level Attention, have been proven to be effective. The Multi-Graph Encoding layer has the most significant impact on performance among all the components, indicating that the sparse user-item bipartite graph is insufficient to learn user/item embedding. When multi-graph information is added, the cross-attention network can effectively improve the recommendation performance compared to processing different graphs individually and then combining them, which also validates our hypothesis: using complementary information between different graphs can get better user/item embeddings. In addition, it can be

Table 2. Ablation studies on Yelp

Architecture	Yelp	
	RMSE	MAE
Remove Multi-Graph Encoding	0.3885	0.1045
Remove Cross-Attention	0.3702	0.0994
Remove Preference-level Attention	0.3581	0.0942
LCAN	**0.3451**	**0.0925**

found that preference-level attention filters out unimportant preferences and leads to significant performance improvements, and combining all three components receives further improvements.

4.3   Hyper-Parameter Studies

In LCAN, the number of user preferences is a key parameter, and we explored its impact on performance. The impact of embedding dimension d on performance is also studied.

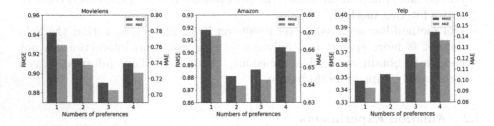

Fig. 2. Impact of preference numbers on three real world datasets.

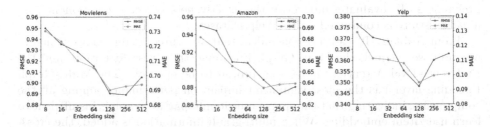

Fig. 3. Effect of different embedding dimensions.

Number of Preferences. To study the preference extraction ability of the model, the preference amount M is changed within the range of $\{1, 2, 3, 4\}$ while keeping other parameters unchanged. Experiments on three datasets and the experimental results are shown in Fig. 2. It can be concluded that the number of preferences is inconsistent for different datasets when the best-recommended performance is reached. On the Yelp dataset, most of the ratings are 1 or 2, which indicates that one preference is enough to model latent semantics. On the Amazon and Movielens datasets, user interaction is more complicated. At this point, the power of multiple performances is more prominent. Increasing M leads to performance improvement, proving that more fine-grained user preferences are hidden in the user-item interaction, which cannot be expressed by edge information. However, LCAN can effectively extract this information. At the same time, as the number of preferences increases, the recommendation performance first increases and then decreases, which may be caused by overfitting.

Impact of Embedding Dimensions. The dimension of embeddings d is also a crucial hyper-parameter to control the complexity and capacity of LCAN. Therefore, the influence of the embedding dimension on the recommendation effect is studied, and the embedding size is set from 8 to 512. The experimental results are shown in Fig. 3. A common trend can be found from the experimental data: with the gradually increasing of embedding dimension, recommendation performance will first increase progressively and reach the peak, and then with the further increase of dimension, the performance will remain stable or even decline. Therefore, an appropriate embedding Dimension d is used to balance the trade-off between performance and complexity.

5   Conclusion

This work focuses on modeling the motivation behind user decision-making through graph convolutional networks. A Light Cross-Attention Network (LCAN) is proposed. To improve the recommendation performance, the additional information beyond the user interaction history is explicitly modeled. First, the preference extractor is used to learn the preference-level embedding of users and items. Second, the cross-attention network is used to capture the complementarity between graphs. Finally, a fusion layer with preference-level attention filters unimportant preferences and generates embeddings for prediction. Comparison experiments and ablation studies have shown that LCAN is superior to existing methods in terms of recommendation accuracy and captures the different potential preferences of users. We will work on further improvement by studying a more suitable combination mechanism of preference-level representation of users and items in the future.

Acknowledgements. This work is supported by National Natural Science Foundation of China (Grant No. 62072060, 72074036), this work is also partly supported by the China Postdoctoral Science Foundation (2020M673145) and Program for Innovation Research Groups at Institutions of Higher Education in Chongqing (CXQT21032).

References

1. van den Berg, R., Kipf, T.N., Welling, M.: Graph convolutional matrix completion. CoRR abs/1706.02263 (2017)
2. Chen, L., Wu, L., Hong, R., Zhang, K., Wang, M.: Revisiting graph based collaborative filtering: a linear residual graph convolutional network approach. In: AAAI 2020, pp. 27–34. AAAI Press (2020)
3. Cheng, Z., Ding, Y., Zhu, L., Kankanhalli, M.S.: Aspect-aware latent factor model: rating prediction with ratings and reviews. In: WWW 2018, Lyon, France, 23–27 April 2018, pp. 639–648. ACM (2018)
4. He, X., Deng, K., Wang, X., Li, Y., Zhang, Y., Wang, M.: LightGCN: simplifying and powering graph convolution network for recommendation. In: SIGIR, pp. 639–648. ACM (2020)
5. Jorro-Aragoneses, J.L., Díaz-Agudo, B., Recio-García, J.A., Jiménez-Díaz, G.: RecoLibry suite: a set of intelligent tools for the development of recommender systems. Autom. Softw. Eng. **27**(1), 63–89 (2020)
6. Koren, Y., Bell, R.M., Volinsky, C.: Matrix factorization techniques for recommender systems. Computer **42**(8), 30–37 (2009)
7. Salakhutdinov, R., Mnih, A.: Probabilistic matrix factorization. In: Advances in Neural Information Processing Systems 20, pp. 1257–1264. Curran Associates, Inc. (2007)
8. Sedhain, S., Menon, A.K., Sanner, S., Xie, L.: AutoRec: autoencoders meet collaborative filtering. In: WWW 2015, pp. 111–112. ACM (2015)
9. Sun, J., et al.: Neighbor interaction aware graph convolution networks for recommendation. In: SIGIR 2020, pp. 1289–1298. ACM (2020)
10. Wang, X., He, X., Wang, M., Feng, F., Chua, T.: Neural graph collaborative filtering. In: SIGIR 2019, pp. 165–174. ACM (2019)
11. Wang, X., Jin, H., Zhang, A., He, X., Xu, T., Chua, T.: Disentangled graph collaborative filtering. In: SIGIR 2020, pp. 1001–1010. ACM (2020)
12. Wang, X., Wang, R., Shi, C., Song, G., Li, Q.: Multi-component graph convolutional collaborative filtering. In: AAAI 2020, pp. 6267–6274. AAAI Press (2020)
13. Xu, Z., et al.: When recommender systems meet fleet management: practical study in online driver repositioning system. In: Huang, Y., King, I., Liu, T., van Steen, M. (eds.) WWW 2020, pp. 2220–2229. ACM/IW3C2 (2020)
14. Xue, H., Dai, X., Zhang, J., Huang, S., Chen, J.: Deep matrix factorization models for recommender systems. In: IJCAI 2017, pp. 3203–3209. ijcai.org (2017)
15. Zhang, Y., et al.: Towards poisoning the neural collaborative filtering-based recommender systems. In: Chen, L., Li, N., Liang, K., Schneider, S. (eds.) ESORICS 2020. LNCS, vol. 12308, pp. 461–479. Springer, Cham (2020). https://doi.org/10.1007/978-3-030-58951-6_23
16. Zhou, J., et al.: Graph neural networks: a review of methods and applications. AI Open **1**, 57–81 (2020)
17. Zhu, H., et al.: Bilinear graph neural network with neighbor interactions. In: IJCAI 2020, pp. 1452–1458. ijcai.org (2020)

Coded Hate Speech Detection
via Contextual Information

Depeng Xu[1], Shuhan Yuan[2(✉)], Yueyang Wang[3], Angela Uchechukwu Nwude[1],
Lu Zhang[1], Anna Zajicek[1], and Xintao Wu[1]

[1] University of Arkansas, Fayetteville, AR 72701, USA
{depengxu,aunwude,lz006,azajicek,xintaowu}@uark.edu
[2] Utah State University, Logan, UT 84322, USA
shuhan.yuan@usu.edu
[3] Paycom Software, Inc., Oklahoma City, OK 73142, USA
Yueyang.wang@paycomonline.com

Abstract. Hate speech on online social media seriously affects the experience of common users. Many online social media platforms deploy automatic hate speech detection programs to filter out hateful content. To evade detection, coded words have been used to represent the targeted groups in hate speech. For example, on Twitter, "Google" is used to indicate African-Americans, and "Skittles" is used to indicate Muslim. As a result, it would be difficult to determine whether a hateful text including "Google" targets African-Americans or the search engine. In this paper, we develop a coded hate speech detection framework, called CODE, to detect hate speech by judging whether coded words like Google or Skittles are used in the coded meaning or not. Based on a proposed two-layer structure, CODE is able to detect the hateful texts with observed coded words as well as newly emerged coded words. Experimental results on a Twitter dataset show the effectiveness of our approach.

Keywords: Coded hate speech · Few-shot learning

1 Introduction

Online social media bring people together and encourage people to share their thoughts freely. However, due to the openness of social media, some users misuse the platforms to promote hateful language. As a result, hate speech, which "expresses hate or encourages violence towards a person or group based on characteristics such as race, religion, sex, or sexual orientation"[1], unfortunately becomes a common phenomenon on online social media. Hate speech on online social media not only seriously affects the experience of regular users, but also could lead to a real-world consequence. Currently, many online social media, such as Facebook and Twitter, have a policy prohibiting hate speech. Users who violate the policy could result in permanent account suspension.

---

[1] https://dictionary.cambridge.org/dictionary/english/hate-speech.

© The Author(s), under exclusive license to Springer Nature Switzerland AG 2022
J. Gama et al. (Eds.): PAKDD 2022, LNAI 13280, pp. 93–105, 2022.
https://doi.org/10.1007/978-3-031-05933-9_8

These companies also deploy programs to automatically filter out hateful content. In early 2016, Google unveiled a tech incubator Jigsaw to reduce online hate and harassment. In response to these programs, trolls started the "Operation Google Campaign", which replaces racial slurs with names of technology brands and products. For example, when racists publish hate speech to attack African-American, instead of using the n-word that can be easily detected, they use the word "Google" to represent African-American, such as "worthless google, kill yourself"[2]. Since then, more coded words are proposed to represent different targeted groups to avoid the censorship of hate speech. Table 1 shows some coded words and their corresponding targeted groups. We call such hate speech using coded words the *coded hate speech*.

Table 1. Commonly-used coded words

Coded word	Targeted group	Coded word	Targeted group
Google	African-American	Butterfly	Gay
Skittles	Muslim	Car Salesman	Liberal
Pepe	Alt-right	Bing	Asian
M&M	Mexican & Muslim	Yahoo	Mexican

By replacing the targeted groups with specific coded words, it becomes an uneasy task to detect whether a text is hateful or not because it is difficult to determine whether these words are in the coded meanings. In this paper, we detect the hateful texts by distinguishing the meanings of coded words based on their contexts. Specifically, we consider coded words as a special case of polysemy, i.e., a single word is associated with two or more different meanings. In our case, each coded word is associated with a regular meaning, e.g., Google is a search engine, and Skittles is a brand of candies, and also a coded meaning, e.g., African-American or Muslin. We consider that if a coded word is in its coded meaning, the text containing the coded word expresses hate.

Given a training dataset $\mathcal{T}$ consisting of regular and hateful texts and collected with a set of coded words $\mathcal{C}$, we aim to build a classifier that is able to: 1) detect the hateful texts with some observed coded words in $\mathcal{C}$; and 2) detect the hateful texts with new coded words. To achieve these goals, there are three challenges: 1) how to represent the coded word with two different meanings; 2) how to detect whether a coded word is used in its coded or regular meaning; 3) how to detect new coded words without collecting new labeled texts.

In this work, we propose a **CO**ded hate speech **DE**tection (CODE) framework, which is able to detect the observed coded words $\mathcal{C}$ as well as newly emerged coded words. CODE makes use of Embeddings from Language Models (ELMo) [10] as the pre-trained word embedding model to derive the contextual embeddings of coded words. ELMo is capable of handling the polysemy by

---

[2] https://knowyourmeme.com/memes/events/operation-google.

representing a word in different texts with different word embeddings. CODE then applies transformations on the derived embeddings via a two-layer structure (general and specific layers) to detect the meaning of a coded word. In particular, the general layer derives a generalized coded meaning anchor based on all the observed coded hate speech and makes the coded words in the coded meanings close to the anchor. We then expect a new coded word in the coded meaning also close to this generalized coded meaning anchor. The specific layer then derives the specialized coded meaning anchor for each observed coded word to detect whether a coded word is used in its coded meaning. Experimental results on a collected Twitter dataset indicate that CODE can detect hate speech by distinguishing the coded and regular meanings of coded words, even the new coded words without collecting new training samples.

2   Related Work

Hate speech detection has attracted increasing attention in recent years [2,14]. Research in [3] develops a semantic dictionary of hate domain and uses a rule-based classifier to detect hate speech. Research in [1,16] uses a bag-of-word model to derive the features of hate speech and then adopt classical machine learning models such as SVM, logistic regression, random forests to detect the hate speech. Recently, deep learning models, such as recurrent neural networks or deep auto-encoder, are also proposed for hate speech detection [9,11–13,15].

There are few studies focusing on coded hate speech detection. Research in [6] composes a balanced dataset and adopt SVM to predict the coded hate tweets. In this work, we consider the problem of coded hate speech detection given both regular and hateful tweets while only a small number of coded hate tweets is available for training, which is closer to the real-world scenario. Research in [7] aims at identifying coded words from hateful tweets with a known coded word. The proposed approach can only detect the coded words that co-occur with the known coded word within the same corpus. Meanwhile, it cannot detect whether a coded word is used in the coded or regular meaning given a new tweet.

3   Method

Given a set of coded words $\mathcal{C}$ and related hateful texts $\mathcal{T}^h$ and regular texts $\mathcal{T}^r$, for each coded word $c \in \mathcal{C}$, there are associated hateful texts $\mathcal{T}_c^h \subset \mathcal{T}^h$ and regular texts $\mathcal{T}_c^r \subset \mathcal{T}^r$. Note that in most cases, the coded word is still used as its regular meaning, i.e., $|\mathcal{T}_c^r| > |\mathcal{T}_c^h|$. In this work, we aim to detect a hateful text by determining whether an observed coded word $c \in \mathcal{C}$ or a new coded word $c \notin \mathcal{C}$ in the text is in the coded meaning based on the training corpus $\{\mathcal{T}^r, \mathcal{T}^h\}$.

We propose a **CO**ded hate speech **DE**tection (CODE) framework as shown in Fig. 1. CODE uses ELMo to derive contextual embeddings for coded words and then applies transformations on the derived embeddings via a two-layer structure. CODE derives generalized coded and regular meaning anchors and makes the coded words with coded meanings close to the coded meaning anchor and away from the regular meaning anchor by training on $\mathcal{T}^r$ and $\mathcal{T}^h$. We expect if a new coded word $c \notin \mathcal{C}$ is in its coded meaning, it should be also close to the generalized coded meaning anchor. For the observed coded word $c \in \mathcal{C}$, we have the specific texts $\mathcal{T}_c^r$ and $\mathcal{T}_c^h$. CODE further derives the specialized coded and regular meaning anchors for each observed coded word c based on $\mathcal{T}_c^r$ and $\mathcal{T}_c^h$ so that CODE can leverage the specialized anchors to detect the hate speech.

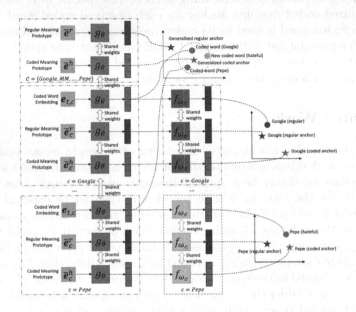

Fig. 1. The proposed CODE framework

3.1 Coded Word Representation

We use word embeddings to represent coded words. However, the traditional word embedding approaches, such as word2vec, usually represent one word by one embedding, which is unable to capture different meanings of coded words [8]. In this work, we use ELMo to derive the coded word embeddings, which addresses the issue of polysemy by representing each word based on its context [10]. As a result, one word has different representations in different contexts. We represent the coded word c in a text t as $\mathbf{e}_{t,c} = \mathbf{ELMo}_t(c)$.

3.2 Transformation Layers

After obtaining $\mathbf{e}_{t,c}$, we aim to detect whether the coded word c is in its coded meaning. CODE consists of one general layer and one specific layer, both of which make transformations to separate the regular and coded meanings of coded words in the embedding space.

General Layer. The general layer of CODE is to identify the generalized coded meanings. In general, for different coded words, although the regular meanings could be different, the coded meanings should be all alike since they all express the social hate towards specific groups. Based on this hypothesis, we first derive prototype representations to represent the regular and coded meanings by using ELMo on $\mathcal{T}^r$ and $\mathcal{T}^h$. Then, we adopt a neural network as the general layer to apply a transformation on the original embedding as well as the prototypes. After we apply the transformation, we consider the transformed prototype representations of regular and coded meanings as the generalized regular and coded meaning anchors, respectively. The objective is to make the transformed coded word embedding in different meanings close to their corresponding anchors. For example, if "Google" is used to indicate the African-American in a text, we aim to make it close to the coded meaning anchor. Because the generalized coded meaning anchor is derived from several coded words $\mathcal{C}$, the anchor should represent the generalized coded meaning. Hence, after training, given a new coded word $c \notin \mathcal{C}$, if the coded word is used in its coded meaning, we expect the transformed embedding has a smaller distance to the coded meaning anchor compared with the distance to the regular meaning anchor.

More specifically, given a set of regular and hateful texts $\mathcal{T}^r$ and $\mathcal{T}^h$, we first adopt ELMo to obtain the word embeddings $\mathbf{e}_{t,c}$ of the coded word c in text t. Note that for ELMo, the coded word has a unique word embedding for each text t. Based on the obtained coded word embedding, we consider the centroid of coded word embeddings in regular or coded meaning as the corresponding prototype representation. Symbolically, the prototype representations of regular meaning and coded meaning of a set of coded words $\mathcal{C}$ are computed by a mean operation, respectively:

$$\bar{\mathbf{e}}^r = \frac{1}{\sum_{c \in \mathcal{C}} |\mathcal{T}_c^r|} \sum_{v \in \mathcal{C}, \iota \in \mathcal{T}_c^r} \mathbf{e}_{t,c} \quad \bar{\mathbf{e}}^h = \frac{1}{\sum_{c \in \mathcal{C}} |\mathcal{T}_c^h|} \sum_{v \in \mathcal{C}, \iota \in \mathcal{T}_c^h} \mathbf{e}_{t,c}, \tag{1}$$

where $\bar{\mathbf{e}}^r$ and $\bar{\mathbf{e}}^h$ indicate the prototype representations of regular and coded meanings, respectively.

Then, we aim to map the coded word embedding derived from each text close to the prototype representation based on its meaning via a neural network. To this end, we adopt a neural network g to make a transformation on the coded word embedding in each text t as well as the prototype representations:

$$\mathbf{g}_{t,c} = g_\theta(\mathbf{e}_{t,c}) \quad \bar{\mathbf{g}}^r = g_\theta(\bar{\mathbf{e}}^r) \quad \bar{\mathbf{g}}^h = g_\theta(\bar{\mathbf{e}}^h), \tag{2}$$

where θ indicates the parameters of the neural network g. We consider $\bar{\mathbf{g}}^r$ and $\bar{\mathbf{g}}^h$ as the generalized regular and coded meaning anchors, respectively.

In order to separate $\bar{\mathbf{g}}^r$ and $\bar{\mathbf{g}}^h$ and to make the coded word embedding $\mathbf{g}_{t,c}$ close to the corresponding anchor, the objective of the neural network g_θ is to make the word embedding in the coded meaning (regular meaning) close to the generalized coded meaning anchor (regular meaning anchor) and far from the regular meaning anchor (coded meaning anchor). To this end, we propose the following triplet loss function to train the neural network g_θ:

$$
\mathcal{L}_{tri} = \frac{1}{\sum_{c \in C} |T_c^h|} \sum_{c \in C, t \in T_c^h} \max\{d(\bar{\mathbf{g}}^h, \mathbf{g}_{t,c}) - d(\bar{\mathbf{g}}^r, \mathbf{g}_{t,c}) + \alpha, 0\}
$$
$$
+ \frac{1}{\sum_{c \in C} |T_c^r|} \sum_{c \in C, t \in T_c^r} \max\{d(\bar{\mathbf{g}}^r, \mathbf{g}_{t,c}) - d(\bar{\mathbf{g}}^h, \mathbf{g}_{t,c}) + \alpha, 0\}, \tag{3}
$$

where α is the margin and $d(\mathbf{x}, \mathbf{y}) = \|\mathbf{x} - \mathbf{y}\|_2$.

Besides the triplet loss, as discussed above, for the coded hate speech, if the coded words are in the coded meanings, they should group together in the embedding space because the contexts of different coded words are similar, i.e., expressing the social hate. To this end, we further aim to minimize the distance between the word embedding in coded meaning $\mathbf{g}_{t,c}$ and the generalized coded meaning anchor $\bar{\mathbf{g}}^h$ based on the mean squared loss:

$$
\mathcal{L}_{mse} = \frac{1}{\sum_{c \in C} |T_c^h|} \sum_{c \in C, t \in T_c^h} (\mathbf{g}_{t,c} - \bar{\mathbf{g}}^h)^2. \tag{4}
$$

Finally, the objective function for the general layer is defined as:

$$
\mathcal{L}_g = \mathcal{L}_{tri} + \lambda \mathcal{L}_{mse}, \tag{5}
$$

where λ is a hyper-parameter.

Specific Layer. The general layer mainly derives the generalized coded meaning anchor to detect the coded hate speech. Because we have the labeled corpus for each coded word $c \in C$, we can leverage the regular and hateful texts related with each coded word, T_c^r and T_c^h, to further improve the performance of specific hate speech detection. To this end, we propose the specific layer to detect whether a coded word $c \in C$ is in its coded meaning. In the specific layer, we derive the prototype representations of regular and coded meanings for each coded word c, respectively, and then adopt another neural network to finetune the coded word representation to make it close to the corresponding prototype.

Given a coded word $c \in C$, we adopt the mean operation to derive prototype representations of regular and coded meanings, $\bar{\mathbf{e}}_c^r$ and $\bar{\mathbf{e}}_c^h$, respectively.

$$
\bar{\mathbf{e}}_c^r = \frac{1}{|T_c^r|} \sum_{t \in T_c^r} \mathbf{e}_{t,c} \quad \bar{\mathbf{e}}_c^h = \frac{1}{|T_c^h|} \sum_{t \in T_c^h} \mathbf{e}_{t,c}. \tag{6}
$$

Then, given $\mathbf{e}_{t,c}$, $\bar{\mathbf{e}}_c^r$ and $\bar{\mathbf{e}}_c^h$, we first use general layer to derive $\mathbf{g}_{t,c}$ by Eq. (2). Similarly, we have $\bar{\mathbf{g}}_c^r = g_\theta(\bar{\mathbf{e}}_c^r)$, $\bar{\mathbf{g}}_c^h = g_\theta(\bar{\mathbf{e}}_c^h)$. After that, we adopt a neural network f_{ω_c} to further derive the specialized representation for $c \in C$.

$$
\mathbf{s}_{t,c} = f_{\omega_c}(\mathbf{g}_{t,c}) \quad \bar{\mathbf{s}}_c^r = f_{\omega_c}(\bar{\mathbf{g}}_c^r) \quad \bar{\mathbf{s}}_c^h = f_{\omega_c}(\bar{\mathbf{g}}_c^h). \tag{7}
$$

It is worth noting that to achieve the specific hate speech detection, we train a unique neural network f_{ω_c} for each coded word c. We consider $\bar{\mathbf{s}}_c^r$ and $\bar{\mathbf{s}}_c^h$ as the specialized regular and coded meaning anchors for the coded word c.

To train the neural network, for each coded word c, we adopt the triplet loss as the objective function to make $\mathbf{s}_{t,c}$ close to the corresponding anchors.

$$
\mathcal{L}_{s,c} = \frac{1}{|\mathcal{T}_c^h|} \sum_{t \in \mathcal{T}_c^h} \max\{d(\bar{\mathbf{s}}_c^h, \mathbf{s}_{t,c}) - d(\bar{\mathbf{s}}_c^r, \mathbf{s}_{t,c}) + \alpha, 0\}
$$
$$
+ \frac{1}{|\mathcal{T}_c^r|} \sum_{t \in \mathcal{T}_c^r} \max\{d(\bar{\mathbf{s}}_c^r, \mathbf{s}_{t,c}) - d(\bar{\mathbf{s}}_c^h, \mathbf{s}_{t,c}) + \alpha, 0\}. \tag{8}
$$

Overall, CODE is trained in an end-to-end manner. The complete objective function is defined as below:

$$
\mathcal{L} = \mathcal{L}_g + \lambda' \sum_{c \in \mathcal{C}} \mathcal{L}_{s,c}, \tag{9}
$$

where λ' is a hyperparameter to balance two layers.

3.3 Coded Hate Speech Detection

CODE can detect the hateful texts with observed and new coded words.

Coded Hate Speech Detection on Observed Coded Words. Given a new text t with an observed coded word $c \in \mathcal{C}$, CODE derives the coded word representation $\mathbf{s}_{t,c}$ from the specific layer and then compares the distances from $\mathbf{s}_{t,c}$ to the specialized regular and coded meaning anchors ($\bar{\mathbf{s}}_c^r$ and $\bar{\mathbf{s}}_c^h$), respectively. If the coded word representation is closer to the specialized coded meaning anchors, i.e., $d(\bar{\mathbf{s}}_c^h, \mathbf{s}_{t,c}) < d(\bar{\mathbf{s}}_c^r, \mathbf{s}_{t,c})$, the text t will be predicted as hate speech.

Coded Hate Speech Detection on New Coded Words. Given a new text t with a new coded word $c \notin \mathcal{C}$, CODE leverages the general layer for coded meaning detection because there is no specialized network built for the new coded word. First, CODE derives the coded word representation $\mathbf{g}_{t,c}$ from the general layer. It then compares the distances from $\mathbf{g}_{t,c}$ to the generalized regular and coded meaning anchors ($\bar{\mathbf{g}}^r$ and $\bar{\mathbf{g}}^h$), respectively. If the coded word representation is closer to the generalized coded meaning anchor, i.e., $d(\bar{\mathbf{g}}^h, \mathbf{g}_{t,c}) < d(\bar{\mathbf{g}}^r, \mathbf{g}_{t,c})$, CODE will predict the text t as hate speech.

4 Experiments

4.1 Experimental Setup

Coded Hate Speech Corpus. In this work, we evaluate our approach by using the coded hate speech on Twitter. We first collect two sets of tweets, i.e., the benign and hateful tweets, for several selected coded words shown in Table 2. In benign tweets, the coded word is used in its regular meaning, while in hateful tweets, the coded word is used in coded meaning targeting a specific group. In order to collect benign and hateful tweets, we combine different sets of keywords with the coded word as query words to search tweets on Twitter, where the query words are chosen based on the regular and coded meanings of the coded word. For example, given the coded word Skittles, for crawling benign tweets, we compose a set of query words, including pack, eat, and kid, while for crawling hateful tweets, we use query words, such as Muslim, refugee, and terrorist combining with Skittles. By using this strategy, we collected 10000 benign tweets and 6342 hateful tweets. The detailed information of collected benign and hateful tweets for each coded word is shown in Table 2.

Table 2. The information of crawled benign and hateful tweets in the collection and the adopted training corpus size for each coded word

c	Benign	Hateful	Time period	$|\mathcal{T}_c^r|$	$|\mathcal{T}_c^h|$
Google	10000	69	09/20/16-04/09/18	300	20
M&M	10000	520	09/20/16-09/26/19	300	50
Pepe	10000	6499	08/16/17-09/26/19	300	50
Butterfly	10000	618	11/12/09-09/21/19	300	50
Skittles	10000	6342	09/20/16-10/01/19	0	0

Evaluation Tasks. We evaluate our model on two tasks. (1) *Coded hate speech detection on observed coded words.* We build a training corpus consisting of the coded words "Google", "M&M", "Pepe", and "Butterfly". We train a model with this corpus and evaluate its performance on new tweets containing these observed words. (2) *Coded hate speech detection on a new coded word.* Using the model built on observed coded words, we detect coded hate speech on new tweets containing an unobserved coded word, "Skittles".

As shown in Table 2, in order to simulate the unbalanced nature of coded hate speech, for the coded word, Google, we adopt 300 regular and 20 hateful tweets as labeled texts in training. For coded words, M&M, Pepe, and Butterfly, we adopt 300 regular and 50 hateful tweets in training. The remaining tweets in our collection for each coded word all serve as the testing set.

Hyperparameters. We adopt the pre-trained ELMo model[3] to derive coded word contextual embeddings. The dimension of coded word embeddings $\mathbf{e}_{t,c}$ is 2048. The dimension of the transformed general embeddings $\mathbf{g}_{t,c}$ is 1024. The dimension of the transformed word specific embeddings $\mathbf{s}_{t,c}$ is 512. The hyperparameters λ and λ' are set to 1 in default unless specified otherwise. Both neural networks g_θ and f_θ are fully connected networks with one hidden layer.

Baselines. We compare CODE with following baselines. (1) **LSTM** [4], which is a deep learning based binary classifier; (2) **ELMo**, which is based on the raw embeddings derived from ELMo and uses the same distance-based approach

---

[3] https://github.com/allenai/allennlp/blob/v0.9.0/tutorials/how_to/elmo.md.

proposed in this work to detect coded hate speech. The purpose is to show the advantage of our two-layer framework; (3) **S model**, which removes the general layer and train the neural network for each coded word separately to detect coded hate speech as a comparison to show the advantage of the general layer.

Table 3. Experimental results on coded hate speech detection for the observed (Google, M&M, Pepe, and Butterfly) and unobserved (Skittles) coded words in terms of precision (P), recall (R) and F1-score (F1)

Model	Google			M&M			Pepe			Butterfly			*Skittles*		
	P	R	F1	P	R	F1	P	R	F1	P	R	F1	P	R	F1
LSTM	0.589	0.524	0.555	0.579	0.392	0.468	0.937	0.489	0.643	0.933	0.997	0.963	0.674	0.548	0.605
ELMo	1.000	0.490	0.658	0.437	0.871	0.582	0.854	0.592	0.699	0.957	1.000	0.978	0.573	0.863	0.689
S model	0.821	0.653	0.727	0.600	0.758	0.670	0.919	0.558	0.694	0.957	1.000	0.978	-	-	-
CODE	0.889	0.653	**0.753**	0.630	0.715	**0.670**	0.871	0.664	**0.754**	0.959	1.000	**0.979**	0.689	0.855	**0.763**

4.2   Experimental Results

Coded Hate Speech Detection on Observed Coded Words. Table 3 shows experimental results on coded hate speech detection for each observed coded word, Google, M&M, Pepe, and Butterfly. For most coded words, our approach significantly outperforms the LSTM model. For LSTM, it usually requires a large amount of training data and a relatively balanced dataset between positive and negative labels. However, it is mostly not the case when we are dealing with coded hate speech. In comparison to ELMo, the post-transformation models (S model and CODE) have improvement on the F1-score. It indicates that training a coded hate speech detection transformation could further separate the coded word embeddings in regular and coded meanings than the original coded word embeddings by the pre-trained ELMo. For the coded word "Butterfly", all models have a nearly perfect performance on coded hate speech detection, which means "Butterfly" in regular meanings and coded meanings are already separated in the original embedding space. However, for other coded words in our experiments, making the transformation on original coded word embeddings can boost the performance on coded hate speech detection with a large margin. In CODE, the transformation on the general layer leverages more training data from multiple coded words, which creates a better context of hateful tweets. The specific layer embedding further combines contexts specifically related to each coded word to further improve the model performance. Hence, CODE has the best performance in terms of the F1-score for all words.

Coded Hate Speech Detection on a New Coded Word. The last column of Table 3 shows experimental results on coded hate speech detection for the unobserved coded word, "Skittles". Tweets containing "Skittles" are not in the training set. The predictions are made by models trained on other words. Similar to the first task, LSTM does not have the best performance due to the same issues. The embeddings by ELMo before any transformation already can achieve 0.689 F1-score. It indicates that there is a similarity in the context of the hateful tweets across different coded words. The general layer transformation in our approach further extracts the hateful context into a general layer embedding, which has the best performance of 0.763 on the new coded hate detection. The S model is not applicable here because it can only detect observed coded words. However, if we compare CODE with the S model trained only on tweets containing Skittles (F1-score: 0.702), our approach, which benefits from more training tweets, even outperforms it.

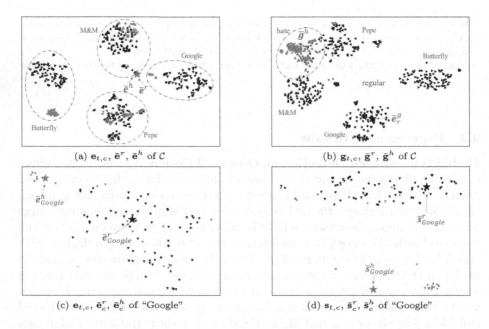

(a) $\mathbf{e}_{t,c}$, $\bar{\mathbf{e}}^r$, $\bar{\mathbf{e}}^h$ of $\mathcal{C}$

(b) $\mathbf{g}_{t,c}$, $\bar{\mathbf{g}}^r$, $\bar{\mathbf{g}}^h$ of $\mathcal{C}$

(c) $\mathbf{e}_{t,c}$, $\bar{\mathbf{e}}^r_c$, $\bar{\mathbf{e}}^h_c$ of "Google"

(d) $\mathbf{s}_{t,c}$, $\bar{\mathbf{s}}^r_c$, $\bar{\mathbf{s}}^h_c$ of "Google"

Fig. 2. The visualizations of coded word embeddings. The red and black colors indicate the coded and regular meanings, respectively. The triangles pointing down, up, left, and right indicate Google, M&M, Pepe, and Butterfly, respectively. The stars indicate the respected prototypes of coded/regular meanings. (Color figure online)

Visualization. We randomly select 300 regular tweets and 50 coded hateful tweets from the testing set for each coded word. We adopt TSNE [5] to project coded word embeddings to a two-dimensional space and visualize regular and coded meanings of each word in the corpus. Figures 2 (a) and (b) show the visualization results of regular and coded meanings as well as two prototype representations before and after the transformations. We can observe that before conducting the general layer transformation on the contextual embeddings, the embeddings of regular and coded meanings are mixed within the same words, and the prototypes of regular and coded meanings are also close to each other.

The embedding space is divided by words. There is no obvious general trend to detect hateful meanings for all words. After conducting the general layer transformation, the coded word embeddings in regular and coded meanings are clearly separated, and the prototypes of the two meanings are also separated. Meanwhile, most of the coded word embeddings in coded meanings are close to the prototype of coded meaning regardless of the regular meaning of the coded words, which meets our assumption that all hateful tweets have similar hateful contexts. The embedding space is divided by hateful or regular meanings. It indicates that our model exploits the common space for all coded words in the direction of hatefulness. We use the coded word "Google" as an example for the coded word specific layer embedding visualization. Figures 2 (c) and (d) show the visualization results of regular and coded meanings before and after two layers of transformations. After the transformations, the regular and coded meanings of "Google" are better separated in the new embedding space.

Table 4. Parameter sensitivity of λ and λ' on observed (average of all observed words) and unobserved (Skittles) coded words

(λ, λ')	Observed			Skittles		
	P	R	F1	P	R	F1
(1, 1)	0.718	0.819	0.763	0.689	0.855	0.763
(0.1, 1)	0.807	0.734	0.768	0.703	0.803	0.750
(10, 1)	0.608	0.874	0.718	0.660	0.911	0.765
(1, 0.1)	0.683	0.834	0.751	0.705	0.874	0.781
(1, 10)	0.783	0.779	0.781	0.685	0.819	0.746

Parameter Sensitivity. We also evaluate the parameter sensitivity of λ and λ' defined in Eqs. (5) and (9), respectively. We evaluate on observed (average of all observed words) and unobserved (Skittles) coded words. The results are shown in Table 4. λ controls the relative weight of MSE loss in comparison to triplet loss in the general layer. We change λ while keeping $\lambda' = 1$. When $\lambda = 0.1$, F1-score of the unobserved word decreases. When $\lambda = 10$, the average F1-score of the observed words decreases.

λ' controls the relative weight of the loss for specific layer in comparison to general layer in the overall objective function. We change λ' while keeping $\lambda = 1$. As λ' controls the trade-off of generalization and specialization of the whole model, the increase of $\lambda' = (0.1, 1, 10)$ increases the average F1-score of the observed words but decreases F1-score of the unobserved word.

5   Conclusion

In this paper, we have developed CODE to achieve coded hate speech detection by determining whether the coded word is in its coded meaning or not. We treat each coded word as a polysemy and use ELMo to map the coded word in each text to an embedding space. Based on that, we have developed a two-layer (general layer and specific layer) transformation approach. The general layer derives a generalized coded meaning anchor and makes the coded word in its coded meaning close to the anchor, while the specific layer derives a specialized coded meaning anchor for each observed coded word and also makes the specific coded word in its coded meaning close to the specialized coded meaning anchor. CODE can detect the hate speech with either observed or newly emerged coded words by comparing the distance to the specialized or generalized anchor, separately. Experimental results on a Twitter dataset show that our approach can effectively detect coded hate speech and significantly outperform the baseline methods. In the future, we plan to study how to identify newly emerged coded words based on the existing coded hate speech corpus.

Acknowledgement. This work was supported in part by NSF grants 1564250, 1946391, and 2103829.

References

1. Davidson, T., Warmsley, D., Macy, M., Weber, I.: Automated hate speech detection and the problem of offensive language. In: ICWSM (2017)
2. Fortuna, P., Nunes, S.: A survey on automatic detection of hate speech in text. ACM Comput. Surv. **51**(4) (2018)
3. Gitari, N.D., Zuping, Z., Damien, H., Long, J.: A lexicon-based approach for hate speech detection. Int. J. Multimed. Ubiquit. Eng. **10**(4), 215–230 (2015)
4. Hochreiter, S., Schmidhuber, J.: Long short-term memory. Neural Comput. **9**(8), 1735–1780 (1997)
5. van der Maaten, L., Hinton, G.: Visualizing data using t-SNE. J. Mach. Learn. Res. **9**, 2579–2605 (2008)
6. Magu, R., Joshi, K., Luo, J.: Detecting the hate code on social media. In: Eleventh International AAAI Conference on Web and Social Media (2017)
7. Magu, R., Luo, J.: Determining code words in euphemistic hate speech using word embedding networks. In: ALW (2018)
8. Mikolov, T., Chen, K., Corrado, G., Dean, J.: Efficient estimation of word representations in vector space. arXiv preprint arXiv:1301.3781 (2013)
9. Mou, G., Ye, P., Lee, K.: SWE2: SubWord enriched and significant word emphasized framework for hate speech detection. In: Proceedings of the 29th ACM International Conference on Information & Knowledge Management, pp. 1145–1154. Association for Computing Machinery, New York, October 2020
10. Peters, M., et al.: Deep contextualized word representations. In: NAACL (2018)
11. Qian, J., ElSherief, M., Belding, E., Wang, W.Y.: Hierarchical CVAE for fine-grained hate speech classification. In: EMNLP (2018)
12. Qian, J., ElSherief, M., Belding, E., Wang, W.Y.: Leveraging intra-user and inter-user representation learning for automated hate speech detection. In: NAACL (2018)
13. Rajamanickam, S., Mishra, P., Yannakoudakis, H., Shutova, E.: Joint modelling of emotion and abusive language detection. In: ACL, May 2020

14. Schmidt, A., Wiegand, M.: A survey on hate speech detection using natural language processing. In: Proceedings of the Fifth International Workshop on Natural Language Processing for Social Media (2017)
15. Tran, T., et al.: HABERTOR: an efficient and effective deep hatespeech detector. In: EMNLP, October 2020
16. Waseem, Z., Hovy, D.: Hateful symbols or hateful people? Predictive features for hate speech detection on twitter. In: Proceedings of the NAACL Student Research Workshop (2016)

Reducing Catastrophic Forgetting in Neural Networks via Gaussian Mixture Approximation

Hoang Phan[1,2], Anh Phan Tuan[1], Son Nguyen[1,2], Ngo Van Linh[1], and Khoat Than[1(✉)] (iD)

[1] Hanoi University of Science and Technology, Hanoi, Vietnam
khoattq@soict.hust.edu.vn
[2] VinAI Research, Hanoi, Vietnam

Abstract. Our paper studies the continual learning (CL) problems in which data comes in sequence and the trained models are expected to be capable of utilizing existing knowledge to solve new tasks without losing performance on previous ones. This also poses a central difficulty in the field of CL, termed as Catastrophic Forgetting (CF). In an attempt to address this problem, Bayesian methods provide a powerful principle, focusing on the inference scheme to estimate the importance of weights. Variational inference (VI), one of the most widely used methods within this vein, approximates the intractable posterior by a factorized distribution, thus offering computational efficiency. Notwithstanding many state-of-the-art performances in practice, this simple assumption about the posterior distribution typically limits the model capacity to some extent. In this paper, we introduce a novel approach to mitigate forgetting in the Bayesian approach via enriching the posterior distribution with mixture models, which intuitively promotes neural networks to acquire knowledge from multiple tasks at a time. Moreover, in order to reduce the model's complexity growth when the number of components increases, we propose a solution that conducts low-rank decomposition on the variance of each component based on neural matrix factorization. Extensive experiments show that our method yields significant improvements compared to prior works on different benchmarks.

Keywords: Continual learning · Catastrophic forgetting · Gaussian mixture

1 Introduction

Despite the fact that artificial intelligent agents have surpassed human beings at many specified tasks, their abilities are still far behind humans in terms of performing well on wide-ranged, disjointed problems. However, this could not be completely comparable, since current systems often fail to retain the previous

---

H. Phan, A. P. Tuan and S. Nguyen—Equal contribution.

© The Author(s), under exclusive license to Springer Nature Switzerland AG 2022
J. Gama et al. (Eds.): PAKDD 2022, LNAI 13280, pp. 106–117, 2022.
https://doi.org/10.1007/978-3-031-05933-9_9

knowledge while learning new tasks [13], whilst humans have a great capability of learning in a continuous manner: accumulating knowledge and avoiding forgetting. Continual learning, therefore, has emerged in recent years as a learning regime, allowing deep learning models to learn sequential tasks efficiently.

A branch of preliminary work in CL [1,5,11,14] capitalizes on the idea of injecting uncertainty into the neural network's parameters, which is known as Bayesian Neural Networks (BNNs) [2]. Unlike the original NN, BNN considers its parameters as random variables drawn from a given prior distribution. Based on this framework, Variational Continual Learning [14], or simply VCL, was one of the first proposals to formulate the continual learning as an approximation Bayesian inference problem in which the combination of online variational inference with an efficient sampling method soon achieved significant results. Uncertainty-based Continual Learning [1] then re-interpreted the KL-divergence term in VCL, defined the concept of uncertainty for hidden nodes and modified the KL-divergence term following the principle: the more uncertain a parameter is, the more likely it will be changed in subsequent tasks. Another advantage this method brings is that the number of parameters stored is less than other earlier works, which placed the importance on weights [11]. By contrast, Variational Generative Replay (VGR) [6] has shown that likelihood-focused methods - those that estimate the likelihood of the preceding tasks with synthetic data rather than directly employ the model's posteriors as prior for successive tasks - can outperform prior-focused methods. In a recent study, Uncertainty-guided Continual Bayesian Neural Networks (UCB) [5] defined a metric for measuring the weight's importance and thereby developed an appropriate training strategy.

In spite of the variational inference's popularity and the advantages it brings, the inference quality is still heavily affected by the parametric family of the posterior approximation distribution. Both recent methods VCL and UCB utilized a simple diagonal covariance Gaussian for posterior approximation, which is likely not flexible enough to match the true posterior, especially in the continual learning context. That being said, the nature of the data stream exhibits a large inconsistency. Because each of them is sampled from a different distribution, the problem of CF, that existing approaches suffer from, could be explained as the well-known mode-seeking in statistics, where the unimodal distribution is unable to capture information in multiple modes. We use an example in Fig. 1 as a simple illustration for this conclusion, a trimodal Gaussian mixture (blue curve) is approximated by a single Gaussian distribution and another mixture of two components. While both of these approximation distributions cover the middle mode, the RHS is covered by the mixture only. Since the overall range and shape of the mixture is much closer to the true distribution, it is obviously a better approximation to the target trimodal mixture.

Meanwhile, choosing the more expressive (i.e. richer representation capacity) variational family could help obtain better inference. With this intuition in mind, recent studies that go beyond mean-field variational inference are: normalizing flows [16], auxiliary variables [12] and mixture models [7]. Firstly, normalizing flows is a powerful framework that allows simple probabilistic density functions

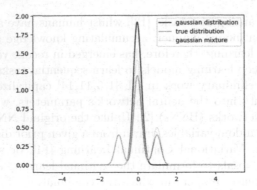

Fig. 1. The approximation abilities of Gaussian distribution and Mixture model. Given the true distribution that is a mixture of 3 Gaussian distributions $0.2\mathcal{N}(-1,0.2) + 0.6\mathcal{N}(0,0.2) + 0.2\mathcal{N}(1,0.2)$, KL-divergence is used to find the approximation of the true distribution. The learned Gaussian distribution is $\mathcal{N}(0,0.208)$ while the learned mixture of 2 Gaussian distributions is $0.754\mathcal{N}(-0.007,0.214) + 0.246\mathcal{N}(1.003,0.197)$. (Color figure online)

(PDFs) to be iteratively transformed into the target PDFs via a chain of invertible mappings. Secondly, auxiliary variables are included in the posterior in order to augment itself into more expressive and structured distributions. Finally, the mixture model is deemed to have the ability to approximate any given distribution with arbitrary closeness. However, all the directions seem to be impractical to apply to BNNs in the continual learning context, since their need is to add a huge number of extra parameters. In this work, we focus on an efficient solution to can exploit Gaussian mixture approximations in CL.

The main contributions of our paper could be depicted as follows: first, we present a novel inference method exploiting the Gaussian Mixture as a posterior approximation distribution and an information-theoretic view on how this could handle the CF. Accordingly, an efficient learning algorithm is also proposed, combining the Gumbel softmax reparameterization trick and the closed-form upper bound of KL diverge between two mixture models. Moreover, we employ a parameter reduction technique using Neural matrix factorization [4], which offers computational complexity benefits required for training. Finally, our inference process is experimentally proved to be widely incorporated into existing Bayesian-based learning methods for CL.

2 Backgrounds

2.1 Bayesian Inference

Consider the Bayesian Inference in the supervised learning setting, given the dataset $\mathcal{D} = \{x_i, y_i\}_i^n$ and a BNN parameterized by θ following the prior distribution $p(\theta)$. Typically, the main goal of Bayesian inference is to derive the posterior distribution over the weights $p(\theta|\mathcal{D})$, which is in an intractable form (i.e. involving integrals). The variational inference (VI) provides an efficient solution in which the true posterior is approximated to a variational distribution $q(\theta|\lambda)$ via minimizing the Kullback-Leibler divergence between these two distributions $\mathrm{KL}(p(\theta|\mathcal{D})||q(\theta|\lambda))$. In the most popular form, both of the prior and variational distributions are assumed to be Gaussian. In short, this optimization is equivalent to maximizing the Evidence Lower Bound (ELBO) w.r.t variational parameter λ:

$$\mathcal{L}(\theta) = \underbrace{E_{q(\theta|\lambda)}\log p(\mathcal{D}|\theta)}_{\text{expected-log likelihood}} - \underbrace{\mathrm{KL}(q(\theta|\lambda)||p(\theta))}_{\text{regularizer}} \qquad (1)$$

The above objective function now can be optimized with the Stochastic Gradient Variational Bayes (SGVB, [9]) estimator. More specifically, this process includes two main steps: Reparameterization Trick and Monte Carlo sampling [9].

2.2 Bayesian Approach in Continual Learning

Variational Continual Learning [14] used the online Bayesian update following the Bayes rule, and the posterior after observing previous tasks would be the prior of the next one. The sequential tasks are denoted as $\mathcal{D} = \{\mathcal{D}_1, \mathcal{D}_2, \ldots, \mathcal{D}_T\}$. Then, the posterior distributions are recursively computed as:

$$p(\theta|\mathcal{D}_{1:t}) = \frac{p(\theta|\mathcal{D}_{1:t-1})\, p(\mathcal{D}_t)}{p(\mathcal{D}_t|\mathcal{D}_{1:t-1})} \quad (0 < t \leq T) \qquad (2)$$

With $p(\theta)$ as the prior distribution placed on θ. Due to the intractable property, $q_t(\theta) \approx p(\theta|\mathcal{D}_{1:t}) \ \forall t$ is the tth task's posterior approximation. Specifically, both the prior q_0 and posterior q_i $(i > 0)$ are chosen to be multivariate diagonal Gaussian distributions to simplify the computation. The VCL's objective function on t^{th} step is:

$$\mathcal{L}_{\mathrm{VCL}}(\theta) = E_{q_t(\theta)} \log(p(\mathcal{D}_t|\theta)) - \mathrm{KL}(q_t(\theta)|q_{t-1}(\theta))$$

In a major advance in regularizing the change of the parameters, UCB [5] estimates the importance of each weight by the multiplicative inverse of its standard deviation $\Omega = \frac{1}{\sigma}$. According to this notion of importance, they controlled the parameter-wise learning rate update at each step as:

$$\Omega_\mu \leftarrow \frac{1}{\sigma} \Rightarrow \alpha_\mu \leftarrow \frac{\alpha_\mu}{\Omega_\mu}$$

$$\Omega_\sigma \leftarrow 1 \Rightarrow \alpha_\sigma \leftarrow \frac{\alpha_\sigma}{\Omega_\sigma}$$

The learning rate scheduler in UCB aimed to lessen the substantial shifting in important parameters, placed on their inherent uncertainty. This also provided a memory-beneficial consequence, since it neither accesses the past data nor stores the quantities associated with previous tasks.

2.3   Gumbel Softmax and Categorical Reparameterization

Normally, training a neural network often involves backpropagation through a chain of continuous-valued and differentiable functions. Even so, the discrete random variables used in stochastic neural networks to represent distributions sometimes yield a more meaningful and interpretable representation. In this section, we briefly summarize some concepts behind the idea of smoothly relaxing discrete distributions with Gumbel-Softmax and the way of training these models with reparameterization trick (path-derivative) gradients.

Gumbel-Softmax Trick [8]: Let α be an n-dimensional vector on simplex $\Delta^{n-1} = \{(x_1, x_2, ...x_n) | \ x_i \in (0,1), \sum_{i=1}^{n} x_i = 1\}$ and $g = \{g_1, g_2, ...g_n\}$ with g_i are i.i.d drawn from the Gumbel distribution $G(0, 1)$. Clearly, sampling from the multinomial distribution with probability vector given by α can be written as $y = \text{softmax}(\frac{\log \alpha + g}{\tau})$ where $\tau > 0$ is the temperature parameter.

3   Gaussian Mixture Approximation in Bayesian Inference for Continual Learning

In this section, we first present our proposal that exploits Gaussian mixture approximation in continual learning. Then we introduce a solution to reduce the number of parameters of the Gaussian mixture.

3.1   Proposed Method

A proper choice of the posterior approximation distribution theoretically expands the searching space for the true posterior. Especially in continual learning settings, neural networks must be learned on data from several tasks, a typically used unimodal Gaussian distribution is not rich and expressive enough to approximate the true posterior of weights. Accordingly, a Gaussian mixture approximation is more suitable to capture the multi-modality in data modeling. We conducted a simple experiment to show how GM can be a better approximation to the true distribution than a single Gaussian distribution (Fig. 1). This result provides convincing evidence to apply Gaussian mixture approximation in learning multiple tasks (i.e. scenario of continual learning).

For continual learning, for each task, we approximate the true posterior of weights to a Gaussian mixture instead of using unimodal Gaussian distribution as in existing studies [1,5,14]. The approximate distribution is presented as follows:

$$q(\theta|\lambda) = \sum_{i=1}^{K} \pi_i \, \mathcal{N}(\mu_i, \sigma_i) \ \ (\pi \in \Delta^{K-1}) \tag{3}$$

Note that the posterior learned from the previous task is often used as the prior in the current task [1,14]. Therefore, both the prior and variational distributions are Gaussian mixtures. They are plugged into Eq. 1. However, optimizing the evidence lower bound (Eq. 1) poses two main challenges. On the one hand, KL divergence between two Gaussian mixtures (the regularization term) does not have a closed-form formula. On the other hand, sampling from such distributions typically involves categorical variables (the expected-log likelihood). The lower bound in this case is thus difficult to optimize.

Algorithm 1: Reparameterization trick for Gaussian mixture

Input: Prior distribution: $p(\theta)$ and posterior: $q(\theta|\lambda) = \sum\limits_{i=1}^{K} \pi_i \, \mathcal{N}(\mu_i, \sigma_i)$

Number of samples: N; temperature : τ

Output: Estimation of the log-likelihood $E_{q(\theta|\lambda)} \log p(\mathcal{D}, \theta)$

Function Log-likelihoodEstimation($p(\theta)$, $q(\theta|\lambda)$, N, τ):

    for $n \leftarrow 1$ **to** N **do**

        for $k \leftarrow 1$ **to** K **do**

            $u_k \leftarrow \mu_k + \sigma_k \odot \epsilon_k$ where $\epsilon_k \sim \mathcal{N}(0,1)$

            $g_k \sim \text{Gumbel}(0,1)$

        end

        $y = (y_1, y_2, \ldots y_K) \leftarrow \text{softmax}(\frac{\log(\pi) + g}{\tau})$ //   Gumbel softmax trick

        $h_n \leftarrow \sum\limits_{k=1}^{K} u_k y_k$

    end

    return $\frac{1}{N} \sum\limits_{n=1}^{N} \log p(h_n)$;

In terms of the first difficulty, the Kullback-Leibler divergence, fortunately, has a closed-form upper bound [3]. Rather than directly maximizing the ELBO, we substitute the KL divergence in Eq. 1 with its upper bound. Theorem 1 presents the Upperbound_KL.

Theorem 1. *Consider 2 mixtures* $f(x) = \sum_{i=1}^{K} \pi_i^a \mathcal{N}(\mu_i^a, \sigma_i^a)$ *and* $g(x) = \sum_{i=1}^{K} \pi_i^b \mathcal{N}(\mu_i^b, \sigma_i^b)(\pi_i^a, \pi_i^b \in \Delta^{K-1})$, *we have the below inequality:*

$$\text{KL}(f|g) \leq \text{KL}(\pi^a|\pi^b) + \sum_{i=1}^{K} \pi_i^\alpha \, \text{KL}(\mathcal{N}(\mu_i^a, \sigma_i^a)|\mathcal{N}(\mu_i^b, \sigma_i^b)) \qquad (4)$$

Regarding the latter challenge, Gaussian mixture is regarded as a consolidation of the categorical and Gaussian distributions. We propose a strategy to approximate the expected-log likelihood (Eq. 1) with Monte Carlo and reparameterization trick. Algorithm 1 presents the approximation of the expected-log likelihood. The discrete mixing coefficients $y = \{y_1, y_2, .., y_K\}$ are reparameterized based on the Gumbel-Softmax trick. Then samples $u = \{u_1, u_2, .., u_K\}$) generated by each Gaussian component are linearly combined: $h = \sum\limits_{k=1}^{K} u_k y_k$ to calculate the expected-log likelihood. Moreover, the mixture coefficient π is reparameterized by a 1-1 transformation via softmax function: $\pi = \text{soft\_max}(0, \hat{\pi}_1, \hat{\pi}_2, \ldots \hat{\pi}_{K-1})$. For brevity, this turns the constrained optimization problem into an unconstrained one that could be optimized by back-propagation gradient.

We emphasize that the proposed technique is model-agnostic, which means that it is compatible with any Bayesian-based approaches. Without loss of generality, we analyze the application of our scheme on VCL [14] and UCB [5]. In Algorithm 2, we derived the training algorithm for-almost-all continual learning methods following the Bayesian principle. For those methods which adapt the learning in each iteration (e.g. UCB, Sect. 2.2), a `LearningRateUpdate` procedure, which takes the current learning rate α and parameter of interest λ and returns the updated learning rate, would be retained as in the original study.

Algorithm 2: Training of proposed method in continual learning scenario

---

Input: A sequence of T datasets $\mathcal{D}_{t=1}^{T} = \{x_t^{(n)}, y_t^{(n)}\}_{n=1}^{N_T}$
Prior distribution $p_t(\theta)$, number of samples : N
Learning rate α_λ, temperature: τ

Output: Update the variational parameter λ for t^{th} task
for $t \leftarrow 1$ to T do
> repeat
>> $L1 = \texttt{Log-likelihoodEstimation}(p_t(\theta), q_t(\theta|\lambda), \text{N}, \tau)$
>> $L2 = \texttt{Upperbound\_KL}(q_t(\theta|\lambda)|p_t(\theta))$
>> $L = L1 - L2$
>> $\lambda = \lambda - \alpha_\lambda \nabla_\lambda L$
>> $\alpha_\lambda = \texttt{LearningRateUpdate}(\alpha_\lambda, \lambda)$ // Optional
> until *convergence*;
end
return Updated variational parameter λ

3.2   Dimension Reduction via Neural Matrix Factorization

Using Gaussian mixture approximation leads to the number of parameters to being multiplied (along with the size of components) in comparison with unimodal Gaussian approximation. Consequently, optimization via a typical algorithm is expensive, and a dimensional reduction technique is necessary in this case. Recently, [17] proposed an idea of decomposing variance of each Gaussian component $\mathcal{N}(. |\mu, \sigma)$ in (3) by: $\sigma = \text{diag}(H) = \text{diag}(UV^T)$ for some matrix $H \in \mathbb{R}^{M \times N}, U \in \mathbb{R}^{M \times K}, V \in \mathbb{R}^{N \times K}$ and K is often set equal to a small positive integer. As a result, the number of parameters decreases from $M \times N$ to $(M + N) \times K$ in each component. Experimental results showed that this matrix factorization not only compresses the neural network, but also provides competitive performance.

Table 1. Average accuracy on final task.

Method	Dataset			
	Permuted MNIST	Split MNIST	Fashion MNIST	NotMNIST
VCL-Gauss	73.77	96.9	95.7	92.1
VCL-GMM	75.52	97.77	97.78	93.9

Unfortunately, it would be tough to tune the target rank for factorization in the continual learning scenario, since a large K causes the over-parameterization in some tasks whereas a small K might be underfitting in the others. For a better solution, we alternatively use neural matrix factorization [4] to overcome this shortcoming and describe the amelioration gained in Sect. 4.4. With $H \approx$ MLP(U, V), this dimension reduction is no longer a low-rank approximation. The density function of each Gaussian component in the posterior is: $\mathcal{N}(\, . \, |\mu, \sigma) = \mathcal{N}(\, . \, |\mu, f(U, V))$ where f is defined as the multilayer perceptron parameterized by θ.

4   Experiments

To study the contributions of the above methodologies in the continual learning scenario, we conducted extensive experiments in comparison with earlier baselines, which are also governed by the Bayesian regime. Additionally, in the last subsection, we analyze the effect of the matrix factorized used in dimension reduction. All the results are averaged on five random seeds.

Datasets: The datasets for evaluation are Permuted MNIST, Fashion MNIST, Split MNIST, and notMNIST.

Evaluation Metrics: At the point our model has been trained on i consecutive tasks so far, let $R_{i,j}$ be the accuracy of the achieved on j^{th} task. We use two different benchmark protocols (higher is better) to evaluate the performance at T^{th} task: $\text{ACC}_T = \frac{1}{T} \sum_{i=1}^{T} R_{T,i}$ and $\text{BWT}_T = \frac{1}{T-1} \sum_{i=1}^{T-1} R_{T,i} - R_{i,i}$. Conceptually, the **Average accuracy** (ACC_T) score is to measure model's overall performance, whereas **Backward Transfer** (BWT_T) indicates its knowledge transfer ability on preceding tasks.

4.1   Task-Incremental with Multi-head Architecture

First, we incorporate the introduced techniques into VCL on the four below benchmark datasets:

- **Permuted MNIST:** [10] is a variation of the original MNIST, in which the image's pixels in are randomly shuffled via a random (and fixed) permutation at each task.

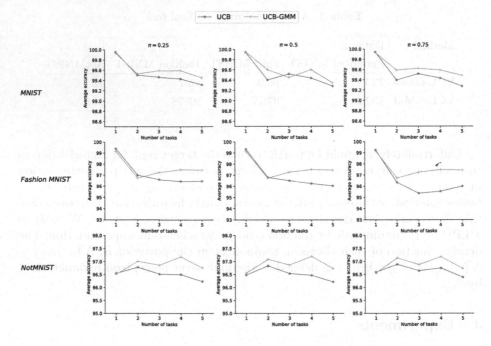

Fig. 2. Gaussian and Gaussian mixture on UCB.

- **Split MNIST:** [18] MNIST is partitioned into five subsets. In particular, each of them comprises images from two different classes, namely 0/1, 2/3, 4/5, 6/7, 8/9.
- **FashionMNIST, NotMNIST:** Similar to the split MNIST, our model would be incrementally trained in five separated binary classification tasks.

   We replicate the model architectures used in [14], which is composed of two fully connected layers (100 units each for permuted MNIST and 256 for split MNIST). The given results in Table 1 show that VCL with mixture posterior significantly outperforms the one with single diagonal Gaussian in terms of Average accuracy (1–2% each dataset).

   Ordinarily, the architecture of a discriminative model can be divided into two parts: classifier (final layer) and extractor (earlier layers), the catastrophic forgetting might occur on two components with different degrees, depending on the data. Various methods [14,15] relied on the additional information about task identification at inference time, as such, to avoid CF in the classifier.

4.2   Task-Incremental with Single-Head Architecture

The VCL's prerequisite about task boundaries, however, is rarely feasible in the use cases. In contrast, UCB implemented a single head network for all tasks, thus becoming an effective baseline. Before digging deeper into the experiment

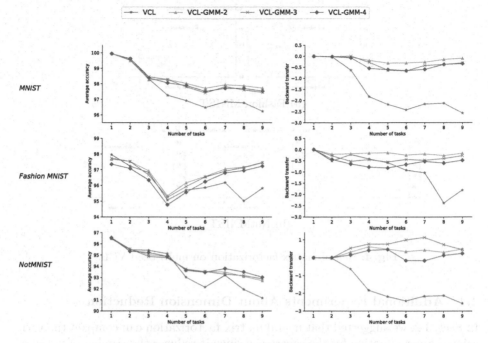

Fig. 3. ACC (left) and BWT (right).

result, we briefly recall a minor difference between UCB and VCL, which lies in the chosen prior. UCB uses a Gaussian mixture of two components (as mentioned in [2]) with the weight factor π.

To ensure a fair comparison, the model architecture again remains as the original UCB implementation[1]. Moreover, we carefully select π on different values $(0.25, 0.5, 0.75)$. We then plot the overall performance on split MNIST, fashion MNIST and notMNIST in Fig. 2.

4.3   Task-Incremental with Data Overlapping

Up to now, many prior works focused on datasets with isolation characteristics, which means there is no class overlapping in two separated tasks. To the best of our knowledge, this is the first experimental setup to simulate the repetition of data at different times. For example, the MNIST dataset now is allocated for nine classification problems: $0/1, 1/2, 2/3, \ldots, 8/9$. This larger correlation allows the learners to selectively transfer information between tasks in a soft way.

From Fig. 3, we observe that our proposed method in this setting produces stable outputs: the performances on a range of hyper-parameters (ACC curves stay closed to each other for all values of num_component $\in \{2, 3, 4\}$) and the knowledge transfer abilities (BWTs almost keep unchanged).

---

[1] https://github.com/SaynaEbrahimi/UCB.

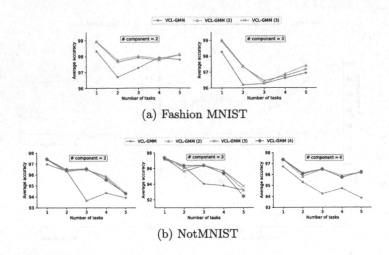

(a) Fashion MNIST

(b) NotMNIST

Fig. 4. Neural matrix factorization on multi-head VCL.

4.4   Additional Experiments About Dimension Reduction

In Sect. 3.2, we suggested that neural matrix factorization can compact the variational approximation for the covariance since it reduces the number of parameters without diminishing the capacity. In the implementation, we employ a two-hidden-layer MLP with the size of $(2K, K, K, 1)$ which decreases the number of trainable parameters for each component from $2MN$ to $MN+K(M+N+3K+1)$ and offers comparatively less resource in training when $M, N \gg K$. With similar strategies, Fig. 4 quantify the impact of this technique in Fashion MNIST and NotMNIST datasets. We observe ~1% increase in average accuracy on almost all experiments, especially $2, 4\%$ (from $93, 8 \rightarrow 96, 4$) in the case of 4 components GMM on NotMNIST.

5   Conclusion and Future Work

In this paper, we propose a framework, which is applicable to any Bayesian-based approach for continual learning. Our methodology unifies several recent proposals in variational inference and latent feature models and entails significant gains in the model's performance. We found that it is beneficial for incorporating these enhancements into VCL and UCB via distinct setups. The future work should take the initialization of mixtures into account (e.g. Iterated Laplace Approximations). In addition, we plan to properly find the number of components of the Gaussian mixture according to each task or even cast it into a trainable parameter while ensuring computation time and memory.

Acknowledgements. This work was funded by Gia Lam Urban Development and Investment Company Limited, Vingroup and supported by Vingroup Innovation Foundation (VINIF) under project code VINIF.2019.DA18.

References

1. Ahn, H., Cha, S., Lee, D., Moon, T.: Uncertainty-based continual learning with adaptive regularization. arXiv preprint arXiv:1905.11614 (2019)
2. Blundell, C., Cornebise, J., Kavukcuoglu, K., Wierstra, D.: Weight uncertainty in neural network. In: International Conference on Machine Learning, pp. 1613–1622. PMLR (2015)
3. Do, M.N.: Fast approximation of Kullback-Leibler distance for dependence trees and hidden Markov models. IEEE Signal Process. Lett. **10**(4), 115–118 (2003)
4. Dziugaite, G.K., Roy, D.M.: Neural network matrix factorization. arXiv preprint arXiv:1511.06443 (2015)
5. Ebrahimi, S., Elhoseiny, M., Darrell, T., Rohrbach, M.: Uncertainty-guided continual learning with Bayesian neural networks. In: International Conference on Learning Representations (ICLR) (2020)
6. Farquhar, S., Gal, Y.: A unifying Bayesian view of continual learning. arXiv preprint arXiv:1902.06494 (2019)
7. Jaakkola, T.S., Jordan, M.I.: Improving the mean field approximation via the use of mixture distributions. In: Jordan, M.I. (ed.) Learning in Graphical Models, pp. 163–173. Springer, Cham (1998). https://doi.org/10.1007/978-94-011-5014-9_6
8. Jang, E., Gu, S., Poole, B.: Categorical reparameterization with Gumbel-Softmax. In: International Conference on Learning Representations, ICLR (2017)
9. Kingma, D.P., Welling, M.: Auto-encoding variational Bayes. In: International Conference on Learning Representations, ICLR (2014)
10. Kirkpatrick, J., et al.: Overcoming catastrophic forgetting in neural networks. Proc. Natl. Acad. Sci. **114**(13), 3521–3526 (2017)
11. Linh, N.V., Nam, L.H., Phan, H., Than, K.: Auxiliary local variables for improving regularization/prior approach in continual learning. In: Pacific-Asia Conference on Knowledge Discovery and Data Mining (PAKDD) (2022)
12. Maaløe, L., Sønderby, C.K., Sønderby, S.K., Winther, O.: Auxiliary deep generative models. In: International Conference on Machine Learning, pp. 1445–1453. PMLR (2016)
13. McCloskey, M., Cohen, N.J.: Catastrophic interference in connectionist networks: the sequential learning problem. In: Psychology of Learning and Motivation, vol. 24, pp. 109–165. Elsevier (1989)
14. Nguyen, C.V., Li, Y., Bui, T.D., Turner, R.E.: Variational continual learning. In: International Conference on Learning Representations (ICLR) (2018)
15. Rebuffi, S.A., Kolesnikov, A., Sperl, G., Lampert, C.H.: iCaRL: incremental classifier and representation learning. In: Proceedings of the IEEE Conference on Computer Vision and Pattern Recognition, pp. 2001–2010 (2017)
16. Rezende, D., Mohamed, S.: Variational inference with normalizing flows. In: International Conference on Machine Learning, pp. 1530–1538. PMLR (2015)
17. Swiatkowski, J., et al.: The k-tied normal distribution: a compact parameterization of Gaussian mean field posteriors in Bayesian neural networks. In: International Conference on Machine Learning, pp. 9289–9299. PMLR (2020)
18. Zenke, F., Poole, B., Ganguli, S.: Continual learning through synaptic intelligence. In: International Conference on Machine Learning, pp. 3987–3995. PMLR (2017)

A Novel Semi-supervised Neural Network for Recognizing Parkinson's Disease

Zhehao Zhang[1], Xiaobo Zhang[1,2,3]($\boxtimes$), Dengmin Wen[1], Lilan Peng[1,3], and Yuxin Zhou[1]

[1] School of Computing and Artificial Intelligence, Southwest Jiaotong University, Chengdu 611756, China
{zhzhang7,llpeng,yuzs}@my.swjtu.edu.cn, {zhangxb,dmwen}@swjtu.edu.cn
[2] Institute of Artificial Intelligence, Southwest Jiaotong University, Chengdu 611756, China
[3] National Engineering Laboratory of Integrated Transportation Big Data Application Technology, Southwest Jiaotong University, Chengdu 611756, China

Abstract. Generative adversarial networks (GAN) have been mainly applied to tasks such as synthesis, segmentation, and reconstruction in the field of medical images since its appearance, and the research results in the field of classification are relatively rare. In the field of Parkinson's disease, the development of deep learning in this field has been limited due to the lack of available data sets and the differences between medical images and natural images. This paper proposes a new neural network for recognizing Parkinson's disease called Triple Progressive Generative Adversarial Networks (TP-GAN). Adding a classifier makes the model change from a two-person game to a three-person game, and introduces a manifold regularization method to guide the direction of classification decisions. The use of progressive networks to replace the traditional convolutional network makes the model perform better than the original network structure when processing large resolution data. The exprimental results demonstrate that our model performs better than the state-of-the-art baselines on the dataset of brain Magnetic Resonance Imaging (MRI).

Keywords: Semi-supervised · Generative adversarial networks · Parkinson's disease · MRI

1 Introduction

Parkinson disease (PD) is the second largest neurodegenerative disease in the world after Alzheimer's disease. At present, the diagnosis of Parkinson's disease mainly relies on medical history, clinical symptoms and physical signs. When clinical symptoms appear, the death of substantia nigra dopaminergic neurons is at least 50%, and the dopamine content of the striatum is reduced by more than 80% [5,12]. Therefore, when Parkinson's disease patients are diagnosed, most of the lesions are already in the middle or advanced stages. So far, there

© The Author(s), under exclusive license to Springer Nature Switzerland AG 2022
J. Gama et al. (Eds.): PAKDD 2022, LNAI 13280, pp. 118–130, 2022.
https://doi.org/10.1007/978-3-031-05933-9_10

is no cure for Parkinson's disease, so all efforts should be devoted to the early diagnosis of Parkinson's disease [22].

In recent years, generative adversarial networks have received wide attention in the field of medical imaging [3,20]. Many models have been applied to modern medical-assisted diagnosis, such as synthesis, segmentation and reconstruction.

However, the current research still has some limitations. Compared with natural images, medical images have higher resolution, lower contrast and fewer samples [24]. These factors make it impossible for GAN to achieve as good results when processing medical samples as when processing natural images [23].

In order to solve above problems, we propose a new Semi-supervised neural network Triple Progressive Generative Adversarial Networks (TP-GAN) for recognizing Parkinson's disease. Compared with the baseline in the experiment, and the experimental results show that TP-GAN shows better performance. The contributions of this paper are as follows:

1. We preprocess the PPMI dataset [11], and obtain a Parkinson's disease brain MRI image dataset containing three stages of control, PD and Prodromal PD at the same size.
2. A new neural network model is proposed to classify the different stages of Parkinson's disease. Used the PPMI dataset to recognizing Parkinson's disease and evaluate the model in the same experimental environment.
3. TP-GAN uses progressive network to replace the traditional convolutional network. Add classifiers and introduce manifold regularization to alleviate problems such as overfitting due to scarcity of medical data.

The rest of this paper is organized as follows. Section 2 introduces the related work of generative confrontation network and semi-supervised classification. Section 3 presents our new model Triple Progressive GAN. Section 4 gives the experimental results. Section 5 summarizes this paper.

2   Related Work

In recent years, various machine learning techniques have been used to identify Parkinson's disease from medical data. Prasuhn et al. proposed to cut the brain expansion image into multiple brain network atlas regions, using a greedy algorithm to filter and perform a weighted average on the selected combination of regions [13]. Gabriel et al. used the first-order and second-order statistical methods to detect the regions of the extracted features, so as to perform VBM on the magnetic resonance image [18]. Bowman et al. used elastic network regression technology to replace the original log loss function with a square loss function, and jointly use the L1 and L2 norm penalty terms for regularization, which can be effective when multimodal features are interconnected [2]. Sahand et al. used the diffusion-based MRI data set to calculate the average apparent diffusion coefficient of each patient, and used this as a feature to classify PD and patients with progressive supranuclear palsy patients with linear kernel support vector machines [21]. Huang et al. solved the problem of ambiguous gray and white matter in MRI images of PD patients by enhancing and segmenting the information of MRI images extracted by discrete wavelet transform and inputting the segmented images into fuzzy median filter, and improved the classification accuracy of PD images [9].

In recent years, deep learning methods have been extensively studied in the computer-aided diagnosis of Parkinson's disease based on MRI images. Based on the T1-weighted MRI image, Zhang et al. added a normalization layer to the fifth layer of the original AlexNet network to improve the overall learning rate and accelerate the convergence of the model [1]. Mohamed et al. used Siamese neural network for the diagnosis of Parkinson's disease to enhance the distribution of similar samples among groups by clustering data sets before applying classification [15]. Suvita et al. proposed the binary versions of Rao algorithms and applied to four publicly available Parkinson's disease datasets [16]. Basnin et al. combining the DenseNet model and LSTM to enhance the ability of model feature selection and used LSTM to discover the relationship between temporal characteristics [7].

3   Methods

In this section, we propose our TP-GAN. First of all, we added a classifier to make GAN have strong classification capabilities. In addition, we use progressive network to replace the traditional convolutional network, which makes our model more stable when processing high-resolution images and train faster. Finally, we added manifold regularization to the classifier to solve problems such as the model's easy overfitting and poor generalization ability. The framework of TP-GAN is shown in Fig. 1.

3.1   The Addition of Classifier

Our model contains three parts: generator, discriminator and classifier. The generator makes the generated sample distribution as similar to the real sample distribution as possible, $p_g(x|y) \approx p_{data}(x|y)$; The discriminator distinguishes whether a pair of data (x, y) is a real sample from the real sample distribution $p_{data}(x|y)$ or a generated sample generated by the generator that conforms to the generated sample distribution $p_g(x|y)$; The classifier makes the distribution of classified samples as similar as possible to the true conditional distribution, $p_c(y|x) \approx p_{data}(y|x)$.

During the training process, the generator maps the input random noise z $\sim p_z(z)$ that conforms to the distribution of the noise variable according to the given label y into a pseudo sample $G(y, z) \sim p_g(x|y)$, and output the corresponding sample-label pair(x_g, y_g). The classifier inputs the sample x, generates a pseudo-label y according to the classification sample distribution $p_c(y|x)$, and outputs a pair of data (x_c, y_c) belongs to the joint distribution $p(x)p_c(y|x)$. The discriminator needs to judge (x, y) from the true conditional distribution $p_{data}(x|y)$, (x_g, y_g) from the generator and (x_c, y_c) from the classifier, as much

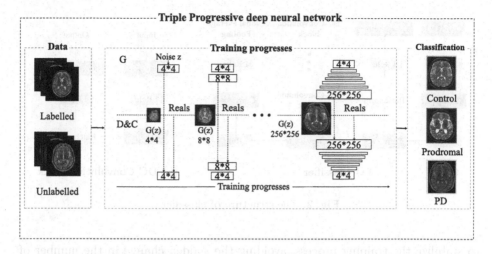

Fig. 1. General framework of TP-GAN

as possible put the sample-label pairs from the generator in the pseudo-sample classes.

Since the generator and classifier have opposite goals to the discriminator, in order to adapt to this situation, the optimal parameters of the model are solved, and the objective function is optimized in a mini-max method, so the objective function of training is:

$$\min_{C,G} \max_{D} U(C,G,D) = E_{(x,y) \sim p_{data}(x,y)}[\log D(x,y)]$$
$$+ (1-\alpha)E_{y \sim p(y), z \sim p_z(z)}[\log(1 - D(G(y,z),y))] \quad (1)$$
$$+ \alpha E_{(x,y) \sim p_c(y|x)}[\log(1 - D(x,y))]$$

where α is a constant adjustment parameter used to control the degree of influence of the classifier and generator on the objective function, $p_{data}(x,y)$ is the true sample distribution, $p_z(z)$ is the noise variable distribution, and $p_c(y \mid x)$ is the categorical sample distribution. The structure of classifier is shown in Fig. 2.

3.2 Progressive Network

Equation (1) is balanced when $p_{data}(x,y) = (1-\alpha)p_g(x,y) + \alpha p_c(y \mid x)$, so when the generated sample distribution is closer to the true sample distribution, the classification sample distribution is also more close to the true sample distribution. We gradually increase the scale of generators, discriminators and classifiers so that the model can better improve the details during the training process.

The generator, discriminator, and classifier have 7 levels, each of which learns images with different resolutions, and the level is the logarithm of the resolution learned by the current level with the base two $(level = log_2(pix))$, during the transition from one level to another, the smooth transition technology is used

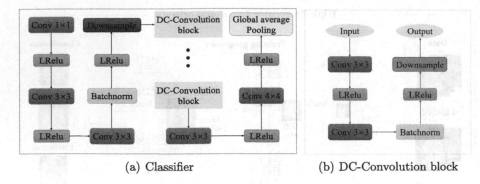

(a) Classifier   (b) DC-Convolution block

Fig. 2. The structure of classifier

to stabilize the training process, avoiding the sudden change in the number of network layers, which makes it take extra time for the model to converge from a turbulent state to a stable state. By gradually increasing the number of network layers, most of the iterations in the training process can be completed at low resolution, which greatly accelerates the training speed and makes the training process more stable, while retaining and outputting low-resolution training information will go to the next level.

3.3 Manifold Regularization and Use of Generated Samples

We added manifold regularization to the classifier to keep the classifier unchanged for the local perturbation of the data manifold, and similarly mark adjacent points on the manifold to enhance the stability of the classifier. In the training process, based on the high-quality samples generated by the generator, we input the generated samples as unlabeled samples into the classifier, which alleviates the problem that the model is easy to overfit due to the scarcity of medical images, and further helps the classifier learn the distribution characteristic information of the sample.

The total loss function of the model can be defined as:

$$L = L_D + L_G + L_C \tag{2}$$

where L_D is the discriminator loss, L_G is the generator loss, L_C is the classifier loss.

$$
\begin{aligned}
\nabla_D = \nabla[\frac{1}{k}(\sum_{(x,y)} \log D(x,y) + \sum_{(x_c,y_c)} \log D(x_c,y_c)) \\
+ \frac{1-\alpha}{m} \sum_{(x_g,y_g)} \log(1 - D(x_g,y_g)) + \frac{\alpha}{n} \sum_{(x_c,y_c)} \log(1 - D(x_c,y_c))]
\end{aligned}
\tag{3}
$$

$$\nabla_G = \nabla[\frac{1-\alpha}{m} \sum_{(x_g,y_g)} \log(1 - D(x_g, y_g))] \tag{4}$$

where ∇_D is the gradient ascend equation when the discriminator updates the parameters during the training process, ∇_G is the gradient descent equation when the generator is updated, k, m, n are the sample numbers of the input discriminator, generator and classifier respectively.

The classifier loss L_c is divided into three parts: supervised loss, unsupervised loss and semi-supervised loss. Supervised loss is the cross-entropy loss between the prediction of the real labeled image and the real label; the unsupervised loss is the real unlabeled image and the generator generates the confrontation loss of the image to help the classifier better learn the distribution feature information of the real sample; the semi-supervised loss is a manifold regularization term that penalizes the change gradient of the classification decision to make the classifier output a locally consistent classification result, thereby improving the classification decision-making ability of the classifier under disturbance.

$$L_C = \rho L_{supervised} + \varphi L_{unsupervised} + \omega L_{MR} \tag{5}$$

where $L_{supervised}$ is the cross-entropy loss, $L_{unsupervised}$ is the confrontation loss function, L_{MR} is the manifold regularization term, ρ, φ and ω are the corresponding weight coefficients, representing the influence degree of different types of data on the classifier.

$$L_{supervised} = -E_{(x,y)\sim p_{data}(x,y)}[\log D(x, y)] \tag{6}$$

$$L_{unsupervised} = E_{y\sim p(y),z\sim p_z(z)}[\log(1 - D(G(y, z), y))] \\ + E_{(x,y)\sim p_c(y|x)}[\log(1 - D(x, y))] \tag{7}$$

$$L_{MR} = \int_{x\in M} ||\nabla_M c||^2 dP_x(x) \tag{8}$$

where M is a submanifold of P_x, ∇_M is the gradient of c along manifold M, and c represents the classifier. Since it is very difficult to directly calculate the manifold regularization term, the graph Laplacian approximation is used based on labeled samples and unlabeled samples.

$$L_{MR} \approx \frac{1}{n} \sum_{i=1}^{n} ||c(g(z^{(i)})) - c(g(z^{(i)}) + \epsilon \bar{r}(z^{(i)}))||_2^2 \tag{9}$$

$$\bar{r}(z) = g(z + \eta\bar{\delta}) - g(z^{(i)}) \tag{10}$$

where $g(z^{(i)})$ represents the classification decision on the manifold, ϵ represents the adjustment parameter of the update step size of the gradient direction of the manifold, $\bar{r}(z)$ represents the approximation of the manifold gradient on z, $\delta \sim N(0,1)$, $\bar{\delta} = \frac{\delta}{||\delta||}$, η is the adjustable step size.

$$L_{MR} = E_{z\sim p_z(z),\delta\sim N(\delta)} ||c(g(z)) - c(g(z + \epsilon\bar{\delta}))||_2^2 \tag{11}$$

$$\nabla_C = \nabla[\alpha(-\frac{\rho}{h}\sum_{(x_c,y_c)}\log D(x_c,y_c)) + \frac{\varphi}{j+l}(\sum_{(x_{un},y_{un})}\log(1 - D(x_{un},y_{un}))$$

$$+ \sum_{(x_g,y_{cg})}\log(1 - D(x_g,y_{cg}))) + \frac{\omega}{l}\sum_{i=1}^{l}||c(g(z^{(i)})) - c(g(z^{(i)} + \epsilon\bar{\delta}))||_2^2]$$

(12)

where ∇_C is the gradient descend equation when the classifier updates the parameters during the training process, h, j, and l are the number of real labeled samples, real unlabeled samples and pseudo samples in the input classifier, respectively.

In summary, the algorithm is shown in Algorithm 1.

Algorithm 1. Algorithm process of TP-GAN

Input: The noises z~ $p_z(z)$, the real labelled samples x_l ~p(x), the unlabelled real samples $x_{un} \sim p(x)$.

Output: The classification result (x_c, y_c), the generate pseudo samples-label pairs (x_g, y_g).

1: **for** number of training iterations **do**
2:      Sample z and y to generate G(y,z) and (x_g, y_g).
3:      Sample x_l to compute y_l and (x_l, y_l).
4:      Sample x_{un} to compute y_{un} and (x_{un}, y_{un}).
5:      Sample x_g and $\{z^{(1)}, z^{(2)}, \cdots, z^{(l)},\}$ to compute y_{cg} and (x_g, y_{cg}).
6:      Concatenate $(x_l, y_l), (x_{un}, y_{un})$ and (x_g, y_{cg}) to (x_c, y_c).
7:      Sampling (x, y) and (x_c, y_c).
8:      Update D by ascending along its stochastic gradient according to Eq. 3
9:      Compute $L_{supervised}$, $L_{unsupervised}$ and L_{MR} aaccording to Eq. 6, 7, 11
10:     Update C by descending along its stochastic gradient according to Eq. 4
11:     Update G by descending along its stochastic gradient according to Eq. 12
12: **end for**

---

4   Experiments and Results

4.1   Dataset

We used brain MRI dataset from the Parkinson Progression Marker Initiative (PPMI) database to evaluate the proposed method. The PPMI is a comprehensive observational, international, multicenter study designed to identify PD progression biomarkers both to improve understanding of disease etiology and course and to provide crucial tools to enhance the likelihood of success of PD modifying therapeutic trials [11].

We downloaded the MRI dataset of PPMI database from the PPMI platform. In our MRI data, there is a total of 2292 samples with three categories: normal samples (Control), confirmed PD (PD), and samples in the incubation period (Prodromal PD). All samples are in axial view.

The dataset is split into training, testing and validation set with 1836:228:228 respectively. The training set contains 228 MRI samples with data labels and 1836 MRI samples without data labels. The dataset allocation is shown in Table 1.

Table 1. The details of dataset allocation

Subject	Training set		Validation set	Testing set	Total
	Labelled	Unlabelled	Labelled	Labelled	
Control	54	381	54	54	543
PD	155	1091	155	155	1556
Prodromal PD	19	136	19	19	193

4.2 Experimental Settings

All experiments are conducted on a platform with ubuntu system and NVIDIA Tesla P100 16G GPU, and tensorflow and keras frameworks were used to build models to complete the experiments. In our experiments, we found the best batch size is 32 in the range of {8,16,32,64}. Also, we found the best epoch is 100, and the best learning rate is 3×10^{-4}. Moreover, we optimize all of our networks with Adam and we found when the first order exponential decay rate for the moment estimate β_1 is equal to 0.3, the epsilon ϵ is 10^{-6}, and the second order exponential decay rate for the moment estimate β_2 is 0.999, training for the best results. In order to view these parameters more intuitively, Table 2 lists the best parameters of our model in the PPMI dataset.

In Table 2, on the PPMI dataset, the trade-off parameter α is set 0.3 as the best parameter.

Table 2. The parameters of TP-GAN based on the PPMI dataset.

Parameters	Optimal value
Batch size	32
epoch	100
Learning rate	10^{-4}
β_1	0.3
β_2	0.999
ϵ	10^{-6}
α	0.3

Specifically, we used k-fold cross-validation with $k = 10$ for choosing appropriate model parameters, and effectively avoid over-fitting that may be caused by too small data sets.

The performance of the classification is measured using three metrics, i.e., Accuracy (ACC), Sensitive (SEN), F1-score (F1).

4.3   Comparison Methods

We compared our model with Categorical Generative Adversarial Networks (Cat-GAN) [19], Improved Techniques for Training GANs (ImprovedGAN) [14], Triple Generative Adversarial Nets(TripleGAN) [10], Deep residual network(Resnet) [6], Densely Connected Convolutional Networks(Densenet) [8], Very Deep Convolutional Networks (VGGnet) [17]. We chose these models because they are used as baselines to achieve better results in image classification tasks.

4.4   Experimental Results and Analysis

In this section, we presented the results of the PPMI dataset under the same experimental environment.

The performance of our model and baselines on the PPMI dataset for different performance indicators is shown in Tables 3, 4 and 5, and the results with the best classification performance in each epoch are marked in bold.

In Tables 3, 4 and 5, the ACC, SEN and F1 of our model reached 86.53%, 85.92% and 86.95%, respectively, achieving classification performance comparable to or better than the current baseline. This shows that our model can learn the true sample distribution better than baselines, and our model is a promising network for classification tasks.

Table 3. Accuracy of the proposed model and six advanced classification models (%)

Model	25 epochs	50 epochs	75 epochs	100 epochs	Average
CatGAN	69.89	72.28	74.75	77.63	73.64
ImprovedGAN	78.26	79.69	80.29	81.57	79.95
TripleGAN	77.03	79.52	81.24	82.84	80.16
Resnet	79.67	81.71	83.08	83.61	82.01
Densenet	70.87	73.29	76.33	79.91	75.10
VGGnet	77.19	79.37	81.73	83.12	80.35
TP-GAN	**79.76**	**83.84**	**85.24**	**86.53**	**83.84**

Table 4. Sensitive of the proposed model and six advanced classification models (%)

Model	25 epochs	50 epochs	75 epochs	100 epochs	Average
CatGAN	73.27	75.67	77.06	81.94	76.99
ImprovedGAN	77.28	78.29	79.37	80.61	78.89
TripleGAN	77.76	79.33	81.91	83.06	80.51
Resnet	78.82	80.26	83.27	84.71	81.77
Densenet	71.27	74.34	78.73	82.29	76.66
VGGnet	78.09	80.02	82.01	84.36	81.12
TP-GAN	**79.63**	**82.39**	**84.26**	**85.92**	**83.05**

Table 5. F1-score of the proposed model and six advanced classification models (%)

Model	25 epochs	50 epochs	75 epochs	100 epochs	Average
CatGAN	69.93	79.28	75.37	77.81	74.10
ImprovedGAN	78.47	79.16	79.83	81.91	79.84
TripleGAN	77.18	79.29	81.87	83.07	80.35
Resnet	79.28	81.75	83.47	85.26	82.44
Densenet	71.44	74.42	77.27	79.97	75.78
VGGnet	77.28	78.92	81.49	83.73	80.36
TP-GAN	**80.17**	**83.69**	**85.63**	**86.95**	**84.11**

In order to compare the experimental results based on the PPMI data set more intuitively, the visualization of Table 3, 4 and 5 above is shown in Fig. 3. It can be clearly seen from Fig. 3 that on the PPMI data set, our model outperforms other baselines in every evaluation index, which further proves that our model can be applied to image classification tasks.

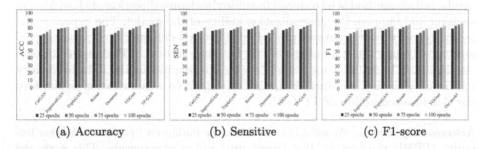

(a) Accuracy   (b) Sensitive   (c) F1-score

Fig. 3. Visualization of comparing results for TP-GAN and other baselines method on PPMI dataset by three metrics.

4.5 Parameters Sensitivity Analysis

For our TP-GAN, we adjusted different batchsize, learning rate, β_1 and β_2 to make the model achieve the best performance [4]. We also designed the trade-off parameter α of TP-GAN, and find the best α in the range of {0.1, 0.2, 0.3, ..., 0.9}. The TP-GAN classification performance with different parameters α on PPMI dataset is shown in Fig. 4. As shown in Fig. 4, how the classification performance of TP-GAN changes with different α. The horizontal axis represents the change of the parameter α, and the vertical axis is an index for evaluating classification performance. In Fig. 4, we can see that when $\alpha = 0.3$, all indicators have achieved good results.

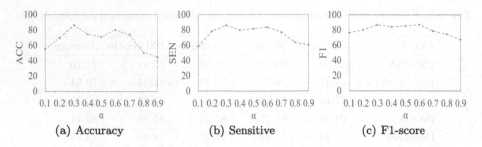

(a) Accuracy   (b) Sensitive   (c) F1-score

Fig. 4. Visualization results of the TP-GAN classification performance with the parameters α on PPMI dataset by three metrics.

5 Conclusion and Future Work

This paper proposes a new neural network for recognizing Parkinson's disease called Triple Progressive Generative Adversarial Networks (TP-GAN). We first add a classifier module to make the training process more stable. In addition, we use progressive networks, which greatly improves the training speed and has better performance when processing high-resolution images. Then, the manifold regularization method is introduced to guide the change direction of classification decision and improve the generalization ability of the model. The experimental results on a real-world dataset prove that TP-GAN demonstrated superior performance against the state-of-the-art six baselines in task of Parkinson's disease classification. In future work, we will try to utilize multi-modal data to identify Parkinson's disease, and further update the manifold regularization to improve the classification ability of TP-GAN.

Acknowledgements. We would like to thank the Parkinson Progression Marker Initiative (PPMI) platform for the dataset used in our experiments. This work was supported by the National Natural Science Foundation of China (No. 62102330), the Key Research and Development Programme in Sichuan Province of China (No. 2021YFS0302) and the Fundamental Research Funds for the Central Universities (Nos. 2682021 ZTPY009 and 2682021CX040).

References

1. Aghzal, M., Mourhir, A.: Early diagnosis of Parkinson's disease based on handwritten patterns using deep learning. In: 2020 Fourth International Conference on Intelligent Computing in Data Sciences (ICDS), pp. 1–6. IEEE (2020)
2. Bowman, F.D., Drake, D.F., Huddleston, D.E.: Multimodal imaging signatures of Parkinson's disease. Front. Neurosci. **10**, 131 (2016)
3. Delannoy, Q., et al.: SegSRGAN: super-resolution and segmentation using generative adversarial networks - application to neonatal brain MRI. Comput. Biol. Med. **120**, 103755 (2020)

4. Ebtehaj, I., Bonakdari, H., Zaji, A.H., Sharafi, H.: Sensitivity analysis of parameters affecting scour depth around bridge piers based on the non-tuned, rapid extreme learning machine method. Neural Comput. Appl. **31**(12), 9145–9156 (2018). https://doi.org/10.1007/s00521-018-3696-6

5. Fayyad, M., et al.: Parkinson's disease biomarkers based on α-synuclein. J. Neurochem. **150**(5), 626–636 (2019)

6. He, K., Zhang, X., Ren, S., Sun, J.: Deep residual learning for image recognition. CoRR, abs/1512.03385 (2015)

7. Hegazy, M., Cho, M.H., Lee, S.Y.: Half-scan artifact correction using generative adversarial network for dental CT. Comput. Biol. Med. **132**, 104313 (2021)

8. Huang, G., Liu, Z., Weinberger, K.Q.: Densely connected convolutional networks. CoRR, abs/1608.06993 (2016)

9. Huang, Y., et al.: Wavelet k-means clustering and fuzzy-based method for segmenting MRI images depicting Parkinson's disease. Int. J. Fuzzy Syst. **23**(6), 1600–1612 (2021)

10. Li, C., Xu, K., Zhu, J., Zhang, B.: Triple generative adversarial nets. CoRR, abs/1703.02291 (2017)

11. Marek, K., et al.: The Parkinson progression marker initiative (PPMI). Progress Neurobiol. **95**(4), 629–635 (2011)

12. Olanow, C.W., Stern, M.B., Sethi, K.: The scientific and clinical basis for the treatment of Parkinson disease. Neurology **72**(21 Suppl. 4), S1–S136 (2009)

13. Prasuhn, J., Heldmann, M., Münte, T.F., Brüggemann, N.: A machine learning-based classification approach on Parkinson's disease diffusion tensor imaging datasets. Neurol. Res. Pract. **2**(1), 1–5 (2020)

14. Salimans, T., Goodfellow, I.J., Zaremba, W., Cheung, V., Radford, A., Chen, X.: Improved techniques for training GANs. CoRR, abs/1606.03498 (2016)

15. Shalaby, M., Belal, N.A., Omar, Y.: Data clustering improves Siamese neural networks classification of Parkinson's disease. Complexity **2021**, 3112771:1–3112771:9 (2021)

16. Sharma, S.R., Singh, B., Kaur, M.: Classification of Parkinson disease using binary Rao optimization algorithms. Expert Syst. J. Knowl. Eng. **38**(4) (2021)

17. Simonyan, K., Zisserman, A.: Very deep convolutional networks for large-scale image recognition. In: 3rd International Conference on Learning Representations, ICLR 2015, San Diego, CA, USA, 7–9 May 2015, Conference Track Proceedings (2015)

18. Solana-Lavalle, G., Rosas-Romero, R.: Classification of PPMI MRI scans with voxel-based morphometry and machine learning to assist in the diagnosis of Parkinson's disease. Comput. Methods Programs Biomed. **198**, 105793 (2021)

19. Springenberg, J.T.: Unsupervised and semi-supervised learning with categorical generative adversarial networks. In: 4th International Conference on Learning Representations, ICLR 2016, San Juan, Puerto Rico, 2–4 May 2016, Conference Track Proceedings (2016)

20. Suh, S., Lee, H., Lukowicz, P., Lee, Y.O.: CEGAN: classification enhancement generative adversarial networks for unraveling data imbalance problems. Neural Netw. **133**, 69–86 (2021)

21. Talai, S., Boelmans, K., Sedlacik, J., Forkert, N.D.: Automatic classification of patients with idiopathic Parkinson's disease and progressive supranuclear palsy using diffusion MRI datasets. In: Medical Imaging 2017: Computer-Aided Diagnosis, Orlando, Florida, United States, 11–16 February 2017. SPIE Proceedings, vol. 10134, p. 101342H. SPIE (2017)

22. Zeighami, Y., et al.: A clinical-anatomical signature of Parkinson's disease identified with partial least squares and magnetic resonance imaging. NeuroImage **190**, 69–78 (2019)

23. Zhao, C., Wang, T., Lei, B.: Medical image fusion method based on dense block and deep convolutional generative adversarial network. Neural Comput. Appl. **33**(12), 6595–6610 (2020). https://doi.org/10.1007/s00521-020-05421-5

24. Zhou, Z., Zhai, X., Tin, C.: Fully automatic electrocardiogram classification system based on generative adversarial network with auxiliary classifier. Expert Syst. Appl. **174**, 114809 (2021)

ADAM: An Attentional Data Augmentation Method for Extreme Multi-label Text Classification

Jiaxin Zhang, Jie Liu$^{(\boxtimes)}$, Shaowei Chen, Shaoxin Lin, Bingquan Wang, and Shanpeng Wang

College of Artificial Intelligence, Nankai University, Tianjin, China
jliu@nankai.edu.cn, {2120200411,wangbq,1811579}@mail.nankai.edu.cn

Abstract. Extreme Multi-label text Classification (XMC) is a fundamental text mining task, which aims to assign multiple labels related to the given text from a large-scale label set. Various models and many data augmentation methods are proposed to improve classification performance. However, the classification performance is limited due to the long tail distribution of labels, which is an essential characteristic of XMC. To address this problem, we propose a novel data augmentation method named **A**ttentional **D**ata **A**ugmentation **M**ethod (**ADAM**) for long tail labels. Specifically, we split each sentence into several segments of equal length and use an attention-based neural network to explore the core segments of long tail labels. The unimportant segments of each instance from the dataset are considered to be replaced by those segments related to the long tail labels. Extensive experiments show that **ADAM** has an improvement based on the XMC method, especially on the prediction of long tail labels.

Keywords: Extreme multi-label text classification · Long tail labels · Data augmentation

1 Introduction

Extreme Multi-label text Classification (XMC) is an important task in the natural language processing field, which has attracted universal interest. Given a text, the XMC task aims to find multiple relevant labels from a pre-defined label set. Different from the classical multi-label classification task, XMC needs to select relevant labels from an extremely large label set. Therefore, existing studies generally focus on two issues, including how to explore the significant semantics associated with labels from the given text and how to reduce the computational complexity.

Considerable deep learning models are proposed, such as AttentionXML [17], X-Transformer [2], and LightXML [6]. They generally utilize the recurrent neural network, Transformer, or pre-trained language model as encoder and classify by the obtained text representation.

J. Gama et al. (Eds.): PAKDD 2022, LNAI 13280, pp. 131–142, 2022.
https://doi.org/10.1007/978-3-031-05933-9_11

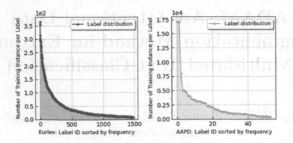

Fig. 1. The labels distribution of EUR-Lex and AAPD datasets.

Although these models have achieved significant progress, they generally neglect the effect of the long tail distribution. As shown in Fig. 1, most instances correspond to few head labels, while a large number of tail labels only contain minimal instances. This imbalance between head labels and tail labels can lead to two problems. First, the associations between instances and tail labels are difficult to capture due to the lack of training data corresponding to tail labels. Second, being dominated by head labels, the model tends to predict these high-frequency labels and omits low-frequency labels, leading to label absence and label hallucination.

To alleviate the above two problems, an intuitive method is to augment instances with tail labels. Recent data augmentation methods [4,11,13] for text classification generally adopt synonym replacement, random insertion, and back-translation mechanism. However, these methods hardly change the semantics of texts and keep the original label set completely for each instance, which limits the effectiveness of data augmentation. Therefore, it is an enormous challenge how to obtain diverse training data with tail labels.

To address this challenge, we propose an Attentional Data Augmentation Method (ADAM). Specifically, ADAM mainly consists of three steps, including relevance learning, segments memory construction, and segments replacement. Compared to a single token, segments of texts generally play a more important role in text classification due to their abundant information. Hence, we first divide the given text into different segments according to a fixed window size in the first step. Then, we devise a label-segment attention mechanism to learn the relevance degree between labels and segments. Due to the long tail distribution problem, the reliability of the correlation degree between the label and segment pairs for the head labels is higher than the tail labels. Thus, in the second step, for each instance, the segments that have a low correlation with the head labels are used as the candidate segments for tail labels. And we assign the candidate segments to each tail label in the instance according to the attention weights. Subsequently, we use long tail labels and their related segments to construct memory. In the last step, we design a random replacement strategy to augment the training data for tail labels, which uses the segments related to a long tail label that is come from the memory to replace the segments that have low relevance with each label in the selected instance, and the corresponding long tail label will be put into the original label set of the selected instance to form a new sample. To verify the effectiveness of our framework, we conduct extensive experiments on three benchmark datasets. The experimental results show that our framework achieves significant improvements.

In summary, the contributions to this paper are three-fold:

- To deal with the long tail distribution problem existing in XMC, we propose an Attentional Data Augmentation Method (ADAM).
- To generate diverse training data for tail labels, we devise a label-segment attention mechanism to learn the relevance between labels and segments and utilize a random replacement strategy to augment the training data for tail labels.
- Our extensive experiments show that our method significantly outperforms the baselines and other data augmentation methods on three benchmark datasets.

2 Related Work

2.1 Extreme Multi-label Text Classification

There are many methods that have been proposed to solve the XMC problem, and they can be broadly divided into two categories: One is traditional machine learning methods that use the raw text features like TF-IDF as input, the other is the neural network methods that extract the semantic level features before classification.

Traditional Methods. The methods that treat each label as a binary classification problem and classification is independent of each other like ProXML [1] are called one-vs-all methods. Although they have improved a certain level of performance, they always suffer from an enormous computational cost, and space complexity is really high. To alleviate this problem, Tree-based methods that use a probabilistic label tree [5] to partition labels are proposed to overcome the high computational complexity. The state-of-the-art tree-based method named Parabel [10] constructs a binary balanced tree by iteratively partitioning nodes into two balanced clusters until the cluster size is less than a value. But there is a problem that some long tail labels will be mistakenly grouped into a cluster by unrelated labels, ultimately affecting the classification performance.

Neural Network Methods. Various neural networks have shown significant improvement in XMC. These methods always extract the semantic features by deep learning and classify based on the representation. XMLCNN [7] constructs a CNN network and a fully connected layer to solve the problem. And AttentionXML [17] is also a typical method that adopts a probabilistic label tree

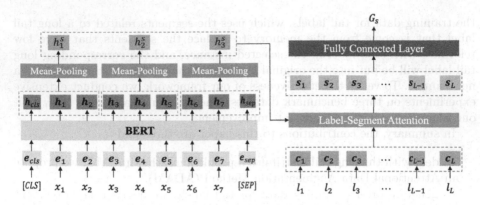

Fig. 2. Relevance learning model

(PLT) and RNN. What's more, LightXML [6] obtains a representation of the original text through the transformer-based model like BERT [3], and adopts the dynamic negative labels sampling during the label recalling part. However, none of the methods mentioned above solve the problem of poor long tail label classification performance.

2.2   Data Augmentation in NLP

Data augmentation has shown its effectiveness in many fields of NLP. **Wei et al.** [13] proposed EDA, which augments data by using Synonym Replacement, Random Insertion, Random Swap, Random Deletion. Unsupervised learning is becoming more and more popular because it reduces the cost of data labeling. UDA [14] trains in an unsupervised way, and it uses back-translation to augment the raw data, which translates raw texts to another language and translates them back. However, all data augmentation methods have no significant improvement in long tail label classification, which affects the overall performance improvement.

3   Method

To effectively augment data for long tail labels, firstly, we split a sentence into several segments, and learn the relationship between each label and each segment through an attention-based neural network, as shown in Fig. 2. Subsequently, we find the segments that have high relevance with the long tail labels and form a memory. Finally, those segments that have low relevance to each label of one instance are replaced by the segments that belong to the memory, and the corresponding long tail label will be added to the label set of the current data.

3.1   Notations

Given an original training set X_N that is composed of $\{(x_i, y_i)\}_{i=1}^{N}$, where x_i is raw text, and $y_i \in \{0,1\}^L$ is a L dimensional multi-hot vectors and stands for the label of x_i. Each x_i can be split into T segments $\{w_{ik}\}_{k=1}^{T}$ of equal length. The dataset that has been augmented can be represented as $\hat{X}_M$, which is composed of $\{(\hat{x}_i, \hat{y}_i)\}_{i=1}^{M}$.

3.2   Relevance Learning

BERT [3] and its related models have shown excellent performance in the various benchmark datasets. To better capture the relationship between segments and labels, in our method, we adopt three pre-trained transformer [12] based models: BERT, Roberta [8] and XLNet [16].

We can obtain word representation $h_t \in R^d$ through various encoders,

$$h_t = encoder(x), \tag{1}$$

where the encoder can be one of BERT, Roberta, XLNet, t is t-th word of the instance x, $t \in [0, z]$, z is the max length of the text.

How to judge a label belong to an instance is usually based on the features of some segments in the text. Therefore, we split each text into T segments of equal length where T is a hyperparameter. The representation of the k-th segment is the mean pooling of all the h_t in w_k:

$$h_k^s = mean(h_{(k-1)l}, h_{(k-1)l+1}, \ldots, h_{kl-1}), \tag{2}$$

where $h_k^s \in R^d$, l is the number of words in the segment, which can be calculated by z/T.

And then, we can get the representation $H \in R^{T \times d}$ of all segments:

$$H = [h_1^s, h_2^s, \ldots, h_T^s], \tag{3}$$

To capture the relationship between labels and segments, we use an attention mechanism to learn the correlation between labels and segments. Because of the different semantic information of each label, a trainable matrix $C \in R^{L \times g}$ is adopted to denote the label's representation, g is the dimension of the label.

And then, we calculate the attention matrix $A \in R^{L \times T}$, whose element $A_{i,j}$ denotes the correlation intensity between the i-th label and the j-th segment:

$$A = softmax(CWH^T), \tag{4}$$

where $W \in R^{g \times d}$ is a model parameter.

We can get the updated representation S of the segments through the attention matrix A:

$$S = AH, \tag{5}$$

where $S \in R^{L \times d}$.

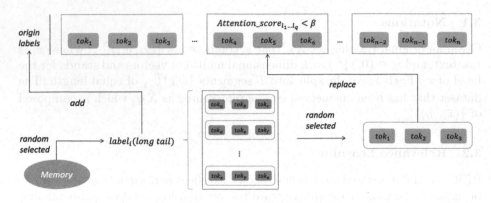

Fig. 3. The process of segments replacement

Subsequently, we feed S into a fully connected layer with sigmoid and obtain a L dimensional vector representation $G(S)$, which is the scores of all labels:

$$G(S) = \sigma(W_s S^T + b_s),\tag{6}$$

where $W_s \in R^d$ and $b \in R^L$ are fully connected layer parameters.

And finally, we can update the model parameters, especially W, which in attention by calculating Cross-Entropy loss:

$$\mathcal{L}(G(S), y) = \sum_{i=0}^{L}(1 - y^i)(-\log(1 - G(S)^i)) + y^i(-log(G(S)^i)),\tag{7}$$

3.3   Segments Memory Construction

Based on the above training, we obtain a model that can learn the relationship between labels and segments of texts, which is used to find the segments related to long tail labels. We call these segments the core segments of long tail labels. However, due to the fact that the data with long tail labels is rarely in the datasets, the model cannot learn the relationship between the long tail labels and segments well. On the opposite, because of the large frequency of head labels in the dataset, the segments related to the head labels have strong confidence.

Therefore, we select the instances with both head labels and long tail labels and filter out the segments whose attention scores with head labels greater than α, the retaining segments can be used as core segments corresponding to the long tail label.[1]

In this way, we can iteratively get the core segments corresponding to each long tail label and form a memory M.

---

[1] If there is more than one long tail label in an instance, segments that have a low attention score with head labels belong to the long tail label that has the highest attention score with them.

3.4 Segments Replacement

In data augmentation, we make a new instance by replacing some segments which have low relevance to all labels of the original instance.

Firstly, we randomly select a raw instance (x_q, y_q) from the dataset and randomly select a long tail label y_a. Secondly, for a segment w_k in x_q, determine whether w_k needs to be replaced is $A(w_k, y_q) < \beta$, if the attention scores between w_k and all labels in y_q are less than β, w_k will be replaced by a segment that is randomly selected from the memory M_{y_a}, where y_a is a long tail label that is randomly selected from M. Finally, if there are r segments in x_q have been replaced, a new pair of instance $(\hat{x}_q, \hat{y}_q)$ is formed, where $\hat{y}_q$ is composed of the original labels y_q and the long tail label y_a. After n repeated operations above, the new training set $\hat{X}_M$ can be obtained. Figure 3 shows the whole process of segments replacement.

4 Experiment

4.1 Datasets

To validate the effectiveness of our method, we use two extreme multi-label text classification datasets (EUR-Lex and Wiki10-31K) and an ordinary multi-label text classification dataset with long tail label problem (AAPD).

- **EUR-Lex**[2] [9] is a European Union legal document dataset containing 15449 documents in the training set and 3865 documents in the test set. And There are 3956 labels in the EUR-Lex.
- **AAPD** [15] collects the abstracts and subjects of papers from arXiv, and this dataset contains 53840 training data, 1000 test data, and 1000 validation data. There are 54 labels in the AAPD.
- **Wiki10-31K**[3] [18] contains 20762 articles from Wikipedia. Among them, there are 14146 articles in the training set and 6616 articles in the test set. The Wiki10-31K has a large label set which includes 30,938 labels totally.

4.2 Evaluation Measures

We use the evaluation metrics that are widely used in extreme multi-label text classification, named $P@k$, which represents the percentage of accuracy labels in top k score labels:

$$P@k = \frac{1}{k} \sum_{i \in \text{rank}_k(\tilde{y})} y_i \tag{8}$$

where $y_i \in \{0, 1\}^L$, $\tilde{y}$ is the score of all labels predicted, i denotes the index of the i-th highest element in $\tilde{y}$.

---

[2] http://www.ke.tu-darmstadt.de/resources/eurlex/eurlex.html.
[3] http://manikvarma.org/downloads/XC/XMLRepository.html.

Table 1. The performance of various data augmentation methods based on XMLCNN and AttentionXML. "BT" stands for the Back Translation.

Dataset		XMLCNN	+EDA	+BT	+ADAM	AttentionXML	+EDA	+BT	+ADAM
EUR-Lex	P@1	66.9	66.2	71.2	**73.1**	**86.5**	85.1	86.0	86.1
	P@3	51.7	53.3	57.2	**58.8**	72.4	72.8	73.4	**73.4**
	P@5	42.0	43.5	47.3	**48.5**	60.4	60.9	61.6	**61.6**
AAPD	P@1	70.8	65.9	71.7	**81.5**	83.0	82.9	84.1	**84.6**
	P@3	50.9	48.0	51	**56.8**	59.8	60.7	60.2	**60.7**
	P@5	35.4	33.9	35.7	**38.0**	41.3	41.3	**41.5**	40.7
Wiki10-31K	P@1	**80.7**	45.7	80.6	66.5	**87.1**	80.7	82.5	80.2
	P@3	51.9	40.5	**53.4**	43.8	**77.4**	61.5	73.5	64.3
	P@5	39.9	36.2	**41.4**	35.0	**68.7**	50.6	65.6	53.5

4.3 Baseline

We apply our method to XMLCNN [7] and AttentionXML [17] which are proposed to solve the XMC problem in recent years. And we compare our method with two widely used data augmentation methods:

- **EDA** [13] is a universal data augmentation method, which uses a small probability to decide whether to do a Synonym Replacement, Random Insertion, Random Swap, Random Deletion for every word.
- **Back Translation** [4,11] was originally used when there is a large amount of monolingual corpus in machine translation. However, in recent years, due to its practicality, it has been widely used in data augmentation for various tasks in the NLP.

4.4 Experiment Details

For each dataset, we count the frequency of all labels and consider labels with a frequency lower than the median frequency of all labels as long tail labels. In relevance learning, we try three different Transformer architecture encoders, namely Bert, RoBerta, XLNet, and we finally choose the Bert as encoder because of its best performance in relevance learning. Due to the need to focus on learning the parameters of the Attention mechanism and classifier, we set two different learning rates during training. The learning rate of the encoder similar to Bert is 0.00001, and the learning rate of the remaining parameters is 0.01, and batch size is 10. The dimension of label embedding is 300, and the dimension of W in attention is 300 too. The number of segments in each sentence is set to 100, the hyper-parameters α and β are both set to 0.01. For EUR-Lex, AAPD, WiKi10-31K, we set the number of data augmentation n is 50000, 25000, 15000, respectively.

For EDA, we have tried to set the number of augmentation for each instance to 1 to 9, respectively. According to the performance of XMLCNN and AttentionXML, we finally augment each instance only once.

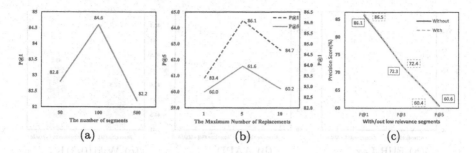

Fig. 4. The performance of different hyper-parameter and the proof of semantic integrity on AttentionXML. (a) represents the performance of the various segments number. (b) represents the performance of diverse maximum replacement numbers in each sentence. (c) represents the performance of the data with those low relevance segments and the data without low relevance segments.

For Back-Translation, we translate each instance into seven different languages and randomly pick one of them into the new dataset.

When we use the augmented data to train AttentionXML and XMLCNN, the experimental settings are the same as training the original data.

4.5 Comparison Results and Discussion

In this section, the whole method ADAM is evaluated on three benchmark datasets by comparing with all the baselines mentioned above. Like most methods in XMC, we use top prediction by varying k at 1, 3 and 5 in $P@k$.

Table 1 shows the prediction results on three datasets and two baselines. Firstly, for EUR-Lex and AAPD, our method almost outperforms all methods on both XMLCNN and AttentionXML. At the same time, both EDA and Back-Translation have improved on the EUR-Lex. Secondly, for AAPD, EDA has some performance degradation on XMLCNN, it is because CNN has a limited ability to capture text semantics compared with LSTM, when there is some noise in the data, it is difficult for CNN to obtain useful information from the text. Unexpectedly, our method and EDA are limited on the Wiki10-31K, on the contrary, the performance of Back-Translation has not decreased much, and even almost better than baseline on XMLCNN. We find the average number of labels per sample is 18.64 in the Wiki10-31K, which is a large number for instance. Hence, almost every segment of the text is related to one of the labels of a sample. When the core segments of the text are replaced or changed by the data augmentation method, the classification of the corresponding label will be affected. Back-Translation can retain the original semantics of the text, but our method and EDA inevitably draw into some noise that affects the semantics of the original data. However, due to our relation learning, we capture the relation between segments and labels, some core segments of the original sample can avoid being replaced. Therefore, our method outperforms EDA in $P@1$ and $P@3$.

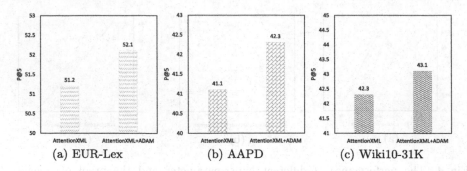

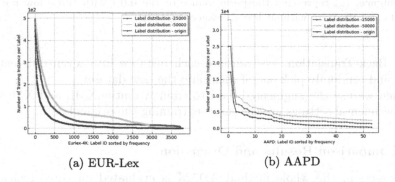

(a) EUR-Lex   (b) AAPD   (c) Wiki10-31K

Fig. 5. The performance of the instances that include long tail labels on EUR-Lex, AAPD and Wiki10-31K.

(a) EUR-Lex   (b) AAPD

Fig. 6. The labels distribution based on various augmentation number on EUR-Lex and AAPD.

4.6   Analysis of Different Segment Numbers

To observe the impact of the number of segments on performance, we set the number of segments on the AAPD to 500, 100, 50, respectively. To eliminate other influences, we fix the maximum number of replacements to 5. Since we set the maximum text length of the input model to 500, thus, for the model that has 500 segments, there is only one word in each segment.

As Fig. 4 (a) has shown, the performance of 100 segments is better than the other two in $P@1$, and we analyze that the model with 500 segments cannot capture the real core segment due to its limited number of words in a segment. On the contrary, since there are a lot of words in a segment for the model with 50 segments, one segment may be associated with multiple labels, resulting in incorrect classification.

4.7   Analysis of Maximum Number of Replacements

We also report the effect of the different maximum number of replacements in Fig. 4 (b). Similarly, we fix the number of segments to 100. As we guessed, the

Digital elevation models \(dem \) are images having terrain information embedded into them using cognitive mapping concepts for dem registration , has evolved from this basic idea of using the mapping between the space to objects and defining their relationships to form the basic landmarks that need to be marked , stored and manipulated in and about the environment or other candidate environments...

And simulate real world complex into them using cognitive mapping estimation of the parameters by complex network, social network idea of using the mapping between the space to objects and defining their relationships to form the basic landmarks that need to be marked, stored and manipulated in and about the environment or other candidate environments...

Fig. 7. The case study of original text and updated text on AAPD. The original labels are "cs.AI" and "cs.CV", and the updated labels are "cs.AI", "cs.CV" and "physics.data-an".

replacement of too few or too many segments will affect performance. When we only replace one segment for each sample, the new segment related to the long tail label may not promote the classification, even because the newly added long tail label increases the difficulty of classification. Replacing ten segments for each instance also obtains limited performance, although the new segments have a correlation with the long tail label, it also causes the loss of some information in the original text, which will disturb the classification of the original labels.

4.8 Analysis of Long Tail Labels

To further verify the effectiveness of our methods, we analyze the results on the long tail labels for different datasets. We select the instances that include long tail labels from the test dataset, the results are shown in Fig. 5. Our method has an improvement on both three datasets. For Eur-Lex, our method outperforms the baseline by almost 1%. Furthermore, on the AAPD, ADAM improves 1.2% from the baseline in $P@5$. Especially, although we obtained limited performance on the Wiki10-31K, we still achieved 0.8% growth on the long tail labels. This also proves that our method can improve the performance of long tail labels.

4.9 Effectiveness of Data Augmentaion

To prove that our data augmentation method has little influence on the semantic integrity of the original sentence, we compared the performance of the original data with the data that have removed low-relevance segments(need to be replaced by the core segments related to the long tail labels) through the relevance learning model, as shown in Fig. 4 (c), their performance on AttentionXML is almost the same. Therefore, it can explain that removing the segments with low relevance to the labels has little impact on classification performance. Fig. 6 shows the distribution of original datasets and updated datasets. For long tail labels, we increase their frequency to make the distribution of the entire label set smoother. This is more significant on the extreme multi-label classification dataset EUR-Lex. The case study of original data and updated data can be seen in Fig. 7.

5   Conclusion

In this paper, we focus on the problem of long tail labels in XMC and propose a simple but effective attention-based data augmentation method. With the relation learning between segments and labels, the segments that have low correlation with labels are replaced by the segments related to one of the long tail labels. Meanwhile, the corresponding long tail labels are put into the original label set of the selected instance. Extensive experiments on three benchmark datasets prove the effectiveness of ADAM by comparing two widely used data augmentation methods based on various baselines.

Acknowledgements. This research is supported by the National Natural Science Foundation of China under the grant No. 61976119 and the Natural Science Foundation of Tianjin under the grant No. 18ZXZNGX00310.

References

1. Babbar, R., Schölkopf, B.: Data scarcity, robustness and extreme multi-label classification. Mach. Learn. **108**(8–9), 1329–1351 (2019)
2. Chang, W., Yu, H., Zhong, K., Yang, Y., Dhillon, I.S.: Taming pretrained transformers for extreme multi-label text classification. In: KDD 2020. ACM (2020)
3. Devlin, J., Chang, M., Lee, K., Toutanova, K.: BERT: pre-training of deep bidirectional transformers for language understanding. In: NAACL-HLT 2019 (2019)
4. Edunov, S., Ott, M., Auli, M., Grangier, D.: Understanding back-translation at scale. In: EMNLP 2018. Association for Computational Linguistics (2018)
5. Jasinska, K., Dembczynski, K., Busa-Fekete, R., et al.: Extreme f-measure maximization using sparse probability estimates. In: ICML 2016 (2016)
6. Jiang, T., Wang, D., Sun, L., et al.: LightXML: transformer with dynamic negative sampling for high-performance extreme multi-label text classification. CoRR (2021)
7. Liu, J., Chang, W., Wu, Y., Yang, Y.: Deep learning for extreme multi-label text classification. In: Proceedings of the 40th International ACM SIGIR (2017)
8. Liu, Y., et al.: Roberta: a robustly optimized BERT pretraining approach
9. Mencía, E.L., Fürnkranz, J.: Efficient pairwise multilabel classification for large-scale problems in the legal domain. In: ECML/PKDD 2008 (2008)
10. Prabhu, Y., Kag, A., Harsola, S., et al.: Parabel: partitioned label trees for extreme classification with application to dynamic search advertising. In: WWW (2018)
11. Sennrich, R., et al.: Improving neural machine translation models with monolingual data. In: ACL 2016 (2016)
12. Vaswani, A., et al.: Attention is all you need. In: NeurIPS 2017 (2017)
13. Wei, J.W., Zou, K.: EDA: easy data augmentation techniques for boosting performance on text classification tasks. In: EMNLP-IJCNLP 2019 (2019)
14. Xie, Q., et al.: Unsupervised data augmentation for consistency training
15. Yang, P., Sun, X., Li, W., Ma, S., Wu, W., Wang, H.: SGM: sequence generation model for multi-label classification. In: COLING 2018 (2018)
16. Yang, Z., Dai, Z., Yang, Y., et al.: XLNet: generalized autoregressive pretraining for language understanding. In: NeurIPS 2019 (2019)
17. You, R., Zhang, Z., Wang, Z., et al.: AttentionXML: label tree-based attention-aware deep model for high-performance extreme multi-label text classification
18. Zubiaga, A.: Enhancing navigation on Wikipedia with social tags. CoRR (2012)

AutoTransformer: Automatic Transformer Architecture Design for Time Series Classification

Yankun Ren, Longfei Li, Xinxing Yang, and Jun Zhou[✉]

Ant Group, Hangzhou, China

{yankun.ryk,longyao.llf,xinxing.yangxx,jun.zhoujun}@antgroup.com

Abstract. Time series classification (TSC) aims to assign labels to time series. Deep learning methods, such as InceptionTime and Transformer, achieve promising performances in TSC. Although deep learning methods do not require manually crafted features, they do require careful manual design of the network structure. The design of architectures heavily relies on researchers' prior knowledge and experience. Due to the limitations of human's knowledge, the designed architecture may not be optimal on the dataset of interest. To automate and optimize the architecture design, we propose a data-driven TSC network architecture design method called AutoTransformer. AutoTransformer designs the suitable network architecture automatically depending on the target TSC dataset. Inspired by the overall architecture of Transformer, we first propose a novel search space tailored for TSC. The search space includes a variety of substructures that are capable of extracting global and local features from time series. Then, with the help of neural architecture search (NAS) technique, a suitable network architecture for the target TSC dataset can be found from the search space. Experimental results show that AutoTransformer finds proper architectures on different TSC datasets and outperforms state-of-the-art methods on the UCR archive. Ablation studies verify the effectiveness of the proposed search space.

Keywords: Time series classification · Neural architecture search · Transformer · Convolutional neural network

1 Introduction

A time series is a sequence of data points chronologically arranged. Time series classification (TSC) is the task that classifies time series as predefined classes. It is considered as an important and challenging problem in data mining [9] and attracts much research interest.

Different from supervised learning for structured data, TSC methods should be able to harness the temporal information in time series. Early methods are mostly traditional feature-based methods. They identify various hand-crafted features that could represent the global/local time series patterns and feed them into classifiers [4,19]. These feature-based methods need hand-engineering and domain-specific knowledge to derive

J. Gama et al. (Eds.): PAKDD 2022, LNAI 13280, pp. 143–155, 2022.
https://doi.org/10.1007/978-3-031-05933-9_12

features. Then, inspired by the successful application of deep learning models in computer vision, researchers apply many deep learning models to TSC [11,24] which are free from hand-crafted features, such as InceptionTime [11] and Transformer [23].

However, existing deep learning models depend on the careful manual design of the model architecture. Manual designed architecture's performance is limited by researchers' prior knowledge and experience, thus the designed architecture may not be the optimal one on the target dataset.

To address the aforementioned problem, we propose a novel deep learning method for TSC called AutoTransformer. First, AutoTransformer adapts the neural architecture search (NAS) algorithm for TSC and searches for the suitable network architecture for each TSC dataset in a data-driven way. For example, one suitable neural architecture is designed automatically for each dataset in the UCR archive instead of one architecture for all datasets. Second, considering both the short-term local features and long-term global features in time series data, we proposed a novel search space inspired by the overall architecture of Transformer [23], which achieves great success in modeling sequence data in natural language processing (NLP). As for the concrete operation in the overall architecture, we retain the convolution operations used in the previous TSC models to extract local features. Besides, we include operations such as RNN and self-attention to capture global long-term dependency in time series. We also proposed an improvement on the search algorithm to enable the search of layer inputs and residual inputs. Our search space and search method can automatically combine multiple operations freely to modeling both local and global features in time series based on the characteristics of each TSC dataset.

Experimental results on the UCR archive show that our proposed AutoTransformer can design suitable architectures tailored for each TSC dataset in the UCR archive and archives the new state of the art. Case study is performed to show the rationality of searched architectures. Moreover, several ablation studies verify the effectiveness of the proposed novel search space.

In summary, the major contributions of this paper are as follows:

- Unlike the existing deep learning methods, we design network architecture automatically in a data-driven way. We adapt the neural architecture search (NAS) for TSC to automatically design the proper architecture tailored for each TSC dataset, while former deep learning methods design only one model architecture based on human's prior knowledge and apply it to all TSC datasets.
- Different from the search space used in image classification domain, we propose a novel search space tailored for TSC tasks. We adopt the overall architecture of Transformer and include operations specialized in capturing local and global features in time series. Also, we propose an improvement to the search algorithm which makes the search process of input choice and residual input choice feasible.
- Extensive experiments are conducted on the UCR time series classification archive (85 datasets), experimental results show that AutoTransformer outperforms state-of-the-art methods. We also conduct case studies to demonstrate AutoTransformer's ability to design the suitable architecture for each TSC task. Ablation studies verify the effectiveness of key components of the proposed search space.

2   Relate Work

Time Series Classification (TSC). Over the last two decades, many efforts have been invested in TSC. Early researches focus on feature-based methods. A variety of hand-crafted features are proposed, such as Dynamic Time Warping (DTW) distance [18]. These approaches transform time series to one or several feature spaces and use an ensemble of discriminant classifiers to classify time series [2,4,6,13,17,19,21].

Then, several deep learning methods free from hand-engineering efforts are proposed and achieve promising results in TSC [4,10,11,24]. These methods adopt convolutions which are commonly used to capture local information in image classification and utilize RNN to extract global features [16]. Different from previous studies which design complex networks manually, we aim to automatically learn suitable neural architectures for different tasks.

Neural Architecture Search. In recent years, many neural architecture search (NAS) methods are proposed [26,27], which make automatic model design feasible. Also, recent methods can finish searching in only a few GPU-hours [8,15], which is similar as training a normal neural model. These methods mainly utilize various convolution operations in their search space to extract local features, as they focus on the image classification domain. In this paper, we make an early exploration to design an search space tailored for TSC tasks and make use of convolution operations and other operations to capture both local and global features in time series.

Transformer. Transformer [23] achieves great success in many tasks of NLP [7]. One important reason is that Transformer have great ability to capture long-term global feature in sequence data. In this paper, we adapt the overall architecture of Transformer and the multi-head self-attention operation to help modeling time series.

3   Methodology

We generalize neural architecture search (NAS) to time series classification (TSC). We propose AutoTransformer to automatically design one TSC network architecture depending on each target TSC dataset.

3.1   Time Series Classification Search Space

The overall architecture of AutoTransformer search space is shown in Fig. 1(a). Inspired by Transformer encoder, we formulate the overall architecture of AutoTransformer as a stack of N layers. For each layer L, there are three components that are searchable: (i) the operation to conduct on the input in layer L; (ii) the input of layer L; (iii) the residual inputs of layer L. One layer of Transformer is shown in Fig. 1(b), we can observe that Transformer is a special case of AutoTransformer. Each layer of Transformer can be viewed as two cascaded layers of AutoTransformer with the searchable components in each layer fixed. In the following, we will introduce the searchable parts in detail.

Operation Choice. In each layer, an operation is selected from the candidate operations to extract features from the input. We construct the candidate operation set based on the following considerations.

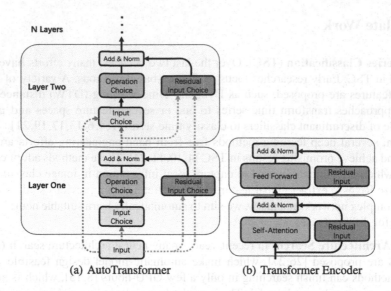

Fig. 1. (a) The whole search space of AutoTransformer. Solid lines are fixed data flows, red dotted line are candidate inputs and blue dotted lines are candidate residual inputs. (b) Architecture of one layer of Transformer encoder. (Color figure online)

First, in TSC tasks, some subsequences of a time series may be discriminating in classification [25]. Thus, it is important to capture local features from subsequences. We adopt convolution operations to model such local correlation. In addition, convolutions of different kernel sizes model the spatial dependency in different ranges, so we include various convolution kernel sizes. We use 1D standard convolutions and dilated convolutions with kernel size $\{1, 3, 5, 7, 9, 11, 13, 17, 21\}$ among which the dilated convolution can be used to enhance the capability of capturing longer subsequence information and each convolution is applied as a Relu-Conv-BatchNorm structure. Note that the 1-D convolution with *kernel size* $= 1$ is equivalent to feed-forward. We also utilize max pooling and average pooling with kernel size 3 to extract local features. To keep the shape of output the same as input, we utilize the convolution of *stride* $= 1$ with SAME padding and the filter dimensions of output is the same as input.

Second, we include LSTM [14] and multi-head self-attention [23] in the operation candidate set, because of their verified excellent performances in capturing global features in sequence data [7].

Input Choice. In InceptionTime [11], convolutions with different kernel sizes are utilized to extract local features in time series in parallel from different aspects, then multiple feature maps are fused (e.g., concatenation or summation) to form the final output, which can improve the model's performance. This technique is shown in Fig. 2(a).

Motivated by InceptionTime, we make the input of each layer searchable. Formally, one layer can select one output from lower layers as its input. For example, Layer 3 can select one from set $\{original\ input,\ layer1\ output,\ layer\ 2\ output\}$ as its input. In this way, the layers can be assembled both in cascade and in parallel. The proposed input choice schema makes the architecture in Fig. 2(b) possible. This architecture is equivalent to the aforementioned InceptionTime technique, which makes the InceptionTime technique a special case of AutoTransformer.

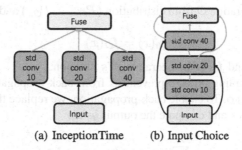

(a) InceptionTime   (b) Input Choice

Fig. 2. (a) InceptionTime extracts multi-aspect features in parallel with different kernel size convolutions. (b) An instance of our search space. Each layer can select one from lower layers' outputs as its input. These two architectures are equivalent.

Residual Input Choice. Low-level features captured by low layers and high-level features encoded by high layers are both important to TSC. Therefore, residual connections are often useful in combining multi-level features. Inspired by ResNet [12] and Transformer, we add residual input choice module to AutoTransformer. A layer can decides whether each one of its former layers' outputs can be its residual input independently, this means that one layer can have multiple residual inputs. For example, the second layer in AutoTransformer can have two residual inputs both from the original input and the first layer.

3.2 Optimization Procedure

The goal of neural architecture search (NAS) is to find an architecture α and the model weights ω_α of α, which together achieve the minimum loss $\mathcal{L}$ on the train data. For example, ω_α can be weights of RNN in the model.

Given an α, optimal model weights ω_α^* can be easily obtained through gradient descent as shown in Eq. 1.

$$\omega_\alpha^{'} = \omega_\alpha - \beta \nabla_{\omega_\alpha} \mathcal{L} \tag{1}$$

where β is the learning rate.

To optimize α through gradient descent, we follow the gradient-based search method used in GDAS [8]. We sample architectures from a distribution characterized by a set of learnable distribution weights. As the architecture α is sampled from a discrete probability distribution, $\mathcal{L}$ is not differentiable with respect to the distribution weights. Thus, we use the Gumbel-Max trick and softmax to make $\mathcal{L}$ differentiable with respect to the distribution weights. The details are as follows.

Operation Choice in Each Layer. In each layer, we sample an operation to extract features from the input. Suppose the candidate operation set is $\{\mathcal{O}_1, ..., \mathcal{O}_n\}$, the corresponding distribution weights of the operations are $\mathbf{o} = [o_1, ..., o_n]$. Then, the probability of sampling operation $\mathcal{O}_i$ is computed as Eq. 2

$$P_{\mathcal{O}_i} = \frac{\exp(o_i)}{\sum_{j=1}^{n} \exp(o_j)} \tag{2}$$

We can sample an operation $\mathcal{O}$ from distribution $(P_{\mathcal{O}_1}, ..., P_{\mathcal{O}_n})$ and compute the output y_0 as Eq. 3.

$$y_0(x) = f_{\mathcal{O}}(x) \tag{3}$$

where x is the input and $f_{\mathcal{O}}$ is the operation $\mathcal{O}$'s function.

However, the sampling process prevent us from back-propagate gradients from $\mathcal{L}$ to distribution weights $\mathbf{o}$. To enable back-propagation, we replace the sampling process with Gumbel-Max trick and compute the output y_0 as:

$$y_0(x) = \sum_{i=1}^{n} h_i \cdot f_{\mathcal{O}_i}(x) = 1 \cdot f_{\mathcal{O}_k}(x)$$

$$\text{s.t. } h = \text{one\_hot}(k), k = \arg\max_i(o_i + g_i) \tag{4}$$

where g_i are samples drawn from Gumbel$(0, 1)$, and h is an n-dim one-hot vector with the k-th item being 1. As argmax is not differentiable, we relax argmax with softmax. Thus, we compute the output y_0 as:

$$P'_{\mathcal{O}_i} = \frac{\exp((o_i + g_i)/\tau)}{\sum_{j=1}^{n}[\exp((o_j + g_j)/\tau)]} \tag{5}$$

$$y_0(x) = \sum_{i=1}^{n} P'_{\mathcal{O}_i} \cdot f_{\mathcal{O}_i}(x) \approx P'_{\mathcal{O}_k} \cdot f_{\mathcal{O}_k}(x)$$

$$\text{s.t. } k = \arg\max_i(P'_{\mathcal{O}_i}) \tag{6}$$

where τ is the temperature. When $\tau \to 0$, then $P'_{\mathcal{O}_k} \to 1$, the architecture distribution approaches one-hot, thus $y_0(x) \to P'_{\mathcal{O}_k} f_{\mathcal{O}_k}(x)$. In practice, we compute y_0 as:

$$y_0(x) = (1 - \text{detached}(P'_{\mathcal{O}_k}) + P'_{\mathcal{O}_k}) * f_{\mathcal{O}_k}(x) \tag{7}$$

where $\text{detached}(a)$ means a does not receive gradients. In this way, the output is $f_{\mathcal{O}_k}(x)$ during forward and y_0 is differentiable with respect to operations' distribution weights $\mathbf{o}$. After trained, we select the operation with the highest weight as the final operation choice in this layer as shown in Eq. 8.

$$\mathcal{O} = \arg\max_{\mathcal{O}_i}(o_i) \tag{8}$$

GDAS [8] does not conduct input choice and residual choice. Here, we propose to formulate these two modules as follows.

Input Choice in Each Layer. Suppose there are n layers' outputs for selection, which form a set $\{\mathcal{U}_1, ..., \mathcal{U}_n\}$. We introduce distribution weights $\mathbf{u} = [u_1, ..., u_n]$, thus the output of the operation $\mathcal{O}$ with input choice can be modified from y_0 (see Eq. 7) to y_1 as Eq. 9:

$$y_1 = (1 - \text{detached}(P'_{\mathcal{U}_k}) + P'_{\mathcal{U}_k}) * y_0(\mathcal{U}_m)$$
$$\text{s.t. } k = \arg\max_i(P'_{\mathcal{H}_i}) \tag{9}$$

where the probability $P'_{\mathcal{U}_i}$ is computed the same as $P'_{\mathcal{O}_i}$ in Eq. 5. After trained, we choose $\mathcal{U} = \arg\max_{\mathcal{U}_i}(u_i)$ as the final input similar as Eq. 8.

Residual Choice in Each Layer. All of one layer's former layers' outputs can be its residual inputs. For one former layer's output x^r, suppose $\mathcal{R}_1$ represents that we keep x^r as a residual input while $\mathcal{R}_0$ means that we discard x^r. Then we utilize weights $\mathbf{r} = [r_0, r_1]$ as distribution weights. The final output y_2 of a layer is computed as:

$$y_2 = y_1 + (1 - \text{detached}(P'_{\mathcal{R}_k}) + P'_{\mathcal{R}_k})(R_k == \mathcal{R}_1) * x^r$$
$$\text{s.t. } k = \arg\max_i(P'_{\mathcal{R}_0}, P'_{\mathcal{R}_1}) \tag{10}$$

where $P'_{\mathcal{R}_i}$ is computed the same as $P'_{\mathcal{O}_i}$ in Eq. 5 and y_1 is from Eq. 9. After trained, if $\arg\max_{\mathcal{R}_i}(r_0, r_1) = \mathcal{R}_1$, this means $\mathcal{R}_1$ has the highest weight, thus we keep the residual input x^r. On the contrary, if $\arg\max_{\mathcal{R}_i}(r_0, r_1) = \mathcal{R}_0$, we discard x^r. For each possible residual input, the aforementioned process is repeated to decide whether we keep it as a residual input.

Finally, all layers' outputs are attentively added up and then passed to a linear layer and a softmax layer to give a probability prediction $\text{Pr}(y \mid x)$ of a possible class. The final loss $\mathcal{L}$ is computed as:

$$\mathcal{L} = \mathbb{E}_{(x,y)\sim\mathbb{D}_{train}} - \log\text{Pr}(y \mid x; \mathbf{o}, \mathbf{u}, \mathbf{r}, w_\alpha) \tag{11}$$

where $\mathbb{D}_{train}$ is train data. In the search stage, we follow DSNAS [15] and optimize model weights w_α and distribution weights ($\mathbf{o}$, $\mathbf{u}$ and $\mathbf{r}$) of architectures α by minimizing $\mathcal{L}$ at the same time. After converge, we infer the optimal architecture α from optimized $\mathbf{o}$, $\mathbf{u}$ and $\mathbf{r}$ as aforementioned. In this way, we get a suitable model architecture α for a specific dataset. Then, in the retrain stage, we fix the model architecture as the inferred architecture α, and we train the model of the inferred architecture α by minimizing $\mathcal{L}$. This is just the same as we train a normal model such as InceptionTime and RNN, as we only optimize model weights w_α by minimizing $\mathcal{L}$. We evaluate the model on the test data to get metrics such as accuracy. In this way, we can get the final performance of the searched architecture on a TSC dataset.

4   Experiments

In this section, we conduct experiments on the UCR Time Series Classification Archive [3] to evaluate AutoTransformer. We compare our method with several TSC methods. Most of them are deep learning models and some of them are traditional methods. We also compare AutoTransformer with GDAS, which is a differentiable neural architecture search method. For a better understanding, we further report several ablation studies.

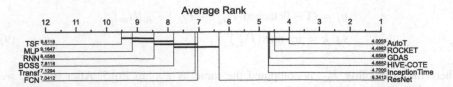

Fig. 3. Critical difference diagram showing the performance of AutoTransformer (AutoT) compared with other methods. Each number on the axis means the average rank on all datasets. If a thick horizontal line connected a group of classifier, these classifiers are not-significantly different in terms of accuracy when performing the Wilcoxon Signed-rank test.

4.1 Experiment Setup and Details

Dataset. Following existing literature [1,11,22], we use the UCR Time Series Classification Archive 2015 [3]. It consists of 85 univariate time series classification datasets, which can be used to evaluate the performances of methods in different scenarios.

Evaluation Measurements. On each dataset of the UCR archive, we report the accuracies. As no single method performs best on all datasets, we follow the recommended evaluation measurement [1,5,11,22]. We first adopt the Friedman test to reject the null hypothesis which claims that all methods have the same performance. Then, we perform the pairwise post-hoc analyse using the Wilcoxon signed-rank test to analyse whether a pair of methods are significantly different. To visualize this comparison we use a critical difference diagram [5], where a thick horizontal line indicates that a group of classifiers are not-significantly different in terms of accuracy.

Model Setting. In all experiments, for AutoTransformer, we set the hidden dimension $d = 128$ and the layer number as 6. We train all models for $1,500$ epochs and pick the model with the lowest train loss.

Table 1. Accuracies on Adiac, CricketX, DiatomSizeReduction (DSR) and DistalPhalanxOutlineAgeGroup (DPOAG) datasets.

Method	Adiac	CricketX	DSR	DPOAG
Transformer	0.6854	0.7769	0.9705	0.8650
ResNet	0.7953	0.8102	0.9477	0.8399
InceptionTime	**0.8235**	0.8153	0.9509	0.8500
HIVE-COTE	0.7962	0.8161	0.9142	0.8239
GDAS	0.7953	0.8179	0.9542	0.8375
ROCKET	0.7720	0.8390	0.9580	0.8115
AutoTransformer	0.8184	**0.8512**	**0.9869**	**0.8675**

4.2 Overall Results

We compare AutoTransformer with several methods.

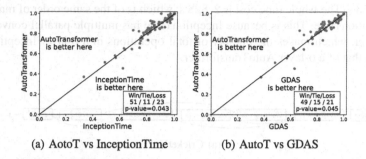

(a) AotoT vs InceptionTime   (b) AutoT vs GDAS

Fig. 4. Accuracy plot for comparisons of pairs of methods. Each axis represent one method. A point in the plot represent the accuracies of the two methods on one dataset.

– **Deep Learning Methods:** We choose several deep learning models, which are shown to be effective in TSC [10]: the multi-layer perceptron (MLP), the fully convolutional neural network (FCN) [20], the residual network (ResNet) [12], InceptionTime [11], the long short-term memory (LSTM) [14] and Transformer (Transf) [23]. Moreover, we choose NAS method GDAS [8] to demonstrate the effectiveness of AutoTransformer search space.
– **Traditional TSC Methods:** Time Series Forest (TSF) [6], Bag-of-SFA-Symbols (BOSS) [21], Hierarchical Vote Collective of Transformation-Based Ensembles (HIVE-COTE) [19] and ROCKET [4]. We obtain the results of these 4 methods from UEA & UCR Time Series Classification Repository[1].

The critical difference diagram of all methods is shown in Fig. 3. Due to space limitations, in Table 1, we only report part of these methods' accuracies on four datasets.

From Fig. 3, we can observe that a thick horizontal line delineates a high-rank group composed of AutoTransformer, ROCKET, GDAS, HIVE-COTE and InceptionTime. Methods in this group perform significantly better than other methods. Among methods in the high-rank group, AutoTransformer achieves the highest average rank. This demonstrates the effectiveness of AutoTransformer.

To further visualize the difference between AutoTransformer and two best performing deep learning methods, we depict the accuracy plot of AutoTransformer against InceptionTime and GDAS for each of the 85 UCR datasets in Fig. 4. From Fig. 4(a), we can see a Win/Tie/Loss of 51/11/23 in favor of AutoTransformer against InceptionTime (p-value $= 0.043$ with Wilcoxon signed-rank test). This implies an advantage of AutoTransformer over the SOTA deep learning TSC method. From Fig. 4(b), we can observe that compared with GDAS, AutoTransformer achieves a Win/Tie/Loss of 49/15/21, this verifies the superiority of proposed search space in TSC tasks (p-value $= 0.045$).

---

[1] http://www.timeseriesclassification.com/.

Training Time Cost Comparison. We compare the training time cost of 6-layer InceptionTime and 6-layer AutoTransformer on the StarLightCurves dataset. StarLightCurves has $1,000$ train samples, each of length $1,024$. Trainig time of InceptionTime is $2,310$ s. While training time of AutoTransformer search and retrain is $1,472$ s and $1,353$ s. The whole time cost is $2,828$ s, which is of the same order of magnitude with InceptionTime. This is because InceptionTime has multiple parallel convolutions in one layer, which makes the number of total operations in a 6-layer InceptionTime more than that of a 6-layer AutoTransformer.

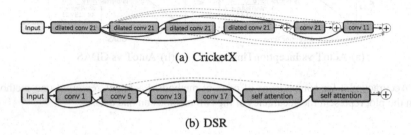

(a) CricketX

(b) DSR

Fig. 5. Searched architectures on two datasets: cricket and DiatomSizeReduction (DSR). Solid black lines represent data input, and dotted blue lines indicate residual connection. Numbers represent convolution filter sizes. Final output is computed by adding all layers' outputs attentively.

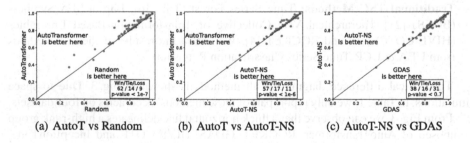

(a) AutoT vs Random   (b) AutoT vs AutoT-NS   (c) AutoT-NS vs GDAS

Fig. 6. Accuracy plot for comparisons of pairs of methods.

4.3   Case Study

In this section, we demonstrate AutoTransformer's ability to design proper architecture for each specific dataset visually. The searched architectures on datasets CricketX and DSR are shown in Fig. 5. We can observe these architectures have 2 main differences.

First, the searched architecture for DSR utilizes self-attention while the searched architecture for CricketX only utilizes convolution operations. As shown in Table 1, Transformer performs better than convolution-based ResNet and InceptionTime on

DSR, this implies the effectiveness of self-attention on DSR. Thus AutoTransformer chooses to include the effective self-attention on DSR.

Second, the searched architecture for DSR has less long filter convolution operations (average filter size is only 9 while CricketX is 19), since encoding local features is enough for DSR which is an easier 4-class classification task, while CricketX is a more complex 12-class classification task.

4.4 Ablation Studies

In order to further prove that the performance increments are independently contributed by each of the key components, we conduct ablation studies from the following aspects.

Effectiveness of Search. To verify the effectiveness of the search process, we adopt a randomly sampled architecture for each dataset, then we retrain the sampled architectures and get the accuracies on test set. Figure 6(a) depicts the accuracy plot of Auto-Transformer against randomly sampled architectures for each of the 85 UCR archive datasets. We can observe a Win/Tie/Loss of 62/14/9 in favor of AutoTransformer against the random method with a p-value $< 1 * 10^{-7}$ after performing the Wilcoxon Signed-rank test. This verifies that the search process can find more suitable architectures in the search space compared with random selection.

Effectiveness of RNN and Self-attention in the Search Space. In order to find out whether it is useful to adopt RNN and self-attention in the search space, we remove RNN and self-attention operations from the search space and repeat experiments. The reduced search space is named AutoTransformer-NoSeq (AutoT-NS). In Fig. 6(b), we can see a Win/Tie/Loss of 57/17/11 in favor of AutoTransformer, which shows that adding RNN and self-attention in the search space makes the found architectures better (p-value $< 1 * 10^{-6}$).

Effectiveness of Overall Architecture of Transformer. Last, we examine the effectiveness of the overall architecture of Transformer. Note that AutoTransformer without RNN and self-attention (AutoT-NS) has the same candidate operation set (convolution and pooling operations) as GDAS. The only difference between them is the overall architecture. The overall architecture of GDAS is the stack of the same searched cells. We compare AutoTransformer without RNN and self-attention with GDAS on the UCR archive datasets, From Fig. 6(c), we can observe that AutoTransformer without RNN and self-attention is slightly better than GDAS with a Win/Tie/Loss of 38/16/31 (p-value < 0.7). This proves that the overall architecture of Transformer is slightly better than GDAS.

5   Conclusion

In this paper, we propose AutoTransformer, which can automatically design suitable neural architecture for a specific TSC task. Experiments show that AutoTransformer is capable of learning proper architectures for TSC tasks and achieves state-of-the-art results in the UCR archive. In addition, it is also demonstrated that the proposed novel search space is effective in modeling time series data.

References

1. Bagnall, A., Lines, J., Bostrom, A., Large, J., Keogh, E.: The great time series classification bake off: a review and experimental evaluation of recent algorithmic advances. Data Min. Knowl. Disc. **31**(3), 606–660 (2016). https://doi.org/10.1007/s10618-016-0483-9
2. Baydogan, M.G., Runger, G., Tuv, E.: A bag-of-features framework to classify time series. PAMI **35**(11), 2796–2802 (2013)
3. Chen, Y., et al.: The UCR time series classification archive, July 2015. www.cs.ucr.edu/~eamonn/time_series_data/
4. Dempster, A., Petitjean, F., Webb, G.I.: Rocket: exceptionally fast and accurate time series classification using random convolutional kernels. Data Min. Knowl. Disc. **34**(5), 1454–1495 (2020)
5. Demšar, J.: Statistical comparisons of classifiers over multiple data sets. J. Mach. Learn. Res. **7**(Jan), 1–30 (2006)
6. Deng, H., Runger, G., Tuv, E., Vladimir, M.: A time series forest for classification and feature extraction. Inf. Sci. **239**, 142–153 (2013)
7. Devlin, J., Chang, M.W., Lee, K., Toutanova, K.: BERT: pre-training of deep bidirectional transformers for language understanding. arXiv preprint arXiv:1810.04805 (2018)
8. Dong, X., Yang, Y.: Searching for a robust neural architecture in four GPU hours. In: CVPR, pp. 1761–1770 (2019)
9. Esling, P., Agon, C.: Time-series data mining. ACM Comput. Surv. (CSUR) **45**(1), 1–34 (2012)
10. Ismail Fawaz, H., Forestier, G., Weber, J., Idoumghar, L., Muller, P.-A.: Deep learning for time series classification: a review. Data Min. Knowl. Disc. **33**(4), 917–963 (2019). https://doi.org/10.1007/s10618-019-00619-1
11. Fawaz, H.I., et al.: Inceptiontime: finding AlexNet for time series classification. arXiv preprint arXiv:1909.04939 (2019)
12. He, K., Zhang, X., Ren, S., Sun, J.: Deep residual learning for image recognition. In: CVPR, pp. 770–778 (2016)
13. Hills, J., Lines, J., Baranauskas, E., Mapp, J., Bagnall, A.: Classification of time series by shapelet transformation. Data Min. Knowl. Disc. **28**(4), 851–881 (2013). https://doi.org/10.1007/s10618-013-0322-1
14. Hochreiter, S., Schmidhuber, J.: Long short-term memory. Neural Comput. **9**(8), 1735–1780 (1997)
15. Hu, S., et al.: DSNAS: direct neural architecture search without parameter retraining. In: CVPR, pp. 12084–12092 (2020)
16. Karim, F., Majumdar, S., Darabi, H., Chen, S.: LSTM fully convolutional networks for time series classification. IEEE Access **6**, 1662–1669 (2017)
17. Kate, R.J.: Using dynamic time warping distances as features for improved time series classification. Data Min. Knowl. Disc. **30**(2), 283–312 (2015). https://doi.org/10.1007/s10618-015-0418-x
18. Lines, J., Bagnall, A.: Time series classification with ensembles of elastic distance measures. Data Min. Knowl. Disc. **29**(3), 565–592 (2014). https://doi.org/10.1007/s10618-014-0361-2
19. Lines, J., Taylor, S., Bagnall, A.: Time series classification with hive-cote: the hierarchical vote collective of transformation-based ensembles. TKDD **12**(5) (2018)
20. Long, J., Shelhamer, E., Darrell, T.: Fully convolutional networks for semantic segmentation. In: CVPR, pp. 3431–3440 (2015)
21. Schäfer, P.: The boss is concerned with time series classification in the presence of noise. Data Min. Knowl. Disc. **29**(6), 1505–1530 (2015)

22. Shifaz, A., Pelletier, C., Petitjean, F., Webb, G.I.: TS-CHIEF: a scalable and accurate forest algorithm for time series classification. Data Min. Knowl. Discov. 1–34 (2020)
23. Vaswani, A., et al.: Attention is all you need. In: NeurIPS, pp. 5998–6008 (2017)
24. Wang, Z., Yan, W., Oates, T.: Time series classification from scratch with deep neural networks: a strong baseline. In: IJCNN, pp. 1578–1585. IEEE (2017)
25. Ye, L., Keogh, E.: Time series shapelets: a new primitive for data mining. In: SIGKDD, pp. 947–956 (2009)
26. Zoph, B., Le, Q.V.: Neural architecture search with reinforcement learning. arXiv preprint arXiv:1611.01578 (2016)
27. Zoph, B., Vasudevan, V., Shlens, J., Le, Q.V.: Learning transferable architectures for scalable image recognition. In: CVPR, pp. 8697–8710 (2018)

Aspect-Based Sentiment Analysis Through EDU-Level Attentions

Ting Lin[1(✉)], Aixin Sun[1], and Yequan Wang[2]

[1] School of Computer Science and Engineering, Nanyang Technological University, Singapore, Singapore
ting005@e.ntu.edu.sg, axsun@ntu.edu.sg
[2] Beijing Academy of Artificial Intelligence, Beijing, China

Abstract. A sentence may express sentiments on multiple aspects. When these aspects are associated with different sentiment polarities, a model's accuracy is often adversely affected. We observe that multiple aspects in such hard sentences are mostly expressed through multiple clauses, or formally known as *elementary discourse units* (EDUs), and one EDU tends to express a single aspect with unitary sentiment towards that aspect. In this paper, we propose to consider EDU boundaries in sentence modeling, with attentions at both word and EDU levels. Specifically, we highlight sentiment-bearing words in EDU through word-level sparse attention. Then at EDU level, we force the model to attend to the right EDU for the right aspect, by using EDU-level sparse attention and orthogonal regularization. Experiments on three benchmark datasets show that our simple EDU-Attention model outperforms state-of-the-art baselines. Because EDU can be automatically segmented with high accuracy, our model can be applied to sentences directly without the need of manual EDU boundary annotation.

1   Introduction

Aspect-based sentiment analysis (ABSA) is challenging because a sentence may express complex sentiments towards multiple aspects. We call these sentences *hard sentences*. For example, *"Despite the waiter's mediocre service, the food is tasty and the bill is never too large."* mentions three aspects: service, food, and price, and they are associated with different sentiment polarities. Because an aspect may not always be explicitly expressed through such representative terms, ABSA has been approached by dividing this challenging task into subtasks, *e.g.,* to identify aspects in a sentence, and to predict sentiment polarities of the identified aspects. In this paper, we focus on the latter, also known as *aspect category sentiment analysis* (ACSA). In ACSA, the aspects expressed in a sentence are given, and the task is to predict the corresponding sentiment on each given aspect. Note that, the aspect in ACSA is an abstractive category label (*e.g.,* price). The name of such an aspect category may not literally appear in a sentence (*e.g.,* bill is never large). Our task is different from aspect term-based sentiment analysis (ATSA), where the aspect indicative term in the input sentence are pre-annotated.

Y. Wang—This work was done when Yequan worked at Institute of Computing Technology, Chinese Academy of Sciences, China.

J. Gama et al. (Eds.): PAKDD 2022, LNAI 13280, pp. 156–168, 2022.
https://doi.org/10.1007/978-3-031-05933-9_13

Table 1. Each of the three 3 EDUs expresses a clear sentiment on one aspect.

Elementary discourse unit (EDU)	Aspect	Polarity
e_1. Despite the waiter's mediocre service	Service	Neutral
e_2. the food is tasty	Food	Positive
e_3. and the bill is never too large	Price	Positive

In the example sentence, sentiments to the three aspects are expressed in three clauses, or more formally *Elementary Discourse Units* (EDUs), shown in Table 1. EDUs are clause-like grammatical units for discourse parsing in rhetorical structure theory (RST) [16]. One EDU carries coherent semantic meaning towards a subtopic [6,7]. Thanks to the development of neural models, EDU segmentation can be achieved automatically with high accuracy [12].

From three benchmark datasets (see Experiments), we observe that, **an EDU tends to express at most one aspect** and **unitary sentiment polarity** towards an aspect. Motivated by this observation, we propose the **EDU-Attention** model. Our model learns aspect-specific representation for each EDU independently, as a part of a full sentence representation. Because of single aspect and unitary sentiment in one EDU, we apply sparse self-attention to select only relevant words to the target aspect in an EDU and ignore irrelevant ones. Considering all EDUs in one sentence, we apply EDU-level sparse self-attention to select the correct EDU(s) for a target aspect. As each EDU only describes a single aspect, we further apply Orthogonal Regularization on EDU-level attention scores to diversify the attention distributions among all aspects, *i.e.,* to ensure that the same EDU is not selected for more than one aspect.

The EDU-Attention model is simple and effective in handling hard sentences in ACSA. Experiments show that our model achieves better accuracy than BERT based models on hard sentences with a much smaller model size and faster inference time.

2 Related Work

Predicting sentiment at aspect level is a fine-grained task. Attention mechanism [2], as a way of extracting important features from an input sentence, has shown its success in previous studies. A line of work uses target aspect as a 'query' on terms in an input sentence, to give more weights to aspect relevant terms [8,22,23,28]. There are also work that try to fuse target aspect representation with each term in the sentence before applying attention [5,15,26,29]. The word-level aspect and term feature fusion makes the input to a model to be more target-specific.

Syntactic dependency between an aspect and its corresponding opinion expression has also been explored [13,18,24]. By utilizing additional syntactic knowledge obtained from external syntax parsers, the relative position in a syntactic tree is used to measure the distance between aspect-related terms and opinion-bearing text span in the sentence. These approaches require the terms that describe target aspect explicitly appearing in the sentence, and are pre-annotated. They are not applicable in our work.

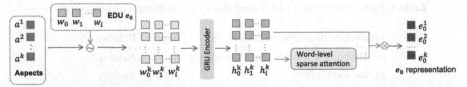

(a) A k-aspect-specific EDU representation is learned for the given k aspects. Word-level sparse attention is applied to give more weights to sentiment-bearing words.

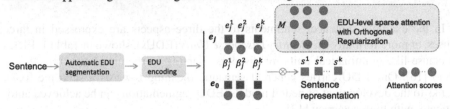

(b) Aspect-specific sentence representation is a linear sum of EDU representations and their respective attention weights, on each aspect. Sentiment is predicted on the corresponding s^k; detecting the existence of the aspect is an additional learning objective.

Fig. 1. The model architecture of EDU-Attention.

Several studies utilize discourse structure/relationship for sentiment analysis. Authors in [10,31] explore discourse relationship between two adjacent EDUs for predicting sentiment polarity. Hand-crafted rules are used to segment text into sentiment expression units (SEUs); a SEU contains either a sentiment, or an aspect, or both [30]. In [4], a full discourse parse tree is utilized to find precise context for a given aspect term. There are also neural network approaches that utilize EDU segments for document-level sentiment prediction [1,11,25]. The overall sentiment polarity becomes an aggregation of sentiment distribution of EDUs in the document.

Different from these models, we do not consider relationships between adjacent EDUs or their locations in a discourse parse tree. Instead, we model each EDU independently and apply word-level sparse attention to give more weights to relevant terms. By assuming one EDU expresses at most one aspect, we use regularisation at EDU-level attention to avoid the same EDU being selected for multiple aspects.

3 The Proposed Model: EDU-Attention

We follow the definition of *aspect category sentiment analysis* (ACSA) in previous studies [8,14,27]. There are k predefined aspect categories $A = \{a^1, \ldots, a^k\}$, and a list of predefined sentiment polarities $P = \{negative, neutral, positive\}$. Given a sentence s and the m aspect label(s) expressed in the sentence $A_s = \{a^1, \ldots, a^m\}, A_s \subseteq A$, we aim to predict the sentiment polarity associated with each aspect label, *i.e.*, all pairs of (a^m, p_o), for $a^m \in A_s, p_o \in P$.

3.1 Model Overview

EDU-Attention is a simple model that takes in an EDU-segmented sentence as input. EDU segmentation of a sentence can be achieved with high accuracy by using off-the-shelf tools.[1] After *automatic EDU segmentation*, an input sentence is denoted by its EDUs $\{e_0, \ldots, e_j\}$, and an EDU is a sequence of words $e_j = \{w_0, \ldots, w_i\}$.

Figure 1a shows the EDU-encoder in our model. It learns an aspect-specific EDU representation for a given EDU. Specifically, an EDU is represented as a k-dimensional feature, one for each of the k aspects. The encoder applies sparse self-attention to words within the EDU, to give sentiment-bearing words more weight for that particular aspect. Shown in Fig. 1b, with all aspect-specific EDUs representations learned in a sentence, we apply EDU-level sparse self-attention for locating the right EDU(s) for the right aspect. We further apply orthogonal regularization to force different aspects to focus on different EDUs. Finally, sentence representation is obtained as a linear sum of the EDUs representations with respective attention scores. We use aspect category prediction as an auxiliary learning objective, in addition to sentiment label prediction.

3.2 EDU Representation

The EDU-encoder learns aspect-specific EDU-representation for an EDU in three steps: word-aspect feature fusion, EDU encoder, and word-level sparse attention, see Fig. 1a.

Word-Aspect Feature Fusion. To promote interaction between aspect and words, we fuse word embedding w_i with aspect embedding a^k to derive aspect-specific word feature w_i^k, shown in Eq. 1. Here, W_e and W_k shared for all aspects are learnable parameters. Aspect embedding a^k can be initialized by using embedding of a matching word for the aspect (*e.g.*, 'food', 'service'), or be initialized randomly, if there is no representative word for the aspect (*e.g.*, 'anecdotes/miscellaneous').

$$w_i^k = \tanh\left(w_i W_e + a^k W_k\right) \tag{1}$$

EDU Encoder. With the fused aspect-word features, we perform EDU encoding by using a bidirectional GRU shown in Eq. 2. The bidirectional GRU learns contextual information within an EDU, for each aspect. Specifically, the encoder takes in an aspect-word feature matrix of the EDU e_j for aspect k, and stacks hidden output from every time step of GRU into H_j^k for further processing.

$$H_j^k = \text{GRU}([w_0^k, w_1^k, \ldots, w_i^k]) \tag{2}$$

---

[1] In our implementation, we use the pretrained SEGBOT tool http://138.197.118.157:8000/segbot/ released by its authors [12]. If an EDU returned by SEGBOT contains conjunction words (*i.e.*, 'but', 'and', 'although', and 'or'), we further split this EDU by using regular expression.

Word-Level Sparse Attention. Ideally, a classifier only needs to extract aspect-specific opinion-bearing words for predicting the sentiment. The remaining words can be ignored. We apply the attention mechanism on H_j^k to highlight such important words.

A straightforward solution is to apply self-attention with softmax normalization [2]. The resulting probability is distributed to all words in an EDU, *i.e.,* the attention score after softmax is not equal to 0 for every word $w_i \in e_j$. This does not well serve our purpose of ignoring irrelevant words in the EDU. Hence, we adopt sparsemax function [17], which returns the euclidean projection of the k element input vector z onto the $(k - 1)$-dimensional simplex $\triangle^{k-1}$ defined as $\{p \in \mathbb{R}^k | 1^T p = 1, p \geq 0\}$ [17]. The projection is likely to hit the boundary of the simplex, in which case the sparsemax(z) becomes sparse. sparsemax produces sparse distribution while retaining the important properties of softmax. Equation 3 shows the sparse attention computation.

$$\text{sparsemax}(z) = \underset{p \in \triangle^{k-1}}{\text{argmin}} ||p - z||^2 \tag{3}$$

By using feature matrix H_j^k obtained earlier, we apply sparsemax to compute a weight vector for all words in EDU e_j, for aspect a^k. The sparse attention computation, where α_i^k is the attention score of word w_i in EDU e_j towards aspect a^k is calculated by:

$$\alpha_0^k, \ldots, \alpha_i^k = \text{sparsemax}(H_j^k W_e), \tag{4}$$

where W_e is a learnable parameter.

Then, for EDU e_j, we derive the k number of EDU representations $[e_j^1, \ldots, e_j^k]$, one for each aspect. For aspect a^k, e_j^k is the weighted sum of the sparse attention scores α_i^k's and h_i^k's for all the words in e_j. We also add position embedding [3] of EDU e_j, denoted by $\text{Emb}_p(e_j)$, for its relative position in the sentence, as shown in Eq. 5.

$$e_j^k = \text{Emb}_p(e_j) + \sum_{w_i \in e_j} \alpha_i^k h_i^k \tag{5}$$

3.3 Sentence Representation and Learning Objective

So far, for an EDU e_j, we obtain its aspect-specific representation $[e_j^1, \ldots, e_j^k]$. As we observe that one EDU tends to express sentiment on one aspect, we now try to identify the right aspect for each EDU in a sentence.

EDU-Level Sparse Attention. For each aspect a^k, we apply sparse attention on the corresponding aspect-specific EDU-representations in the sentence $[e_0^k, \ldots, e_j^k]$, for choosing the right EDU for this aspect. The sparse attention shares the similar process as word-level sparse attention, or formally:

$$\beta_0^k, \ldots, \beta_j^k = \text{sparsemax}([e_0^k, \ldots, e_j^k]W_s), \tag{6}$$

where W_s is a learnable parameter. Then, the aspect a^k specific sentence representation s^k is a linear combination of its aspect-specific EDU representations.

$$s^k = \sum_{e_j \in s} \beta_j^k e_j^k \tag{7}$$

Table 2. Statistics of datasets, with number of sentences expressing single and multiple aspects. We remove sentence with conflict polarities.

Datasets	Rest14		Rest14-Hard	Laptop15		Laptop15-Hard	MAMS-ACSA		
	Train	Test	Test	Train	Test	Test	Train	Val	Test
Single	2,345	595	–	1,174	539	–	–	–	–
Multiple	539	172	25	209	93	20	2,839	710	400
Negative	841	222	20	616	258	14	1,883	460	263
Neutral	501	94	12	58	44	8	2,776	689	393
Positive	2,174	657	21	860	424	19	1,742	428	263

Each s^k is used to predict sentiment label, and also to predict the corresponding aspect as an additional objective.

The Aspect-Level Orthogonal Regularization. As stated earlier, an EDU tends to express a single aspect and a unitary sentiment. In a complex sentence, opinions for different aspects reside in different EDUs. The sparse attention computed in Eq. 6 is for one specific aspect, and attention scores for different aspects are computed independently.

To constraint that one EDU should be attended to a single aspect, we put the EDU-level attention scores β_j^k's computed for the k aspects over the j EDUs in a sentence, into a $j \times k$ attention matrix M. Then we apply orthogonal regularization to force the dot product of attention vectors of each aspect to be orthogonal, as shown in Eq. 8. I is an identity matrix.

$$R_{orth} = \left\| M^T M - I \right\| \tag{8}$$

Learning Objectives. The key objective is to predict sentiment polarities for the given aspects mentioned in a given sentence. In EDU-Attention, we also use aspect prediction as an additional learning objective in addition to the sentiment labels. Specifically, a binary prediction on each aspect existence in the input sentence is added into the loss function of our model. We use cross-entropy loss $J(\theta)$ for sentiment labels prediction, and binary cross-entropy loss $U(\theta)$ for aspect prediction. The aspect-level orthogonal regularization $R(\theta)_{orth}$ is also a part of our learning objectives. The full loss function of our model is as follows:

$$L(\theta) = \lambda_1 J(\theta) + \lambda_2 U(\theta) + \lambda_3 R(\theta)_{orth}, \tag{9}$$

where $\lambda_1, \lambda_2, \lambda_3$ are the scaling parameters set for each loss. The collection of model parameters is θ.

4 Experiments

We evaluate the proposed EDU-Attention on three benchmark datasets, with a focus on hard sentences.

4.1   Datasets and Baselines

Table 2 summarizes the three datasets in our experiments. Following previous studies [23, 26], we remove samples with conflict polarities. **Rest14** is from SemEval-2014 Task-4 Restaurant Review [20]. **Rest14-Hard** is a collection of hard sentences sampled from Rest14 test set [29]. Each sentence in Rest14-Hard contains at least two aspects, and the aspects have different sentiment polarities. **Laptop15** is from SemEval-2015 Task-12 Laptop Review [20]. To be consistent with other datasets, we keep aspects only and ignore attributes. Accordingly, we update sentiment labels of aspects following the original annotation guideline [20].[2] We choose to keep the aspects that contain at least one sentence in every sentiment class (positive, neutral, and negative). In total, there are 22 aspects.[3] **Laptop15-Hard** is a collection of hard sentences sampled from Laptop15 test set, following the same sampling strategy as Rest14-Hard. **MAMS-ACSA** is a restaurant review dataset [8]. *All sentences in MAMS-ACSA are hard sentences*; Note that, MAMS-ACSA's *annotation scheme is different* from that of other datasets. We observe that a sentence is annotated with an aspect if the sentence mentions a matching keyword. The differences in annotation scheme results in majority of the sentiment labels being 'neutral' in MAMS-ACSA, making the dataset challenging.

We evaluate the following baselines. **ATAE-LSTM** [26], is a strong baseline where aspect embeddings are concatenated with word vectors. **MemoryNet** [23], employs two LSTMs and an interactive attention mechanism to learn representations of sentence and aspect. **HAN** [25] is a hierarchical attention network built on word, clause (EDU), and sentence for aspect-specific sentence representation. **GCAE** [29] uses CNNs to extract features and then employs two Gated Tanh-Relu units to selectively output the sentiment information flow towards the aspect, for predicting sentiment labels. **ATAE-CAN-2R$_o$** [5] uses aspect detection as an auxiliary task. **AS-Capsule** [27] is a capsule alike network. Each capsule encloses a set of computations for one aspect. **CapsNet** [8] is a capsule-network based model. It learns the association between aspect and context. **AC-MIMLLN** [14] is a multi-instance learning model. It has two separate encoders to learning aspect- and sentiment- representations.

These baselines use different approaches of learning aspect-specific sentence representations. ATAE-LSTM, ATAE-CAN, GCAE, HAN fuse aspect features with term features, then apply attention on the fused features. MemoryNet and CapsNet use carefully designed attention mechanisms on term features only. In addition, AC-MIMLLN, ATAE-CAN, and AS-Capsule use aspect category prediction as an auxiliary task for learning the interaction between aspect representation and sentiment representation.

We also evaluated Bert-based models. The simple **Bert-baseline** and its 'distillation' versions (DistillBERT [21], TinyBERT[9]) encode an aspect-specific sentence representation with this input format: [CLS] words in sentence [SEP] aspect category [SEP] [28]. The sentiment prediction is done by softmax with a linear layer. In the Bert variant of our model, we replace the EDU representation learning (*i.e.,* Fig. 1a) by Bert encoding with: [CLS] words in EDU [SEP] aspect category [SEP]. For each EDU, we enumerate all aspect categories to obtain its aspect-specific representation.

---

[2] https://alt.qcri.org/semeval2014/task4.
[3] The re-processed Laptop15 dataset can be found at: https://github.com/Ting005/EDU_Attentions.

Table 3. Accuracy and Macro-F1 of all models on all datasets. We reproduce results of baselines by using authors' implementation except two models. ATAE-CAN-2R_o is by our own implementation following authors' paper. Results of AC-MIMLLN model are reported in its original paper. We run a model 5 times with random seeds and report the average. The best scores are in boldface and second best underlined, among non-Bert and Bert models respectively.

Model	Rest14		Rest14-Hard		Laptop15		Laptop15-Hard		MAMS-ACSA	
	Acc.	F_1	Acc.	F_1	Acc.	F_1	Acc.	F_1	Acc.	F_1
MemoryNet	81.29	70.79	54.72	46.65	71.80	52.03	36.59	26.09	64.04	62.57
AC-MIMLLN*	81.60	–	65.28	–	–	–	–	–	76.64	–
HAN	81.74	71.49	58.49	49.67	73.21	53.64	48.78	41.05	73.64	72.42
AS-Capsule	82.03	71.55	59.24	51.99	76.31	55.65	48.78	41.05	75.44	74.37
GCAE	82.32	72.08	56.13	51.07	76.08	54.00	55.29	42.78	70.59	69.01
ATAE-CAN-2R_o	82.43	71.18	64.62	53.74	76.24	53.00	51.22	43.66	76.42	75.23
ATAE-LSTM	83.10	73.32	59.91	53.08	76.86	56.88	46.34	38.35	75.02	73.93
CapsNet	83.10	72.58	53.78	44.50	75.48	52.33	48.78	33.65	72.92	71.86
Ours	**83.97**	**73.96**	**70.28**	**65.59**	**77.83**	56.39	**56.10**	**46.35**	**77.14**	**76.00**
Ours w/o reg.	82.88	72.69	68.68	60.79	76.37	55.05	51.22	41.49	75.65	74.50
Ours w/o aux.	82.99	73.02	58.51	54.49	77.05	55.23	53.66	45.08	75.03	73.91
DistilBERT	65.57	38.36	41.51	26.72	59.09	32.39	53.66	33.73	59.16	47.17
TinyBERT	67.52	26.87	39.62	18.92	57.44	32.58	56.10	38.02	59.49	47.45
Bert-baseline	87.82	80.07	66.98	62.83	83.47	63.94	58.40	24.58	78.86	78.06
CapsNet-Bert	87.80	79.81	50.94	38.66	84.07	57.25	48.78	33.65	77.42	76.65
BERT-pair-QA-B	87.25	78.09	52.83	46.58	83.88	69.77	46.34	39.19	79.35	78.89
Ours-Bert	**87.94**	**80.74**	**72.95**	**70.71**	**84.85**	65.07	**58.54**	**41.56**	**79.64**	**79.02**

CapsNet-Bert is a variation of CapsNet; **BERT-pair-QA-B** [22] constructs an auxiliary sentence for each aspect and transforms the task into a sentence-pair classification. For CapsNet-Bert and BERT-pair-QA-B, we use authors' implementation.

4.2 Implementation and Parameter Setting

All models are implemented by using Pytorch[4] with CUDA 11.1 on RTX3090 GPU in Windows OS. Models' parameters are optimized by using Adam. For non-Bert models, we set a learning rate of $1e-3$ for model parameters and $1e-4$ for word embedding adjustment. The word embeddings are initialized by Glove [19] with 300 dimensions, and randomly initialize positional embedding for each EDU with dimension of 300. We set the mini-batch size to 32 and evaluate every 16 mini-batches, and use a dropout rate of 0.5 during model training. We use 'bert-base-uncased'[5] for fine-tuning models use BERT, 'distilbert-base-uncased'[6] for DistillBERT model, and 'Tiny-BERT_General_6L_768D'[7] for TinyBERT model. For fine-tuning, we keep dropout probability at 0.1, learning rate at $3e-5$. We set the scaling parameter λ_1, λ_2 to 1, and λ_3 to 0.1 for EDU-Attention (see Eq. 9). In the Bert variation of EDU-Attention, we set $\lambda_1, \lambda_2, \lambda_3$ to 0.5, 0.4, and 0.1 respectively for the best performance. Models'

---

[4] https://pytorch.org/.

[5] https://huggingface.co/bert-base-uncased.

[6] https://huggingface.co/distilbert-base-uncased.

[7] https://huggingface.co/huawei-noah/TinyBERT_General_6L_768D.

parameters are tuned on validation set. The MAMS-ACSA dataset comes with a validation set. For Rest14 and Laptop15 datasets, we randomly sample 20% of training data as validation set. We run the models for 5 times with random seed initialization, and report the average metric on test sets.

4.3 Comparison with Baselines

Non-Bert Models. Reported in Table 3, among non-Bert models, our EDU-Attention performs the best on almost all metrics, except Macro-F1 on Laptop15, which is the second best. Large improvements are achieved on Rest14-Hard and Laptop15-Hard.

The Rest14 dataset contains 19% multi-aspect sentences (see Table 2). Models that use aspect and term fused features (*e.g.,* ATAE-CAN, ATAE-LSTM, GCAE) generally perform better than others. In particular, ATAE-LSTM performs the second-best among non-bert models by both accuracy and Macro-F1. We produce results on Rest14-Hard by using the models trained on Rest14 training dataset. Our EDU-Attention outperforms all baselines by a large margin on both metrics, showing the effectiveness of modeling EDU contextual boundaries in handling hard sentences.

The Laptop15 dataset contains 15% multi-aspect sentences. All models are less affected by noise introduced by sentiment terms of non-target aspects, compared to other datasets. On hard sentences, our model shows clear advantage on Laptop15-Hard.

All sentences in MAMS-ACSA are hard sentences. A different annotation scheme is adopted in MAMS-ACSA as described in the Dataset section. The annotation based on appearance of surface terms, makes the dataset challenging with many neutral labels. Our model is the best performer, demonstrating its effectiveness in handling hard sentences.

Bert Models. As expected, Bert-baseline outperforms all non-Bert models and brings in big improvements on Rest14 and Laptop15 datasets. The improvement over EDU-Attention on the MAMS dataset, however, is relatively small. On Rest14-Hard, Bert-baseline performs poorer than EDU-Attention by about 3 points for both Accuracy and Macro-F1. Recall that MAMS, Rest14-Hard, and Laptop15-Hard datasets only contain hard sentences; the comparable performance between EDU-Attention and Bert-baseline suggests that modeling EDU contextual boundary is beneficial to aspect category sentiment analysis. By using Bert for EDU representation learning, our EDU-Attention-Bert model outperforms all models. In particular, on Rest14-Hard, our EDU-Attention-Bert outperforms Bert-baseline by 6 to 7 points on both Accuracy and Macro-F1. We also compare the 'distillation' version of BERT (DistillBERT, TinyBERT) with the same input format and model training strategy, the performance of both models does not come close to the rest of the baselines.

4.4 Ablation Study

The orthogonal regularization in our model makes the distribution of EDU-level attention scores diverse by aspects. We also predict the existence of an aspect (*i.e.,* $U(\theta)$ in Eq. 9) as an additional learning objective to enable the model to concentrate more on the aspect relevant EDU(s).

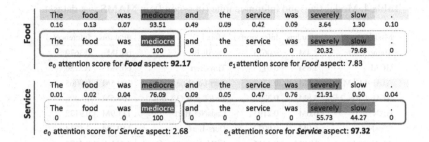

Fig. 2. Attention scores from aspect-specific representations of ATAE-LSTM and EDU-Attention, on two aspects food and service. Outputs from ATAE-LSTM is shown in one sentence for each aspect. EDU-Attention has both word- and EDU-level attentions, shown with EDU boundaries.

We conduct ablation study to analyze effectiveness of the orthogonal regularization (reg.) and the auxiliary aspect prediction (aux.) in EDU-Attention. Reported in Table 3, removing either orthogonal regularization or auxiliary aspect prediction leads to performance drop on all datasets and on all measures. The amount of performance drop is comparable on Rest14, Laptop15, and MAMS-ACSA datasets. Large drops are observed on Rest14-Hard. On hard sentences, our model relies on orthogonal regularization to spread out the EDU-level attention scores. The additional objective function guides the model to recognize the aspect expressed in an EDU. The amount of drop on the MAMS-ACSA dataset is slightly larger than that on Rest14 and Laptop15. On the one hand, all sentences in MAMS-ACSA are hard sentences. On the other hand, its annotation scheme is different from the other datasets. The model might have captured the association between term appearance and aspects, and the model performance is heavily affected by the large number of neutral labels (see Table 2).

4.5   Analysis of Sparse Attention

As a case study, we compare attention scores computed by ATAE-LSTM [26], and the attention scores of our model at both word- and EDU- levels. Figure 2 shows an example sentence from Rest14. The sentence expresses sentiments towards two aspects: food and service, with indicative words. ATAE-LSTM computes two aspect-specific representations, one for each aspect. In both representations, the model highlights 'mediocre' and 'severely' with larger scores. Nevertheless, the score of 'mediocre' dominates in both representations. Even though ATAE-LSTM also correctly highlights 'severely' for aspect 'service', the larger score of 'mediocre' would interfere with the correct prediction of sentiment label for service.

In contrast, the word-level attention scores computed by EDU-Attention is confined within an EDU contextual boundary, and normalized within the EDU. Opinion words 'mediocre' and 'severely slow' are given large attention scores. In fact, all the rest non-opinion words are assigned 0 scores, thanks to sparse attention (see Eq. 3). The association between the opinion words and their corresponding aspects is through the EDU-level sparse attention. Recall that, for each EDU, we learn an aspect-specific representation. As shown in Fig. 2, for the first clause (*i.e.*, EDU e_0), its attention score

Table 4. Model size, and inference time (in second) on MAMS test dataset.

Model	#Params	Inference(s)	Model	#Params	Inference(s)
MemoryNet	2.4M	0.0673	Distill-Bert	66.3M	0.3679
GCAE	2.8M	0.0775	Tiny-Bert	66.9M	0.3736
ATAE-LSTM	3.8M	0.1298	Bert-baseline	109.5M	0.9065
ATAE-CAN-2R_0	3.8M	0.3702	BERT-pair-QA-B	109.5M	17.496
HAN	6.0M	0.2195	CapsNet-Bert	111.9M	0.9122
CapsNet	6.0M	0.9072	Ours-Bert	110.7M	3.5734
AS-Capsule	10.0M	0.4215	EDU-Attention (**Ours**)	3.5M	0.4625

for 'food' aspect is 92.17, compared to 2.68 for 'service' aspect. Note that, EDU-level attention scores are normalized across all EDUs in the same sentence on each aspect. Similarly, the second clause, EDU e_1, receives high score 97.32 for 'service', and a very small score 7.83 for 'food' aspect. In short, sparse attentions at both EDU- and word- levels guide our model to correctly identify aspect-relevant EDU in a sentence, and opinion words in an EDU.

4.6   Model Size and Inference Time

Table 4 summarizes the model size in number of parameters, and reports the inference time for processing the 400 sentences in MAMS-ACSA test dataset. Among non-Bert models, our model has relatively small size. The slightly longer inference time is due to aspect-specific EDU representation computation. A hard sentence often contains multiple EDUs. Compared to Bert-baseline, our model is much smaller and only takes about half of its inference time. The Bert variant of our model has a longer inference time as our model needs to encode an EDU at a time using Bert, instead of encoding a full sentence at a time as in Bert-baseline. Overall, our EDU-Attention has a small model size, and achieves good performance with reasonable inference time.

5   Conclusion

We observe that text span in an EDU tends to express a single aspect and unitary sentiment towards the aspect. Hence, we propose a simple EDU-Attention model to learn aspect-specific representations of EDUs in a sentence. Based on our observation, we apply sparse attention at both word-level and EDU-level, to highlight sentiment-bearing words, and to constraint an EDU to one aspect. Our model shows improvement over strong baselines on three benchmark datasets. The detailed ablation study also shows that the model behaves as expected. Interestingly, the prediction of aspect labels of EDUs, based on EDU-level attention scores, can be beneficial to other applications like review summarization.

References

1. Angelidis, S., Lapata, M.: Multiple instance learning networks for sentiment analysis. Trans. Assoc. Comput. Linguistics **6**, 17–31 (2018)

2. Bahdanau, D., Cho, K., Bengio, Y.: Neural machine translation by jointly learning to align and translate. In: ICLR (2015)
3. Devlin, J., Chang, M., Lee, K., Toutanova, K.: BERT: pre-training of deep bidirectional transformers for language understanding. In: NAACL-HLT, pp. 4171–4186 (2019)
4. Hoogervorst, R., et al.: Aspect-based sentiment analysis on the web using rhetorical structure theory. In: Bozzon, A., Cudre-Maroux, P., Pautasso, C. (eds.) ICWE 2016. LNCS, vol. 9671, pp. 317–334. Springer, Cham (2016). https://doi.org/10.1007/978-3-319-38791-8_18
5. Hu, M., et al.: CAN: constrained attention networks for multi-aspect sentiment analysis. In: EMNLP-IJCNLP, pp. 4600–4609 (2019)
6. Huber, P., Carenini, G.: From sentiment annotations to sentiment prediction through discourse augmentation. In: COLING, pp. 185–197 (2020)
7. Jiang, F., Fan, Y., Chu, X., Li, P., Zhu, Q., Kong, F.: Hierarchical macro discourse parsing based on topic segmentation. In: AAAI, pp. 13152–13160 (2021)
8. Jiang, Q., Chen, L., Xu, R., Ao, X., Yang, M.: A challenge dataset and effective models for aspect-based sentiment analysis. In: ACL, pp. 6279–6284 (2019)
9. Jiao, X., et al.: TinyBERT: distilling BERT for natural language understanding. In: EMNLP, pp. 4163–4174 (2020)
10. Lazaridou, A., Titov, I., Sporleder, C.: A Bayesian model for joint unsupervised induction of sentiment, aspect and discourse representations. In: ACL, pp. 1630–1639 (2013)
11. Li, J., Chiu, B., Shang, S., Shao, L.: Neural text segmentation and its application to sentiment analysis. TKDE **34**, 828–842 (2020)
12. Li, J., Sun, A., Joty, S.R.: SegBot: a generic neural text segmentation model with pointer network. In: IJCAI, pp. 4166–4172 (2018)
13. Li, R., Chen, H., Feng, F., Ma, Z., Wang, X., Hovy, E.H.: Dual graph convolutional networks for aspect-based sentiment analysis. In: ACL/IJCNLP, pp. 6319–6329 (2021)
14. Li, Y., Yin, C., Zhong, S., Pan, X.: Multi-instance multi-label learning networks for aspect-category sentiment analysis. In: EMNLP, pp. 3550–3560 (2020)
15. Lin, P., Yang, M., Lai, J.: Deep mask memory network with semantic dependency and context moment for aspect level sentiment classification. In: IJCAI, pp. 5088–5094 (2019)
16. Mann, W.C., Thompson, S.A.: Assertions from discourse structure. In: Annual Meeting of the Berkeley Linguistics Society, pp. 245–258 (1985)
17. Martins, A.F.T., Astudillo, R.F.: From softmax to sparsemax: a sparse model of attention and multi-label classification. In: ICML, pp. 1614–1623 (2016)
18. Oh, S., et al.: Deep context- and relation-aware learning for aspect-based sentiment analysis. In: ACL/IJCNLP, pp. 495–503 (2021)
19. Pennington, J., Socher, R., Manning, C.D.: GloVe: global vectors for word representation. In: EMNLP, pp. 1532–1543 (2014)
20. Pontiki, M., Galanis, D., Pavlopoulos, J., Papageorgiou, H., Androutsopoulos, I., Manandhar, S.: SemEval-2014 task 4: Aspect based sentiment analysis. In: SemEval@COLING, pp. 27–35 (2014)
21. Sanh, V., Debut, L., Chaumond, J., Wolf, T.: DistilBERT, a distilled version of BERT: smaller, faster, cheaper and lighter. CoRR abs/1910.01108 (2019)
22. Sun, C., Huang, L., Qiu, X.: Utilizing BERT for aspect-based sentiment analysis via constructing auxiliary sentence. In: NAACL-HLT, pp. 380–385 (2019)
23. Tang, D., Qin, B., Liu, T.: Aspect level sentiment classification with deep memory network. In: EMNLP, pp. 214–224 (2016)
24. Tian, Y., Chen, G., Song, Y.: Aspect-based sentiment analysis with type-aware graph convolutional networks and layer ensemble. In: NAACL-HLT, pp. 2910–2922 (2021)
25. Wang, J., et al.: Aspect sentiment classification with both word-level and clause-level attention networks. In: IJCAI, pp. 4439–4445 (2018)

26. Wang, Y., Huang, M., Zhu, X., Zhao, L.: Attention-based LSTM for aspect-level sentiment classification. In: EMNLP, pp. 606–615 (2016)
27. Wang, Y., Sun, A., Huang, M., Zhu, X.: Aspect-level sentiment analysis using as-capsules. In: WWW, pp. 2033–2044 (2019)
28. Xu, H., Liu, B., Shu, L., Yu, P.S.: BERT post-training for review reading comprehension and aspect-based sentiment analysis. In: NAACL-HLT, pp. 2324–2335 (2019)
29. Xue, W., Li, T.: Aspect based sentiment analysis with gated convolutional networks. In: ACL, pp. 2514–2523 (2018)
30. Zhang, Z., Singh, M.P.: Limbic: author-based sentiment aspect modeling regularized with word embeddings and discourse relations. In: EMNLP, pp. 3412–3422 (2018)
31. Zirn, C., Niepert, M., Stuckenschmidt, H., Strube, M.: Fine-grained sentiment analysis with structural features. In: IJCNLP, pp. 336–344 (2011)

Interconnected Neural Linear Contextual Bandits with UCB Exploration

Yang Chen[1]([⊠])[iD], Miao Xie[2][iD], Jiamou Liu[1][iD], and Kaiqi Zhao[1][iD]

[1] The University of Auckland, Auckland, New Zealand
{yang.chen,jiamou.liu,kaiqi.zhao}@auckland.ac.nz
[2] Kuaishou Technology Co. Ltd., Beijing, China

Abstract. Contextual multi-armed bandit algorithms are widely used to solve online decision-making problems. However, traditional methods assume linear rewards and low dimensional contextual information, leading to high regrets and low online efficiency in real-world applications. In this paper, we propose a novel framework called *interconnected neural-linear UCB* (INLUCB) that interleaves two learning processes: an offline representation learning part, to convert the original contextual information to low-dimensional latent features via non-linear transformation, and an online exploration part, to update a linear layer using upper confidence bound (UCB). These two processes produce an effective and efficient strategy for online decision-making problems with non-linear rewards and high dimensional contexts. We derive a general expression of the finite-time cumulative regret bound of INLUCB. We also give a tighter regret bound under certain assumptions on neural networks. We test INLUCB against state-of-the-art bandit methods on synthetic and real-world datasets with non-linear rewards and high dimensional contexts. Results demonstrate that INLUCB significantly improves the performance on cumulative regrets and online efficiency.

Keywords: Contextual bandits · Upper confidence bound · Neural networks · Regret bound

1 Introduction

Contextual multi-armed bandit algorithms are powerful solutions to online sequential decision making problems such as influence maximisation [17] and recommendation [20]. In its setting, an agent sequentially observes a feature vector associated with each arm (action), called the *context*. Based on the contexts, the agent selects an arm which provides a random reward that is assumed to follow some distribution. Since the underlying distribution is unknown and the reward can only be observed at run-time, the agent should balance exploration and exploitation to maximise total rewards or, equivalently, to minimise *regret*.

Well-established contextual bandit methods, e.g., linear upper confidence bound (LINUCB) [7] and linear Thompson sampling (LINTS) [2], were effective

© The Author(s), under exclusive license to Springer Nature Switzerland AG 2022
J. Gama et al. (Eds.): PAKDD 2022, LNAI 13280, pp. 169–181, 2022.
https://doi.org/10.1007/978-3-031-05933-9_14

assuming that the reward is linear and the contexts are low-dimensional. However, these methods face two important challenges when applied to real-world scenarios such as online image classification [19] and online audio recognition [23]: First, these applications involve reward distributions that are non-linear w.r.t. the contexts. This violates the linear-reward assumption which is needed to achieve non-trivial regret bounds. Thus it is possible that these methods could result in high regrets. Second, most of these existing methods involve inverting a matrix online [4] whose dimension coincides with that of the contexts. However, the applications above usually involve high-dimensional contexts. Thus the existing methods will suffer from poor online efficiency. Although some recent methods relax the liner-reward assumption, they still rely on relatively restrictive modelling assumptions on rewards and/or cannot provide acceptable online efficiency. For instance, KERNELUCB [16] relaxes the linear reward assumption by asserting that the reward function belongs to a reproducing kernel Hilbert space, but it incurs an even higher computation cost on matrix inversions as the dimension of the kernel matrix increases with time.

Recently, several new methods under the name of *neural contextual bandits* [24] are proposed to extend classical contextual bandit algorithms. Leveraging the expressive power of neural networks, these methods aim to learn richer non-linear reward function and latent features through representation learning. So far, two major neural contextual bandit paradigms have been proposed: NEURAL-LINEAR and NEURALUCB. The former uses neural networks to extract a dimension-reduced latent feature (representation learning) and conduct exploration on top of the latent features [15,22], while the latter uses neural networks as a reward predictor and use UCB for exploration [24]. Despite showing promise in certain empirical tasks, these methods still suffer from some significant shortcomings. (1) While NEURAL-LINEAR is time-efficient, the method often incurs high regrets. This is because that it trains networks end-to-end, failing to use the result of exploration to boost representation learning. Worse yet, its regret bound is still unknown [15]. (2) NEURALUCB, in contrast, can provide the theoretical guarantee on regret bound. But updating the entire network every step results in low online efficiency, which makes it infeasible in practice.

Contributions. This paper addresses the need for an *efficient* contextual bandit algorithm applicable to *non-linear rewards* and *high-dimensional contexts*. We summarise our main contributions as follows: **(1)** We propose a new neural contextual bandit framework, called *interconnected neural-linear upper confidence bound* (INLUCB) (see Sect. 4). To our knowledge, INLUCB is the first contextual bandit method that achieves high online efficiency with a theoretical guarantee on its regret bound. INLUCB uses neural networks with two parts: the lower layers transform raw contexts to a low-dimensional latent feature space; and the last linear layer represents a linear model that fits the observed reward in terms of the latent features. The key novelty of INLUCB lies in an *interconnected offline-online update mechanism* to train the two parts. The offline process (*representation learning*) updates lower layers subject to the current linear model, simplifying the task at hand. The online process (*exploration*) follows

UCB-based exploration to update only the last linear layer based on the proposed representation, thereby guaranteeing online efficiency. (2) We derive a general expression of the regret bound of INLUCB by decomposing the total regrets into regrets caused by representation learning and online exploration (see Theorem 1). Specifically, we present a tighter regret bound under certain assumptions on neural networks (see Corollary 1). (3) We test INLUCB against state-of-the-art contextual, non-linear contextual, non-parametric contextual and neural contextual bandit methods on synthetic dataset with high-dimensional contexts and non-linear rewards as well as on real-world datasets with audio and images as contextual information (see Sect. 6). Results demonstrate that INLUCB achieves much lower cumulative regrets than linear contextual bandit baselines and higher online efficiency than neural contextual bandit baselines.

2 Related Work

Classical Contextual Bandits. Both classical multi-armed bandits and contextual bandits have been studied extensively along with their variants. Classical bandit algorithms such as Upper Confidence Bound (UCB) and Thompson Sampling (TS) [1] achieve $\widetilde{O}(\sqrt{KT})$ regret, where K is the number of candidate arms, T is the number of steps, and $\widetilde{O}(\cdot)$ hides the logarithmic factors. Since this regret bound depends on K, they are inefficient in real-world applications when K is large. To alleviate this problem, one can assume that the reward of each arm is a function of some observed features (i.e., contexts), yielding the family of methods called the *contextual bandits* [9,13]. As two widely-adopted contextual bandit algorithms with the linear-reward assumption, LINUCB [7] and LINTS [2] have a regret bound of $\widetilde{O}(\sqrt{dT})$ and $\widetilde{O}(d\sqrt{T})$, respectively, which depends on the dimension d of features rather than K. However, contextual bandits may result in high regrets when the reward function is non-linear or d is large.

High-Dimensional Contexts and Non-linear Rewards. Despite works exist that attempt to extend contextual bandits to the setting of either high-dimensional contexts or non-linear rewards [6,11,18,21], no method so far can resolve these two challenges simultaneously with acceptable efficiency. Lasso regression is investigated for the sparse contexts [6,11,18]. Although its regret bound [11] is superior to LINUCB, optimising a lasso regression problem online makes it too time-consuming to be used in practice. CBRAP [21] adopts random projection to map the high-dimensional contexts onto a low-dimensional space. Although it improves efficiency, its performance heavily relies on a good initial projection matrix, leading to poor robustness. KERNELUCB [16] adopts kernel functions to handle non-linear rewards but it uses matrix inversion which incurs high computation costs. NEURAL-LINEAR [22] and NEURALUCB [24] adopt neural networks to model rewards, each with issues mentioned above. In INLUCB, we propose a novel interconnected-update framework that makes our method unique and allows our method to overcome the shortcomings of the existing neural contextual bandit methods.

3   Problem Formulation and Background

Problem Setting. We consider the stochastic contextual bandit problem with K *arms* (actions) and T *steps*. At each step $t \leq T$, the agent observes the *context* (feature) $\mathbf{x}_{t,a} \in \mathbb{R}^d$ of each arm a with $\|\mathbf{x}_{t,a}\|_2 \leq 1$, where the *contextual dimension d* is usually very **large** in applications. An algorithm selects an action $a_t \in [K]$ at step t and receives a *reward* $r_{t,a_t} \in [0,1]$, where $[K]$ denotes the set $\{1, 2, \ldots, K\}$. The reward r_{t,a_t} is an independent random variable conditioned on context $\mathbf{x}_{t,a_t}$. The *regret* of the algorithm is defined as:

$$R_T \triangleq \sum_{t=1}^{T} r_{t,a_t^*} - \sum_{t=1}^{T} r_{t,a_t}, \tag{1}$$

where $a_t^* = \arg\max_{a \in [K]} \mathbb{E}[r_{t,a}|\mathbf{x}_{t,a}]$ is the optimal action at step t that maximises the expected reward. The goal is to find an algorithm to minimise R_T.

We focus on the cases where the reward function is **non-linear** in terms of contexts. To capture this fact, for each step t, we assume that the reward is generated by:

$$r_{t,a} \triangleq g(\mathbf{x}_{t,a}) + \xi_t, \tag{2}$$

where $g \colon \mathbb{R}^d \to \mathbb{R}$ is an unknown non-linear function satisfying $g(\mathbf{x}) \in [0,1]$ for any $\mathbf{x}$, and ξ_t is a sub-Gaussian noise satisfying $\mathbb{E}[\xi_t] = 0$. The sub-Gaussian noise is a standard assumption in the stochastic bandit literature, which can represent any bounded noise [14].

Neural Contextual Bandits. Neural contextual bandit methods [15, 22, 24] compute the rewards using a neural network. In this way, the method handles high dimensional contexts and non-linear rewards. Formally, the function g in Eq. (2) is realised by:

$$g(\mathbf{x}_{t,a}) = f_\star^\top(\mathbf{x}_{t,a})\boldsymbol{\theta}_\star, \tag{3}$$

where $f_\star \colon \mathbb{R}^d \to \mathbb{R}^p$ represents all layers except the last that satisfies $\|f_\star(\mathbf{x}_{t,a})\|_2 \leq 1$, $\boldsymbol{\theta}_\star$ represents the weights of the last linear layer that satisfies $\|\boldsymbol{\theta}_\star\|_2 \leq 1$, and $p \ll d$. We call f and $\boldsymbol{\theta}$ the *dimension reduction mapping* and *latent weight vector*, respectively. Intuitively, f serves as a non-linear transformation that converts raw contexts of a large dimension d to latent features of a much lower dimension p, and the reward function is linear in latent features. Since a neural network with suitable size and activation functions is a global function approximator [5], it is reasonable to assume that Eq. (3) expresses the underlying reward function, i.e., there exists a pair $(f_\star, \boldsymbol{\theta}_\star)$ that fulfils Eq. (3).

We introduce the two major neural contextual bandits: (1) NEURAL-LINEAR [15, 22] trains $\boldsymbol{\theta}$ by applying linear contextual bandit methods (e.g., UCB or TS) on top of f for exploration. The training of f (representation learning) and $\boldsymbol{\theta}$ (exploration) are executed at different time-scales. Whenever the exploration is terminated, we turn to representation learning by training the entire model (both f and $\boldsymbol{\theta}$) end-to-end. Although online exploration quantifies uncertainties over rewards, end-to-end training makes NEURAL-LINEAR ignore this important information in representation learning. This may lead to a low convergence speed and thereby result in high regrets. Notably, the regret bound of NEURAL-LINEAR is still unknown. (2) and NEURALUCB [24] provides a regret bound of $\widetilde{O}(\tilde{d}\sqrt{T})$ through leveraging the *neural tangent kernel* (NTK) [10] to characterise a fully connected neural network, where $\tilde{d}$ is the effective dimension of a NTK matrix. However, reformulating a neural network as a NTK matrix requires updating all parameters (both f and θ) of a neural network at once after each step of online decision-making, making NEURALUCB too inefficient to be used in practice.

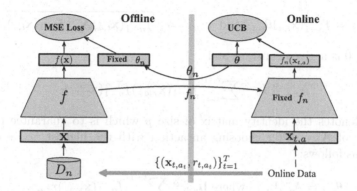

Fig. 1. The process flow of INLUCB framework. Solid and dashed arrows represent input/output and sampling, respectively.

4 The Interconnected Neural-Linear UCB Framework

To address the need for novel contextual bandit methods with non-linear rewards and high-dimensional contexts, we propose a new contextual bandit framework called *interconnected neural-linear UCB* (INLUCB). Following the neural contextual bandits regime, INLUCB alternates between the training of f and θ.

The key to INLUCB is an interconnected online-offline mechanism rather than end-to-end training. Fixing f, the online process tunes θ using UCB to balance exploration and exploration. In turn, freezing θ, the offline process updates f based on samples collected by online exploration. Figure 1 depicts this mechanism. This interconnected update mechanism overcomes the shortcoming of NEURAL-LINEAR in the sense that representation learning and online exploration are alternatively performed to *boost each other*. Besides, the method has two extra advantages: (1) online exploration is an effective way to sample data since initially data is often too scarce to train the entire model offline, i.e., the *cold-start problem*; (2) moving the heavy workload of updating hidden layers offline can significantly improve online efficiency. Formally, let $n \in [N]$ denote the index of iterations, and denote by θ_n and f_n the values of θ and f after the nth iteration, respectively. Let D_n denote the offline dataset at the nth iteration. Initially, we assume $D_0 = \varnothing$. We next formally introduce the two processes.

Online Exploration. Each iteration starts from the online exploration. In the nth iteration, we fix f_{n-1} to extract a latent feature $f_{n-1}(\mathbf{x}_{t,a})$ for each context $\mathbf{x}_{t,a}$. As for updating $\boldsymbol{\theta}$, we apply LINUCB on top of the extracted latent features for exploration. The basic idea is to maintain a *reward predictor* (i.e., predicted expected reward) $\hat{r}_{t,a}$ and a confidence interval around it with *width* $w_{t,a}$ that captures the variance of rewards. Then, at each step t, we choose the action with the highest upper confidence bound $\hat{r}_{t,a} + w_{t,a}$. Formally, we use $\boldsymbol{\theta}_{n,t}$ to denote the estimation of $\boldsymbol{\theta}$ at the tth step of the nth iteration. For each action a, the reward predictor and the width of the confidence interval are given by

$$\hat{r}_{t,a} \triangleq f_{n-1}^{\top}(\mathbf{x}_{t,a})\boldsymbol{\theta}_{n,t}, \text{ and } w_{t,a} \triangleq \alpha\sqrt{f_{n-1}^{\top}(\mathbf{x}_{t,a})\mathbf{A}_{n,t}^{-1}f_{n-1}(\mathbf{x}_{t,a})}, \qquad (4)$$

where $\alpha > 0$ is a given constant and

$$\mathbf{A}_{n,t} \triangleq \mathbf{I}_p + \sum_{\tau=1}^{t-1} f_{n-1}(\mathbf{x}_{\tau,a_\tau})f_{n-1}^{\top}(\mathbf{x}_{\tau,a_\tau}). \qquad (5)$$

Here, $\mathbf{I}_p$ denotes the identity matrix of size p which is to guarantee the non-singularity of $\mathbf{A}_{n,t}$. After choosing an action with the highest $\hat{r}_{t,a} + w_{t,a}$, we update $\boldsymbol{\theta}$ as follows:

$$\boldsymbol{\theta}_{n,t} = \mathbf{A}_{n,t}^{-1}\mathbf{b}_{n,t}, \text{ where } \mathbf{b}_{n,t} \triangleq \sum_{\tau=1}^{t-1} f_{n-1}(\mathbf{x}_{\tau,a_\tau})r_{\tau,a_\tau}. \qquad (6)$$

Online training terminates after T steps (T is a predefined constant). Then, accumulated online samples $\{(x_{t,a_t}, r_{t,a_t})\}_{t=1}^{T}$ are appended to the offline dataset D_{n-1}, yielding D_n.

Offline Representation Learning. In the nth iteration of offline learning, we fix $\boldsymbol{\theta}_n$ and train f_n on D_n by minimising the mean square error (MSE) loss:

$$\mathcal{L}_{D_n}(f; \boldsymbol{\theta}_n) \triangleq \mathbb{E}_{(\mathbf{x},r)\sim D_n}\left[(f^{\top}(\mathbf{x})\boldsymbol{\theta}_n - r)^2\right], \qquad (7)$$

Since the sub-Gaussian noise on rewards has zero mean, the minimiser of Eq. (7) is an unbiased estimator of the optimal f w.r.t. $\boldsymbol{\theta}_n$. Algorithm 1 presents the pseudocode of INLUCB.

5   Regret Analysis

This section studies the regret bound of INLUCB. By Eq. (1), the total regret of INLUCB with N iterations, each of T online steps, can be written as

$$R_{N,T} = \sum_{n=1}^{N} R_{n,T} \triangleq \sum_{n=1}^{N}\left[\sum_{t=1}^{T} r_{n,t,a_t^*} - \sum_{t=1}^{T} r_{n,t,a_t}\right], \qquad (8)$$

where $R_{n,T}$ denotes the regret at the nth iteration. We study the per-iteration regret $R_{n,T}$. The total regret can then be obtained by summing $R_{n,T}$ over all N iterations. For simplicity, we will omit the iteration index n in some notations.

Algorithm 1. INLUCB

Input: $\alpha \in \mathbb{R}^+, N, T, K, d \in \mathbb{N}, p < d \in \mathbb{N}$
Output: $f : \mathbb{R}^d \to \mathbb{R}^p$ and $\theta \in \mathbb{R}^p$.
Initialisation: $D_0 = \varnothing$, random f_0, $\mathbf{A}' \leftarrow \mathbf{I}_p$ and $\mathbf{b}' \leftarrow \mathbf{0}_p$

1: **while** $n = 1, 2, \ldots, N$ **do**
2:     ▷ *Online Exploration*
3:         $\mathbf{A} \leftarrow \mathbf{A}'$, $\mathbf{b} \leftarrow \mathbf{b}'$
4:         **for** $t = 1, 2, \ldots, T$ **do**
5:             $\theta_{n,t} \leftarrow \mathbf{A}^{-1}\mathbf{b}$
6:             Observe K features $\mathbf{x}_{t,1}, \mathbf{x}_{t,2}, \ldots, \mathbf{x}_{t,K} \in \mathbb{R}^d$
7:             **for** each arm $a = 1, 2, \ldots, K$ **do**
8:                 Compute $p_{t,a} \leftarrow f_{n-1}^\top(\mathbf{x}_{t,a})\theta_{n,t} + \alpha\sqrt{f_{n-1}^\top(\mathbf{x}_{t,a})\mathbf{A}^{-1}f_{n-1}(\mathbf{x}_{t,a})}$
9:             **end for**
10:             Choose action $a_t \leftarrow \arg\max_{a \in [K]} p_{t,a}$
11:             Observe payoff $r_{t,a_t} \in [0, 1]$
12:             $\mathbf{A} \leftarrow \mathbf{A} + f_{n-1}(\mathbf{x}_{t,a})f_{n-1}^\top(\mathbf{x}_{t,a})$
13:             $\mathbf{b} \leftarrow \mathbf{b} + f_{n-1}(\mathbf{x}_{t,a})r_{t,a_t}$
14:         **end for**
15:         $\theta_n \leftarrow \theta_{n,T}$, $D_n \leftarrow D_{n-1} \cup \{(\mathbf{x}_{t,a_t}, r_{t,a_t})\}_{t=1}^T$
16:     ▷ *Offline Representation Learning*
17:         Fix θ_n, train $f_{(n)}$ on D_n through gradient descent on the loss defined in Eq. (7)
18: **end while**

Recall that the agent always pulls the arm with the highest UCB which is a sum of the reward predictor $\hat{r}_{t,a}$ and a width term $w_{t,a}$. Therefore, to bound $R_{n,T}$, we need to know the *error in reward prediction*:

$$|\hat{r}_{t,a} - f_\star^\top(\mathbf{x}_{t,a})\theta_\star| = |f_{n-1}^\top(\mathbf{x}_{t,a})\theta_{n,t} - f_\star^\top(\mathbf{x}_{t,a})\theta_\star|.$$

Same as LINUCB, the reward predictors $\hat{r}_{t,a}$ in INLUCB are sums of *dependent* variables since predictions in later steps are made using previous outcomes, which prevents us from applying Azuma-Hoeffding inequality to control the error in reward prediction. Thus, directly analysing regret bound of INLUCB is intractable. To sidestep this problem, we use the construction in [3] to modify the online learning of INLUCB into BASEINLUCB which assumes statistical independence among samples. We then use a master algorithm SUPINLUCB to pull arms in a way that ensures this assumption holds. The pseudocode for both algorithms can be found in Appendix A. In the literature of contextual bandits, due to the intractability of the regret bound of the original algorithm, the convention is to instead analyse the regret bound of the master algorithm [3,7,16], which can be viewed as an appropriate modification of the original algorithm. Following this convention, we next analyse the regret bound of SUPINLUCB.

However, although the above technique ensures independence among samples, directly calculating the error in reward prediction is still intractable due to the coupling between the estimation errors of $\theta_{n,t}$ and f_{n-1} in the total error of reward prediction. One of our main contributions is proposing a method to separate them by defining the *offline error*:

$$\epsilon_n \triangleq \max_{\mathbf{x} \in \mathbb{R}^d} |f_{n-1}^\top(\mathbf{x})\theta_\star - f_\star^\top(\mathbf{x})\theta_\star| \in [0, 1], \tag{9}$$

and the *online error*: $\gamma_n(\mathbf{x}_{t,a}) \triangleq |f_{n-1}^\top(\mathbf{x}_{t,a})\theta_{n,t} - f_{n-1}^\top(\mathbf{x}_{t,a})\theta_\star|$.

Intuitively, the offline error and online error capture the effects of the estimation error of f_{n-1} and $\boldsymbol{\theta}_{n,t}$ arising from representation learning and exploration, respectively. By applying a triangle inequality, we derive an upper bound of the error in reward prediction by splitting it into online and offline errors:

$$|\hat{r}_{t,a} - f_\star^\top(\mathbf{x}_{t,a})\boldsymbol{\theta}_\star| \leq \gamma_n(\mathbf{x}_{t,a}) + \epsilon_n.$$

Our main result given below is derived by bounding $\gamma_n(\mathbf{x}_{t,a})$, and leaving ϵ_n as a factor in the total regret. The proof is given in Appendix B.

Theorem 1. *If* SUPINLUCB *is run with* $\alpha = \sqrt{\frac{1}{2}\ln\frac{2NTK}{\delta}}$, *with probability at least* $1 - \delta$, *the regret of the algorithm is*

$$O\left(\left(N + T\sum_{n=1}^{N}\epsilon_n\right)\sqrt{Tp\ln^3(NTK\ln(T)/\delta)}\right). \tag{10}$$

Remark 1. Theorem 1 provides a general expression of the regret bound which implies that the rate of convergence of the sequence of offline errors $\{\epsilon_n\}_{n=1}^N$ determines the order of the regret bound. Generally, we know $\sum_{n=1}^N \epsilon_n \leq N$ as $\epsilon_n \leq 1$. But substituting N for $\sum_{n=1}^N \epsilon_n$ leads to a loose bound $\tilde{O}(NT\sqrt{Tp})$. In general, the bound of ϵ_n depends on the complexity of the underlying dimension reduction mapping $f_\star$ and the error of estimating $f_\star$ using the neural network. Thus, we cannot derive a universal non-trivial upper bound for ϵ_n as we cannot guarantee that the neural network attains global minimum. While, if we discard the error of neural networks and assume the latent feature is in a simple form (e.g., linear in raw contexts), we can derive a tighter regret bound by further bounding ϵ_n (see Corollary 1). Also, empirically, we show that ϵ_n decreases fast with the number of iteration n increases (see next section).

Remark 2. We relate our regret analysis of INLUCB to that of LINUCB [7]. As for INLUCB, if we known in davance that the reward function degenerates to a linear mapping, offline representation learning is no longer needed, which means that only online exploration remains (i.e., $N = 1$) and the offline error would be zero (i.e., $\epsilon_n = 0$). Then in this case, the regret bound in Theorem 1 reduces to $\sqrt{Td\ln^3(TK\ln(T)/\delta)}$, which is the same as that of LINUCB [7, Theorem 1]. This suggests that INLUCB recovers LINUCB as a special case.

Corollary 1. *Assume that* $f_\star$ *is linear, i.e.,* $f_\star(\mathbf{x}) = \mathbf{Q}_\star\mathbf{x}$ *where* $\mathbf{Q}_\star \in \mathbb{R}^{p\times d}$. *Let* INLUCB *use a fully connected network of three layers, each of size* d, p *and* 1. *Also assume for all* $n \in [N]$, $(f_n, \boldsymbol{\theta}_n)$ *minimises* $\mathcal{L}_{D_n}(f;\boldsymbol{\theta})$. *If* SUPINLUCB *is run with* $\alpha = \sqrt{\frac{1}{2}\ln\frac{2NTK}{\delta}}$, *then there exist constants* $\sigma \in [0,1]$ *and* $C_\sigma \geq 0$ *such that with probability at least* $(1 - \delta)(1 - \sigma)$, *the regret is*

$$O\left(\left(N + T + C_\sigma\sqrt{NT}\right)\sqrt{Tp\ln^3(NTK\ln(T)/\delta)}\right).$$

Remark 3. The regret of random selection is $O(NT)$. Thus, the regret bound in Corollary 1 is non-trivial as the magnitude of $\tilde{O}((N + T + C_\sigma\sqrt{NT})\sqrt{Tp})$ (p is constant) is smaller than that of $O(NT)$. On the other hand, the non-trivial regret bound of other neural contextual bandit methods, e.g., NEURALUCB [24], also relies on the assumption that the error of neural networks can be bounded.

6   Experiments

We empirically evaluate the accuracy (cumulative regret) and efficiency (run-time per step) of INLUCB on both high-dimensional synthetic and real-world datasets with non-linear rewards. We adopt eight bandit methods as baselines: **(1) LinUCB** [7], a linear contextual bandit method using UCB for exploration. Its regret bound is $\tilde{O}(\sqrt{dT})$; **(2) LinTS** [2], a linear contextual bandit method using Thompson sampling (TS) for exploration. It has a regret bound of $\tilde{O}(d\sqrt{T})$. **(3) CBRAP** [21], a method that uses random projection to do dimension reduction and UCB for exploration. **(4) KernelUCB** [16], a method that ultilises kernel functions for handling non-linear rewards and uses UCB for exploration. Its regret bound is $\tilde{O}(\sqrt{\tilde{d}T})$, where $\tilde{d}$ is the effective dimension of kernel matrixes. **(5) NeuralUCB** [24], it uses a fully connected neural network for reward prediction, uses UCB for online exploration, and updates the whole neural network at each step. It has a regret bound of $\tilde{O}(\tilde{d}\sqrt{T})$; **(6) Neural-Linear** [22], a method that extracts latent features using NN and use TS on the top of the last linear for exploration. The regret bound is not given by authors. **(7) EXP3** [4], a representative adversarial bandits algorithm that pulls arms with probabilities and adjusts such probabilities based on received rewards; **(8) ε-Greedy**: a classic exploration method; with high probability $1 - \varepsilon$ pulling the arm with highest average reward in history and with small probability ε pulling an arm randomly.

6.1   Experimental Setting

For UCB-based methods, we tune the constant α through a grid search over $\{0.01, 0.1, 1\}$. For TS-based methods, we do grid search over $\{0.01, 0.01, 1\}$ for the hyper-parameter that controls the covariance of the prior and posterior distributions. For KERNELUCB, we adopt radial basis function (RBF) kernel and empirically set the parameters with best results. For EXP3 and ε-GREEDY, we do grid search for the exploration parameter over $\{0.01, 0.1, 1\}$. For INLUCB, NEURALUCB and NEURAL-LINEAR, we use the same neural network structure: a fully connected network of four layers of size d, d, p and 1, respectively. For CBRAP, the dimension after projection is also p. We vary p from 10 to 100 with step size 10, and vary T from 100 to 1000 with step size 100. For all grid-searched parameters, we choose the best setting for comparisons. For all contextual bandit methods we test their efficiency with respect to the context dimension and the number of steps. Results are averaged over 10 independent runs.

Table 1. Real-world Datasets statistics.

Dataset	Feature dimension	Number of classes	Number of samples
MUSIC	518	193	106,574
FONT	409	20	100,000
MNIST	784	10	60,000

Synthetic Datasets. We generate synthetic datasets with contextual dimension $d = 500$ and $K = 200$ arms. Contextual vectors are chosen uniformly at random from the unit ball. We use three artificial non-linear reward functions: $g(\mathbf{x}) = \cos(3\mathbf{x}^\top \mathbf{a})$ (shorten as COS), $g(\mathbf{x}) = 10(\mathbf{x}^\top \mathbf{a})^2$ (SQU), and $g(\mathbf{x}) = \exp(\mathbf{x}^\top \mathbf{a})$ (EXP), where $\mathbf{a}$ is randomly generated from uniform distribution over unit ball. These typical functions cover a wide range of non-linear mappings [24].

Real-World Datasets. We use three real-world classification datasets: MUSIC and FONT from the UCI Machine Learning Repository [8] and the MNIST dataset [12]. Table 1 lists key statistics of the datasets. Following [15], we transform classification tasks into bandit tasks: each step we randomly select one sample; the agent gets reward 1 if it classifies the sample correctly, and 0 otherwise.

6.2 Results

Figure 2 and Table 2 report results of cumulative regrets and runtime per step, respectively. Overall, INLUCB exhibits the lowest regret and superior efficiency in all cases. Specifically, only INLUCB shows convergence in synthetic datasets, which indicates the fast decrease of offline errors with the number of iterations grows. For real-world datasets, although we do not observe convergence in some cases, INLUCB achieves the lowest regret on all tasks.

Table 2. Results for runtime per step (in milliseconds).

Algorithm	COS	SQU	EXP	MUSIC	FONT	MNIST
LINUCB	0.9	0.6	0.6	0.8	0.5	0.5
LINTS	6.6	6.5	6.6	6.5	4.7	4.2
CBRAP	0.7	0.9	0.9	0.9	0.1	0.1
KERNELUCB	2089.4	2114.8	1850.7	2128.1	2100.6	2228.0
NEURALUCB	1121.2	1075.2	1003.1	1027.3	1193.5	1239.6
NEURAL-LINEAR	8.7	6.1	6.0	7.1	1.8	2.3
EXP-3	0.3	0.6	0.6	0.5	0.08	0.08
ε-GREEDY	0.01	0.01	0.01	0.009	0.04	0.004
INLUCB (ours)	4.3	5.6	5.7	9.5	3.8	5.0

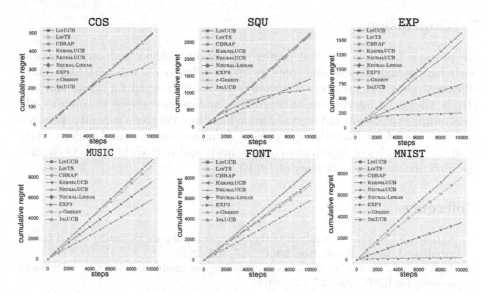

Fig. 2. Results for cumulative regrets.

LINUCB and LINTS show low regrets in some cases but fail to converge, as they are able to take use of complete contextual information but cannot handle non-linear reward functions.

KERNELUCB costs more than 10s after 1500 steps since it needs to invert a matrix whose dimension is proportional to the number of steps, which gives the evidence that it is inefficient for practical use. Although NEURAL-LINEAR is efficient, it suffers from high regrets since the end-to-end training framework prevents the online exploration from effectively boosting representation learning. CBRAP is relatively efficient but has high regret since empirically we find that it is really sensitive to the initial value of the projection matrix. The classical probability-based exploration techniques EXP3 and ε-GREEDY have the highest regret although they are most efficient. The reason is that they lack the capability of modeling environments with contextual information.

Results of sensitivity test (see Appendix C.1) show that NEURALUCB is not applicable to environments with high dimensional contexts. Thus, the result of NEURALUCB in Fig. 2 is reported using a selected subset of features that result in the best performance. Even though NEURALUCB runs on a subset of original contexts, it has extremely high runtime cost. In contrast, INLUCB is three orders of magnitude faster than NEURALUCB for online decision making. We also show that INLUCB has comparable cumulative regrets to NEURALUCB on the same subset of features in Appendix C.2. Thus, we conclude that INLUCB achieves a better balance between the accuracy and online efficiency.

7 Conclusion

We propose INLUCB, the first contextual bandit method that can simultaneously handle high dimensional contexts and non-linear rewards with high online efficiency. INLUCB uses neural networks to model reward functions and creatively adopts an interleaving online/offline update mechanism to combine efficient online exploration and representation learning. We give a general expression of regret bound for INLUCB and present a tighter regret bound under certain conditions. Results of experiments on synthetic and real-world datasets confirm the high accuracy and efficiency of INLUCB.

References

1. Agrawal, S., Goyal, N.: Analysis of thompson sampling for the multi-armed bandit problem. J. Mach. Learn. Res. **23**(4), 357–364 (2011)
2. Agrawal, S., Goyal, N.: Thompson sampling for contextual bandits with linear payoffs. In: ICML, pp. 127–135 (2013)
3. Auer, P.: Using confidence bounds for exploitation-exploration trade-offs. J. Mach. Learn. Res. **3**(Nov), 397–422 (2002)
4. Auer, P., Cesa-Bianchi, N., Fischer, P.: Finite-time analysis of the multiarmed bandit problem. Mach. Learn. **47**(2–3), 235–256 (2002)
5. Barron, A.R.: Universal approximation bounds for superpositions of a sigmoidal function. IEEE Trans. Inf. Theory **39**(3), 930–945 (1993)
6. Bastani, H., Bayati, M.: Online decision making with high-dimensional covariates. Oper. Res. **68**(1), 276–294 (2020)
7. Chu, W., Li, L., Reyzin, L., Schapire, R.: Contextual bandits with linear payoff functions. In: AISTATS, pp. 208–214 (2011)
8. Dua, D., Graff, C.: UCI machine learning repository (2017). http://archive.ics.uci.edu/ml
9. Ghosh, A., Chowdhury, S.R., Gopalan, A.: Misspecified linear bandits. In: AAAI, pp. 3761–3767 (2017)
10. Jacot, A., Gabriel, F., Hongler, C.: Neural tangent kernel: convergence and generalization in neural networks. In: NIPS, pp. 8580–8589 (2018)
11. Kim, G.S., Paik, M.C.: Doubly-robust lasso bandit. In: NIPS, pp. 5869–5879 (2019)
12. LeCun, Y., Bottou, L., Bengio, Y., Haffner, P.: Gradient-based learning applied to document recognition. Proc. IEEE **86**(11), 2278–2324 (1998)
13. Li, L., Chu, W., Langford, J., Schapire, R.E.: A contextual-bandit approach to personalized news article recommendation. In: WWW, pp. 661–670 (2010)
14. Li, L., Lu, Y., Zhou, D.: Provably optimal algorithms for generalized linear contextual bandits. In: ICML, pp. 2071–2080 (2017)
15. Riquelme, C., Tucker, G., Snoek, J.: Deep Bayesian bandits showdown. In: ICLR (2018)
16. Valko, M., Korda, N., Munos, R., Flaounas, I., Cristianini, N.: Finite-time analysis of kernelised contextual bandits. In: UAI, pp. 654–663 (2013)
17. Vaswani, S., Kveton, B., Wen, Z., Ghavamzadeh, M., Lakshmanan, L.V., Schmidt, M.: Model-independent online learning for influence maximization. In: ICML, pp. 3530–3539 (2017)
18. Wang, X., Wei, M., Yao, T.: Minimax concave penalized multi-armed bandit model with high-dimensional covariates. In: ICML, pp. 5200–5208 (2018)

19. Weng, J., Hwang, W.S.: Online image classification using IHDR. Int. J. Doc. Anal. Recogn. **5**(2-3), 118-125 (2003)
20. Xie, M., Yin, W., Xu, H.: AutoBandit: a meta bandit online learning system. In: Proceedings of the Thirtieth International Joint Conference on Artificial Intelligence, IJCAI-21, pp. 5028-5031 (2021)
21. Yu, X., Lyu, M.R., King, I.: CBRAP: contextual bandits with random projection. In: AAAI (2017)
22. Zahavy, T., Mannor, S.: Deep neural linear bandits: overcoming catastrophic forgetting through likelihood matching. arXiv preprint arXiv:1901.08612 (2019)
23. Zeng, Z., Li, X., Ma, X., Ji, Q.: Adaptive context recognition based on audio signal. In: 2008 19th International Conference on Pattern Recognition, pp. 1-4. IEEE (2008)
24. Zhou, D., Li, L., Gu, Q.: Neural contextual bandits with UCB-based exploration. In: ICML, pp. 11492-11502 (2020)

Node Information Awareness Pooling for Graph Representation Learning

Chuan Sun, Feihu Huang, and Jian Peng$^{(\boxtimes)}$

College of Computer Science, Sichuan University, Chengdu, China
sunchuan@stu.scu.edu.cn, {huangfh,jianpeng}@scu.edu.cn

Abstract. Graph neural networks (GNNs), which generalize deep neural network models to graph structure data, have attracted increasing attention and achieved state-of-the-art performance in graph-related tasks such as graph classification, link prediction, and node classification. To adapt GNNs to graph classification, existing works aim to define the graph pooling method to learn graph-level representation by down-sampling and summarizing the information present in the nodes. However, most existing pooling methods lack a way of obtaining information about the entire graph from both the local and global aspects of the graph. Moreover, in these pooling methods, the difference features between nodes and their neighbors are usually ignored, which is crucial in obtaining graph information in our opinions. In this paper, we propose a novel graph pooling method called Node Information Awareness Pooling (NIAPool), which addresses the limitations of previous graph pooling methods. NIAPool utilizes a novel self-attention framework and a new convolution operation that can better capture the difference features between nodes to obtain node information in the graph from both local and global aspects. Experiments on five public benchmark datasets demonstrate the superior performance of NIAPool for graph classification compared to the state-of-the-art baseline methods.

Keywords: Graph pooling · Graph classification · Graph representation learning · Graph neural network

1 Introduction

Graph neural networks (GNNs) are a class of deep learning models that operate on data represented as graphs with arbitrary topological structures such as body skeletons [22], brain networks [14], molecules [6], and social networks [12]. Unlike some regular grid data (e.g., images and texts), the inputs of GNN are permutation-invariant variable-size graphs consisting of rich information. By passing, transforming, and aggregating node features across the graph, GNNs can capture graph information effectively [9] and demonstrate strong ability on related tasks such as text classification [10], mental illness analysis [32], drug discovery [24], relation extraction [25], and particle physics analysis [23]. Some

J. Gama et al. (Eds.): PAKDD 2022, LNAI 13280, pp. 182–193, 2022.
https://doi.org/10.1007/978-3-031-05933-9_15

of the existing methods focus on node-level representation learning to perform tasks such as link prediction [11,18] and node classification [12,27]. Others focus on learning graph-level representations for tasks like graph classification [7,34] and graph regression [21]. In this paper, we focus on graph-level representation learning for the task of graph classification.

In brief, the task of graph classification is to generate the graph-level representation of the entire graph to predict the label of the input graph by utilizing the given graph structure information and node representation. The majority of existing GNNs usually generate the graph-level representation by applying simple global pooling strategies [15,28,35] (i.e., a summation over the final learned node representations). These methods are inherently "flat" [36] and lack the capability of aggregating node information in a hierarchical manner since they treat all the nodes equivalently when generating graph representation using the node representations. Furthermore, the structure information of the entire graph is almost neglected during this process. For example, to prove a molecule is toxic or not, which depends on not only the features of atoms but also the structure information in atoms' interaction networks.

To address this problem, hierarchical pooling architectures have been proposed, which have the ability to coarsen the graph in an adaptive, data-dependent manner within a GNN pipeline, analogous to image downsampling within convolutional neural networks. The first end-to-end learnable hierarchical pooling operator is DiffPool [33]. DiffPool groups nodes into super-nodes by computing soft clustering assignments of nodes. Since the cluster assignment matrix is dense, its ability to scale to large graphs is limited. TopKPool [7] uses a simple scalar projection score for each node to select top-k nodes in a sparse pooling manner, so the computing limitation of DiffPool is overcome. Following that, SAGPool [13], a variant of TopKPool, uses self-attention to incorporate global structure information, and EdgePool [4] learns a localized and sparse hard pooling transform by edge contraction. Nevertheless, a critical limitation of these pooling operations is the lack of a way to combine the local and global information in graphs simultaneously. To address the above limitations, ASAP [17] devises a cluster scoring procedure to select nodes depending on the feature-based fitness scores and self-attention network. [8] uses a two-stage attention voting process that selects more important nodes in a graph. Although [17] and [8] have proved the ability of their methods to capture graph information, we argue that self-attention mechanisms in these methods may render mutual exclusion effect on node importance, which means these methods focus too much on the nodes with high similarity and almost ignore the node information with high differences. In addition, we find that the difference features of nodes are vital for the entire graph. For example, even if two molecules have the same structure and atomic number, as long as there is a pair of different types of atoms, there will be huge differences between the molecules. However, most of the existing methods cannot effectively capture the difference features between nodes or even do not take them into consideration. Therefore, hierarchical graph pooling methods that capture the difference features between nodes effectively while reasonably considering both local and global graph information currently do not exist.

In this work, we propose a new pooling operator called Node Information Awareness Pooling (NIAPool) which overcomes the problems mentioned above. It uses a novel Difference2Token attention framework (D2T), which balances important node selection reasonably, to enhance the node information representation locally, and then utilizes a new convolution operation called Neighbor Feature Awareness Convolution (NFAConv) to capture the difference features between neighbor nodes and perform global node scoring.

Our contributions can be summarized as follows:

- We propose D2T to evaluate each node's information given its neighborhood and then enhance local node information.
- We propose a new convolution operator NFAConv. Compared with state-of-the-art convolution operations, NFAConv is more powerful in extracting difference features between nodes.
- We conduct extensive experiments on five public datasets to demonstrate NIAPool's effectiveness as well as superiority compared to a range of state-of-the-art methods.

2 Preliminaries

2.1 Notations and Problem Formulation

Given a set of graph data $D = \{(G_1, y_1), (G_2, y_2), \ldots, (G_n, y_n)\}$, where the number of nodes and edges in each graph might be different. For an arbitrary graph $G_i = (\mathcal{V}_i, \mathcal{E}_i, X_i)$ with $N = |\mathcal{V}_i|$ nodes and $|\mathcal{E}_i|$ edges. Let $A \in \mathbb{R}^{N \times N}$ be the adjacent matrix describing its edge connection information, and $X \in \mathbb{R}^{N \times d}$ represents the node feature matrix assuming each node has d features. Each graph is also associated with a label y_i indicating the class it belongs to. The goal of graph classification is to learn a mapping function $f : \mathcal{G} \to \mathcal{Y}$ where $\mathcal{G}$ is the set of graphs and $\mathcal{Y}$ is the set of labels. A pooled graph and its adjacency matrix are denoted by $G^p = (\mathcal{V}^p, \mathcal{E}^p, X^p)$ and A^p, respectively. For each node v_i, we use $\mathcal{N}(v_i)$ to represent its 1-hop neighbors.

2.2 Graph Convolutional Neural Network

Graph convolutional neural network (GCN) [12] is a powerful tool for handling graph-structured data and has shown promising performance in various challenging tasks. Thus, we choose GCN as the building block to design the framework for graph classification. For the l-th layer in GCN, it takes both the adjacent matrix A and hidden representation matrix X^l of the graph as input, and then the new node embedding matrix will be generated as follows:

$$X^{l+1} = \sigma \left(\tilde{D}^{-\frac{1}{2}} \tilde{A} \tilde{D}^{-\frac{1}{2}} X^l W^l \right) \qquad (1)$$

Here, $\sigma(\cdot)$ is the non-linear activation function, $\tilde{A} = A + I$ is the adjacency matrix with self-loops. $\tilde{D}$ is the diagonal degree matrix of $\tilde{A}$, and $W^l \in R^{d_l \times d_{l+1}}$ is the trainable weight matrix.

2.3 Self-attention Mechanism

Self-attention is used to discover the dependence of input on itself [26]. The attention coefficient $\alpha_{i,j}$ is computed to map the importance of candidates h_i on target query h_j, where h_i and h_j are obtained from input entities $\boldsymbol{h} = \{h_1, \ldots, h_n\}$. We introduce three variants of self-attention mechanisms.

Token2Token (T2T) [19] explores the dependency between the target and candidates from the input set $\boldsymbol{h}$. The attention coefficient $\alpha_{i,j}$ is computed as:

$$\alpha_{i,j} = \text{softmax}\left(\vec{v}^T \sigma\left(W h_i \| W h_j\right)\right) \tag{2}$$

where $\|$ is the concatenation operator.

Source2Token (S2T) [19] drops the target query term to explore the dependency between each candidate and the entire input set $\boldsymbol{h}$.

$$\alpha_{i,j} = \text{softmax}\left(\vec{v}^T \sigma\left(W h_j\right)\right) \tag{3}$$

Master2Token (M2T) [17] is a self-attention mechanism that works on graph data. M2T utilizes intra-cluster information by using a master function to generate a query vector within the node and its 1-hop neighbors. Compared with other self-attention mechanisms, M2T can capture graph information better. Formally, M2T is defined as:

$$\alpha_{i,j} = \text{softmax}\left(\vec{v}^T \sigma\left(W m_i \| h_j\right)\right) \tag{4}$$

$$m_i = \max_{h_i \in \mathcal{N}(h_j)} (h_i) \tag{5}$$

Here, h_j refer to node representation in graph and $\mathcal{N}(h_j)$ are the 1-hop neighbors of h_j.

3 NIAPool (Proposed Method)

In this section, we give an overview of the proposed method NIAPool. As shown in Fig. 1, NIAPool initially focuses on the local structure of the given graph, considering all nodes and their 1-hop neighbors, and utilizes a new self-attention to enhance node information by aggregating neighbor node features. These enhanced nodes are then globally scored using a modified GCN. Further, a fraction of the top-scoring nodes will be saved. Below, we discuss the modules of NIAPool in detail.

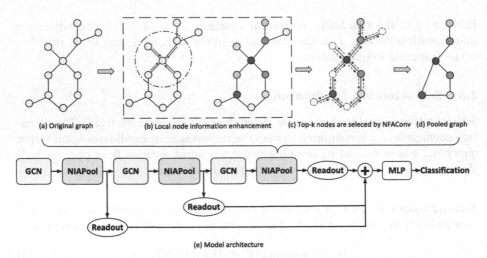

(a) Original graph   (b) Local node information enhancement   (c) Top-k nodes are seleced by NFAConv (d) Pooled graph

(e) Model architecture

Fig. 1. The illustration of the proposed NIAPool: (a) Input graph to NIAPool. (b) Local node information enhancement based on D2T attention. (c) We utilize NFAConv to capture node difference features and then score nodes. (d) A fraction of top-scoring nodes are kept in the pooled graph. (e) The overview of hierarchical graph classification architecture

3.1 Local Node Information Enhancement

Initially, we consider each node and its 1-hop neighbors. We learn the attention coefficient between each node in the graph and its neighbor nodes through the self-attention mechanism. Further, the learned attention coefficient is used to fuse information from v_i and its neighbors $\mathcal{N}(v_i)$. The task here is how to learn the enhancement representation of nodes effectively by attending to the relevant nodes. We observe that the self-attention mechanisms mentioned in Sect. 2.3 may have a node importance mutual exclusion problem (Please refer to Sect. 4.3 or Sect. 1 for more details). To address this problem, we propose a new variant of self-attention, called **Difference2Token (D2T)** to balance the attention procedure. D2T is defined as:

$$\alpha_{i,j} = \text{softmax}\left(\vec{v}^T \sigma\left(W_1\left(h_i - h_j\right) \| W_2 h_j\right)\right) \tag{6}$$

Here, h_i is node representation and $h_j \in \mathcal{N}(h_i)$.

Our motivation for designing D2T is that if a node's information can be reconstructed or inferred by its neighborhood information, it means this node can probably be deleted in the pooled graph with almost no information loss. In general, the nodes with similar information can be substituted for each other. That is, the more similar the nodes are, the less important they are, which is different from the self-attentions in Sect. 2.3. In Eq. (6), the difference $h_i - h_j$ will show significant differences when a node representation h_i is similar to one representation of its neighbor h_j or not, so D2T can balance the attention procedure reasonably by taking $h_i - h_j$ into consideration.

After obtaining the attention coefficient, node information can be enhanced as follows:

$$h_i^e = h_i + \sum_{j=1}^{|\mathcal{N}(v_i)|} \alpha_{i,j} h_j \tag{7}$$

where h_i^e is the enhanced node representation.

3.2   Global Node Scoring Using NFAConv

The difference feature between nodes is critical for generating graph-level representation. For example, even if the graph has the same topology, different node features can make the whole graph greatly different. How to effectively capture the difference features between nodes becomes a new problem, which is ignored by most graph classification methods. To solve this problem, inspired by [17] and [30], we propose **Neighbor Feature Awareness Convolution (NFAConv)**, a powerful variant of GCN which is aware of node features:

$$\theta_i = h_i \cdot W + \phi \left(\sum_{h_j \in \mathcal{N}(h_i)} (h_i - h_j) \parallel (h_i \odot h_j) \right) \tag{8}$$

where W is learnable parameter matrix and ϕ denotes a neural network. $\odot$ is broadcasted hadamard product.

NFAConv utilizes $h_i - h_j$ and $h_i \odot h_j$ to obtain difference features between nodes and their neighbors, and then a neural network ϕ is applied to extract useful information from these features. In this paper, ϕ is a Multilayer Perceptron (MLP) with three linear layers. It is worth noting that ϕ can also be Convolutional Neural Network (CNN) like the one used in [35], or other neural networks. We keep this for future work.

After enhancing local node information, we sample nodes based on the global node fitness score θ_i which calculated by NFAConv. For a given pooling ratio $k \in [0, 1)$, the top $\lceil kN \rceil$ nodes are saved in the pooled graph G^p.

3.3   Graph Coarsening

Following the graph coarsening procedure in [33], we make global node fitness vector $\Theta = [\theta_1, \theta_2, \ldots, \theta_N]^T$ learnable by multiplying it to the node feature matrix X^l. The indices of selected nodes $\hat{i}$ are obtained by choosing the top $\lceil kN \rceil$ nodes based on X^l:

$$\hat{X}^l = \Theta \odot X^l, \quad \hat{i} = \text{TOP}_k \left(\hat{X}^l, \lceil kN \rceil \right) \tag{9}$$

The node feature matrix $X^p \in \mathbb{R}^{\lceil kN \rceil \times d}$ and the pruned cluster assignment matrix $\hat{S} \in \mathbb{R}^{N \times \lceil kN \rceil}$ are given by:

$$X^p = \hat{X}^l(\hat{i}, :), \quad \hat{S} = S(:, \hat{i}) \tag{10}$$

Here, $S \in \mathbb{R}^{N \times N}$ is the cluster assignment matrix, where $S_{i,j}$ represents the membership of node $v_i \in \mathcal{V}$ in $\mathcal{N}(v_j)$. Finally, the pooled new adjacency matrix A^p can be obtained as follows:

$$A^p = \hat{S}^T (A + I) \hat{S} \tag{11}$$

3.4 Model Architecture

The architecture used in SAGPool [13] is adopted in our experiments. We use JK-net [31] as our readout layers to aggregate pooled node features. Figure 1 shows the details of the model architecture.

4 Experiments

In this section, we present our experimental setup and results. Our experiments are designed to answer the following questions:

- **Q1** How does NIAPool perform compared to other state-of-the-art pooling methods at the task of graph classification?
- **Q2** Is local node information enhancement by D2T attention more powerful compared to other self-attentions?
- **Q3** Compared with some state-of-the-art GCN variants, can NFAConv effectively capture the features of nodes in the graph?

4.1 Experimental Setting

Datasets: Five graph datasets are selected from the public benchmark graph data collection. D&D [5,20] and Proteins [3,5] are two datasets containing graphs of protein structures. NCI1 [29] and NCI109 are biological datasets used for anticancer activity classification. Frankenstein [16] is a set of molecular graphs representing the molecules with or without mutagenicity. Table 1 summarizes the statistics of all datasets.

Table 1. Statistics of the datasets.

Dataset	# $|G|$	# $|c|$	Avg.# $|\mathcal{V}|$	Avg.# $|\mathcal{E}|$
D&D	1178	2	284.32	715.66
Proteins	1113	2	39.06	72.82
NCI1	4110	2	29.87	32.30
NCI109	4127	2	29.68	32.13
Frankenstein	4337	2	16.90	17.88

Baselines: We compare NIAPool with previous state-of-the-art graph pooling methods, including Set2Set [28], GlobalAttention [15], SortPool [35], DiffPool [33], TopKPool [7], SAGPool [13], ASAP [17], EdgePool [4], GSAPool [34], MinCUT-Pool [2], HGP-SL [36], and GMT [1].

Training Procedures: In our experiments, we use 10-fold cross-validation to evaluate the pooling methods, where each time we split each dataset into three parts: 80% as the training set, 10% as the validation set, and the remaining 10% as the test set. The average performance with standard derivation is reported. For NIAPool, we choose $k = 0.5$ and set the dimension of node representations 128 for all datasets. For baseline algorithms, we use source codes released by authors and follow the experimental setup that is mentioned in their manuscript. Adam optimizer with a learning rate of 0.001 is adopted as our optimizer.

4.2 Q1: Comparison with Baseline Models

We compare the performance of NIAPool with baseline methods on five datasets. The graph classification results are reported in Table 2. We also show the different pooling ratios based on the NIAPool architecture in Table 3.

Table 2. Comparison with the state-of-the-art graph pooling methods.

Methods	D&D	Proteins	NCI1	NCI109	Frankenstein
Set2Set	71.60 ± 0.87	72.16 ± 0.43	66.97 ± 0.74	61.04 ± 2.69	61.46 ± 0.47
GlobalAttention	71.38 ± 0.78	71.87 ± 0.60	69.00 ± 0.49	67.87 ± 0.40	61.31 ± 0.41
SortPool	71.87 ± 0.96	73.91 ± 0.72	68.74 ± 1.07	68.59 ± 0.67	63.44 ± 0.65
DiffPool	66.95 ± 2.41	68.20 ± 2.02	62.32 ± 1.90	61.98 ± 1.98	60.60 ± 1.62
TopKPool	75.01 ± 0.86	71.10 ± 0.90	67.02 ± 2.25	66.12 ± 1.60	61.46 ± 0.84
SAGPool	76.45 ± 0.97	71.86 ± 0.97	67.45 ± 1.11	67.86 ± 1.41	61.73 ± 0.76
ASAP	76.87 ± 0.7	74.19 ± 0.79	71.48 ± 0.42	70.07 ± 0.55	**66.26 ± 0.47**
EdgePool	70.37 ± 3.81	73.67 ± 4.18	73.65 ± 2.20	70.66 ± 2.01	65.28 ± 1.88
GSAPool	74.95 ± 4.51	73.22 ± 3.37	72.21 ± 1.35	70.11 ± 2.39	60.99 ± 2.08
MinCUT-Pool	77.50 ± 4.89	74.85 ± 3.72	75.38 ± 1.46	73.73 ± 1.56	63.57 ± 2.26
HGP-SL	76.66 ± 3.07	74.03 ± 3.21	74.59 ± 1.46	72.09 ± 2.06	62.81 ± 1.36
GMT	77.75 ± 3.48	75.01 ± 2.89	74.57 ± 1.59	73.25 ± 2.63	64.35 ± 1.41
NIAPool (ours)	**79.28 ± 4.30**	**75.25 ± 3.71**	**78.08 ± 1.92**	**75.56 ± 2.32**	65.35 ± 2.58

Overall, a general observation we can draw from the results is that our model obtains the highest accuracy on most of the datasets compared with baseline models. In particular, NIAPool achieves approximate 2.70% higher accuracy over the best baseline on the NCI1 dataset and 2.31% on the NCI109 dataset, respectively. This superiority of NIAPool may be attributed to its advanced mechanism for effectively capturing both local and global node information in pool operation. Although ASAP exhibit the best performance among all baseline methods and is even better than ours on the Frankenstein dataset, NIAPool has an average improvement of 3.89% over ASAP on the other four datasets.

Table 3. The evaluation of different pooling ratios based on the NIAPool architecture.

Pooling ratio	D&D	Proteins	NCI1	NCI109	Frankenstein
0.25	77.67 ± 4.13	74.03 ± 2.18	76.35 ± 1.81	74.63 ± 2.38	62.39 ± 3.28
0.50	**79.28 ± 4.30**	75.25 ± 3.71	**78.08 ± 1.92**	75.56 ± 2.32	**65.35 ± 2.58**
0.75	76.05 ± 3.40	**75.73 ± 4.14**	77.38 ± 1.17	**75.87 ± 1.95**	64.46 ± 2.71

4.3   Q2: Effectiveness of D2T Attention

To show the effectiveness of D2T attention, We replace the D2T attention module in NIAPool with previously proposed S2T, T2T, and M2T attention techniques, respectively. The results are shown in Table 4. We observe that D2T attention achieves better performance than other attentions, which indicates that the proposed D2T attention framework can reasonably select important nodes.

T2T models the membership of a node by generating a query based only on the medoid nodes. S2T attention scores each node for a global task. M2T extends T2T by using a master function to utilize intra-cluster information. There is a disadvantage causing the node importance mutual exclusion problem when these methods are used for local node information enhancement. That is, they pay too much attention to the nodes with high similarity, resulting in the loss of other node information. From the results, we can prove that D2T deal with the above problem well by taking $h_i - h_j$ into consideration.

Table 4. Effectiveness of different attention framework.

Attention module	D&D	Proteins	NCI1	NCI109	Frankenstein
T2T	78.18 ± 4.44	74.71 ± 3.12	77.13 ± 3.23	75.07 ± 2.02	65.18 ± 1.53
S2T	78.35 ± 4.43	74.82 ± 4.02	76.55 ± 1.89	74.11 ± 1.95	65.30 ± 2.73
M2T	79.03 ± 4.40	74.93 ± 3.90	77.10 ± 0.85	74.23 ± 2.24	65.29 ± 2.23
D2T	**79.28 ± 4.30**	**75.25 ± 3.71**	**78.08 ± 1.92**	**75.56 ± 2.32**	**65.35 ± 2.58**

4.4   Q3: Effectiveness of NFAConv

In this section, we analyze the impact of NFAConv as a fitness scoring function in NIAPool. We use three famous graph convolutional operations, including GCN [12], GAT [27], and GraphSAGE [9] as our baselines. In Table 5, we can see that NFAConv performs significantly better than others. In particular, our method has obvious advantages in datasets NCI1 and NCI109, which achieve approximate 3.24% and 2.67% accuracy improvement over the best baseline, respectively.

GCN can be viewed as a procedure that computes a score for each node followed by a weighted average operation over neighbors and a nonlinearity operation. If some of the nodes get a high score, it may increase the scores of its neighbors indirectly, which biases the pooling operator to select nodes. GraphSAGE directly averages the neighbor features of central nodes, ignoring the feature diversity among nodes. GAT is an example of T2T attention in graphs. GAT utilizes attention coefficient to weight and sum node features but lacking a way to capture specific node difference features. NFAConv addresses the above problems by focusing on capturing node difference features between nodes. The results from Table 5 verify the effectiveness of NFAConv.

Table 5. Effectiveness of different graph convolutional operations as fitness scoring function.

Fitness function	D&D	Proteins	NCI1	NCI109	Frankenstein
GCN	78.89 ± 3.54	74.48 ± 3.66	74.14 ± 1.49	72.21 ± 2.25	62.44 ± 3.06
GraphSAGE	78.61 ± 4.03	74.39 ± 3.38	74.84 ± 2.68	72.89 ± 2.10	64.65 ± 3.84
GAT	77.08 ± 2.67	74.11 ± 3.23	74.44 ± 1.50	71.97 ± 2.14	61.24 ± 3.50
NFAConv	**79.28 ± 4.30**	**75.25 ± 3.71**	**78.08 ± 1.92**	**75.56 ± 2.32**	**65.35 ± 2.58**

5 Conclusion

In this paper, we introduce NIAPool, a novel graph pooling operator for the graph classification task. NIAPool is aware of the node information from both the local and global aspects of the graph. For the local aspect, we propose a Difference2Token self-attention framework to better capture the membership between each node and its 1-hop neighbors. For the global aspect, we propose NFAConv, a novel GCN variant that focuses on capturing node difference features and uses it to score nodes. We validate the effectiveness of the components of NIAPool empirically. Through extensive experiments, we demonstrate that NIAPool achieves state-of-the-art performance on multiple graph classification datasets.

Acknowledgments. This work was supported by the National Key R&D Program of China (2017YFB0202403) and the Key R&D Program of Sichuan Province, China (2017GZDZX0003, 2020YFG0089, 2020YFG0308, 2020YFG0304).

References

1. Baek, J., Kang, M., Hwang, S.J.: Accurate learning of graph representations with graph multiset pooling. In: International Conference on Learning Representations. OpenReview.net (2021)
2. Bianchi, F.M., Grattarola, D., Alippi, C.: MinCUT pooling in graph neural networks. ArXiv abs/1907.00481 (2019)
3. Borgwardt, K.M., Ong, C.S., Schönauer, S., Vishwanathan, S.V.N., Smola, A., Kriegel, H.P.: Protein function prediction via graph kernels. Bioinformatics 21(Suppl 1), i47–56 (2005)
4. Diehl, F.: Edge contraction pooling for graph neural networks. CoRR abs/1905.10990 (2019)

5. Dobson, P.D., Doig, A.J.: Distinguishing enzyme structures from non-enzymes without alignments. J. Mol. Biol. **330**(4), 771–83 (2003)
6. Duvenaud, D., et al.: Convolutional networks on graphs for learning molecular fingerprints. In: Advances in Neural Information Processing Systems, pp. 2224–2232 (2015)
7. Gao, H., Ji, S.: Graph U-nets. In: International Conference on Machine Learning. Proceedings of Machine Learning Research, vol. 97, pp. 2083–2092. PMLR (2019)
8. Gao, H., Liu, Y., Ji, S.: Topology-aware graph pooling networks. IEEE Trans. Pattern Anal. Mach. Intell. **PP** (2021)
9. Hamilton, W.L., Ying, Z., Leskovec, J.: Inductive representation learning on large graphs. In: Advances in Neural Information Processing Systems, pp. 1024–1034 (2017)
10. Hu, L., Yang, T., Shi, C., Ji, H., Li, X.: Heterogeneous graph attention networks for semi-supervised short text classification. In: Empirical Methods in Natural Language Processing, pp. 4820–4829. Association for Computational Linguistics (2019)
11. Kipf, T.N., Welling, M.: Variational graph auto-encoders. CoRR abs/1611.07308 (2016)
12. Kipf, T.N., Welling, M.: Semi-supervised classification with graph convolutional networks. In: International Conference on Learning Representations. OpenReview.net (2017)
13. Lee, J., Lee, I., Kang, J.: Self-attention graph pooling. In: Chaudhuri, K., Salakhutdinov, R. (eds.) International Conference on Machine Learning. Proceedings of Machine Learning Research, vol. 97, pp. 3734–3743. PMLR (2019)
14. Li, X., et al.: BrainGNN: interpretable brain graph neural network for fMRI analysis. Med. Image Anal. **74**, 102233 (2021)
15. Li, Y., Tarlow, D., Brockschmidt, M., Zemel, R.S.: Gated graph sequence neural networks. In: International Conference on Learning Representations (2016)
16. Orsini, F., Frasconi, P., Raedt, L.D.: Graph invariant kernels. In: Yang, Q., Wooldridge, M.J. (eds.) International Joint Conference on Artificial Intelligence, pp. 3756–3762. AAAI Press (2015)
17. Ranjan, E., Sanyal, S., Talukdar, P.P.: ASAP: adaptive structure aware pooling for learning hierarchical graph representations. In: AAAI Conference on Artificial Intelligence, pp. 5470–5477. AAAI Press (2020)
18. Schlichtkrull, M., Kipf, T.N., Bloem, P., van den Berg, R., Titov, I., Welling, M.: Modeling relational data with graph convolutional networks. In: Gangemi, A., et al. (eds.) ESWC 2018. LNCS, vol. 10843, pp. 593–607. Springer, Cham (2018). https://doi.org/10.1007/978-3-319-93417-4_38
19. Shen, T., Zhou, T., Long, G., Jiang, J., Pan, S., Zhang, C.: DiSAN: directional self-attention network for RNN/CNN-free language understanding. In: Conference on Artificial Intelligence, pp. 5446–5455. AAAI Press (2018)
20. Shervashidze, N., Schweitzer, P., van Leeuwen, E.J., Mehlhorn, K., Borgwardt, K.M.: Weisfeiler-Lehman graph kernels. J. Mach. Learn. Res. **12**, 2539–2561 (2011)
21. Shi, J.Y., Huang, H., Zhang, Y.N., Long, Y.X., Yiu, S.: Predicting binary, discrete and continued lncRNA-disease associations via a unified framework based on graph regression. BMC Med. Genom. **10**, 55–64 (2017)
22. Shi, L., Zhang, Y., Cheng, J., Lu, H.: Two-stream adaptive graph convolutional networks for skeleton-based action recognition. In: Proceedings of the IEEE/CVF Conference on Computer Vision and Pattern Recognition, pp. 12026–12035 (2019)
23. Shlomi, J., Battaglia, P.W., Vlimant, J.: Graph neural networks in particle physics. Mach. Learn. Sci. Technol. **2**(2), 21001 (2021)

24. Stokes, J.M., et al.: A deep learning approach to antibiotic discovery. Cell **180**, 688–702.e13 (2020)
25. Vashishth, S., Joshi, R., Prayaga, S.S., Bhattacharyya, C., Talukdar, P.P.: RESIDE: improving distantly-supervised neural relation extraction using side information. In: Empirical Methods in Natural Language Processing, pp. 1257–1266. Association for Computational Linguistics (2018)
26. Vaswani, A., et al.: Attention is all you need. In: Advances in Neural Information Processing Systems, pp. 5998–6008 (2017)
27. Velickovic, P., Cucurull, G., Casanova, A., Romero, A., Liò, P., Bengio, Y.: Graph attention networks. CoRR abs/1710.10903 (2017)
28. Vinyals, O., Bengio, S., Kudlur, M.: Order matters: sequence to sequence for sets. In: International Conference on Learning Representations (2016)
29. Wale, N., Watson, I.A., Karypis, G.: Comparison of descriptor spaces for chemical compound retrieval and classification. Knowl. Inf. Syst. **14**, 347–375 (2006)
30. Xu, K., Hu, W., Leskovec, J., Jegelka, S.: How powerful are graph neural networks? In: International Conference on Learning Representations. OpenReview.net (2019)
31. Xu, K., Li, C., Tian, Y., Sonobe, T., Kawarabayashi, K., Jegelka, S.: Representation learning on graphs with jumping knowledge networks. In: International Conference on Machine Learning. Proceedings of Machine Learning Research, vol. 80, pp. 5449–5458. PMLR (2018)
32. Yang, H., et al.: Interpretable multimodality embedding of cerebral cortex using attention graph network for identifying bipolar disorder. In: Shen, D., et al. (eds.) MICCAI 2019. LNCS, vol. 11766, pp. 799–807. Springer, Cham (2019). https://doi.org/10.1007/978-3-030-32248-9_89
33. Ying, Z., You, J., Morris, C., Ren, X., Hamilton, W.L., Leskovec, J.: Hierarchical graph representation learning with differentiable pooling. In: Advances in Neural Information Processing Systems, pp. 4805–4815 (2018)
34. Zhang, L., et al.: Structure-feature based graph self-adaptive pooling. In: International World Wide Web Conference, pp. 3098–3104. ACM/IW3C2 (2020)
35. Zhang, M., Cui, Z., Neumann, M., Chen, Y.: An end-to-end deep learning architecture for graph classification. In: AAAI Conference on Artificial Intelligence, pp. 4438–4445. AAAI Press (2018)
36. Zhang, Z., et al.: Hierarchical graph pooling with structure learning. CoRR abs/1911.05954 (2019)

dK-Personalization: Publishing Network Statistics with Personalized Differential Privacy

Masooma Iftikhar(✉), Qing Wang, and Yang Li

The Australian National University, Canberra, Australia
{masooma.iftikhar,qing.wang,kelvin.li}@anu.edu.au

Abstract. Preserving privacy of an individual in network structured data while enhancing utility of published data is one of the most challenging problems in data privacy. Moreover, different individuals might have different privacy levels based on their own preferences, thereby personalization needs to be considered to achieve personal data protection. In this paper, we aim to develop a privacy-preserving mechanism to publish network statistics, particularly *degree distribution*, and *joint degree distribution*, which guarantees personalized (edge or node) differential privacy while enhancing network data utility. To this extend we propose four approaches to handle personal privacy requirements of individuals in a differentially private computation. We have empirically verified the utility enhancement and privacy guarantee of our proposed approaches on four real-world network datasets. To the best of our knowledge, this is the first study to publish network data distributions under personalized differential privacy, while enhancing network data utility.

Keywords: Privacy-preserving graph data publishing · Personalized differential privacy · Network data distributions · Graph data utility

1 Introduction

Network analysis provides unique insights about social network activities, disease transmission, consumer behaviour, communication patterns, and recommendations [22]. However, given the private nature of data about individuals stored in networks, releasing network data raises privacy concerns, and there has been much interest to devise privacy preserving mechanisms for network data analysis [2,11]. The current focus of privacy is around differential privacy (DP) [5], because of its provable mathematical privacy guarantee. DP ensures that the output of a computation undergoes enough perturbation to mask whether an individual is present or not in the output. The magnitude of random noise for perturbation is determined by the sensitivity of the computation (i.e., the maximum impact that one individual can have on the output), and a global privacy parameter $\varepsilon \in [0, \infty)$, also called privacy level, where a smaller value of ε implies a stronger privacy guarantee and requires larger noise.

© The Author(s), under exclusive license to Springer Nature Switzerland AG 2022
J. Gama et al. (Eds.): PAKDD 2022, LNAI 13280, pp. 194–207, 2022.
https://doi.org/10.1007/978-3-031-05933-9_16

One fundamental shortcoming of DP is that, a uniform privacy level (i.e., ε) is assigned to each individual while performing perturbation; however, in practice, different individuals have different privacy levels based on their own preferences subject to their data [6,10]. For instance, in social networks an individual (user) tends to share their personal information with their close ones and only share obscured data with acquaintances or strangers. Therefore, DP may lead to provide insufficient protection for some individuals, while over-protecting others.

Early studies [1,6,10,16] which consider personalization under DP framework, such as personalized DP (PDP) [6,10] or heterogeneous DP (HDP) [1] are limited to relational databases. Later some works [13,14,24] explore graph data perturbation with PDP, however, these works are also limited to publish single queries. Additionally, DP has two variants when applying to network data, i.e., edge-DP [11] and node-DP [2], and to the best of our knowledge there is no work which considers these variations before under personalization.

In this paper, we aim to publish higher-order network statics such as *degree distribution* and *joint degree distribution* under personalized edge and node DP, where individuals (nodes) in a network can specify their own privacy preferences.

Undertaking the problem of releasing network data distributions under PDP brings up two challenges: (i) each node in a network has its own privacy preference in personalized settings whereas each data point in data distribution reflects information about more than one node, and (ii) network data is highly sensitive to structural changes under DP. To address these challenges, we propose *four* PDP mechanisms and introduce *degree queries* for controlled sensitivity.

Contributions. To summarize, our work makes the following contributions: (1) We show how to publish network data distributions in a personalized differentially private manner under edge and node DP. (2) We analyse the sensitivity of *degree queries* for publishing *degree distribution*, and *joint degree distribution* under edge and node DP. (3) We introduce four approaches for generating personalized differentially private network data distributions while enhancing utility. (4) We conduct comprehensive experiments over four real-world networks, and the results demonstrate that our proposed approaches can effectively enhance the utility of differentially private network data distributions with personalization.

2   Related Work

Privacy issues have received much attention in recent years, due to the growing popularity of social networking, and recommendation platforms providing personalized services for enhancing user experience and anticipating individual needs. In recent years, privacy-persevering techniques based on personalized differential privacy (PDP) [5] have been proposed [1,6,10]. These techniques can be broadly categorized into three areas: user-grained [6,10,16], distance-grained [13,14] and item-grained [1,24]. User-grained PDP approaches generalized the classic definition of DP [5] to provide freedom to each individual to have a personalised privacy preference [6,10] in relational database settings. Distance-grained PDP approaches [13,14] consider social network and scale individuals' privacy

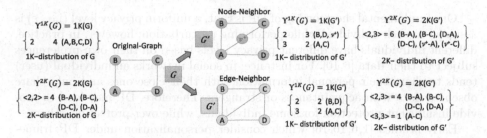

Fig. 1. An illustrative example of two (edge or node) neighboring graphs $G \sim G'$ with their corresponding *dK-distributions*, for $d = 1$, and $d = 2$.

based on distance between individuals in a network (e.g., length of a shortest path between two nodes). Item-grained PDP approaches for relational databases include heterogeneous DP (HDP) [1], which allows different privacy levels for different data items (e.g. salary, age, etc.). Later for social networks, a fine-grained approach based on both distance and item-grained has been presented in [24]. Despite considerable progress being achieved in understanding PDP, these works have either limited to relational database setting [6,10,16] or only studied the release of single queries in a network setting [1,13,14,24].

Different from existing work, we aim to release network data distributions, i.e., *degree distribution* and *joint degree distribution* under edge-DP [11] and node-DP [2] with personalization while enhancing network data utility.

3   Problem Formulation

Throughout, a network is an undirected graph $G = (V, E)$, where V is the set of nodes and $E \subseteq V \times V$ is the set of edges. Let $N_G(v) = \{u \in V | (u, v) \in E\}$ be the set of neighbors of a vertex v in G. Denote v's degree by $deg(v)$ and G's maximum degree by $deg(G) = max\{deg(v) | v \in V\}$.

Definition 1 (NEIGHBORING GRAPHS). *Two graphs* $G = (V, E)$ *and* $G' = (V', E')$ *are* edge neighboring graphs $G \overset{e}{\sim} G'$ *if* $V = V'$, $E \subset E'$ *and* $|E| + 1 = |E'|$, *and are* node neighboring graphs $G \overset{n}{\sim} G'$ *if* $V' = V \cup \{v^+\}$ *and* $E' = E \cup E^+$, *where* E^+ *is the set of all edges incident to* $v^+ in$ G'.

Personalization allows each $v \in V$ to specify its own *privacy preference* ε in G. However, given two (edge or node) neighboring graphs $G \sim G'$ where G' is obtained from G by adding (or deleting) an edge (or node) can affect more than one node. Thus, PDP should be formalized in terms of all affected nodes to guarantee ε-indistinguishability. In a privacy specification $\Phi = \{\varepsilon_1, \dots \varepsilon_n\}$, denote Φ^v the privacy preference ε of a node v. For $G \overset{e}{\sim} G'$, adding (or deleting) an edge affects exactly two nodes u and v; likewise, for $G \overset{n}{\sim} G'$, adding (or deleting) a node v^+ affects $|E^+|$ nodes incident to v^+ and v^+ itself. We formally define edge and node PDP below.

Definition 2 (EDGE Φ-PDP). *A randomized mechanism $\mathcal{K}$ satisfies edge Φ-personalized differential privacy, if for each pair of edge neighboring graphs $G \overset{e}{\sim} G'$, and all possible outputs $\mathcal{O} \subseteq range(\mathcal{K})$, it holds:*

$$Pr[\mathcal{K}(G) \in \mathcal{O}] \leq e^{min\{\Phi^u, \Phi^v\}} \times Pr[\mathcal{K}(G') \in \mathcal{O}].$$

Definition 3 (NODE Φ-PDP). *A randomized mechanism $\mathcal{K}$ satisfies node Φ-personalized differential privacy, if for each pair of node neighboring graphs $G \overset{n}{\sim} G'$, and all possible outputs $\mathcal{O} \subseteq range(\mathcal{K})$, it holds:*

$$Pr[\mathcal{K}(G) \in \mathcal{O}] \leq e^{min\{\Phi^v | (v^+, v) \in E^+\}} \times Pr[\mathcal{K}(G') \in \mathcal{O}] \quad and$$

$$Pr[\mathcal{K}(G) \in \mathcal{O}] \leq e^{\Phi^{v^+}} \times Pr[\mathcal{K}(G') \in \mathcal{O}]$$

The dK-graph model [20] offers a systematized method to extract subgraph degree distributions from a given graph, i.e. *dK-distributions*. For d nodes $v_1, \ldots, v_d$ of degrees $g_1, \ldots, g_d$, where $d \in [1, |V|]$, let D^d be the set of all possible degree tuples $d_t = (g_1, \ldots, g_d)$ in a graph G.

Definition 4 (DK-DISTRIBUTION). *A dK-distribution over a graph $G = (V, E)$, denoted as $dK(G)$, is a probability distribution $p : D^d \rightarrow \mathbb{N}$ defined by $p(d_t) = n$, where n is the number of the subgraphs with degrees $g_1, \ldots, g_d$.*

The degree distribution of a graph is captured by $1K$-distribution. The joint degree distribution, i.e., the number of edges between nodes of different degrees, is captured by $2K$-distribution. This paper studied the private release of $1K$ and $2K$-distributions. Figure 1 depicts the $1K$ and $2K$-distributions of the graph G. In $1K(G)$ we have $p(2) = 4$ which corresponds to 4 nodes A, B, C and D. In $2K(G)$ we have $p(2, 2) = 4$ which corresponds to 4 edges with joint degree $(2, 2)$.

To show how a dK-distribution is extracted from a graph, we define dK function. Let $\mathcal{G}_n$ be the set of graphs with n nodes and $\mathcal{D}$ be the set of dK-distributions over a graph.

Definition 5 (DK-FUNCTION). *A dK function $\gamma^{dK} : \mathcal{G}_n \rightarrow \mathcal{D}$ defined by $\gamma^{dK}(G) = dK(G)$ maps a graph $G \in \mathcal{G}_n$ to its dK-distribution.*

Conceptually, $\gamma^{dK}(G)$ queries the dK-distribution of G. Figure 1 provides an illustrative example of two (edge or node) neighboring graphs with their corresponding dK-distribution for $d = 1$, and 2. In our work, for each dK-distribution D, we want to generate D_Φ that is an anonymized version of D satisfying (edge or node) Φ-PDP. Thus, we view the response to γ^{dK} for $d = 1$ and 2 as a collection of responses to degree queries, one for each tuple (entry) in a dK-distribution.

Definition 6 (DEGREE QUERY). *A degree query $\gamma_q : \gamma^{dK}(G) \rightarrow \mathbb{N}$ maps a degree tuple $d_t \in \gamma^{dK}(G)$ to a frequency value in $\mathbb{N}$ s.t. $(d_t, \gamma_q(G)) \in \gamma^{dK}(G)$.*

As DP implies PDP [10], to guarantee (edge or node) Φ-PDP, we perform perturbation over real responses of γ_q by adding controlled noise from Laplace stochastic process [5]. The response to each γ_q can be combined into a complete dK-distribution using the parallel composition property of DP [21]. More specifically, we add random noise into the real response $\gamma_q(G)$, yielding a randomized response $\gamma_q(G) + Lap(\Delta(\gamma_q)/\varepsilon)$, where $\Delta(\gamma_q)$ denotes the sensitivity of γ_q and $Lap(\Delta(\gamma_q)/\varepsilon)$ denotes random noise drawn from a Laplace distribution. Here, ε refers to a personalized privacy preference. We will discuss in detail how to utilize ε to perform personalized perturbation in Sect. 5.

4   Sensitivity Analysis

The key challenge of releasing dK-distributions under PDP is to determine the right amount of noise that guarantees both privacy and accuracy. Unlike previous studies [2,7,9,22,23] that analyzed the entire dK-distribution sensitivity, we focus on the sensitivity of a single dK-distribution entry, i.e., *degree query γ_q*.

Sensitivity of γ_q for 1K-Distribution Under Node-DP. Suppose that a node v^+ is added to G with a set E^+ of new edges. Each edge in E^+ can cause at most one change in a current 1K-distribution entry, so the total change in an entry is at most $|E^+|$. Furthermore, the new node v^+ can cause an additional change in the same entry. Thus, the maximum change in γ_q is at most $|E^+| + 1$.

Lemma 1. *Let $G \overset{n}{\sim} G'$ be two node neighboring graphs. When $d = 1$, $\gamma_q(G)$ and $\gamma_q(G')$ differ by at most $|E^+| + 1$.*

In the worst case, $|E^+|$ can be $|V|$. Thus, the total number of changes in $\gamma_q(G)$, for $d = 1$, by adding a node is upper bounded by $|V| + 1$.

Sensitivity of γ_q for 2K-Distribution Under Node-DP. For a 2K-distribution, when it is connected to a maximum degree node, each new edge (v^+, v_i) in E^+ can affect at most $deg(G)$ edges. If all the v_is incident to v^+ have the same degree, then the maximum increase in an entry is $deg(G) \times |E^+|$. This can be further increased if v^+ has the same degree as all the v_is. Hence, the maximum change in γ_q is at most $(deg(G) + 1) \times |E^+|$.

Lemma 2. *Let $G \overset{n}{\sim} G'$ be two node neighboring graphs. When $d = 2$, $\gamma_q(G)$ and $\gamma_q(G')$ differ by at most $(deg(G) + 1) \times |E^+|$.*

Prior studies [15,22] have shown that, in large social networks, $deg(G)$ is upper bounded by $O(\sqrt{|V|})$. Thus, for such networks, the sensitivity of γ_q for $d = 2$ of two node neighboring graphs is upper bounded by $O(|V|^{3/2})$.

Sensitivity of γ_q for 1K-Distribution Under Edge-DP. This is straightforward, because each new edge can affect at most two nodes.

Lemma 3. *Let $G \overset{e}{\sim} G'$ be two edge neighboring graphs. When $d = 1$, $\gamma_q(G)$ and $\gamma_q(G')$ differ by at most 2.*

Sensitivity of γ_q for $2K$-Distribution Under Edge-DP. This is similar to the node-DP case. But since only one edge is added, it can affect at most $2 \times deg(G)$ existing edges. The new edge itself can further increase the entry by 1 when its end nodes have the same degrees as the those of the affected edges.

Lemma 4. *Let $G \overset{e}{\sim} G'$ be two edge neighboring graphs. When $d = 2$, $\gamma_q(G)$ and $\gamma_q(G')$ differ by at most $2 \times deg(G) + 1$.*

Prior studies [15,22] have shown that, in large social networks, $deg(G)$ is upper bounded by $O(\sqrt{|V|})$. Thus, for such networks, the sensitivity of γ_q for $d = 2$ is upper bounded by $O(\sqrt{|V|})$ too.

We have observed that, by adding a single node or edge in G, the maximum change induced in the entries of G's corresponding dK-distribution is two times greater than the maximum change induced in a single entry of a dK-distribution. Thus the sensitivity of *degree query* γ_q, is half as compared to dK-*function* γ^{dK}. This shows that our analysis as compared to analysis in existing studies [2,8,9, 22,23] significantly enhances utility of published dK-distribution.

5   Proposed Approaches

In this section, we present *four* mechanisms for generating personalized differentially private dK-distribution for $d = 1$ and $d = 2$.

5.1   Local Least Based Personalized Perturbation

A straightforward perturbation can be done over dK-distribution D by invoking the Laplace mechanism with the strongest privacy preference in G. This will, however, overprotect some individuals and degrade the output utility. To address this issue, we propose *Local Least (LL-dK)* mechanism to perturb the entries of D with the strongest *local* privacy preference. More specifically, LL-dK perturbs each entry $x_i \in D$ with the smallest privacy preference ε_{x_i} associated the corresponding nodes for x_i. For instance, in Fig. 2 the frequency $p(1) = 2$ in $1K(G)$, and the frequency $p(2,4) = 3$ in $2K(G)$ are perturbed with the privacy preference $\varepsilon = min(\Phi^B, \Phi^F)$, and $\varepsilon = min(\Phi^A, \Phi^C, \Phi^D, \Phi^E)$, respectively.

In LL-dK, the perturbation is conducted using the Laplace mechanism based on the sensitivity of γ_q, and personalized privacy preferences ε_{x_i} Thus, by the parallel composition property of DP [21] we have the following lemma.

Lemma 5. *LL-dK generates $(\max_i \varepsilon_{x_i})$-personalized differentially private dK-distributions.*

As LL-dK uses the local strongest privacy preferences, which may lead to adding excessive noise when nodes with strong privacy preference have high centrality. To control the amount of random noise and improve personalization, another option is to simply discard high degree privacy conscious nodes.

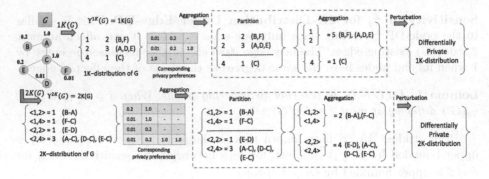

Fig. 2. An illustrative example of aggregation based perturbation over dK-*distributions*, for $d = 1$, and $d = 2$.

5.2 Threshold Projection Based Personalized Perturbation

The *threshold projection (TP-dK)* approach (Algorithm 1) first transforms a graph G to a θ-bounded graph G^θ with $\theta < deg(G)$ [2,3], removes all nodes in G^θ with $\Phi^v < \tau$ to get $G^{\theta,\tau}$, where $min(\Phi^v) < \tau < max(\Phi^v)$ is a global threshold, and then perturbs entries of $\gamma^{dK}(G^{\theta,\tau})$. Since we have $deg(G) \leq \theta$, the sensitivity of γ_q is reduced; likewise, with threshold τ all nodes with high privacy concerns are removed which results in adding less noise to release D of $G^{\theta,\tau}$.

Algorithm 1: *TP-dK algorithm*

Input: A graph $G = (V, E)$;
  a value d in 1, 2;
  a threshold τ;
  a projection parameter θ;
  a projection algorithm $\mathcal{P}$

Output: a perturbed $\widehat{dK}$

1 $G^\theta \leftarrow$ Project G by $\mathcal{P}$ w.r.t. θ
2 $G^{\theta,\tau} \leftarrow$ Truncate G^θ w.r.t. τ
3 $dK^{\theta,\tau} \leftarrow$ Query $G^{\theta,\tau}$ with γ^{dK}
4 $\widehat{dK} \leftarrow$ Perturb $dK^{\theta,\tau}$
5 **Return** $\widehat{dK}$

In Algorithm 1, the perturbation is conducted using the Laplace mechanism (Line 4) based on the sensitivity of γ_q, and personalized privacy preference τ. Thus, by the parallel composition property of DP [21] we have the following lemma.

Lemma 6. *TP-dK generates τ-personalized differentially private dK-distributions.*

The main challenge in this approach is to select good thresholds (i.e., θ, and τ) that can lower the sensitivity and preserve as much topological structure of a graph as possible [2]. To address this limitation, we now propose a personalized mechanism based on *sampling*.

5.3   Sampling Based Personalized Perturbation

Sampling is shown to be a powerful tool that can be integrated into DP to amplify privacy protection [10,17]. The *sampling (ST-dK)* approach (Algorithm 2) first splits each entry $x_i \in D$ such that frequency count of each x_i is equal to one. Then samples each x_i with probability $p_i = 1$ if $\varepsilon_{x_i} \geq \tau$, and samples other entries i.i.d. with probability $p_i = \frac{e^{\varepsilon_{x_i}} - 1}{e^{\tau} - 1}$ if $\varepsilon_{x_i} < \tau$, where

---
Algorithm 2: *ST-dK algorithm*

Input: A graph $G = (V, E)$;
       a value d in $1, 2$;
       a threshold τ;

Output: a perturbed $\widehat{dK}$

1 $dK \leftarrow$ Query G with γ^{dK}
2 $dK^\tau \leftarrow$ Split and Sample dK
3 $\widehat{dK} \leftarrow$ Perturb dK^τ
4 **Return** $\widehat{dK}$

$min(\Phi^v) < \tau < max(\Phi^v)$, and ε_{x_i} is the smallest privacy preference associated the corresponding nodes for x_i. The inclusion probability for each x_i depends on the corresponding ε_{x_i}, and the global threshold τ.

In Algorithm 2, the perturbation is conducted using the Laplace mechanism (Line 3) over sampled dK-distribution based on the sensitivity of γ_q, and personalized privacy preference τ. Thus, by the parallel composition property of DP [21] we have the following lemma.

Lemma 7. *ST-dK generates τ-personalized differentially private dK-distributions.*

Our sampling mechanism is inspired by the results from [10,17], where sampling prior to DP is shown to benefit privacy by combining two sources of randomness; however, introducing two kinds of errors: sampling error, and perturbation error. We observe that, smaller τ increases perturbation error, and larger τ increases sampling error, and vice versa. To reduce overall error, we now propose a personalized mechanism based on *aggregation*.

5.4   Aggregation Based Personalized Perturbation

The technique of aggregation is shown to reduce noise and enhance the utility of publishing differentially private histogram [2,22]. Since degree distribution is obtained from normalizing the degree histogram [2], the idea of aggregation can be equivalently applied. Our aggregation mechanism AG-dK (Algorithm 3) computes a table $\mathcal{T}$ by combining privacy preferences ε of corresponding nodes associated with each $x_i \in D$ as a sorted list. Then, performing aggregation over $\gamma^{dK}(G)$ consists of

---
Algorithm 3: *AG-dK algorithm*

Input: A graph $G = (V, E)$;
       a value d in $1, 2$;
       a partitioning parameter k;
       a partitioning algorithm $\mathcal{M}$

Output: a perturbed $\widehat{dK}$

1 $dK \leftarrow$ Query G with γ^{dK}
2 $\mathcal{T} \leftarrow$ Compute corresponding ε
3 $dK_i^p \leftarrow \mathcal{M}(dK)$ with k and $\mathcal{T}$
4 $dK_i^a \leftarrow$ Aggregate dK_i^p
5 $\widehat{dK} \leftarrow$ Perturb dK_i^a
6 **Return** $\widehat{dK}$

---

two steps: (i) entries with similar degrees and privacy preferences are partitioned into the same group; (ii) the frequency values of entries in the same group are aggregated. For instance in Fig. 2, a dK-distribution with $d = 1$, and $d = 2$ is partitioned into multiple groups, then the frequency values of entries in each group are aggregated by an aggregation process. The perturbation is performed over each aggregated frequency value with the smallest privacy preference ε_{p_i} associated with the corresponding nodes for each partition p_i.

In Algorithm 3, the perturbation is conducted using the Laplace mechanism (Line 5) over aggregated dK-distribution based on the sensitivity of γ_q, and personalized privacy preferences ε_{p_i} associated to each partition. Thus, by the parallel composition property of DP [21] we have the following lemma.

Lemma 8. *AG-dK generates* $(\max_i \varepsilon_{p_i})$*-personalized differentially private dK-distributions.*

AG-dK significantly reduces the total amount of noise to achieve PDP, particularly when the number of partitions in aggregated dK-distribution is smaller.

6   Experiments

In this section we evaluate our proposed approaches, and discuss the experimental results.

6.1   Experimental Setup

Datasets. We use four real-world network datasets from different domains including social, citation, and email networks: (1) *Facebook*[1] contains 4,039 nodes and 88,234 edges. (2) *Wiki-Vote* (see footnote 1) contains 7,115 nodes and 103,689 edges. (3) *Ca-HepPh* (see footnote 1) contains 12,008 nodes and 118,521 edges. (4) *Email-Enron* (see footnote 1) contains 36,692 nodes and 183,831 edges.

Privacy Specifications. Following [10], we randomly divide nodes in a network into three groups: conservative with fraction $f_C = 0.54$, moderate with fraction $f_M = 0.37$, and liberal with fraction $f_L = 1.0 - (f_C + f_M)$, having high (randomly drawn from range $[\varepsilon_C, \varepsilon_M]$), medium (randomly drawn from range $[\varepsilon_M, \varepsilon_L]$), and low (fixed at $\varepsilon_L = 1.0$) privacy preferences, respectively. We use $\varepsilon_C, \varepsilon_M, \varepsilon_L \in [0.01, 1.0]$, which cover the range of DP levels used in the literature [8–10], where default values of $\varepsilon_C = 0.01$, and $\varepsilon_M = 0.2$ [10].

Utility Metrics. Following [2,3,7], we use two utility metrics to measure the difference between the original dK-distribution D and its private version D_Φ: (1) *L1 distance* measures the network structural error by calculating $\|D - D_\Phi\|_1 = \sum_{i=1}^{deg(G)} |D_i - D_{\Phi_i}|$. We pad a distribution entry with 0 if it does not exist in D or D_Φ; (2) *Kolmogorov-Smirnov (KS) distance* measures the closeness between

---

[1] Network datasets are available at http://snap.stanford.edu/data/index.html .

the cumulative distribution functions of D and D_Φ by calculating $KS(D, D_\Phi)$ = $max_i|CDF_{D_i} - CDF_{D_{\Phi_i}}|$.

Baseline Methods. We compare the utility of the following methods for generating personalized differentially private dK-distributions: (1) PDP_{min} is a standard DP algorithm using Laplace mechanism [5] with the minimum privacy preference in a network [10]. (2) $LL\text{-}dK$ is our proposed *local least* approach. (3) $TP\text{-}dK$ is our *threshold projection* approach which extends the projection algorithm Stable-Edge-Removal [7] for graph projection. (4) $ST\text{-}dK$ is our *sampling* approach. Additionally, we investigate two variations of $TP\text{-}dK$, and $ST\text{-}dK$ with threshold [10] $\tau = max(\Phi^v)$, and $\tau = \frac{1}{n}\sum \Phi^v$ (i.e., the average privacy preference in a network). We denote these as $TP\text{-}dK_{avg}$, and $ST\text{-}dK_{avg}$. 5) $AG\text{-}dK$ is our proposed *aggregation* based approach which extends the algorithm MDAV-dK [9] for partitioning dK-distributions.

Parameter Settings and Others. We choose $\theta \in \{1, 2, 4, \ldots, 2^{\lfloor 2log_2(|V|)\rfloor}\}$ [2,3], for projection in $TP\text{-}dK$. We vary sample size $m \in [30\%, 70\%]$ [18] to study the impact of sample size in $ST\text{-}dK$. Following [4,8,9], we choose $k \in \{2, 4, 6, \ldots, 100\}$ for partitioning in $AG\text{-}dK$. For each method, we ran 3 times and took the average result [8]. We use *Orbis* [19] to generate dK-distributions for $d = 1$ and $d = 2$.

6.2   Results and Discussion

Evaluating Personalized Differentially Private $1K$-Distributions. Figure 3 presents our experimental results. For all four datasets, w.r.t. L1 distance, under both edge-PDP and node-PDP, our methods yield less network structural error than PDP_{min} for every value of k, θ, and m, except for the largest network *Email-Enron*, where PDP_{min} performs better than $ST\text{-}dK$ and $ST\text{-}dK_{avg}$, when sample size is greater than 30%. This is because, by increasing the sample size for larger dataset, perturbation error dominates the impact on the utility. Overall, our methods $AG\text{-}dK$, $TP\text{-}dK$, and $TP\text{-}dK_{avg}$ outperform all other methods, because aggregation reduces the overall noise and projection reduces the sensitivity, thus enhancing output utility significantly. When measured by the KS distance, for all four datasets, under edge-PDP our method $AG\text{-}dK$, and under node-PDP our method $LL\text{-}dK$ outperforms PDP_{min} by generating more similar $1K$-distributions after perturbation. Overall, $TP\text{-}dK$, and $TP\text{-}dK_{avg}$ yield higher values of KS distance under both edge-PDP and node-PDP because graph transformation may change the topological structure of an original graph which results in generating less similar dK-distributions.

Evaluating Personalized Differentially Private $2K$-Distributions. Figure 4 presents our experimental results. For all four datasets, w.r.t. L1 distance, under both edge-PDP and node-PDP, our methods yield less network structural error than PDP_{min} for every value of k, θ, and m. Also, unlike to $1K$-distribution, in $2K$-distribution $ST\text{-}dK$, and $ST\text{-}dK_{avg}$ perform better than $LL\text{-}dK$ because sensitivity of $2K$-distributions is high as compared to $1K$-distribution, thus overall more noise is added into $2K$-distribution as compared

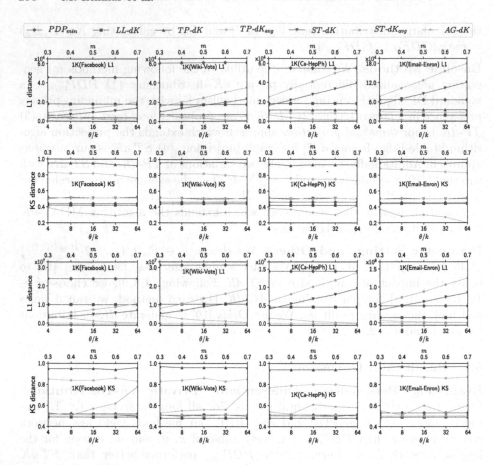

Fig. 3. The L1 and KS distance of $1K$-distributions on four datasets varying θ, k, and m: (upper half) under edge-PDP and (lower half) under node-PDP.

to sampled $2K$-distribution. Overall, our methods $TP\text{-}dK$, and $TP\text{-}dK_{avg}$ outperform all other methods, w.r.t. L1 distance, and the results of $AG\text{-}dK$, for larger k values, are close to them, because aggregation and projection enhance the overall utility. On the other hand, w.r.t. KS distance, for all four datasets, under edge-PDP our method $AG\text{-}dK$, and under node-PDP our method $LL\text{-}dK$ outperforms PDP_{min} by generating more similar $2K$-distributions after perturbation. In addition, under edge-PDP, the results of $LL\text{-}dK$, $ST\text{-}dK$, $ST\text{-}dK_{avg}$ and PDP_{min} are close, which reflects both sampling error and perturbation error contribute towards the results. Contrary, under node-PDP, the results of $ST\text{-}dK$, $ST\text{-}dK_{avg}$ and PDP_{min} are almost same which reflect the impact of high sensitivity, where perturbation error dominates. Overall, $TP\text{-}dK$, and $TP\text{-}dK_{avg}$ yield higher values of KS distance under both edge-PDP, and node-PDP because of graph transformation.

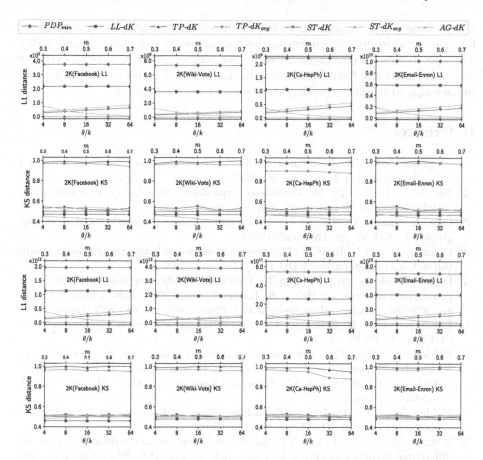

Fig. 4. The L1 and KS distance of $2K$-distributions on four datasets varying θ, k, and m: (upper half) under edge-PDP and (lower half) under node-PDP.

Discussion. We analyse the trade-offs between utility and privacy of dK-distributions under PDP generated using our proposed approaches. We have noticed that, the error caused by random noise which depends on the sensitivity and the personalized privacy preference ε dominates the impact on output utility. Increasing ε and decreasing sensitivity can help to reduce error, though, generating more similar dK-distributions to their original dK-distribution is challenging due to noise addition. Reducing sensitivity can increase the output utility without compromising privacy; however, it is more challenging for node-PDP than for edge-PDP as graph data is highly sensitive under node-PDP.

7   Conclusions and Future Work

In this paper, we have studied the problem of publishing *degree distribution* and *joint degree distribution* under PDP. We have theoretically analyzed the sensitivity of these distributions under edge-PDP and node-PDP. We have proposed four personalized privacy-preserving mechanisms while enhancing output utility. The effectiveness of our proposed work has been empirically verified over four real-world networks. Future extensions to this work will consider local differential privacy [12] to release network statistics under personalization.

References

1. Alaggan, M., Gambs, S., Kermarrec, A.-M.: Heterogeneous differential privacy. J. Priv. Confidentiality **7**(2), 127–158 (2016)
2. Day, W.Y., Li, N., Lyu, M.: Publishing graph degree distribution with node differential privacy. In: SIGMOD, pp. 123–138 (2016)
3. Ding, X., Zhang, X., Bao, Z., Jin, H.: Privacy-preserving triangle counting in large graphs. In: CIKM, pp. 1283–1292 (2018)
4. Domingo-Ferrer, J., Torra, V.: Ordinal, continuous and heterogeneous k-anonymity through microaggregation. KDD **11**(2), 195–212 (2005)
5. Dwork, C., McSherry, F., Nissim, K., Smith, A.: Calibrating noise to sensitivity in private data analysis. In: Halevi, S., Rabin, T. (eds.) TCC 2006. LNCS, vol. 3876, pp. 265–284. Springer, Heidelberg (2006). https://doi.org/10.1007/11681878_14
6. Ebadi, H., Sands, D., Schneider, G.: Differential privacy: now it's getting personal. POPL **50**(1), 69–81 (2015)
7. Iftikhar, M., Wang, Q.: dK-projection: publishing graph joint degree distribution with node differential privacy. In: Karlapalem, K., et al. (eds.) PAKDD 2021. LNCS (LNAI), vol. 12713, pp. 358–370. Springer, Cham (2021). https://doi.org/10.1007/978-3-030-75765-6_29
8. Iftikhar, M., Wang, Q., Lin, Y.: Publishing differentially private datasets via stable microaggregation. In: EDBT, pp. 662–665 (2019)
9. Iftikhar, M., Wang, Q., Lin, Yu.: dK-microaggregation: anonymizing graphs with differential privacy guarantees. In: Lauw, H.W., Wong, R.C.-W., Ntoulas, A., Lim, E.-P., Ng, S.-K., Pan, S.J. (eds.) PAKDD 2020. LNCS (LNAI), vol. 12085, pp. 191–203. Springer, Cham (2020). https://doi.org/10.1007/978-3-030-47436-2_15
10. Jorgensen, Z., Yu, T., Cormode, G.: Conservative or liberal? Personalized differential privacy. In: ICDE, pp. 1023–1034. IEEE (2015)
11. Jorgensen, Z., Yu, T., Cormode, G.: Publishing attributed social graphs with formal privacy guarantees. In: SIGMOD, pp. 107–122 (2016)
12. Kasiviswanathan, S.P., Lee, H.K., Nissim, K., Raskhodnikova, S., Smith, A.: What can we learn privately? SICOMP **40**(3), 793–826 (2011)
13. Koufogiannis, F., Han, S., Pappas, G.J.: Gradual release of sensitive data under differential privacy. J. Priv. Confidentiality **7**(2), 23–52 (2016)
14. Koufogiannis, F., Pappas, G.J.: Diffusing private data over networks. TCNS **5**(3), 1027–1037 (2017)
15. Kwak, H., Lee, C., Park, H., Moon, S.: What is twitter, a social network or a news media? In: WWW, pp. 591–600 (2010)
16. Li, H., Xiong, L., Ji, Z., Jiang, X.: Partitioning-Based Mechanisms Under Personalized Differential Privacy. In: Kim, J., Shim, K., Cao, L., Lee, J.-G., Lin, X., Moon, Y.-S. (eds.) PAKDD 2017. LNCS (LNAI), vol. 10234, pp. 615–627. Springer, Cham (2017). https://doi.org/10.1007/978-3-319-57454-7_48
17. Li, N., Qardaji, W., Su, D.: On sampling, anonymization, and differential privacy or, k-anonymization meets differential privacy. In: ASIACCS, pp. 32–33 (2012)

18. Lin, B.R., Wang, Y., Rane, S.: On the benefits of sampling in privacy preserving statistical analysis on distributed databases. CoRR, abs/1304.4613 (2013)
19. Mahadevan, P., Hubble, C., Krioukov, D., Huffaker, B., Vahdat, A.: Orbis: rescaling degree correlations to generate annotated internet topologies. In: SIGCOMM, vol. 37, pp. 325–336 (2007)
20. Mahadevan, P., Krioukov, D., Fall, K., Vahdat, A.: Systematic topology analysis and generation using degree correlations. In: SIGCOMM, pp. 135–146 (2006)
21. McSherry, F.D.: Privacy integrated queries: an extensible platform for privacy-preserving data analysis. In: SIGMOD, pp. 19–30 (2009)
22. Sala, A., Zhao, X., Wilson, C., Zheng, H., Zhao, B.Y.: Sharing graphs using differentially private graph models. In: SIGCOMM, pp. 81–98 (2011)
23. Wang, Y., Wu, X.: Preserving differential privacy in degree-correlation based graph generation. TDP 6(2), 127–145 (2013)
24. Yan, S., Pan, S., Zhao, Y., Zhu, W.-T.: Towards privacy-preserving data mining in online social networks: distance-grained and item-grained differential privacy. In: Liu, J.K., Steinfeld, R. (eds.) ACISP 2016. LNCS, vol. 9722, pp. 141–157. Springer, Cham (2016). https://doi.org/10.1007/978-3-319-40253-6_9

Residual Vector Product Quantization for Approximate Nearest Neighbor Search

Zhi Xu[1]($\boxtimes$), Lushuai Niu[1], Ruimin Meng[1], Longyang Zhao[1], and Jianqiu Ji[1,2]

[1] School of Computer Science and Information Security,
Guilin University of Electronic Technology, Guilin 541004, China
xuzhi@guet.edu.cn
[2] Doodod Technology Co. Ltd., Guilin, China
jijianqiu@doodod.com

Abstract. Product Quantization is popular for approximate nearest neighbor search, which decomposes the vector space into Cartesian product of several subspaces and constructs separately one codebook for each subspace. The construction of codebooks dominates the quantization error that directly impacts the retrieval accuracy. In this paper, we propose a novel quantization method, residual vector product quantization (RVPQ), which constructs a residual hierarchy structure consisted of several ordered residual codebooks for each subspace. The proposed method minimizes the quantization error by jointly optimizing all the codebooks in each subspace using the efficient mini-batch stochastic gradient descent algorithm. Furthermore, an efficient encoding method, based on H-variable Beam Search, is also proposed to reduce the computation complexity of encoding with negligible loss of accuracy. Extensive experiments show that our proposed method outperforms the-state-of-the-art on retrieval accuracy while retaining a comparable computation complexity.

Keywords: Vector quantization · Approximate nearest neighbor search · Residual encoding

1 Introduction

Nearest neighbor search plays an important role in many fields, such as information retrieval, computer vision, pattern recognition and machine learning, etc. Given a data point $x \in \mathcal{R}^D$, The aim of nearest neighbor search is to find an element in a finite set, which is the closest point to x under certain distance measurement. However, due to the curse of dimensionality [4] of high-dimensional data, the nearest neighbor search is impractical because of the high computation complexity. Therefore, lots of attentions have been paid to more practical approximate nearest neighbor (ANN) search methods, which achieves efficient retrieval through compressing data and simplifying distance calculation between vectors.

© The Author(s), under exclusive license to Springer Nature Switzerland AG 2022
J. Gama et al. (Eds.): PAKDD 2022, LNAI 13280, pp. 208–220, 2022.
https://doi.org/10.1007/978-3-031-05933-9_17

Vector quantization [14] is a popular way for ANN search. It maps the data space into a finite set named codebook, and then the data points are presented as short binary indices of codewords from the codebook and hence reducing the memory cost. In addition, the approximate distance between vectors can be efficiently computed using table lookups. Because of the advantages, lots of vector quantization based methods have been proposed for large scale ANN search [17]. These methods can be divided into two main categories according to the generation way of codebooks, which are Cartesian product based methods, such as Product Quantization (PQ) [8] and linear addition based methods, such as Addition Quantization (AQ) [3].

PQ [8] is a pioneering Cartesian product based quantization method for ANN search. PQ decomposes the vector space into several subspaces, and for each subspace generates one sub-codebook. The Cartesian product of all the sub-codebooks can generate enormous codewords so that a large scale dataset can be indexed exactly. In order to improve the performance of PQ, OPQ [6] finds the best partition of subspaces through an orthonormal matrix R, and OCKM [15] generates two codebooks for each subspace to further reduce the quantization error. DCPQ [13] constructs dual PQ quantizers to encode a vector and selects the one with lower quantization error as the codes of a vector. LOPQ [9] locally optimizes an individual PQ for per inverted list and use it to encode the residuals of the corresponding list.

AQ [3] is the representative method of linear addition. AQ generates several codebooks without any constraint, and then vectors are quantized using the sum of several codewords coming from different codebooks. Despite that the quantization error of AQ is smaller than PQ, the encoding method is very computationally expensive due to the high level of freedom of codewords combination. In addition, the distance computation is also less efficient since the codeword dimension is higher than PQ. Residual Vector Quantization (RVQ) [5] is another linear addition based method, which defines a hierarchy structure for the residual codebooks, and each layer encodes the residuals of the previous layers. However, due to the isolated training and encoding, the quantization error is high. And improvements based on RVQ have been proposed [1,12,16] to improve the encoding of vectors or the training of residual codebooks.

In summary, compared to the Cartesian product based methods, because of the more complex codebooks structure, linear addition based methods achieve lower quantization error and hence acquire higher retrieval accuracy, while they suffer from higher complexity of distance computation, which significantly decreases the search efficiency. In this paper, we propose a novel quantization method which utilizes the advantages of both the categories. The followings are the key contributions of this work:

- A novel Cartesian product based quantization method Residual Vector Product Quantization (RVPQ) is proposed, which achieves better retrieval accuracy for ANN search than the state-of-the-art methods.

- All the residual codebooks for each subspace are jointly optimized by minimizing the quantization error, using an efficient training algorithm to converge into a satisfying quantization error.
- An efficient encoding algorithm based on H-Variable Beam Search is proposed, which improves the encoding efficiency without increasing the quantization error.

The rest of this paper is organized as follows: In Sect. 2, the related work and some notations of vector quantization are introduced. In Sect. 3, the proposed residual vector product quantization is described, and the theoretical analysis is provided. In Sect. 4, the experiment setup and the datasets are described, and then the experiment results are shown and analyzed. Finally, Sect. 5 concludes this paper.

2   Related Work

Vector quantization [7] is a function Q that maps a space $X \subset \mathcal{R}^D$ into a finite codebook C of size K. Therefore, a vector x is presented by a binary indices of the nearest codeword c_k in codebook C and hence reducing the memory load. And Q is formally defined as $Q(x) = \underset{k \in \{1, \cdots, K\}}{argmin} \; dis\,(x, c_k)$, where dis is a distance function, which usually is the L_2 distance between the vector x and codewords c_k coming from C. The construction of the codebook is crucial, and the quality of vector quantization is measured by quantization error that is the mean squared error (MSE) between the vectors and their corresponding codewords. Therefore, vector quantization can be viewed as an optimization problem which optimizes the codebook by minimizing the MSE. And the MSE of N vectors is formally formulized as $\frac{1}{N} \sum_{i=1}^{N} \|x_i - c_{k_i}\|_2^2$, $k_i = Q(x_i)$. However, traditional vector quantization methods, such as K-means, only construct a single codebook, and thus they are impractical due to the high computation complexity when K is large, and hence are not suitable for ANN search which requires significant codewords to index a large scale dataset.

PQ [8] divides the space $X \subset \mathcal{R}^D$ into J subspaces and generates one sub-codebook of size K for each subspace. Then the Cartesian product of these J sub-codebooks, denoted as $\{C^1 \times C^2 \times \cdots \times C^J\}$, constructs the codebook C of PQ, which increases the codewords number from JK to K^J. When K is large, the amount of codewords is enough for ANN search. Each sub-vector x^j of x is quantized separately by the corresponding sub-codebook C^j. Therefore the MSE of PQ can be formulized as:

$$MSE_{PQ} = \frac{1}{N} \sum_{i=1}^{N} \sum_{j=1}^{J} \left\| x_i^j - c_{k_i}^j \right\|_2^2.$$

In the search stage of PQ, the distance between a query vector and any vectors in the database can be approximated by asymmetric distance computation

(ADC) [8], which computes the distances between the query and the codewords of the database vectors. And this can be efficiently done by table lookups, all distances between any sub-vector of the query and all the codewords from the corresponding sub-codebook are precomputed and stored in J tables of size $1 \times K$. Then the distance is obtained by the sum of J lookups.

RVQ [5] defines a residual hierarchy structure for M ordered residual code-books. The residual $r = x - \sum_{i=1}^{m-1} C_{x_k}^i$ obtained by the previous $m - 1$ layers is quantized as the nearest codeword coming from the corresponding residual codebook C^m. RVQ optimizes the quantization error by solving M subprob-lems, which are training the M residual codebooks successively. However, RVQ only gets the local optimal solution of codebooks and codes of vectors due to the interrelationship between residual codebooks and thus results in high quan-tization error.

3   Residual Vector Product Quantization

The framework of RVPQ is that the vector space is divided into J subpaces and then we construct a residual hierarchy structure consisting of M residual code-books, denoted as $C^{(j)} = \left\{ C_1^j, \cdots, C_M^j \right\}$, for each subspace. Subspace partition has a crucial role for minimizing quantization error and can be presented by an orthonormal matrix R initialized by Eigenvalue Allocation [6]. We optimize RVPQ using an alternative iteration strategy, thus all residual hierarchy struc-tures are trained independently by solving J subproblems that minimizing the quantization error with R fixed, and vice versa. Given N training vectors, we formulate the MSE of the jth subspace as follows:

$$\underset{RVPQ^{(j)}}{MSE} = \frac{1}{N} \sum_{i=1}^{N} \left\| Rx_i^{(j)} - \sum_{m=1}^{M} c_{k_i}^{m\,(j)} \right\|_2^2, \tag{1}$$

where $Rx_i^{(j)}$ represents the sub-vectors in the jth subspace and $c_{k_i}^{m\,(j)}$ represents the codeword selected from the mth residual codebook of the jth residual hier-archy structure. Therefore, the MSE of RVPQ is the sum of the quantization error of all the J subspaces. The training algorithm of RVPQ is described in the Algorithm 1, which consists of two sections. The function of the first section, from line 3 to 10, is to update the residual hierarchy structure for all the J sub-spaces, and the function of the second section, from line 10 to 12, is to optimize R. In the subsequent subsections, we describe in detail the learning for RVPQ.

3.1   Learning for Residual Hierarchy Structure

Recall that RVQ obtains the residual codebook for each layer separately by clustering the residual r computed by previous layers. However, the codebooks in the residual hierarchy structure are interrelated, even though the optimal

---

Algorithm 1: Training Procedure for RVPQ

Input: Training Set $X = \{x_1, \cdots, x_N\}$; Learning Rate γ; Mini-Batch Size S;
         The training times $Epoch$, the number of inner iteration I
Output: residual hierarchy structure $C = \left\{ C^{(1)}, C^{(2)}, \cdots, C^{(J)} \right\}$,
         transformation matrix R;

1 Initialize the residual hierarchy structure C and rotation matrix R;
2 **while** $Epoch \geq 1$ **do**
3   Project the training set X to obtain $\hat{X} = RX$;
4   partition the training set $\hat{X}$ into L mini-batches $\{MB_1, \cdots, MB_L\}$;
5   **for** *each subspace* $j \in [1, J]$ **do**
6   **for** *each mini-batch* MB_l **do**
7   encode the samples $\hat{x}_i^j \in MB_l$ by the H- Variable Beam Search;
8   **while** $I \geq 1$ **do**
9   calculate the gradient by formula (2), and update
         $C^{(j)} = \left\{ C_1^j, \cdots C_M^j \right\}$;
10                  $I - -$;

11   calculate the approximation Y of $\hat{X}$;
12   apply SVD to obtain $\hat{X}Y^T = U\Sigma V^T$;
13   update R by $R = UV^T$;
14   decrease the learning rate γ;
15      $Epoch - -$;
16 **return** $C = \left\{ C^{(1)}, C^{(j)}, \cdots, C^{(J)} \right\}$ and R;

---

residual codebook can be obtained for each layer, the whole residual hierarchy structure generally gets a local optimal solution which leads to high quantization error.

Therefore, in order to reduce the quantization error, the codebooks from a residual hierarchy structure should train jointly. This is a challenging optimization problem due to the large number of unknown parameters. We propose an efficient solution for this problem using mini-batch stochastic gradient descent. The formula (2) calculates the gradient of (1) about $C_m^{(j)}$, where b_m^j is a K-dimensional unit vector which denotes the indices of a codeword, and thus the code of sub-vector $Rx^{(j)}$ is presented as $\left[b_1^j, \cdots, b_m^j, \cdots b_M^j \right]$, and the codeword indexed by b_m^j can be obtained through the product $C_m^{(j)} b_m^j$. The training dataset is divided into L batches with size S. For the jth subproblem (1), all the training vectors from a mini-batch is utilized to calculate the gradients $\left\{ \nabla \left(C_1^{(j)} \right), \cdots, \nabla \left(C_M^{(j)} \right) \right\}$ using formula (2). Then all the residual codebooks $\left\{ C_1^{(j)}, \cdots, C_M^{(j)} \right\}$ of the jth residual hierarchy structure are updated simultaneously using formula (3), where γ is the learning rate and t represents the epoch number.

$$\nabla\left(C_m^{(j)}\right) - \frac{\partial}{\partial C_m^{(j)}}\left(\sum_{i=1}^{N}\left\|Rx_i{}^{(j)} - \sum_{m=1}^{M} C_m^{(j)} b_m^j\right\|_2^2\right) = \sum_{i=1}^{N} 2\left(\sum_{m=1}^{M} C_m^{(j)} b_m^j - Rx_i{}^{(j)}\right)(b_m^j)^T \quad (2)$$

$$C_m^{(j)}(t+1) = C_m^{(j)}(t) - \gamma_j(t)\ \nabla\left(C_m^{(j)}\right) \quad (3)$$

Hence, each epoch updates the residual codebooks L times. The larger S is, the less update frequency is, and hence reducing the communication cost. However, as S increases, the convergence level of quantization error rises because of the over-training for each mini-batch. This problem can be solved using EM-SGD [10] which sets a large S to improve the computation efficiency while retaining a satisfying convergence level. Thus we add a regularization term to prevent the codebooks from over-training in a single mini-batch for RVPQ. And hence we reformulate formula (3) as (4), where $\eta(t)$ is used to restrain the extent of variation of residual codebooks in each mini-batch.

$$C_m^{(j)}(t+1) = C_m^{(j)}(t) - \gamma_j(t)\ \left(\nabla\left(C_m^{(j)}\right) + \eta(t)\left(C_m^{(j)}(t+1) - C_m^{(j)}(t)\right)\right) \quad (4)$$

3.2 Encoding Algorithm

The quantization code of d-dimensional subvector $Rx^{(j)}$ consists of M codewords selected from the residual hierarchy structure of the jth subspace, where $d = D/J$. In fact, the encoding for $Rx^{(j)}$ is a combinatorial optimization problem which aims to find the best codewords combination from the corresponding residual hierarchy structure. We solve this problem by using a heuristic method which is similar to Beam Search used in AQ [3]. More specifically, we remain a number of candidates of codewords combinations with the minimum quantization error for each layer. In the first layer, we choose the $H = H_0$ nearest codewords to subvector $Rx^{(j)}$ from the corresponding codebook to construct H_0 candidates, and then these H_0 corresponding residuals are computed for the next layer. For the subsequent layers, one way proposed in [12] still remains the best H_0 candidates from H_0K combinations which are generated by adding the K codewords of the corresponding codebook to the H_0 candidates. After finishing the quantization for all the M layers, the candidate with the best quantization error is selected as the codes of subvector $Rx^{(j)}$. We name this method as H-Fixed Beam Search since the H for each layer is always H_0. Based on the hierarchy structure, this encoding method is more efficient than AQ which needs to search for the best H_0 candidates from $(M - m + 1)H_0K$ combinations for the mth layer.

However, the contribution of each layer to the quantization error is not equal, and the next layer contributes much less than the previous one. For instance, at the mth layer, adding an identical codeword to the H candidates, the candidate with a smaller quantization error is probably still smaller than the others. The deeper the layer is, the greater the probability is. Therefore, H-Fixed Beam Search involves many unnecessary computation at the deep layers, and this property suggests that it is not necessary to remain the fixed number of candidates for each layer. Since the upper layer has a more important contribution to the quantization error, we initialize H to a big value H_0 in order to remain enough candidates. With the deepening of the layer, we set $H = H_0/2$ to reduce the computation of quantization error from dH_0K to $dH_0K/2$, and hence improves the encoding efficiency. We name this method as H-Variable Beam Search, due to the variable number H of candidates for different layers.

When encoding the residual $r_m^{(j)}$ using the codebook of the mth layer, the quantization error between $r_m^{(j)}$ and codewords $c_{m,k}$ of the mth layer presented in (5) needs to be computed, where $r_m^{(j)} = Rx^{(j)} - \sum_{l=1}^{m-1} c_{l,k}$, $r_1^{(j)} = Rx^{(j)}$.

$$e\left(r_m^{(j)}, c_{m,k}\right) = \left\|r_m^{(j)}\right\|_2^2 - 2\left\langle Rx^{(j)}, c_{m,k}\right\rangle + 2\sum_{l=1}^{m-1}\left\langle c_{l,k}, c_{m,k}\right\rangle + \left\|c_{m,k}\right\|_2^2 \quad (5)$$

The first term is the length of the subvector $r_m^{(j)}$, which is calculated by the previous layer, and thus no extra computation is required. The second term is the inner product between $r_m^{(j)}$ and the codewords $c_{m,k}$ of the mth layer, which costs $O(dK)$. The third and the fourth terms are the inner product between $c_{l,k}$ and the codewords $c_{m,k}$ of the mth layer, cost $(m-1)dKH/2^{m-2}$. Based on the H-variable beam search, the best $H/2^{m-1}$ candidates are selected from $KH/2^{m-2}$ candidates, and it costs $\left(KH/2^{m-2}\right)\log\left(H/2^{m-1}\right)$. After finishing the quantization of all the M layers, the best codewords combination for $Rx^{(j)}$ is retained. The total computation complexity of J subvectors is presented in Table 1, and the computation complexity of H-fixed Beam Search and Beam Search are also presented. As the Table 1 shows, our encoding method is the most efficient.

Table 1. Computation complexity for different encoding algorithm

Method	Computation complexity for encoding Rx
H-variable BS	$O\left(MDK + K\log H + \sum_{m=2}^{M}\left((m-1)DKH/2^{m-2} + \left(KH/2^{m-2}\right)\log\left(H/2^{m-1}\right)\right)\right)$
H-fixed BS	$O\left(MDK + K\log H + \sum_{m=2}^{M}((m-1)DKH + KH\log H)\right)$
Beam search	$O\left(MDK + MK\log H + \sum_{m=2}^{M}(M-m+1)(DKH + KH\log H)\right)$

3.3   Learning for Transformation Matrix R

PQ [8] involves decomposing the D-dimensional data space into J subspaces. The experiment results of PQ suggest that the way of data space decomposition has a great impact on quantization error. For instance, vectors such as SIFT and GIST are structured descriptors, using a random order to divide the vectors instead of the natural order will lead to poor results, and using the structured order which groups together dimensions that are related can significantly improve the performance. However, the structured information of data is not always available in advance. Therefore, we view the space decomposition as an unkonwn parameter to be solved, which can be obtained by minimizing the objective function of quantization error. Specifically, we take a similar method proposed in [6]. Because of any reordering of the dimensions can be represented by an identity matrix with rows rearranged that is an orthonomal matrix, thus any space decomposition can be accomplished by an orthonormal matrix R. Therefore, the D-dimensional data space is first transformed by R, then the dimensions of transformed space are grouping in natural order into J subspaces.

Given the J residual hierarchy structures of RVPQ fixed, we optimize the orthonormal matrix R. The decoding of $Rx^{(j)}$ is the sum of codewords selected from the jth residual hierarchy structure, and then the decoding of Rx is presented as a vector $y = [\sum_{m=1}^{M} C_m^{(1)} b_m^1, \cdots, \sum_{m=1}^{M} C_m^{(j)} b_m^j, \cdots, \sum_{m=1}^{M} C_m^{(J)} b_m^J]$, which is the concatenation of the approximations of J subvectors.

We use a matrix $X \in \mathcal{R}^{D \times N}$ to represent the training set, in which each column is a training vector, and a matrix $Y \in \mathcal{R}^{D \times N}$ to represent the decoding of the training vectors. Thus the matrix form of MSE of RVPQ can be formulated as follows:

$$\min_{R} \underset{RVPQ}{MSE} = \|RX - Y\|_F^2 \text{ s.t. } R^T R = I. \qquad (6)$$

where $\|\cdot\|_F^2$ is the Frobenius norm, and this optimization problem is an Orthogonal Procrusters Problem which has a closed-form solution that can be solved by SVD, $XY^T = U\Sigma V^T$, so $R = VU^T$.

4 Experiment Results and Analysis

In this section, we evaluate our approach on three datasets[1], and the descriptions of these datasets are listed in Table 2. We use the *Recall@R* as the performance measurement of ANN search, it means the proportion of query vectors for which the nearest neighbor is ranked within the first R positions. Hence *Recall@R* represents the ratio of queries for which the nearest neighbor is retrieved correctly.

Table 2. Detail of the SIFT-1M and GIST-1M datasets

Dataset	SIFT-1M	GIST-1M	SIFT-100M
Descriptor dimension	128	960	128
Training samples	100000	500000	5000000
Database samples	1000000	1000000	100000000
Query samples	10000	1000	10000

4.1 Parameters Setting for RVPQ

We set $J = 2, M = 4, K = 256$ for 64bits codes and train the RVPQ using the Algorithm 1, which has several hyper-parameters including learning rate γ,

[1] All of datasets used in this section are available at http://corpus-texmex.irisa.fr/.

regularization coefficient η and mini-batch size S. And these hyper-parameters altogether decide the convergence level of quantization error. For epoch t, we use the $O(1/\sqrt{t})$ decay learning rate for EM-SGD, and hence $\gamma_l(t) = \gamma_l(0)\sqrt{\frac{\alpha}{t+\alpha}}$, where $\gamma_l(0)$ is the initial learning rate used to update the codebook of the lth layer for a subspace, and α corresponds to the decaying speed. Note that as the layer deepens, the corresponding magnitude of codebook decreases, and thus we set different initial learning rates for different layers of a subspace. Therefore, we define a total value $\gamma_{total} = \sum_{l=1}^{m} \gamma_l(0)$ and let $\gamma_l(0) = \frac{1}{\log_2 l+1}\gamma_0$. Then we search the range of $\gamma_{total} \in \{10^{-1}, \cdots, 10^{-5}\}$, $S \in \{10^2, 10^3, 10^4\}$, $\alpha \in \{10^0, \cdots, 10^4\}$ and $\eta \in \{10^0, \cdots, 10^5\}$.

Experimental results demonstrate that the quantization error is the best when $\gamma_{total} = 0.05, S = 10^3, \alpha = 10^4, \eta = 50$. We also compare the EM-SGD with SGD and the curves of quantization error are shown in Fig. 1. This figure shows that the convergence rate of EM-SGD is slower than SGD. However, as epoch increases, the quantization error of SGD converged to a higher level after about $150th$ epoch, while the error of EM-SGD still has a significant downward trend and eventually converged to a lower level after 350 epochs.

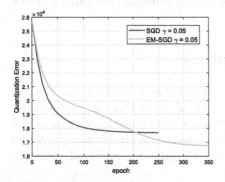

Fig. 1. Convergence curve of EM-SGD and SGD in SIFT-1M.

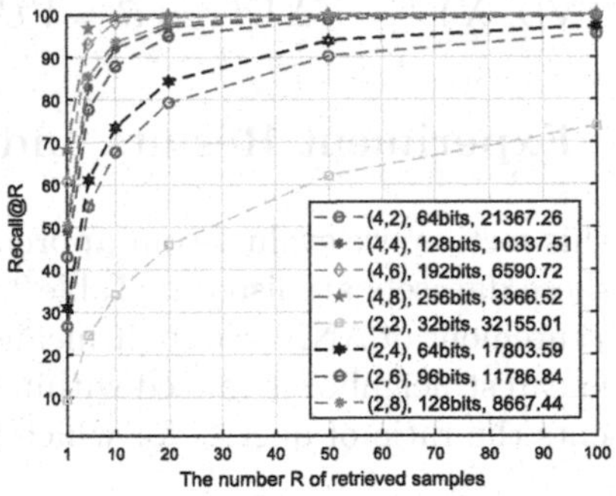

Fig. 2. The recall@R of exhaustive search on SIFT-1M.

4.2 Comparison of Encoding Algorithms

In this section, we compare our encoding method to the Beam Search [3] and the H-fixed Beam Search [12]. We encode SIFT-1M using the RVPQ based on different encoding methods, and then we record the quantization error and time consumption respectively. We test the performance of these encoding methods for $H_0 = 32$ and $H_0 = 64$, and the results are shown in Table 3, which shows that H-fixed Beam Search achieves the best quantization error. Although the encoding time of Beam Search is the longest, the quantization error is still higher than H-fixed Beam Search. It suggests that the high level freedom of codewords combination is not suitable for residual hierarchy structure. Note that our encoding method is the most efficient with only 1.3% and 0.8% loss of accuracy compared to the H-fixed Beam Search.

Table 3. The error and time cost for encoding the SIFT-1M (64bits)

	$H_0 = 32$		$H_0 = 64$	
	Time (s)	Error	Time (s)	Error
Beam search	14872.46	18405.48	38833.47	18325.38
H-fixed beam search	3906.60	**18375.01**	15023.68	**18304.68**
H-variable beam search	**2438.10**	18621.38	**4524.73**	18427.71

4.3 Exhaustive ANN Search

We perform exhaustive search for all query vectors of SIFT-1M and GIST-1M. We set $J = 2, M = 2$ for 32$bits$ codes and $J = 2, M = 4$ for 64$bits$ codes to train the RVPQ. And then the performance is compared to the state-of-the-art methods that are the Cartesian product based methods including PQ [8], OPQ [6], APQ [3] OCKM [15], DCPQ [13] and the linear addition based methods including RVQ [5], AQ [3] and CompQ [12]. The results of *Recall@R* are listed in Table 4, where the results of the other methods are taken from the original publications. Note that the *Recall@R* results of all methods are based on the identical datasets partition. In other words, the results are based on identical training set, database, and query set.

As the Table 4 shows, RVPQ outperforms all the state-of-the-art Cartesian product based methods. Furthermore, the recall of our method is comparable to AQ and Compq which have much higher distance computation complexity than RVPQ. Note that our method performs better than RVQ which suggests that the training for residual codebooks of residual hierarchy structure is not a separately optimal problem and should be solved jointly.

We also evaluate the impact of the number of subspaces J and the number of codebooks M in a subspace on quantization error and accuracy of ANN search. We set $J \in \{2, 4\}$ and $M \in \{2, 4, 6, 8\}$ to train the RVPQ respectively, and then we encode SIFT-1M to obtain the codes of diverse lengths ranging from 32 bits to 256 bits. We perform the exhaustive search on SIFT-1M using these codes of different lengths. Figure 2 shows the recall@R of different settings and the corresponding quantization error. As can be seen from Fig. 2, the longer the code length, the smaller the error and the higher the recall rate of the retrieval. In addition, for the same length of codes, $J = 2, M = 4$ performs better than $J = 4, M = 2$, while $J = 4, M = 4$ performs better than $J = 2, M = 8$. It suggests that given a code length, we should find a proper trade-off between J and M.

Table 4. Exhaustive search on SIFT-1M and GIST-1M

(a) *Recall@R* on SIFT-1M, 32bits

	recall@1	recall@10	recall@100
PQ	0.055	0.225	0.592
OPQ	0.068	0.263	0.658
OCKM	*NA*	0.349	0.741
RVPQ	**0.102**	**0.347**	**0.745**
AQ	*0.105*	*0.414*	*0.823*
CompQ	*0.135*	*0.435*	*0.818*

(b) *Recall@R* on GIST-1M, 32bits

	recall@1	recall@10	recall@100
PQ	0.024	0.056	0.155
OPQ	0.059	0.152	0.394
OCKM	*NA*	0.173	0.466
RVPQ	**0.067**	**0.177**	**0.471**
AQ	*0.069*	*0.188*	*0.465*
CompQ	*0.072*	*0.200*	*0.504*

(c) Recall@R on SIFT-1M, 64bits

	recall@1	recall@10	recall@100
PQ	0.220	0.605	0.921
OPQ	0.239	0.634	0.939
OCKM	0.279	0.683	0.945
APQ	0.297	0.740	0.972
DCPQ	0.254	0.656	0.943
RVPQ	**0.303**	**0.738**	**0.974**
RVQ	*0.259*	*0.658*	*0.954*
CompQ	*0.327*	*0.773*	*0.984*

(d) *Recall@R* on GIST-1M, 64bits

	recall@1	recall@10	recall@100
PQ	0.048	0.114	0.336
OPQ	0.117	0.336	0.713
OCKM	0.131	0.355	0.722
DCPQ	0.049	0.121	0.382
RVPQ	**0.146**	**0.397**	**0.765**
RVQ	*0.116*	*0.325*	*0.676*
CompQ	*0.155*	*0.419*	*0.801*

4.4 Non Exhaustive Search Based on Inverted Multi-index

The inverted multi-index (IMI) [2,11] is an efficient non-exhaustive search framework. IMI uses a PQ consisting of J sub-codebooks of size K to build the inverted file system. Each codeword of PQ is assigned a short list that contains all the vectors belonging to this codeword. For each query, IMI only reranks the vectors from a number of short lists instead of the whole database, and hence significantly improving the search efficiency. Based on the IMI, we perform the non-exhaustive search using the codes quantized by RVPQ, and we name this method as Multi-RVPQ-ADC.

During the non-exhaustive search, for each query, at first, a number of nearest codewords from PQ codebooks are found. Since the distance between the query and codewords of PQ can be obtained by table lookups, we can efficiently find the corresponding short lists. The total number of vectors extracted from these short lists is denoted as T. We only need to rerank the T vectors to retrieve the nearest neighbor for each query.

We set $J = 2, M = 1, K = 256$ for PQ to build the IMI, and $J = 2, M = 4, K = 256$ for 64*bits* codes for RVPQ on SIFT-1M. Table 5 shows the recall versus the T top re-ranked vectors. As can be seen, Multi-RVPQ-ADC significantly decreases the number of comparisons with a negligible drop in the Recall, and when $T = 10000$, the *Recall@R* is close to that of exhaustive search. In addition, we also evaluate the performance of Multi-RVPQ-ADC on a larger database SIFT-100M and compare it to the Multi-OPQ-ADC [6]. We also set $J = 2, M = 1, K = 256$ for PQ to build the IMI, and then we set $J = 8, M = 1, K = 256$ for OPQ, $J = 2, M = 4, K = 256$ for RVPQ. The results of *Recall@R* are listed in Table 5 which shows that RVPQ outperforms OPQ at the same IMI for non exhaustive search.

Table 5. Non-exhaustive search based on inverted multi-index

(a) Results of *Recall@R* for SIFT-1M

	T	recall@1	recall@10	recall@100
Multi-RVPQ-ADC	10000	30.88	72.05	91.54
Multi-RVPQ-ADC	30000	31.09	73.88	96.76
Multi-RVPQ-ADC	100000	31.04	73.99	97.46
Exhaustive	1000000	31.04	73.95	97.46

(b) Results of *Recall@R* for SIFT-100M

	T	recall@1	recall@10	recall@100
Multi-RVPQ-ADC	100000	18.680	53.250	74.070
Multi-RVPQ-ADC	1000000	20.150	60.630	91.420
Multi-OPQ-ADC	100000	8.870	30.810	62.110
Multi-OPQ-ADC	1000000	8.800	32.040	69.520

5   Conclusion

In this paper, a novel vector quantization method RVPQ is proposed to effectively solve ANN search tasks. The training algorithm of RVPQ achieves high training efficiency and satisfying quantization error. In addition, our encoding method improves the encoding efficiency with negligible loss of accuracy using the H-variable Beam Search. Due to these novel improvements, RVPQ achieves the state-of-the-art performance with comparable computation for ANN search.

Acknowledgement. The work was supported by the National Natural Science Foundation of China (No. 61702130), Guangxi Natural Science Foundation (Nos. 2020GXNSFAA297186, 2020GXNSFAA159137), Guangxi Project of technology base and special talent (No. AD19110022), Guangxi Science and Technology Major Project (No. 2018AA32001).

References

1. Ai, L., Yu, J., Wu, Z., He, Y., Guan, T.: Optimized residual vector quantization for efficient approximate nearest neighbor search. Multimed. Syst. **23**(2), 169–181 (2015). https://doi.org/10.1007/s00530-015-0470-9

2. Babenko, A., Lempitsky, V.S.: The inverted multi-index. In: 2012 IEEE Conference on Computer Vision and Pattern Recognition, Providence, RI, USA, 16–21 June 2012, pp. 3069–3076. IEEE Computer Society (2012)
3. Babenko, A., Lempitsky, V.S.: Additive quantization for extreme vector compression. In: 2014 IEEE Conference on Computer Vision and Pattern Recognition, CVPR 2014, Columbus, OH, USA, 23–28 June 2014, pp. 931–938. IEEE Computer Society (2014)
4. Böhm, C., Berchtold, S., Keim, D.A.: Searching in high-dimensional spaces: index structures for improving the performance of multimedia databases. ACM Comput. Surv. 33(3), 322–373 (2001)
5. Chen, Y., Guan, T., Wang, C.: Approximate nearest neighbor search by residual vector quantization. Sensors 10(12), 11259–11273 (2010)
6. Ge, T., He, K., Ke, Q., Sun, J.: Optimized product quantization for approximate nearest neighbor search. In: 2013 IEEE Conference on Computer Vision and Pattern Recognition, Portland, OR, USA, 23–28 June 2013, pp. 2946–2953. IEEE Computer Society (2013)
7. Gray, R., Neuhoff, D.: Quantization. IEEE Trans. Inf. Theory 44(6), 2325–2383 (1998)
8. Jégou, H., Douze, M., Schmid, C.: Product quantization for nearest neighbor search. IEEE Trans. Pattern Anal. Mach. Intell. 33(1), 117–128 (2011)
9. Kalantidis, Y., Avrithis, Y.: Locally optimized product quantization for approximate nearest neighbor search. In: 2014 IEEE Conference on Computer Vision and Pattern Recognition, CVPR 2014, Columbus, OH, USA, 23–28 June 2014, pp. 2329–2336. IEEE Computer Society (2014)
10. Li, M., Zhang, T., Chen, Y., Smola, A.J.: Efficient mini-batch training for stochastic optimization. In: Macskassy, S.A., Perlich, C., Leskovec, J., Wang, W., Ghani, R. (eds.) The 20th ACM SIGKDD International Conference on Knowledge Discovery and Data Mining, KDD 2014, New York, NY, USA, 24–27 August 2014, pp. 661–670. ACM (2014)
11. Noh, H., Kim, T., Heo, J.P.: Product quantizer aware inverted index for scalable nearest neighbor search. In: Proceedings of the IEEE/CVF International Conference on Computer Vision (ICCV), pp. 12210–12218, October 2021
12. Ozan, E.C., Kiranyaz, S., Gabbouj, M.: Competitive quantization for approximate nearest neighbor search. IEEE Trans. Knowl. Data Eng. 28(11), 2884–2894 (2016)
13. Pan, Z., Wang, L., Wang, Y., Liu, Y.: Product quantization with dual codebooks for approximate nearest neighbor search. Neurocomputing 401, 59–68 (2020)
14. Sloane, N., Wyner, A.: Coding theorems for a discrete source with a fidelity criterioninstitute of radio engineers, international convention record, vol. 7, p. 1959 (1993)
15. Wang, J., Wang, J., Song, J., Xu, X., Shen, H.T., Li, S.: Optimized cartesian k-means. IEEE Trans. Knowl. Data Eng. 27(1), 180–192 (2015)
16. Wei, B., Guan, T., Yu, J.: Projected residual vector quantization for ANN search. IEEE Multim. 21(3), 41–51 (2014)
17. Wu, Z., Yu, J.: Vector quantization: a review. Front. Inf. Technol. Electron. Eng. 20(4), 507–524 (2019). https://doi.org/10.1631/FITEE.1700833

Data Removal from an AUC Optimization Model

Jie Li, Jun-Qi Guo, and Wei Gao[✉]

National Key Laboratory for Novel Software Technology, Nanjing University,
Nanjing 210023, China
{lijie,guojq,gaow}@lamda.nju.edu.cn

Abstract. Our learned model may be required to make some dynamic adjustments owing to data removals in privacy, adversarial learning, etc. Previous studies on this issue mostly focus on the standard classification accuracy. This work takes one step on data removal for AUC optimization, where previous methods can not be applied directly since AUC is measured by a sum of losses defined over pairs of instances from different classes. We develop the Data Removal algorithm for AUC optimization (DRAUC), and the basic idea is to adjust the trained model according to the removed data, rather than retrain another model again from the scratch. Our algorithm only needs to maintain some data statistics, without storing the training data in memory. For high-dimensional data, we utilize the frequent direction algorithm to approximate the second-order statistics, and solve the numerical solution based on gradient descent so as to avoid calculating the inverse of Hessian matrix. We verify the effectiveness of the proposed DRAUC both theoretically and empirically.

Keywords: Machine learning · AUC optimization · Data removal

1 Introduction

Sometimes, it is necessary to make some dynamic adjustments for our learned model due to data removal or accesses right in real applications. For example, we may realize certain deceptive training data after getting a well-trained model in data-poisoning attack [16], and removing such deceptive data may yield better performance. Moreover, machine learning model may leak information on training data, and the data owners may demand the removal of their data from a trained model [20]. Actually, data removal has become some basic right, which has been written into the General Data Protection Regulation by European Union [21]. This remains an important problem on how to 'remove' some data from a trained model, rather than retrain another model from the scratch with expensive costs on computation and storage.

Much attention has been paid to data removal in recent years. Data removal is first formally proposed as a forgetting system that could forget certain data quickly [3]. In addition, Ginart et al. [9] devise a notion of removal efficiency and

J. Gama et al. (Eds.): PAKDD 2022, LNAI 13280, pp. 221–235, 2022.
https://doi.org/10.1007/978-3-031-05933-9_18

give two algorithms for data removal from k-means clustering models. Recently, the hessian-based methods are designed for removing certain data from linear model [10,13]. However, these proposed methods need access to the training data when updating the model, which needs expensive costs on computation and storage. To this end, Wu et al. [23] propose the DeltaGrad which stores all the gradients when training and approximates Hessian matrix using LBFS algorithm. Izzo et al. [12] also proposed an algorithm, which calculates and stores the Hessian of training data ahead, and updating it when removing data. The storage of all the gradients and Hessian both cost a lot, making it unsuitable for high-dimensional tasks. Previous studies on data removal focus on classification accuracy. However, the classification accuracy may be not appropriate for some applications with unevenly-distributed classes; for example, some categories may have more instances than others in class-imbalanced tasks such as face detection and collaborative filtering, and the level of imbalance can be as high as 10^6 [22], where the model is almost determined by the data from the majority class.

AUC (Area Under the ROC Curve) has been a preferable measure for class-imbalanced learning, cost-sensitive learning, information retrieval, etc., [1,5,18]. It is measured by a sum of losses defined over pairs of instances from different classes, which is different from accuracy defined on a single instance, making it challenging to develop algorithms for large-scale datasets. Various algorithms have been developed for AUC optimization. Herschtal and Raskutti [11] uses gradient descent to optimize an approximate objective of pairwise loss. Gao et al. [6] develop an one-pass AUC maximization method that maintains and updates the mean and covariance of training data with $O(d^2)$ space and per-iteration time complexity. In recent years, Ying et al. [24] formulates the minimization of the AUC loss into an equivalent min-max optimization problem and use primal-dual SGD algorithm, which has $O(d)$ space and per-iteration time complexity. The followed works [14,17] also use the min-max formulation and give faster convergence rate. However, previous works mainly focus on AUC maximization, and it remains open for data removal from an AUC optimization model.

This work takes one step for data removals from an AUC optimization model, and the main contributions can be summarized as follows:

- We develop an algorithm on Data Removal from an AUC optimization model (DRAUC) and the basic idea is to adjust the trained model using the removed data, rather than retrain another model from scratch, which only needs to maintain some data statistics, without storing the training data.
- For high-dimensional tasks, we introduce the frequent direction algorithm to approximate the second-order statistics with low-rank matrix, and look for the numerical solution based on gradient descent so as to avoid the direct calculation and storage of the inverse of Hessian matrix.
- Theoretically, we show that our DRAUC method, with exact data statistics, could recover the model retrained from the scratch, and further present the convergence analysis for high-dimensional tasks, where we approximate the data statistics and obtain numerical solution.

- We finally provide extensive experiments to support the effectiveness of the proposed DRAUC method.

2   Preliminaries

Let $\mathcal{X} \subseteq [0,1]^d$ and $\mathcal{Y} = \{+1, -1\}$ denote the instance and label space, respectively. Suppose that $\mathcal{D}$ is an underlying (unknown) distribution over the product space $\mathcal{X} \times \mathcal{Y}$. Note that the distribution $\mathcal{D}$ is unknown in practice, and what we can observe is a training sample $S_n = \{(\mathbf{x}_1, y_1), (\mathbf{x}_2, y_2), \dots, (\mathbf{x}_n, y_n)\}$, where each element is drawn independently and identically (i.i.d.) from distribution $\mathcal{D}$. We could divide S_n into $S_n^+ = \{(\mathbf{x}_1^+, +1), (\mathbf{x}_2^+, +1), \dots, (\mathbf{x}_{n_+}^+, +1)\}$, and $S_n^- = \{(\mathbf{x}_1^-, -1), (\mathbf{x}_2^-, -1), \dots, (\mathbf{x}_{n_-}^-, -1)\}$, where n_+ and n_- are the cardinalities of positive and negative instances, respectively, and $n = n_+ + n_-$. We further derive their first- and second-order statistics as follows:

$$\mathbf{c}^+ = \sum_{i=1}^{n_+} \frac{\mathbf{x}_i^+}{n_+}, \quad \mathcal{S}^+ = \sum_{i=1}^{n_+} \frac{\mathbf{x}_i^+ [\mathbf{x}_i^+]^\top - \mathbf{c}^+ [\mathbf{c}^+]^\top}{n_+},$$

$$\mathbf{c}^- = \sum_{j=1}^{n_-} \frac{\mathbf{x}_j^-}{n_-}, \quad \mathcal{S}^- = \sum_{j=1}^{n_-} \frac{\mathbf{x}_j^- [\mathbf{x}_j^-]^\top - \mathbf{c}^- [\mathbf{c}^-]^\top}{n_-},$$

that is, the mean and covariance matrices of positive and negative training instances, which is easy to obtain when training the model.

Let $f: \mathcal{X} \to \mathbb{R}$ be a score function, and the AUC of f over S_n is given by

$$\text{AUC}(f; S_n) = \sum_{i=1}^{n_+} \sum_{j=1}^{n_-} \frac{\mathbb{I}[f(\mathbf{x}_i^+) > f(\mathbf{x}_j^-)] + \mathbb{I}[f(\mathbf{x}_i^+) = f(\mathbf{x}_j^-)]/2}{n_+ n_-},$$

where $\mathbb{I}[\cdot]$ is the indicator function which returns 1 if the argument is true; and 0 otherwise. This work focuses on linear hypothesis space $\mathcal{W} = \{\mathbf{w} \in \mathbb{R}^d\}$ for simplicity, and it can be extended to deep neural networks by treating the final layer as a linear model and fixing the other layers [12].

Direct optimization of AUC often yields an NP-hard problem since it can be cast into a combinatorial optimization problem. A feasible solution in practice is to optimize some pairwise surrogate losses as follows:

$$\mathcal{L}(\mathbf{w}; S_n) = \sum_{i=1}^{n_+} \sum_{j=1}^{n_-} \frac{\ell(\mathbf{w}^\top (\mathbf{x}_i^+ - \mathbf{x}_j^-))}{2n_+ n_-} + \frac{\lambda}{2} \|\mathbf{w}\|^2, \tag{1}$$

where ℓ is a convex loss function, and λ denotes the regularized parameter.

A function $f : \mathcal{W} \to \mathbb{R}$ is said to be L-Lipschitz continuous for constant $L > 0$, if it holds that $|f(\mathbf{w}) - f(\mathbf{w}')| \leq L \|\mathbf{w} - \mathbf{w}'\|$ for every $\mathbf{w}, \mathbf{w}' \in \mathcal{W}$. A function $f : \mathcal{W} \to \mathbb{R}$ is said to be μ-strongly convex for constant $\mu > 0$, if it holds that $f(\mathbf{w}) - f(\mathbf{w}') \geq \nabla f(\mathbf{w}')^\top (\mathbf{w} - \mathbf{w}') + \mu \|\mathbf{w} - \mathbf{w}'\|^2 / 2$ for every $\mathbf{x}, \mathbf{x}' \in \mathcal{W}$. A function $f : \mathcal{W} \to \mathbb{R}$ is said to be β-smooth for constant $\beta > 0$, if it holds that $\|\nabla f(\mathbf{w})) - \nabla f(\mathbf{w}')\| \leq \beta \|\mathbf{w} - \mathbf{w}'\|$ for every $\mathbf{w}, \mathbf{w}' \in \mathcal{W}$.

Let $|A|$ be the cardinality of set A, and denote by $[n] = \{1, 2, \ldots, n\}$ for integer $n > 0$. Let $\mathbf{I}_d$ and $\mathbf{0}_d$ denote the identity matrix of size $d \times d$ and the zero vector of size d, respectively. Let $\|\cdot\|$ denote the Euclidian L_2 norm of a vectors: $\|\mathbf{x}\| = \sqrt{\sum_i x_i^2}$, and $\lfloor r \rfloor$ is the largest integer which is no more than r.

3   Data Removal for AUC Optimization

This section proposes the data removal framework for AUC optimization. We consider square loss $\ell(\mathbf{w}^\top(\mathbf{x}_i^+ - \mathbf{x}_j^-)) = (1 - \mathbf{w}^\top(\mathbf{x}_i^+ - \mathbf{x}_j^-))^2$ for AUC optimization, and the surrogate loss of Eq. (1) over sample S_n can be given by

$$\mathcal{L}(\mathbf{w}; S_n) = \sum_{i=1}^{n_+} \sum_{j=1}^{n_-} \frac{(1 - \mathbf{w}^\top(\mathbf{x}_i^+ - \mathbf{x}_j^-))^2}{2n_+ n_-} + \frac{\lambda}{2}\|\mathbf{w}\|^2.$$

Notice that square loss has been shown to be statistically consistent with AUC [7], and has been used for a series of studies on AUC optimization [15,19,24].

Based on empirical risk minimization, we can get the optimal linear model as $\mathbf{w}^* = \arg\min_{\mathbf{w}}\{\mathcal{L}(\mathbf{w}; S_n)\}$ over the full training data S_n. It is possible to make some necessary adjustments in real applications after we train the model $\mathbf{w}^*$. For example, it is necessary to remove some noisy and incorrect training data, which are detected after training; and we may be required to delete partial Internet data from the data-owners due to privacy.

Let $R \subset S_n$ be the removed data from training sample S_n, and the updated classifier $\mathbf{w}_R^*$ can be given by $\mathbf{w}_{-R}^* = \arg\min_{\mathbf{w}}\{\mathcal{L}(\mathbf{w}; S_n\backslash R)\}$. We could simply and directly retrain the linear model $\mathbf{w}_{-R}^*$ from training data $S_n\backslash R$, while this requires to keep the entire training dataset, and take the same computational cost on retraining of linear model $\mathbf{w}_{-R}^*$ as that of $\mathbf{w}^*$. Therefore, the goal of this work is to get the model $\mathbf{w}_{-R}^*$ directly from the trained model $\mathbf{w}^*$ and removed data R, rather than retrain the model $\mathbf{w}_{-R}^*$ from training data $S_n\backslash R$.

For simplicity, we first consider the data removal of one single instance, and assume that some positive instance $(\mathbf{x}_i^+, +1)$ is removed from training data S_n without loss of generality. We then try to solve the following problem:

$$\mathbf{w}_{-i}^* = \arg\min_{\mathbf{w}} \left\{\mathcal{L}(\mathbf{w}; S_n\backslash(\mathbf{x}_i^+, +1))\right\} = \arg\min_{\mathbf{w}} \left\{\frac{n_+\mathcal{L}(\mathbf{w}; S_n) - \mathcal{L}_i^+(\mathbf{w})}{n_+ - 1}\right\}, \quad (2)$$

with $\mathcal{L}_i^+(\mathbf{w}) = \sum_{j=1}^{n_-}(1 - \mathbf{w}^\top(\mathbf{x}_i^+ - \mathbf{x}_j^-))^2/n_- + \lambda\|\mathbf{w}\|^2/2$. Motivated from [4,13], our basic idea is to first minimize $n_+\mathcal{L}(\mathbf{w}; S_n) + \epsilon\mathcal{L}_i^+(\mathbf{w})$ with some perturbation ϵ, and then approximate $\mathbf{w}_{-i}^*$ according to Taylor expansion with the selection of $\epsilon = -1$. Specifically, we have

$$\hat{\mathbf{w}}_{-i} = \mathbf{w}^* + \left[H_{\mathbf{w}^*}^{-i}\right]^{-1}\nabla\mathcal{L}_i^+(\mathbf{w}^*)/(n_+ - 1),$$

where $H_{\mathbf{w}^*}^{-i} = \left[\partial^2\mathcal{L}(\mathbf{w}; S_n\backslash(\mathbf{x}_i^+, +1))/\partial\mathbf{w}^2\right]_{\mathbf{w}=\mathbf{w}^*}$. It is easy to calculate

$$H_{\mathbf{w}^*}^{-i} = \mathcal{S}_{-i}^- + (\mathbf{c}_{-i}^- - \mathbf{c}_{-i}^+)(\mathbf{c}_{-i}^- - \mathbf{c}_{-i}^+)^\top + \mathcal{S}_{-i}^+ + \lambda\mathbf{I}_d,$$

$$\nabla\mathcal{L}_i^+(\mathbf{w}^*) = \lambda\mathbf{w}^* - \mathbf{x}_i^+ + \mathbf{c}_{-i}^- + (\mathbf{x}_i^+ - \mathbf{c}_{-i}^-)(\mathbf{x}_i^+ - \mathbf{c}_{-i}^-)^\top\mathbf{w}^* + \mathcal{S}_{-i}^-\mathbf{w}^*,$$

Algorithm 1. Data Removal algorithm for AUC optimization (DRAUC)

Input: Trained model $\mathbf{w}^*$, some data statistics n_+, n_-, $\mathbf{c}^-$, $\mathbf{c}^+$, $\mathcal{S}^-$ and $\mathcal{S}^+$, and removed data R

1: Count r_+ and r_- the cardinalities of positive and negative instances in R, resp.
2: Compute the mean $\mathbf{c}^+_{-R}$ and $\mathbf{c}^-_{-R}$ from Eq. (3) and (4) , resp.
3: Compute the covariance matrices $\mathcal{S}^+_{-R}$ and $\mathcal{S}^-_{-R}$ from Eqs. (5) and (6), resp.
4: Compute the model $\hat{\mathbf{w}}_{-R}$ from Eq. (7)

Output: Model $\hat{\mathbf{w}}_{-R}$

---

where $\mathbf{c}^+_{-i}$, $\mathbf{c}^-_{-i}$, $\mathcal{S}^+_{-i}$ and $\mathcal{S}^-_{-i}$ denote the mean and covariance matrices of positive and negative instances in training data $S_n \backslash (\mathbf{x}^+_i, +1)$, respectively.

We now consider the data removal of a subset $R \subset S_n$. Let r_+ and r_- be the cardinalities of positive and negative instances in R, respectively. Denote by $\mathbf{c}^+_{-R}$, $\mathbf{c}^-_{-R}$, $\mathcal{S}^+_{-R}$ and $\mathcal{S}^-_{-R}$ the mean and covariance matrices of positive and negative instances in data $S_n \backslash R$, respectively. It is easy to derive

$$\mathbf{c}^+_{-R} = \mathbf{c}^+ - \sum_{(\mathbf{x}_i, y_i) \in R} \frac{(\mathbf{x}_i - \mathbf{c}^+)\mathbb{I}[y_i = +1]}{n_+ - r_+}, \tag{3}$$

$$\mathbf{c}^-_{-R} = \mathbf{c}^- - \sum_{(\mathbf{x}_j, y_j) \in R} \frac{(\mathbf{x}_j - \mathbf{c}^-)\mathbb{I}[y_j = -1]}{n_- - r_-}, \tag{4}$$

and

$$\mathcal{S}^+_{-R} = \frac{n_+(\mathcal{S}^+ + \mathbf{c}^+[\mathbf{c}^+]^\top)}{n_+ - r_+} - \mathbf{c}^+_{-R}[\mathbf{c}^+_{-R}]^\top - \sum_{(\mathbf{x}_i, y_i) \in R} \frac{\mathbf{x}_i \mathbf{x}_i^\top \mathbb{I}[y_i = +1]}{n_+ - r_+}, \tag{5}$$

$$\mathcal{S}^-_{-R} = \frac{n_-(\mathcal{S}^- + \mathbf{c}^-[\mathbf{c}^-]^\top)}{n_- - r_-} - \mathbf{c}^-_{-R}[\mathbf{c}^-_{-R}]^\top - \sum_{(\mathbf{x}_j, y_j) \in R} \frac{\mathbf{x}_j \mathbf{x}_j^\top \mathbb{I}[y_j = -1]}{n_- - r_-}. \tag{6}$$

We aim to solve $\mathbf{w}^*_{-R} = \arg\min_{\mathbf{w}} \mathcal{L}(\mathbf{w}; S_n \backslash R)$ with

$$\mathcal{L}(\mathbf{w}; S_n \backslash R) = \sum_{\substack{y_i = +1 \\ (\mathbf{x}_i, y_i) \in S_n \backslash R}} \sum_{\substack{y_j = -1 \\ (\mathbf{x}_j, y_j) \in S_n \backslash R}} \frac{\ell(\mathbf{w}^\top (\mathbf{x}^+_i - \mathbf{x}^-_j))}{2(n_+ - r_+)(n_- - r_-)} + \frac{\lambda}{2}\|\mathbf{w}\|^2.$$

Denote $\Delta(\mathbf{w}) = n_+ n_- \mathcal{L}(\mathbf{w}, S_n) - (n_+ - r_+)(n_- - r_-)\mathcal{L}(\mathbf{w}; \mathcal{S} \backslash R)$. We first solve $\min_{\mathbf{w}} \{n_+ n_- \mathcal{L}(\mathbf{w}; S_n) + \epsilon \Delta(\mathbf{w})\}]$ for some perturbation ϵ, and then approximate $\mathbf{w}^*_{-R}$ according to Taylor expansion with the selection of $\epsilon = -1$. Specifically, we have

$$\hat{\mathbf{w}}_{-R} = \mathbf{w}^* + \frac{[\partial^2 \mathcal{L}(\mathbf{w}; S_n \backslash R)/\partial \mathbf{w}^2]^{-1}_{\mathbf{w}=\mathbf{w}^*} \times [\partial \Delta(\mathbf{w})/\partial \mathbf{w}]_{\mathbf{w}=\mathbf{w}^*}}{(n_+ - r_+)(n_- - r_-)}, \tag{7}$$

where

$$\left[\frac{\partial^2 \mathcal{L}(\mathbf{w}; S_n \backslash R)}{\partial \mathbf{w}^2}\right]_{\mathbf{w}=\mathbf{w}^*} = \mathcal{S}^-_{-R} + (\mathbf{c}^-_{-R} - \mathbf{c}^+_{-R})(\mathbf{c}^-_{-R} - \mathbf{c}^+_{-R})^\top + \mathcal{S}^+_{-R} + \lambda \mathbf{I}_d$$

Algorithm 2. Frequent Directions

Input: instance $\mathbf{x}$, sketch $Z \in \mathbb{R}^{d \times m}$

1: Insert x into a zero-valued column of Z
2: **if** Z has no zero-valued column **then**
3:     $[U, \Sigma, V] \leftarrow \text{SVD}(Z)$
4:     $\delta = \Sigma_{\lfloor m/2 \rfloor, \lfloor m/2 \rfloor}$
5:     $Z = U\sqrt{\max(\Sigma^2 - \mathbf{I}_{d \times m}\delta^2, \mathbf{0}_{d \times m})}$ (max($\cdot$) denotes an element-wise maximum)
6: **end if**

Output: Z

---

and $[\partial \Delta(\mathbf{w})/\partial \mathbf{w}]_{\mathbf{w}=\mathbf{w}^*}$ equals to

$$n_- \sum_{\substack{y_i=+1 \\ (\mathbf{x}_i, y_i) \in R}} \left(\lambda \mathbf{w}^* - \mathbf{x}_i^+ + \mathbf{c}^- + (\mathbf{x}_i^+ - \mathbf{c}^-)(\mathbf{x}_i^+ - \mathbf{c}^-)^\top \mathbf{w}^* + \mathcal{S}^- \mathbf{w}^* \right)$$

$$+ (n_+ - r_+) \sum_{\substack{y_j=-1 \\ (\mathbf{x}_j, y_j) \in R}} \left(\lambda \mathbf{w}^* + \mathbf{x}_j^- - \mathbf{c}_{-R}^+ + (\mathbf{x}_j^- - \mathbf{c}_{-R}^+)(\mathbf{x}_j^- - \mathbf{c}_{-R}^+)^\top \mathbf{w}^* + \mathcal{S}_{-R}^+ \mathbf{w}^* \right).$$

Algorithm 1 presents the detailed description of the Data Removal algorithm for AUC optimization, short for DRAUC. As can be seen, our algorithm does not retrain the model $\mathbf{w}^*_{-R} = \arg\min_{\mathbf{w}} \mathcal{L}(\mathbf{w}; S_n \backslash R)$ from training data $S_n \backslash R$, but to learn from the trained model $\mathbf{w}^*$ and removed data R, and merely need to store the data statistics instand of the entire training data S_n.

3.1 Deal with High Dimension

The limitations of Algorithm 1 include the $O(d^2)$ storage cost for covariance and Hessian matrices, as well as the $O(d^3)$ computational cost for the inverse of Hessian matrix, making it unsuitable for high-dimensional data.

For the storage of covariance matrices, we take a deterministic sketching method from [8] to approximate the covariance matrices by low-rank matrices. Let X_n^+ and X_n^- denote the matrices of positive and negative instances in S_n, respectively. For covariance matrix, we have

$$\mathcal{S}^+ = X_n^+ [X_n^+]^\top / n_+ - \mathbf{c}^+[\mathbf{c}^+]^\top \quad \text{and} \quad \mathcal{S}^- = X_n^- [X_n^-]^\top / n_- - \mathbf{c}^-[\mathbf{c}^-]^\top.$$

To represent the high-dimensional matrix $\mathcal{S}^+$, we approximate $X_n^+[X_n^+]^\top$ by $Z^+[Z^+]^\top$ of low-rank $d \times m$ sketch matrix Z^+ ($m \ll d$). More precisely, we initialize the sketch matrix $Z^+ = \mathbf{0}_{d \times m}$, and insert instance $\mathbf{x}_i$ into a zeros valued column of sketch matrix Z^+ every time. If Z^+ has no zero-valued column, then we utilize SVD to rotate Z^+ with orthogonal columns and descending magnitude order, and the sketch columns norms are "shrunk" so that at least half of them is set to zero. We could approximate $X_n^-[X_n^-]^\top$ similarly. Algorithm 2 presents the detailed description of frequent direction.

The covariance matrices S^+ and S^- can be approximated, respectively, by

$$\hat{S}^+ = Z^+[Z^+]^\top/n_+ - \mathbf{c}^+[\mathbf{c}^+]^\top \quad \text{and} \quad \hat{S}^- = Z^-[Z^-]^\top/n_- - \mathbf{c}^-[\mathbf{c}^-]^\top .$$

Given the removed dataset R, we could further approximate the covariance matrix S^+_{-R} and S^-_{-R} of data $S_n \backslash R$, respectively, by

$$\hat{S}^+_{-R} = \frac{Z^+[Z^+]^\top}{n_+ - r_+} - \mathbf{c}^+_{-R}[\mathbf{c}^+_{-R}]^\top - \sum_{\substack{(\mathbf{x}_i,y_i) \in R}} \frac{\mathbf{x}_i\mathbf{x}_i^\top \mathbb{I}[y_i = +1]}{n_+ - r_+}, \tag{8}$$

$$\hat{S}^-_{-R} = \frac{Z^-[Z^-]^\top}{n_- - r_-} - \mathbf{c}^-_{-R}[\mathbf{c}^-_{-R}]^\top - \sum_{\substack{(\mathbf{x}_j,y_j) \in R}} \frac{\mathbf{x}_j\mathbf{x}_j^\top \mathbb{I}[y_j = -1]}{n_- - r_-}. \tag{9}$$

This follows that we can approximate $[\partial\Delta(\mathbf{w})/\partial\mathbf{w}]_{\mathbf{w}=\mathbf{w}^*}$ by

$$n_- \sum_{\substack{y_i=+1 \\ (\mathbf{x}_i,y_i) \in R}} \left(\lambda\mathbf{w}^* - \mathbf{x}_i^+ + \mathbf{c}^- + (\mathbf{x}_i^+ - \mathbf{c}^-)(\mathbf{x}_i^+ - \mathbf{c}^-)^\top\mathbf{w}^* + \hat{S}^-\mathbf{w}^*\right)$$

$$+ (n_+ - r_+) \sum_{\substack{y_j=-1 \\ (\mathbf{x}_j,y_j) \in R}} \left(\lambda\mathbf{w}^* + \mathbf{x}_j^- - \mathbf{c}^+_{-R} + (\mathbf{x}_j^- - \mathbf{c}^+_{-R})(\mathbf{x}_j^- - \mathbf{c}^+_{-R})^\top\mathbf{w}^* + \hat{S}^+_{-R}\mathbf{w}^*\right),$$

and

$$\left[\frac{\partial^2 \mathcal{L}(\mathbf{w}; S_n \backslash R)}{\partial\mathbf{w}^2}\right]_{\mathbf{w}=\mathbf{w}^*} \approx \hat{S}^-_{-R} + (\mathbf{c}^-_{-R} - \mathbf{c}^+_{-R})(\mathbf{c}^-_{-R} - \mathbf{c}^+_{-R})^\top + \hat{S}^+_{-R} + \lambda\mathbf{I}_d.$$

where $\Delta(\mathbf{w}) = n_+ n_- \mathcal{L}(\mathbf{w}, S_n) - (n_+ - r_+)(n_- - r_-)\mathcal{L}(\mathbf{w}, S\backslash R)$. We obtain the final model

$$\hat{\mathbf{w}}_{-R} = \mathbf{w}^* + \frac{[\partial^2\mathcal{L}(\mathbf{w}; S_n\backslash R)/\partial\mathbf{w}^2]^{-1}_{\mathbf{w}=\mathbf{w}^*} \times [\partial\Delta(\mathbf{w})/\partial\mathbf{w}]_{\mathbf{w}=\mathbf{w}^*}}{(n_+ - r_+)(n_- - r_-)}. \tag{10}$$

Notice that we do not require to calculate and store $\hat{S}^+$, $\hat{S}^-$, $\hat{S}^+_{-R}$ and $\hat{S}^-_{-R}$, but to maintain Z^+, Z^- and removed data R in memory. We take the $O(md)$ computational cost instead of $O(d^2)$ by using the trick $A[A]^\top\mathbf{w} = A([A]^\top\mathbf{w})$, where $A \in \mathbb{R}^{d\times 1}$ or $A \in \mathbb{R}^{d\times m}$.

Another challenge is to calculate $[\partial^2\mathcal{L}(\mathbf{w}; S_n\backslash R)/\partial\mathbf{w}^2]^{-1}_{\mathbf{w}=\mathbf{w}^*}$, i.e., the inverse of Hessian. Here, we introduce an auxiliary function as follows:

$$f(\mathbf{w}) = \mathbf{w}^\top A\mathbf{w}/2 + \mathbf{b}^\top\mathbf{w} + c \quad \text{(for positive semi-definite matrix } A\text{)}, \tag{11}$$

and minimizing $f(\mathbf{w})$ gives the closed-form solution $\mathbf{w}^* = A^{-1}\mathbf{b}$. In practice, it is difficult to calculate A^{-1} directly, especially for high-dimensional matrix. We could instead calculate the numerical solution based on gradient descent.

Precisely, we set $A = [\partial^2\mathcal{L}(\mathbf{w}; S_n\backslash R)/\partial\mathbf{w}^2]_{\mathbf{w}=\mathbf{w}^*}$ and $b = [\partial\Delta(\mathbf{w})/\partial\mathbf{w}]_{\mathbf{w}=\mathbf{w}^*}$ in Eq. (11), and calculate the gradient $\nabla f(\mathbf{w}) = A\mathbf{w} + \mathbf{b}$ with $O(md)$ computational cost. We finally solve the numerical solution of $\mathbf{w}^* = A^{-1}\mathbf{b}$ base on gradient descent with several iterations.

To implement this approach, we just maintain $\mathbf{c}^+$, $\mathbf{c}^-$, Z^+ and Z^- in memory, and calculate $\mathbf{c}^+_{-R}$ and $\mathbf{c}^-_{-R}$ from Eq. (3) and (4). We finally obtain the output model $\hat{\mathbf{w}}_{-R}$ from Eq. (10) based on gradient descent with T iterations.

Table 1. Benchmark datasets

Datasets	# instance	# feature	Datasets	# instance	# feature	Datasets	# instance	# feature
vehicle	846	18	letter	15,000	16	acoustic	78,823	50
splice	3,175	60	a9a	32,561	123	aloi	108,000	128
satimage	4,435	36	w8a	49,749	300	ijcnn1	141,691	22
usps	7,291	256	cod-rna	59,535	8	skin	245,057	3
phishing	11,055	68	connect-4	67,557	126	covtype	581,012	54

4   Theoretical Results

This section presents the main theoretical results for our proposed algorithm. We first present the theoretical analysis on Algorithm 1 as follows:

Theorem 1. *Let $\hat{\mathbf{w}}_{-R}$ be the output model of Algorithm 1, and let $\mathbf{w}^*_{-R}$ be the retrained model from data $S_n \backslash R$. We have $\hat{\mathbf{w}}_{-R} = \mathbf{w}^*_{-R}$.*

This can be obtained from that the hessian-based method gives an exact local quadratic approximation for square loss. From this theorem, we can see that the output model $\hat{\mathbf{w}}_{-R}$ of Algorithm 1 is exactly the same as the model $\mathbf{w}^*_{-R}$, which is retrained from remaining training data $S_n \backslash R$. In contrast, our algorithm merely stores the data statistics in memory and does not need access to the remaining training data $S_n \backslash R$.

We now present the convergence analysis for high-dimensional tasks, where the covariance matrices are appproximated by low-rank matrices.

Theorem 2. *Suppose $\|\mathbf{w}\| \leq B$. Let $\hat{\mathbf{w}}_{-R}$ be the output of Eq. (10) from trained model $\mathbf{w}^*$ and removed data R, and let $\mathbf{w}^*_{-R}$ be the retrained model from data $S_n \backslash R$. We have*

$$\|\hat{\mathbf{w}}_{-R} - \mathbf{w}^*_{-R}\| \leq \frac{2\tau B}{\lambda m} + \left(\frac{2}{\lambda+2}\right)^T \left(\frac{2\tau B}{\lambda m} + \frac{4B(4+\lambda)(r_+n_- + r_-n_+ - r_+r_-)}{\lambda n_+ n_-}\right),$$

where T is the iteration number of gradient descent, m is the sketch size of frequent direction, $\tau = \max(rank(X_n^+[X_n^+]^\top), rank(X_n^-[X_n^-]^\top))$, and λ is the regularization parameter.

The detailed proof is presented in the Appendix. This theorem shows that the convergence of approximation algorithm depends on the iteration number t of gradient descent, sketch size m, the rank τ of training data, and the proportion of removed data $(r_+n_- + r_-n_+ - r_+r_-)/(n_+n_-)$.

5   Experiment

In this section, we evaluate the performance of DRAUC on benchmark datasets in Sect. 5.1, and present an evaluation and parameter influence on high-dimensional sparse datasets in Sects. 5.2. The program code for replicating our experiments is available on Github[1].

---

[1] https://github.com/sven-lijie/DRAUC.

Table 2. Distance, Test AUC and Running Time for Retraining and DRAUC with 10% removal rate over benchmark datasets

Dataset	Distance		Test AUC		Running time		
	$\|\mathbf{w}^* - \mathbf{w}^*_{-R}\|$	$\|\hat{\mathbf{w}}_{-R} - \mathbf{w}^*_{-R}\|$	Retraining	DRAUC	Retraining	DRAUC	Speed-up
vehicle	7.3×10^{-3}	1.0×10^{-6}	$.7872 \pm .0050$	$.7872 \pm .0050$	0.061	0.002	$30.5\times$
splice	3.3×10^{-2}	2.7×10^{-6}	$.9239 \pm .0056$	$.9239 \pm .0056$	0.023	0.001	$23.0\times$
satimage	2.4×10^{-2}	4.5×10^{-6}	$.9123 \pm .0088$	$.9123 \pm .0088$	0.027	0.001	$27.0\times$
usps	4.0×10^{-2}	1.2×10^{-4}	$.9164 \pm .0031$	$.9164 \pm .0031$	0.967	0.009	$107.4\times$
phishing	2.1×10^{-2}	4.0×10^{-5}	$.9751 \pm .0033$	$.9751 \pm .0033$	0.375	0.009	$41.6\times$
letter	1.7×10^{-2}	4.6×10^{-5}	$.7900 \pm .0073$	$.7900 \pm .0073$	0.290	0.008	$37.2\times$
a9a	2.0×10^{-2}	7.6×10^{-5}	$.8958 \pm .0015$	$.8958 \pm .0015$	0.457	0.012	$38.7\times$
w8a	6.0×10^{-2}	1.3×10^{-4}	$.9553 \pm .0058$	$.9553 \pm .0058$	1.010	0.031	$32.6\times$
cod-rna	6.9×10^{-3}	5.2×10^{-4}	$.9561 \pm .0059$	$.9560 \pm .0059$	0.295	0.007	$43.4\times$
connect-4	1.8×10^{-2}	6.8×10^{-5}	$.8410 \pm .0030$	$.8410 \pm .0030$	0.528	0.016	$32.4\times$
acoustic	8.9×10^{-3}	2.4×10^{-5}	$.7907 \pm .0027$	$.7907 \pm .0027$	0.187	0.015	$12.6\times$
aloi	1.7×10^{-2}	1.7×10^{-6}	$.6123 \pm .0046$	$.6123 \pm .0046$	0.173	0.007	$24.7\times$
ijcnn1	1.1×10^{-2}	1.7×10^{-5}	$.9177 \pm .0030$	$.9177 \pm .0030$	0.128	0.010	$12.8\times$
skin	1.3×10^{-3}	5.2×10^{-7}	$.9464 \pm .0007$	$.9464 \pm .0007$	0.035	0.002	$17.5\times$
covtype	4.6×10^{-3}	4.1×10^{-4}	$.8572 \pm .0064$	$.8573 \pm .0064$	0.953	0.033	$29.1\times$

5.1   Benchmark Results

We conduct our experiments on ten benchmark datasets[2] as summarized in Table 1, which have been used in previous studies on AUC optimization. We transform all multi-class datasets into binary ones by randomly partitioning classes into two groups, where each group contains the same number of classes, and the features of all dataset have been scaled to $[-1, +1]$.

The regularized parameter λ is set to 0.1 in all experiments as done in [23], since it is insensitive to the final AUC scores for relatively-small parameter. Let $\mathbf{w}^*$ be the model trained from the training dataset S_n, and store data statistics simultaneously. Then, 10% of the training data, denoted by R, are removed from S_n randomly and independently without replacement. Let $\hat{\mathbf{w}}_{-R}$ be the output model of DRAUC (Algorithm 1) from trained model $\mathbf{w}^*$ and removed data R.

Let $\mathbf{w}^*_{-R}$ denote the model which is retrained from data $S_n \backslash R$. For faster convergence, we use gradient descent with step-sizes $1/(\gamma t + 10)$ to get $\mathbf{w}^*_{-R}$, where the parameter $\gamma \in 10^{[-7,0]}$ is selected in advance with 5-fold cross-validation.

Table 2 shows the performance between our DRAUC model $\hat{\mathbf{w}}_{-R}$ and the retained model $\mathbf{w}^*_{-R}$ from data $S_n \backslash R$. We first observe that the distance $\|\mathbf{w}^* - \mathbf{w}^*_{-R}\|$ is large, since the removed data could yield the changes of training model. However, it is relatively small for the distance $\|\hat{\mathbf{w}}_{-R} - \mathbf{w}^*_{-R}\|$, which shows our model $\hat{\mathbf{w}}_{-R}$, trained from $\mathbf{w}^*$ and removed data R, have no significantly difference from the model $\mathbf{w}^*_{-R}$, which is retained from $S_n \backslash R$. This is in accordance with our theoretical analysis in Theorem 1.

It is also observable that, from Table 2, our DRAUC algorithm takes the same AUC on test data sets most times as the retraining model $\mathbf{w}^*_{-R}$, since

---

[2] http://www.csie.ntu.edu.tw/~cjlin/libsvmtools/.

Table 3. Distance, Test AUC and Running Time for retraining and DRAUC$_h$ with 10% removal rate over High-dimensional dataset

Dataset		smallNORB	real-sim	siam-competition2007	rcv1.binary
Instance		24,300	72,309	21,519	20,242
Feature		18,432	20,958	30,438	47,236
Distance	$\|\mathbf{w}^* - \mathbf{w}^*_{-R}\|$	3.8×10^{-2}	1.6×10^{-1}	3.3×10^{-1}	2.5×10^{-1}
	$\|\hat{\mathbf{w}}_{-R} - \mathbf{w}^*_{-R}\|$	5.8×10^{-3}	2.5×10^{-2}	4.0×10^{-2}	3.6×10^{-2}
Test AUC	Retraining	$.9199 \pm .0041$	$.9828 \pm .0008$	$.7505 \pm .0075$	$.9852 \pm .0015$
	DRAUC$_h$	$.9199 \pm .0041$	$.9828 \pm .0008$	$.7504 \pm .0075$	$.9853 \pm .0015$
Running time	Retraining	102.511	102.641	158.401	380.680
	DRAUC$_h$	4.680	6.535	7.305	13.846
	Speed-up	21.9×	15.7×	21.7×	27.5×

our DRAUC model $\hat{\mathbf{w}}_{-R}$ is close to $\mathbf{w}^*_{-R}$. However, our DRAUC method takes much smaller running time than the retraining method on $S_n \backslash R$; for example, our DRAUC achieves significant 10~100× speed-up compared to the retraining method. In addition, our DRAUC only needs some data statistics, rather than store the entire training data S_n in memory.

5.2 High-Dimensional Results

We consider four high-dimensional datasets[3] as shown in Table 3, and all multi-class datasets are transformed randomly into binary ones with the same number of classes, and the features of all dataset have been scaled to $[-1, +1]$.

Let $\mathbf{w}^*$ be the original model trained from the entire training dataset S_n. 10% of the training data, denoted by R, are removed from S_n randomly and independently without replacement. We set the regularized parameter λ as 0.01, the iteration numbers T as 200 and the sketch size m as 2000 in all experiments. The low-rank sketch matrices are generated by frequent direction (Algorithm 2) when we train $\mathbf{w}^*$, and calculate the first-order statistics simultaneously. We refer our algorithm as DRAUC$_h$ for high-dimensional data, and let $\hat{\mathbf{w}}_{-R}$ be our output model of DRAUC$_h$ from the trained model $\mathbf{w}^*$ and removed data R. Let $\mathbf{w}^*_{-R}$ denote the model which is retrained from data $S_n \backslash R$. We use gradient descent with learning rate $1/(\gamma t + 10)$ to get $\mathbf{w}^*_{-R}$, similarly to Sect. 5.1.

Table 3 shows the comparisons of different measure between our DRAUC$_h$ model $\hat{\mathbf{w}}_{-R}$ and the model $\mathbf{w}^*_{-R}$ retrained from data $S_n \backslash R$. We first observe that the original model $\mathbf{w}^*$ is quite different from $\mathbf{w}^*_{-R}$ due to data removal, and our DRAUC$_h$ model $\hat{\mathbf{w}}_{-R}$ is a relatively-good approximation to the retrained model $\mathbf{w}^*_{-R}$ because of low-rank approximation on covariance matrices and numerical solution. Despite of the low-rank approximation, our DRAUC$_h$ takes almost the same AUC on test data sets as retraining. In addition, our DRAUC$_h$ method takes smaller running time than retraining (e.g., about 15–28× speed-up).

---

[3] http://www.csie.ntu.edu.tw/~cjlin/libsvmtools/.

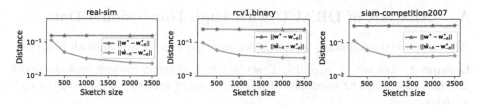

Fig. 1. Distance with varied sketch size m

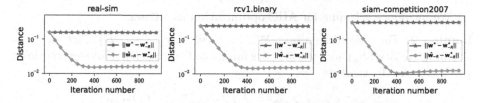

Fig. 2. Distance with varied iteration number T

We finally analyze the influence of parameters for high-dimensional datasets. Figure 1 shows that we can make good approximation when the sketch size is greater than 500, and there is no significantly improvement when the sketch size reaches 2000. Figure 2 shows the convergence of our approach on finding the numerical solution, which convergences after about 300 iterations.

6   Conclusion

In this work, we develop an efficient algorithm on data removal from an AUC optimization model. The main challenge lies in that AUC is measured by a sum of losses defined over pairs of instances from difference classes, while most previous works focus on univariate loss. Our algorithm maintains the data statistics, rather than store the entire training data. For high-dimensional data, we approximate the second-order statistics with low-rank matrices by frequent direction, and obtain the numerical solution with gradient descent so as to avoid the calculation of Hessian and its inverse explicitly. We verify the effectiveness of our algorithm both theoretically and empirically.

Acknowledgement. The authors want to thank the anonymous reviewers for helpful comments and suggestions. This research is supported by National Science Foundation of China (61921006, 61876078).

A   Analysis of DRAUC with High-Dimensional Data

We introduce the following lemma [2] for strongly convex and smooth function:

Lemma 1. *Let* $\mathbf{w}^* = \arg\min_{\mathbf{w}} \mathcal{L}(\mathbf{w})$ *w.r.t.* μ-*strongly convex and* β-*smooth function* $\mathcal{L}(\mathbf{w})$. *After* t *iterations of gradient descent with step size* $\eta_t = 2/(\beta+\mu)$, *we have*

$$\|\mathbf{w}_t - \mathbf{w}^*\| \leq \left(\frac{\beta - \mu}{\beta + \mu}\right)^t \|\mathbf{w}_0 - \mathbf{w}^*\|.$$

We introduce a lemma for AUC optimization as follows:

Lemma 2. *For bounded space* $\mathcal{W} = \{\mathbf{w}: \|\mathbf{w}\| \leq B\}$, *let* $\mathbf{w}^* = \arg\min_{\mathbf{w} \in \mathcal{W}} \mathcal{L}(\mathbf{w}; S_n)$ *and* $\mathbf{w}^*_{-R} = \arg\min_{\mathbf{w} \in \mathcal{W}} \mathcal{L}(\mathbf{w}; S_n \backslash R)$. *For regularization parameter* $\lambda > 0$, *we have*

$$\|\mathbf{w}^* - \mathbf{w}^*_{-R}\| \leq 4B \frac{4 + \lambda}{\lambda} \left(\frac{r_+}{n_+} + \frac{r_-}{n_-} - \frac{r_+ r_-}{n_+ n_-}\right).$$

Proof. From the definition of $\mathcal{L}(\mathbf{w}; S_n)$, we have

$$\|\nabla \mathcal{L}(\mathbf{w}_1; S_n) - \nabla \mathcal{L}(\mathbf{w}_2; S_n)\| \leq (4 + \lambda)\|\mathbf{w}_1 - \mathbf{w}_2\|,$$

where $\mathbf{w}_1, \mathbf{w}_2 \in \mathcal{W}$, $\|\mathbf{x}_i\| \leq 1$ and $\|\mathbf{x}_j\| \leq 1$, and thus $\mathcal{L}(\mathbf{w}; S_n)$ is $(4+\lambda)$-smooth. From Cauchy's mean-value theorem, we have

$$|\mathcal{L}(\mathbf{w}_1; S_n) - \mathcal{L}(\mathbf{w}_2; S_n)| = |\nabla \mathcal{L}^\top(\kappa \mathbf{w}_1 + (1 - \kappa)\mathbf{w}_2; S_n)(\mathbf{w}_1 - \mathbf{w}_2)|$$
$$\leq \|\nabla \mathcal{L}(\kappa \mathbf{w}_1 + (1 - \kappa)\mathbf{w}_2; S_n)\|\|\mathbf{w}_1 - \mathbf{w}_2\| \quad (12)$$

where $\kappa \in [0, 1]$ and $\kappa \mathbf{w}_1 + (1 - \kappa)\mathbf{w}_2 \in \mathcal{W}$. We also have $\nabla \mathcal{L}(\mathbf{w}^*; S_n) = 0$, and it holds that

$$\|\nabla \mathcal{L}(t\mathbf{w}_1 + (1 - t)\mathbf{w}_2; S_n)\| \leq \max_{\mathbf{w} \in \mathcal{W}} \|\nabla \mathcal{L}(\mathbf{w}; S_n) - \nabla \mathcal{L}(\mathbf{w}^*; S_n)\| \leq 2B(4 + \lambda)$$

which yields that $\mathcal{L}(\mathbf{w}; S_n)$ is $2B(4 + \lambda)$-Lipschitz from Eq. (12). Recall that $\Delta(\mathbf{w}) = n_+ n_- \mathcal{L}(\mathbf{w}; S_n) - (n_+ - r_+)(n_- - r_-)\mathcal{L}(\mathbf{w}; S_n \backslash R)$, and we have

$$n_+ n_- \mathcal{L}(\mathbf{w}^*_{-R}; S_n) = (n_+ - r_+)(n_- - r_-)\mathcal{L}(\mathbf{w}^*_{-R}; S_n \backslash R) + \Delta(\mathbf{w}^*_{-R})$$
$$\leq (n_+ - r_+)(n_- - r_-)\mathcal{L}(\mathbf{w}^*; S_n \backslash R) + \Delta(\mathbf{w}^*_{-R})$$
$$= n_+ n_- \mathcal{L}(\mathbf{w}^*; S_n) + (\Delta(\mathbf{w}^*_{-R}) - \Delta(\mathbf{w}^*))$$
$$\leq n_+ n_- \mathcal{L}(\mathbf{w}^*; S_n)$$
$$+ 2B(n_+ n_- - (n_+ - r_+)(n_- - r_-))(\lambda + 4)\|\mathbf{w}^*_{-R} - \mathbf{w}^*\| \quad (13)$$

where the first inequality holds from the optimal solution of $\mathcal{L}(\mathbf{w}^*_{-R}; S_n \backslash R)$, and the last inequality follows from the $2B(n_+ n_- - (n_+ - r_+)(n_- - r_-))(\lambda + 4)$-Lipschitzness of $\Delta(\mathbf{w})$. For λ-strongly convex function $\mathcal{L}(\mathbf{w}; S_n)$, we have

$$\mathcal{L}(\mathbf{w}^*_{-R}; S_n) - \mathcal{L}(\mathbf{w}^*; S_n) \geq \frac{\lambda}{2}\|\mathbf{w}^*_{-R} - \mathbf{w}^*\|^2 \quad (14)$$

from $\nabla \mathcal{L}(\mathbf{w}^*; S_n) = 0$. Combining Eq. (13) and (14) completes the proof.

It is necessary to introduce the following lemma from [8]:

Lemma 3. *Let Z be the sketch matrix of X using frequent direction. We have*

$$\left\| X[X]^\top - Z[Z]^\top \right\| \le 2tr(X[X]^\top)/m$$

where m is the sketch size.

Proof of Theorem 2.

Proof. Let $\hat{\mathcal{L}}(\mathbf{w}, S_n \backslash R)$ be the loss by replacing covariance matrices $\mathcal{S}^+_{-R}$ and $\mathcal{S}^-_{-R}$ with $\hat{\mathcal{S}}^+_{-R}$ and $\hat{\mathcal{S}}^-_{-R}$, and $\hat{\mathbf{w}}^*_{-R} = \arg\min_\mathbf{w} \{ \hat{\mathcal{L}}(\mathbf{w}; S_n \backslash R) \}$, we have

$$\left\| \hat{\mathbf{w}}_{-R} - \mathbf{w}^*_{-R} \right\| \le \left\| \hat{\mathbf{w}}_{-R} - \hat{\mathbf{w}}^*_{-R} \right\| + \left\| \hat{\mathbf{w}}^*_{-R} - \mathbf{w}^*_{-R} \right\|. \tag{15}$$

Combining with Lemma 1, this follows that

$$\left\| \hat{\mathbf{w}}_{-R} - \mathbf{w}^*_{-R} \right\| \le \left(\frac{2}{\lambda + 2} \right)^T \left\| \mathbf{w}^* - \mathbf{w}^*_{-R} \right\| + \left(1 + \left(\frac{2}{\lambda + 2} \right)^T \right) \left\| \hat{\mathbf{w}}^*_{-R} - \mathbf{w}^*_{-R} \right\|. \tag{16}$$

To bound $\left\| \hat{\mathbf{w}}^*_{-R} - \mathbf{w}^*_{-R} \right\|$, we first rewrite $\mathcal{L}(\mathbf{w}; S_n \backslash R)$ as

$$\mathcal{L}(\mathbf{w}; S_n \backslash R) = \mathbf{w}^\top (A_1 + A_2) \mathbf{w} + \mathbf{w}^\top \mathbf{a} + 1/2$$

where $\mathbf{a} = \mathbf{c}^-_{-R} - \mathbf{c}^+_{-R}$, $A_1 = \mathcal{S}^+_{-R} + \mathcal{S}^-_{-R} + \lambda \mathbf{I}_d$ and $A_2 = (\mathbf{c}^-_{-R} - \mathbf{c}^+_{-R})(\mathbf{c}^-_{-R} - \mathbf{c}^+_{-R})^\top$. Similarly, we rewrite $\hat{\mathcal{L}}(\mathbf{w}; S_n \backslash R)$ as

$$\hat{\mathcal{L}}(\mathbf{w}; S_n \backslash R) = \mathbf{w}^\top (\hat{A}_1 + A_2) \mathbf{w} + \mathbf{w}^\top \mathbf{a} + 1/2 \quad \text{with} \quad \hat{A}_1 = \hat{\mathcal{S}}^+_{-R} + \hat{\mathcal{S}}^-_{-R} + \lambda \mathbf{I}_d.$$

Minimizing $\mathcal{L}(\mathbf{w}; S_n)$ and $\hat{\mathcal{L}}(\mathbf{w}, S_n)$ gives

$$\mathbf{w}^*_{-R} = (A_1 + A_2)^{-1} \mathbf{a} \quad \text{and} \quad \hat{\mathbf{w}}^*_{-R} = (\hat{A}_1 + A_2)^{-1} \mathbf{a}, \quad \text{respectively.}$$

It is easy to get

$$\left\| (A_1 + A_2)^{1/2} (\hat{A}_1 + A_2)^{-1} (A_1 + A_2)^{1/2} - \mathbf{I}_d \right\|$$
$$= \left\| (\hat{A}_1 + A_2)^{-1/2} (A_1 - \hat{A}_1)(\hat{A}_1 + A_2)^{-1/2} \right\| \le \left\| A_1 - \hat{A}_1 \right\| \left\| (\hat{A}_1 + A_2)^{-1} \right\| \le \frac{2\tau}{\lambda m},$$

where $\tau = \max(rank(X^+_n [X^+_n]^\top), rank(X^-_n [X^-_n]^\top))$, and the inequality comes from Lemma 3. Denote $\Omega = (A_1 + A_2)^{1/2} (\hat{A}_1 + A_2)^{-1} (A_1 + A_2)^{1/2} - \mathbf{I}_d$, we have

$$\left\| \hat{\mathbf{w}}^*_{-R} - \mathbf{w}^*_{-R} \right\| = \left\| \left((\hat{A}_1 + A_2)^{-1} - (A_1 + A_2)^{-1} \right) \mathbf{a} \right\|$$
$$= \left\| (A_1 + A_2)^{-1/2} \Omega (A_1 + A_2)^{-1/2} \mathbf{a} \right\| \le \frac{2\tau}{\lambda m} \left\| (A_1 + A_2)^{-1} \mathbf{a} \right\| \le \frac{2\tau}{\lambda m} B,$$

which completes the proof by combining with Eq. (16) and Lemma 2.

References

1. Brzezinski, D., Stefanowski, J.: Prequential AUC: properties of the area under the ROC curve for data streams with concept drift. Knowl. Inf. Syst. **52**(2), 531–562 (2017). https://doi.org/10.1007/s10115-017-1022-8
2. Bubeck, S.: Convex optimization: algorithms and complexity. Found. Trends® Mach. Learn. **8**(3–4), 231–357 (2015)
3. Cao, Y., Yang, J.: Towards making systems forget with machine unlearning. In: 2015 IEEE Symposium on Security and Privacy, pp. 463–480. IEEE (2015)
4. Cook, R., Weisberg, S.: Residuals and Influence in Regression. Chapman and Hall, New York (1982)
5. Flach, P., Hernández-Orallo, J., Ramirez, C.: A coherent interpretation of AUC as a measure of aggregated classification performance. In: Proceedings of the 28th International Conference on Machine Learning, pp. 657–664 (2011)
6. Gao, W., Jin, R., Zhu, S., Zhou, Z.H.: One-pass AUC optimization. In: Proceedings of the 30th International Conference on Machine Learning, pp. 906–914 (2013)
7. Gao, W., Zhou, Z.H.: On the consistency of AUC pairwise optimization. In: Proceedings of the 24th International Joint Conference on Artificial Intelligence, pp. 939–945 (2015)
8. Ghashami, M., Liberty, E., Phillips, J., Woodruff, D.: Frequent directions: simple and deterministic matrix sketching. SIAM J. Comput. **45**(5), 1762–1792 (2016)
9. Ginart, A., Guan, M., Valiant, G., Zou, J.: Making AI forget you: data deletion in machine learning. In: Advances in Neural Information Processing Systems, vol. 32, pp. 3518–3531 (2019)
10. Guo, C., Goldstein, T., Hannun, A., van der Maaten, L.: Certified data removal from machine learning models. In: Proceedings of the 37th International Conference on Machine Learning, pp. 3832–3842 (2019)
11. Herschtal, A., Raskutti, B.: Optimising area under the ROC curve using gradient descent. In: Proceedings of the 21st International Conference on Machine Learning, p. 49 (2004)
12. Izzo, Z., Anne Smart, M., Chaudhuri, K., Zou, J.: Approximate data deletion from machine learning models. In: Proceedings of The 24th International Conference on Artificial Intelligence and Statistics, pp. 2008–2016 (2021)
13. Koh, P., Liang, P.: Understanding black-box predictions via influence functions. In: Proceedings of the 34th International Conference on Machine Learning, pp. 1885–1894 (2017)
14. Liu, M., Yuan, Z., Ying, Y., Yang, T.: Stochastic AUC maximization with deep neural networks. In: Proceedings of the 8th International Conference on Learning Representations (2019)
15. Liu, M., Zhang, X., Chen, Z., Wang, X., Yang, T.: Fast stochastic AUC maximization with $o(1/n)$-convergence rate. In: Proceedings of the 35th International Conference on Machine Learning, pp. 3189–3197 (2018)
16. Mozaffari-Kermani, M., Sur-Kolay, S., Raghunathan, A., Jha, N.: Systematic poisoning attacks on and defenses for machine learning in healthcare. IEEE J. Biomed. Health Inform. **19**(6), 1893–1905 (2014)
17. Natole, M., Ying, Y., Lyu, S.: Stochastic proximal algorithms for AUC maximization. In: Proceedings of the 35th International Conference on Machine Learning, pp. 3710–3719. PMLR (2018)
18. Provost, F., Fawcett, T., Kohavi, R.: The case against accuracy estimation for comparing induction algorithms. In: Proceedings of the 15th International Conference on Machine Learning, pp. 445–453 (1998)

19. Shen, S.Q., Yang, B.B., Gao, W.: AUC optimization with a reject option. In: Proceedings of the 34th AAAI Conference on Artificial Intelligence, pp. 5684–5691 (2020)
20. Shokri, R., Stronati, M., Song, C., Shmatikov, V.: Membership inference attacks against machine learning models. In: 2017 IEEE Symposium on Security and Privacy, pp. 3–18 (2017)
21. Voigt, P., Von dem Bussche, A.: The EU General Data Protection Regulation (DGPR), vol. 10. Springer, Cham (2017). https://doi.org/10.1007/978-3-319-57959-7
22. Wu, J., Brubaker, S., Mullin, M., Rehg, J.: Fast asymmetric learning for cascade face detection. IEEE Trans. Pattern Anal. Mach. Intell. **30**(3), 369–382 (2008)
23. Wu, Y., Dobriban, E., Davidson, S.: DeltaGrad: rapid retraining of machine learning models. In: Proceedings of the 37th International Conference on Machine Learning, pp. 10355–10366 (2020)
24. Ying, Y., Wen, L., Lyu, S.: Stochastic online AUC maximization. In: Advances in Neural Information Processing Systems, vol. 29, pp. 451–459 (2016)

Distributed Differentially Private Ranking Aggregation

Baobao Song[1], Qiujun Lan[1], Yang Li[2], and Gang Li[3]([✉])

[1] Hunan University, Changsha 410082, China
bbsong@tulip.academy, lanqiujun@hnu.edu.cn
[2] The Australian National University, Canberra, ACT 2600, Australia
kelvin.li@anu.edu.au
[3] Deakin University, Geelong, VIC 3216, Australia
gang.li@deakin.edu.au

Abstract. Ranking aggregation is commonly adopted in cooperative decision-making to assist combining multiple rankings into a single representative. To protect the actual ranking of each individual, some privacy-preserving strategies, such as differential privacy, are often used. This, however, does not consider the scenario where the curator, who collects all rankings from individuals, is untrustworthy. This paper proposed a mechanism to solve the above issue using the *distribute differential privacy* framework. The proposed mechanism collects locally differential private rankings from individuals, then randomly permutes pairwise rankings using a shuffle model to further amplify the privacy protection. The final representative is produced by hierarchical rank aggregation. The mechanism was theoretically analysed and experimentally compared against existing methods, and demonstrated competitive results in both the output accuracy and privacy protection.

Keywords: Ranking aggregation · Distributed differential privacy ·
HRA algorithm, Shuffle model

1 Introduction

Cooperative decision-making [14] is pervasive in business management, because of its superiority in providing information from different aspects for better decision-making. As an essential step in cooperative decision-making, aggregation combines all individual preferences into a representative output. In daily life, individuals often rank all available alternatives to reveal the preference relation of multiple alternatives, hence *ranking aggregation* has become essential for society, and many researchers focus on its two requirements, which are hard to be satisfied simultaneously: *privacy* and *utility*. Preference data in ranking has sensitive information, and the leaking of it may make individuals susceptible to coercion. *Utility* represents whether the aggregation result stands for the majority preference. Consequently, the ability to effectively aggregate private ranking into a representative result is important in ranking aggregation. In the past few years, substantial research efforts have been devoted to ranking aggregation.

Traditional anonymizing methods such as anonymization hardly solves the problem. For example, Hugo Awards 2015 incident [8] shows that the anonymized preferences could result in the re-identification of individuals because the adversary with background knowledge is able to launch a *linkage attack*. According to the weakness of the traditional anonymizing method, many researchers resort to *differential privacy* (DP) [5].

DP is an effective method to provide a rigorous privacy guarantee, and it can defend against various attacks, no matter how much background knowledge the adversary has [19]. As a lightweight methodology to protect privacy, many current works address the ranking aggregation problems with DP. Shang et al. [16] designed a privacy-preserving rank aggregation algorithm, and whatever the ranking rules, the algorithm adds noise to votes and returns the histogram of rankings. Based on `Quicksort` [10], Hay et al. [9] proposed three differentially private rank aggregation algorithms includes `P-SORT`, a pairwise comparison method about private ranking aggregation. The benefit of using DP to protect individual sensitive information is that the adversary is unlikely to obtain sensitive information by observing the releasing results. In the meantime, the results have high availability. Nevertheless, DP is not without limitations: in real-world applications, the curator may collude with the adversary to leak some information before perturbing.

Local differential privacy (LDP) [7] alleviates this issue through adding noise locally and uploading noisy data to the untrusted curator. Yan et al. [18] proposed the `LDP-KwikSort` algorithm, and they use the number of queries K to trade-off between utility and privacy. Besides privacy and utility, Wang et al. [17] studied another property, soundness, and then proposed the weighted sampling mechanism and the additive mechanism to improve the ranking utility. Unfortunately, LDP needs a large amount of data to achieve an acceptable utility. Moreover, existing approaches that use pairwise comparison information to rank [9, 18] share a common limitation: they introduce additional errors through the random pivot selection. In conclusion, two obstacles need to be overcome simultaneously. Firstly, the ranking algorithm needs to output an aggregation result with utility as high as possible. Secondly, in order to protect individuals' sensitive data, the untrusted curator should not receive the original raw preferences.

With the increased awareness of privacy protection, many researchers are interested in *distributed differential privacy* (DDP) [15] to amplify the privacy. DDP builds on LDP but further protects privacy using an intermediate node. This may mitigate the problem of poor utility in LDP. Besides, recently advances in ranking aggregation such as the algorithm HRA [4], which takes advantage of pairwise comparisons to aggregate ranking, and provides one way to eliminate the errors of random selecting pivot. However, as this algorithm applies *Borda count* [2] and pairwise comparison method, it costs too much privacy budget if it is directly combined with DP, and results in very poor performance.

In this paper, we propose a novel algorithm DDP-Helnaksort to meet the requirements of *privacy* and *utility* in ranking aggregation. The contributions of this algorithm are two-fold:

- DDP-Helnaksort employs a new ranking aggregation that avoids the random pivot selection as in quicksort-based LDP methods. Moreover, Borda count in HRA was replaced by a new method that scores the alternatives to reduce the noise effect caused by small privacy budget. Experiments show that it outperforms some pairwise comparison-based DP rank aggregations.
- We firstly adopt DDP to deal with the ranking aggregation problem. This was achieved by combining LDP with a shuffle model [1] that randomly permutes the preferences in order to amplify the privacy before submitting rankings to the untrusted curator. This provides a stronger DP guarantee, which can be measured by calculating the amplification bound.

The rest of this paper is organized as follows. Section 2 provides the preliminary of ranking aggregation, differential privacy and shuffle model. Section 3 presents the DDP-Helnaksort algorithm and gives the privacy guarantee. Section 4 reports the comparison results with baseline algorithms and analyzes the effect of adjusting parameters. Final conclusions and future directions are shown in Sect. 5.

2 Preliminaries

2.1 Ranking Aggregation

2.1.1 Conception and Measurement

In a ranking scenario, an agent u is asked to rank a set $A = \{a_1, a_2, ..., a_m\}$ of alternatives and to provide the order of preferences, denoted by $p_u = [x_1, x_2, ..., x_m]$, where x_i is the ranking index of a_i, and $x_i = 1$ means that agent u's favourite alternative is a_i. A curator then collects the order from each agent and uses a ranking aggregation algorithm to output a representative ranking R based on $\{p_u\}$. In this paper, $a_i \succ a_j$ means that a_i is preferred than a_j.

Ranking aggregation aims to find the most representative ranking R^*. *Kenmeny optimal aggregation* (KOA) [6] is used to find R^* by minimising the *average Kendall tau distance* $\overline{K}$. *Kendall tau* distance [11] measures the distance between two rankings by counting the number of inconsistent pairs among all pairs of alternatives: $K(R, p_u) = \frac{1}{\binom{m}{2}} \sum_{i \neq j, i, j \in [m]} \kappa_{ij}(R, p_u)$, $\kappa_{ij}(R, p_u)$ is 1 when the pair a_i and a_j is ordered differently in rankings R and p_u, otherwise it is 0. The average *Kendall tau distance* is then computed over the rankings $\{p_u\}$ from all agents: $\overline{K}(R, p_u) = \frac{1}{n} \sum_{u \in [n]} K(R, p_u)$.

Hierarchical Ranking Aggregation. Hierarchical ranking aggregation (HRA) [4] algorithm can consolidate all agents' rankings into a total order. It is a recursive process like Quicksort, but does not rank alternatives based on the random pivot selections as in `Quicksort`, which is likely to reduce the utility of a private ranking besides the impact of additive noise such as in DP.

Given m ranking alternatives, the `HRA` algorithm first computes an m by m *pairwise comparison matrix* (PCM) M, where each entry $M(i,j) = \frac{1}{n}\sum_{u=1}^{n} l_{ij}^{u}$ is a comparison score for (a_i, a_j) over all n rankings. $l_{ij}^{u} = 1$ and 0 when $a_i \succ a_j$ and $a_j \succ a_i$ respectively and 0.5 otherwise. Then, the algorithm computes an m by m *pairwise preference relation* (PPR) matrix D. Each entry $D(i,j) = 1$ or 0 when $M(i,j)$ is greater or smaller than $M(i,j)$ respectively, and 0.5 for equality. Third, every alternative is allocated to a different level according to its score $L(i)$ that is the row sum of $D(i,j)$. As multiple alternatives can be allocated at the same level, they are further compared and allocated into sublevels using a sub-PCM that only includes the rankings of the corresponding alternatives. If alternatives have the same score in sublevel, `Borda count` is used to select a winner. Finally, the algorithm finishes when each level contains only one alternative.

2.2 Differential Privacy

Differential privacy (DP) [5] is a privacy protection model that adds calibrated noise to query outputs to ensure an adversary having negligible chance of guessing the sensitive information in a database. Formally, DP can be defined as:

Definition 1. (ϵ, δ)-***Differential Privacy***. *A random algorithm $\mathcal{M}$ provides (ϵ, δ)-differential privacy if for any two datasets D and D' that differ in at most a single record, and for all outputs $A \in Range(O)$: $Pr[\mathcal{M}(D) \in O)] \leq e^{\epsilon} Pr[\mathcal{M}(D') \in O)] + \delta$.*

The parameter ϵ is defined as the privacy budget, which controls the privacy guarantee level of the mechanism. Another parameter δ is responsible for the probability that ϵ does not hold. DP assumes that there is a trusted curator, but in reality, the adversary has possibility to collect information from the curator. Hence, the local differential privacy (LDP) [7] has been utilized. In the LDP model, each agent uses an algorithm $\mathcal{M}$ to perturb data locally and then upload the noisy one to the untrusted curator. The definition of LDP is as follows:

Definition 2. (ϵ, δ)-***Local Differential Privacy***. *A local algorithm $\mathcal{M}$ provides (ϵ, δ)-local differential privacy if for any two value x and x', and for every output y: $Pr[\mathcal{M}(x = y)] \leq e^{\epsilon} Pr[\mathcal{M}(x' = y)] + \delta$.*

Although LDP solves the problem that the curator may disclose information, it requires a huge amount of data to achieve a satisfactory utility [1]. Based on LDP, *distributed differential privacy* (DDP) [15] can improve the data utility. In DDP, every agent adds noise locally, and uploads the data to a trusted intermediate node to protect privacy further, finally sends the results to the curator. On the one hand, we do not need to worry about the privacy leakage from the curator in DDP. On the other hand, it has a higher utility than LDP. In this paper, we apply DDP model to aggregate ranking. And we use Gaussian mechanism [7] to perturb preferences. An application of Gaussian mechanism satisfies (ϵ, δ)-DP, if variables drawn from the Gaussian distribution with $\mu = 0$ and $\sigma = \frac{\Delta_g f \sqrt{2 \log(\frac{1.25}{\delta})}}{\epsilon}$, where $\Delta_g f$ is the global sensitivity of the function f.

2.3   Shuffle Model

Shuffle model can be used in intermediate nodes to realise DDP, and the protocol P was proposed in [1]. The protocol has three components: a randomizer $\mathcal{R}$, a shuffler $\mathcal{S}$ and an analyzer $\mathcal{A}$. First, $\mathcal{R}$ applies LDP to perturb data to get (ϵ, δ) protection. Then, $\mathcal{S}$ chooses a random permutation π to shuffle the data, and cut the connection between the outputs and their sources. Finally, $\mathcal{A}$ analyses the data and gets the query result. The shuffle step can amplify the privacy, and the following theorem [3] quantifies the amplification bound of shuffling:

Theorem 1. *If every agent sends a message to the shuffle model, and the randomizer R satisfies $(\epsilon + \ln n, \delta)$-local differential privacy, then the protocol $P = (R, S, A)$ satisfies both (ϵ, δ)-differential privacy and $(\epsilon + \ln n, \delta)$-local differential privacy, where ϵ' is smaller than ϵ, and n is the number of agents.*

3   Ranking Aggregation Algorithm Under DDP

In this section, we propose an algorithm `DDP-Helnaksort` to solve the private ranking aggregation problem. It can be formalised as follows: given m alternatives to be ranked by n agents, a curator need to present a final ranking that represents most agents' preferences. In addition, each agent u's ranking p_u must not reveal to the curator his true preferences over the alternatives.

The `DDP-Helnaksort` algorithm consists of three steps, as shown in Fig. 1. These steps are discussed respectively in Sect. 3.1, Sect. 3.2 and Sect. 3.3. The first step (①–④) is ranking preference collection, in which each agent, before

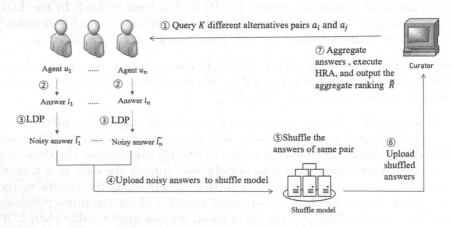

Fig. 1. Overview of `DDP-Helnaksort`

submitting the answers to the curator, adds the Gaussian noise to the rank of each pair (a_i, a_j) that being queried. The second step (⑤–⑥) is a shuffling process, which collects the ranking of (a_i, a_j) from the corresponding agents that answered the query, in order to further reduce the risk of privacy breach. The third step (⑦) aggregates to generate a final ranking of all m alternatives.

3.1   Ranking Preference Collection

The first step collects K private pairwise rankings from each agent, where K is an input parameter. A larger K leads to a more accurate aggregated ranking, because each pair (a_i, a_j) will be answered by more agents. The drawback is the partition of the privacy budget into a tiny piece for each query, which results in adding large noise that diminishes the utility. A smaller K can guarantee the utility, but the curator may end up with a less representative final ranking. We explore the optimal K in Sect. 4.2. The ranking preference collection step is shown in Algorithm 1. $l_{ij} \leftarrow p_u$ represents the preference in agent u's ranking of a randomly selected pair (a_i, a_j). This algorithm uses the Gaussian mechanism for noise addition (other mechanisms can be used too).

Algorithm 1. Ranking Preference Collection

Input: Agent u's ranking p_u, K queries, privacy parameter ϵ and δ
Output: Private pairwise preferences Q
 1: $Q_u = \emptyset$
 2: **for** $k \in [K]$ **do**
 3:    $l_{ij} \leftarrow p_u$
 4:    $\widetilde{l}_{ij} = l_{ij} + Gau(\frac{K\Delta f \sqrt{2\ln\frac{1.25}{\delta}}}{\epsilon})$
 5:    **if** $\widetilde{l}_{ij} > 0.5$ **then**
 6:       $\widetilde{l}_{ij} = 1$
 7:    **else**
 8:       $\widetilde{l}_{ij} = 0$
 9:    **end if**
10:    $Q_u = Q_u \cup \{\widetilde{l}_{ij}\}$
11: **end for**

3.2   Shuffling

Shuffling before aggregation can amplify privacy without affecting the output utility. In DDP-Helnaksort, each pair (a_i, a_j)'s answers from the corresponding agents are collected and randomly permuted at an intermediate node, so that when the private rankings are submitted, the curator is unable to guess the source of an answer with a non-negligible probability. The shuffle model finally provides a protection of DP with a smaller ϵ, which is further discussed in Sect. 3.4.

3.3   Ranking Aggregation

Once all the private rankings are submitted, the DDP-Helnaksort algorithm goes into the final stage, ranking aggregation. This step is based on the HRA algorithm, but with a different fallback to sort equal alternatives in sublevels in order to reduce the noise effect. The algorithm is shown in Algorithm 2. M is the number of alternatives in the unsorted sublevel, and $C_{a_i a_j}$ is the number of agents who voted $a_i \succ a_j$. The method uses pairwise preference to calculate a score for a_j

$$C_{a_j} = \sum_{j \in [M]} (C_{a_j a_i} - C_{a_i a_j}), \tag{1}$$

hence avoids splitting some privacy budgets as in *Borda count*.

This RA(ranking aggregation) algorithm mainly adopts a separate-layer ranking thought to generate the aggregation ranking, which uses the information about $C_{a_i a_j}$ and $C_{a_j a_i}$. The calculations of PCM and PPR matrix happen at *Line* 6–8 and *Line* 9, respectively. After that, we can count the scores of every alternative in M (*Line* 10–12). And if the scores are same in two rounds, we calculate the C_{a_j}, and then put the highest one in a high level and others at a low level to do the next round (*Line* 13–16). The algorithm iterates until $M = 1$ in each level, and finally the aggregated ranking $\widetilde{R}$ is generated (*Line* 17–20).

Algorithm 2. RA

Input: Agents pairwise aggregation $C_{a_i a_j}$ and $C_{a_j a_i}$

Output: Aggregate ranking $\widetilde{R}$

1: M = number of alternatives needed to rank
2: $L = [0] * M$
3: **if** $M = 1$ **then**
4:    **return**
5: **end if**
6: **for** each $i, j \in [m]$ **do**
7:    Calculate $PCM(i, j) = \frac{C_{a_i a_j}}{C_{a_i a_j} + C_{a_j a_i}}$
8: **end for**
9: Calculate PPR according to PCM
10: **for** $j, i \in [M]$ **do**
11:    Calculate alternatives' level score $L(j) += PPR(i, j)$
12: **end for**
13: **if** $L(1) = L(2) = ... = L(M)$ **then**
14:    put the C_{a_j} winner into a high-ranking level and others into a low level
15: **end if**
16: **for** $l = 1$ to the number of different levels **do**
17:    ranking of l-th level = RA (input ranking about the alternatives in l-th level)
18: **end for**
19: Rank the alternatives according to their levels to get aggregate ranking $\widetilde{R}$

3.4   Privacy Guarantee

Theorem 2. *DDP-Helnaksort satisfies (ϵ, δ)-local differential privacy and $(\epsilon - \ln \frac{n}{\binom{m}{2}}, \delta)$-differential privacy when $K = 1$.*

Proof. In the ranking preference collection phase, Gaussian mechanism is used to add noise into every agent's answers. Because are K rounds, $\epsilon_k = \frac{\epsilon}{K}$ in each round. In Gaussian mechanism, we set

$$\delta = \frac{\Delta_g f \sqrt{2 \ln \frac{1.25}{\delta}}}{\epsilon_k} = \frac{K \Delta_g f \sqrt{2 \ln \frac{1.25}{\delta}}}{\epsilon} \tag{2}$$

And DDP-Helnaksort executes the post-processing procedure after applying Gaussian mechanism, hence it satisfies $(\epsilon, \delta) - LDP$. Besides, $K = 1$ means that every agent answers once and uploads a single message (latter experiments confirm the algorithm utility is the highest when $K = 1$). In the shuffling phase, there are $\binom{m}{2}$ pairs of alternatives, so the number of same pair and the size of set S in shuffle model is

$$n' = \frac{n}{\binom{m}{2}} \tag{3}$$

Therefore, by using Theorem 1, the algorithm DDP-Helnaksort satisfies $(\epsilon - \ln \frac{n}{\binom{m}{2}}, \delta)$-DP when $K = 1$.

4   Experiments

In this section, we evaluate the performance of DDP-Helnaksort, and compare it with benchmark methods on both real and synthetic datasets. All algorithms were implemented in Python and executed 300 times to get the result.

4.1   Experiment Settings

Datasets. The experiments were conducted on synthetic datasets and a real-world dataset **TurkDots** [13]. By using R package PerMallows 1.13, we obtained four synthetic datasets with $n \in \{100, 1000, 2500, 5000\}$, $\theta = 0.25$, and $m = 15$ from Mallows model [12]. The dispersion parameter θ represents the distance between the generated ranking and ground truth ranking. The generated ranking is closer to the ground truth ranking when θ is larger. **TurkDots** is from *Amazon Mechanical Turk*, and it contains $m = 4$ alternatives rankings.

Baseline Algorithms

- LDP-Kwiksort [18]. It has K rounds' interactions between every agent and the untrusted curator. In each round, the curator random selects paired alternatives to ask agents preference and receives noisy answers from agents (queries to an agent are not the same), then uses the Kwiksort algorithm to get the aggregate ranking. Its utility is the highest when $K = 1$.

- LDP-Quicksort. Compared with LDP-Kwiksort, it only differs in when a new pivot random chosen in Quicksort, the curator queries the preference between the pivot and other alternatives. This setting is only to collect preference used in Quicksort, and avoid the waste of privacy budget for other pairs. K in this algorithm represents the times of the agent's answers. Finally, when the Quicksort algorithm is finished, the curator gets an aggregated ranking.

Utility Metric - Average Kendall Tau Distance. We use the *average Kendall tau distance* to measure the accuracy of the aggregated ranking. The larger the *average Kendall tau distance*, the worse the algorithm performance. We normalise this distance by $m(m-1)/2$ because we can directly compare it with different number of alternatives. Hence, the *average Kendall tau distance* can be calculated as $\overline{K}(R, R_u) = \frac{2}{nm(m-1)} \sum_{u \in N} K(R, R_u)$.

4.2   Performance of DDP-Helnaksort

Comparison between DDP-Helnaksort and Baseline Algorithms. We ran three algorithms LDP-Quicksort, LDP-Kwiksort, DDP-Helnaksort with Gaussian noise. Here ϵ is the parameter in LDP. We set $K \in \{1, m, max\}$ to observe the performance of different algorithms in different K, and m is the number of alternatives. When $K = max$, the maximum value of K in LDP-Kwiksort and DDP-Helnaksort is $\binom{m}{2}$, but in LDP-Quicksort, the value is according to the chosen pivot, and it is $(m-1)\log m$ in general. We did the experiment on **TurkDots** with $n = 100$. With $\epsilon = 1$, $\delta = 10^{-4}$ in local differential privacy, the *average Kendall tau distance* of LDP-Quicksort, LDP-Kwiksort and DDP-Helnaksort are shown in Fig. 2.

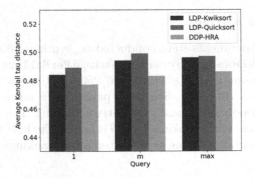

Fig. 2. Comparison of algorithms according to *average Kendall tau distance* on **Turk-Dots** across different K

The results in Fig. 2 proves our algorithm outperforms others across different K. When we add the same scale of noise to these algorithms, the *average Kendall tau distance* of DDP-Helnaksort is the shortest. The cause is when adding same scale of noise, DDP-Helnaksort uses more pairwise alternatives' information (the comparison information provided by pairwise comparisons) to rank, which leads to a more accuracy result. Besides, it keeps away from the error of pivot random selection, which can not be avoided by the other algorithms.

Impact of Query Amount to Every Agent. Different number of the queries has different ranking aggregation results. More information can be obtained when increasing the number of queries, but at the same time, the privacy budget of each round becomes smaller, and the larger noise is added to every answer. In order to get the best performance with the best K, we ran DDP-Helnaksort on dataset **TurkDots** and the synthetic dataset with 100 agents. We set $\delta = 10^{-4}$, $\epsilon \in \{0.5, 1\}$ (this ϵ is the parameter of DP, also means that it is the amplification result of local randomizer, and ϵ in following experiment is the same) as well as varying the number of queries K to observe the performance of DDP-Helnaksort. The results are shown in Fig. 3.

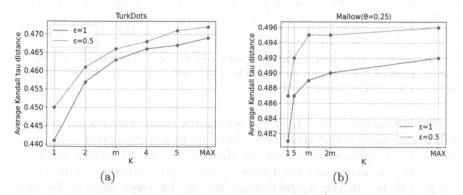

Fig. 3. Performance of DDP-Helnaksort: *Average Kendall tau distance* on **TurkDots** (a) and a synthetic dataset (b) across different K when $\delta = 10^{-4}$, $\epsilon \in \{0.5, 1\}$

It is apparent that as the decreasing of K, the performance of DDP-Helnaksort is better. The *average Kendall tau distance* reaches the minimum when $K = 1$. This experiment result is the same as [18], which reveals the best performance is achieved when $K = \frac{\epsilon}{2}$. The reason of this phenomenon is large K leads to a small ϵ in each round, and large scale of noise has a great impact on results. Although some information about agents' preferences is lost when K is small, a small noise is added to each answer, and the impact is smaller than large noise with more information. The result also implies if we want to further improve performance of the algorithm, we can do some works about handling ϵ such as implementation of personalised differential privacy which can release some needless privacy.

Ablation Study: Impact of Shuffle Model and Privacy Budget. As seen in Sect. 3.4, shuffle model turns LDP to DDP and amplifies the privacy. When every agent gives his noisy answers, a shuffling mechanism used before aggregation can offer another protection. After using the shuffle model, the algorithm satisfies DP with a smaller ϵ than before. In order to demonstrate the privacy amplification of shuffling, we compared the algorithm with and without shuffle model in a same ϵ. Besides, ϵ reflects the level of privacy protection of every agent. We varied ϵ to observe the changes in *average Kendall tau distance*. We set $k = 1$, and other experimental setup is unchanged.

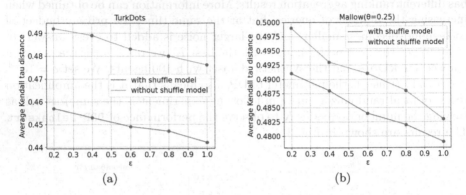

(a)   (b)

Fig. 4. Comparison of DDP-Helnaksort with and without shuffle model according to *average Kendall tau distance* on **TurkDots** (a) and a synthetic dataset (b) across different ϵ when $K = 1$ and $n = 100$

We can conclude from Fig. 4 that adding the shuffle step results in a better utility. The reason is that shuffling is equivalent to adding another noise on data. Consequently, when we compared the algorithm with and without shuffling at a certain ϵ, the second one has a large ϵ locally, so it perturbs less on data and performs better. In Fig. 4, the distance average increases more in **TurkDots** than the synthetic dataset from with shuffling to without shuffling, and this mainly relates to different number of alternatives m. The synthetic dataset has more alternatives than **TurkDots**, thus the synthetic dataset has more alternative pairs and it has fewer collected preferences about a certain pair. Therefore, the shuffle model offers a smaller amplification on the synthetic dataset. This phenomenon is consistent with the Theorem 2 that the amplification bound is proportional to the amount of data about a certain pair. Moreover, when decreasing the privacy budget, the *average Kendall tau distance* increases due to large scale of noise, which make the final aggregation ranking further to the representative ranking. Furthermore, in DDP, we can choose alternative methods, such as some cryptography tools, to amplify the privacy.

5   Conclusions

In order to improve the utility of private ranking aggregation, we proposed a new algorithm `DDP-Helnaksort`, which avoids the issue of random pivot selection which appears in other private ranking algorithm using the pairwise method.

We designed a new method to give alternatives' score according to preference in pairs, which can save some privacy budget and lead to a higher utility. Experimental results indicate that our algorithm achieves a better performance. Besides, We're first applying the DDP mechanism shuffle model to amplify the privacy. Theoretical analysis of amplification bound of shuffle model and experimental results all confirm that the shuffle model is valid.

In the future, we will further improve the ranking utility, such as using some cryptography tools. Besides, this algorithm can be further optimised if it could apply personalised DP, which can release some redundant privacy budget to achieve a higher utility.

Acknowledgement. This research is supported by the National Natural Science Fund of China (Project No. 71871090), and Hunan Provincial Science & Technology Innovation Leading Project (2020GK2005).

References

1. Bittau, A., et al.: Prochlo: strong privacy for analytics in the crowd. In: Proceedings of the 26th Symposium on Operating Systems Principles, pp. 441–459 (2017)
2. Black, D., et al.: The theory of committees and elections (1958)
3. Cheu, A., Smith, A., Ullman, J., Zeber, D., Zhilyaev, M.: Distributed differential privacy via shuffling. In: Ishai, Y., Rijmen, V. (eds.) EUROCRYPT 2019. LNCS, vol. 11476, pp. 375–403. Springer, Cham (2019). https://doi.org/10.1007/978-3-030-17653-2_13
4. Ding, J., Han, D., Dezert, J., Yang, Y.: A new hierarchical ranking aggregation method. Inf. Sci. **453**, 168–185 (2018)
5. Dwork, C.: Differential privacy. In: Bugliesi, M., Preneel, B., Sassone, V., Wegener, I. (eds.) ICALP 2006. LNCS, vol. 4052, pp. 1–12. Springer, Heidelberg (2006). https://doi.org/10.1007/11787006_1
6. Dwork, C., Kumar, R., Naor, M., Sivakumar, D.: Rank aggregation methods for the web. In: Proceedings of the 10th International Conference on World Wide Web, pp. 613–622 (2001)
7. Dwork, C., Roth, A., et al.: The algorithmic foundations of differential privacy. Found. Trends Theor. Comput. Sci. **9**(3–4), 211–407 (2014)
8. Eppstein, D.: Instability vs anonymization in e pluribus hugo (2015). https://11011110.github.io/blog/2015/09/09/instability-vs-anonymization.html
9. Hay, M., Elagina, L., Miklau, G.: Differentially private rank aggregation. In: Proceedings of the 2017 SIAM International Conference on Data Mining, pp. 669–677. SIAM (2017)
10. Hoare, C.A.: Quicksort. Comput. J. **5**(1), 10–16 (1962)
11. Kendall, M.G.: Rank correlation methods (1948)
12. Mallows, C.L.: Non-null ranking models. I. Biometrika **44**(1/2), 114–130 (1957)
13. Mao, A., Procaccia, A., Chen, Y.: Better human computation through principled voting. In: Proceedings of the AAAI Conference on Artificial Intelligence, vol. 27 (2013)
14. Meinhardt, H.I.: Cooperative Decision Making in Common Pool Situations, vol. 517. Springer, Heidelberg (2012)
15. Narayan, A.: Distributed differential privacy and applications. University of Pennsylvania (2015)

16. Shang, S., Wang, T., Cuff, P., Kulkarni, S.: The application of differential privacy for rank aggregation: privacy and accuracy. In: 17th International Conference on Information Fusion (FUSION), pp. 1–7. IEEE (2014)
17. Wang, S., et al.: Aggregating votes with local differential privacy: usefulness, soundness vs. indistinguishability. arXiv preprint arXiv:1908.04920 (2019)
18. Yan, Z., Li, G., Liu, J.: Private rank aggregation under local differential privacy. Int. J. Intell. Syst. **35**(10), 1492–1519 (2020)
19. Zhu, T., Li, G., Zhou, W., Philip, S.Y.: Differentially private data publishing and analysis: a survey. IEEE Trans. Knowl. Data Eng. **29**(8), 1619–1638 (2017)

Semantics-Guided Disentangled Learning for Recommendation

Dianer Yu[1], Qian Li[2(✉)], Xiangmeng Wang[1], Zhichao Wang[3], Yanan Cao[4], and Guandong Xu[1(✉)]

[1] Data Science and Machine Intelligence Lab, School of Computer Science,
University of Technology Sydney, Sydney, Australia
{Dianer.Yu-1,Xiangmeng.Wang}@student.uts.edu.au, Guandong.Xu@uts.edu.au
[2] School of Electrical Engineering, Computing and Mathematical Sciences,
Curtin University, Perth, Australia
qli@curtin.edu.au
[3] University of New South Wales, Sydney, Australia
[4] Institute of Information Engineering, Chinese Academy of Sciences, Beijing, China
caoyanan@iie.ac.cn

Abstract. Although traditional recommendation methods trained on observational interaction information have engendered a significant impact in real-world applications, it is challenging to disentangle users' true interests from interaction data. Recent disentangled learning methods emphasize on untangling users' true interests from historical interaction records, which however overlook auxiliary information to correct bias. In this paper, we design a novel method called **SeDLR** (**Se**mantics **D**isentangled **L**earning **R**ecommendation) to bridge this gap. Particularly, by leveraging rich heterogeneous information networks (HIN), SeDLR is capable of untangling high-order user-item relationships into multiple independent components according to their semantic user intents. In addition, SeDLR offers reliable explanations for the disentangled graph embeddings by the designed Monte Carlo edge-drop component. Finally, we conduct extensive experiments on two benchmark datasets and achieve state-of-the-art performance compared against recent strong baselines.

Keywords: Semantic-aware representation · Disentangled learning · Monte Carlo edge-drop · Explainable recommendation

1 Introduction

Recommendation systems (RS) have become popular personalization tools to assist users in sorting through the ever-growing corpus of content and discovering contents in which they would be interested [3,11,13,22]. Early work mainly used collaborative filtering methods to simply learn user/item ID representation based on historical interactions [1,20,23]. More effective methods exploit interaction as graph-structured data and aggregate feature information from high-order

J. Gama et al. (Eds.): PAKDD 2022, LNAI 13280, pp. 249–261, 2022.
https://doi.org/10.1007/978-3-031-05933-9_20

neighborhoods using neural networks [7,14,20]. Despite effectiveness, modeling user-item relationships via embedding functions fails to differentiate user intents on different items, which could easily lead to suboptimal representations [5,9,21]. The disentangled learning emerges as the state-of-the-art and aims to explore the diverse user-item relationships and learn disentangled representation for users' true interests [4,11,12,21,25].

The principal motivation of disentangled learning is to separate users' intents behind each interaction in order to achieve a robust recommendation. Although disentangled learning has made promising improvements for distilling users' intents, a deficiency is that they emphasize historical interaction records and overlook auxiliary information to correct bias in the recommendation. As shown in Fig. 1, there are four interactions between u_1 and movies (i.e., i_1, i_2, i_3 and i_4). With context information of user and item, we may infer that u_1 prefers to watch a movie with type and director. More importantly, the interaction between u_1 and i_4 might be due to the conformity bias that u_1 tends to watch a movie i_4 that is strongly recommended by a friend u_2, even if this goes against u_1's own preference. Merely using interactions without contextual information fails to capture users' pure interests that are independent of conformity. Therefore, exploiting the contextual information of users (e.g., social relationship) and movie (i.e., director and type) is crucial for distinguishing the conformity bias from users' true interests.

Towards this end, we empower disentangled learning with contextual information, with the aim of discovering users' true interests from the biased interactions and offering explainable recommendations. Overall, the three main contributions of this work are summarized as followings:

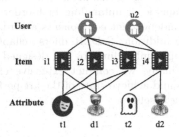

Fig. 1. An illustration of interactions between users and items with contexts.

- To the best of our knowledge, we are the first to incorporate heterogeneous information networks (HIN) into disentangled learning. Our SeDLR model can exploit high-order user-item relationships at the finer granularity and learn disentangled representations towards different semantic-aware aspects.
- We design a Monte Carlo edge-drop strategy, which modifies the HIN structure and drops users' intents-irrelevant semantic information, with the aim of facilitating the explainability of our SeDLR model.

- We conduct extensive experiments on two benchmark datasets to show the superiority and explainability of our SeDLR model.

2   Preliminary and Related Work

In this section, we will introduce recent works that are highly related to ours includes HIN-based learning, graph-based entangled learning and disentangled learning for the recommendation.

Heterogeneous information networks (HIN) include multiple node types and connection relationships, which can flexibly use rich objects and information to model heterogeneous data effectively [18]. HIN enhanced methods leverage meta-path based social relationships derived from rich HIN information, which greatly improve the Top-K recommendation performance. Many HIN-based recommendations have proven the effectiveness of using HIN. For example, IF-BPR [24] propose meta-path based social relations derived from a HIN, then capture the similarity between users for the recommendation. While MCRec [8] uses rich meta-path context representation and attention mechanism.

Graph-based entangled methods learn user/item embeddings by linearly propagating with neighborhood aggregation in the Graph Convolution Network (GCN) component, such as NeuMF [7] and NGCF [20]. NeuMF [7] combines traditional matrix factorization and neural network, which can extract low and high dimensional features at the same time, then concatenate multiple neural network layers with matrix factorization layer to gain the final likelihood score [10]. While NGCF [20] refine the embedding vector from high-order connection information, and integrates by three Graph Neural Network (GNN) layers, then trains by optimizing losses to gain the affinity score of the pair of user and item.

Disentanglement recommendation methods learn users' intents by disentangling users' latent factors, which is more effective to recommend items by knowing the intent rather than the historical records [21]. For instance, DGCF [21] is a state-of-the-art disentanglement recommendation method, which disentangles latent factors of user intents by the neighbor routing and embedding propagation, then applies an independent module to separate intents. M-VAE [16] achieves the macroscopic entanglement by inferring the high-level concepts associated with user intentions, and simultaneously captures user preferences for different items. However, neither M-VAE nor DGCF is able to associate learned intent with real-world users' aspects which can be seen as pre-defined intent.

3   Methodology

The architecture of the proposed SeDLR's framework is shown in Fig. 2. Our method takes the holistic user-item interaction graph with a HIN as the input, and passes through a graph disentangling network (top-left) to divide the holistic interaction graph into q intent-aware sub-graphs for learning the separated user

intent representations, while the HIN embedding network (bottom-left) leverages meta-path schemes retained in the HIN to construct expressive representations of context (i.e., aspects). The learned context representations are then
incorporated into user intent representations to derive the semantic-aware intent
representation for the later recommendation task. Finally, to better explain the
disentangled learning based recommendation, we use Monte Carlo edge-drop
strategy to select the important aspects as explanations (right).

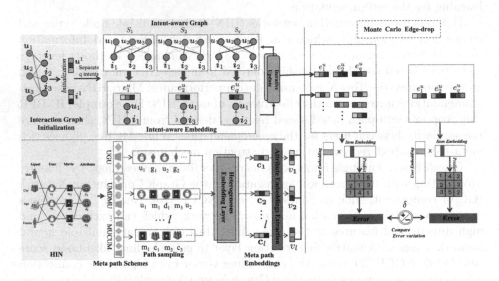

Fig. 2. The overview of the proposed SeDLR framework.

3.1   Graph Disentangling for Users Intents

Our first target is to disentangle q intents of the user as initialization, hence we
divide user/item embedding into q chunks and associate each with a potential
intent as follows:

$$u = (u_1, u_2, \ldots, u_q), \quad i = (i_1, i_2, \ldots, i_q) \tag{1}$$

where u_q and i_q illustrate chunked representation for q-th intent on interaction
of user/item. Additionally, we employ random initialization for each chunk representation to ensure the difference before the training stage. We then adopt a
score vector to explore the relationships between intent and interaction as:

$$S(u, i) = (S_1(u, i), S_2(u, i), \cdots, S_q(u, i)) \tag{2}$$

where $S_q(u, i)$ represents the score vector over q-th intent on interaction, which
is the possibility of adopting interaction is due to q-th intent. Accordingly, a

set of score vectors can be initialized as the same values to indicate the same contribution on interaction before training. Hence, this score vector can be seen as an adjacency matrix for an intent-aware graph.

Next, we design a graph disentangling layer to explore valuable information via the high-order connectivity, we employ a graph disentangling layer that consists of embedding propagation mechanisms and neighbor routing as follows:

$$e_q^{u(1)} = g\left(u_q, \{i_q \mid i \in \mathcal{N}_u\}\right) \tag{3}$$

where $e_q^{u(1)}$ collects information from neighbors of u, and index 1 represents the first-order neighbor. $\mathcal{N}_u$ is the historical interacted items, more formally, it is the first-hop neighbor. We then perform an iterative update rule which is to update the intent-aware embedding by embedding propagation in the intent-aware graph, then use it to refine the graph. Consequently, score vector S_q for each interaction after n iteration can be generated. To get its distribution across all intents, we normalized by softmax as:

$$\tilde{S}_q^n(u,i) = \frac{\exp S_q^n(u,i)}{\sum_{q'=1}^{q} \exp S_{q'}^n(u,i)} \tag{4}$$

to illustrate the importance of each intent. Accordingly, we can obtain normalized adjacency matrix $\tilde{S}_q^n$ for each intent. The Laplacian matrix of $\tilde{S}_q^n$ is adopted as:

$$\mathcal{M}_q^n(u,i) = \frac{\tilde{S}_q^n(u,i)}{\sqrt{D_q^n(u) \cdot D_q^n(i)}} \tag{5}$$

where $D_q^n(u) = \sum_{i' \in \mathcal{N}_u} \tilde{S}_q^n(u,i')$ and $D_q^n(i) = \sum_{u' \in \mathcal{N}_i} \tilde{S}_q^n(u',i)$ are the degrees of u and i, respectively. Besides, the embedding propagation for each graph can encode the information influenced to the interaction and the sum aggregator is defined as:

$$u_q^n = \sum_{i \in N_u} \mathcal{M}_q^n(u,i) \cdot i_q^0 \tag{6}$$

where u_q^n illustrates the sum of historical items and importance weighting in q-th aspect at n-th iteration, and i_q^0 is the input representation for the historical item. It can temporarily memorize the information collected from neighbors $\mathcal{N}_u$.

Thereafter, we iteratively the update intent-aware graph. Intuitively, interacted items driven by the same intent tend to have similar chunked representations, encouraging stronger relationships between them can achieve this purpose. Thus, we iteratively update the interaction score vector $S_q^n(u,i)$ to adjust the degree of u and neighbor i as follows:

$$S_q^{n+1}(u,i) = S_q^n(u,i) + u_q^{n\top} \tanh\left(i_q^0\right) \tag{7}$$

where $u_q^{n\top} \tanh\left(i_q^0\right)$ represents the affinity between u_q^n and i_q^0 in Eq. (6), while $tanh$ is a nonlinear activation function can improve the representation ability.

Finally, output a graph disentangling layer after n iterations, that contains disentangled representation $e_q^{u(1)} = u_q^n$ and intent-aware graph $\tilde{S}_q^n$.

We then combine multiple layers to gather rich semantics from high-order connectivity. While the first-order neighbors have been used above, hence we can stack r layers to obtain influence signals from r-th high-order connectivity as:

$$e_q^{u(r)} = g\left(e_q^{u(r-1)}, \left\{e_q^{i(r-1)} \mid i \in \mathcal{N}_u\right\}\right) \tag{8}$$

where $e_q^{u(r-1)}$ and $e_q^{i(r-1)}$ serve as the representations of u and i on q-th intent which save the propagated information from $(r-1)$-hop neighbors. Every disentangled representation is associated with explanatory graph serve as weighted adjacency matrix $\tilde{S}_q^r$. We can sum up the intent-aware representations after r layers as $e_q^u = \left(e_q^{u(0)}, e_q^{u(1)}, \cdots, e_q^{u(r)}\right)$ and $e_q^i = \left(e_q^{i(0)}, e_q^{i(1)}, \cdots, e_q^{i(r)}\right)$ for u and i, respectively. Furthermore, we summarized user/item representations as $e^u = (e_1^u, \cdots, e_q^u)$, $e^i = (e_1^i, \cdots, e_q^i)$, respectively.

3.2   Semantic-Aware Intent Representation Learning

In this section, we aim to extract aspect embeddings from meta-paths of the rich HIN context. The HIN, which records different types of relationships between users and items, carries diverse semantics and is beneficial to intent representation learning. Specifically, such semantics can be reflected in the meta-path schemes of the given HIN, which is some paths defined composites of different node types with diverse edge types. By characterizing meta-path schemes, the complex relations of the involved nodes can be captured, reflecting higher-level semantics to augment user intent learning. Taking the UMU as an example, the path sequence $U_{u_1} - M_{m_1} - U_{u_2}$ defined under such a meta-path can reflect the behavior similarity of u_1 and u_2, while the social influence of u_2 to u_1 is the important aspect that may affect the intent of u_1. This motivates us to leverage the aspect embeddings modeled from meta-paths as the context to refine the user intent representations.

Formally, given the pre-defined meta-path $\mathbf{p}$, we should firstly generate a series of high-quality path instances $\rho = \{u_1, u_2, \cdots, u_l\}$. Here we resort to *Meta-path Based Random Walks* [2], which is a wildly used path sampling strategy that generates path instances that constitute multiple types of nodes, under a specific meta-path $\mathbf{p}$ to further capture both the semantics and structural correlations between various types of nodes. Then we learn the embeddings of the acquired path instances ρ by a Convolution Neural Network (CNN) [6,15] parameterized by Θ, then adopt the max-pooling operation to derive the final embedding for a meta-path $\mathbf{p}$ by aggregating the embeddings of L selected path instances:

$$\mathbf{c_p} = \text{max-pooling}\left(\{CNN(\{\mathbf{X}_i^\rho\}; \Theta)\}_{i=1}^L\right) \tag{9}$$

where $\{\mathbf{X}_i^\rho\}$ means the set of embeddings for L path instances of meta-path $\mathbf{p}$. Each $\mathbf{X}_i^\rho$ illustrates the embedding matrix.

The meta-paths carry important semantic meanings, which can guide the intent learning of users. We propose to extract the semantic embedding from meta path embedding c_p, serving as the context information that waits to be incorporated into the latter semantic-aware intent learning. Specifically, the semantic representation of user u can be derived by the embedding lookup operation as:

$$v_p = c_p^\top \cdot u \tag{10}$$

where $u \in \mathbb{R}^{1 \times |\mathcal{U}|}$ is the one-hot encoding of user u. The learned v_p is then serves as the aspect embeddings for all user $u \in \mathcal{U}$ under meta-path p. Therefore, we can extract and generate all the aspect embeddings under different meta-paths.

We then perform the semantic-aware intent learning from intent representation e^u and e^i for users and items in Eq. (3), and aim to incorporate semantics as retained in v_p for users and items to learn the semantic-aware intent representations for the latter recommendation. Towards this end, we design a Factorization Machine (FM) operator to instantiate semantics-aware intent representation h_p, which denotes the user intent towards different aspects under meta-path p. Formally, we now have obtained $v_p \in \mathbb{R}^{1 \times d}$ as the semantics-aware representation and the intent-aware representations $e^u = (e_1^u, \cdots, e_q^u) \in \mathbb{R}^{1 \times d}$ for user u. Then h_p can be calculated by a FM module:

$$h_p = e^u \odot v_p \tag{11}$$

where $\odot$ denotes the element-wise product.

Lastly, we perform optimization for model parameters. In detail, the semantics-aware intent representation h_p can be incorporated into recommendation models as one additional user representation. Formally, we use the collaborative filtering to calculate the prediction score $\hat{y}_{ui}$ given user and item ID representations as follows:

$$\hat{y}_{ui} = \alpha u^\top i + (1 - \alpha) h_p^\top i \tag{12}$$

where u and i are the ID embeddings given by id mapping techniques in Eq. (1), such as Multi-OneHot [17] and α is the coefficient that describes how much each component contributes to the prediction score. After obtaining the final representation for user/item, we optimize the parameters for h_p in Eq. (11) by using Bayesian Personalized Ranking (BPR) loss, which encourages the prediction of an observed to be higher than its unobserved counterparts user:

$$\mathcal{L}_{BPR} = \sum_{u,i,j \in \mathcal{D}} -\ln \sigma(\hat{y}_{ui} - \hat{y}_{uj}) + \lambda \|E\|_2^2 \tag{13}$$

where $\mathcal{D} = \{(u, i) : u \in U, i \in I, j \in I\}$ is the training set and E is the embedding matrix of all users and items.

3.3   Monte Carlo Edge-Drop for Explainability

To further explain the recommendation we propose a novel strategy namely Monte Carlo edge-drop, which aims to provide explainable semantics for the recommendation. By optimizing Eq. (13), we finally produce our prediction model denoted as $f(\cdot)$. We further conduct an inference with a HIN with an edge b removed from meta-path $\mathbf{p}$, i.e., removing the influence of attribute b, which generates the prediction as $\hat{y}_{ui}^s$. Thereafter, we define a criterion, which denotes the absolute error variation between $\hat{y}_{ui}^s$ and the original prediction $\hat{y}_{ui}$, to determine the importance of attribute b. If the variation is greater than a threshold δ, we then claim this aspect is influential since it has a significant impact on the prediction.

4   Experiments

In order to thoroughly evaluate and analyze the proposed methodology, we conducted extensive experiments to answer the following research questions:

- **(RQ1)** How does our method compare with other state-of-the-art models?
- **(RQ2)** How does the threshold δ in Monte Carlo edge-drop strategy improve Top-K recommendation?
- **(RQ3)** How does our method explain users' aspects and provide semantic information for the recommendation?

4.1   Settings

We conduct extensive experiments on two publicly available datasets: *Walmart Recruit*[1], and *Douban Book*[2]. *Walmart Recruit* contains historical retail data from 2011 to 2013 as HIN context includes price, discount, user, gender, category type and city and has been widely used for recommendation related research [19]. The ratings of *Walmart Recruit* are the user's rating number of transactions. *Douban Book* includes rich HIN information such as 3 attributes for the user and 4 attributes for the book. The ratings of *Douban Book* are the user's rating number of books. For both two datasets, we binarize the feedback data (i.e., ratings) by interpreting ratings of 5 or higher as positive feedback (i.e., $r = 1$) or lower as negative feedback (i.e., $r = 0$). Moreover, we use negative sampling to randomly sample unobserved items and pair them with the user as negative instances. The statistics detail are summarized in Table 1.

---

[1] https://www.kaggle.com/c/walmart-recruiting-store-sales-forecasting.
[2] https://github.com/librahu/HIN-Datasets-for-Recommendation-and-Network-Embedding/tree/master/Douban%20Book.

Table 1. Statistic details: density is $\#Connections/(\#Users \cdot \#Items)$, relation is connection number and Avg.Degree of A is $\#Relation/\#A$.

Dataset (density)	Node	Relation A-B	Avg.Degree of A/B
Walmart Recruit (0.11%)	#User(U): 5,647	#U-G: 5,645	#U/G: 1/2822.5
	#Gender(G): 2	#U-C: 5,645	#U/C: 1/564.5
	#City(C): 10	#U-T: 23,053	U/T: 4.1/1.1
	Transaction(T): 20,878	#U-U: 0	U/U: 0/0
	#Category Type(CT): 5	#T-A: 23,053	#T/A: 1.1/4.0
	#Amount(A): 5,764	#T-CT: 23,053	#T/CT: 1.1/4610.6
Douban Book (0.27%)	#User(U): 13,024	#U-Bo: 792,062	#U/Bo: 60.8/35.4
	#Book(Bo): 22,347	#U-U: 169,150	#U/U: 13.0/13.0
	#Group(Gr): 2,936	#U-Gr: 1,189,271	#U/Gr: 91.3/405.1
	#Author(Au): 10,805	#Bo-Au: 21,907	#Bo/Au: 1.0/2.0
	#Publisher(P): 1,815	#Bo-P: 21,773	#Bo/P: 1.0/12.0
	#Year(Y): 64	#Bo-Y: 21,192	#Bo/Y: 1.0/331.1

All experiments are conducted on a Linux server with RTX3070 GPU. We adopt three popular metrics including Recall@K, NDCG@K, and Precision@K, where K is set as 1, 10, 20 and 40 in Table 2. Both two datasets are split as a proportion of 80%/10%/10%, train/test/validate set, respectively. A grid search is used to find the best parameter settings. The embedding size is initialized with Xavier and searched in $\{16, 32, 64, 128\}$, and learning rate is in $\{0.001, 0.01, 0.05, 0.1\}$. The maximum epoch is set as 1000 with an early stopping strategy. Default hyperparameters of SeDLR are: embedding size 128, disentangled layer iteration number $n = 3$, latent intent number $q = 4$, learning rate 0.01. We compare our proposed SeDLR with three kinds of state-of-the-art recommendation methods: (1) HIN-based methods including IF-BPR [24] and MCRcc [8]; (2) Graph-based entangled methods including NeuMF [7] and NGCF [20]; (3) Disentangled-based methods including DGCF [21] and M-VAE [16][3].

4.2 RQ1 Performance Comparison

To understand the performance of SeDLR, we adopt deep comparison with multiple state-of-the-art models on Top-K recommendations. The overall statistical outcomes can be found in Table 2. On both two datasets, our SeDLR consistently outperforms all other approaches. Especially, SeDLR improves over the strongest baselines at NDCG@20 by 27.7% and 15.2% on *Walmart Recruit*, and *Douban Book*, respectively. Specifically, most improvements are more than 10%, which validates the Monte Carlo edge-drop has critical effects on improving

---

[3] Refer to related work for more details of baselines.

Table 2. Overall performance comparison: the best results are marked as bold, strongest baselines are marked with underline.

Datasets	Metrics	NeuMF	NGCF	DGCF	M-VAE	IF-BPR	MCRec	SeDLR	Improv.
Walmart Recruit	Recall@1	0.0376	0.0299	0.0421	0.0391	0.0385	0.0381	**0.0476**	13.1%
	Recall@10	0.0401	0.0387	0.0447	0.0472	0.0419	0.0437	**0.0512**	8.5%
	Recall@20	0.0451	0.0430	0.0516	0.0509	0.0479	0.0448	**0.0552**	7.0%
	Recall@40	0.0612	0.0582	0.0572	0.0519	0.0556	0.0622	**0.0672**	8.0%
	Precision@1	0.0301	0.0315	0.0357	0.0322	0.0316	0.0351	**0.0417**	16.8%
	Precision@10	0.0457	0.0385	0.0477	0.0369	0.0399	0.0426	**0.0516**	8.2%
	Precision@20	0.0528	0.0497	0.0519	0.0489	0.0462	0.0512	**0.0556**	5.3%
	Precision@40	0.0609	0.0599	0.0712	0.0603	0.0591	0.0621	**0.0776**	9.0%
	NDCG@1	0.0201	0.0315	0.0362	0.0288	0.0291	0.0343	**0.0415**	14.6%
	NDCG@10	0.0341	0.0392	0.0448	0.0429	0.0409	0.0422	**0.0512**	14.3%
	NDCG@20	0.0396	0.0499	0.0513	0.0489	0.0502	0.0511	**0.0591**	15.2%
	NDCG@40	0.0670	0.0689	0.0711	0.0676	0.0709	0.0712	**0.0823**	15.6%
Douban Book	Recall@1	0.0267	0.0205	0.0333	0.0301	0.0329	0.0324	**0.0387**	16.2%
	Recall@10	0.0311	0.0377	0.0411	0.0339	0.0362	0.0401	**0.0458**	11.4%
	Recall@20	0.0339	0.0252	0.0431	0.0309	0.0396	0.0478	**0.0515**	7.7%
	Recall@40	0.0641	0.0707	0.0749	0.0691	0.0628	0.0481	**0.0801**	6.9%
	Precision@1	0.0302	0.0344	0.0351	0.0325	0.0281	0.0294	**0.0401**	14.2%
	Precision@10	0.0391	0.0402	0.0415	0.0378	0.0356	0.0352	**0.0476**	14.7%
	Precision@20	0.0420	0.0495	0.0538	0.0322	0.0376	0.0309	**0.0541**	0.6%
	Precision@40	0.0599	0.0618	0.0725	0.0425	0.0564	0.0468	**0.0745**	2.8%
	NDCG@1	0.0301	0.0295	0.0327	0.0341	0.0205	0.0202	**0.0395**	15.8%
	NDCG@10	0.0356	0.0441	0.0457	0.0401	0.0398	0.0268	**0.0552**	20.8%
	NDCG@20	0.0391	0.0301	0.0502	0.0425	0.0463	0.0294	**0.0641**	27.7%
	NDCG@40	0.0682	0.0691	0.0663	0.0645	0.0601	0.0507	**0.0813**	19.2%

recommendation performance. Additionally, disentangled methods achieved better results than the other two methods in most cases, which justifies the disentangled representation has a better performance by separating intents, therefore our SeDLR adopts it in our method.

4.3   RQ2 Aspect Threshold Influence

We conduct extensive experiments to explore the influence of aspect threshold δ in Monte Carlo edge-drop strategy for three popular metrics Recall@K, NDCG@K, and Precision@K on the recommendation. The empirical results can be found in Fig. 3. Through the comparison, we observed the highest accuracy existing in the δ value 0.6 for all three metrics on both two datasets with K@20 and K@40. Then the accuracy drops dramatically later, which is reasonable since the Monte Carlo edge-drop start to filter aspects from 0 and leads to improvement. But the accuracy has decreased when dropped too many aspects, which is a lack of inputs. Accordingly, we summarized with 0.6 is the best aspect threshold on HIN-based disentangled network recommendation.

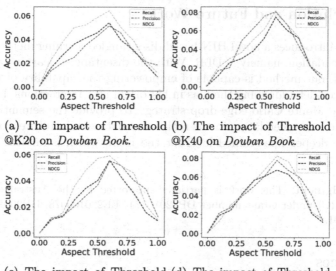

(a) The impact of Threshold @K20 on *Douban Book*.

(b) The impact of Threshold @K40 on *Douban Book*.

(c) The impact of Threshold @K20 on *Walmart Recruit*.

(d) The impact of Threshold @K40 on *Walmart Recruit*.

Fig. 3. The influence of aspect threshold δ in Monte Carlo edge-drop strategy on Top-K recommendation evaluated by Recall@K, NDCG@K and Precision@K.

4.4   RQ3 Model Explainability and Visualization

We visualize two case studies include two users and one item from *Walmart Recruit* to gain a deeper understanding of SeDLR's explainability in Fig. 4. By jointly analyzing interaction and aspect, we find the aspect differ across each interaction, which is reflected by score values. For instance, user 268136 only keeps male aspect scoring 0.78, and man store aspects scoring 0.62, under thresh old δ value 0.6. It can provide semantics meaning that a male user is likely to interact with an item laptop at the man store. These results show SeDLR not only can effectively untangle users' intents but also add semantic supplements for learned intents.

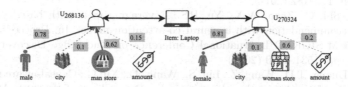

Fig. 4. A visualization of two case studies from *Walmart Recruit*, the bold lines represent retained aspects after Monte Carlo edge-drop strategy with threshold δ value 0.6.

5   Conclusion and Future Work

This paper introduces a novel HIN-based disentangled learning method for Top-K recommendation, namely SeDLR. With the disentangled learning augmented by the HIN, our method is capable of empowering the capability of the recommendation model addressing the bias in historical user interactions. In addition, we resort to Monte Carlo edge-drop strategy to provide the semantic explanations for the recommendation in the real-world datasets. In future work, we will explore the deeper fine-grained level for the item, which is another promising direction.

Acknowledgment. This work is partially supported by the Australian Research Council (ARC) under Grant number DP22010371, LE220100078, DP200101374, and LP170100891.

References

1. Berg, R.V.D., Kipf, T.N., Welling, M.: Graph convolutional matrix completion. arXiv preprint arXiv:1706.02263 (2017)
2. Dong, Y., Chawla, N.V., Swami, A.: metapath2vec: scalable representation learning for heterogeneous networks. In: Proceedings of the 23rd ACM SIGKDD International Conference on Knowledge Discovery and Data Mining. pp. 135–144 (2017)
3. Duong, T.D., Li, Q., Xu, G.: Stochastic intervention for causal effect estimation. In: 2021 International Joint Conference on Neural Networks (IJCNN), pp. 1–8. IEEE (2021)
4. Duong, T.D., Li, Q., Xu, G.: Stochastic intervention for causal effect estimation. arXiv preprint arXiv:2105.12898 (2021)
5. Duong, T.D., Li, Q., Xu, G.: Stochastic intervention for causal inference via reinforcement learning. arXiv preprint arXiv:2105.13514 (2021)
6. He, X., Deng, K., Wang, X., Li, Y., Zhang, Y., Wang, M.: LightGCN: simplifying and powering graph convolution network for recommendation. In: Proceedings of the 43rd International ACM SIGIR Conference on Research and Development in Information Retrieval, pp. 639–648 (2020)
7. He, X., Liao, L., Zhang, H., Nie, L., Hu, X., Chua, T.S.: Neural collaborative filtering. In: Proceedings of the 26th International Conference on World Wide Web, pp. 173–182 (2017)
8. Hu, B., Shi, C., Zhao, W.X., Yu, P.S.: Leveraging meta-path based context for top-n recommendation with a neural co-attention model. In: Proceedings of the 24th ACM SIGKDD International Conference on Knowledge Discovery & Data Mining, pp. 1531–1540 (2018)
9. Li, Q., Duong, T.D., Wang, Z., Liu, S., Wang, D., Xu, G.: Causal-aware generative imputation for automated underwriting. In: Proceedings of the 30th ACM International Conference on Information & Knowledge Management, pp. 3916–3924 (2021)
10. Li, Q., Niu, W., Li, G., Cao, Y., Tan, J., Guo, L.: Lingo: linearized grassmannian optimization for nuclear norm minimization. In: Proceedings of the 24th ACM International on Conference on Information and Knowledge Management, pp. 801–809 (2015)

11. Li, Q., Wang, X., Xu, G.: Be causal: de-biasing social network confounding in recommendation. arXiv preprint arXiv:2105.07775 (2021)
12. Li, Q., Wang, Z., Li, G., Pang, J., Xu, G.: Hilbert Sinkhorn divergence for optimal transport. In: Proceedings of the IEEE/CVF Conference on Computer Vision and Pattern Recognition, pp. 3835–3844 (2021)
13. Li, Q., Wang, Z., Liu, S., Li, G., Xu, G.: Causal optimal transport for treatment effect estimation. IEEE Trans. Neural Netw. Learn. Syst. (2021)
14. Li, Q., Wang, Z., Liu, S., Li, G., Xu, G.: Deep treatment-adaptive network for causal inference. arXiv preprint arXiv:2112.13502 (2021)
15. Lo, S.C.B., Chan, H.P., Lin, J.S., Li, H., Freedman, M.T., Mun, S.K.: Artificial convolution neural network for medical image pattern recognition. Neural Netw. 8(7–8), 1201–1214 (1995)
16. Ma, J., Zhou, C., Cui, P., Yang, H., Zhu, W.: Learning disentangled representations for recommendation. arXiv preprint arXiv:1910.14238 (2019)
17. Rodríguez, P., Bautista, M.A., Gonzalez, J., Escalera, S.: Beyond one-hot encoding: lower dimensional target embedding. Image Vis. Comput. **75**, 21–31 (2018)
18. Shi, C., Hu, B., Zhao, W.X., Philip, S.Y.: Heterogeneous information network embedding for recommendation. IEEE Trans. Knowl. Data Eng. **31**(2), 357–370 (2018)
19. Taghizadeh, E.: Utilizing artificial neural networks to predict demand for weather-sensitive products at retail stores. arXiv preprint arXiv:1711.08325 (2017)
20. Wang, X., He, X., Wang, M., Feng, F., Chua, T.S.: Neural graph collaborative filtering. In: Proceedings of the 42nd International ACM SIGIR Conference on Research and Development in Information Retrieval, pp. 165–174 (2019)
21. Wang, X., Jin, H., Zhang, A., He, X., Xu, T., Chua, T.S.: Disentangled graph collaborative filtering. In: Proceedings of the 43rd International ACM SIGIR Conference on Research and Development in Information Retrieval, pp. 1001–1010 (2020)
22. Xu, G., Duong, T.D., Li, Q., Liu, S., Wang, X.: Causality learning: a new perspective for interpretable machine learning. arXiv preprint arXiv:2006.16789 (2020)
23. Ying, R., He, R., Chen, K., Eksombatchai, P., Hamilton, W.L., Leskovec, J.: Graph convolutional neural networks for web-scale recommender systems. In: Proceedings of the 24th ACM SIGKDD International Conference on Knowledge Discovery & Data Mining, pp. 974–983 (2018)
24. Yu, J., Gao, M., Li, J., Yin, H., Liu, H.: Adaptive implicit friends identification over heterogeneous network for social recommendation. In: Proceedings of the 27th ACM International Conference on Information and Knowledge Management, pp. 357–366 (2018)
25. Zheng, Y., Gao, C., Li, X., He, X., Li, Y., Jin, D.: Disentangling user interest and conformity for recommendation with causal embedding. In: Proceedings of the Web Conference 2021, pp. 2980–2991 (2021)

Multi-task Knowledge Graph Representations via Residual Functions

Adit Krishnan[1]([⊠]), Mahashweta Das[2], Mangesh Bendre[2], Fei Wang[2], Hao Yang[2], and Hari Sundaram[1]

[1] University of Illinois at Urbana-Champaign, Champaign, USA
{aditk2,hs1}@illinois.edu
[2] Visa Research, Palo Alto, CA, USA
{mahdas,mbendre,feiwang,haoyang}@visa.com

Abstract. In this paper, we propose MuTATE, a Multi-Task Augmented approach to learn Transferable Embeddings of knowledge graphs. Previous knowledge graph representation techniques either employ task-agnostic geometric hypotheses to learn informative node embeddings or integrate task-specific learning objectives like attribute prediction. In contrast, our framework unifies multiple co-dependent learning objectives with knowledge graph enrichment. We define co-dependence as multiple tasks that extract covariant distributions of entities and their relationships for prediction or regression objectives. We facilitate knowledge transfer in this setting: tasks→graph, graph→tasks, and task-1→task-2 via task-specific residual functions to specialize the node embeddings for each task, motivated by domain-shift theory. We show 5% relative gains over *state-of-the-art* knowledge graph embedding baselines on two public multi-task datasets and show significant potential for cross-task learning.

Keywords: Knowledge graphs · Knowledge graph embedding · Graph neural networks · Multi-task learning · Residual learning

1 Introduction

Knowledge graphs enable versatile storage, visualization, interpretation, and manipulation of large volumes of contextual information across interacting entities (nodes) via relations (links) in diverse domains such as linguistics (Wang et al. (2013)), biomedicine (Ernst et al. (2015)) and finance (Cheng et al. (2020)). The transitive entity association structure enhances inferencing applications involving entity attribute prediction and *entity-to-entity* relation prediction. However, the persistent challenges with knowledge graphs are two-fold, *link sparsity and its* task-unaware inflexible structure (Huang et al. (2019); Wang et al. (2014)). To overcome these challenges, a popular direction is to embed knowledge graphs in dense vector spaces (Bordes et al. (2013); Wang et al. (2014); Sun et al. (2019)) via path-based patterns such as *symmetry, anti-symmetry, composition* and *analogy* (Sect. 3.1). However, these learned patterns are static

J. Gama et al. (Eds.): PAKDD 2022, LNAI 13280, pp. 262–275, 2022.
https://doi.org/10.1007/978-3-031-05933-9_21

and not task-specific. To address this, the second direction integrates knowledge graph embeddings with specific learning tasks (Huang et al. (2019); Wang et al. (2019a)). In this case, the node/link embeddings are optimized for a single-task, but cannot combine or benefit multiple tasks.

Unlike these two directions, our approach unifies multi-task learning, graph enrichment, and embedding learning. We specifically focus on co-dependent tasks, i.e., tasks depending on shared aspects of the graph structure. As an example, we consider two well-defined prediction objectives in Fig. 1, book recommendation and book genre prediction. We consider a collaborative recommender model on the user-book links and a prediction model on the book-genre links.

These two task-models extract task-biased views of the knowledge graph depending on their *inductive biases*. However, both tasks (recommendation, genre prediction) require accurate book embeddings, i.e., shared subspace of the joint (*User, Item, Genre*) latent distribution. Further, each model can address link sparsity in the graph by predicting new links of the same type, thus transferring the extracted knowledge back to the graph. These newly predicted links represent the task-biased distribution learned by each model. Combining multiple tasks in this manner jointly enriches the graph as well as the other tasks through their shared subspaces. In summary, our contributions are as follows:

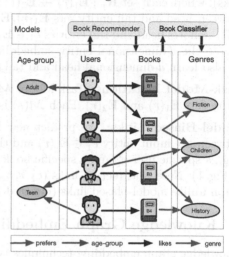

Fig. 1. Toy knowledge graph with four entity types: *users, books, age-groups, genres*. Entities are linked via user *prefers* genre, user *in* age-group, user *likes* book, book *is genre* relations. *Sample task-models include recommendation and book genre prediction.*

Merging Multi-task Learning and Knowledge Graph Embedding/Knowledge Graph Enrichment: We propose a holistic view of knowledge graphs and multi-task learning to enable bidirectional knowledge transfer between the graph and multiple co-dependent learning objectives.

Generalizability: The proposed framework makes no assumptions about the data-domain or learning tasks. We validate this empirically.

Modeling Multi-task Embedding Updates via Residuals: We identify the connection between multi-task knowledge graph updates and covariate shift (Johansson et al. (2016)) to unify multiple task distributions over shared node embeddings via task-specific residual functions.

Strong Experimental Results: We demonstrate strong experimental results on knowledge graphs constructed from two large public datasets, *Google Local*

Reviews[1] (He et al. (2017); Pasricha and McAuley (2018)) and *Yelp Challenge*[2] and using two co-dependent task-models, word2vec (Mikolov et al. (2013)) and a context-aware recommender (Krishnan et al. (2020)).

2   Problem Definition

Knowledge Graph Notations: We consider a heterogeneous directed knowledge graph with multiple entity (node) types, $\mathcal{E} = \{\mathbf{E}_1, \mathbf{E}_2 \cdots \mathbf{E}_{|\mathcal{E}|}\}$.

Factual Links: $\mathcal{R} = \{\mathbf{R}_1, \mathbf{R}_2 \cdots \mathbf{R}_{|\mathcal{R}|}\}$ is the set of all links (called factual links), where each set $\mathbf{R}_r : \mathbf{E}_1(r) \to \mathbf{E}_2(r)$ is a specific relation $r \in \{1, 2, \cdots |\mathcal{R}|\}$ between head and tail entity sets $\mathbf{E}_1(r), \mathbf{E}_2(r) \in \mathcal{E}$. Each factual link $(e_1, r, e_2) \in \mathbf{R}_r$ denotes head and tail entities $e_1 \in \mathbf{E}_1(r)$, $e_2 \in \mathbf{E}_2(r)$ with relation r. $\vec{e}_1, \vec{e}_2$ denote the d-dimensional entity embeddings of e_1 and e_2. For each relation r, we also learn d-dimensional head and tail embedding projectors $(\vec{p}_1(r), \vec{p}_2(r))$.

Task-Model Notations: Task-Model $\mathcal{M}(r)$ predicts relation-r links between entity sets $\mathbf{E}_1(r)$ and $\mathbf{E}_2(r)$. Each $\mathcal{M}(r)$ is trained with factual links $\mathbf{R}_r$.

Model-Biased Links: We predict new links (e_1', r, e_2') via task-model $\mathcal{M}(r)$ between the input entity $e_1' \in \mathbf{E}_1(r)$ and the model predicted output $e_2' \in \mathbf{E}_2(r)$ (e.g., a specific user e_1' and a specific book e_2' from the recommender task-model in Fig. 1). Note that factual links $(e_1, r, e_2) \in \mathbf{R}_r$ exist apriori in the knowledge graph unlike model-biased links (denoted $(e_1', r, e_2') \in \mathbf{R}'_r$).

3   Knowledge Graph Embeddings

Knowledge graph embedding techniques typically encode static node connectivity pattens to mitigate link-sparsity (Sun et al. (2019)) such as:

- **Symmetry:** $(e_1, r_a, e_2) \implies (e_2, r_a, e_1)$
- **Anti-Symmetry:** $(e_1, r_a, e_2) \implies not\ (e_2, r_a, e_1)$
- **Inversion:** $(e_1, r_a, e_2) \implies (e_2, r_b, e_1)$
- **Composition:** (e_1, r_a, e_2) *and* $(e_2, r_b, e_3) \implies (e_1, r_c, e_3)$
- **Analogy:** (e_1, r_a, e_2) *and* $(e_3, r_a, e_4) \implies (e_1, r_b, e_3)/(e_2, r_c, e_4)$

None of these first-cut patterns are task-specific. Prior approaches in this vein do not provide mechanisms for task-adaptation or multi-task learning. We formalize task-to-task knowledge transfer as follows:

- How do we leverage links (e_1, r_a, e_2) for link predictions of the form (e_1, r', e'), (e_2, r', e'), (e'', r'', e_1), (e'', r'', e_2)?

Note that the solution to the above transfer learning is specific to the relation types r_a, r', r'' as well the entity nodes e_1 and e_2, and thus can be combined with task-models $\mathcal{M}(r)$ involving these entities or relations. We thus propose a two-step solution where we first leverage the static patterns to generate first-cut embeddings and then augment them with task-specific residual functions (Sect. 3.2) to enable adaptation to the respective task-models.

---

[1] http://cseweb.ucsd.edu/~jmcauley/datasets.html.
[2] https://www.yelp.com/dataset/challenge.

3.1   Link Embedding Model

Parallelizable embedding learning is critical for knowledge graph applications owing to their massive sizes. DistMult (Yang et al. (2014)) describes a block-optimizable bilinear form with a learnable diagonal embedding projector ($\mathbf{P}_r$) for each relation type r Lerer et al. (2019). Under this approach, the likelihood of a link (e_1, r, e_2) is given by:

$$\mathcal{L}(\vec{e}_1, r, \vec{e}_2) = \vec{e}_1^T \mathbf{P}_r \vec{e}_2 \tag{1}$$

However, due to the symmetric nature of the above transformation, it cannot encode anti-symmetry and inversion patterns (Sun et al. (2019)). In contrast, other methods that do not have a symmetric objective wrt. head and tail entities (e.g., Sun et al. (2019)) pose block optimization constraints. To overcome these limitations, we break the symmetry in Eq. (1) by describing two projectors (for the head and tail entity embeddings) for each relation type. Our form adds twice as many relation-specific projectors. However, the number of relation-types is typically orders of magnitude less than the number of nodes so that the overhead is insignificant. We now define the likelihood of a link (e_1, r, e_2):

$$\mathcal{L}(\vec{e}_1, r, \vec{e}_2) = cosine\text{-}sim\Big(\vec{e}_1 \otimes \vec{p}_1(r), \ \vec{e}_2 \otimes \vec{p}_2(r)\Big) \tag{2}$$

The above modification enables composition, inversion, and anti-symmetry:

- **Anti-Symmetry:** Consider relations r_a to be anti-symmetric, so that, $(e_1, r_a, e_2) \implies not \ (e_2, r_a, e_1)$ We can encode this in our likelihood term with orthogonal projectors for the head and tail, i.e., $\vec{p}_1(r) \perp \vec{p}_2(r)$ so that we take the orthogonal projections of the head and tail entity when the direction of the relation is reversed.
- **Inversion:** Consider relations r_a, r_b to be inversions of each other, so that, $(e_1, r_a, e_2) \implies (e_2, r_b, e_1)$ We can encode this in our likelihood term by switching the head and tail projectors, i.e., $\vec{p}_1(r_a) = \vec{p}_2(r_b)$ and $\vec{p}_2(r_a) = \vec{p}_1(r_b)$. It is easy to verify that this would result in $\mathcal{L}(\vec{e}_1, r_a, \vec{e}_2) = \mathcal{L}(\vec{e}_2, r_b, \vec{e}_1)$ which results in the desired inversion.
- **Composition:** Relation r_c composes r_a and r_b if $(e_1, r_a, e_2), (e_2, r_b, e_3) \implies (e_1, r_c, e_3)$. We can encode this in our likelihood terms with the following simple switch, i.e., $\vec{p}_1(r_c) = \vec{p}_1(r_a)$ and $\vec{p}_2(r_c) = \vec{p}_2(r_a)$. This would transitively align the composed relation with the head and tail entities e_1 and e_3.

Finally, we also add a scale factor to Equation (2) ($sim = cosine\text{-}similarity$):

$$\mathcal{L}(\vec{e}_1, r, \vec{e}_2) = sim\Big(\vec{e}_1 \otimes (\vec{p}_1(r) + s\mathbb{I}), \ \vec{e}_2 \otimes (\vec{p}_2(r) + s\mathbb{I})\Big) \tag{3}$$

In the next subsection, we describe task-specific embedding adaptation and link-sparsity mitigation on the first-cut factual embeddings from Eq. (3).

3.2   Embedding Augmentation via Model-Biased Links

Consider the prediction task for relation r between entity sets $\mathbf{E}_1(r)$, $\mathbf{E}_2(r)$. Task-model $\mathcal{M}(r)$ predicts model-biased links $(e_1'(r), r, e_2'(r))$ where $e_1'(r) \in \mathbf{E}_1(r)$, $e_2'(r) \in \mathbf{E}_2(r)$, from its inferred co-occurrence distribution. In this manner, each $\mathcal{M}(r)$ generates model-biased links $\mathbf{R}'_r$ different from the factual links $\mathbf{R}_r$ of the same relation type. Under Eq. (3), the likelihood of each factual link $(e_1, r, e_2) \in \mathbf{R}_r$ is given by:

$$\mathcal{L}(\vec{\mathbf{e}}_1, r, \vec{\mathbf{e}}_2) = sim\Big(\vec{\mathbf{e}}_1 \otimes (\vec{\mathbf{p}}_1(r) + s\mathbb{I}), \quad \vec{\mathbf{e}}_2 \otimes (\vec{\mathbf{p}}_2(r) + s\mathbb{I}) \Big) \tag{4}$$

Upon optimization, we obtain the first-cut factual embedding space $\vec{\mathbf{E}}$ with the latent factual embedding distribution $P(\vec{\mathbf{E}})$. However, each task-model $\mathcal{M}(r)$ represents a co-occurrence distribution between entity sets $\mathbf{E}_1(r)$, $\mathbf{E}_2(r)$ which differs from those in $P(\vec{\mathbf{E}})$, depending on the specific task and the model-architecture (inductive bias). We thus learn model-specific embedding spaces $\vec{\mathbf{E}}'_r$ by optimizing Eq. (3) over the model-biased links $\mathbf{R}'_r$ instead of $\mathbf{R}_r$ (Fig. 2).

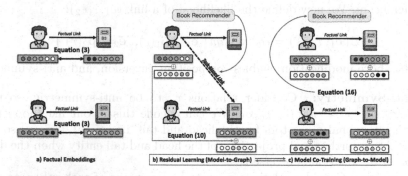

Fig. 2. (a) We learn the facutal entity embeddings via Eq. (3), (b) we then generate model-biased links with the *Book Recommender* model to train residual functions (Eq. (7)), (c) improve the task-model with the residual functions from step (b) in Eq. (12). Steps (b), (c) can be iteratively optimized.

Thus for pairs of entities $e_1 \in \mathbf{E}_1(r)$, $e_2 \in \mathbf{E}_2(r)$, we obtain both factual and model-biased embeddings ($\vec{\mathbf{e}'}$ denotes the model-biased embedding of entity e):

$$\vec{\mathbf{e}}_1, \vec{\mathbf{e}}_2 \sim P(\vec{\mathbf{E}}); \quad \vec{\mathbf{e}'}_1, \vec{\mathbf{e}'}_2 \sim P(\vec{\mathbf{E}}'_r) \tag{5}$$

We learn the divergence $\boldsymbol{\Delta}(r)$ between distributions $P(\vec{\mathbf{E}})$ and each $P(\vec{\mathbf{E}}'_r)$ so that the knowledge graph embeddings can be adapted to each task-model:

$$\boldsymbol{\Delta}(r) = \mathbf{KL}(P(\vec{\mathbf{E}}), \ P(\vec{\mathbf{E}}'_r)) \tag{6}$$

We encode $\boldsymbol{\Delta}(r)$ for each task-model $\mathcal{M}(r)$ via embedding residual shifts motivated by covariate domain-shift theory (He et al. (2016), Johansson et al. (2016)). In the next subsection, we show how this enables task$\rightarrow$graph and graph$\rightarrow$task embedding conversion via task-specific residual functions.

3.3 Residual Shift

The factual and model-biased embedding distributions $(P(\vec{\mathbf{E}}),\ P(\vec{\mathbf{E}}'_r))$ represent different covariate-shifts in the node embedding space depending on the biases of each task-model $\mathcal{M}(r)$. We model each of these shifts with a task-specific residual function $\boldsymbol{\delta}_r$ to translate between the spaces $\vec{\mathbf{E}}$ and $\vec{\mathbf{E}}'_r$:

$$\vec{e'}_1 = \vec{e}_1 + \boldsymbol{\delta}_r(\vec{e}_1); \quad \vec{e'}_2 = \vec{e}_2 + \boldsymbol{\delta}_r(\vec{e}_2) \tag{7}$$

where $\vec{e}_1$ denotes the factual embedding of the entity $e_1(r)$ and each residual function $\boldsymbol{\delta}_r$ is given by,

$$\boldsymbol{\delta}_r(\vec{e}) = tanh(\ \boldsymbol{W}_r(\vec{e}) + \mathbf{b}_r\ ) \tag{8}$$

We learn the weights $\boldsymbol{W}_r$ and biases $\mathbf{b}_r$ to optimize the likelihoods of the model-biased links $(\mathcal{L}(\vec{e'}_1, r, \vec{e'}_2)\ \forall\ (e'_1, r, e'_2) \in \mathbf{R}'_r)$ by placing the residual shifted entity embeddings $\vec{e'}_1,\ \vec{e'}_2$ in Eq. (3).

4 Training Methods

4.1 Learning the Task-Specific Residual Functions

We generate the model-biased links $(e'_1, r, e'_2) \in \mathbf{R}'_r$ for each $e'_1 \in \mathbf{E}_1(r)$ via $\mathcal{M}(r)$. We then learn the residual function $\boldsymbol{\delta}_r$ via alternating optimization of the following likelihoods:

$$\mathcal{L}(\mathbf{R}_r) = \sum_{(e_1, r, e_2) \in \mathbf{R}_r} \log \mathcal{L}(\vec{e}_1, r, \vec{e}_2) \tag{9}$$

$$\mathcal{L}(\mathbf{R}'_r) = \sum_{(e'_1, r, e'_2) \in \mathbf{R}'_r} \log \mathcal{L}(\vec{e'}_1, r, \vec{e'}_2) \tag{10}$$

with notations following from Eq. (3), Eq. (7) and Table 1.

4.2 Graph and Model Co-training

We now describe our training approach to concurrently learn entity embeddings and task-models with continuous differentiable objective functions. In Eq. (10), the task-model is held constant, i.e., we only learn the entity embeddings and residual functions. For co-training, we apply the same residual transformations to the factual links in the graph; and add them to the task-model's optimization objective as soft-criteria.

Table 1. Residual function notations

Symbol	Description
$\vec{e}_1, \vec{e}_2$	Factual embeddings
$\boldsymbol{\delta}_r(.)$	Residual function for $\mathcal{M}(r)$
$\boldsymbol{W}_r, \mathbf{b}_r$	Weight, bias for $\boldsymbol{\delta}_r$
$\vec{e'}_1, \vec{e'}_2$	Residual shifted embeddings $\vec{e'}_1 = \vec{e}_1 + \boldsymbol{\delta}_r(\vec{e}_1)$ $\vec{e'}_2 = \vec{e}_2 + \boldsymbol{\delta}_r(\vec{e}_2)$

For each factual link $(e_1, r, e_2) \in \mathbf{R}_r$, we estimate the residual shifted likelihood as follows:

$$SA(e_1, e_2) = \mathcal{L}(\vec{e'}_1, r, \vec{e'}_2) \tag{11}$$

where $\mathcal{L}$ follows from Eq. (3). We now add the following regularization term to the objective function $\mathcal{O}(r)$ of $\mathcal{M}(r)$:

$$\lambda(r)\Big(\sum_{\mathbf{R}_r} SA\,(e_1, e_2) - \mathcal{M}(r)\,(e_1, e_2) \Big) \tag{12}$$

Here, $\mathcal{M}(r)\,(e_1, e_2)$ indicates the confidence score assigned by $\mathcal{M}(r)$ to link e_2 to e_1 and $\lambda(r)$ is the regularization strength.

4.3   Model to Model Cross-Training

Let us consider the following direction of transfer, $\mathcal{M}(r_1) \to \mathcal{M}(r_2)$ (teacher-model $\to$ student-model). To cross-train $\mathcal{M}(r_2)$ with $\mathcal{M}(r_1)$, we need at least one entity set to be shared across the two models. Let us denote a shared entity set $\mathbf{E}$ with factual embeddings $\vec{e}, e \in \mathbf{E}$ obtained via Eq. (3). We then learn the residual function δ_{r_1} corresponding to the teacher-model $\mathcal{M}(r_1)$, and update the entity embeddings for $\mathbf{E}$ with Eq. (10), while holding δ_{r_1} constant. Finally, we perform the graph-to-model updates described in Sect. 4.2 to train student-model $\mathcal{M}(r_2)$ with the updated embeddings.

5   Experimental Results

Here, we present our experimental analyses on diverse multi-domain datasets and validate our framework. First, we show that counterfactual enrichment with effective task-models can significantly improve node embedding quality with sparse connections, by evaluating the updated embeddings on the held-out link completion task. Next, we show that co-training a context-aware neural recommendation model with the knowledge graph leads to simultaneous embedding updates and better model performance for nodes with lower degrees. We also notice a small degradation in the performance for high-degree nodes. Additionally, we exhibit that we can significantly improve the above context-aware neural recommendation model by leveraging a distributed word embedding model using the illustrated cross-training method. Finally, we do a scalability analysis against publicly available baseline implementations and conclude with limitations and discussion.

5.1   Data Description, Setup

Google Local Reviews Dataset: He et al. (2017); Pasricha and McAuley (2018): Users rate businesses on a 0–5 scale with temporal, spatial, and textual context features in each review. We filter this dataset for at least 10 users per business and 5 businesses per user recursively and eliminate all reviews with less than a 3-star rating. The resulting dataset has 38,614 users and 26,922 businesses, and contextual node types - Review Words (5000 nodes), Business Name Words (2000 nodes), Categories of the Business (650 nodes), Pricey-ness (4 nodes), Location (312 nodes) - states, cities, and Time (23 nodes) - time (binned into 6-h chunks), month, day.

We create our knowledge graph by connecting all users to the businesses they rated, business name and review words to each business, review words, categories of visits, and business names to users who rated them, the pricey-ness, locations, and times to businesses and users. On each of these links, we associated a 1–4 level depending on the strength of the associations (measured statistically on a per-user and per-business basis). These levels constitute our relation types. The total number of nodes and links in the graph is 73,525 and 7,325,614 respectively.

Yelp Challenge Dataset: Users rate businesses on a 0–5 scale with temporal, spatial, and textual context features for each review. We filter this dataset for at least 30 users per business and 10 businesses per user recursively and eliminate all reviews with less than a 3-star rating. The resulting dataset has 25,3695 users and 69,738 businesses. We obtain the following contextual nodes - Review Words (2000 nodes), Business Attributes (200 nodes), Location (1062 nodes) - states, cities, lat-long (binned using a KD-tree), Time (23 nodes) - time (binned by 6-h chunks), month, day.

We create our knowledge graph by connecting all users to the restaurants they rated, the review words and attributes of the restaurants to each restaurant, the location nodes, the associated time nodes, and likewise for the users as well. On each of these links, we associated a 1–4 level depending on the strength of the associations (measured statistically on a per-user and per-business basis). These levels constitute our relation types. The total number of nodes and links in the graph is 99,906 and 10,102,877 respectively.

Baselines: We choose a broad array of diverse knowledge graph embedding baselines as a representative set to evaluate the edge completion task: TransE Bordes et al. (2013), DistMult Yang et al. (2014), ComplEx Trouillon et al. (2016), Rotate Sun et al. (2019), RotH Chami et al. (2020) and GAAT Wang et al. (2019b). We used the OpenKE implementations[3] in Tensorflow/PyTorch with default parameter settings, wherever applicable.

5.2   Task-Models

For both datasets, we used a pair of task models that both have the same input entity-set (users), and different output entity sets (business category and businesses respectively).

We train the distributional word2vec word-embedding model Mikolov et al. (2013) on the set of review text words, business names, and all the business attributes text over all the reviews in the dataset. We use the basic version (non-transfer) of the context-aware recommender proposed in Krishnan et al. (2020)

---

[3] http://139.129.163.161//.

Table 2. Overall Link Prediction Results. Bold-font denotes statistically significant gains over all baselines at the 0.05 significance-level under *paired t-tests*, while * denotes the second-best performer.

Link type	User to business		User to category	
Metric	R @ 5	R @ 10	R @ 5	R @ 10
TransE [Bordes et al. 13]	0.43	0.60	0.52	0.68
RotatE [Sun et al. 19]	0.59*	0.72	0.64	0.80
RotH [Chami et al. 20]	0.58	0.76*	0.65*	0.79
DistMult [Yang et al. 14]	0.56	0.70	0.63	0.77
CompleX [Trouillon et al. 15]	0.57	0.70	0.61	0.76
GAAT [Wang et al. 19]	0.59*	0.74	0.63	0.82*
MutatE-F	0.58	0.73	0.64	0.79
MutatE-CF	**0.62**	**0.80**	**0.68**	0.84

with the non-textual categorical links of the users and businesses (as above) forming the context of each review. To predict business category/attribute words for each user, we take an average of their review word set embeddings, and map the average to the closest business category words as learned by the model.

Parameters: In both the above datasets, for the context-aware recommendation model Krishnan et al. (2020), we use the author recommended parameters with 200-dimensional embeddings, while we use the gensim[4] implementation of word2vec with a maximum 10-length window. The additional parameters of our model, such as the discrepancy scaling in Eq. (10) were tuned with an exponential grid-search approach (e^{-5} to e^0). The knowledge graph and counterfactual residuals were also trained with 200-dimensional embeddings, and implemented in Tensorflow, and run on a Tesla K80 GPU.

Metrics for Link Prediction: In both the datasets, we attempt to predict held-out links using the embeddings learned by our models, as well as the embedding baselines. For each held-out link of the form (e_1, r, e_2), we create several negative samples of the form $(e_1, r, \tilde{e}_2)$ and $(\tilde{e}_1, r, e_2)$, i.e., with the same relation type and head and tail entity types, however a randomly sampled entity for either the head or tail. We then rank the entire list of negative samples against the true link (e_1, r, e_2) under each embedding model and measure the **Recall@K** metric for the respective ranked lists. Specifically, we measure the **Recall@5**, **Recall@10** for two types of held-out links - *User → Business* and *User → Category-word* (Attribute in case of yelp), for a 100-length ranked list.

5.3   Primary Results - Link Prediction

We evaluate the above two knowledge graphs on the link completion task. We randomly tag 20% of the user nodes as held-out nodes. We then held out two

---

[4] https://pypi.org/project/gensim/.

types of links for these users - we held out half of their user-business links and half of their user-business attribute/category word links. These two link types directly correspond to the two task models we used: The word2vec model predicts user-business category word links while the context-aware recommender predicts the user-business links.

For our model, we present two variants - MUTATE-F, which only uses the factual nodes, and MUTATE-CF, which uses counterfactual enrichment for the held-out user set. Specifically, we use the top-5 words predicted by the word2vec model, and the top-5 businesses predicted by the recommender to form counterfactual user-business and user-word links. We also trained all the baseline embedding models on the same knowledge graphs and attempted to predict the same set of held-out links using their trained embeddings.

Key Observations from Table 2: The relative order of performance of the baselines is as expected, DistMult Yang et al. (2014) performs moderately owing to the inverse nature of some relation-types in our graphs across user-context-business paths. In contrast, our base model can overcome this challenge and perform comparably to the other baselines.

We also observe that our MUTATE-CF model strongly outperforms all the competing models on the User-Word link prediction and User-Business link prediction tasks. The two external task models, namely word2vec and the context-aware recommender, can better predict the missing links and enrich the graph compared to the heuristic or path-based link completion approach in the other baselines. It is easy to see how we can leverage the inductive biases of the specific models. While the word2vec model can interpret the review text's distributional properties, the context-aware recommender leverages the multiplicative predictors from the context features. Also, note that these two models use the same data as the Knowledge Graphs and do not depend on any external sources.

5.4   Co-training Model with Graph

In this section, we describe our co-training approach for the recommender model with the knowledge graph. Specifically, we make predictions from these models for users and use these counterfactual links to update knowledge graph embeddings, as described in Eq. (9). Simultaneously, we make predictions from the updated embeddings for users and use these to augment the recommendation loss function as described in Eq. (11).

Table 3. Co-training performance gains against the information-flow parameter λ^j

λ^j	e^{-5}	e^{-4}	e^{-3}	e^{-2}	e^{-1}
Word2Vec	-5.6%	-1.3%	$\mathbf{+8.1\%}$	-4.9%	-18.6%
Context recommender	$+2.8\%$	-1.03%	$\mathbf{+5.4\%}$	-8.6%	-28.9%

Although we did not achieve a dramatic performance difference, we observe that overregularizing the model or under-regularizing the model is suboptimal. In other words, the co-training proceeds best when we set the regularizer λ^j to an optimal balance. The numbers in Table 3 indicate the best performance improvements we were able to achieve for the recommender model under different settings of λ^j. A higher value of λ^j meant that the recommender was more constrained by the knowledge graph, while a lower value meant that more information flows from the model to the graph. Thus, we need an ideal trade-off between the forward and reverse information flow.

5.5   Cross-Training Across Tasks

Next, we describe our cross-training approach for the recommender model by leveraging the word2vec model.

Table 4. Cross-training performance gains for the context-recommender with word2vec, $\mathcal{M}^{word2vec} \rightarrow Knowledge\ Graph \rightarrow \mathcal{M}^{context\text{-}aware\text{-}recommender}$, parameter λ^j is set to varying values as in Eq. (10), percentages relative to isolated performance

λ^j	e^{-5}	e^{-4}	e^{-3}	e^{-2}	e^{-1}
Context recommender	−1.2%	**+6.4%**	**+12.9%**	−10.3%	−22.1%

We first train the word2vec model on the base data, then use it to update the knowledge graph embeddings using the model to graph knowledge transfer method from Sect. 4.3. We then use the reverse direction to regularize the recommender model as in Eq. (12), i.e., knowledge now flows from the updated graph to the recommender model. Thus, the overall direction of knowledge flow is as follows:

$$\mathcal{M}^{word2vec} \rightarrow Knowledge\ Graph \rightarrow \mathcal{M}^{context\text{-}aware\text{-}recommender}$$

Since the review text is informative of both user and business embeddings owing to their shared link structure, we were able to achieve noticeable performance gains for the recommender model (Table 4) after leveraging the sequence of steps described in Sect. 4.3.

Table 5. Cross-training performance gains for the word2vec model, $\mathcal{M}^{context\text{-}aware\text{-}recommender} \rightarrow Knowledge\ Graph \rightarrow \mathcal{M}^{word2vec}$, parameter λ^j is again set to varying values as in Eq. (10), percentages relative to isolated performance

λ^j	e^{-5}	e^{-4}	e^{-3}	e^{-2}	e^{-1}
Word2vec	−7.9%	**−2.1%**	**−1.6%**	−4.1%	−18.3%

However, we observe that the reverse transfer direction, i.e. context-aware recommender to word2vec model, does not result in noticeable performance gains (Table 5), indicating the importance of choosing the more informative model to enrich the knowledge graph.

5.6 Sparsity Analysis

In this subsection, we study the impact of counterfactual updates on sparse and non-sparse nodes. Specifically, for both the tasks, user-word link prediction, and user-business link prediction, we study the relative gains obtained by counterfactual updates, i.e., the difference in the performance of MUTATE and MUTATE-F for the different sparsity sets.

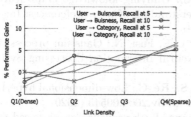

Q_1, Q_2, Q_3 and Q_4 denote the four sparsity quartiles for each respective user node, and we then measure the average performance difference between MUTATE and MUTATE-F for each quartile in Fig. 3. As expected, we obtain the strongest gains for sparse users, i.e., users in quartiles Q3/Q4, since they lack the word-associations to help us learn better embeddings. Thus, the distributional knowledge encoded in the word2vec model bridges this gap in the knowledge graph and enriches the corresponding node embeddings.

Fig. 3. The gains of MUTATE-CF relative to MUTATE-F on the two types of link prediction. In each case, we measure the performance gains across 4 quartiles of users, arranged by the density of that specific type of link for the user.

5.7 Limitations and Discussion

The two primary limitations of our work are the non-exchangeability of cross-training and homoscedastic embedding assumption in each entity set. This results from our assumption that a single residual function, conditioned on the node embeddings, can encode the distributional differences introduced by the task-models. Alternatives such as Gaussian mixture embedding spaces (Casale et al. (2018)) can encode heteroscedastic node embeddings. However, they are quite hard to implement efficiently within a knowledge graph neural network optimization framework. We plan to study the trade-offs between generalizability and overall exchangeability in future work.

6 Conclusion

We propose a holistic view of knowledge graphs and multi-task learning to enable task-enhancement and graph enrichment. Our framework unifies co-dependent task distributions with the underlying knowledge graph via residual learning. The key strength of our approach lies in delegating the extraction of task-specific distributions to the respective task-models while enabling cross-task knowledge transfer. While the current work primarily demonstrates empirical applications of such a framework, we intend to study the theoretical exchangeability of the proposed method for future work.

References

Ai, Q., Azizi, V., Chen, X., Zhang, Y.: Learning heterogeneous knowledge base embeddings for explainable recommendation. Algorithms **11**, 9 (2018)

Bordes, A., Usunier, N., Garcia-Duran, A., Weston, J., Yakhnenko, O.: Translating embeddings for modeling multi-relational data. In: Advances in Neural Information Processing Systems (NIPS) (2013)

Cao, Y., Wang, X., He, X., Hu, Z., Chua, T.-S.: Unifying knowledge graph learning and recommendation: towards a better understanding of user preferences. In: The World Wide Web Conference (WWW) (2019)

Casale, F.P., Dalca, A., Saglietti, L., Listgarten, J., Fusi, N.: Gaussian process prior variational autoencoders. In: Advances in Neural Information Processing Systems (2018)

Chami, I., Wolf, A., Juan, D.-C., Sala, F., Ravi, S., Ré, C.: Low-dimensional hyperbolic knowledge graph embeddings. arXiv preprint arXiv:2005.00545 (2020)

Cheng, D., Yang, F., Wang, X., Zhang, Y., Zhang, L.: Knowledge graph-based event embedding framework for financial quantitative investments. In: International Conference on Research and Development in Information Retrieval (SIGIR) (2020)

Daume III, H., Marcu, D.: Domain adaptation for statistical classifiers. J. Artif. Intell. Res. **26**, 101–126 (2006)

Ernst, P., Siu, A., Weikum, G.: KnoWlife: a versatile approach for constructing a large knowledge graph for biomedical sciences. BMC Bioinform. **16**, 1 (2015)

Guo, S., Wang, Q., Wang, L., Wang, B., Guo, L.: Knowledge graph embedding with iterative guidance from soft rules. arXiv preprint arXiv:1711.11231 (2017)

He, K., Zhang, X., Ren, S., Sun, J.: Deep residual learning for image recognition. In: Proceedings of the IEEE Conference on Computer Vision and Pattern Recognition (2016)

He, R., Kang, W.-C., McAuley, J.: Translation-based recommendation. In: Proceedings of the Eleventh ACM Conference on Recommender Systems (2017)

Huang, X., Zhang, J., Li, D., Li, P.: Knowledge graph embedding based question answering. In: International Conference on Web Search and Data Mining (WSDM) (2019)

Ji, G., He, S., Xu, L., Liu, K., Zhao, J.: Knowledge graph embedding via dynamic mapping matrix. In: Association for Computational Linguistics and International Joint Conference on Natural Language Processing (ACL-IJCNLP) (2015)

Ji, G., Liu, K., He, S., Zhao, J.: Knowledge graph completion with adaptive sparse transfer matrix. In: Schuurmans, D., Wellman, M.P. (eds.) International Conference on Artificial Intelligence (AAAI) (2016)

Jia, Y., Wang, Y., Lin, H., Jin, X., Cheng, X.: Locally adaptive translation for knowledge graph embedding. In: International Conference on Artificial Intelligence (AAAI) (2016)

Johansson, F., Shalit, U., Sontag, D.: Learning representations for counterfactual inference. In: International Conference on Machine Learning (ICML) (2016)

Krishnan, A., Das, M., Bendre, M., Yang, H., Sundaram, H.: Transfer learning via contextual invariants for one-to-many cross-domain recommendation. arXiv preprint arXiv:2005.10473 (2020)

Krishnan, A., Sharma, A., Sundaram, H.: Insights from the long-tail: learning latent representations of online user behavior in the presence of skew and sparsity. In: International Conference on Information and Knowledge Management (CIKM) (2018)

Lerer, A., et al.: PyTorch-BigGraph: a large-scale graph embedding system. arXiv preprint arXiv:1903.12287 (2019)

Li, W.: Zipf's law everywhere. Glottometrics **5**, 14–21 (2002)

Lin, Y., Liu, Z., Sun, M., Liu, Y., Zhu, X.: Learning entity and relation embeddings for knowledge graph completion. In: International Conference on Artificial Intelligence (AAAI) (2015)

Mansour, Y., Mohri, M., Rostamizadeh, A.: Domain adaptation: learning bounds and algorithms. arXiv preprint arXiv:0902.3430 (2009)

Mikolov, T., Sutskever, I., Chen, K., Corrado, G.S., Dean, J.: Distributed representations of words and phrases and their compositionality. In: Advances in Neural Information Processing Systems (NIPS) (2013)

Vu, T., Nguyen, T.D., Nguyen, D.Q., Phung, D.: A capsule network-based embedding model for knowledge graph completion and search personalization. In: North American Chapter of the Association for Computational Linguistics: Human Language Technologies (NAACL-HLT) (2019)

Nickel, M., Tresp, V.: An analysis of tensor models for learning on structured data. In: Blockeel, H., Kersting, K., Nijssen, S., Železný, F. (eds.) ECML PKDD 2013. LNCS (LNAI), vol. 8189, pp. 272–287. Springer, Heidelberg (2013a). https://doi.org/10.1007/978-3-642-40991-2_18

Nickel, M., Tresp, V.: Tensor factorization for multi-relational learning. In: Blockeel, H., Kersting, K., Nijssen, S., Železný, F. (eds.) ECML PKDD 2013. LNCS (LNAI), vol. 8190, pp. 617–621. Springer, Heidelberg (2013b). https://doi.org/10.1007/978-3-642-40994-3_40

Pasricha, R., McAuley, J.: Translation-based factorization machines for sequential recommendation. In: International Conference on Recommender Systems (RecSys) (2018)

Sun, Z., Deng, Z.-H., Nie, J.-Y., Tang, J.: RotatE: knowledge graph embedding by relational rotation in complex space. arXiv preprint arXiv:1902.10197 (2019)

Sun, Z., Yang, J., Zhang, J., Bozzon, A., Huang, L.-K., Xu, C.: Recurrent knowledge graph embedding for effective recommendation. In: International Conference on Recommender Systems (RecSys) (2018)

Trouillon, T., Welbl, J., Riedel, S., Gaussier, É., Bouchard, G.: Complex embeddings for simple link prediction. In: International Conference on Machine Learning (ICML) (2016)

Wang, R., Li, B., Hu, S., Du, W., Zhang, M.: Knowledge graph embedding via graph attenuated attention networks. IEEE Access **8**, 5212–5224 (2019)

Wang, X., He, X., Cao, Y., Liu, M., Chua, T.-S.: KGAT: knowledge graph attention network for recommendation. In: International Conference on Knowledge Discovery & Data Mining (SIGKDD) (2019)

Wang, Z., et al.: XLore: a large-scale English-Chinese bilingual knowledge graph. In: International Semantic Web Conference (ISWC) (2013)

Wang, Z., Zhang, J., Feng, J., Chen, Z.: Knowledge graph embedding by translating on hyperplanes. In: International Conference on Artificial Intelligence (AAAI) (2014)

Yang, B., Yih, W., He, X., Gao, J., Deng, L.: Embedding entities and relations for learning and inference in knowledge bases. arXiv preprint arXiv:1412.6575 (2014)

Adaptive Feature Generation for Online Continual Learning from Imbalanced Data

Yingchun Jian[1], Jinfeng Yi[2], and Lijun Zhang[1(✉)]

[1] National Key Laboratory for Novel Software Technology, Nanjing University,
Nanjing 210023, China
{jianyc,zhanglj}@lamda.nju.edu.cn
[2] JD AI Research, Beijing, China

Abstract. Online continual learning (OCL) is the setting where deep neural network (DNN) incrementally learns new tasks with online data streams. The major problem in OCL is *catastrophic forgetting*, that DNN forgets the acquired knowledge on previous tasks quickly. Recently emerged studies tackle a more realistic problem that the data follows an imbalanced class distribution in OCL by storing particular exemplars. However, preserving exemplars causes memory burden and privacy issues. In this paper, we propose a non-exemplar based method—Adaptive Feature Generation (AdaFG) for OCL from imbalanced data, which tackles the class imbalance and catastrophic forgetting problems simultaneously. Specifically, we argue that one common reason for these problems is the decision boundaries of minority or old classes with few or no samples are affected by majority classes. Therefore, we first maintain a representative prototype for each class in the feature space, which dynamically changes with the streaming data to approximate the class mean feature. Then, we generate new features adaptively for old and minority classes based on their prototypes and train the DNN's classifier to adjust the decision boundaries. Experiments on three popular datasets demonstrate AdaFG's effectiveness in consolidating previous knowledge and addressing the class imbalance problem without preserving exemplars.

Keywords: Online continual learning · Imbalanced learning · Data augmentation

1 Introduction

In the last decade, Deep Neural Network (DNN) has achieved human-level or even better performance in many individual tasks [8,20,26]. When applying the DNN to practice, a typical paradigm is training the DNN sufficiently on a prepared dataset, then fixing the model parameters to deploy on various devices. However, the well-trained model can only tackle a specific task, lacking the capacity of continually learning from data when the environment changes, e.g., new

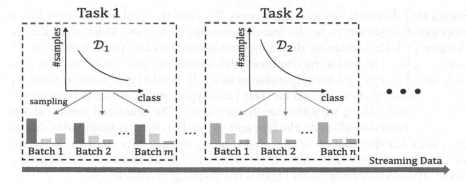

Fig. 1. Online continual learning from imbalanced data. Each batch is sampled from an imbalanced distribution.

classes not seen in the prepared dataset occur. Motivated by human's lifelong-learning ability [25], *continual learning* has been proposed to incrementally learn new tasks without access to previous data while maintaining the acquired knowledge on old tasks at the same time [15].

In this paper, we consider classification tasks and focus on the *online continual learning (OCL)* setting where the task data are coming in an online manner [18]. Different from the offline continual learning (Off-CL) [21] setting where the entire data of the new task are accessible and can be processed numerous times, we can only obtain a tiny batch of data at a time in OCL, which resembles the way humans learn more closely [3]. The major problem in OCL is *catastrophic forgetting* [19], i.e., DNN forgets previous knowledge quickly when learning a new task. Existing methods in OCL can be divided into two categories: exemplar based methods, which keep previous knowledge by storing observed samples (i.e., *exemplars*), and non-exemplar based methods that remember important parameters. However, most studies implicitly assume that the data follows a balanced distribution in each task [12,16,22,30], ignoring many realistic scenarios of imbalanced distributions, e.g., fraud detection and spam classification.

In this paper, we tackle the problem of online continual learning from imbalanced data (OCL-Imb) that each batch is sampled from an imbalanced distribution, as shown in Fig. 1. Besides catastrophic forgetting, we also need to solve the *class imbalance* problem in OCL-Imb, i.e., minority classes that have few samples in the new task are hard to learn [10]. Recently emerged studies address these problems in OCL-Imb by storing particular exemplars, e.g., Class-Balancing Reservoir Sampling [5] and Partitioning Reservoir Sampling [11]. However, these exemplar based methods bring memory burden for resource-constrained devices and cause privacy issues that arouse wide attention nowadays. Inspired by the non-exemplar based method protoAug [31] in Off-CL, we propose Adaptive Feature Generation (AdaFG) to address the class imbalance and catastrophic forgetting problems in OCL-Imb without preserving any exemplars. Specifically, we argue that one common reason of these problems is the decision boundaries for minority or old classes with few or no samples are affected by majority classes that dominate the learning process. To solve this problem, we maintain a representative prototype in the feature space for each class, which dynamically changes with the streaming data to approximate the class mean feature on all seen samples. The prototype contains rich information and can be used to generate new features by adding gaussian noises [31]. AdaFG generates features for old and minority classes based on their prototypes, which are used to train the DNN's classifier along with the new coming data. The generated feature shows to be very helpful to alleviate the bias and adjust the decision boundaries. Moreover, since the distribution of observed samples dynamically changes, we adopt an adaptive strategy that controls the number of generated features according to the dynamic distribution to balance the learning process better.

To verify the effectiveness of the proposed AdaFG, we construct imbalanced tasks on three popular datasets (i.e., CIFAR-100 [13], Food-101 [2] and Mini-ImageNet [27]), and compare it with the state-of-the-art exemplar and non-exemplar based methods. Our empirical results show that AdaFG outperforms previous methods by large margins.

2   Related Works

In this section, we briefly review the related works of the OCL-Imb problem.

2.1   Online Continual Learning (OCL)

The existing OCL methods can be divided into two categories: exemplar based and non-exemplar based methods. Exemplar based methods use a memory buffer to store exemplars selected from previous tasks, which are retrieved to train the model along with the new coming data. Experience Replay [23] takes a naive approach that updates the memory with reservoir sampling and randomly retrieves the exemplars. Various memory updating and retrieving strategies are proposed to improve the performance, such as diversifying the gradients of the exemplars in the memory update [1] and leveraging Shapley Value adversarially in the memory retrieval [24]. As for the OCL-Imb problem, recent works design balanced schemes to make the memory updating and retrieving processes friendly to minority classes, e.g., Class-Balancing Reservoir Sampling [5] and Partitioning Reservoir Sampling [11]. Exemplar based methods try to maintain previously acquired knowledge by exploiting the information of exemplars as more as possible, but they may bring memory burden and cause privacy issues in many resource-restricted and data-sensitive applications.

Non-exemplar based methods usually use various regularization terms to consolidate the acquired knowledge on previous tasks. As representative methods, EWC [12] uses Fisher information matrix to estimate the importance of model parameters and penalizes the drastic changes of important parameters, and LwF [16] adopts knowledge distillation terms to prevent the feature drift of old classes. In addition to designing regularization terms, protoaug [31], a pioneering work in the field of Off-CL, maintains prototypes for old classes in the feature space and generates prototype-based features to keep and expand the decision boundaries of old classes. The class-representative prototype shows to be very effective in keeping previous knowledge. Moreover, Lange and Tuyte-laars [14] propose to use the prototype for nearest neighbor classification in OCL-Imb. However, this method needs to store exemplars to update the prototype with online data streams. In this paper, we extend the prototype-based method with no exemplars preserved to tackle the catastrophic forgetting problem in OCL-Imb.

2.2   Online Imbalanced Learning

Different from OCL that incrementally learns new tasks, online imbalanced learning focuses on learning a single task, and the streaming data is sampled from an imbalanced distribution. Due to the lack of samples, minority classes are highly under-represented and harder to learn compared with majority classes [7]. Many re-sampling strategies are proposed to solve this problem. For instance, Wang and Pineau [28] introduce online bagging techniques for online binary classification by randomly oversampling and undersampling samples of minority and majority classes, respectively. Furthermore, Wang et al. [29] extend online bagging techniques to tackle the multi-class imbalance problem. Besides re-sampling samples, data augmentation strategies are used to address the class imbalance problem. For instance, Generative Adversarial Networks [6] produce virtual samples by approximating the distribution of minority classes, and SMOTE [4] generates new samples for minority classes around the neighbor of original data. In this paper, we tackle the class imbalance problem in OCL-Imb also from the perspective of data augmentation.

3   Method

In this section, we first introduce the OCL-Imb setting and analyze the major problems in OCL-Imb, then illustrate the framework and details of our proposed method AdaFG.

3.1   Problem Analysis

Firstly, we consider the OCL setting. When learning the t-th task T_t, the model receives a tiny batch of samples of size b at a time, which is denoted as $\mathcal{B}_t^i$ for the i-th step. The entire data of T_t are $D_t = \{\mathcal{B}_t^0, \mathcal{B}_t^1, \cdots, \mathcal{B}_t^\tau\}$, where τ is the number of the totally received batches. When it comes to the OCL-Imb problem, T_t is not a balanced task and $\mathcal{B}_t^i$ are sampled from an imbalanced distribution $\mathcal{D}_t$, which is unknown in advance. As for the DNN model, we divide it into two parts: the feature extractor $\mathcal{G}$ and unified classifier $\mathcal{F}$. The goal is to minimize the statistical risk incurred by all seen tasks, which is formulated as:

$$\min \sum_{n=1}^{t} \mathbb{E}_{(x,y) \in D_n} \left[\ell \left(\mathcal{F}_t \left(\mathcal{G}_t \left(x; \phi_t \right); \theta_t \right), y \right) \right], \tag{1}$$

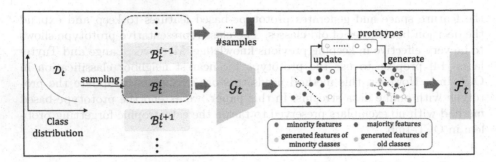

Fig. 2. The framework of Adaptive Feature Generation (AdaFG).

where ϕ_t and θ_t are the parameters of $\mathcal{G}_t$ and $\mathcal{F}_t$ after learning T_t, $\mathcal{G}_t(x;\phi_t)$ extracts the feature of a sample (x,y), $\mathcal{F}_t(\mathcal{G}_t(x;\phi_t);\theta_t)$ gets the outputs of the classifier, and ℓ is the loss function. Note that the data D_n of previous tasks $T_n(n < t)$ are not accessible, and we can only receive b samples of T_t at a time. The statistical risk of T_t can be approximated by the empirical risk [15]

$$\frac{1}{\tau}\sum_{i=1}^{\tau}\sum_{(x,y)\in\mathcal{B}_t^i}\ell\left(\mathcal{F}_t\left(\mathcal{G}_t(x;\phi_t^i);\theta_t^i\right),y\right),\tag{2}$$

where ϕ_t^i and θ_t^i are parameters after the model trained on the i-th batch $\mathcal{B}_t^i$.

There are two major problems in OCL-Imb. One is catastrophic forgetting, which is caused by the drastic changes of decision boundaries for old classes when learning new classes in T_t [31]. Moreover, the learning process mainly focuses on majority classes, which contribute most to the changes of decision boundaries. The other is class imbalance that the decision boundaries are close to minority classes, because the optimization process of minimizing Eq. (2) is dominated by majority classes. The catastrophic forgetting and class imbalance problems are correlated, and one common reason of them is the decision boundaries for both old and minority classes are affected by the majority classes.

3.2   Adaptive Feature Generation (AdaFG)

In this subsection, we propose Adaptive Feature Generation (AdaFG) to adjust the decision boundaries by generating new features for old and minority classes, which are used to train the DNN's classifier.

Firstly, we handle the class imbalance problem by generating different numbers of features for classes in the current task T_t. For the k-th class in T_t (denoted as $c_{t,k}$), we maintain a prototype $\mu_{t,k}^i$, which is a feature approximating the class mean on all observed samples and dynamically changes with the incoming batch $\mathcal{B}_t^i$ (illustrated in Sect. 3.3). The new feature is generated based on $\mu_{t,k}^i$:

$$g_{t,k}^i = \mu_{t,k}^i + e \cdot r,\tag{3}$$

where e is a Gaussian noise with the same dimension as $\mu^i_{t,k}$, and r is a predefined value (e.g., 0.1). The generated feature can be seen as a certain disturbance of the prototype, which is used to train the classifier $\mathcal{F}_t$ to consolidate the just learned knowledge and expand the decision boundaries [31]. When receiving a new batch $\mathcal{B}^i_t$, the number of observed samples of $c_{t,k}$ is denoted as $s^i_{t,k}$. Since the distribution of $s^i_{t,k}$ ($k = 1, 2, \cdots, C_t$, C_t is the number of classes in T_t) changes with incoming batches, we adopt an adaptive strategy to control the number of generated features for each class:

$$a^i_{t,k} = \left[N_1 \cdot \left(1 - \frac{s^i_{t,k}}{s^i_{max}} \right) \right], \tag{4}$$

where s^i_{max} is the maximum number of $s^i_{t,k}$ ($k = 1, 2, \cdots, C_t$), N_1 is a predefined positive value (e.g., 10), and $[a]$ returns the integer closest to a. $a^i_{t,k}$ is close to N_1 if $s^i_{t,k} \ll s^i_{max}$ and close to 0 if $s^i_{t,k} \approx s^i_{max}$. In this way, minority classes have superiority in the number of generated features, making the decision boundaries far from minority classes, thus improving their performance.

Furthermore, we tackle the catastrophic forgetting problem by generating features for old classes in a similar way as Eq. (3) to keep previous decision boundaries. For the k-th class in the old task T_n ($n < t$), the maintained prototype is $\mu_{n,t}$ and the new feature is generated as:

$$g_{n,k} = \mu_{n,k} + e \cdot r. \tag{5}$$

To generate features in a balanced way, we randomly select an old class to generate one feature by Eq. (5) and repeat N_2 times when receiving a new batch. To reduce the computation, N_2 is set to be a constant (e.g., 10) instead of a variable proportional to the number of seen classes.

The framework of AdaFG is illustrated in Fig. 2. When training on a new batch sampled from an imbalanced distribution, we update the number of observed samples for each class and the prototypes by current features. Then, we generate new features for old and minority classes in the ways mentioned above, which are used for training the classifier $\mathcal{F}_t$ to adjust the decision boundaries.

3.3 Online Prototype Update

For each class, the class mean in the feature space contains rich information and can be used for data augmentation [17]. Recent work [31] in Off-CL generates features for old classes to alleviate forgetting based on the maintained class mean features, which are computed until the feature extractor $\mathcal{G}_t$ is sufficiently trained on T_t. However, since only the current batch $\mathcal{B}^i_t$ are accessible and feature extractor parameters change over time in OCL-Imb, the class mean feature on all observed samples can't be computed directly. To overcome this problem, we adopt a *moving average* strategy to update the maintained prototype online to approximate the true class mean feature. When receiving $\mathcal{B}_t^i$, the previously maintained prototype for $c_{t,k}$ is $\mu_{t,k}^{i-1}$ and will be updated by:

$$\mu_{t,k}^i = (1 - \alpha_{t,k}^i) \cdot \mu_{t,k}^{i-1} + \alpha_{t,k}^i \cdot \tilde{\mu}_{t,k}^i,$$

$$\tilde{\mu}_{t,k}^i = \frac{1}{|X_{t,k}^i|} \sum_{x \in X_{t,k}^i} \mathcal{G}_t(x; \phi_t^{i-1}), \tag{6}$$

where $X_{t,k}^i$ are samples of $c_{t,k}$ in $\mathcal{B}_t^i$, $|X_{t,k}^i|$ is the size of $X_{t,k}^i$, and ϕ_t^{i-1} is the parameters of $\mathcal{G}_t$ after training on last batch. $\alpha_{t,k}^i$ is a factor controlling the prototype update, which can be adopted as:

$$\alpha_{t,k}^i = \frac{|X_{t,k}^i|}{\sum_{j=1}^i |X_{t,k}^j|}.$$

The prototype $\mu_{t,k}^i$ will not be updated if $|X_{t,k}^i|$ is 0. After learning the last batch $X_{t,k}^\tau$, the obtained prototype $\mu_{t,k}^\tau$ (also denoted as $\mu_{t,k}$) will be maintained for continually learning later tasks. In Sect. 4.3, we conduct experiments to demonstrate that the prototype updated by Eq. (6) is a good approximation to the true class mean on previously observed samples.

3.4   Training Process

When learning from the streaming data of a new task T_t, the training process can be divided into two parts. The first part is training on the new coming batch $\mathcal{B}_t^i$. For a new sample (x, y) in $\mathcal{B}_t^i$, the outputs of the current and last model are ξ_t and ξ_{t-1}, respectively. Typically, we adopt the cross-entropy loss $\mathcal{L}_{ce}(\xi_t, y)$ for classification and use the well-known knowledge distillation loss $\mathcal{L}_{kd}(\xi_t, \xi_{t-1})$ [16,22,30] to mitigate forgetting by making outputs of the current model close to those of the last model, which are defined as:

$$\mathcal{L}_{ce}(\xi_t, y) = - \sum_{c=1}^C y_c \log(\sigma(\xi_t)_c),$$

$$\mathcal{L}_{kd}(\xi_t, \xi_{t-1}) = - \sum_{c=1}^C \sigma(\xi_{t-1})_c \log(\sigma(\xi_t)_c), \tag{7}$$

where C is the number of classes seen so far, $y \in \mathbb{R}^C$ is a label vector, and $\sigma(\cdot)$ is a softmax function. The overall loss of learning from the new data can be defined as previous works [22,30]:

$$\mathcal{L}_{new} = \frac{1}{t}\mathcal{L}_{ce}(\xi_t, y) + \left(1 - \frac{1}{t}\right)\mathcal{L}_{kd}(\xi_t, \xi_{t-1}). \tag{8}$$

With the growth of tasks, the proportion of $\mathcal{L}_{kd}$ increases to remember more and more knowledge.

The second part is training the classifier on the generated features by AdaFG. For the generated data (g_o, y_o) and (g_m, y_m) of old and minority classes, we adopt $\mathcal{L}_{fgt}$ and $\mathcal{L}_{imb}$ to train the classifier $\mathcal{F}_t$:

$$
\begin{aligned}
\mathcal{L}_{fgt} &= \mathcal{L}_{ce}(\mathcal{F}_t(g_o; \theta_t^{i-1}), y_o), \\
\mathcal{L}_{imb} &= \mathcal{L}_{ce}(\mathcal{F}_t(g_m; \theta_t^{i-1}), y_m) + \mathcal{L}_{kd}(\mathcal{F}_t(g_m; \theta_t^i), \mathcal{F}_{t-1}(g_m; \theta_{t-1})).
\end{aligned}
\tag{9}
$$

$\mathcal{L}_{fgt}$ and $\mathcal{L}_{imb}$ focus on mitigating forgetting and learning minority classes, respectively. Notice that in the first batch of each new task, the prototypes of new classes are not available, and only $\mathcal{L}_{fgt}$ is calculated.

The total loss is comprised of the above terms:

$$
\mathcal{L} = \mathcal{L}_{new} + \eta \mathcal{L}_{fgt} + \gamma \mathcal{L}_{imb},
\tag{10}
$$

where η and γ are coefficients that control the impact of corresponding terms.

4 Experiments

In this section, we compare AdaFG with several state-of-the-art methods and analyze the results to validate our approach. Furthermore, we visualize the generated features to verify the effectiveness of the maintained prototype in AdaFG.

4.1 Setup

Datasets. CIFAR-100, Food-101 and Mini-ImageNet are used in our experiments, which are both balanced datasets. CIFAR-100 contains 50k training images and 10k test images in 100 classes. Food-101 contains 75k training images and 25k test images in 100 classes. Mini-ImageNet contains 60k images in 100 classes, and we split them into 50k training images and 10k test images.

OCL-Imb Settings. Similar to previous works [9,31], we divide the whole classes into two parts: base classes (20 classes for CIFAR-100 and Mini-ImageNet, and 21 classes for Food-101) and rest classes (80 classes). The base classes are used to train a base feature extractor offline, which is beneficial for the DNN model to cope with the streaming data. The rest classes are divided

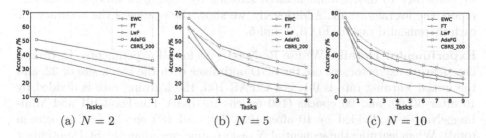

(a) $N = 2$ (b) $N = 5$ (c) $N = 10$

Fig. 3. Accuracy for each incremental task on CIFAR-100 when $N = 2$, 5, and 10.

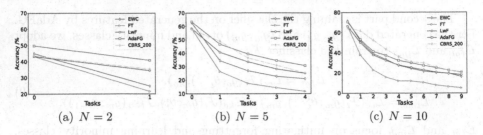

Fig. 4. Accuracy for each incremental task on Food-101 when $N = 2$, 5, and 10.

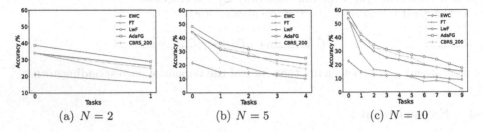

Fig. 5. Accuracy for each incremental task on Mini-ImageNet when $N = 2$, 5, and 10.

into N tasks, and N can be 2, 5, or 10, which means each task contains 40, 16, and 8 classes, respectively. Following Chrysakis and Moens [5], we select a random percentage p of instances in the original dataset for each rest class to construct imbalanced streams. p is randomly selected from a retention set $\{1, 10^{-r}, 10^{-2r}, 10^{-3r}, 10^{-4r}\}$. In this paper, we use $r = 0.25$ for all experiments, i.e., the maximum imbalance between two classes is 10. When learning the sequential tasks, the number of samples received at a time is set to 10 (i.e., $b = 10$), and each sample can only be processed once.

Evaluation. After learning a new task, the performance is evaluated on test images of the observed rest classes by computing the average accuracy. We use two popular criteria to measure the ability to incrementally learn new tasks [22,30]. One is the *last accuracy*, which is the performance after learning the last task. The other is the *average incremental accuracy*, which computes the mean value of the performance over all incremental tasks. For each task division N, we construct 15 different imbalanced streams by setting 15 random seeds and report the average result. Additionally, we show the results of the accuracy for each incremental task in Fig. 3, 4 and 5.

Experimental Details. We use ResNet-18 [8] for all experiments. To train a base feature extractor, we use the SGD optimizer with the batch size of 32, and the initial learning rate is 0.1. For CIFAR-100, the learning rate is divided by 10 after 30, 60, and 90 epochs (100 epochs in total). For Food-101 and Mini-ImageNet, it is divided by 10 after 100, 150, and 180 epochs (200 epochs in total). When learning the sequential N tasks online, we adopt the SGD optimizer with the learning rate of 0.1. The hyper-parameters used in AdaFG are set to: $N_1 = N_2 = 10$, $\eta = \gamma = 1.0$, and $r = 0.1$.

Table 1. Last accuracy (Last) and average incremental accuracy (Aver) on CIFAR-100.

CIFAR-100	$N = 2$		$N = 5$		$N = 10$	
	Last	Aver	Last	Aver	Last	Aver
FT	21.9 ± 1.2	32.9 ± 1.4	12.0 ± 0.8	26.9 ± 0.7	2.0 ± 3.3	19.9 ± 0.4
LwF	30.2 ± 1.8	37.0 ± 2.0	23.6 ± 1.3	37.0 ± 1.6	17.6 ± 1.1	32.2 ± 1.4
EWC	20.2 ± 1.0	29.3 ± 1.6	16.7 ± 1.4	27.6 ± 1.5	13.4 ± 2.0	28.3 ± 1.7
AdaFG	**36.1 ± 1.1**	**43.5 ± 1.4**	**31.5 ± 1.2**	**44.1 ± 0.8**	**23.4 ± 1.8**	**39.3 ± 1.0**
CBRS-50	25.8 ± 0.9	34.7 ± 1.4	17.3 ± 0.7	32.5 ± 0.9	9.8 ± 3.6	30.3 ± 1.2
CBRS-100	28.5 ± 1.0	36.0 ± 1.5	20.8 ± 0.9	36.0 ± 0.8	13.6 ± 3.5	35.3 ± 0.8
CBRS-200	30.9 ± 1.0	37.2 ± 1.5	25.1 ± 0.6	39.4 ± 0.6	19.1 ± 3.1	**39.9 ± 1.0**

Table 2. Last accuracy (Last) and average incremental accuracy (Aver) on Food-101.

Food-101	$N = 2$		$N = 5$		$N = 10$	
	Last	Aver	Last	Aver	Last	Aver
FT	21.5 ± 2.0	32.4 ± 2.5	11.7 ± 0.9	25.7 ± 1.0	5.2 ± 1.7	19.7 ± 0.7
LwF	34.4 ± 2.5	38.8 ± 2.9	25.1 ± 1.3	37.2 ± 1.6	18.1 ± 1.2	32.0 ± 2.0
EWC	24.8 ± 1.9	34.9 ± 2.0	21.0 ± 1.9	33.6 ± 1.5	16.3 ± 1.5	32.5 ± 1.9
AdaFG	**40.8 ± 1.6**	**45.3 ± 2.4**	**30.8 ± 1.9**	**43.0 ± 2.0**	**20.6 ± 1.8**	**36.0 ± 1.9**
CBRS-50	29.3 ± 1.2	36.3 ± 2.0	16.7 ± 1.1	32.2 ± 0.9	11.4 ± 0.7	28.3 ± 0.9
CBRS-100	32.0 ± 1.4	37.9 ± 2.2	20.0 ± 1.4	35.4 ± 1.2	15.2 ± 1.3	32.6 ± 0.9
CBRS-200	35.3 ± 1.8	39.5 ± 2.5	25.3 ± 1.3	39.9 ± 0.9	20.2 ± 1.1	**38.7 ± 0.7**

Compared Methods. We compare our proposed method AdaFG with several state-of-the-art non-exemplar and exemplar based methods:

- **FT:** (non-exemplar) A naive but important method that fine-tunes the model on the receiving data without any approach for avoiding forgetting.
- **LwF** [16]: (non-exemplar) Learning without Forgetting trains the model with the classification loss $\mathcal{L}_{ce}$ and knowledge distillation loss $\mathcal{L}_{kd}$.
- **EWC** [12]: (non-exemplar) Elastic Weight Consolidation uses a regularization term to constrain the updates of important parameters.
- **CBRS** [5]: (exemplar) Class-Balanced Reservoir Sampling uses a memory buffer to solve the OCL-Imb problem by storing particular samples. The buffer size is set to 50, 100 and 200 in our experiments.

4.2 Results

The results are reported in Tables 1, 2 and 3. For different task divisions ($N = 2, 5$ or 10), FT gets poor results, e.g., only 2.0% last accuracy on CIFAR-100 when $N = 10$. Since the importance of model parameters is hard to estimate,

Table 3. Last accuracy (Last) and average incremental accuracy (Aver) on Mini-ImageNet.

Mini-ImageNet	$N = 2$		$N = 5$		$N = 10$	
	Last	Aver	Last	Aver	Last	Aver
FT	19.8 ± 0.9	26.9 ± 1.4	9.9 ± 0.8	21.7 ± 0.7	2.3 ± 2.7	16.2 ± 0.8
LwF	26.2 ± 1.3	30.1 ± 1.8	20.8 ± 1.5	29.4 ± 1.5	15.1 ± 1.3	26.0 ± 1.6
EWC	15.9 ± 1.0	18.5 ± 1.4	12.4 ± 1.2	15.3 ± 2.0	8.9 ± 1.4	12.5 ± 2.1
AdaFG	**28.8 ± 1.2**	**33.8 ± 1.4**	**25.2 ± 1.2**	**33.8 ± 1.2**	**17.3 ± 2.3**	**30.9 ± 1.1**
CBRS-50	22.3 ± 0.6	28.0 ± 1.4	12.3 ± 0.5	24.4 ± 0.6	6.5 ± 2.9	21.0 ± 0.6
CBRS-100	23.3 ± 0.8	28.6 ± 1.5	14.3 ± 1.0	26.5 ± 0.8	8.7 ± 2.9	23.8 ± 0.9
CBRS-200	24.8 ± 0.7	29.4 ± 1.6	17.2 ± 0.9	28.8 ± 0.9	11.4 ± 2.3	27.6 ± 0.7

the results of EWC are not good compared with the distillation-based method LwF. In contrast, AdaFG achieves superior performance over all compared non-exemplar based methods whether on CIFAR-100, Food-101 or Mini-ImageNet. This is because these methods don't consider the class imbalance problem when designing the algorithms, thus shows unsatisfactory performance in the OCL-Imb setting. On Mini-ImageNet, AdaFG outperforms LwF by 3.7%, 4.4%, and 4.9% on the average incremental accuracy when $N = 2$, 5, and 10, respectively. On Food-101, the gaps are 6.5%, 6.2%, and 4.0%. As for CIFAR-100, the gaps are increased to 6.5%, 7.1%, and 7.1%. The performance of AdaFG on CIFAR-100 is much better than that on Mini-ImageNet, because the data in Mini-ImageNet are more complex (84×84 pixels v.s. 32×32 pixels) and the representation learning of the DNN model becomes more difficult.

As for the exemplar-based method, CBRS heavily depends on the memory size, and the performance gain of increasing the size is quite obvious. For instance, CBRS achieves 9.3% improvements on the last accuracy on CIFAR-100 ($N = 10$) when enlarging the memory size from 50 to 200. Compared with CBRS, AdaFG achieves comparable or even better results even if the memory size is 200. Specifically, AdaFG outperforms CBRS by large margins on Mini-ImageNet, e.g., 8.0% improvements on the last accuracy when $N = 5$, demonstrating the effectiveness of AdaFG to cope with complex data.

Performance of the Largest and Smallest Classes. After learning a new task, we compute the average accuracy of the largest and smallest observed classes (i.e., $p = 1.0$ and 0.1). As shown in Fig. 6, compared with other methods, AdaFG achieves comparable results on the largest classes but surpasses others greatly on the smallest classes, which shows that AdaFG can improve the learning of minority classes effectively at a slight expense of performance degradation on majority classes.

Ablation Study. We analyze the impact of hyper-parameter r that controls the scope of the generated features in AdaFG, and the results of the average incremental accuracy on CIFAR-100 are shown in Fig. 7. When $r > 0.4$, the accuracy drops rapidly with the increasing value of r. In this paper, we set $r = 0.1$ as it obtains good performance when learning a long sequence of tasks ($N = 10$).

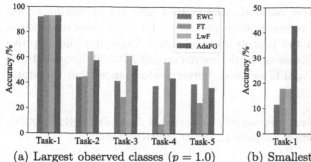
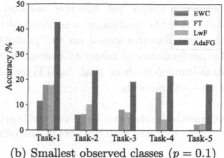

(a) Largest observed classes ($p = 1.0$)   (b) Smallest observed classes ($p = 0.1$)

Fig. 6. Performance of specific classes when continually learning 5 tasks on CIFAR-100 ($N = 5$). (a) The mean accuracy of the observed largest classes ($p = 1.0$). (b) The mean accuracy of the observed smallest classes ($p = 0.1$).

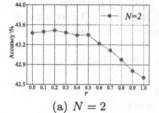
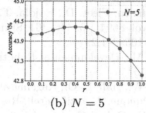
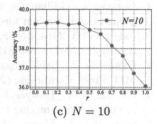

(a) $N = 2$ (b) $N = 5$ (c) $N = 10$

Fig. 7. Average incremental accuracy on CIFAR-100 w.r.t. r when $N = 2, 5,$ and 10.

4.3 Visualization of Generated Features

To verify the quality of generated features by AdaFG, we visualize the generated features based on the prototypes and the real features of classes with different percentages p by collecting all seen samples. As shown in Fig. 8, the generated features are in the core of real features and can almost cover the real features of minority classes (e.g., $p = 0.10$ and 0.18), which demonstrates the prototype computed by Eq. (6) is close to the true class mean feature.

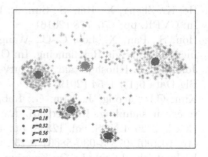

Fig. 8. Real (light color) and generated (dark color) features (Color figure online)

5   Conclusion

In this paper, we tackle the realistic problem of online continual learning from imbalanced data (OCL-Imb) and analyze the catastrophic forgetting and class imbalance problems encountered in OCL-Imb. To address these problems, we propose a simple yet effective method AdaFG that gets rid of storing exemplars. AdaFG maintains a representative prototype for each class and generates new features based on the prototype to mitigate forgetting and improve the performance of minority classes. Experiments on CIFAR-100, Food-101 and Mini-ImageNet show that AdaFG achieves better performance than the state-of-the-art methods.

Acknowledgements. This work was partially supported by JiangsuSF (BK20200064), and the Open Research Projects of Zhejiang Lab (NO. 2021KB0AB02).

References

1. Aljundi, R., Lin, M., Goujaud, B., Bengio, Y.: Gradient based sample selection for online continual learning. In: NeurIPS, pp. 11816–11825 (2019)
2. Bossard, L., Guillaumin, M., Van Gool, L.: Food-101 – mining discriminative components with random forests. In: Fleet, D., Pajdla, T., Schiele, B., Tuytelaars, T. (eds.) ECCV 2014. LNCS, vol. 8694, pp. 446–461. Springer, Cham (2014). https://doi.org/10.1007/978-3-319-10599-4_29
3. Cangelosi, A., Schlesinger, M.: Developmental Robotics: From Babies to Robots. MIT Press, Cambridge (2015)
4. Chawla, N.-V., Bowyer, K.-W., Hall, L.-O., Kegelmeyer, W.-P.: SMOTE: synthetic minority over-sampling technique. J. Artif. Intell. Res. **16**(1), 321–357 (2002)
5. Chrysakis, A., Moens, M.-F.: Online Continual Learning from Imbalanced Data. In: ICML, pp. 1952–1961 (2020)
6. Douzas, G., Bacao, F.: Effective data generation for imbalanced learning using conditional generative adversarial networks. Expert Syst. Appl. **91**, 464–471 (2018)
7. He, H., Garcia, E.-A.: Learning from imbalanced data. TKDE **9**(21), 1263–1284 (2009)
8. He, K., Zhang, X., Ren, S., Sun, J.: Deep residual learning for image recognition. In: CVPR, pp. 770–778 (2016)
9. Hou, S., Pan, X., Loy, C.-C., Wang, Z., Lin, D.: Learning a Unified Classifier Incrementally via Rebalancing. In: CVPR, pp. 831–839 (2019)
10. Johnson, J., Khoshgoftaar, T.: Survey on deep learning with class imbalance. J. Big Data **6**(1), 1–54 (2019)
11. Kim, C.D., Jeong, J., Kim, G.: Imbalanced continual learning with partitioning reservoir sampling. In: Vedaldi, A., Bischof, H., Brox, T., Frahm, J.-M. (eds.) ECCV 2020. LNCS, vol. 12358, pp. 411–428. Springer, Cham (2020). https://doi.org/10.1007/978-3-030-58601-0_25
12. Kirkpatrick, J., et al.: Overcoming catastrophic forgetting in neural networks. PNAS **114**(13), 3521–3526 (2017)
13. Krizhevsky, A., Hinton, G.: Learning multiple layers of features from tiny images. Technical report (2009)
14. Lange, M., Tuytelaars, T.: Continual prototype evolution: learning online from non-stationary data streams. In: ICCV, pp. 8250–8259 (2021)
15. Lange, M., et al.: A continual learning survey: defying forgetting in classification tasks. arXiv preprint arXiv:1909.08383 (2019)
16. Li, Z., Hoiem, D.: Learning without forgetting. In: ECCV, pp. 614–629 (2016)

17. Liu, J., Sun, Y., Han, C., Dou, Z., Li, W.: Deep representation learning on long-tailed data: a learnable embedding augmentation perspective. In: CVPR, pp. 2970–2979 (2020)
18. Mai, Z., et al.: Online continual learning in image classification: an empirical survey. arXiv preprint arXiv:2101.10423 (2021)
19. McCloskey, M., Cohen, J.-N.: Catastrophic interference in connectionist networks: the sequential learning problem. Psychol. Learn. Motiv. **24**, 109–165 (1989)
20. Mnih, V., et al.: Playing atari with deep reinforcement learning. In: NeurIPS Workshop (2013)
21. Prabhu, A., Torr, P.H.S., Dokania, P.K.: GDumb: a simple approach that questions our progress in continual learning. In: Vedaldi, A., Bischof, H., Brox, T., Frahm, J.-M. (eds.) ECCV 2020. LNCS, vol. 12347, pp. 524–540. Springer, Cham (2020). https://doi.org/10.1007/978-3-030-58536-5_31
22. Rebuffi, S., Kolesnikov, A., Sperl, G., Lampert, C.: iCaRL: incremental classifier and representation learning. In: ICCV, pp. 5533–5542 (2017)
23. Rolnick, D., Ahuja, A., Schwarz, J., Lillicrap, T., Wayne, G.: Experience replay for continual learning. In: NeurIPS, pp. 350–360 (2019)
24. Shim, D., Mai, Z., Jeong, J., Sanner, S., Kim, H., Jang, J.: Online class-incremental continual learning with adversarial shapley value. In: AAAI, pp. 9630–9638 (2021)
25. Tani, J.: Exploring Robotic Minds: Actions, Symbols, and Consciousness as Self-Organizing Dynamic Phenomena. Oxford University Press, Oxford (2016)
26. Vaswani, A., et al.: Attention is all you need. In: NeurIPS, pp. 6000–6010 (2017)
27. Vinyals, O., Blundell, C., Lillicrap, T., Kavukcuoglu, K., Wierstra, D.: Matching networks for one shot learning. In: NeurIPS, pp. 3637–3645 (2016)
28. Wang, B., Pineau, J.: Online bagging and boosting for imbalanced data streams. TKDE **28**(12), 3353–3366 (2016)
29. Wang, S., Minku, L.-L., Yao, X.: Dealing with multiple classes in online class imbalance learning. In: IJCAI, pp. 2118–2124 (2016)
30. Zhao, B., Xiao, X., Gan, G., Zhang, B., Xia, S.-T.: Maintaining discrimination and fairness in class incremental learning. In: ICCV, pp. 13205–13214 (2020)
31. Zhu, F., Zhang, X.-Y., Wang, C., Yin, F., Liu, C.-L.: Prototype augmentation and self-supervision for incremental learning. In: CVPR, pp. 5871–5880 (2021)

Order-Aware Graph Neural Network for Sequential Recommendation

Xinlei Zhang, Wendi Ji, Jiahao Yuan, and Xiaoling Wang(✉)

School of Computer Science and Technology, East China Normal University, Shanghai 200062, China
jhyuan@stu.ecnu.edu.cn, xlwang@cs.ecnu.edu.cn

Abstract. Graph neural networks (GNNs) have gained impressive success in the task of sequential recommendation due to their advantage in obtaining the complex transition patterns of items. However, existing GNN-based sequential recommenders still face some problems: (1) The global order information is lost when converting a sequence into a graph. (2) The long-term dependencies in a sequence are ignored due to the over-smoothing problem in GNNs. In this paper, we propose an order-aware GNN with long-range connections (OAG-LC) for sequence modeling. To capture the global order of a sequence, a novel graph update mechanism is proposed, which evolves the graph embedding recurrently over time rather than concurrently for order preservation. And a novel gate is used to incorporate both order and structural information in the update phase. To model the long-term dependencies of user behaviors, we convert the sequence into a graph via reachability and apply the attention mechanism for information propagation through the long-range connections. Furthermore, the proposed graph construction method differentiated repeated items with their positions for information lossless encoding. We conduct extensive experiments on four public datasets, and the experimental results demonstrate the effectiveness of our proposed model.

Keywords: Graph neural network · Recommendation · Sequence model

1 Introduction

Recommender system is an important information filtering tool to solve the information overload problem. The chronological order of users' behavior plays an important role in inferring users' interests. Therefore, many sequential recommendation models are proposed to utilize user history in a sequential manner for future behavior prediction and recommendation. Early works use Markov chains to learn a transition graph over time and predict users' next action based on their recent actions [11]. Recent studies have highlighted the importance of using deep neural networks in sequential recommendation tasks, e.g. RNNs [3,13,16] and

ⓒ The Author(s), under exclusive license to Springer Nature Switzerland AG 2022
J. Gama et al. (Eds.): PAKDD 2022, LNAI 13280, pp. 290–302, 2022.
https://doi.org/10.1007/978-3-031-05933-9_23

attentive networks [4,7]. Furthermore, due to their remarkable advantage to learn the complex transition patterns, GNN-based methods [14,15] are proposed in recent years to convert a sequence into a graph and encode sequential behaviors through information aggregation and updating. Although the GNN-based methods enable sequential recommendation to capture the graph-structured dependencies, there are still some drawbacks.

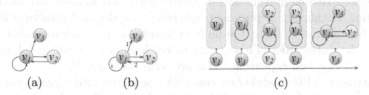

$$(a) \qquad\qquad (b) \qquad\qquad\qquad\qquad (c)$$

Fig. 1. Lossy sequence encoding problem of GNN. Two different sequences $(v_1, v_1, v_2, v_1, v_3)$ and $(v_1, v_2, v_1, v_1, v_3)$ are converted into the same graph (a), EOP multigraph solves this problem by assigning an order to each edge (b), the proposed OAG module solves this problem by updating graph representation recurrently (c).

First, when converting a sequence into a graph, the order information is usually lost. The existing GNN-based methods [14,15] usually convert a sequence to a graph using the unique items as nodes and transitions between items as edges, which will cause the order unawareness problem. Take the sequence $s_1 = (v_1, v_1, v_2, v_1, v_3)$ and $s_2 = (v_1, v_2, v_1, v_1, v_3)$ as an example. As shown in Fig. 1(a), graphs constructed from these two different sequences are the same and it fails to reconstruct the two sequences from this graph. However, the order of items in a sequence embodies critical information for the sequence encoding task. To solve the problem, recent work [2] designs EOP multigraph that assigns orders to edges and aggregates information following these orders (Fig. 1(b) is the EOP multigraph converted from s_1). However, EOP multigraph can only keep the local order (the relative order of a pair of items) and miss the global order (the absolute position of items) of the sequence. Take the sequence $s_1 = (v_1, v_1, v_2, v_1, v_3)$ as an example. Without loss of generality, when updating v_1's embedding, the EOP takes its neighbors (v_1, v_2) sequentially as the input of GRU. Therefore, even assign order numbers to edges, the EOP multigraph can only tell that as v_1's neighbors, v_1 appears before v_2 (local order). It can not distinguish the positions of v_1 or v_2 (global order) in the sequence s_1.

Secondly, GNNs cannot capture the long-term dependencies in the sequence. Because of the over-smoothing problem, the optimal number of layers for these GNN models is usually no larger than 3 [2], which is not enough for finding the long-term dependencie. To solve this problem, researches [2] propagate information along long-range connections. However, repeated items are treated equally and the position information of each item are ignored when propagating messages from the neighbors to a target item. As a word with different contexts in a sentence are represented differently, different representations should be learned from repeated items according to their positions and neighbors.

To overcome the aforementioned problems, we propose an order-aware graph neural network with long-range connections (OAG-LC for short). OAG-LC consists of two modules: an order-aware graph update module for the lossy sequence encoding problem (OAG for short) and an attentive graph aggregation module for long-term dependency capturing problem (LPG for short). As shown in Fig. 1(c), the general idea of the OAG module is to recurrently encode a sequence into a graph and the snapshots of the graph-structured sequence are different at each state. To take advantages of GNN-based and RNN-based methods, a novel gating mechanism is proposed, where both order and structure information can be unitized to learn the node embeddings. In this way, OAG module maintains the global order of a sequence by the lossless recurrent graph encoding and mainly focuses on capturing the short-term interests.

Furthermore, LPG module first converts a sequence into a graph via reachability where edges exist between all the reachable pairs. To further streathen the different importance for repeated items at different positions, we treat repeated items as different nodes in the graph and distinguish them with their positions. Then, it propagates information using an attention mechanism and repeated items at different positions learn different representation with their context. Finally, we apply a readout function to generate the representation of users' next interests for Top-K recommendation by taking both short-term and long-term preferences into account. We will show that the OAG module and the LPG module are crucial in the sequential recommendation task on four real-world datasets in experiment.

In summary, we conclude the main contributions of this paper as follows:

- To handle the lossy global order problem of GNN, we propose an order-aware graph update module that recurrently encodes a sequence into a graph, and uses a gating mechanism to obtains both order and structure information.
- To efficiently capture the long-term dependencies in a sequence, we construct a sequence into a directed graph via reachability and treat the repeated items in a sequence differently in the graph aggregation phase.
- We conduct extensive experiments on four public datasets. The experimental results demonstrate that the proposed OAG-LC model outperforms the state-of-the-art methods.

2   Related Work

User preferences dynamically evolve with time. The order information of behavior sequences is crucial to capture interests of users. Compared with the general recommenders that model the static preferences of users [12], the sequential recommendation mainly focuses on the order of user behaviors. Early works use Markov chains [11] to learn a transition graph over items, which usually suffered from the computation complexity problem of exponentially growing state space.

Inspired by neural language models, deep neural recommenders are proposed to capture the dynamic preferences from users' sequential behaviors. GRU4Rec

[3] is the first to employ RNN-based models in sequential recommendation. The NARM model [5] further incorporates an attention mechanism into RNN to emphasize the user's main purpose in a sequence. Some researchers adopt self-attention in user modeling, e.g. SASRec [4] seeks to identify the correlations of items by the dot-product attention. However, these attentive and recurrent models mainly focus on modeling the sequential patterns without the consideration of the complex transitions of items.

In recent years, to model the complex transition patterns in sequential behaviors, some GNNs has been applied to sequential recommendation. SRGNN [14] first converts a sequence to graph-structured data, then learns item embedding using gated GNN. Though the GNN-based methods show promising potential to model the complex transition in a sequence, they still face some problems. The first one is that GNNs fail to capture the long-term dependencies in a behavior sequence due to the over-smoothing problem [6]. GCSAN [15] adds an additional self-attention layer after GNN to preserve the long-term dependencies, but does not change the inherent disadvantage of GNN. LESSR [2] uses an SGAT layer to learn the global dependency by propagating information along long-range edges. However, repeated items in a sequence are treated equally, so that absolute order of items in the sequence is also ignored. The second problem of GNN is the lossy graph construction of the sequential order information. Recent research [2] attempts to solve the lossy order problem by assigning an order to each edge and aggregating latent features following the edge order. LESSR focuses on retaining the local order, but how to maintain the global order information of a sequence in GNN-based methods is still not explored.

In this paper, we develop a graph neural network, which can preserve the global order and the long-term dependency information.

3 Methodology

In this section, We will first give a formal description of the sequential recommendation task (Sect. 3.1). Then we describe our OAG-LC model in detail. The complete framework is demonstrated in Fig. 2. At first, an input sequence is converted to a graph and the snapshots of the graph-structured sequence are different at each state (Sect. 3.2). Then, for each snapshots, the LPG module uses attention mechanism to learn the long-term dependencies in the sequence (Sect. 3.3). After that, the OAG module is applied to learn item representation recurrently (Sect. 3.4). Finally, a readout function is used to learn the sequence representation (Sect. 3.5).

3.1 Problem Definition

Suppose there are N items. A behaviour sequence is denoted as $s = (v_1, v_2, ..., v_{l_s})$, where $v_i \in V$ is the i-th item in the sequence and l_s is the sequence length. As shown in Fig. 2, we focus on the candidate generation stage

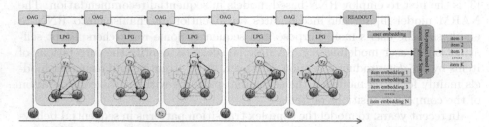

Fig. 2. The overview of the OAG-LC. A sequence is first converted to a graph with long-range connections. The Aggregation stage captures long-term dependency with attention mechanism. Meanwhile, our model learns graph-structured sequence representation recurrently to remain the order information at the update stage. Then a readout function is used to computes the sequence representation from both short-term and long-term aspects. Finally, we apply dot-product based k-nearest neighbor search method to retrieval the top-K items for next item recommendation.

of an industrial recommendation system [1], of which the sequential recommendation task is to recommend the top K items for the next click based on the previous behavior sequence s. Let $\boldsymbol{E} \in R^{d \times N}$ be the item projection matrix, where d is the dimension of the embedding size. We first project the behavior sequence $s = (v_1, v_2, ..., v_{l_s})$ into the embedding space $\boldsymbol{s} = (\boldsymbol{x}_1, \boldsymbol{x}_2, ..., \boldsymbol{x}_{l_s})$. The proposed OAG-LC model aims to learn a preference embedding $\boldsymbol{x}_s \in R^{1 \times d}$ based on the behavior sequence $\boldsymbol{s}$, which is formulated as: $\boldsymbol{x}_s = OAG\text{-}LC(\boldsymbol{s})$. Then we calculate the scores for all items as the dot-product between the preference embedding $\boldsymbol{x}_s$ and the item embedding matrix $\boldsymbol{E}$: $\boldsymbol{y}_s = \boldsymbol{x}_s \boldsymbol{E}$, where $\boldsymbol{y}_s \in R^{1 \times N}$ can be used for ranking the top K items. Finally, we use the cross-entropy loss as the optimization target: $L(\boldsymbol{y}', \hat{\boldsymbol{y}}) = \sum_{s \in S} -\boldsymbol{y}'_s log \hat{\boldsymbol{y}}_s{}^T$. $\boldsymbol{y}'_s$ is the one-hot coding of ground truth item, $\hat{\boldsymbol{y}}_s = \text{sotfmax}(\boldsymbol{y}_s)$ is the predicted probability distribution. S is the sequence set (Table 1).

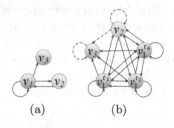

(a)   (b)

Fig. 3. The adjacent graph (a), and our proposed graph (b) of the sequence $[v_1, v_1, v_2, v_1, v_3]$.

Table 1. Dataset statistics

Statistics	#user	#item	#interaction	Avg. length
Elec	41861	87128	1325678	31.67
MT	31609	34587	1430261	45.25
HK	165420	102425	2657933	16.07
Tmall	101118	36648	1572610	15.55

3.2 Graph Construction

To learn the representation of sequential data with a GNN, a sequence should be converted into a graph. Here we take the sequence $s = (v_1, v_1, v_2, v_1, v_3)$ as an example, as shown in Fig. 3 (The solid line refers to the one-hop relationship, while the dotted line refers the self-loop and multi-hop relationship). A common way is to construct s into an adjacent graph $G_s = (V_s, E_s)$, where V_s is a set of unique items in sequence s and E_s is a set of weighted [10] or unweighted [14,15] edges (Fig. 3(a)). We call it an adjacent graph because the edge set contains edge $e_{i,j} = (v_i, v_j)$ iff v_i and v_j are adjacent items in sequences s.

However, the one-hop relationship fails to capture the long-term dependencies between items, which is necessary for sequential recommendation. And unique items in the graph will brings order-losing problem of the repeated times. To solve these problems, we develop a lossless graph construction method.

For each sequence s we convert it to a directed graph $G_s = (V_s, E_s)$ (Fig. 3(b)), where V_s stands for all the items and E_s are the reachable item pairs in the sequence s. Inspired by [2], we set an item pair (v_i, v_j) reachable iff there exists a path from v_i to v_j. We also add self-loops to all of the items in s. Besides, repeated items with different positions in a sequence may contains different information. Therefore, we differentiate repeated items with their positions.

3.3 Aggregation Stage

In this subsection, we introduce a long-term dependency preserved graph aggregation mechanism (LPG for short) which aims to learn the long-term dependencies of graph-structured sequences. Since we construct an lossless graph $G_s = (V_s, E_s)$ from sequence s where the edges exist between all the reachable pairs, the information from multi-hop relationships can be captured without going through intermediate nodes. This construction method enables us to learn long-term dependency information in one hop.

When aggregating information from the neighbors, the position information is necessary for finding the most related items. And our repeated item setting in the graph makes it convenient for position incorporation. Therefore, we add items' position encoding into their original embedding before message propagation: $z_i = x_i + p_i$, where $p_i \in R^{1 \times d}$ is the position embedding of item v_i, x_i is the initial embedding of v_i, and the output z_i is used for information propagation. We apply the attention mechanism [2,10] for message passing and the aggregation function can be formulated as:

$$b_i^l = \sum_{(v_j, v_i) \in E_s} e_{ij} z_j^l W_v, \alpha_{ij} = \frac{(z_i^l W_q)(z_j^l W_k)^T}{\sqrt{d}}, e_{ij} = \frac{exp(\alpha_{ij})}{\sum_{(v_j, v_i) \in E_s} \alpha_{ij}} \quad (1)$$

where $W_* \in R^{d \times d}$ are the learnable parameters. z_j^l is the embedding of item v_j at layer l with $z_j^0 = x_j^0 + p_j$ as its initial embedding. b_i^l is the neighbors' information of v_i at layer l.

To further capture the complex structure information in sequence s, we can stack multiple layers. After we stack L layers, the final neighbors' embedding of item v_i is b_i^L. These embeddings will serve as the input of the OAG module that we will introduce in detail below.

3.4   Update Stage

After aggregating information from items' neighbors, the common GNN methods usually update items' embeddings with their neighbor information concurrently, which only maintain the structure information like the neighborhood relationship and ignores the order of each item in the sequence. To overcome this defect, we develop an order-aware graph update mechanism (OAG for short) which updates the item embedding recurrently based on the structure information and the order information. We use b_i^L as the structure information of item v_i, which is denoted as g_i in this section.

The general idea of the OAG module is to encode the sequence into a history state which preserves the order information at each step, and then update the items' embedding and the new history state recurrently following the order of the sequence. Formally, we encode the sequence at state $i - 1$ as a hidden vector h_{i-1}, which preserves the order up to $i - 1$. At the i_{th} step, we update hidden vector h_i and the node embedding x_i with a recurrent updating function f_{OAG}, where the inputs are the history state h_{i-1}, its structure information g_{i-1} and node representation x_i. The update function can be formulated as:

$$x_i', h_i = f_{OAG}(x_i, h_{i-1}, g_i), \tag{2}$$

where x_i' is the updated node embedding. In this way, the order is preserved by both the history state h_i and the recurrent update mechanism.

Inspired by the gating mechanism in the gated GNN model [14], we design a gate to update the sequence representation in the OAG module, which aims to take both order information and structure information into account. The function f_{OAG} is formulated as:

$$r_i = \sigma([x_i, h_{i-1}]W_r + b_r), \tag{3}$$

$$z_i = \sigma([x_i, h_{i-1}]W_z + b_z), \tag{4}$$

$$h_i' = \phi([x_i, h_{i-1} \odot r_i]W_h + b_h), \tag{5}$$

$$u_i = \sigma([x_i W_1 + b_1, g_i W_2 + b_2]W_u + b_u), \tag{6}$$

$$h_i = z_i \odot h_{i-1} + (1 - z_i) \odot h_i' \odot u_i. \tag{7}$$

$$x_i' = h_i W_t + b_t, \tag{8}$$

where $W_r, W_h, W_z, W_u \in R^{2d \times d}$, $W_1, W_2, W_t \in R^{d \times d}$ and $b_* \in R^{1 \times d}$ are the learnable parameters, r_i, z_i, u_i are the reset gate, update gate and our proposed structure gate resepectively, h_i' is the candidate state, h_{i-1} is the history state. $[\cdot, \cdot]$ is the concatenate operation. $\odot$ denote the element-wise multiplication. σ is the sigmoid function and ϕ is the tanh function. The outputs of update function f_{OAG} are h_i and x'_i, where h_i is the output hidden state and x'_i is the new representation of node v_i.

3.5 Generating Sequence Embedding

After the information propagation through all the modules, we then apply a read-out function to generate the sequence representations. The neighbors' embedding of the last item considers the interaction between all the previous items with the last item. Therefore, $b^L_{l_s}$ can be used to represent the long-term preference. Besides, the gating update mechanism makes the OAG module focus more on the short-term information. By combing the long-term and short-term preference, the sequence representation of current interest can be computed as: $x_s = x'_{l_s} + b^L_{l_s}$.

4 Experiments

In this section, we first describe the datasets, compared methods, and evaluation metrics used in our experiments. Then we will give a detailed analysis of the experimental results.

4.1 Experimental Settings

Datasets. We evaluate different recommenders on four public real-world datasets, including three review datasets from Amazon[1]: Movies & TV (MT), Home & Kitchen (HK), and Electronics (Elec) and one transaction dataset from Tmall[2].

Baselines. To evaluate the proposed OAG-LC model, we compare it with the following representative methods. Top Pop/P-Pop recommends items of the most interactions with all users/the target user. SKNN/VSKNN [8] recommend items based on the nearest neighbors of the previous behaviors. VSKNN emphasizes more on recent items. GRU4Rec [3] recommends items with GRU. NARM [5] employs attention mechanism in the recurrent neural recommender to capture users' main purpose. STAMP [7] captures users' long-term and short-term interests with an attentive method. SASRec [4] uses the self-attention methods for sequential recommendation. SRGNN [14] learns item embeddings through message passing in graph. GCSAN [15] modifies SRGNN by adding a self-attention module. LESSR [2] uses two graphs to retains the order and long-term dependencies. SGNN-HN [9] applies a star node to model transitions between items. The implementation of all the baseline models is based on Recbole [17]. The source code of our model and the baselines is available online[3].

---

[1] http://deepyeti.ucsd.edu/jianmo/amazon/index.html.
[2] https://tianchi.aliyun.com/dataset/dataDetail?dataId=53.
[3] https://drive.google.com/file/.

Evaluation Metrics. We use the commonly adopted HR@K (Hit Ratio) and NDCG@K (Normalized Discounted Cumulative Gain) as the metrics for all the models. We report the result when $K = \{5, 10\}$.

4.2   Overall Performance

To demonstrate the effectiveness of the proposed model, we compare it with the state-of-the-art sequential recommendation methods. The overall performance is illustrated in Table 2, and the best results are highlighted in boldface.

The Top-Pop and P-Pop methods are not competitive because they solely base on the occurence frequency of items. The VSKNN outperform SKNN shows that the recent interacted items are important for finding users' intention. All of the neural networks methods significantly outperform conventional models, which proves the powerful ability of deep learning models in this task.

The RNN-based models (GRU4Rec and NARM) outperform GNN-based models (SRGNN, GCSAN, and LESSR, SGNN-HN) on two of four datasates including Elec and MT, and get a comparable result on the other two datasets, suggesting that the order of items in a sequence is important to model users' preference. When it comes to whether long-term information is necessary in sequence modeling, we may conduct several comparisons between models with and without long-term information. (1) NARM model outperforms the GRU4Rec model on two of four datasets. (2) The LESSR defeats the other GNN-based models on two datasets including MT and HK, and gets competitive results on Elec. Such results prove that users' long-term preferences are also important in the sequential recommendation task. The proposed OAG-LC model captures both global order and long-term dependencies information in a sequence. Thus it outperforms all the baseline models significantly.

Table 2. Performance comparison of different methods

Datasets	Metrics	T-Pop	P-Pop	SKNN	VSKNN	GRU4Rec	NARM	STAMP	SASRec	SRGNN	GCSAN	LESSR	SGNN-HN	OAG-LC	Improv	p-val
Elec	HR@5	0.00095	0.00070	0.01035	0.01450	0.04985	0.04790	0.04575	0.04755	0.04805	0.04795	0.04860	0.04525	**0.05285**	6.0%	8.9e−4
	NDCG@5	0.00042	0.00055	0.00592	0.00828	0.04488	0.04386	0.04258	0.04260	0.04294	0.04071	0.04245	0.04078	**0.04665**	3.9%	4.9e−3
	HR@10	0.00275	0.00115	0.01710	0.02360	0.05760	0.05455	0.05065	0.05380	0.05435	0.05385	0.05580	0.05265	**0.06025**	4.6%	1.3e−2
	NDCG@10	0.00100	0.00069	0.00810	0.01124	0.04737	0.04597	0.04418	0.04460	0.04496	0.04260	0.04478	0.04316	**0.04904**	3.5%	8.0e−3
MT	HR@5	0.00320	0.00165	0.03135	0.03940	0.18585	0.18845	0.16865	0.17275	0.17590	0.17645	0.18660	0.17655	**0.19305**	2.4%	1.0e−3
	NDCG@5	0.00156	0.00136	0.01661	0.02110	0.17133	0.17345	0.15455	0.15547	0.16220	0.15771	0.17229	0.16297	**0.17590**	1.4%	1.2e−2
	HR@10	0.00640	0.00245	0.06000	0.07045	0.20690	0.20970	0.18420	0.19630	0.19495	0.19535	0.20750	0.19610	**0.21630**	3.1%	5.3e−5
	NDCG@10	0.00257	0.00160	0.02586	0.03111	0.17811	0.18030	0.15956	0.16303	0.16831	0.16382	0.17901	0.16924	**0.18338**	1.7%	1.4e−3
HK	HR@5	0.00340	0.00115	0.02305	0.02855	0.04670	0.04775	0.04425	0.04420	0.04610	0.04615	0.04790	0.04660	**0.05030**	5.0%	5.7e−4
	NDCG@5	0.00271	0.00076	0.01231	0.01536	0.04294	0.04333	0.04038	0.04087	0.04213	0.04122	0.04312	0.04254	**0.04588**	5.9%	2.4e−7
	HR@10	0.00595	0.00775	0.03500	0.04150	0.05080	0.05245	0.04745	0.04900	0.05010	0.04995	0.05265	0.05100	**0.05570**	5.8%	1.3e−4
	NDCG@10	0.00357	0.00277	0.01615	0.01957	0.04426	0.04485	0.04141	0.04240	0.04342	0.04245	0.04467	0.04398	**0.04763**	6.2%	1.3e−8
Tmall	HR@5	0.00015	0	0.02310	0.02420	0.08660	0.08285	0.08890	0.08155	0.08725	0.07610	0.08335	0.08340	**0.09215**	3.7%	2.3e−2
	NDCG@5	0.00015	0	0.01090	0.01138	0.06645	0.06250	0.06779	0.06156	0.06621	0.05393	0.06333	0.06397	**0.07040**	3.9%	1.0e−2
	HR@10	0.00030	0	0.03740	0.04205	0.11095	0.11000	0.11245	0.10705	0.11090	0.09925	0.10695	0.10615	**0.11770**	4.7%	6.7e−4
	NDCG@10	0.00020	0	0.01547	0.01706	0.07425	0.07128	0.07533	0.06977	0.07388	0.06141	0.07091	0.07131	**0.07855**	4.3%	7.9e−4

4.3 Ablation Studies

In this section, we conduct some ablation studies to demonstrate the contribution of each module. To better compare the influence of order and long-term information, we compare the proposed methods with two representative GNN-based models: SRGNN and LESSR. We also report the significance test results between our OAG-LC model and the best ablation module on each metric of all the datasets. $-$, $*$ and $**$ means not significance, significance, and strong significance. Due to the space limit, we only report the results on the Elec and HK datasets. Results on the other datasets follow similar patterns.

OAG stands for the variant that removes the LPG aggregation phase, which is degenerated into a normal gated recurrent unit. This variant only considers the order information. LPG only keeps the long-term dependency preserved aggregation phase, and we use the neighbors' representation as the target item's representation. As shown in Fig. 4, OAG-LC defeated LPG and OAG modules consistently, suggesting that both the order and long-term information are important for sequence modeling. Not surprisingly, LESSR performs better than SRGNN on most datasets because the former can preserves items' order and the long-term dependencies. Compared with LESSR, which only keeps the local order around a target item, our proposed model performs better.

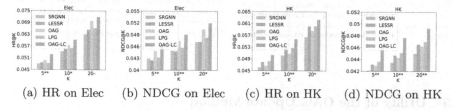

(a) HR on Elec   (b) NDCG on Elec   (c) HR on HK   (d) NDCG on HK

Fig. 4. Ablation analysis on four datasets. OAG stands for the original model removes the LPG aggregation phase, which is degenerated into a normal gated recurrent unit. LPG refers to the model only keeps the long-term dependency preserved aggregation phase, and uses the neighbors' representation as the target item's representation.

4.4 Utility of the LPG Aggregation Method

We perform some discussion studies to show the effectiveness of our LPG module. First, to show the repeated nodes in the graph construction phase help capture the long-term dependencies in a sequence, we remove the repeated nodes in our proposed graph and still add edges with the reachability of the sequence. We call this variant as *no-repeated-oag*. Furthermore, to evaluate the effectiveness of the long range information captured by the LPG module, we compare it with the aggregation methods from other GNN-based methods, including gated GNN from SRGNN [14], weighted graph attention method from FGNN [10], and a heuristic average aggregation method. These variants are called *gated-oag*, *wgat-oag* and *heur-long-oag* accordingly. For a fair comparison, we keep the same update and readout function for each variant as our original model. From Table 3 (Agg is short for aggregation), although *no-repeated-oag* converts a sequence into a graph via the reachability, it performs worse than the *OAG-LC* model. Such results demonstrate that repeated items with different contexts should be represented with different embeddings. When comparing *no-repeated-oag* with other variants that use different graph construction and message aggregation method, the former performs better. This is because all the other variants construct a graph with one-hop connections, making them fail to aggregate information from long-range connections. Though *heur-long-oag* averages the embedding of items that can reach the target item, it is unaware of relevant and irrelevant items.

Table 3. Aggr methods comparison

	Elec	MT	HK	Tmall
HR@10				
no-repeated-oag	0.05765	0.21305	0.05550	0.11750
gated-oag	0.05770	0.20225	0.05220	0.10825
wgat-oag	0.05785	0.21320	0.05485	0.11260
heur-long-oag	0.05495	0.20815	0.05230	0.10940
OAG-LC	**0.06025**	**0.21630**	**0.05570**	**0.11770**
NDCG@10				
no-repeated-oag	0.04801	0.18069	0.04716	0.07790
gated-oag	0.04730	0.17206	0.04498	0.07135
wgat-oag	0.04797	0.18182	0.04684	0.07524
heur-long-oag	0.04568	0.17692	0.04502	0.07223
OAG-LC	**0.04904**	**0.18338**	**0.04763**	**0.07855**

Table 4. Update methods comparison

	Elec	MT	HK	Tmall
HR@10				
lpg-gru	0.05545	0.19440	0.05445	0.11585
lpg-gated	0.05515	0.19525	0.05475	**0.11900**
lpg-linear	0.05585	0.19805	0.05365	0.11850
lpg-weak-order	0.05560	0.19200	0.05430	0.11040
OAG-LC	**0.06025**	**0.21630**	**0.05570**	0.11770
NDCG@10				
lpg-gru	0.04509	0.16385	0.04636	0.07607
lpg-gated	0.04462	0.16149	0.04584	0.07242
lpg-linear	0.04524	0.16555	0.04634	**0.07936**
lpg-weak-order	0.04477	0.16002	0.04620	0.07338
OAG-LC	**0.04904**	**0.18338**	**0.04763**	0.07855

4.5 Utility of the OAG Update Method

To show the effectiveness of our OAG update module, we replace it with other update functions from the existing GNN-based methods, including gated update function from SRGNN [14] and linear update function from LESSR [2]. We call these two variants as *lpg-gated* and *lpg-linear*. Besides, we also design two variants called *lpg-weak-order* and *lpg-gru*, based on our OAG update method. *lpg-weak-order* uses the average embedding before the target item as the order information. *lpg-gru* removes the structure gate in the OAG module, which makes the update function degenerated to a normal GRU. For a fair comparison, for all the variants, the graph construction, aggregation and readout function are kept the same as our original model. From Table 4, we can see that after replacing our OAG update function with other methods, the performance decreases dramatically. The reason might be that the gated or linear update method can only preserve the structure of the sequence, and they fail to retain the order information of the sequence. Though *lpg-weak-order* takes the average item embeddings that appear before the target item as the order information, it fails to credit the position of each item. Such results show that the order is necessary for sequence modeling. *lpg-gru* gets competitive results by preserving the sequence order using a gated recurrent unit. However, it uses the addition of neighbor and item embeddings as the input, making it can not balance the order and structure information well.

5 Conclusion

In this paper, to capture the global order and long-term dependency information of sequences with the GNN-based models, we propose a novel graph neural network named OAG-LC. This model first converts a sequence to a directed graph where edges exist between connected items and repeated items are distinguished with their positions for long-term dependency capturing, and then a recurrent update module is applied to retain the global order information in the sequence. Extensive experimental results on four real benchmark datasets show the effectiveness of our model.

Acknowledgements. This work was supported by NSFC grants (No. 62136002 and 61972155), the Science and Technology Commission of Shanghai Municipality (20DZ1100300) and the Open Project Fund from Shenzhen Institute of Artificial Intelligence and Robotics for Society, under Grant No. AC01202005020, Shanghai Knowledge Service Platform Project (No. ZF1213), Shanghai Trusted Industry Internet Software Collaborative Innovation Center.

References

1. Beutel, A., Covington, P., et al.: Latent cross: making use of context in recurrent recommender systems. In: Proceedings of the 11th WSDM, pp. 46–54 (2018)
2. Chen, T., Wong, R.C.W.: Handling information loss of graph neural networks for session-based recommendation. In: Proceedings of the 26th ACM SIGKDD, pp. 1172–1180 (2020)
3. Hidasi, B., Karatzoglou, A., et al.: Session-based recommendations with recurrent neural networks. In: Proceedings of the 3rd ICLR (2016)
4. Kang, W.C., McAuley, J.: Self-attentive sequential recommendation. In: Proceedings of the 18th ICDM, pp. 197–206. IEEE (2018)
5. Li, J., Ren, P., Chen, Z., et al.: Neural attentive session-based recommendation. In: Proceedings of the 27th CIKM, pp. 1419–1428 (2017)
6. Li, Q., Han, Z., Wu, X.M.: Deeper insights into graph convolutional networks for semi-supervised learning. In: Proceedings of the 32nd AAAI, vol. 32 (2018)
7. Liu, Q., Zeng, Y., et al.: STAMP: short-term attention/memory priority model for session-based recommendation. In: Proceedings of the 24th SIGKDD, pp. 1831–1839 (2018)
8. Ludewig, M., Jannach, D.: Evaluation of session-based recommendation algorithms. User Model. User-Adap. Inter. **28**(4–5), 331–390 (2018)
9. Pan, Z., Cai, F., Chen, W., et al.: Star graph neural networks for session-based recommendation. In: Proceedings of the 29th CIKM, pp. 1195–1204 (2020)
10. Qiu, R., Li, J., Huang, Z., et al.: Rethinking the item order in session-based recommendation with graph neural networks. In: Proceedings of the 28th CIKM, pp. 579–588 (2019)

11. Shani, G., Heckerman, D., Brafman, R.I.: An MDP-based recommender system. J. Mach. Learn. Res. **6**(Sep), 1265–1295 (2005)
12. Song, Q., et al.: Incremental matrix factorization via feature space re-learning for recommender system. In: Proceedings of the 9th Recsys, pp. 277–280 (2015)
13. Sun, K., Qian, T., Yin, H., Chen, T., Chen, Y., Chen, L.: What can history tell us? In: Proceedings of the 28th CIKM, pp. 1593–1602 (2019)
14. Wu, S., Tang, Y., et al.: Session-based recommendation with graph neural networks. In: Proceedings of the 33rd AAAI, vol. 33, pp. 346–353 (2019)
15. Xu, C., Zhao, P., Liu, Y., et al.: Graph contextualized self-attention network for session-based recommendation. In: Proceedings of the 28th IJCAI, vol. 19, pp. 3940–3946 (2019)
16. Yu, Z., Lian, J., Mahmoody, A., et al.: Adaptive user modeling with long and short-term preferences for personalized recommendation. In: Proceedings of the 28th IJCAI, pp. 4213–4219 (2019)
17. Zhao, W.X., Mu, S., Hou, Y., et al.: RecBole: towards a unified, comprehensive and efficient framework for recommendation algorithms. arXiv preprint arXiv:2011.01731 (2020)

Domain-Level Pairwise Semantic Interaction for Aspect-Based Sentiment Classification

Zhenxin Wu[1]📧, Jiazheng Gong[1], Kecen Guo[1], Guanye Liang[1],
Qingliang Chen[1,2](✉)📧, and Bo Liu[1]

[1] College of Information Science and Technology, Jinan University,
Guangzhou 510632, China
wuzhenxin@stu2020.jnu.edu.cn, tpchen@jnu.edu.cn
[2] Guangzhou Xuanyuan Research Institute Company, Ltd., Guangzhou 510006,
China

Abstract. Aspect-based sentiment classification (ABSC) is a very challenging subtask of sentiment analysis (SA) and suffers badly from the class-imbalance. Existing methods only process sentences independently, without considering the domain-level relationship between sentences, and fail to provide effective solutions to the problem of class-imbalance. From an intuitive point of view, sentences in the same domain often have high-level semantic connections. The interaction of their high-level semantic features can force the model to produce better semantic representations, and find the similarities and nuances between sentences better. Driven by this idea, we propose a plug-and-play Pairwise Semantic Interaction (PSI) module, which takes pairwise sentences as input, and obtains interactive information by learning the semantic vectors of the two sentences. Subsequently, different gates are generated to effectively highlight the key semantic features of each sentence. Finally, the adversarial interaction between the vectors is used to make the semantic representation of two sentences more distinguishable. Experimental results on four ABSC datasets show that, in most cases, PSI is superior to many competitive state-of-the-art baselines and can significantly alleviate the problem of class-imbalance.

Keywords: Aspect-based sentiment classification · Pairwise semantic interaction · Class-imbalance

1 Introduction

Aspect-based sentiment classification (ABSC) is a fine-grained sentiment classification subtask of sentiment analysis [10], which aims to identify the sentiment polarity of each aspect in a sentence (positive, negative or neutral). It is widely used in different domains, such as online comments (e.g., movie and restaurant reviews [9]), data mining and e-commerce customer service. For example, sentence 1 in Fig. 1 shows that the customer enjoys the restaurant's food but thinks

J. Gama et al. (Eds.): PAKDD 2022, LNAI 13280, pp. 303–314, 2022.
https://doi.org/10.1007/978-3-031-05933-9_24

the ambience is just not bad. For this sentence, ABSC needs to recognize that the two aspects "ambience" (A1) and "food" (A2) contained in the sentence are "neutral" and "positive", respectively.

In fact, people's comments often have obvious emotional preferences, which means that they may suffer the problem of class-imbalance. Since there are far more comments with "positive" and "negative" in the same domain than those with "neutral", "neutral" comments are always marginalized and thus misjudged. At present, the commonly used ABSC methods, whether they are traditional methods [1,19] or deep learning models [5,22], none of them has solved the problem of class-imbalance. Moreover, the similarity of semantic contexts between sentences in the same domain has not been fully utilized. In our paper, we define "semantic" as a highly abstract coding vector of sentences extracted by the information extractor, e.g. BERT. If we can make interactive learning of two similar sentences in the same domain, they can learn more domain semantic information from each other and enrich the high-level semantic encoding of sentences. It will also help to find the similarities and nuances between sentences, which can reduce misjudgments due to class-imbalance.

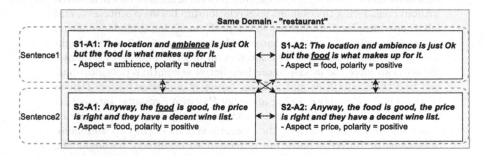

Fig. 1. The two sentences have different aspects, but they all belong to the same domain "restaurant" and have similar semantic context.

For example, making interactive learning of S1-A1 and S1-A2 in Fig. 1, can help to distinguish different sentiment polarities of a sentence which contains different aspects. The interaction between S1-A1 ("neutral") and S2 (both aspects are "positive") can make the semantic encoding of "neutral" more discriminative, by comparing it with the strong "positive" sentence S2. At the same time, the interaction between S1-A2, S2-A1, and S2-A2 can also enrich the features of the same sentiment polarity in different aspects or different sentences.

Based on this intuition, we propose a domain-level plug-and-play Pairwise Semantic Interaction (PSI) module for ABSC. For the construction of sentence pairs, it is worth emphasizing that we consider that the sentences in the same dataset belong to the same domain. We do not limit that two sentences must have the same aspect, and encourage richer interactions between sentences (refer to Sect. 3.4 for detailed sentence pair construction strategy). For PSI module, firstly, we extract the semantic vectors of the two sentences by semantic extractors (e.g., BERT [5]), respectively. Subsequently, through a gating mechanism, sentences can learn each other's high-level semantic information adaptively, which enriches the semantic representation of a single sentence. Finally, we additionally use a design similar to the adversarial network [6] to help promote the model to distinguish the nuances between similar semantic representations. In summary, the contributions of this paper are as follows:

- We introduce a Pairwise Semantic Interaction (PSI) module for the interaction between sentences, help to find the similarities and nuances of sentences, and can significantly reduce the misjudgments due to class-imbalance. Through the interaction between sentences in the same domain, the sentences get better and more discriminative semantic representations.
- The PSI module is plug-and-play and can be easily combined with most mainstream semantic extractors such as BERT.
- The experiments on four prestigious ABSC datasets have justified the efficacy of PSI, achieving or approaching SOTA results.

2   Related Work

Existing ABSA researches focus on the use of deep neural networks, such as target dependent LSTM models [18] and Attention-based LSTM [21] for aspect-level sentiment classification. In recent years, the pre-trained language model BERT [5], which has been very successful in many Natural Language Processing tasks, has been applied in ABSA and achieved significant results such as [11,17,23]. However, all of the above studies have ignored semantic relationship between sentences in the same domain. Recently, contrastive learning has achieved great success in both Computer Vision (CV) and Natural Language Processing (NLP). Its main purpose is to make the features of the same category closer to each other, while the distance between the features of different categories is farther. In [25], through an attention interaction, the network can adaptively find delicate clues from two fine-grained images in pairs. In ABSA, Chen et al. [4] proposed a Cooperative Graph Attention Networks (CoGAN) method for cooperatively learning the aspect-related sentence representation in document level. However, their model is based on transformer combined with Graph Networks, which has high computational overhead. By contrast, our proposed PSI is a plug-and-play module, which can achieve decent performance with a little extra overhead.

3   The Proposed Method

In this section, we will describe our PSI module. PSI compares two similar sentences together to find the common semantic representation and semantic differences between them, rather than studying the semantic representation of a single sentence alone.

The module PSI will take two similar sentences as input and go through three carefully designed sub-modules i.e., mutual vector learning, semantic gate generation, and adversarial interaction. The entire structure of PSI is shown in Fig. 2.

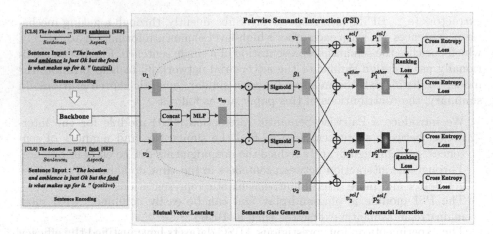

Fig. 2. The structure of PSI. We simply use S1-A1 and S1-A2 in Fig. 1 as an example of pairwise sentences (*Sentence₁, Sentence₂*). In this case, PSI can promote the model to distinguish different sentiment polarities of different aspects in the sentence. It is worth emphasizing that the PSI is a plug-and-play module, i.e., PSI can be combined with most mainstream semantic extraction backbones (e.g., BERT) during the training phase, and flexibly unload it for single-input test sentence.

3.1   Mutual Vector Learning

Before this sub-module, the semantic of these two sentences are extracted by backbone, and two D-dimensional semantic vectors i.e., v_1 and $v_2 \in \mathbb{R}^D$ are generated, respectively. Then, in this sub-module, we learn a mutual vector $v_m \in \mathbb{R}^D$ from individual v_1 and v_2,

$$v_m = f_m([v_1, v_2]).\tag{1}$$

where $[\ ]$ is concatenation operation and $f_m(\cdot)$ is a mapping function of $[v_1, v_2]$. Specifically, we use the multi-layer perceptron (MLP) as the mapping function. By summarizing the two feature vectors v_1 and v_2, the mutual vector v_m is produced accordingly, which contains common high-level semantic information and discriminative semantic clues of two sentences.

3.2   Semantic Gate Generation

After producing the mutual vector v_m, we can use v_m to activate v_1 and v_2. In order to generate more discriminative information for later comparisons, the dot product of v_m with two feature vectors v_1 and v_2 is carried out according to channels to locate the contrastive information in the two vectors. Then, the gate vectors, i.e., g_1 and g_2 are generated by a sigmoid function,

$$g_i = sigmoid(v_m \odot v_i), i \in \{1, 2\}.\tag{2}$$

Therefore, g_i can be used as an attention vector to highlight the important semantic representations belonging to individual v_i. For example, the previous example of S1-A1 and S1-A2 in Fig. 1, the gates will help to highlight two different key words "ambience" and "food", respectively. This helps to distinguish different sentiment polarities belonging to different aspects.

3.3 Adversarial Interaction and Model Training

When comparing two sentences, humans not only focus on the salient parts of one sentence, but also focus on the salient parts of the other one. Based on this, we introduce an adversarial interaction mechanism through residual attention. As shown in Fig. 2, two feature vectors v_1 and v_2 and two gate vectors g_1 and g_2 are combined in pairs, we could then get four attentive semantic vectors with

$$
\begin{aligned}
v_1^{self} &= v_1 + v_1 \odot g_1, \\
v_2^{self} &= v_2 + v_2 \odot g_2, \\
v_1^{other} &= v_1 + v_1 \odot g_2, \\
v_2^{other} &= v_2 + v_2 \odot g_1.
\end{aligned}
\tag{3}
$$

Intuitively, the semantic vector $v_i(i \in \{1,2\})$, is guided by the attention of the gate vector $g_j(j \in \{1,2\})$, strengthens or weakens certain semantic information, and then adds to itself to get the output. While $v_i^{self} \in \mathbb{R}^D$ reinforces the feature region belonging to its own gate vector, and $v_i^{other} \in \mathbb{R}^D$ reinforces the feature region belonging to another gate vector.

Then v_i^j (where $i \in \{1,2\}$, $j \in \{self, other\}$) are feed into the softmax classifier by $p_i^j = softmax(W v_i^j + b)$, where $p_i^j \in \mathbb{R}^C$ represents the score vector of prediction, C indicates the number of polarities, and $\{W, b\}$ is the parameter set of the softmax classifier. In order to effectively train the entire PSI module, we define the following loss function as

$$
J = J_{ce} + \mu J_{rk}.
\tag{4}
$$

Among them, J_{ce} is the cross-entropy loss, and J_{rk} is the score ranking regularization loss with a coefficient of μ. Specifically we choose the hinge loss function as the score ranking regularization J_{rk},

$$
J_{rk} = \sum_{i \in \{1,2\}} \max(0, p_i^{other}(y_i) - p_i^{self}(y_i) + \varepsilon).
\tag{5}
$$

where $p_i^j(y_i) \in \mathbb{R}$ represents the score got in the predicted vector p_i^j, and y_i denotes the index of the true polarity of sentence i, and ε is the penalty term. The motivation of this design is that, v_i^{self} is activated by its own gate vector. Hence, compared to v_i^{other}, it should be more discriminative to the corresponding label. That is, the score difference $p_i^{self}(y_i)$-$p_i^{other}(y_i)$ should be larger than a margin ϵ, which means that $p_i^{self}(y_i)$ should be larger than $p_i^{other}(y_i)$ and must keep a distance with $p_i^{other}(y_i)$. At the same time, when cross-entropy loss J_{ce} is optimized, due to having the same label, $p_i^{self}(y_i)$ and $p_i^{other}(y_i)$ will tend to be closer. Therefore, J_{rk} and J_{ce} will be optimized adversarially. As a result, v_i^{other} can learn the semantic information shared by the two sentences, and v_i^{self} can learn its own unique information which will be more discriminative and reduce the noise of sentence pairs.

3.4   Sentence Pair Construction

Next, we'll provide an explanation on how to construct multiple sentence pairs in a batch for end-to-end training. Specifically, we randomly sample N_p polarities in a batch (there are 3 polarities in total, i.e. positive, negative, neutral). For each polarity, we randomly sample N_s training sentences. Consequently, there are $N_p \times N_s$ different sentences in each batch (we set the same sentence to express different aspects, belonging to different sentences). After getting a batch of sentences, we input these sentences into the backbone to generate their respective semantic vectors. For every sentence, we compare its semantic vector with the different sentences in the batch in accordance to Euclidean distance. We do not limit that two different sentences must have the same aspect, and encourage richer interactions between sentences (the following ablation study proves our point). Then, we can construct the inter/intra-pairs in a batch. The inter-pairs are the following sentence pairs which contains two situations. 1) The current sentence and itself (with different aspect and different polarity), e.g., S1-A1 & S1-A2 in Fig. 1; 2) The current sentence and the most similar sentence with different polarities from the current sentence, e.g., S1-A1 & S2 (A1/A2). On the contrary, intra-pairs refer to the following sentence pairs. 1) The current sentence and itself (with different aspect and same polarity), e.g., S2-A1 & S2-A2; 2) The current sentence and the most similar sentence with the same polarity from the current sentence, e.g., S1-A2 & S2 (A1/A2). This design permits the PSI to learn to distinguish between truly similar and highly overlapping pairs.

3.5   Model Testing

Because PSI is a practical plug-and-play module. In the training phase, the backbone and the PSI module can summarize the comparative clues from sentence pairs, and step by step improve the discriminant capacity of backbone representation for sentences. Therefore, in the testing phase, only the backbone model with updated parameters is used, but not the PSI module, so that the generalization ability of the model can be guaranteed without losing the performance of the model. To be specific, in testing phase, we input a sentence into the backbone, extract its semantic vector $X_* \in \mathbb{R}^D$, and then directly input X_* into the softmax classifier. It is worth emphasizing that the softmax classifier are shared between the training phase and the testing phase. The score vector $P_* \in \mathbb{R}^C$ is applied to label prediction. Thus, our test scheme is the same as a regular backbone, which demonstrates the strong applicability of PSI.

Table 1. Statistics of the datasets.

Polarity	Res14		Lap15		Res16		Lap16	
	Train	Test	Train	Test	Train	Test	Train	Test
Positive	839	222	765	329	749	204	1084	274
Neutral	500	94	106	79	101	44	188	46
Negative	2179	657	1103	541	1657	611	1637	481
Sum	**3518**	**973**	**1974**	**949**	**2507**	**859**	**2909**	**801**

4 Experiments

4.1 Datasets and Metrics

We have carried out experiments on four datasets to verify the performance of our proposed model, PSI. Restaurant14 is from Semeval-2014 Task 4 [16], Laptop15 is from Semeval-2015 Task 12 [15], and the other two datasets (Restaurant16 and Laptop16) are from Semeval-2016 Task 5 [14]. The statistics for these datasets are shown in Table 1. And we use Accuracy (Acc.) and Macro-F1 (F1) as performance metrics.

4.2 Implementation Details

Unless stated otherwise, we implement PSI as follows. For each aspect of a sentence, we concat the corresponding aspect at the end of the sentence. Then we adjust the length of each sentence to 85 (the maximum sentence length after the tokenizer tokenizes is 85). If it is not enough, fill it with zero, and then use it as the input of backbone. Firstly, we extract the semantic vector $v_i \in \mathbb{R}^{786}$ by BERT. Secondly, for all the datasets, we randomly sample 3 polarities, i.e., $N_p = 3$. And for each polarity, we randomly sample 4 sentences to form a batch, i.e., $N_s = 4$. For each sentence, we find its most similar sentence from its own polarity and the rest polarities, according to Euclidean distance between their semantic vectors. As a result, we obtain an intra-pair and an inter-pair for each sentence in the batch. For each pair, we concatenate v_1 and v_2 as input to a two-layer MLP, i.e., FC $(1572 \rightarrow 512)$, FC $(512 \rightarrow 786)$. Consequently, this operation generates the mutual vector $v_m \in \mathbb{R}^{786}$. All of our models are implemented by Pytorch with a single NVIDIA GTX 2080Ti GPU with 11G Memory. For all datasets, the coefficient μ in Eq. 4 is 1, while the margin ϵ is 0.05 in score ranking regularization. Among them, BERT is optimized by Adam optimizer with $\beta_1 = 0.9$, and the initial learning rate is 0.0001. For our PSI (backbone is BERT or BERT-Large) method, we use another Adam optimizer for training, and the initial learning rate is 0.00002 with $\beta_1 = 0.9$. There are a total of 20 training epochs, and if the loss does not decrease for 5 consecutive epochs, it will invoke early-stop. In addition, we set a fixed seed when training the model to ensure the reproducibility of the results.

Table 2. The comparative results, with data for non-BERT models from [4], kumaGCN from [2], RepWalk from [24], and IMN from [8]. The data for BERT-QA from [17], AC-MIMLLN from [11], and CoGAN from [4]. The experimental configuration for standard BERT and PSI is shown in implementation details. "–" denotes no data available yet. The best result of each dataset is bolded, and the second-best result is underlined.

Models	Res14		Lap15		Res16		Lap16	
	Acc.	F1	Acc.	F1	Acc.	F1	Acc.	F1
TC-LSTM	0.781	0.675	0.745	0.622	0.813	0.629	0.766	0.578
ATAE-LSTM	0.772	–	0.747	0.637	0.821	0.644	0.781	0.591
RAM	0.802	0.708	0.759	0.639	0.839	0.661	0.802	0.627
IAN	0.793	0.701	0.753	0.625	0.836	0.652	0.794	0.622
Clause-Level ATT	–	–	0.816	0.667	0.841	0.667	0.809	0.634
LSTM+synATT+TarRep	0.806	0.713	0.822	0.649	0.846	0.675	0.813	0.628
kumaGCN	0.814	0.736	–	–	0.894	0.732	–	–
RepWalk	0.838	0.769	–	–	0.896	0.712	–	–
IMN	0.839	0.757	0.831	0.654	0.892	0.71	0.802	0.623
BERT	0.867	0.764	0.818	0.699	0.884	0.755	0.817	0.665
BERT-QA	–	–	0.827	0.595	0.896	0.715	0.812	0.596
AC-MIMLLN	0.893	–	–	–	–	–	–	–
CoGAN	–	–	0.851	0.745	**0.920**	0.816	0.842	0.707
PSI (BERT)	0.916	0.857	0.860	0.756	0.901	0.788	0.839	0.723
PSI (BERT-Large)	**0.924**	**0.863**	**0.868**	**0.760**	0.913	**0.828**	**0.87**	**0.737**

4.3 Comparison with SOTA Methods

To fully evaluate the performance of our method, we apply PSI based on BERT or BERT-Large. We compare it with the state-of-the-art (SOTA) baselines including (1) ABSA models without BERT: TC-LSTM [18], ATAE-LSTM [21], RAM [3], IAN [12], Clause-LevelATT [20], LSTM+synATT+TarRep [7], kumaGCN [2], RepWalk [24] and IMN [8]. (2) BERT-based models for ABSA: BERT [5], BERT-QA [17], AC-MIMLLN [11] and CoGAN [4]. Table 2 shows the results of our experiments on four datasets.

From Table 2 we can come to the following conclusion. The performance of PSI (Based on BERT or BERT-Large) on Res14, Lap15 and Lap16 is better than those of all baselines. And in Res16, our PSI module approaches SOTA results. The experiments justify that PSI is a very powerful plug-and-play module, showing the effectiveness of our method.

Table 3. Comparison of accuracy (%) of different polarities between PSI and BERT.

Model	Negative (%)	Neutral (%)	Positive (%)	Overall (%)
BERT	92.5	47.9	86.0	86.7
PSI (BERT)	**97.6** (+5.1)	**60.6** (+12.7)	**86.9** (+0.9)	**91.6** (+4.9)

4.4 Alleviating the Problem of Class-Imbalance

As shown in Table 1, the mainstream datasets have the problem of class-imbalance. For example, in Res14, the "negative" comments is significantly more than the data of other polarities ("negative" accounts for 62%), while the "neutral" comments is far lower for other polarities ("neutral" accounted for 14%). In order to illustrate the advantages of the our method, we compared PSI (Based on BERT) with the standard BERT model on the Res14 dataset, for each polarity (positive, negative, and neutral). Table 3 shows the comparison results of the accuracy of different sentiment polarities.

It can be seen that the "positive" and "neutral" accuracy of our model (PSI) is better than that of BERT model. Specifically, PSI significantly improves the performance of "neutral" classification. And the overall accuracy of our model was greatly improved compared with BERT model. It indicates that the sentence pair interaction learning in PSI module can make the semantics between sentences complement each other and effectively learn the nuances between sentences. This can make the semantic representation of different polarities of sentences more distinguishable, and can effectively alleviate the problem of class-imbalance.

4.5 Ablation Study

Our proposed method encourages richer interactions between sentences and do not limit that two different sentences must have the same aspect. In order to study the impact of different sample extraction methods (sentiment polarity and aspect) on ABSC, we conducted the following ablation experiments. We use BERT as semantic vector extractor to evaluate different sample extraction methods on Res14.

As mentioned above, our proposed sample extraction method is that Intra/inter pairs are constructed from the same/different sentiment polarities without limiting the range of aspect (**Interacting Polarity, I_P**). And there are three other sample extraction methods, including **1) Interacting Aspect (I_A)**. Intra/inter pairs are constructed from the same/different aspects without limiting the range of sentiment polarity. **2) Interacting Polarity and Limiting Aspect (I_P & L_A)**. Intra/inter pairs are constructed from the same/different polarities by limiting the same aspect. **3) Interacting Aspect and Limiting Polarity (I_A & L_P)**. Intra/inter pairs are constructed from the same/different aspects by limiting the same polarity.

From Table 4, in all the four datasets, the results of the other three ablation experiments are worse than I_P (Ours). For I_P & L_A and I_A & L_P, results demonstrate that we should not limit aspects (in order to better distinguish different aspects of a sentence) and should also allow different polarities to interact in pairs (in order to better distinguish different sentiment polarities). For I_A, due to class-imbalance, if we do not limit the range of sentiment polarity (by constructing the inter-pair of different sentiment polarities), there will be a lot of interactions between sentences belonging to the same majority class

Table 4. Different sample extraction method.

Methods	Res14		Lap15		Res16		Lap16	
	Acc.	F1	Acc.	F1	Acc.	F1	Acc.	F1
I_P (Ours)	**0.916**	**0.857**	**0.860**	**0.756**	**0.901**	**0.788**	**0.839**	**0.723**
I_A	0.914	0.854	0.834	0.699	0.896	0.753	0.830	0.680
I_P & L_A	0.909	0.852	0.840	0.699	0.893	0.787	0.819	0.656
I_A & L_P	0.895	0.826	0.836	0.689	0.873	0.738	0.820	0.641

("negative"), while the interaction between different sentiment polarities will be insufficient. Finally, for I_P, we explicitly construct inter-pairs (belonging to different polarities) in each batch to ensure that the number of interactions between different polarities is sufficient. In this way, the nuances between different sentiment polarities can be better learned by the model. Therefore, we choose I_P as the sample extraction method for ABSC.

4.6   Visualization Analysis

In order to understand the discriminability of our method, we use UMAP [13] to visualize the polarity separability and compactness in the semantic features extracted from a standard BERT and the PSI (based on BERT) in Res14. In Fig. 3, it is evident that when using our PSI module, the clusters are farther apart and more compact, leading to a more clear distinction of various clusters representing different polarities. This also proves that adding PSI module can promote the model to learn better semantic representation of sentences and make them more distinguishable.

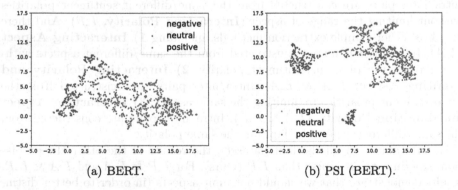

(a) BERT.   (b) PSI (BERT).

Fig. 3. Discriminability using UMAP to visualize polarity separability and compactness.

5   Conclusion

In this paper, we proposed a domain-level Pairwise Semantic Interaction (PSI) for ABSC. Through the interactions between sentences, PSI can effectively enrich the semantic encoding of sentences and produce better semantic representations. Meanwhile, PSI is plug-and-play module and can further help the model distinguish the nuances between similar sentences and effectively alleviate the problem of class-imbalance. Finally, the empirical results on four prestigious ABSC datasets justified the power of PSI that has achieved SOTA performance in most cases. In future work, we will consider integrating some advanced attention mechanisms into this method.

Acknowledgements. This research is supported by Qinghai Provincial Science and Technology Research Program (grant No. 2021-QY-206), Guangdong Provincial Science and Technology Research Program (grant No. 2020A050515014), and National Natural Science Foundation of China (grant No. 62071201).

References

1. A novel lexicalized hmm-based learning framework for web opinion mining. In: ICML. ACM International Conference Proceeding Series, vol. 382, pp. 465–472. ACM (2009, withdrawn)
2. Chen, C., Teng, Z., Zhang, Y.: Inducing target-specific latent structures for aspect sentiment classification. In: EMNLP, pp. 5596–5607. Association for Computational Linguistics (2020)
3. Chen, P., Sun, Z., Bing, L., Yang, W.: Recurrent attention network on memory for aspect sentiment analysis. In: EMNLP, pp. 452–461. Association for Computational Linguistics (2017)
4. Chen, X., et al.: Aspect sentiment classification with document-level sentiment preference modeling. In: ACL, pp. 3667–3677. Association for Computational Linguistics (2020)
5. Devlin, J., Chang, M., Lee, K., Toutanova, K.: BERT: pre-training of deep bidirectional transformers for language understanding. In: NAACL-HLT (1), pp. 4171–4186. Association for Computational Linguistics (2019)
6. Goodfellow, I.J., et al.: Generative adversarial nets. In: NIPS, pp. 2672–2680 (2014)
7. He, R., Lee, W.S., Ng, H.T., Dahlmeier, D.: Effective attention modeling for aspect-level sentiment classification. In: COLING, pp. 1121–1131. Association for Computational Linguistics (2018)
8. He, R., Lee, W.S., Ng, H.T., Dahlmeier, D.: An interactive multi-task learning network for end-to-end aspect-based sentiment analysis. In: ACL, pp. 504–515. Association for Computational Linguistics (2019)
9. Kiritchenko, S., Zhu, X., Cherry, C., Mohammad, S.: NRC-Canada-2014: detecting aspects and sentiment in customer reviews. In: SemEval@COLING, pp. 437–442. The Association for Computer Linguistics (2014)
10. Li, S., Huang, C., Zhou, G., Lee, S.Y.M.: Employing personal/impersonal views in supervised and semi-supervised sentiment classification. In: ACL, pp. 414–423. The Association for Computer Linguistics (2010)

11. Li, Y., Yin, C., Zhong, S., Pan, X.: Multi-instance multi-label learning networks for aspect-category sentiment analysis. In: EMNLP, pp. 3550–3560. Association for Computational Linguistics (2020)

12. Ma, D., Li, S., Zhang, X., Wang, H.: Interactive attention networks for aspect-level sentiment classification. In: IJCAI, pp. 4068–4074. Ijcai.org (2017)

13. McInnes, L., Healy, J.: UMAP: uniform manifold approximation and projection for dimension reduction. CoRR abs/1802.03426 (2018)

14. Pontiki, M., et al.: SemEval-2016 task 5: aspect based sentiment analysis. In: SemEval@NAACL-HLT, pp. 19–30. The Association for Computer Linguistics (2016)

15. Pontiki, M., Galanis, D., Papageorgiou, H., Manandhar, S., Androutsopoulos, I.: SemEval-2015 task 12: aspect based sentiment analysis. In: SemEval@NAACL-HLT, pp. 486–495. The Association for Computer Linguistics (2015)

16. Pontiki, M., Galanis, D., Pavlopoulos, J., Papageorgiou, H., Androutsopoulos, I., Manandhar, S.: SemEval-2014 task 4: aspect based sentiment analysis. In: SemEval@COLING, pp. 27–35. The Association for Computer Linguistics (2014)

17. Sun, C., Huang, L., Qiu, X.: Utilizing BERT for aspect-based sentiment analysis via constructing auxiliary sentence. In: NAACL-HLT (1), pp. 380–385. Association for Computational Linguistics (2019)

18. Tang, D., Qin, B., Feng, X., Liu, T.: Effective LSTMs for target-dependent sentiment classification. In: COLING, pp. 3298–3307. ACL (2016)

19. Wagner, J., et al.: DCU: aspect-based polarity classification for SemEval task 4. In: SemEval@COLING, pp. 223–229. The Association for Computer Linguistics (2014)

20. Wang, J., et al.: Aspect sentiment classification with both word-level and clause-level attention networks. In: IJCAI, pp. 4439–4445. Ijcai.org (2018)

21. Wang, Y., Huang, M., Zhu, X., Zhao, L.: Attention-based LSTM for aspect-level sentiment classification. In: EMNLP, pp. 606–615. The Association for Computational Linguistics (2016)

22. Wu, Z., Ong, D.C.: Context-guided BERT for targeted aspect-based sentiment analysis. In: AAAI, pp. 14094–14102. AAAI Press (2021)

23. Xu, H., Liu, B., Shu, L., Yu, P.S.: BERT post-training for review reading comprehension and aspect-based sentiment analysis. In: NAACL-HLT (1), pp. 2324–2335. Association for Computational Linguistics (2019)

24. Zheng, Y., Zhang, R., Mensah, S., Mao, Y.: Replicate, walk, and stop on syntax: an effective neural network model for aspect-level sentiment classification. In: AAAI, pp. 9685–9692. AAAI Press (2020)

25. Zhuang, P., Wang, Y., Qiao, Y.: Learning attentive pairwise interaction for fine-grained classification. In: AAAI, pp. 13130–13137. AAAI Press (2020)

Real-Time Skill Discovery in Intelligent Virtual Assistants

Preeti Gopal[✉], Sunil Gupta, Santu Rana, Vuong Le, Trong Nguyen, and Svetha Venkatesh

Applied Artificial Intelligence Institute, Deakin University, Geelong, Australia
{p.gopal,sunil.gupta,santu.rana,vuong.le,trong.nguyen,
svetha.venkatesh}@deakin.edu.au, gopal.preeti@gmail.com

Abstract. Solution to many real-world problems often involve the use of expert-level knowledge from various specializations. Such interdisciplinary problems are usually divided into tasks which are then assigned to a set of bots, each specialized in a particular skill. Supervised selection of the right bot each time is cumbersome and not scalable. Hence there is a need for an AI system that identifies the type of task and assigns it to a suitably trained bot. Challenges arise in nonstationary environments when the cost of choosing different bots vary or the bots themselves might evolve in their skills. In this paper, as in Conversational AI, a number of bots are at our disposal, each of which is trained to handle (i.e., answer) a specific type of question in a conversation. We develop a meta-algorithm that learns about capabilities (*Skill Discovery*) of the available bots in real-time and appropriately selects a relevant bot for the question at hand. We present contextual bandits as a solution in this setting and introduce gradual finetuning of query information to improve Skill Discovery. Using two popular datasets from conversational AI: CoQA and SQuAD, we show promising results of our method on non-stationary environments.

Keywords: Bandits · Conversational AI · Nonstationary agents

1 Introduction

Intelligent Virtual Assistants (VA) are being increasingly used to provide relevant information to end-users pertaining to varied everyday activities [6,11]. Whenever a new task is assigned to a VA, it needs to allocate appropriate resources to complete it with the highest possible accuracy and least possible cost. Figure 1 shows how often a single interaction with a VA may invoke different bots, each specialized to perform a specific kind of task. Hence real-time identification of the type of query asked and the selection of an appropriate bot is essential for a seamless conversation.

However, to begin with, any VA must first identify the capabilities or *skills* of each of its available bots. Such *Skill Discovery* involves mapping these bots to

J. Gama et al. (Eds.): PAKDD 2022, LNAI 13280, pp. 315–327, 2022.
https://doi.org/10.1007/978-3-031-05933-9_25

VA: {Bot-1} Hi, How may I help you?
User: Is a root-canal treatment covered in my employer's insurance?
VA: {Bot-1} Yes
User: What's the cap for claims I can make for this treatment?
VA: {Bot-1} 2000 per year
User: Ok. Where's the nearest in-network dental clinic?
VA: {Bot-2} Pulling up the covering app...
VA: {Bot-2} < Shows map with selection tools >
User: Can you connect me to clinic A to book an appointment?
VA: {Bot-3 } Connecting...
VA: {Bot-3 }...

Fig. 1. Example of a conversation with a virtual personal assistant which can involve using multiple specialized sub-bots under the hood. Here, bot-1 is trained to pull-up user specific information from employer's database, bot-2 is skilled in searching and presenting geographical information, and bot-3 is skilled in connecting the user to external agencies, if needed.

their respective skills. Performing this manually will be cumbersome, especially when the number of bots is large. This process is further complicated when skills of the bots evolve (i.e., we have evolving or nonstationary bots). Hence, dynamic Skill Discovery must be automated. In addition to this, the VA needs to identify the type and complexity of each task from a query (*Context*) coming from an end-user and assign it to an appropriate bot. Most existing work on Virtual Assistants focus on algorithms for incorporating various Skills into the AI system and much less has been studied on the *Discovery* of the skills. First, identification of the skill required to answer a query is traditionally done based on the presence of a 'trigger' word or 'skill-invocation-name' within the query. This requires end-users to get trained to know and update themselves with a list of skill-invocation-words. This can be confusing in the long-run when the system's skill-set changes [13]. In order to move towards more natural conversations, some systems use 'intent detection' to detect the skill referred to in the query [2]. In such techniques, there needs to be an internal mapping between commonly used words and 'intent' within the machine. In [8], clustering is performed on all available unlabeled data, and a unique encoding function is learnt for each cluster. Once this is done offline, every new incoming data is mapped to its cluster and encoded using the corresponding encoding function. This new encoded data is used by a contextual bandit to choose relevant action. The received reward from the bandit is then used to update the encoding functions of all clusters. This method assumes availability of some training data to begin with.

In many real-life scenarios, there might not be any training data to perform an offline Skill Discovery for a given application. The VA might receive its first stream of tasks (*query data*) directly *in the field* and hence needs to perform Skill Discovery in real-time. It also needs to keep updating its knowledge of the skills of each of its bots in an online manner. This is analogous to a human manager discovering the strengths of her/his new human employees by assigning them new tasks and observing their performances on the job over time. Thus the problems of dynamically discovering skills in an online fashion and using context of the query to assign the task remain open.

In this work, we present contextual bandits as a solution for dynamic *Skill Discovery* of bots. We also discuss a new method to improve the accuracy of this task by finetuning the representation of Contexts fed to the bandit. We test the efficacy of these methods on two popular datasets in the domain of conversational AI: CoQA and SQuAD. Though the proposed algorithm has been tested on conversational AI datasets, it is inherently domain-agnostic and can be applied to any AI system which needs to discover and update its knowledge of the strengths of its available bots in real-time.

2 Preliminaries

Since our proposed method for Skill Discovery uses contextual bandits, we first describe them briefly here.

Contextual Bandits: A contextual bandit [9] is a sequential decision-making machine. As its input, it receives a sequence of d-dimensional vectors, known as 'contexts' $\{c_j \in \mathbb{R}^d | 1 \leq j \leq T\}$ from outside world (or 'environment') over time T. It has at its disposal a finite number of arms $\{a_i | 1 \leq i \leq n\}$. At a time instant t (round t), the bandit needs to choose (activate) one of the arms $a(t)$ based on the context c_t. Depending on its choice, it receives a 'reward' $r(t)$ from its environment. A high reward implies a good decision, i.e., an appropriate arm being chosen for the given context. Therefore, over time, the bandit selects an arm that has maximum likelihood of giving the best reward for the given context. It does this by learning a reward function for each of its arms. After each round, it trains and updates its reward function of *all* arms based on the history of contexts, arms chosen, and rewards received: $\{(c_1, a(1), r(1)), (c_2, a(2), r(2)), \ldots, (c_t, a(t), r(t))\}$. After sufficient number of rounds, the bandit is trained to take nearly accurate actions for every context it receives. If rewards are binary, a binary classifier such as logistic regression is used as a reward function for each arm. If rewards are real-valued (as will be in our experiments), a regression model is used to learn the reward function, and linear, ridge regressions are popular choices. This process is illustrated in Fig. 2. In addition to choosing the best arm based on past experience ('exploitation'), a bandit also has a mechanism to choose previously unexplored arms ('exploration'). Different exploitation-exploration mechanisms exist in literature, each leading to a unique type of bandit-algorithm. Two of them are Epsilon-Greedy and Upper Confidence Bound (UCB).

- **Epsilon-Greedy**: At round t, an epsilon-greedy bandit chooses a random arm with a probability ϵ and the arm with highest reward-estimate with probability $1 - \epsilon$. Hence, if $s_i(t)$ denotes the reward-estimate of arm i for the current context c_t, and ϵ denotes the probability of exploration, then the selected arm at time t is given by

$$a(t) = \begin{cases} \arg\max_i \{s_i(t)\} & \text{with probability } 1 - \epsilon \\ \text{random arm } i & \text{with probability } \epsilon. \end{cases} \tag{1}$$

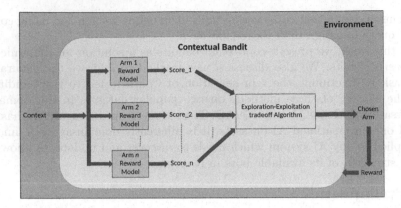

Fig. 2. Framework for selection of an arm using contextual bandits. A 'score' denotes a reward-estimate of an arm for the received context.

After each round, ϵ diminishes by a fraction *decay*. This enables more exploration in the beginning and relies on exploitation later on. ϵ and *decay* are design parameters that decide the exploitation-exploration trade-off. Pseudocode of this technique is illustrated in Algorithm 1.

Algorithm 1: Epsilon Greedy Contextual Bandit

Initialize Ridge regressor for each arm, exploration probability ϵ, *decay* rate
while *Incoming context c_t, and query number t* **do**
    Infer reward-estimates $\{s_1(t), \ldots, s_n(t)\}$ from ridge regressors of each arm.
    Choose a uniform random number r between 0 and 1.
    if $r > \epsilon$ **then**
        | Select arm as $a(t) = \arg\max_i\{s_i(t)\}$.
    else
        | Choose a random arm.
    end
    $\epsilon = \epsilon * decay$
    Get reward $r(t)$ from environment.
    Based on the current and all previous sets of contexts, arms chosen and
        rewards received, $\{c_j, a(j), r(j)|1 \le j \le t\}$, learn a new ridge regression
        model for each arm.
end

- **Linear Upper Confidence Bound**: This is a more principled approach [3,7] than Epsilon-Greedy. Here, selection of an arm $a(t)$ at round t is based on a score that is a sum of two components: expected mean reward of ith arm, $s_i(t)$ for the current context c_t, and variance in expected reward $\sqrt{\frac{\log(t)}{N(i)}}$, where $N(i)$ denotes the number of times the ith arm was chosen in the past. Specifically, $a(t)$ is chosen such that

$$a(t) = \arg\max_i \left\{ s_i(t) + e.\sqrt{\frac{\log(t)}{N(i)}} \right\}. \tag{2}$$

e is a design parameter denoting confidence in exploration. In the beginning, N is small for any arm, and hence the uncertainty term has more weight. After an arm is chosen many times, its selection is done primarily based on its expected mean reward, as illustrated in Algorithm 2.

Algorithm 2: LinUCB Contextual Bandit

Initialize: Ridge regressor for each arm, regularization strength α, matrix
  $\mathbf{A} \longleftarrow \mathbf{I}_d$ (d-dimensional identity matrix).
while *Incoming context* $\mathbf{c}_t$, *and query number* t **do**
  Infer reward-estimates $\{s_1(t), \ldots, s_n(t)\}$ from ridge regressors of each arm.
  Select arm $a(t)$, where $a(t) = \arg\max_i\{s_i(t) + \alpha.\sqrt{\mathbf{c}_t^T \mathbf{A}_i \mathbf{c}_t}\}$.
  Get reward $r(t)$ from environment.
  Update $\mathbf{A}_a \longleftarrow \mathbf{A}_a + \mathbf{c}_t \mathbf{c}_t^T$
  Update reward model: Based on the current and all previous sets of
    contexts, arms chosen and rewards received, $\{\mathbf{c}_j, a(j), r(j) | 1 \leq j \leq t\}$,
    learn a new ridge regression model for each arm.
end

In this work, 'arms' of a bandit will be referred to as 'agents' to appropriately refer to skilled bots in an application.

3 The Proposed Skill Discovery Framework

In conversational AI, a Context is a vector representation of the question together with a reference paragraph and some history of the conversation. This is traditionally embedded using one of the many available text transformations, such as BERT (Bidirectional Encoder Representations from Transformers) [5]. Given this Context, we perform Skill-Discovery in two independent ways:

- Feeding an off the shelf pre-trained BERT-embedded Context directly to a contextual bandit.
- Feeding finetuned BERT+(Conversational-AI network)-embedded Context to a contextual bandit. The network for finetuning the Context (see Fig. 3) consists of a cascade of two networks: a BERT-base network and a task-specific conversational-AI network. The first one outputs a generic BERT-embedded vector[1] and the latter further tunes the BERT-embedded vector to suit to the specific conversational-AI application at hand. At the end of each round (query received and answered), this network is incrementally trained based on

---

[1] We used 'BERT-base-uncased' model from Hugging Face.

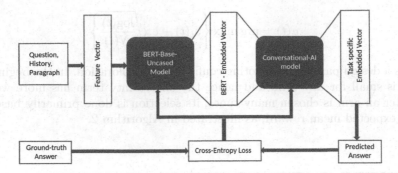

Fig. 3. Pipeline for finetuning the Context. The BERT-base network provides a generic vector representation of the query. This is fed to a task specific Conversational-AI network to generate a task specific vector (Context). At the end of each round (query received and answered), the weights of conversational AI network are incrementally trained or finetuned. This cascade of BERT and conversational AI network is referred to as 'finetuned model' in Fig. 4.

the current query and its true answer. We call this process as 'finetuning the Context'. After sufficient number of rounds, the finetuned Context is better suited for predicting a conversational answer.

Figure 4 shows the overall framework for Skill Discovery using the second method. Once a new query is received, its finetuned Context is computed and fed to a bandit (as shown earlier in Fig. 2) in order to select the right agent which is skilled enough to answer the current query. Pseudocode of this technique is presented in Algorithm 3. Note that in the presence of a cost-factor, we aim to maximize cost-normalized reward-estimates, i.e., if p_i is the cost of accessing agent a_i and $s_i(t)$ is its reward-estimate for query number t, then we aim to maximize $s_i(t)/p_i$ over all $1 \le i \le n$ agents. As will be seen later in the experimental results, finetuning significantly assists the bandit to select the right agent over time. The accuracy of the predicted answers is measured as F1 score of word overlap, where F1 score is the harmonic mean of precision and recall.

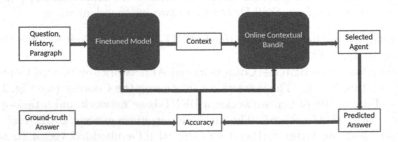

Fig. 4. The Overall framework for Skill Discovery. Each query is passed through the finetuned model (shown in Fig. 3) and the resulting Context is fed to a contextual bandit which selects the appropriate agent to answer the query.

Algorithm 3: Proposed method for Online Skill Discovery in Conversational AI.

Initialize Ridge regressor for each agent, cost p for each agent , bandit algorithm B_A and its hyper-parameters.

while *Incoming Query Q, and query number t* **do**

    if *First batch* **then**

        | Context c_t = BERT embedding of Q.

    else

        | Context c_t = Finetuned BERT+(Conversational-AI) embedding of Q.

    end

    For c_t, compute score (reward-estimate) for each of n agents: $\{s_1(t), s_2(t), \ldots, s_n(t)\}$ from ridge regressors fit on each agent.

    Let cost of each agent a_i be p_i.

    Compute cost-normalized scores for n agents: $\left\{ s_1(t)/p_1, \ldots, s_n(t)/p_n \right\}$.

    Select the best agent based on the exploitation-exploration strategy of B_A.

    Next, the selected agent $a(t)$ predicts an answer A for the query Q.

    Compute reward $r(t)$ = F1-score of the answer A.

    Based on the current and all previous sets of contexts, agents chosen and rewards received, $\{c_j, a(j), r(j) | 1 \le j \le t\}$, learn a new ridge regression model for each agent.

    Based on all previous ground-truth answers and queries, incrementally train the (BERT+Conversational-AI) model.

end

This is the recommended evaluation metric in the popular CoQA [12] dataset. However, one may also choose to use other metrics such as Bilingual Evaluation Understudy Score (BLEU) and exact-match metric.

4 Experimental Design and Datasets

We study Skill Discovery under the presence of agents which have evolving skills and unequal costs. Table 1 shows our design of scores to emulate gradually increasing skill-levels of such agents. An additional agent, called the 'Oracle' is included. 'Oracle' is assumed to be fully skilled for all query-types, but is twice as expensive when compared to other agents. The other agents have mutually-exclusive skills and equal costs. Their skill-level (as indicated by their scores in Table 1) denotes the average accuracy of conversational answers predicted by them. We assume that skill-level of agents improve with time and reaches maximum accuracy after a sufficient number of rounds. This is indicated by their scores slowly increasing from 0 (min. accuracy) to 1 (max. accuracy) in steps of 0.1 after every 1000 queries. This scheme emulates the scenario of increasing the skill-level of an agent by externally training it with more data when it becomes available. Table 1 also shows the right agent ('optimal choice') that the bandit is expected to choose for a particular query. Until the low-cost agents become sufficiently skilled (with score $>= 0.5$), the 'Oracle' is the optimal choice because its normalized-score-to-cost ratio is the maximum. For scores > 0.5, other individual agents are best suited to answer the queries. Table 2 shows examples of scores of low-cost agents once their skills are fully evolved.

Table 1. Design of scores for agents and Oracle in order to simulate cost-sensitivity and evolution of skills over time.

Query number n	Score of agent (Cost = 1)	Score of oracle (Cost = 2)	Maximum (Score/Cost)	Optimal choice
$n < 1000$	0	1	0.5	Oracle
$1000 \leq n < 2000$	0.1	1	0.5	Oracle
$2000 \leq n < 3000$	0.2	1	0.5	Oracle
$3000 \leq n < 4000$	0.3	1	0.5	Oracle
$4000 \leq n < 5000$	0.4	1	0.5	Oracle
$5000 \leq n < 6000$	0.5	1	0.5	Oracle or agent
$6000 \leq n < 7000$	0.6	1	0.6	Agent
$7000 \leq n < 8000$	0.7	1	0.7	Agent
$8000 \leq n < 9000$	0.8	1	0.8	Agent
$9000 \leq n < 10000$	0.9	1	0.9	Agent
$10000 \leq n$	1	1	1	Agent

Note that in a real application, the evolution of skills and costs of different agents will vary from our design. For example, a few agents might have overlapping skills or different agents might evolve at different rates. However, in order to have a predictable trajectory of skill-evolution to enable our study of effects of finetuning Contexts, we choose the design discussed above.

Table 2. Example scores for agents after their skills are fully evolved.

Query	Score of agent 1 (Cost = 1)	Score of agent 2 (Cost = 1)	Score of agent 3 (Cost = 1)	Optimal choice
Type 1	0	0	1	Agent 3
Type 2	1	0	0	Agent 1
Type 3	0	1	0	Agent 2

Datasets: Our dataset consists of 20000 queries from Conversational Question Answering dataset (CoQA) [12] and Stanford Question-Answering Dataset (SQuAD) [10]. For this work, each of these datasets was further used to create sub-datasets as shown in Tables 3 and 4. From the CoQA dataset, we created four sub-datasets: Question Category, Question Domain, Question Difficulty and Question Length. The Question-Category sub-dataset consists of 3 types of queries: those which begin with 'What', those which begin with 'Who','Where' or 'When', and those that begin with other words such as 'How','Why', etc. Here, Skill Discovery denotes mapping of agents to the Question-Category which they are skilled at, and selection of the right agent for answering a given query. In the Question-Domain sub-dataset, each query is labelled with its domain (Wikipedia or Literature or Others) based on its content. This domain name was available to us from the CoQA dataset. The third sub-dataset, Question-Difficulty, is based on the difficulty-level of a query. Every query came along with its reference paragraph and a region-of-interest within the paragraph which potentially has clues for answer to the query. Hence, we created this sub-dataset based on the presence or absence of a lexical-match between words in the query and those in the region of interest within its reference paragraph. We assumed that queries with no lexical match are more difficult to answer than queries with some lexical match. In this case, one of our agents is assumed to be skilled in answering easy questions and the other in answering difficult questions. In the fourth sub-dataset, Question-Length, each query was divided into two types based on the number of words in the question: small-length questions ($<=4$ words) and big-length questions (>4 words). In any conversation, smaller length questions generally refer to some information from the previous query or answer. Hence, any agent that specializes in answering small questions must have a good understanding of the history of the query and this is a specialized skill. The choice of the number 4 was made because average length of queries in CoQA is 5. From the SQuAD dataset, we created two sub-datasets: Question Category and Question Length. The rationale in creating these is the same as the one used in CoQA, the only difference being–SQuAD questions were split into small and big length questions based on question length being $<=9$ or >9. This follows due to the fact that average length of a query in SQuAD is 10 words. Each of the above sub-datasets: four from CoQA and two from SQuAD consist of 20000 queries and form six independent datasets for our experiments. Tables 3 and 4 also show the fraction of questions in each type of query.

Table 3. Sub-datasets created from CoQA

Question Category	% Composition	Question Domain	% Composition
What	29	Wikipedia	24
When, Where, Who	22	Literature	23
Others	49	Others	53

Question Difficult y	% Composition	Question Length	% Composition
Lexical Match	41	$<=$ 4 words	38
No Lexical Match	59	$>$4 words	62

Table 4. Sub-datasets created from SQuAD

Question Category	% Composition	Question Length	% Composition
What	42	<=9 words	49
When, Where, Who	20	>9 words	51
Others	38		

5   Results and Discussion

We use 'cumulative number of correct agents chosen' as our evaluation metric to measure efficacy of Skill Discovery. If an algorithm has better ability to discover skills of its agents, it will eventually choose the right ones over time. Additionally, we assume the right agent will always answer the given query with a mean accuracy indicated as 'scores' in Tables 1 and 2. We use the following three types of algorithms and compare their performances for the task of Skill Discovery: Random-Selection, Epsilon-Greedy and Sliding Window-UCB (SW-UCB). Random-Selection chooses an agent randomly irrespective of Context and skills of the agents. For Epsilon-Greedy, the exploration probability ϵ and decay-rate *decay* were chosen to be 0.2 and 0.99 respectively. In SW-UCB, the agents are updated based on the statistics computed from the query-rounds within the sliding window alone. The window width was set to 500, since the agents evolve after every 1000 rounds in our design (Table 1). Compute infrastructure consisted of Intel(R) Xeon(R) 80-core CPU (E5-2698 v4) running at 2.20 GHz with 500 GB RAM together with NVIDIA's Tesla-V100 series GPU with 32 GB memory. In this machine, completing 20000 turns of queries took 6.5 h when gradual finetuning was performed and 2.5 h when finetuning was not done. All experiments were run 10 times, each with a different shuffle pattern of the query-set i.e., the sequence of questions differed. The code repository is in [1]. Figures 5 and 6 show results for all datasets. The solid lines denote mean number of cumulative right agents chosen and the shaded band represents ±1 standard deviation. In all cases, the deviation from mean was negligible. To interpret these plots, it is helpful to again refer to query-numbers in Table 1. All plots plateau around query 5000, since it is at this stage when the optimal agent shifts from being the Oracle to any of the other low-cost agents. The algorithm takes some time to adapt to this change. Finetuning the Context helps only after the skill-levels have stabilized. Until then, performance using the generic BERT Context is as good as the finetuned Context. This indicates that Bert Context alone may be preferred until skills stabilize and this will offer an added benefit of saving time that is otherwise required for finetuning. However, once all agents are fully evolved, finetuning the Context offers a superior representation enabling better decisions. These experiments also indicate superior performance of SW-UCB over Epsilon-Greedy and Random-Selection. As expected, random choice of agents is the least accurate method.

One of the future directions of work is to study cases when skills of all agents are not mutually exclusive or are evolving at different rates. The possibility of using restricted contexts [4] and neural bandits [14] may also be explored.

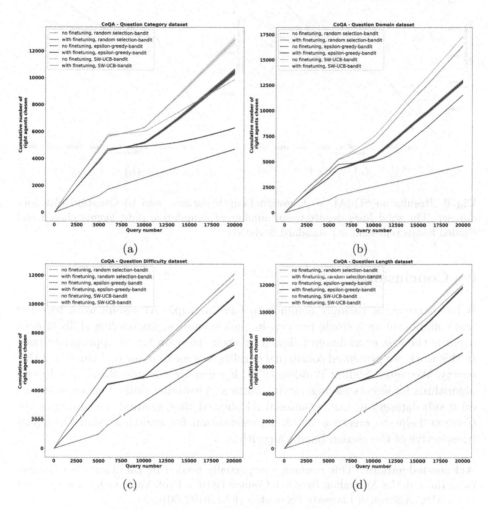

Fig. 5. Results on CoQA a) Question Category dataset, b) Question Domain dataset, c) Question Difficulty dataset, and d) Question Length dataset. The solid lines denote mean number of cumulative right agents chosen and shaded bands represent ±1 standard deviation.

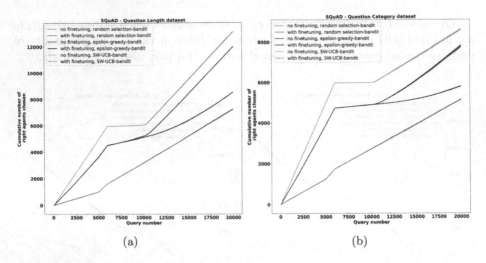

Fig. 6. Results on SQuAD a) Question Length dataset, and b) Question Category dataset. The solid lines denote mean number of cumulative right agents chosen and shaded bands represent ±1 standard deviation.

6   Conclusions

It has become increasingly common to have multiple AI agents work together and collaborate on a single project. In such scenarios, discovering skills of each agent is the first crucial step followed by their selection for an appropriate task. In this work, we presented contextual bandits as a solution for real-time Skill Discovery. We tested Sliding Window UCB, Epsilon-Greedy and Random selection algorithms for agents having varying costs and evolving skills. Our experiments on 6 sub-datasets in Conversational AI showed that gradual finetuning of the Context helps to create a better representation for real-time Skill Discovery, irrespective of the chosen bandit algorithm.

Acknowledgement. This research was partially funded by the Australian Government through the Australian Research Council (ARC). Prof. Venkatesh is the recipient of an ARC Australian Laureate Fellowship (FL170100006).

References

1. Code repository for the work presented in this paper. https://github.com/preetigopal/Skill-Discovery
2. Amazon: How users invoke custom-skills in Alexa (2021). https://developer.amazon.com/en-US/docs/alexa/custom-skills/understanding-how-users-invoke-custom-skills.html, Accessed 07 June 2021
3. Auer, P., Cesa-Bianchi, N., Fischer, P.: Finite-time analysis of the multiarmed bandit problem. Mach. Learn. **47**(2), 235–256 (2002)

4. Bouneffouf, D., Rish, I., Cecchi, G., Féraud, R.: Context attentive bandits: contextual bandit with restricted context. In: Proceedings of International Joint Conference on Artificial Intelligence, IJCAI 2017, pp. 1468–1475 (2017)
5. Devlin, J., Chang, M., Lee, K., Toutanova, K.: BERT: pre-training of deep bidirectional transformers for language understanding (2018). CoRR abs/1810.04805 http://arxiv.org/abs/1810.04805
6. Lau, T.T.: When AI becomes a part of our daily lives (2019). https://hbr.org/2019/05/when-ai-becomes-a-part-of-our-daily-lives, Accessed 30 Oct 2021
7. Li, L., Chu, W., Langford, J., Schapire, R.E.: A contextual-bandit approach to personalized news article recommendation. In: Proceedings of the 19th International Conference on World Wide Web - WWW 2010 (2010)
8. Lin, B., Bouneffouf, D., Cecchi, G., Rish, I.: Contextual bandit with adaptive feature extraction. arXiv (2020)
9. Lu, T., Pal, D., Pal, M.: Contextual multi-armed bandits. In: Proceedings of the Thirteenth International Conference on Artificial Intelligence and Statistics. Proceedings of Machine Learning Research, vol. 9, pp. 485–492. JMLR Workshop and Conference Proceedings (May 2010)
10. Rajpurkar, P., Zhang, J., Lopyrev, K., Liang, P.: SQuAD: 100,000+ questions for machine comprehension of text. In: Proceedings of the 2016 Conference on Empirical Methods in NLP, pp. 2383–2392. ACL, Austin (2016)
11. Rawassizadeh, R., et al.: Manifestation of virtual assistants and robots into daily life: vision and challenges. CCF Trans. Perv. Comput. Interact. 1(3), 163–174 (2019). https://doi.org/10.1007/s42486-019-00014-1
12. Reddy, S., Chen, D., Manning, C.: CoQA: a conversational question answering challenge. TACL 7, 249–266 (2019)
13. White, R.W.: Skill discovery in virtual assistants. Commun. ACM 61(11), 106–113 (2018)
14. Xu, P., Wen, Z., Zhao, H., Gu, Q.: Neural contextual bandits with deep representation and shallow exploration (2020). ArXiv abs/2012.01780 http://arxiv.org/abs/2012.01780

Few-Shot Knowledge Graph Entity Typing

Guozhen Zhu, Zhongbao Zhang[✉], and Sen Su

State Key Laboratory of Networking and Switching Technology, Beijing University
of Posts and Telecommunications, Beijing 100876, People's Republic of China
{gzzhu,zhongbaozb,susen}@bupt.edu.cn

Abstract. Knowledge graph entity typing, which is an important way
to complete knowledge graphs (KGs), aims at predicting the associating
type of certain given entities without any external knowledge. However,
previous methods suppose that many (entity, entity type) pairs (ETPs)
can be obtained for each entity type, performing poorly on entity types
that only have a few associative entities and do not fully utilize the
internal information in KGs. In this work, we propose a novel model
named Meta Entity Typing (MET) for few-shot knowledge graph entity
typing. In MET, we achieve knowledge graph entity typing by meta-
learning with three sub-tasks formed by the hierarchical entity type tree
in its meta-training stage. In this way, MET can focus on transferring
type-specific meta information to learn the most important knowledge for
entity typing. Besides, to fully employ the internal information in KGs
given limited ETPs, inspired by Factorization Machines, we design a
novel Relation To Relation Graph Convolutional Networks (R2R-GCN),
in which we consider different relation combinations could have distinct
influence on its corresponding entity, R2R-GCN can explicitly model the
interactions between different relations. Empirically, our model achieves
state-of-the-art results on few-shot entity typing KG benchmarks.

Keywords: Knowledge graph entity typing · Few-shot learning ·
Graph convolutional networks

1 Introduction

Knowledge bases (KBs), such as DBpedia [10] and YAGO [11], have incorporated
large-scale multi-relational data. These KBs store Knowledge Graphs (KGs) that
can be categorized as two views: (i) the instance-view knowledge graphs that
contain huge amount of triples in the form of (head entity, relation, tail entity)
and (ii) the ontology-view knowledge graphs that constitute semantic meta-
relations of abstract concepts (types).

Z. Zhang—This work was supported in part by: National Natural Science Foundation
under Grant U1936103 and 61921003.

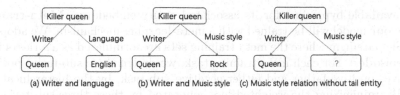

Fig. 1. Different relation combinations for the same entity.

The entity in instance-view KG and its type in ontology-view KG is linked by cross-view link. The cross-view links are incomplete and hard to complete manually, jeopardizing their usefulness in tasks such as relation extraction [8], coreference resolution [6], entity linking [5]. In this work, we focus on knowledge graph entity typing, which aims to infer missing entity types (cross-view links) formed as (entity, entity type) in KGs. It is also an important sub-problem for knowledge graph completion (KGC).

To automatically mine entity types, recent researches mainly focus on inferring missing entity types based on the internal information of KGs, their methods can be divided into embedding and graph convolutional networks (GCNs) methods. Embedding-based methods [7,12,22], obtaining entity embeddings, then use them to infer missing type. GCN methods [16,17,19], infer the type of entity by aggregating the information from its neighbors for multi-relational graph.

Previous knowledge graph entity typing methods usually suppose that many ETPs can be obtained for each entity type. However, the frequency distributions of entity types in real data sets often have long tails, which means a large portion of entity types have only few associated entities in KGs. The long tail scenario incurs the infeasibility of previous models which assume available and sufficient training instances for all entity types. Therefore, it is crucial for models to be able to complete entity types with limited ETPs.

Besides the above-mentioned shortcoming, recent GCN-based methods ignore the interactions between different relations, i.e., they believe that one relation would not be influenced by another relation when aggregating the information of entity neighbors, which is not correct. For instance, in Fig. 1(a) there are two triples, we can infer from (killer queen, writer, queen) that killer queen is a song or a book, thus the writer relation provides more information than the language relation. However, when the writer relation combines with the music style relation, we should pay more attention to the later because it is almost certain that killer queen is a song by music style relation. In short, previous methods have not considered the distinct influence by different relation combinations.

To overcome these shortcomings, we propose a novel method for knowledge graph entity typing in few-shot scenarios, named MET (Meta Entity Typing). MET can be divided into three stages: **Embedding stage:** we design a Relation to Relation Graph Convolutional Networks (R2R-GCN) to obtain the vector representations of entities and entity types. In R2R-GCN, we model the different relation combinations explicitly, and the model could perceive the distinct influence of all kinds of these combinations when aggregating the information of central entity. In this way, we can get the embedding of each kind of relation, which is an available byproduct for its associated entity embedding. **Meta-training stage:** our model will be trained with a meta-learning mechanism. We adopt the episodic paradigm where the meta-training sets are formulated as a series of tasks (i.e., episodes). For each meta-training task, we design three sub-tasks, positive-negative, positive-father and father-negative sub-task, by the hierarchical type tree. By minimizing the weighted loss generated by these three-sub tasks, our model can fully utilize the structure information in the ontology-view KGs. In other words, our model can transfer the most important information for entity typing from types with sufficient ETPs to types with few-shot ETPs. **Meta-testing stage:** when facing a new type which is never seen in meta-training sets, our model can fast adapt to current task and make accurate prediction with limited ETPs. And the model will be tested at this stage.

We summarize our contributions as follows:

- To the best of our knowledge, this is the first attempt to use the meta-learning based methods to alleviate the few-shot problem in knowledge graph entity typing task, and we propose three sub-tasks in meta-training stage to utilize the hierarchical information in type tree.
- We propose R2R-GCN, which can explicitly model the interactions between different relations, thus the central entity can aggregate the information of its connected relations and neighbor entities in a more precise way.
- We evaluate our framework's performance against different baselines in few-shot knowledge graph entity typing task and get state-of-the-art results.

2   Related Work

Knowledge graph entity typing is an important sub-problem for KGC research filed. Recent researches mainly focus on inferring missing entity types based on the information of KGs. In this section, we first introduce the embedding-based methods, and then cover the mainstream methods based on GCNs. At last, we will introduce the meta-learning.

2.1   Embedding-Based Methods

Embedding-based methods aim to obtain the low-dimensional vector representations of entities semantics and then use the vectors to infer the missing types of entities. Neelakantan [13] proposed projection embedding model (PEM), Moon [12] proposed to learn type embedding by combining triple knowledge and entity type instances. Zhao [22] claimed that previous methods lack expressive ability due to its simplicity and proposed the more advanced model ConnectE which includes two mechanisms. One is E2T, which uses a mapping matrix to project the entity from entity space to entity type space. The other is TRT, which uses the types of neighbors to infer the types of central entities. However, these above-mentioned methods assume that there are enough instances for every entity type, ignoring the long tail scenario in knowledge graph entity typing task.

2.2   GCN-Based Methods

GCNs operate on local graph neighborhoods to large-scale relational data. To fully utilize the information in multi-relational graphs like knowledge graph, Schlichtkrull [16] proposed R-GCN, which is an extension of GCNs for multi-relational graphs. R-GCN aggregate the information with relation-specific aggregator. Shang [17] proposed Weighted Graph Convolutional Network, which utilizes learnable relational-specific weights to aggregate the information of neighbors. Vashishth [19] considered that the information of neighbor entity will be influenced by its relation. However, these GCN-based methods all ignore the different relation combinations will have distinct influence when aggregating the information for the central entity.

2.3   Meta-learning

Meta-learning has been widely used in various domains to address few-shot problems. Recent meta learning models have two categories: (1) metric based approaches [18, 20]; (2) meta-optimizer based approaches [4, 9]. The former one learns an effective metric and corresponding matching function among a set of training tasks. The later one aims to learn the global optimal initialization parameters and the model can lead to fast learning on a new task.

3   Problem Formulation

In this section, we introduce the problem definition and episodic meta-learning.

3.1   Few-Shot Entity Typing

We use G_I and G_O to denote the instance-view KG and ontology-view KG respectively. G_I is formed with ε, the set of entities, and R_I, the set of relations. G_O is formed with τ, the set of types, and R_O, the set of meta-relations. S is used to denote the set of known cross-view links (ETPs) in KG.

In few-shot entity typing, we have $\tau = t_b \cup t_n$, where t_b and t_n are the set of base and novel types respectively, and they are disjoint. Every t_b have amount of ETPs. However, for t_n, we only have a few ETPs. Given any subset of m novel types, we aim to train a model with only k ETPs for each novel type, in order to predict the types of the remaining unlabeled entities among these m types. This is called m-way k-shot entity typing.

3.2   Episodic Meta-learning

We adopt the episodic paradigm, which has shown great effectiveness in few-shot learning problem. For episodic paradigm, the problem will be divided into the meta-training stage and the meta-testing stage. We employ t_b and t_n as the meta-training and meta-testing sets respectively. Both of these two stages are formulated as many independent tasks (i.e., episodes).

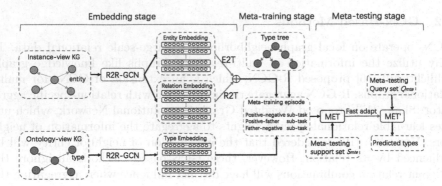

Fig. 2. The architecture of our approach. The blue dotted boxes belong to embedding stage, which will produce entity, relation and type embedding. These embeddings will be input to meta-training stage, which is denoted as yellow dotted boxes. Our model will be fine-tuned at meta-testing stage with red dotted boxes. E2T and R2T mean transferring the entity and relation from the entity space and the relation space to the type space respectively. (Color figure online)

4   Our Approach

The architecture of our approach is given in Fig. 2. Our model mainly consists of three stages: embedding stage, meta-training stage and meta-testing stage. We obtain representations of entity and type at the embedding stage. At the meta-training stage, we use ETPs of each entity type to train our model based on the embedding adopting the episodic paradigm. At the meta-testing stage, we will test our model on novel types with limited ETPs.

4.1   Embedding Stage

R2R-GCN: There are two main inputs to the embedding stage. One is the triples in G_I, which are formed as $(\varepsilon, r, \varepsilon)$, and the other is the triples in G_O, which are formed as (τ, r, τ). We propose Relation to Relation Graph Conventional Networks (R2R-GCN), an extension of R-GCN, to obtain the embeddings of entities and entity types, which are inputs for meta-training stage.

R-GCN operates on local graph neighbors to large-scale multi-relational data and aggregates the information with relation-specific aggregator. However, R-GCN ignores the different combinations of relations could have distinct and non-negligible influence when aggregating the information for the central entity. In other words, R-GCN can not perceive the mutual information of neighbor entities under their particular relation combinations.

Inspired by factorization machines [15], our R2R-GCN uses factored parameters to model all interactions between neighboring entities under different relations as follows:

$$h_i^{(l+1)} = \sigma \left(\sum_{r \in \mathcal{R}} \sum_{j \in \mathcal{N}_i^r} \frac{1}{c_{i,r}} W_r^{(l)} h_j^{(l)} + W_0^{(l)} h_i^{(l)} + \sum_{i=1}^{n} \sum_{j=i+1}^{n} \langle \mathbf{r}_i, \mathbf{r}_j \rangle h_i^{(l)} h_j^{(l)} \right) \quad (1)$$

where N_i^r denotes the set of neighbor nodes (entities) i under relation r. $W_r^{(l)}$ is the relation weight matrix of relation r. $h_i^{(l+1)} \in R^{d(l)}$ is the hidden state of node (entity) in the l-th layer, and $d(l)$ is the dimensionality of representations of this layer. The symbol $\sigma(\cdot)$ denotes a nonlinear activation function (such as ReLU$(\cdot)$). $c_{i,r}$ is a problem normalization constant which is always set as $|N_i^r|$. $\langle \mathbf{r}_i, \mathbf{r}_j \rangle := \sum_{f=1}^{k} r_{i,f} \cdot r_{j,f}$ is the dot product of the relation of the i-th and the j-th entity with k factors. $k \in \mathbf{N}_0^+$ is a hyper-parameter that defines the dimensionality of the factorization.

Generally, $\langle \mathbf{r}_i, \mathbf{r}_j \rangle$ models the interaction between the relations of the i-th and the j-th neighbor entity. It can also be regarded as relation combination specific parameters. Instead of using independent parameter $w_{i,j} \in \mathbf{R}$ for each interaction between relation i and relation j, the R2R-GCN models the interaction by factorizing it. In this way, the parameters can be estimated in sparse settings.

Besides, to build the the embeddings of entities and entity types into two different spaces, we use two R2R-GCN models without sharing their parameters.

Aggregating the Information of Relation: In instance-view KGs, the relations can be considered as predicates in triples, which contain huge amount of information about the type of its head entity. For example, in Fig. 1(c), we remove the tail entity of (killer queen, Music style, rock) and this triple degenerates to (killer queen, Music style, ?). Although we do not know its tail entity, we can still figure out that the head entity is a song based on the Music style relation.

In R2R-GCN, the embeddings of relations can be considered as its byproducts, which are very useful to model the embedding of each head entity for entity typing task. Firstly, the different embeddings of relations can be added to generate the aggregated relation embedding (ARE) as follows:

$$\overrightarrow{\mathrm{ARE}} = \frac{\sum_{i=0}^{|R|} \overrightarrow{r_i}}{|R|} \quad (2)$$

where R means the set of relations connected with a specific entity and $r_i \in R$. For aggregating the relation information when classifying the head entity, we project the relation embedding from relation space to type space and get the relation embedding in type space (R2T). Specifically, we achieve it by a nonlinear affine transformation:

$$\overrightarrow{\mathrm{R2T}} = \sigma \left(\mathbf{W}_{\mathrm{ct}} \cdot \overrightarrow{\mathrm{ARE}} + \mathbf{b}_{\mathrm{ct}} \right) \quad (3)$$

where $\mathbf{W}_{\mathrm{ct}} \in \mathbf{R}^{d(r) \times d(t)}$, $d(r)$ and $d(t)$ is the dimensionality of the relation and type embedding respectively. $\mathbf{b}_{\mathrm{ct}}$ is a bias vector. $\sigma(\cdot)$ is a non-linear activation function, for which we adopt tanh$(\cdot)$. The $\overrightarrow{\mathrm{R2T}}$ will be added as a part of entity embedding when comparing the similarity between entities and types.

Entity Embedding: We can get the embedding of each entity $\overrightarrow{E} \in R^{d(e)}$ and entity type $\overrightarrow{ET} \in R^{d(t)}$. The entity embedding and entity type embedding exist in two completely different embedding spaces. To compare the similarity between entities and entity types, we project the entities and entity types into the same vector space. Specifically, the entity will be mapped to an embedding in the entity type space in the same way as Eq. (3), which is close to the embedding of its corresponding type. We denote the outcome as $\overrightarrow{E2T}$.

To take relation embeddings into consideration, we add $\overrightarrow{E2T}$ and $\overrightarrow{R2T}$ with a hyperparameter α:

$$\overrightarrow{ER} = \alpha\overrightarrow{E2T} + (1 - \alpha)\overrightarrow{R2T} \tag{4}$$

$\alpha \in [0, 1]$ can be used to control the importance of relation and neighbor entity: the smaller the value of α, the lower the importance of neighbor entity when classifying the central entity, while increasing α decreases the importance of relation.

4.2 Meta-training Stage

To overcome the few-shot problem in knowledge graph entity typing, we adopt the episodic paradigm. At meta-training stage, our model will be trained at t_b. Under m-way k-shot setting, we aim to classify the positive type among the m types with k ETPs in each meta-training task.

Training Objectives: Given a distance function $d : \mathbf{R}^M \times \mathbf{R}^M \to [0, +\infty)$, our model produces a distribution for a query entity $\overrightarrow{ER}$ based on a softmax over distances to the types in the embedding space:

$$p(y = k \mid \overrightarrow{ER}) = \frac{\exp\left(-d\left(\overrightarrow{ER}, \overrightarrow{ET}_k\right)\right)}{\sum_{k'} \exp\left(-d\left(\overrightarrow{ER}, \overrightarrow{ET}_{k'}\right)\right)} \tag{5}$$

Learning proceeds by minimizing the negative log-probability $J = -\log p(y = k \mid \overrightarrow{ER})$ of the positive type k via SGD. The training tasks are formed by selecting a subset of types from the training set with some special rules.

Rules of Sub-tasks Construction: The types in the ontology-view KG will form some hierarchical trees rooted by a top-level super type (e.g., /person, /place, etc.). And these (sub-type, father-type) pairs in the forest are constructed by triples in ontology-view KG with *is-a* relation.

We suppose y denotes the positive type. $y \in y_f$ denotes type y is a sub-type of type y_f. The sibling types sharing the same parent type with y are denoted as $Sb(y)$. To fully utilize the hierarchical information, inspired by Chen [3], we propose three sub-tasks for every episode in meta-training stage:

i) **positive-negative sub-task**: The similarity between entity and its corresponding positive type should be higher than other sibling negative types:

$$y > y', \quad \forall y' \in \text{Sb}(y) \tag{6}$$

We define this subset of types as st_1, the log-probability of this sub-task as $J_1 = -\log p(y = k \mid st_1, \overrightarrow{\text{ER}})$.

ii) **positive-father sub-task**: The entity should be more similar with its positive type compared with its father type:

$$y > y_f \tag{7}$$

We define this subset of types as st_2, and $J_2 = -\log p(y = k \mid st_2, \overrightarrow{\text{ER}})$.

iii) **father-negative sub-task**: The father type should rank higher than sibling negative types of y:

$$y_f > y', \quad \forall y' \in \text{Sb}(y) \tag{8}$$

We define this subset of types as st_3, and $J_3 = -\log p(y = k \mid st_3, \overrightarrow{\text{ER}})$.

Besides, different type levels will lead to different penalties. Intuitively, the types in lower level should be harder to classify (e.g., /pop singer vs /classical singer should be more difficult than /singer vs /painter). Therefore, we introduce a loss coefficient to give greater penalties for higher levels. Therefore, the overall learning objective function is given as:

$$J = \frac{1}{l}(\lambda_1 J_1 + \lambda_2 J_2 + (1 - \lambda_1 - \lambda_2)J_3) \tag{9}$$

where l denotes the level of the positive type. λ_1 and λ_2 is set to 0.7 and 0.15 respectively.

4.3   Meta-testing Stage

We adopt the episodic meta-learning framework of MAML [4] for MET. MAML is flexible to work with any model with gradient-based optimization, and learns a set of initial parameters to enable rapid adaptation to new tasks with only a few gradient updates. Since the hierarchical information has been fully utilized at meta-learning stage, for convenience of comparison with other methods, we do not follow the sub-tasks settings and select negative types from t_n randomly. To verify the effectiveness of our model, we perform our model on meta-testing task t_{mte}. The model will fast adapt to each task given meta-testing support set S_{mte}, which consists of a few labeled data. Then we use the meta-testing query set Q_{mte} to calculate the classification accuracy.

5   Experiments and Analysis

In this section, we first introduce our datasets and experimental setup in Sect. 5.1, and then present the experimental results and analysis in Sect. 5.2. Finally, we present the ablation study in Sect. 5.3.

Table 1. Statistics of datasets.

Datasets	Instance-view KG G_I			Ontology-view KG G_O			Type links S
	Entities	Relations	Triples	Types	Meta-relations	Triples	
YAGO26K-906	26,078	34	390,738	906	30	8,962	9,962
DB111K-174	111,762	305	863,643	174	20	763	99,748

5.1   Datasets and Experimental Setup

Datasets: We adopt YAGO26K-906 and DB111K-174 proposed by Hao [7] as our datasets. They are extracted from YAGO and DBpedia respectively. Table 1 provides the statistics of both datasets.

Evaluation Protocols: We separate the types of each dataset into meta-training and meta-testing sets with the ratio of 70% and 30% flowing the method described in Sect. 3. Each type in the support set of meta-training and meta-testing tasks has only three or five ETPs (i.e., $K = 3$ or $K = 5$) for all datasets.

The task of entity typing aims to predict the most probable missing type in ontology-view KG for every entity in instance-view KG. We set sibling and father types of the positive type as negative types for each task. As for evaluation metrics, we use *MRR* (that is the mean reciprocal rank over all the positive types), *Hit@1* (i.e., accuracy) and *Hit@3* (the proportion of the positive types ranked in top 3) on the Q_{mte} for every meta-testing task.

Parameter Settings: In our experiments, we use three-layer R2R-GCN for G_I but one-layer for G_O. We set $\alpha = 0.7$ to make the entity pay more attention to its neighbor entities. And we set $d(e) = d(r) = 200$, $d(t) = 100$, since the type embedding is more general compared with entity and relation. For meta-learning process, we set learning rates for base class training and novel class fine-tuning to 0.001 and 0.01 specifically. We set $m = 10$ for meta-learning stage.

5.2   Compared with the State-of-the-Art Methods

We compare MET with the competitors i.e., TransE [1], DistMult [21], HoIE [14], MTransE [2], etc. For JOIE and ConnetE, we refer to the original paper [7,22] to set the parameters. For TransE, DistMult and HoLE, we convert ETPs to triples $(e, type\text{-}of, t)$. Therefore, entity typing task is equivalent to the triple completion task. For R-GCN and MTransE, we treat types and entities as different views.

Some models are not trained under episodic meta-learning framework. For fair comparison, the ETPs at meta-training stage and the S_{mte} at meta-testing stage will be set as training sets, and the Q_{mte} at meta-testing stage will be set as testing sets. We report results of our model MET on YAGO26K-906 and DB111K-174 in Table 2 for comparing above-mentioned methods.

We observe that all MET methods perform better than other competitors. The original MET outperforms the previous state-of-the-art method JOIE by 2.0% w.r.t accuracy and 1.6% w.r.t MRR on YAGO26K-906, 2.2% and 2.0% on

Table 2. Experimental results of entity typing under the 3-shot setting.

Datasets	YAGO26K-906			DB111K-174		
Metrics	MRR	Acc.	Hit@3	MRR	Acc.	Hit@3
TransE	0.435	38.41	55.34	0.452	39.36	57.63
DistMult	0.662	58.71	75.34	0.673	59.87	77.32
HolE	0.628	53.63	72.35	0.648	55.24	74.31
MTransE	0.753	63.31	85.28	0.768	64.32	86.55
R-GCN	0.762	63.60	86.89	0.776	65.45	87.36
ConnectE	0.721	61.08	83.36	0.735	62.45	84.02
JOIE	0.813	71.32	89.36	0.785	68.35	90.55
MET	**0.829**	**73.35**	**91.25**	**0.805**	**70.57**	**91.87**
MET-E	0.822	72.82	90.84	0.792	69.83	91.14
MET-S	0.818	72.23	89.79	0.802	70.25	91.26
MET-R2R	0.815	71.35	89.54	0.795	69.46	91.17

DB111K-174. R-GCN and MTransE outperform other traditional embedding-based methods significantly, which reflects the advantage of performing different models for two views separately. Surprisingly, ConnectE has not shown competitive results. One possible reason is that ConnectE introduces some hypothetical meta-relations, which may damage the performance of the model.

5.3   Ablation Study

The Effectiveness of Our Proposed Modules: To better understand each module, we implement three variants based on MET. The MET-E only utilizes entity embedding when aggregating the information for central entity without relation embedding. The MET-S removes sub-tasks settings at meta-training stage. And the MET-R2R only applies R2R-GCN without episodic paradigm.

As shown in Table 2, these models adopting episodic paradigm are better than MET-R2R. It demonstrates that meta-learning paradigm brings great benefits for few-shot knowledge graph entity typing task. Compared with MET, MET-E has 0.7% and 0.8% performance degradation w.r.t accuracy and MRR respectively on YAGO26K-906. The performance degradation on MRR and accuracy is 1.1% and 1.6% on DB111K-174. It demonstrates that for the datasets with more kinds of relations, the relation embedding is more important for entity typing task. When compared with MET, the performance of MET-S drops more on YAGO26K-906 compared with DB111K-174. The reason is that YAGO26k-906 has more *is-a* relations in ontology-view KG, and this kind of datasets is more sensitive to sub-tasks setting which can extract hierarchical information.

Influence of the Number of ETPs: We use the same datasets in Sect. 5.2 and explore the influence of the number of ETPs (shot) for every novel type. We increase the ETPs for every novel type from 3 to 5, and choose R-GCN,

Table 3. Experimental results of entity typing under the 5-shot setting.

Datasets	YAGO26K-906			DB111K-174		
Metrics	MRR	Acc.	Hit@3	MRR	Acc.	Hit@3
TransE	0.543	46.34	64.18	0.564	48.52	67.53
R-GCN	0.786	64.60	87.91	0.792	67.55	89.91
ConnectE	0.753	64.25	85.32	0.762	64.86	86.22
JOIE	0.824	72.62	90.89	0.796	69.42	91.30
MET	**0.835**	**73.63**	**91.77**	**0.812**	**71.15**	**92.37**

ConnectE, JOIE and MET for comparison. The results are shown in Table 3. When increasing the ETPs, for TransE, the accuracy and MRR increase 20.64% and 24.82% respectively on YAGO26k-906. For JOIE, the increment is 1.82% and 1.35%. However, the increment drops to 0.5% and 0.7% for MET. MET can fully study the transferable information contained in few-shot ETPs. Therefore, it gets the low growth. But for these methods without meta-learning paradigm, their performances strongly depends on the number of ETPs, thus they can perform better significantly on 5-shot setting compared with 3-shot setting.

6   Conclusion

In this paper, we propose a novel knowledge graph entity typing model named MET, which is particularly designed for few-shot setting. MET adopts meta-learning paradigm and can fully utilize the hierarchical information contained in type tree. In addition, MET includes an GCN-based model R2R-GCN which can explicitly model the interactions between different relations. Our experiments performing on two real-world KGs show that our approach is superior to the most advanced models in most cases.

Appendix

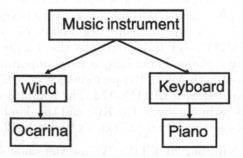

Fig. 3. positive-father sub-task.

We use an example to illustrate the effectiveness of positive-father sub-task in Fig. 3. The entity piano belongs to the entity type keyboard, but also is a kind of music instrument. The entity type keyboard is the positive type of piano, and the entity type music instrument is the father type of piano. Our model want to figure out the fine-grained entity type of each entity, so the probability of positive type should be higher than the probability of father type.

References

1. Bordes, A., Usunier, N., Garcia-Duran, A., Weston, J., Yakhnenko, O.: Translating embeddings for modeling multi-relational data. Adv. Neural Inf. Process. Syst. **26**, 1–9 (2013)
2. Chen, M., Tian, Y., Yang, M., Zaniolo, C.: Multilingual knowledge graph embeddings for cross-lingual knowledge alignment. In: Proceedings of the 26th International Joint Conference on Artificial Intelligence, pp. 1511–1517 (2017)
3. Chen, T., Chen, Y., Van Durme, B.: Hierarchical entity typing via multi-level learning to rank. In: Proceedings of the 58th Annual Meeting of the Association for Computational Linguistics, pp. 8465–8475 (2020)
4. Finn, C., Abbeel, P., Levine, S.: Model-agnostic meta-learning for fast adaptation of deep networks. In: International Conference on Machine Learning, pp. 1126–1135. PMLR (2017)
5. Gupta, N., Singh, S., Roth, D.: Entity linking via joint encoding of types, descriptions, and context. In: Proceedings of the 2017 Conference on Empirical Methods in Natural Language Processing, pp. 2681–2690 (2017)
6. Hajishirzi, H., Zilles, L., Weld, D.S., Zettlemoyer, L.: Joint coreference resolution and named-entity linking with multi-pass sieves. In: Proceedings of the 2013 Conference on Empirical Methods in Natural Language Processing, pp. 289–299 (2013)
7. Hao, J., Chen, M., Yu, W., Sun, Y., Wang, W.: Universal representation learning of knowledge bases by jointly embedding instances and ontological concepts. In: Proceedings of the 25th ACM SIGKDD International Conference on Knowledge Discovery & Data Mining, pp. 1709–1719 (2019)
8. Jain, P., Kumar, P., Chakrabarti, S., et al.: Type-sensitive knowledge base inference without explicit type supervision. In: Proceedings of the 56th Annual Meeting of the Association for Computational Linguistics, vol. 2: Short Papers, pp. 75–80 (2018)
9. Lee, Y., Choi, S.: Gradient-based meta-learning with learned layerwise metric and subspace. In: International Conference on Machine Learning, pp. 2927–2936. PMLR (2018)
10. Lehmann, J., et al.: Dbpedia-a large-scale, multilingual knowledge base extracted from wikipedia. Semant. Web **6**(2), 167–195 (2015)
11. Mahdisoltani, F., Biega, J., Suchanek, F.: Yago3: a knowledge base from multilingual wikipedias. In: 7th Biennial Conference on Innovative Data Systems Research. CIDR Conference (2014)
12. Moon, C., Jones, P., Samatova, N.F.: Learning entity type embeddings for knowledge graph completion. In: Proceedings of the 2017 ACM on Conference on Information and Knowledge Management, pp. 2215–2218 (2017)
13. Neelakantan, A., Chang, M.W.: Inferring missing entity type instances for knowledge base completion: new dataset and methods. arXiv preprint arXiv:1504.06658 (2015)

14. Nickel, M., Rosasco, L., Poggio, T.: Holographic embeddings of knowledge graphs. In: Proceedings of the AAAI Conference on Artificial Intelligence, vol. 30 (2016)
15. Rendle, S.: Factorization machines. In: 2010 IEEE International Conference on Data Mining, pp. 995–1000. IEEE (2010)
16. Schlichtkrull, M., Kipf, T.N., Bloem, P., van den Berg, R., Titov, I., Welling, M.: Modeling relational data with graph convolutional networks. In: Gangemi, A., et al. (eds.) ESWC 2018. LNCS, vol. 10843, pp. 593–607. Springer, Cham (2018). https://doi.org/10.1007/978-3-319-93417-4_38
17. Shang, C., Tang, Y., Huang, J., Bi, J., He, X., Zhou, B.: End-to-end structure-aware convolutional networks for knowledge base completion. In: Proceedings of the AAAI Conference on Artificial Intelligence, vol. 33, pp. 3060–3067 (2019)
18. Snell, J., Swersky, K., Zemel, R.: Prototypical networks for few-shot learning. In: Proceedings of the 31st International Conference on Neural Information Processing Systems, pp. 4080–4090 (2017)
19. Vashishth, S., Sanyal, S., Nitin, V., Talukdar, P.: Composition-based multi-relational graph convolutional networks. arXiv preprint arXiv:1911.03082 (2019)
20. Vinyals, O., Blundell, C., Lillicrap, T., Wierstra, D., et al.: Matching networks for one shot learning. Adv. Neural Inf. Process. Syst. **29**, 3630–3638 (2016)
21. Yang, B., Yih, S.W.t., He, X., Gao, J., Deng, L.: Embedding entities and relations for learning and inference in knowledge bases (2015)
22. Zhao, Y., Zhang, A., Xie, R., Liu, K., Wang, X.: Connecting embeddings for knowledge graph entity typing. In: Proceedings of the 58th Annual Meeting of the Association for Computational Linguistics, pp. 6419–6428 (2020)

Improving Entity Disambiguation Using Knowledge Graph Regularization

Zhi-Rui Tam⑩, Yi-Lun Wu, and Hong-Han Shuai(✉)⑩

National Yang Ming Chiao Tung University, Taipei, Taiwan
{ray.eed08g,w86763777.eed08g,hhshuai}@nctu.edu.tw

Abstract. Entity disambiguation plays the role on bridging between words of interest from an input text document and unique entities in a target Knowledge Base (KB). In this study, to address the challenges of global entity disambiguation, we proposed Conditional Masked Entity Model Using Knowledge Graph Regularization (CMEM-KG), based on a conditional masked language model, in which multiple mentions in a context can be disambiguated in one forward pass. In addition, to address the long-tailed distribution of global entity disambiguation, we proposed a link prediction regularization, in which the entity embeddings were jointly learned to predict knowledge graph links to prevent the model from overfitting. Compared to other global entity disambiguation models, the model proposed in this study exhibited improved performance on six public datasets without an iterative decoding.

Keywords: Entity disambiguation · Parallel decoding

1 Introduction

Entity disambiguation, which is also known as entity linking, refers to the task of assigning entities (such as famous locations, brands, or companies) mentioned in a sentence to a unique identity in a given knowledge base. For example, "A jaguar is cruising on the highway" and "A jaguar is hiding in the jungle" are both mentions of "jaguar", but they refer to two different entity labels based on the contexts. The former refers to a Jaguar_Car, whereas the latter refers to a Jaguar_Animal. Entity disambiguation is an essential task across multiple natural language processing (NLP) applications, such as coreference resolution [24], canonicalization [8] and question answering [17].

Generally, entity disambiguation can be classified into local entity disambiguation and global entity disambiguation. Local entity disambiguation [14,16] focuses on disambiguating mentions based on their local contexts. However, the locality limits the performance when the information provided in a local context is not sufficient. In contrast, global entity disambiguation [4,11,26] overcomes

Supplementary Information The online version contains supplementary material available at https://doi.org/10.1007/978-3-031-05933-9_27.

J. Gama et al. (Eds.): PAKDD 2022, LNAI 13280, pp. 341–353, 2022.
https://doi.org/10.1007/978-3-031-05933-9_27

the limitation of the local entity disambiguation by maximizing the coherence between entities and the context of a document. However, despite the progress in global entity disambiguation, the increase in computing complexity with an increase in the number of mentions in a given document has limited its applications.[1]

Although tremendous efforts have been devoted to overcome the challenges of global entity disambiguation, three issues have restricted the effective application of this task. 1) **Topic Coherence**: When a given context is with multiple mentions, global entity disambiguation should choose entities that belongs to the same topic as the document. For example, "Clark Kent is Superman cover up identity" and the "Superman" mention is known to link to the comic character, global entity disambiguation model should choose the fictional character in the comic for the mention of "Clark_Kent" rather than the music producer since the former fits under the same topic as "Superman". 2) **Slow Decoding Speed**: To achieve an optimal coherence between multiple mentions, entity disambiguation models must explore all possible entity combinations. Existing state-of-the-art methods [15,32] adopt iterative methods, of which the computation complexity increases linearly with an increase in the number of mentions. Although iterative method is better than the exponential complexity from previous works [23,25], a more efficient entity disambiguation is still desirable. 3) **Long-tailed Distribution.** The distribution of entities usually follows a long-tailed distribution. A dataset is majorly composed of the tail of a distribution. Thus, many entities are only contained by a small number of labeled documents, which cannot be disambiguated well owing to overfitting by the model. Although this issue can be alleviated by filtering out infrequent entities, this approach significantly reduces the amount of entity vocabulary; thus, limiting the number of entity predictions.

To address these challenges, in this study we proposed a novel framework, namely Conditional Masked Entity Model Using Knowledge Graph Regularization (CMEM-KG). Particularly, to address the topic coherence challenge, a bidirectional transformer-based decoder was used to model the coherence between mentions. The decoder can implicitly learn to determine the coherent entity via a self-attention between mentions. To address the slow decoding speed, we formulated an entity disambiguation task as a conditional entity disambiguation problem, which is similar to the masked language model [7], to facilitate the parallel prediction of all the unknown entities. To overcome the long-tailed distribution, CMEM-KG utilized a knowledge graph link prediction as a regularization term to improve the disambiguation of long-tailed entities. As a frequently-mentioned entity may share a common relation with less-frequently mentioned entities, updating the parameters of frequently-mentioned entities can also update those of the less-frequent entities as they are linked through a common relation. In contrast to the previous approaches [21,28], the method proposed in the study learned knowledge graph links separately. Furthermore, we proposed an entropy-based filtering method to trim excessive nodes in a knowledge graph. The entropy-based filtering method implicitly represented the

---

[1] Previous work [9] proves the exact solution for the coherence of a global entity is considered as an NP-hard problem.

number of edge degrees and link probability in one value; thus, simplifying the hyperparameters used to remove unwanted nodes. The performance of the model on six public datasets was investigated and the results revealed that the performance of the proposed approach on a CoNLL testing dataset without finetuning was at least 24% better than those of all baseline models. Our implementation is available at: https://github.com/basiclab/CMED.

2 Related Work

Neural network-based entity disambiguation achieves a state-of-the-art performance [12,15,16] owing to the utilization of multiple training objectives to model the relation between an entity and its surrounding context (i.e., words and entities within the same documents). However, this approach relies on finding the co-occurrence matrix between entity-entity and entity-words. Consequently, the application of this approach is limited owing to the sparse matrix. To address this issue, additional features such as entity types [32] and Wikipedia link information, are often required to further improve the performance of this approach [9,30].

Recently, pre-trained transformer-based models (such as BERT [7]) have attracted significant attention as a promising alternative for entity disambiguation. [2] reported that the use of pre-trained bi-directional transformers, such as BERT, can easily achieve competitive scores. In addition, [15] modified the BERT model by iteratively disambiguating one entity with the highest confidence score and feeding the entity back into the model as inputs to disambiguate the next entity until all entities are resolved. However, the use of the same model to encode both context and entities exhibits a sub-optimal performance as they follow two different distributions. In contrast, the model proposed in this study utilized two modules to encode context and entity separately.

As knowledge graphs store interlinked descriptions of entities, recent studies have incorporated knowledge graph into entity disambiguation [21,28]. For example, [21] utilized a local knowledge graph triplets as part of the input sequence. However, this approach does not solve the long-tailed problem as the knowledge graph relations are still part of the model inputs, which falls under the same long-tailed distribution. [28] concatenated the knowledge graph embedding of the entity candidate with the mention embedding as the auxiliary information. Nevertheless, this approach only tackles local entity disambiguation, while our model focuses on global entity disambiguation, which is more challenging.

3 Conditional Masked Entity Model Using Knowledge Graph Regularization (CMEM-KG)

3.1 Architecture Overview

Given a set of M mentions $m_1, m_2, m_3, ..., m_M$ found in a context C and a vocabulary of V predefined entities $e_1, e_2, ..., e_V$, the entity prediction task aims

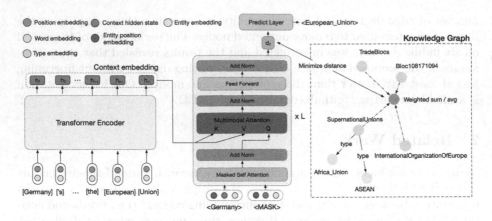

Fig. 1. The model architecture of the proposed CMEM-KG. The decoder predicts the missing entity marked by [MASK].

to predict which entity should be linked to each of the mentions. To tackle this problem, we proposed Conditional Masked Entity Model Using Knowledge Graph Regularization (CMEM-KG). The architecture of CMEM-KG consistsed of a text encoder and bidirectional Transformer-based [29] entity decoder. First, we masked out the entity linked for each mention. Thus, the model must predict the masked entity conditioned to the context features and mention features. Figure 1 shows the overall architecture of the model proposed in this study. The text encoder encodes the mentioned context into a contextual representation, which consisted of position embeddings and word token embeddings, similar to that of the Transformer. Thereafter, the entity decoder utilized the hidden states, h, of the mention tokens and entity embedding (denoted as: $E \in R^{V_e \times H}$), and outputs a set of hidden states d representing the mention features. If the mention tokens span over more than one-word tokens, the average of the hidden states is utilized instead. Next, we passed the mention features to a linear model to predict the entity of the mention. It is worth noting that the entity decoder attends to both left and right entities because the optimal order of entity disambiguation was unknown. Moreover, similar to the sequence-to-sequence transformer model, the entity decoder subjects the outputs from the text encoder to a cross self-attention.

3.2   Entity Prediction

After deriving the mention representation (d), the entity linked to the mention was predicted. Based on the study of [10], which utilized a parallel decoding model, we used a conditional masked loss by randomly replacing half of the entity labels with a mask token $[MASK]$ after which the probability strategy proposed by [10] was utilized. The decoder aims to predict the correct entity label for every masked token conditioned to the mention embedding and context representation. The prediction of each entity label can be learned parallelly, as the probability that an entity is masked is independent to other entities.

To predict the original entity of the masked token, the decoder output of the i-th mention, denoted by $d_i \in R^H$, was multiplied by the full entity embedding matrix E before the softmax function was applied:

$$y_{entity-i} = softmax(E \cdot d_i). \tag{1}$$

To ensure that both spaces were aligned in the same entity space without any bias shift, we did not include the commonly-used bias term in the prediction model of the masked language model.

3.3 Link Prediction Regularization

Although the proposed model could predict entities based on context and mention features, the prediction still suffers from overfitting the entities with the small number of mentions in the Wikipedia paragraphs because neural networks can easily memorize the mentions linked to an entity without learning the surrounding contexts. This indicates that the performance of the model for unseen mentions within similar contexts may be poor. Therefore, in this study, we proposed a novel knowledge graph regularization to prevent overfitting by the model.

Particularly, the knowledge graph enables the linking of a commonly-used entity to a rare entity as a triplet relation (i.e., (e_i, r, e_j)), where r is the relation between e_i and e_j. For example, Jaguar_Car (commonly-used) can be linked to William_Lyons (rare) through a foundedBy relation. Entities in a triplet relation share the same entity vocabulary found in the entity disambiguation task. Moreover, the knowledge graph also contains types (e.g., Jaguar_Car and William_Lyons have the types of Car and AutomotivePioneers, respectively), the type of the entity can be extracted as type sets $t_1, t_2 \dots, t_K$ with a total vocabulary size K. Accordingly, each entity was linked to a various subsets of types to represent the characteristics of an entity. To leverage the information from the knowledge graph, one possible approach is to use the graph embedding [3,22] and margin distance loss similar to TransE [1], which is able to learn one-to-one relations of (e_i, r_{ij}, t_j). Let ϕ_{e_i}, ϕ_{t_j}, and $\phi_{r_{ij}}$ respectively represent the embeddings of the i-th entity, the j-th type, and the relation between them. Moreover, $(\hat{e}_i, r_{ij}, \hat{t}_j)$ represents the corrupted entity triplet pair, i.e., either the entity or type is replaced by a random entity or type. The triplet loss minimizes the distance for (e_i, r_{ij}, t_j) and maximize the distance for $(\hat{e}_i, r_{ij}, \hat{t}_j)$, and can be calculated as follows.

$$L_{triplet} = \gamma + d(\phi_{e_i} + \phi_r, \phi_{t_j}) - d(\phi_{\hat{e}_i} + \phi_r, \phi_{\hat{t}_j}), \tag{2}$$

where γ is the margin hyperparameter. However, the relation between entity to types is a one-to-many mapping (e_i, r, T_k), i.e., an entity e_k has a set of valid type $T_k = \{t_{k,1}, t_{k,2}, \cdots, t_{k,N_k}\}$ with size N_k. According to a previous study [19], this approach cannot learn one-to-many mapping results. Therefore, in this study, the set of types was initially passed through an aggregation function f, after which model the output embedding was modelled as one-to-one relations. Let r_{type} denote the relation between entity and type. Thus, the loss functions of links[2] and types were be represented using:

$$L_{type}(e_i, T_i) = \gamma + d(\phi_{e_i} + \phi_{r_{type}}, f(T_i))$$
$$-d(\phi_{\hat{e}_i} + \phi_{r_{type}}, f(\hat{T}_i)), \tag{3}$$
$$L_{link} = \frac{1}{N}\sum L_{triplet} + \frac{1}{N}\sum L_{type}.$$

The first approach of f is to use the average sum of all the type embeddings for aggregation:

$$f_{avg}(T_k) = \frac{1}{N}\sum_{i=1}^{N}(\phi_{T_{ki}}). \tag{4}$$

However, the average function may not be optimal because not all types share equal contribution to form the entity representation. Therefore, we proposed an attention-weighted sum, in which the query vector was replaced with the representation of an entity, *i.e.*,

$$Q = F_Q(\phi_t), K = F_K(e)$$
$$\alpha = softmax(\frac{Q^T K}{\sqrt{d}}) \tag{5}$$
$$f_{attn}(T_k) = \sum_{i}^{T} \alpha_i \, \phi_{T_{k,i}},$$

Using the attention-weighted sum in Eq. 5, the model dynamically assigned a weight value α to each type. Subsequently, the normalized weights were computed by calculating the dot product of the entity representation entity and the type embeddings, which were scaled using the dimension of entity representation $\sqrt{d}$ after which the normalized weights were passed through the softmax function. The normalized weights were used to calculate the weighted contribution corresponding to each of their type embeddings.

The final objective function of CMEM-KG was:

$$-\sum y \, log(p(x)) + \lambda_1 L_{link} + \lambda_2 \sum L_{type}(d_i, r_{type}, T_i), \tag{6}$$

where the first term is the cross entropy loss for entity disambiguation, λ_1 and λ_2 are the weighted terms for the regularization term, and $d \in \mathbb{R}^h$ is the output embedding from the decoder. The key idea here is that the type relations were uniformly sampled; thus, ensuring that each entity embedding was uniformly updated to alleviate the problem caused by the long-tailed distribution.

---

[2] Only entity to entity relations are minimized in link loss.

3.4   Entropy-Based Type Filtering

During the inference stage of entity disambiguation, the relation and type from the knowledge graph are not required. As the vocabulary size for both relation and type should be sufficiently small to reduce the total training parameters, a good filtering method is required to remove commonly-used and rare types linked to only one entity. For example, `Person` contains lesser information compared to `Male_characters_in_film` as the former type can be found in most entities. Therefore, we proposed an entropy-based filtering to reduce the size of type "vocabulary". As the entropy value of types indicates the amount of the information that a linked entity can provide, we calculated the sum of entity entropy of a type based on the entity conditional probability on types using:

$$H_{t_j}(e, R) = -\frac{1}{N} \sum_{(e_i, t_j) \in R} p(e_i|t_j) log\ p(e_i|t_j), \tag{7}$$

where R is the set of valid entity-type pairs found in the knowledge graph, N is the total numbers of entity-type pairs, and $p(e_i|t_j)$ is the conditional probability of observing the known entity e_i given a type t_j. The conditional probability is equal to the reciprocal of the edge degree of a given type node. Hence, the entropy increases with an increase in the edge degree of a type node. It is worth noting that [5] uses a similar conditional entropy-based filtering to build a diverse conversation data. In contrast to the study of [5], we added a denominator N to the entropy for penalizing common type such as `Person`, `Thing`. To the best of our knowledge, this is the first study that applies an entropy-based filtering for trimming knowledge graph nodes.

4   Experimental Results

4.1   Experimental Settings

Datasets. We evaluated the performance of the proposed model on six benchmark datasets, including AIDA-CoNLL [13], MSNBC [6], AQUIANT [20], ACE2004 [26], WNED-ClueWEB [11] and WNED-Wiki datasets [11]. In addition, according to the study of [9], we utilized the top-30 entity candidates for all datasets evaluated using the prior $p(e|m)$ computed from Wikipedia and Yago datasets. The statistics for all six datasets can be found in the appendix.

Implementation Details. We choose to use the transformer architecture of Roberta $_{base}$[18] as the encoder model while using a significantly smaller Transformer for the decoder since the mentions size is small compared to their document length. We initialized the encoder parameters by Roberta $_{base}$ and entity embeddings parameters by the parameters pre-trained on link prediction task of Dbpedia knowledge graph.[3] Meanwhile, the decoder parameters are initialized

---

[3] The statistics of the knowledge graph can be found in the appendix.

following the previous work [29]. All models are trained on Wikipedia datasets for 250,000 iterations. Mixed precision training is used to speed up training while retaining a large batch size during each feed-forward pass. We implement our models with Pytorch 1.7 and run experiments on an RTX 3090 GPU. The source code and trained models will be publicly available.

Table 1. Micro-F1 on the CoNLL validation dataset. Train marks the model is trained on the CoNLL training dataset.

	Train	Micro-F1
DCA-SL	✓	89.34 ± 0.59
DCA-RL	✓	88.72 ± 0.32
roberta-GCN	✓	87.34 ± 0.05
LUKE+IT	✓	87.06 ± 0.23
LUKE+IT+KG	✓	87.64 ± 0.15
CMEM-KG (no-kg)	✓	84.23 ± 0.19
CMEM-KG (avg)	✓	90.57 ± 0.04
CMEM-KG (attn)	✓	**90.70** ± 0.14
roberta-GCN		68.39
LUKE		62.27
LUKE+IT+KG		66.87
CMEM-KG (no-kg)		68.97
CMEM-KG (avg)		85.66
CMEM-KG (attn)		86.46

Baselines. We compare CMEM-KG with the following state-of-the-art methods. 1) **DCA** [32]: a global entity disambiguation model which constructs a dynamic context of entities through an iterative process, containing two versions of supervised learning (**DCA-SL**) and reinforcement learning (**DCA-RL**) to explore all possible linking results. 2) **Roberta+GCN**, which extends [15] by adding graph convolution network (GCN) [27] to generate features for the knowledge graph. 3) **LUKE+IT** [15], which modifies [31] to tackle global entity disambiguation through an iterative decoding algorithm maximizing the prediction confidence score, together with the pretrained Roberta $_{base}$ encoder. 4) **LUKE+IT+KG**, a variant of **LUKE+IT** that includes our proposed KG regularization.

To ensure a fair comparison, we re-implement all baselines under the same version of English Wikipedia dump (2020-04-20) to prevent domain gap advantage according to [14]. Moreover, a set of entity vocabularies consisting of 274,409 entities is used in all experiments. By sharing the same set of entity vocabularies, all baselines have the same top candidate sets for each mention. The embedding dimension is set to 300, and the number of training iterations is 250,000. Adam is adopted as the optimizer with a learning rate of 1e-4 during 10,000-step warm-up and, followed by a linear decay.

4.2 Quantitative Results

The performance of the CMEM-KG on CoNLL-test dataset was compared to those of other entity disambiguation models, and the results are shown in Table 1 in terms of the micro-F1 scores. CMEM-KG with link prediction regularization (*i.e.*, CMEM-KG (avg) and CMEM-KG (attn)) outperformed other disambiguation models. Furthermore, the performance of the CMEM-KG trained only on Wikipedia-based annotations without training on the in-domain training set of the CoNLL dataset was significantly higher than those of other models by at least 26.4% in terms of micro-F1. This could be attributed to the fact that the link prediction regularization had a positive influence on the architecture of the CMEM-KG; thus, preventing the overfitting of the model at the initial stage.

Table 2. Average Micro-F1 scores on WNED-ClueWEB (CWEB), MSNBC, ACE 2004, WNED-Wiki (Wiki), AQUAINT, each score is averaged from five different runs with different random seeds. The best results are shown in bold.

	CWEB	MSNBC	**ACE2004**	Wiki	AQUAINT	Avg
Prior $p(e\|m)$	**74.81**	75.06	76.15	73.72	76.40	75.23
DCA-RL	70.67	93.50	88.29	70.39	83.36	81.24
DCA-SL	72.75	93.04	87.81	75.21	84.81	82.74
roberta-GCN	69.81	91.10	86.87	71.85	81.15	80.16
LUKE+IT	72.57	90.59	89.90	76.21	84.81	82.82
LUKE+IT+KG	72.71	90.64	90.10	76.40	84.81	82.93
CMEM-KG (no-kg)	67.85	87.47	91.37	75.41	83.34	81.09
CMEM-KG (avg)	69.33	**93.58**	90.19	73.156	83.69	81.99
CMEM-KG (attn)	72.01	91.25	**92.69**	**76.44**	**88.30**	**84.14**

Table 2 shows the comparison of the performance of CMEM-KG and those of the other entity disambiguation models on the other five datasets. All the models listed in Table 2 were fine-tuned based on CoNLL datasets, and their hyper-parameters were tuned using the CoNLL validation set. The results revealed that the proposed CMEM-KG outperformed the baselines; particularly, on ACE2004 and AQUAINT datasets. This could be attributed to the fact that CMEM-KG with a link regularization prevented the overfitting of long-tailed mentions. However, the performance of CMEM-KG on CWEB was slightly poor which could be attributed to the fact that the automatic generation of the label resulted in noisy mentions. As pointed out by [32], CWEB also contains cases where none of the candidates are the actual answer which cause our model to predict the wrong answers. Moreover, adding link prediction regularization to LUKE+IT only slightly improves the overall average micro-F1, since it does not separate entity and word domain using different encoders.

4.3   Ablation Study

Impact of Aggregate Function. We evaluate the performance of two proposed aggregation functions (avg and attn) against the original TransE model by the link prediction on Dbpedia test dataset. Table 3 shows that adding aggregation function to model one-to-many relations improves link prediction with TransE+avg and TransE+attn performs better than TransE baseline in terms of Mean Reciprocal Rank, Hit@3, and Hit@10. The results indicate that our proposed attention mechanism performs better than the average function in all evaluation metrics, since they are able to learn distinct entity embedding with the help of fine-grained types.

Table 3. Comparison of link prediction results on Dbpedia test dataset.

Dbpedia	MRR	Hit@3	Hit@10
TransE	16.5	13.9	29.2
TransE+avg	16.8	14.8	31.0
TransE+attn	19.3	17.4	40.7

Impact of Mention Frequency. To investigate the performance of our model on rare entities, following the setting of previous work [15] method, we first calculate the frequency for each entity found in Wikipedia dataset and split into 4 bins according to the frequency. Afterward, we calculate the Micro-F1 scores of CoNLL test-B split by bins. Table 4 shows the proposed CMEM-KG outperforms LUKE+IT, especially in predicting rare entities.

Table 4. Micro-F1 performance of the CMEM-KG on CoNLL test dataset split using Wikipedia entity frequency. All models were finetuned based on CoNLL dataset.

# Annotations	CMEM-KG		LUKE+IT	
	Attn	Avg	Confidence	Natural
0	**68.37**	58.95	54.90	53.78
1–10	**93.42**	93.61	91.92	91.03
11–50	**89.66**	91.30	89.54	89.54
≥51	**91.26**	90.13	89.91	89.51

4.4 Evaluations on Iterative Entity Disambiguation

Table 5. The Micro-F1% difference compared to parallel entity disambiguation (N = 1) of CMEM-KG on all 6 datasets when decoding using N iterations. Results are averaged from 5 runs of different random seeds.

Iterations (N)	CoNLL-test	CWEB	MSNBC	ACE2004	Wiki	AQUIANT
3	17.36	−1.18	−6.01	10.44	−4.76	15.98
5	23.64	12.08	0.23	25.09	0.85	11.03
M	23.84	10.70	0.02	25.09	0.46	11.03

Following a similar approach by [10], we extend parallel entity disambiguation to N decoding iterations. For each iteration, the model only disambiguates the top-K confidence mentions, where $K = \frac{M}{N}$ and M is the total number of mentions. Each step is repeated until all mentions are resolved. Note that $N = 1$ equals to the parallel entity disambiguation used by our model while $N = M$ equals to the same confidence order proposed by [15].

Table 6. Micro-F1 performance of the CMEM-KG model on CoNLL dataset based on the number of mentions for decoding iterations

# mentions in context	N = 1	N = 3	N = 5	N = 7
1	86.10	86.10	86.10	86.10
$2 \leq M < 5$	86.54	**86.69**	86.69	86.69
$5 \leq M < 10$	88.46	88.46	**88.65**	88.65
$10 \leq M$	78.88	79.26	**79.63**	79.63

Table 5 shows that splitting the parallel decoding into K iterations only provides slight improvements (less than 0.3%) as compared to our original parallel entity disambiguation ($N = 1$). Since most documents only have one mention to resolve, the iterative decoding leads to a minor improvement even when $N = M$. This suggested that our iterative decoding was more effective when there was more than one mention in the documents. To verify this, Table 6 shows the comparison of the performance of CMEM-KG on CoNLL dataset with different number of mentions in a document and different number of iterations. The results indicate that increasing the number of iterations improved the performance of the model on documents with multiple mentions, suggesting that five iterations are sufficient to handle documents with a large number of mentions.

5 Conclusion

In this study, we proposed a new global entity disambiguation model (CMEM-KG) and a new regularization method to train CMEM-KG. The proposed

CMEM-KG only required one feed-forward pass to disambiguate mentions found in a given text. Moreover, the proposed regularization method prevented entity overfitting during training using a link prediction. Furthermore, we employed an entropy-based filtering to reduce the additional computation introduced by the regularization method to prune the nodes in the knowledge graph. Experimental results revealed that the proposed CMEM-KG is effective for different architectures and datasets. In the future, we plan to use a knowledge graph constructed by using relations between mentions from context instead of a predefined knowledge graph.

Acknowledgements. This work is supported in part by the Ministry of Science and Technology (MOST) of Taiwan under the grants MOST-109-2221-E-009-114-MY3 and MOST-110-2221-E-001-001. This work was also supported by the Higher Education Sprout Project of the National Yang Ming Chiao Tung University and Ministry of Education (MOE), Taiwan. We are grateful to the National Center for High-performance Computing for computer time and facilities.

References

1. Bordes, A., Usunier, N., Garcia-Durán, A., Weston, J., Yakhnenko, O.: Translating embeddings for modeling multi-relational data. In: NIPS (2013)
2. Broscheit, S.: Investigating entity knowledge in BERT with simple neural end-to-end entity linking. In: CoNLL (2019)
3. Chen, H.W., Shuai, H.H., Wang, S.D., Yang, D.N.: Quality-aware streaming network embedding with memory refreshing. In: PAKDD (2020)
4. Cheng, S.Y., Chen, Y.L., Yeh, M.Y., Lin, B.T.: Exploiting relevant hyperlinks in knowledge base for entity linking. In: PAKDD (2021)
5. Csáky, R., Purgai, P., Recski, G.: Improving neural conversational models with entropy-based data filtering. In: ACL (2019)
6. Cucerzan, S.: Large-scale named entity disambiguation based on wikipedia data. In: EMNLP-CoNLL (2007)
7. Devlin, J., Chang, M.W., Lee, K., Toutanova, K.: Bert: pre-training of deep bidirectional transformers for language understanding. In: NAACL-HLT (2019)
8. Fatma, N., rVijay Choudhary, Sachdeva, N., Rajput, N.: Canonicalizing knowledge bases for recruitment domain. In: PAKDD (2020)
9. Ganea, O.E., Hofmann, T.: Deep joint entity disambiguation with local neural attention. In: EMNLP (2017)
10. Ghazvininejad, M., Levy, O., Liu, Y., Zettlemoyer, L.: Mask-predict: parallel decoding of conditional masked language models. In: EMNLP-IJCNLP (2019)
11. Guo, Z., Barbosa, D.: Robust named entity disambiguation with random walks. Semant. Web **9**, 459–479 (2018)
12. He, Z., Liu, S., Li, M., Zhou, M., Zhang, L., Wang, H.: Learning entity representation for entity disambiguation. In: ACL (2013)
13. Hoffart, J., et al.: Robust disambiguation of named entities in text. In: EMNLP (2011)
14. van Hulst, J.M., Hasibi, F., Dercksen, K., Balog, K., de Vries, A.P.: Rel: an entity linker standing on the shoulders of giants. In: SIGIR (2020)
15. Yamada, I., Washio, K., Shindo, H., Matsumoto, Y.: Global entity disambiguation with pretrained contextualized embeddings of words and entities. arXiv (2019)

16. Le, P., Titov, I.: Improving entity linking by modeling latent relations between mentions. In: ACL (2018)
17. Li, B.Z., Min, S., Iyer, S., Mehdad, Y., Yih, W.T.: Efficient one-pass end-to-end entity linking for questions. In: EMNLP (2020)
18. Liu, Y., et al.: Roberta: a robustly optimized bert pretraining approach. ArXiv (2019)
19. Fan, M., Zhou, Q., Chang, E., Zheng, T.F.: Transition-based knowledge graph embedding with relational mapping properties. In: PACLIC (2014)
20. Milne, D., Witten, I.H.: Learning to link with wikipedia. In: CIKM (2008)
21. Mulang, I.O., Singh, K., Prabhu, C., Nadgeri, A., Hoffart, J., Lehmann, J.: Evaluating the impact of knowledge graph context on entity disambiguation models. In: CIKM (2020)
22. Ning, Z., Qiao, Z., Dong, H., Du, Y., Zhou, Y.: Lightcake: a lightweight framework for context-aware knowledge graph embedding. In: PAKDD (2021)
23. Pershina, M., He, Y., Grishman, R.: Personalized page rank for named entity disambiguation. In: NAACL-HLT (2015)
24. Pham, M.T.X., Cao, T.H., Huynh, H.M.: Candidate searching and key coreference resolution for wikification. In: IMCOM (2016)
25. Phan, M.C., Sun, A., Tay, Y., Han, J., Li, C.: Pair-linking for collective entity disambiguation: two could be better than all. In: TKDE (2019)
26. Ratinov, L.A., Roth, D., Downey, D., Anderson, M.: Local and global algorithms for disambiguation to wikipedia. In: ACL (2011)
27. Schlichtkrull, M., Kipf, T.N., Bloem, P., van den Berg, R., Titov, I., Welling, M.: Modeling relational data with graph convolutional networks (2018)
28. Sevgili, Ö., Panchenko, A., Biemann, C.: Improving neural entity disambiguation with graph embeddings. In: ACL (2019)
29. Vaswani, A., et al.: Attention is all you need. In: NIPS (2017)
30. Wainwright, M.J., Jordan, M.I.: Graphical models, exponential families, and variational inference. Found. Trends Mach. Learn. 1, 1–305 (2008)
31. Yamada, I., Asai, A., Shindo, H., Takeda, H., Matsumoto, Y.: LUKE: deep contextualized entity representations with entity-aware self-attention. In: EMNLP (2020)
32. Yang, X., et al.: Learning dynamic context augmentation for global entity linking. In: EMNLP-IJCNLP (2019)

PASTA: PArallel Spatio-Temporal Attention with Spatial Auto-Correlation Gating for Fine-Grained Crowd Flow Prediction

Chung Park[1,3], Junui Hong[1,3], Cheonbok Park[2], Taesan Kim[1], Minsung Choi[1],

and Jaegul Choo[3(✉)]

[1] SK Telecom, Seoul, Republic of Korea
{skt.cpark,skt.juhong,ktmountain,ms.choi}@sk.com
[2] Naver Corp, Seongnam, Republic of Korea
cbok.park@navercorp.com
[3] Kim Jaechul Graduate School of AI, KAIST, Daejeon, Republic of Korea
{cpark88kr,secondrun3,jchoo}@kaist.ac.kr

Abstract. Understanding the movement patterns of objects (e.g., humans and vehicles) in a city is essential for many applications, including city planning and management. This paper proposes a method for predicting future city-wide crowd flows by modeling the spatio-temporal patterns of historical crowd flows in fine-grained city-wide maps. We introduce a novel neural network named PArallel Spatio-Temporal Attention with spatial auto-correlation gating (PASTA) that effectively captures the irregular spatio-temporal patterns of fine-grained maps. The novel components in our approach include spatial auto-correlation gating, multi-scale residual block, and temporal attention gating module. The spatial auto-correlation gating employs the concept of spatial statistics to identify irregular spatial regions. The multi-scale residual block is responsible for handling multiple range spatial dependencies in the fine-grained map, and the temporal attention gating filters out irrelevant temporal information for the prediction. The experimental results demonstrate that our model outperforms other competing baselines, especially under challenging conditions that contain irregular spatial regions. We also provide a qualitative analysis to derive the critical time information where our model assigns high attention scores in prediction.

Keywords: Fine-grained city-wide prediction · Spatial auto-correlation · Temporal attention module

1 Introduction

Location-based services with fine-grained maps are being widely introduced as a result of the continuing development of positioning technology [8]. These services provide city-wide crowd flow prediction to aid urban management, for use by the general public and policy makers [6,15]. However, this prediction task turns out to be challenging, because the spatio-temporal dependencies are too complicated in the fine-grained map [7,10].

J. Gama et al. (Eds.): PAKDD 2022, LNAI 13280, pp. 354–366, 2022.
https://doi.org/10.1007/978-3-031-05933-9_28

Spatio-temporal prediction has been actively studied in many previous studies [8,12,14]. They partitioned a city into an $N \times M$ grid map based on latitude and longitude where the grid represents specific region. With this grid map, they predict the number of individuals (e.g., people or taxis) moving in each grid in the future. For example, ST-ResNet [14] proposed a method based on the residual convolution to predict crowd flows. In addition, STDN [12] designed a model of integrating convolution layers and long-short term memory (LSTM) to reflect both spatial and temporal dependencies jointly. The long-term prediction of spatio-temporal data has also been accomplished using a spatial-attention module [8]. However, these previous approaches are still inaccurate and inefficient in practice for predicting a fine-grained city-wide map because of the following three major factors:

(1) Irregular spatial patterns: According to the first law of geography, similar spatial attributes such as crowd flows have a tendency to be located near each other [10]. However, as the spatial pattern becomes more dynamic and irregularly distributed as the resolution increases, cases that widely deviate from the first law of geography are often observed. Figure 1 describes the spatial distribution of the number of moving individuals of Seoul in South Korea. In the coarse map, a specific grid (A) has a value similar to its neighbors. However, grid (B) in the fine-grained map is significantly different from its neighbors, creating spatially irregular patterns. These irregular patterns are caused by a large value grid surrounded by small value grids (i.e., high-low), or vise-versa (i.e., low-high). These high-low or low-high grids are difficult to predict and likely to be regions of practical importance, such as a key commercial hub.

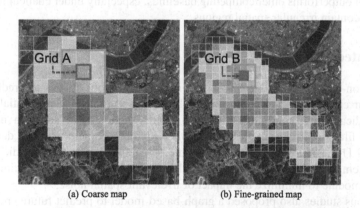

(a) Coarse map   (b) Fine-grained map

Fig. 1. Illustration of the spatial distribution of crowd flows in coarse and fine-grained spatial maps. The larger the volume of crowd flows, the darker the color. The variance in the crowd flow in the fine-grained map is larger than in the coarse map. In the fine-grained map, grid B includes a shopping complex where numerous people gather, and is a region of practical importance.

Therefore, we apply a spatial normalization method that can reflect the spatially irregular patterns by leveraging the concept of local *Moran's I* statistics representing spatial auto-correlation [1]. Spatial auto-correlation indicates the degree to which similar values are located spatially nearby. The high-low and low-high grids have negative statistics with large absolute values in the local *Moran's I* statistics. Therefore, *Moran's I* statistics can be used to identify irregular spatial grids.

(2) Multi-range spatial dependency: Another difficulty comes from multi-range spatial dependency in the fine-grained map. In other words, a specific grid has strong correlations from adjacent to distant grids in terms of connection distance. In addition, spatial distribution is naturally resolution-sensitive, and the interrelation of attributes varies according to the level of resolution [13]. For example, a fine-grained map may contain several geographical hierarchies such as districts and cities, and has multiple interrelations between regions. For this reason, a residual convolution-based architecture with multi-scale filters is adopted to effectively enlarge the receptive field of filters to incorporate the multi-range context in the fine-grained map.

(3) Irrelevant temporal information introducing noise: In a historical observation, a considerable amount of irrelevant temporal information makes noise in the prediction. With a fine-grained map, this phenomenon becomes more pronounced, as the spatial pattern over time changes dynamically. Therefore, we introduce a temporal attention gating module [11], which weights core temporal information for the prediction.

In short, to address these three issues, we propose a PArallel Spatio-Temporal Attention with spatial auto-correlation gating (PASTA) which consists of spatial auto-correlation gating (SAG) to reflect spatial auto-correlation, a multi-scale residual block (MSR) to handle multi-range spatial dependency, and a temporal attention gating module (TAG) to capture significant temporal features for predicting city-wide crowd flows. Our model outperforms other competing baselines, especially under challenging conditions that contain irregular spatial regions.

2 Related Work

Convolution-based models were proposed in many previous studies to predict future population crowds [12, 14]. Those convolution-based models learn spatial patterns which indicate high correlations between nearby regions. An attention module is applied to filter out irrelevant information when predicting spatio-temporal data [8, 12].

STDN [12] employs an attention module to capture long-term periodic information and temporal shifting. DSAN [8] implements long-term flow prediction with an attention module to minimize the impact of irrelevant spatial noise.

Previous studies also proposed a graph-based model to predict future crowd flows [4, 9]. For example, the spatio-temporal graph structure is employed, which connects all grids at the previous and next timestamps, to model spatial and temporal adjacency simultaneously [9]. In addition, spatio-temporal attention, to capture the traffic network's dynamic spatial and temporal correlations, is proposed using an attention based spatial-temporal graph convolutional network [4]. However, we encounter several issues when predicting a fine-grained map, such as irregular spatial patterns. For this reason, this study employs a module to reflect irregular spatial patterns using spatial statistics. In addition, graph-based models are inefficient for fine-grained maps because of the computational cost required to process the large number of nodes in the spatio-temporal graph. Therefore, this paper uses residual convolution with temporal attention to efficiently train the model.

On the other hand, the spatio-temporal prediction task is similar to the semantic segmentation task which predicts a class label to every pixel in the image, in that input and output of the same size are used, and multi-range spatial dependency should be considered. Therefore, to design a model suitable for our problem, we explored various architectures used in semantic segmentation task. Among them, the DeepLab series [2,3] encourages the active utilization of dilated convolution to solve the semantic segmentation task. In particular, atrous spatial pyramid pooling is proposed to apply multi-scale context in DeepLab V2 [2] and DeepLab V3 [3], which adopts parallel dilated convolutional filters to handle multi-scale objects.

In summary, most semantic segmentation tasks use the multi-filters strategy to cope with various scales. This study also devises a structure that uses convolution filters of different sizes in parallel to handle the multi-range spatial dependency.

3 Proposed Method

3.1 Problem Definition

Problem Setting. This study partitions a city into $N \times M$ equal-sized grids based on latitude and longitude, where a grid represents a specific region. The grid is a rectangle corresponding to a small region in a specific city. The actual grid resolution varies from $100\,\text{m} \times 100\,\text{m}$ to $1\,\text{km} \times 1\,\text{km}$. We handle a fine-grained map whose grid map size is under 500 m. Each grid contains flow volume with timestamps.

We use $x_t^{i,j} \in \mathbb{R}$ to denote the value of the (i, j) grid ($i \in \{1, \cdots, N\}$, $j \in \{1, \cdots, M\}$) at timestamp t, and $X_t \in \mathbb{R}^{N \times M}$ denotes the values of all grid maps at timestamp t. We set $\mathcal{X} = (X_{t-T+1}, X_{t-T+2}, ..., X_t) \in \mathbb{R}^{N \times M \times T}$ as the value of all grids over T timestamps.

T timestamps consist of three fragments, denoting recent (*closeness*), daily-periodic (*periodic*), and weekly-periodic (*trend*) segments. Therefore, intervals of $T_{closeness}$, $T_{periodic}$, and T_{trend} are an hour, a day, and a week respectively. From this, the timestamps of *closeness* refers to the timestamps immediately before the target timestamp to be predicted. For example, if we predict the target timestamp t, then the timestamps of the *closeness* in the input sequences can be $\{t - 1, t - 2, t - 3\}$. The timestamps of *periodic* means the identical daily past timestamps of the target timestamp and *trend* refers to the same weekly past timestamps of the target timestamp. The examples of the timestamps of *periodic* and *closeness* are $\{t - 1 \times 24, t - 2 \times 24, t - 3 \times 24\}$ and $\{t - 7 \times 24, t - 14 \times 24\}$, respectively, given the target timestamp t. The number of timestamps in each fragment is represented as T_k, where $k \in (closeness, period, trend)$ indicates each fragment. The total number of timestamps T is equal to $T_{closeness} + T_{periodic} + T_{trend}$.

This study set $T_{closeness}$, $T_{periodic}$, and T_{trend} as five, six, and four respectively. The T_k-channel crowd flows map of each time fragment is then concatenated with the channel axis. In addition, $Y_{t+1} = X_{t+1} \in \mathbb{R}^{N \times M}$ represents the number of individuals in all grids to be predicted at timestamp $t + 1$.

Problem. Given the historical sequence of all the grids over past T slices $\mathcal{X} = (X_{t-T+1}, X_{t-T+2}, ..., X_t) \in \mathbb{R}^{N \times M \times T}$, we aim to predict the future flows $Y_{t+1} \in \mathbb{R}^{N \times M}$.

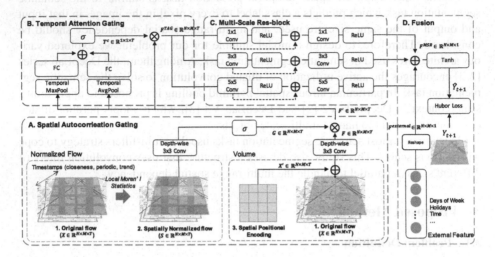

Fig. 2. Architecture of PASTA. There are three major components: spatial auto-correlation gating, temporal attention gating, and multi-scale residual block module. FC indicates fully-connected layers and $Conv$ is a convolution filter. σ is a sigmoid function. $\oplus$ denotes the element-wise summation, and $\otimes$ element-wise multiplication.

3.2 Parallel Spatio-Temporal Attention with Spatial Auto-Correlation Gating

Overview. Figure 2 describes the architecture of PASTA, which is composed of three components including spatial auto-correlation gating (SAG), temporal attention gating (TAG), and multi-scale residual block (MSR) module. First of all, we modify flow volumes throughout a city with T timestamps into an $N \times M$ image-like matrix $\mathcal{X} \in \mathbb{R}^{N \times M \times T}$. We element-wise sum them with their corresponding spatial positional encodings (SPE) before entering the TAG module. The SPE is a particular bias to represent the location of grids [8]. The SAG module for local spatial normalization is applied to $\mathcal{X}$ to produce the indicator of the spatial auto-correlation $\mathcal{S} \in \mathbb{R}^{N \times M \times T}$. Consequently, the normalized outputs $\mathcal{S}$ are fed into the depth-wise convolution layer and a sigmoid activation to produce gating values. The output of SAG is fed into the TAG module and then the MSR module. Such structure captures significant channel-wise features for temporal information and the multi-range dependency between nearby and distant regions. In addition, the external features, such as weather, are fed into two-layer fully-connected neural networks to extract latent features. These features are further element-wise summed with the outputs of the MSR. Lastly, a tanh activation function is adopted to map the aggregation into $[-1, 1]$. We adopts Huber loss [5] as the loss function, which is less sensitive to outliers than the mean squared loss.

SPE: Spatial Positional Encoding. Each grid has its own address. This is a character-istic that differentiates the location data from the image. To consider relative positions in the grid map, we equip the spatial positional encoding (SPE) as a bias to represent the position of a location [8]. Given the coordinate matrix, we calculate the SPE as follows

$$SPE_{i,j}^l = \begin{cases} sin(i/10000^{2l/d}) & , if \ l = 2n, \\ cos(j/10000^{2l/d}) & , if \ l = 2n+1, \end{cases} \tag{1}$$

where $SPE_{i,j}^l \in \mathbb{R}$ is the l-th dimension element of encoding vector of position (i,j) in the grid map. We element-wise sum the SPE encoding with the sequence of grid map $\mathcal{X} = (X_{t-T+1}, X_{t-T+2}, ..., X_t) \in \mathbb{R}^{N \times M \times T}$ to produce $\mathcal{X}' = (X'_{t-T+1}, X'_{t-T+2}, ..., X'_t) \in \mathbb{R}^{N \times M \times T}$.

SAG: Spatial Auto-correlation Gating. In general, spatial information has the prop-erty of spatial auto-correlation, where spatial attributes such as crowd flows in nearby regions tend to be similar on a map. The spatial auto-correlation is denoted as the first law of geography [10], which makes the proposition that *"Everything is related with everything else, but near things are more related than distance things"*.

However, as the map becomes finer, the opposite cases often appear, resulting in a large volume region surrounded by a small volume region (i.e., high-low), or vise-versa (i.e., low-high). These high-low or low-high region produce irregular spatial patterns contrary to the first law of geography. Various indices have been proposed to mea-sure the degree of spatial auto-correlation quantitatively. In particular, *Local Moran's I* statistic [1] is a popular statistics to quantify local spatial auto-correlation.

Suppose that $x_t^{i,j}$ is the value of the (i,j) grid among overall grids in timestamp t, and i and j are the index of horizontal and vertical axis in the whole grid map respec-tively. In addition, N and M are the total number of grids in the horizontal and vertical axis of the map, respectively. Then the *Local Moran's I* statistics $s_t^{i,j}$ of (i,j) grid in timestamp t is calculated as follows

$$s_t^{i,j} = \frac{(x_t^{i,j} - \bar{x}_t)}{P_t} \sum_{z \in W_{ij}} \frac{(x_t^z - \bar{x}_t)}{P_t}, \tag{2}$$

where $\bar{x}_t = \frac{\sum_{i=1}^{N} \sum_{j=1}^{M} x_t^{i,j}}{NM}$ and $P_t = \sqrt{\frac{\sum_{i=1}^{N} \sum_{j=1}^{M} (x_t^{i,j} - \bar{x}_t)}{NM-1}}$. W_{ij} is the set of prede-fined neighbor grids, which is all grids that share an edge or a corner with the (i,j) grid. Therefore, x_t^z is the *Local Moran's I* statistics of neighbor grids with the (i,j) grid.

The statistics represent the relative value of a particular grid to its neighbors. If both a grid (i,j) and its neighbors are greater than the overall average $\bar{x}_t$ together, then a positive $s_t^{i,j}$ is produced in the (i,j) grid. Conversely, if the value of a grid (i,j) is greater than $\bar{x}_t$ and its neighbors are less than $\bar{x}_t$, then the region (i,j) may have a negative $s_t^{i,j}$ value. This region is denoted as high-low, and in the opposite case, low-high. That is, a negative $s_t^{i,j}$ indicates that grid (i,j) has a relatively large or small value compared to its neighbors. These grids with negative $s_t^{i,j}$ are challenging to predict in that their spatial patterns are highly irregular.

The local *Moran's I* statistics are calculated for all grids $\mathcal{X}$. In Fig. 2A, the illustrative calculation process of local *Moran's I* statistics is described. The output of local *Moran's I* statistics $\mathcal{S} \in \mathbb{R}^{N \times M \times T}$ has the same shape as the original flow volumes $\mathcal{X}$. Then, the normalized output $\mathcal{S}$ is passed by depth-wise convolutional layer and sigmoid activation function to produce $\mathcal{G} = (G_{t-T+1}, G_{t-T+2}, ..., G_t) \in \mathbb{R}^{N \times M \times T}$.

Meanwhile, the original flow volume $\mathcal{X}$ is also applied to the depth-wise convolutional layer to extract the unique spatial pattern of each timestamp without sharing temporal information. It is represented as $\mathcal{F} = (F_{t-T+1}, F_{t-T+2}, ..., F_t) \in \mathbb{R}^{N \times M \times T}$. Then, $\mathcal{G}$ are element-wise multiplied by $\mathcal{F}$ to produce a gated output $\mathcal{F}' \in \mathbb{R}^{N \times M \times T}$. This indicates that the normalized value $\mathcal{G}$ explicitly controls the volume information associated with neighbor regions.

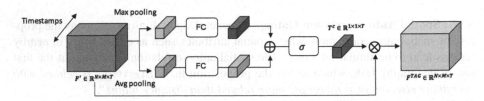

Fig. 3. Diagram of the TAG module. This module assigns high attention scores to meaningful timestamps of input sequences for prediction. Input $\mathcal{F}' \in \mathbb{R}^{N \times M \times T}$ (red tensor) is the output of the SAG module and the output $F^{TAG} \in \mathbb{R}^{N \times M \times T}$ (grey tensor), is fed into the MSR module. FC indicates a fully-connected layer and σ is a sigmoid function. The element-wise summation and element-wise multiplication are denoted as $\oplus$ and $\otimes$, respectively.

TAG: Temporal Attention Gating. Some past historical information may be irrelevant for future prediction. For this reason, inspired by the channel attention module [11], we proposes the TAG module, which filters out temporal features that are irrelevant for prediction. The TAG module is described in Fig. 2B and Fig. 3. It produces a temporal attention map to reflect the inter-time relationship. The channel indicates the timestamp. This temporal attention focuses on the influential timestamps for prediction.

Specifically, we implement the max-pooling and average-pooling operation to $\mathcal{F}'$, the outputs of SAG, and the both outputs F_{max}^c and F_{avg}^c are then forwarded to the two fully connected layers. After the fully connected layer is applied to each feature, we merge them using element-wise summation with a sigmoid activation function. The temporal attention map $T^c \in \mathbb{R}^{1 \times 1 \times T}$ is described as

$$T^c = \sigma(\text{FC}(\text{AvgPool}(\mathcal{F}')) + \text{FC}(\text{MaxPool}(\mathcal{F}'))) \\ = \sigma(\mathbf{W}_1 \mathbf{W}_0 F_{avg}^c + \mathbf{W}_3 \mathbf{W}_2 F_{max}^c), \tag{3}$$

where σ is the sigmoid function. In addition, $\mathbf{W}_0$, $\mathbf{W}_1$, $\mathbf{W}_2$, and $\mathbf{W}_3$ are the trainable weights in the fully connected layers. Then, the final output of the TAG module, $F^{TAG} \in \mathbb{R}^{N \times M \times T}$ is derived as follows

$$F^{TAG} = T^c \otimes \mathcal{F}', \tag{4}$$

where $\otimes$ denotes element-wise multiplication. Through this process, each channel is multiplied by weight between 0 and 1, so that the model can only take critical temporal information. A high attention weight will be given to timestamps essential for prediction.

MSR: Multi-Scale Residual Block. It is crucial to consider multi-range spatial dependence when dealing with a fine-grained map. Therefore, we designed an MSR module composed of shallow layers using parallel convolution filters of different sizes.

The multiple filters are adopted with a parallel scheme as shown in Fig. 2C. The output feature map $F^{TAG} \in \mathbb{R}^{N \times M \times T}$ from the TAG module is fed into parallel convolutional layers with 1×1, 3×3, and 5×5 filters, respectively, followed by skip-connections. Then, the outputs are element-wise summed in time axis. The input and output in the same shape (i.e., $N \times M$) are generated by using padding in a convolution layer. The output feature map $F^{MSR} \in \mathbb{R}^{N \times M \times 1}$ is derived as follows

$$
\begin{aligned}
F^{MSR} = &\sigma(f^{1\times1}(\sigma(f^{1\times1}(F^{TAG})) + f^{1\times1}(F^{TAG})) \\
&+\sigma(f^{3\times3}(\sigma(f^{3\times3}(F^{TAG})) + f^{3\times3}(F^{TAG})) \\
&+\sigma(f^{5\times5}(\sigma(f^{5\times5}(F^{TAG})) + f^{5\times5}(F^{TAG})),
\end{aligned} \tag{5}
$$

where σ is the ReLU function and $f^{l \times l}$ indicates a convolution operation with the filter size $l \times l$.

External Features. Numerous complex external factors significantly affect crowd flows. We adopt time-of-day (24/48 dimensional variables for every 1 h/30 min), day-of-week (7), and holiday (1) as the external features. The embeddings of external features from two fully-connected layers are concatenated with the outputs of the MSR module. In two full-connected layers of external features, the first layer is a fully-connected embedding layer followed by ReLU activation, and the second layer map low to high dimensions with the same shape as F^{MSR}. This output is denoted as $F^{external}$. Then, F^{MSR} and $F^{external}$ are summed in a channel-wise manner. Finally, the predicted value at the $t + 1$ timestamp, denoted by $\hat{Y}_{t+1}$, is defined as

$$
\hat{Y}_{t+1} = \sigma(F^{MSR} + F^{external}), \tag{6}
$$

where σ is a tanh activation function.

4   Experiment

4.1   Experiment Settings

Dataset. We evaluated our model on two public real-world datasets from New York City (NYC). As shown in Table 1, we used three large datasets, NYC-Taxi and NYC-Bike. NYC-Taxi contains the taxi demands in 16×12 grids in New York City every 30 min from January 1, 2016 to February 29, 2016. The training set is set to data from January 1, 2016 to February 15, 2016, and the test set is set to the remaining data.

NYC-Bike contains the flows of bikes in 14×8 grids in New York City every 30 min from August 1, 2016 to September 29, 2016. We set the data from August 1, 2016 to September 15, 2016 as the training set and remaining data as the test set.

In addition, we used fine-grained real-world data from the major cellular network operator in South Korea. This data is collected from the base stations. When users access the base stations for communication or data access, the logs including location records are generated in the base stations. The location data of about three million customers who agreed to collect and analyze their location information in Seoul were collected from April to June 2021. We set the data from April 1, 2021 to June 15, 2021 as the training set and the remaining data as the test set in Seoul-Crowd. We normalized those datasets by min-max scale.

Table 1. Summary of the dataset used in our experiments.

Dataset	NYC-Taxi	NYC-Bike	Seoul-Crowd
Data type	Taxi GPS	Bike rent	Mobile signal
Location	New York	New York	Seoul
Time span	1/1–2/29, 2016	8/1–9/29, 2016	4/1–6/30, 2021
Time interval	30 mins	30 mins	1 h
Grid map size	(16, 12)	(14, 8)	(68, 92)

Compared Algorithms. To evaluate the accuracy of our predictive model, we compared the proposed model with several competitive methods:

- ST-ResNet[14]: It consists of simple architectures with residual block with CNN for spatio-temporal data.
- STDN[12]: This research is based on the local CNN and LSTM to capture the complex spatial dependencies and temporal dynamics.
- DSAN[8]: They focus on the long-term prediction task with an attention module to deal with the dynamic correlation of spatio-temporal data. It is a state-of-the-art model in NYC-Taxi and NYC-Bike dataset.

4.2 Results

Fine-grained Map Prediction. We experimented with fine-grained crowd flow data as shown in Table 2. We measured performance for resolutions of 12×16, 24×32, 48×64, and 68×92. Typically, the higher the resolution, the lower the model performance. This is because the finer the map, the more spatio-temporal irregularity occurs. However, our model showed robust performance in the fine-grained map with 68×92 resolution and better performance than other models. From the result, we discovered that reflecting spatial auto-correlation and multi-range spatial dependencies in the model can reduce error rate in fine-grained map prediction. In addition, we determined that filtering out irrelevant temporal information was critical to achieving reliable performance.

In order to better understand the effectiveness of our model, we carried out additional experiments as shown in Table 3. In this experiment, we used our Seoul-Crowd dataset of 68×92. We observed that our model outperformed other models in high-low and low-high grids. These results demonstrate that our model makes more robust predictions for spatially irregular regions than other competitive baselines. This is because spatially irregular regions are discriminated in our model, using our SAG module, to prevent smoothing of the predicted values by surrounding regions.

Table 2. Fine-grained map predicion results (Seoul-Crowd)

Resolution	12×16		24×32		48×64		68×92	
	RMSE	MAPE	RMSE	MAPE	RMSE	MAPE	RMSE	MAPE
ST-ResNet[14]	115.94	22.47	114.43	22.23	118.11	21.73	132.20	23.92
STDN[12]	101.11	17.85	97.43	17.22	114.53	17.78	119.30	21.67
DSAN[8]	98.12	16.11	93.43	15.98	99.10	16.22	113.44	20.22
PASTA	**96.57**	**15.49**	**91.79**	**14.55**	**96.21**	**14.58**	**108.55**	**18.92**

Coarse Map Prediction. We evaluated the effectiveness of PASTA and other baselines using a coarse map dataset. Table 4 shows the baseline performances. In some cases, our model significantly outperformed other baselines, achieving the lowest RMSE or MAPE even on the coarse map dataset.

Table 3. Result of spatial irregular regions prediction with fine-grained map

Model	High-Low		Low-High	
	RMSE	MAPE	RMSE	MAPE
ST-ResNet[14]	139.77	33.65	60.76	26.88
STDN[12]	135.98	31.10	57.73	25.08
DSAN[8]	135.33	30.41	58.81	25.31
PASTA	**132.12**	**28.78**	**56.12**	**24.78**

Table 4. Coarse map prediction results

Model	Dataset	NYC-Taxi		NYC-Bike		Crowd-Flow	
	Metric	RMSE	MAPE	RMSE	MAPE	RMSE	MAPE
ST-ResNet[14]		23.82	18.87	9.43	22.03	115.94	22.47
STDN[12]		22.98	17.88	9.41	19.94	101.11	17.83
DSAN[8]		20.73	**16.09**	**8.03**	18.33	98.12	16.17
PASTA		**19.89**	16.12	8.26	**17.76**	**96.57**	**15.49**

4.3 Ablation Study

We investigated the effectiveness of each module of our model. Table 5 shows the experimental results, depending on whether each module was applied or not. From Rows (C) and (F), we observe that the SAG module improved the performance of our model. This result indicates that the SAG module helps improve the robustness in the fine-grained map prediction by correcting the spatially irregular pattern. Rows (D) and (F) show the effect of the MSR module. We found that the model with the MSR module was better than the model without it. This is because the MSR module allows the model to cope with multi-scale spatial dependency. The effect of the TAG module is illustrated in Rows (E) and (F). This result demonstrates that filtering out irrelevant temporal feature is indispensable to derive more accurate prediction.

Table 5. Component analysis

Model	SAG	TAG	MSR	RMSE	MAPE
(A)		✓		116.12	24.11
(B)			✓	117.30	24.98
(C)		✓	✓	114.43	21.28
(D)	✓	✓		113.12	21.09
(E)	✓		✓	113.56	21.14
(F)	✓	✓	✓	**108.55**	**18.92**

4.4 Visualization of Temporal Attention Gating

We visualized the temporal attention map $T^c \in \mathbb{R}^{1 \times 1 \times T}$ of the TAG module to verify the weight of temporal information. Each input sequence $\mathcal{X} \in \mathbb{R}^{N \times M \times T}$ produces T^c respectively. We averaged the temporal attention map T^c from our test dataset of Seoul-Crowd for the T timestamps. As shown in Fig. 4, the model assigns large weights to the feature map of an hour ago. This shows that the timestamp of an hour ago is critical to the prediction. However, if the timestamp is close to the target timestamp but not the same period, low weights are given to that timestamp. The attention weights of 2,3,4, and 5 h ago are below 0.3. In addition, the model sometimes gives low weights for feature maps in the far distant past, where the attention weights of 2 and 4 weeks ago are below 0.3. This visualization demonstrates that not all time information is required for prediction.

5 Conclusion

In this work, we proposed PASTA for predicting future city-wide crowd flows. Our model consists of a SAG module to reflect spatial auto-correlation, an MSR module to handle multi-range spatial dependency and a TAG module to filter out irrelevant temporal features for prediction. This study discovered that reflecting spatial relativity and

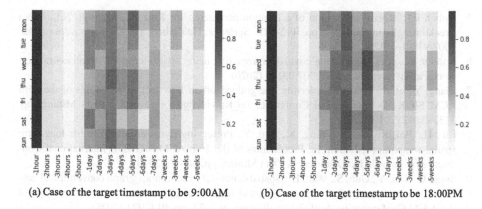

(a) Case of the target timestamp to be 9:00AM   (b) Case of the target timestamp to be 18:00PM

Fig. 4. An illustrative example of temporal attention maps. The higher the attention weights, the darker the color. The y-axis indicates the day of the week and the x-axis is the timestamp. Each grid denotes the average attention weights of the input. (a) is an attention map where the target timestamp to be predicted is 9:00 AM. (b) is the attention map where the target timestamp to be predicted is 18:00 PM.

multi-range spatial dependencies in the model can reduce error rates in fine-grained map prediction. In addition, we determined that filtering out irrelevant temporal information was critical to achieving reliable performance. We extensively evaluated our model on future flow prediction tasks using real-world datasets of fine-grained maps. Our model outperformed other competing baselines in both fine-grained and coarse maps cases. In addition, the results also showed that our model performed better predicting regions with spatially irregular patterns.

Acknowledgment. This work was supported by Institute of Information & communications Technology Planning & Evaluation (IITP) grant funded by the Korea government(MSIT) (No. 2020-0-00368, A Neural-Symbolic Model for Knowledge Acquisition and Inference Techniques, No.2019-0-00075, Artificial Intelligence Graduate School Program(KAIST)).

The authors would like to thank T3K center of SK Telecom for providing GPU cluster support to conduct massive experiments.

References

1. Anselin, L.: Local indicators of spatial association-lisa. Geogr. Anal. **27**(2), 93–115 (1995)
2. Chen, L.C., Papandreou, G., Kokkinos, I., Murphy, K., Yuille, A.L.: Deeplab: semantic image segmentation with deep convolutional nets, atrous convolution, and fully connected crfs. IEEE Trans. Pattern Anal. Mach. Intell. **40**(4), 834–848 (2017)
3. Chen, L.C., Papandreou, G., Schroff, F., Adam, H.: Rethinking atrous convolution for semantic image segmentation. arXiv preprint arXiv:1706.05587 (2017)
4. Guo, S., Lin, Y., Feng, N., Song, C., Wan, H.: Attention based spatial-temporal graph convolutional networks for traffic flow forecasting. In: Proceedings of the AAAI Conference on Artificial Intelligence, vol. 33, pp. 922–929 (2019)

5. Huber, P.J.: Robust estimation of a location parameter. In: Kotz, S., Johnson, N.L. (eds.) Breakthroughs in Statistics, pp. 492–518. Springer, Heidelberg (1992). https://doi.org/10. 1007/978-1-4612-4380-9_35

6. Kapoor, A., et al.: Examining covid-19 forecasting using spatio-temporal graph neural networks. arXiv preprint arXiv:2007.03113 (2020)

7. Liang, Y., et al.: Urbanfm: inferring fine-grained urban flows. In: proceedings of the 25th ACM SIGKDD International Conference on Knowledge Discovery & Data Mining, pp. 3132–3142 (2019)

8. Lin, H., Bai, R., Jia, W., Yang, X., You, Y.: Preserving dynamic attention for long-term spatial-temporal prediction. In: Proceedings of the 26th ACM SIGKDD International Conference on Knowledge Discovery & Data Mining, pp. 36–46 (2020)

9. Song, C., Lin, Y., Guo, S., Wan, H.: Spatial-temporal synchronous graph convolutional networks: a new framework for spatial-temporal network data forecasting. In: Proceedings of the AAAI Conference on Artificial Intelligence, vol. 34, pp. 914–921 (2020)

10. Tobler, W.R.: A computer movie simulating urban growth in the detroit region. Econ. Geogr. **46**(sup1), 234–240 (1970)

11. Woo, S., Park, J., Lee, J.Y., So Kweon, I.: Cbam: convolutional block attention module. In: Proceedings of the European Conference on Computer Vision (ECCV), pp. 3–19 (2018)

12. Yao, H., Tang, X., Wei, H., Zheng, G., Li, Z.: Revisiting spatial-temporal similarity: a deep learning framework for traffic prediction. In: Proceedings of the AAAI Conference on Artificial Intelligence, vol. 33, pp. 5668–5675 (2019)

13. Zhang, J., Atkinson, P., Goodchild, M.F.: Scale in Spatial Information and Analysis. CRC Press, Boca Raton (2014)

14. Zhang, J., Zheng, Y., Qi, D.: Deep spatio-temporal residual networks for citywide crowd flows prediction. arXiv preprint arXiv:1610.00081 (2016)

15. Zhu, S., Bukharin, A., Xie, L., Santillana, M., Yang, S., Xie, Y.: High-resolution spatio-temporal model for county-level covid-19 activity in the us. arXiv preprint arXiv:2009.07356 (2020)

Partially Relaxed Masks for Knowledge Transfer Without Forgetting in Continual Learning

Tatsuya Konishi[1]([⊠]), Mori Kurokawa[1], Chihiro Ono[1], Zixuan Ke[2],
Gyuhak Kim[2], and Bing Liu[2]

[1] KDDI Research, Inc., Fujimino, Japan
{tt-konishi,mo-kurokawa,ono}@kddi-research.jp
[2] University of Illinois at Chicago, Chicago, USA
{zke4,gkim87,liub}@uic.edu

Abstract. The existing research on continual learning (CL) has focused mainly on preventing catastrophic forgetting. In the task-incremental learning setting of CL, several approaches have achieved excellent results, with almost no forgetting. The goal of this work is to endow such systems with the additional ability to transfer knowledge when the tasks are similar and have shared knowledge to achieve higher accuracy. Since the existing system HAT is one of most effective task-incremental learning algorithms, this paper extends HAT with the aim of both objectives, i.e., overcoming catastrophic forgetting and transferring knowledge among tasks without introducing additional mechanisms into the architecture of HAT. The current study finds that task similarity, which indicates knowledge sharing and transfer, can be computed via the clustering of task embeddings optimized by HAT. Thus, we propose a new approach, named "partially relaxed masks" (PRM), to exploit HAT's masks to not only keep some parameters from being modified in learning subsequent tasks as much as possible to prevent forgetting but also enable remaining parameters to be updated to facilitate knowledge transfer. Extensive experiments demonstrate that PRM performs competitively compared with the latest baselines while also requiring much less computation time.

Keywords: Continual learning · Task similarity · Catastrophic forgetting

1 Introduction

Continual learning has recently received substantial attention with the increasing popularity of AI-embedded systems, but these systems still struggle to maintain performance without retraining the model from scratch, which consumes a large amount of time. The main issue in continual learning is *catastrophic forgetting*, which refers to the phenomenon in which once a model has learned a new task, its performance is likely to decline drastically on the previously learned data

J. Gama et al. (Eds.): PAKDD 2022, LNAI 13280, pp. 367–379, 2022.
https://doi.org/10.1007/978-3-031-05933-9_29

[10]; thus, many studies have proposed approaches that address this issue [7,20]. In particular, HAT [25] proposes a mechanism called hard attention that blocks the gradients of parameters, which are important for previous tasks to overcome forgetting, and it achieves learning with almost no forgetting. However, consideration of only the forgetting issue is not sufficient for practical applications. There must sometimes be similar tasks that can be exploited for other tasks and dissimilar tasks that are sensitive to forgetting issues at the same time; however, conventional approaches that have focused mainly on forgetting issues have not fully considered task similarity, which can enhance performance. Therefore, another research theme has become the transfer of knowledge into a newly coming task from previous tasks where, as a matter of course, forgetting should be restrained. These challenges have been represented as *task incremental learning* (TIL), which aims at learning a mixed sequence of similar and dissimilar tasks.

Additionally, looking ahead to the realistic use of continual learning, where AI-embedded edge devices that learn continuously but do not have substantial computational resources are commonly utilized, several studies have focused mainly on the efficiency of learning [3,22]. To advance to the next stage, continual learning methods also need to be as efficient as possible. CAT [13], for instance, is the first approach for tackling a mixed sequence of similar and dissimilar tasks. CAT extends HAT by introducing attention mechanisms across similar tasks to enhance knowledge transfer; however, it is very inefficient and takes a much longer time for task similarity detection because it tries every previous task one by one to judge whether each task is worth being transferred to the current learning task. Although it succeeds in task similarity detection and outperforms HAT, it still faces enormous problems in terms of its efficiency and scalability.

To address these issues, the current study extends HAT so that it can enhance knowledge transfer without another mechanism, such as attention, and even with much shorter computation time. Our contributions to this challenge are as follows. First, the current study discovers that task similarity can be computed from task embeddings that are optimized by a HAT-like approach. Second, we propose a new approach named "partially relaxed masks" (PRM) that employs the masks that are accumulated only for dissimilar tasks so it maintains parameters that are important for the dissimilar tasks as much as possible to prevent forgetting, while keeping the remaining parameters, which are useful for the similar tasks, free to be updated for knowledge transfer. Extensive experiments demonstrate that our approach achieves equal or greater performance than state-of-the-art methods and requires much less computation time.

2   Related Work

Continual learning methods are categorized into three main types: regularization-based methods [14,16,28], which add another penalty so as not to change the important parameters for previous tasks; replay-based methods [4,5,18,23], which keep the small size of previous tasks' samples and exploit them to alleviate forgetting; and parameter isolation-based [11,19,24,26] methods, which create new

---

Algorithm 1. Learning procedure for PRM

Input: $x_{1:T}, y_{1:T}$, Model M with L layers
  for $t = 1 \cdots T$ **do**
    # *Dissimilar task detection (DTD) phase*
    state $\leftarrow$ copy(M)
    1st optimization: Freeze feature extractor, train only classifier of M with x_t, y_t
    2nd optimization: Train whole M with x_t, y_t
    Task embeddings $\{e_l^{1:t}\} \leftarrow M$
    Set of dissimilar tasks $\mathcal{D}_l^t \leftarrow$ Clustering $\left(\{e_l^{1:t}\}\right)$
    Load back: $M \leftarrow$ state
    # *Learning with partially relaxed masks (LwPRM) phase*
    Train whole M with $x_t, y_t, \mathcal{D}_l^t$

---

branches for new tasks, which are defined by new parameters. EWC [14] is one of the most popular regularization-based methods. It computes the Fisher information matrix that represents the importance of each parameter and adds a regularization term that corresponds to the matrix to prevent forgetting. A-GEM [5] is a typical replay-based method, and it uses an efficient approach to select samples of previous tasks that will be learned together in a current task. A major issue with these approaches is that they require an additional memory buffer for saving past samples. To address this issue, many approaches exploit a data generator inside the model, and the generated samples are used with current learning. One of the latest parameter isolation-based methods is CCLL [26], which prevents forgetting with few additional parameters by introducing calibration modules that convert activation maps for previous tasks to a current task. Recently, several approaches have been combined with the meta-learning paradigm to select more effective samples or parameters [4,12].

Although conventional approaches have focused mainly on catastrophic forgetting, most do not have any mechanism for knowledge transfer across similar tasks, which has become another important topic in TIL. In particular, HAT [25] achieves learning with almost no forgetting by introducing hard attention to block the updating of parameters that are important for previous tasks; however, the mechanism no longer enhances knowledge transfer. CAT [13] is the first approach to deal with a mixed sequence of dissimilar tasks and similar tasks at the same time by extending HAT. CAT introduces additional attention operations into classifiers and judges similar tasks using another network separately. Although CAT achieves state-of-the-art performance in this scenario, it requires substantial computational resources for task similarity detection because it tries to build and train the reference and transfer models per previous task and check whether this transfer actually improves its performance. Therefore, the more previous tasks there are, the longer CAT takes to learn a new task.

3 Proposed PRM

The structure and procedure of our proposed method are presented in Fig. 1 and Algorithm 1. The procedure is composed of the two phases: 1) dissimilar task

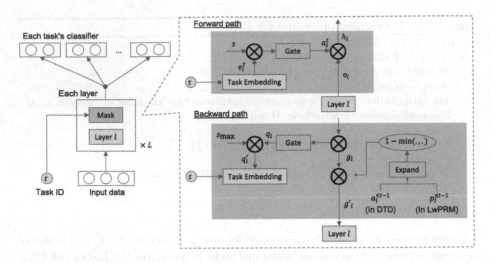

Fig. 1. Structure of PRM for learning task t. PRM follows almost the same procedure as HAT, except for part of the backward path. In dissimilar task detection (DTD), the model uses $a_l^{\leq t-1}$ to block all the parameters. In learning with partially relaxed masks (LwPRM), it uses $p_l^{\leq t-1}$ instead, which blocks only the parameters that are important for the previous dissimilar tasks. In other words, LwPRM aims to relax the masks for some parameters that are not important for previous tasks.

detection (DTD) and 2) learning with partially relaxed masks (LwPRM). As the proposed method basically follows HAT, we refer to HAT first, and then explain the proposed mechanism in detail.

3.1   Mechanism of HAT

HAT requires every layer to have a task embedding, e_l^t, to control the gradient of the layer's parameters. Each layer's mask, a_l^t, is computed from $a_l^t = \sigma(s e_l^t)$, and each layer's output, o_l, is replaced with $h_l = o_l \otimes a_l^t$, where σ is an activation function (e.g., sigmoid), s is a positive scaling parameter, and $\otimes$ denotes element-wise multiplication. To preserve the information obtained in previous tasks, after learning task t, HAT computes an accumulated mask, $a_l^{\leq t}$, as follows:

$$a_l^{\leq t} = \max\left(a_l^t, a_l^{\leq t-1}\right), \tag{1}$$

using elementwise maximum and the all-zero vector for $a^{\leq 0}$. In learning task $(t+1)$, the gradients of parameters, including e_l^t, are computed by a standard back-propagation, and then reduced based on the accumulated mask:

$$g'_{l,ij} = \left[1 - \min\left(a_{l,i}^{\leq t}, a_{l-1,j}^{\leq t}\right)\right] g_{l,ij}, \tag{2}$$

where unit indices i and j denote the l-th and $(l-1)$-th layer outputs, respectively. $g_{l,ij}$ denotes its gradient. Additionally, HAT utilizes two more tricks to stabilize learning. First, in learning, scaling parameter s is linearly annealed as follows:

$$s = \frac{1}{s_{\max}} + \left(s_{\max} - \frac{1}{s_{\max}}\right)\frac{b-1}{B-1},\tag{3}$$

where $s_{\max}$ is a hyper parameter, the value of which is a large positive number, and b and B denote the batch index and the total number of batches, respectively. In testing, $s_{\max}$ is used instead of s. Second, to alleviate the side effect on embedding gradient compensation, the formula below is used:

$$q'_{l,i} = \frac{s_{\max}\left[\cosh(se^t_{l,i}) + 1\right]}{s\left[\cosh(e^t_{l,i}) + 1\right]}q_{l,i},\tag{4}$$

where $q_{l,i}$ denotes the gradient that corresponds to $e^t_{l,i}$ and is replaced with $q'_{l,i}$.

3.2 Mechanism of PRM

First, the DTD phase aims to obtain the task similarity from the task embeddings that are optimized in the same way as with HAT. Once the optimized task embeddings, which are represented by $\{e^{1:t}_l\}$, where l and t denote the indices of the layer and task, respectively, are obtained, the set of dissimilar tasks, $\mathcal{D}^t_l$, can be computed via a clustering on the embeddings. The reason that we focus on task embeddings to measure task similarity is that since they are used as the basis for masking the output of each layer, if two tasks emphasize similar parameters and try to pass them without blocking (masking), their task embeddings should be similar. Since HAT utilizes accumulated masks to block the gradients of the model's parameters, the task embeddings are more flexibly updated than the model's parameters; thus, we expect the task embeddings to provide an informative representation for task similarity. Second, in the LwPRM phase, the model's parameters are optimized again using $\mathcal{D}^t_l$ with the intention of not only blocking some parameters to overcome forgetting, as with HAT, but also making remaining parameters free for updating to transfer knowledge.

Dissimilar Task Detection (DTD). To balance knowledge transfer and the prevention of forgetting, it is important to determine which tasks are similar and can be transferred to the current task and which tasks are dissimilar and should be blocked so that they are not forgotten. To address this issue, we focus on the task embeddings that are learned through the HAT mechanism. Since the mechanism employs an accumulated mask that reduces the gradients of the model's parameters, the task embeddings can be more easily updated than the model's parameters. Therefore, the task embeddings are expected to provide an informative representation of the tasks and their relations with one another. Specifically, we adopt an unsupervised clustering method for judging task similarity.

First, the model follows the approach of HAT in learning task t by reducing the gradients of the parameters according to (1) and (2), namely, as illustrated in Fig. 1 at the bottom, $a_l^{\leq t-1}$ is used to reduce the gradients for all the previous tasks in the DTD phase. After learning task t, optimized $\{e_l^{1:t}\}$ are obtained. Using a clustering method, the set of previous tasks with embeddings that do not belong to the same cluster as task t are regarded as dissimilar tasks, which are represented by $\mathcal{D}_l^t$. Although any clustering method can be used, we exploit X-means[21] since it does not require the number of clusters as input; instead, it searches for the optimal number of clusters based on the Bayesian information criterion by applying K-means recursively.

Two Stage Optimization. We introduce a new optimization that proceeds in two stages on the DTD phase. The model consists of two parts: the first is the feature extractor that is to be shared across tasks, and the other is the classifier that is built for each task. Therefore, we hypothesize that if both the feature extractor and classifier are optimized simultaneously, the information that represents the difference across tasks and can be used as a clue for task similarity comparison may be dispersed both into not only the task embeddings but also the classifier, which may degrade the performance of our approach. To ensure maximum sharing in the task embedding inside the feature extractor, which will facilitate similarity comparison, we first freeze the feature extractor and learn only the classifier (i.e., depicted as "1st optimization" in Algorithm 1); then, we optimize both the feature extractor and classifier simultaneously, from which task embeddings are obtained for the clustering (i.e., "2nd optimization").

Learning with Partially Relaxed Masks (LwPRM). Although the accumulated mask, $a_l^{\leq t}$, in HAT plays a large role in preventing forgetting, it may also restrain knowledge transfer among similar tasks because the gradients of the parameters are reduced, regardless of task similarity. Thus, following HAT, we extend it so that it can promote knowledge transfer by introducing a new mechanism named "partially relaxed masks" (PRM).

PRM employs masks per previous task like HAT; however, not all masks are used to reduce the gradients of the parameters according to the task similarity. Instead, PRM accumulates only the masks that belong to previous dissimilar tasks so that it can prevent forgetting only for dissimilar tasks while maintaining opportunities for improvement for other tasks at the same time. Namely, the accumulated mask focuses only on dissimilar tasks and is partially relaxed to keep the parameters for other tasks updatable, which can enhance knowledge transfer. Given that we know which tasks are dissimilar to the current new task by clustering, (1) and (2) are replaced as follows:

$$p_l^{\leq t} = \max\left(\{a_l^i \mid i \in \mathcal{D}_l^t, i \leq t\}\right), \quad g'_{l,ij} = \left[1 - \min\left(p_{l,i}^{\leq t}, p_{l-1,j}^{\leq t}\right)\right] g_{l,ij} \quad (5)$$

Table 1. The statistics of each task.

Dataset	# Tasks	# Classes	# Trainings	# Validations	# Tests
CIFAR100-10T	10	10	4500	500	1000
EMNIST-10T	10	5 (Last three: 4)	500	200	200
F-CelebA-10T	10	2	400	40	80
F-EMNIST-10T	10	62	1240	310	310

4 Experiments

4.1 Datasets

We use the following four kinds of datasets to evaluate the performance in terms of both prevention of forgetting and knowledge transfer. The datasets are split into multiple tasks, and the statistics of each task are presented in Table 1.

Dissimilar Tasks: CIFAR100[15], which contains 100 classes, is split into 10 tasks, each of which has 10 classes; the dataset is named CIFAR100-10T. EMNIST[6], which contains 47 classes, is split into 10 tasks, each of which has 5 (the last three tasks have 4) classes; the dataset is named EMNIST-10T. These datasets are expected to be sensitive to forgetting as each task has different classes and there are few relations or similarities across tasks.

Similar Tasks: F-CelebA[17] is a dataset that contains face images of celebrities and labels that indicate whether or not they are smiling. Different celebrities correspond to different tasks, and 10 celebrities are used in the experiments; the dataset is named F-CelebA-10T. F-EMNIST[17] is a dataset that contains 62 classes of character images handwritten by different users. We use the images that correspond to 10 writers; the dataset is named F-EMNIST-10T. These tasks are supposed to have shared knowledge across tasks as each task has the same set of labels and the data that are naturally similar.

We conduct experiments with three kinds of sequences that combine at most two different tasks in random order, as presented in Table 2, Table 3, and Table 4: **only dissimilar tasks - #1 and #2, only similar tasks - #3 and #4**, and **mixed of dissimilar and similar tasks - #5 and #6.**

4.2 Baselines

We compare PRM with classic and latest continual learning methods that can work as TIL systems, namely, EWC[14] in the HAT package (EHAT), ACL[8], PathNet[9] (PNT), SupSup[27] (SS), HyperNet[19] (HYP), HAT[25] and CAT[13]. Since HAT focuses only on preventing forgetting and does not have any mechanism for knowledge transfer, it is expected to perform poorly on similar tasks. To the best of our knowledge, CAT is the only approach that focuses on a mixed sequence of similar and dissimilar tasks; however, CAT takes much longer to learn. Also, we prepare two reference methods: naive continual learning (NCL) and single-task learning (STL). NCL learns a new task without considering previous tasks; thus, severe forgetting is expected to occur for dissimilar tasks. STL learns all the tasks at once. Although it does not follow the continual learning scenario, it is expected to be the upper bound, only for dissimilar tasks.

4.3   Implementation Details

The input go through two fully connected layers passing ReLU and dropout layers. The networks are optimized by minimizing the last classifier's cross-entropy loss using SGD. The learning rate starts from 0.025 and is gradually reduced until it reaches 0.001. With no improvement in the validation loss for 5 epochs, the training stops. The batch size and s_{max} are set to 64 and 400, respectively. Other hyper parameters, such as the dropout rate and the weight of regularization, are searched over 20 trials using Tree-structured Parzen Estimator (TPE)[2], which is implemented in Optuna[1]. The baselines are evaluated on the original code with modifications while aligning with our setting as much as possible.

4.4   Metrics

Accuracy (Acc): The average accuracy for all tasks after learning them, where the model is optimized with the best hyper parameters that are found in the search. **Parameter Sensitivity (PS):** The standard deviation of "Acc" with varied hyper parameters over 20 searches. **Forward Transfer (FWT):** The test accuracy of task i just after learning task i is compared to the accuracy for task i by STL, which is expressed as $1/T \sum_t^T (\alpha_t^t - \tilde{\alpha}_t)$, where α_i^j denotes the evaluated accuracy of task i after learning task j, $\tilde{\alpha}_i$ denotes the test accuracy for task i by STL, and T denotes the total number of tasks. **Backward Transfer (BWT):** The average of improvement from each task's initial accuracy to the final accuracy, which is represented by $1/T \sum_t^T (\alpha_t^T - \alpha_t^t)$[18]. Negative values represent that forgetting occurs. **Computation Time (CT):** The total computation time for learning all the tasks, which is measured in seconds.

4.5   Results

The results are presented in Table 2, Table 3, and Table 4. Each row presents to the average results over the same set of three random task sequences.

Only Dissimilar Tasks (#1 and #2): NCL causes severe forgetting (−3.4% and −5.5%, as presented in BWT). PRM achieves competitive accuracy compared to the baselines (especially PNT and CAT) without much forgetting. Also, PRM requires only one-tenth the computation time of CAT in #2. Among the baselines, PRM achieves almost the best score in the second shortest time.

Table 2. Results for only dissimilar tasks sequences.

	(#1) EMNIST-10T				(#2) CIFAR100-10T			
	Acc	PS	CT	FWT/BWT	Acc	PS	CT	FWT/BWT
(STL)	0.929	0.6%	45	-/-	0.612	1.4%	142	-/-
NCL	0.879	0.3%	26	-1.7%/-3.4%	0.534	1.1%	123	-2.3%/-5.5%
ACL	0.902	0.5%	916	-2.6%/-0.1%	0.508	0.6%	6328	-10.3%/-0.1%
PNT	**0.910**	0.3%	43	-2.0%/0.0%	0.571	0.5%	394	-4.1%/0.0%
SS	0.829	11.9%	172	-5.6%/-4.4%	0.462	11.2%	2800	-10.8%/-4.2%
HYP	0.822	10.9%	1105	-10.7%/-0.1%	0.219	3.4%	10825	-38.8%/-0.5%
HAT	0.905	0.3%	101	-2.4%/0.0%	0.582	0.8%	912	-3.0%/0.0%
EHAT	0.899	0.3%	187	-3.1%/0.0%	0.578	0.6%	1011	-3.4%/0.0%
CAT	0.907	0.3%	1276	-2.2%/0.0%	**0.587**	0.7%	6018	-2.5%/0.0%
PRM	0.897	0.4%	80	-2.3%/-1.0%	0.582	0.9%	635	-2.8%/-0.1%

Table 3. Results for only similar tasks sequences.

	(#3) F-EMNIST-10T				(#4) F-CelebA-10T			
	Acc	PS	CT	FWT/BWT	Acc	PS	CT	FWT/BWT
(STL)	0.717	3.2%	126	-/-	0.823	0.6%	15	-/-
NCL	0.654	10.0%	48	-5.3%/ 0.9%	0.820	1.2%	14	-3.7%/3.4%
ACL	0.043	0.9%	1633	-66.4%/-0.9%	0.695	2.4%	628	-14.3%/1.5%
PNT	0.572	15.7%	178	-14.5%/0.0%	0.716	1.1%	32	-10.6%/0.0%
SS	0.456	14.6%	1022	-14.3%/-11.7%	0.780	9.5%	440	-5.0%/0.8%
HYP	0.060	1.5%	3313	-65.5%/-0.2%	0.525	1.5%	1495	-26.6%/-3.2%
HAT	0.655	3.7%	245	-6.2%/0.0%	0.759	1.1%	70	-6.3%/0.0%
EHAT	0.655	3.8%	497	-6.1%/0.0%	0.769	0.8%	110	-5.3%/0.0%
CAT	0.643	2.1%	2171	-7.4%/0.0%	0.781	0.7%	707	-4.1%/0.0%
PRM	**0.657**	3.3%	176	-5.7%/-0.2%	**0.796**	1.4%	54	-5.5%/2.8%

Only Similar Tasks (#3 and #4): As the tasks are similar, NCL does not cause much forgetting, and even improves task by task, as presented in BWT. In both sequences, PRM outperforms all baselines without forgetting and even with significant backward transfer, as shown in #4. As in the case of only dissimilar tasks, which is presented in Table 2, PRM's efficiency is only behind PNT, but PNT's performance is markedly lower than PRM in these only similar tasks.

Mixed of Dissimilar and Similar Tasks (#5 and #6): While PRM achieves the best performance in #6, it underperforms CAT in #5. However, the performance of CAT is highly sensitive to the hyper parameters, according to the PS value of 7.2%, while PRM has 3.4% PS, which is the most stable among the baselines. Moreover, CAT requires a long computation time as shown in Fig. 2, where the computation time for each task and the accumulated time

Table 4. Results for mixed of dissimilar and similar tasks sequences.

	(#5) EMNIST-10T & F-EMNIST-10T				(#6) CIFAR100-10T & F-CelebA-10T			
	Acc	PS	CT	FWT/BWT	Acc	PS	CT	FWT/BWT
(STL)	0.791	1.4%	113	-/-	0.629	1.1%	254	-/-
NCL	0.728	5.0%	87	-5.6%/-0.6%	0.540	1.0%	154	-3.1%/-5.9%
ACL	0.370	9.3%	5437	-41.6%/-0.5%	0.521	0.4%	7515	-10.9%/0.0%
PNT	0.628	4.7%	281	-16.3%/0.0%	0.556	0.7%	425	-7.3%/0.0%
SS	0.572	16.2%	586	-7.6%/-14.3%	0.461	10.5%	1865	-11.1%/-5.7%
HYP	0.322	6.0%	6439	-46.2%/-0.7%	0.199	1.7%	19973	-42.2%/-0.8%
HAT	0.721	3.6%	357	-7.0%/0.0%	0.572	0.5%	891	-5.8%/0.0%
EHAT	0.728	6.9%	854	-6.3%/0.0%	0.572	0.6%	1801	-5.8%/0.0%
CAT	**0.737**	7.2%	9851	-5.3%/0.0%	0.581	0.8%	22810	-4.8%/0.0%
PRM	0.720	3.4%	310	-6.8%/-0.3%	**0.582**	0.6%	642	-4.2%/-0.6%

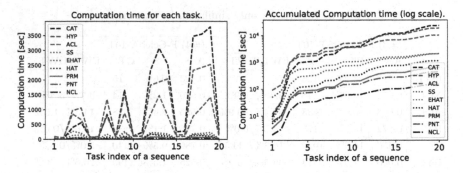

Fig. 2. Computation times for one sequence of #6. The right figure is plotted in log scale. CAT, HYP and ACL take more time, while PRM requires less time.

for all tasks in #6 are plotted. According to these figures, CAT takes much longer than the others, and learning more tasks requires more time. Based on these observations, in practice, it is difficult to use CAT when there is a large amount of data or many tasks, and due to its unstable performance, tuning is essential. In contrast, PRM needs much less computation time, e.g., 1/35 that of CAT when learning 20 tasks, and its performance is also stable with varied hyper parameters.

4.6   Ablation Study

We check how much the DTD phase influences the total performance, as LwPRM completely depends on its behavior. The results are presented in Table 5. PRM(S) and PRM(D) indicate the cases in which all previous tasks are regarded as similar and dissimilar, respectively, instead of actual clustering. PRM(T) represents the cases in which the types of tasks are given and used as a replacement for clustering (e.g., when learning a new task of CIFAR100-10T, the model can

Table 5. Results of ablation experiments.

	PRM(S)	PRM(D)	PRM(T)	PRM	w/o 2SO
(#1) EMNIST-10T	0.882	**0.904**	–	0.897	0.897
(#2) CIFAR100-10T	0.562	0.581	–	**0.582**	0.579
(#3) F-EMNIST-10T	0.633	0.654	–	**0.657**	0.658
(#4) F-CelebA-10T	**0.825**	0.759	–	0.796	0.775
(#5) EMNIST-10T & F-EMNIST-10T	0.713	**0.731**	0.720	0.720	0.721
(#6) CIFAR100-10T & F-CelebA-10T	0.568	0.571	0.567	**0.582**	0.586

tell which previous tasks come from CIFAR100-10T, and only these tasks are regarded as similar). When the sequence of tasks consists only of tasks from the same dataset (#1 to #4), the behavior of PRM(T) is the same as that of PRM(S). Additionally, we check the effect of the two stage optimization, as shown in the column of "w/o 2SO", where both the feature extractor and classifier are optimized simultaneously in the DTD phase.

Notably, PRM(T), where the types of tasks are given, does not always show the best performance. Instead, PRM, which employs task embeddings through clustering, can utilize not explicit but implicit relation across tasks, thereby resulting in similar or better performance than PRM(T). Conversely, it is reasonable that PRM(D) has the best performance in #1 and PRM(S) performs the best in #4. Although PRM(D) performs the best in #5, the performance differences among other the types of PRM are small. Generally, PRM treats well task embeddings via clustering to handle various types of data sequences. Moreover, it is demonstrated by comparing "PRM" and "w/o 2SO" that the two stage optimization contributes PRM's performance especially in #4, which is consistent with our hypothesis that it can facilitate the similarity comparison.

4.7   Limitations

While PRM tries to open up masks for knowledge transfer and achieves better transfer in similar tasks as shown in Table 3, there is room to employ another explicit mechanism, such as attention. However, it is still unclear which combinations perform best. Additionally, as presented in Table 5, the current PRM is sometimes outperformed by PRM(S) and PRM(D), which may indicate that with a more effective method for utilizing task embeddings, it will be possible to improve its performance. The effectiveness of exploiting task embeddings is proven in most cases; however, a more effective approach needs to be developed.

5   Conclusion

To extend HAT so that it can not only overcome catastrophic forgetting but also transfer knowledge, the current study makes two contributions. First, we discover that the task embeddings optimized by parameter masking approaches, such as

HAT, provide an informative representation for the task similarity. Second, we propose a new approach, namely PRM, that controls which parameters should be blocked or relaxed based on the task similarity that is obtained via clustering on the task embeddings. The experiments show that PRM achieves at least competitive performance in terms of both prevention of forgetting and knowledge transfer compared to the latest baselines with much less computation time.

References

1. Akiba, T., Sano, S., Yanase, T., Ohta, T., Koyama, M.: Optuna: a next-generation hyperparameter optimization framework. In: Proceedings of SIGKDD (2019)
2. Bergstra, J., Bardenet, R., Bengio, Y., Kégl, B.: Algorithms for hyper-parameter optimization (2011)
3. Borsos, Z., Mutný, M., Krause, A.: Coresets via bilevel optimization for continual learning and streaming. In: Proceedings of NeurIPS (2020)
4. Chaudhry, A., Gordo, A., Dokania, P.K., Torr, P., Lopez-Paz, D.: Using hindsight to anchor past knowledge in continual learning. In: Proceedings of AAAI (2021)
5. Chaudhry, A., Ranzato, M., Rohrbach, M., Elhoseiny, M.: Efficient lifelong learning with A-GEM. In: Proceedings of ICLR (2019)
6. Cohen, G., Afshar, S., Tapson, J., van Schaik, A.: EMNIST: extending MNIST to handwritten letters. In: Proceedings of IJCNN (2017)
7. Delange, M., et al.: A continual learning survey: defying forgetting in classification tasks. IEEE Trans. Pattern Anal. Mach. Intell. (2021)
8. Ebrahimi, S., Meier, F., Calandra, R., Darrell, T., Rohrbach, M.: Adversarial continual learning. In: Proceedings of ECCV (2020)
9. Fernando, C., et al.: PathNet: evolution channels gradient descent in super neural networks (2017)
10. Goodfellow, I.J., Mirza, M., Xiao, D., Courville, A., Bengio, Y.: An empirical investigation of catastrophic forgetting in gradient-based neural networks. In: Proceedings of ICLR (2014)
11. Hu, W., et al.: Overcoming catastrophic forgetting for continual learning via model adaptation. In: Proceedings of ICLR (2018)
12. Javed, K., White, M.: Meta-learning representations for continual learning. In: Proceedings of NeurIPS (2019)
13. Ke, Z., Liu, B., Huang, X.: Continual learning of a mixed sequence of similar and dissimilar tasks. In: Proceedings of NeurIPS (2020)
14. Kirkpatrick, J., et al.: Overcoming catastrophic forgetting in neural networks. Proc. Natl. Acad. Sci. **114**, 3521–3526 (2017)
15. Krizhevsky, A., Hinton, G., et al.: Learning multiple layers of features from tiny images. Technical report (2009)
16. Li, Z., Hoiem, D.: Learning without Forgetting. In: Proceedings of ECCV (2016)
17. Liu, Z., Luo, P., Wang, X., Tang, X.: Deep learning face attributes in the wild. In: Proceedings of ICCV (2015)
18. Lopez-Paz, D., Ranzato, M.: Gradient episodic memory for continual learning. In: Proceedings of NeurIPS (2017)
19. von Oswald, J., Henning, C., ao Sacramento, J., Grewe, B.F.: Continual learning with hypernetworks. In: Proceedings of ICLR (2020)
20. Parisi, G.I., Kemker, R., Part, J.L., Kanan, C., Wermter, S.: Continual lifelong learning with neural networks: a review. Neural Netw. **113**, 54–71 (2019)

21. Pelleg, D., Moore, A.W., et al.: X-means: extending K-means with efficient estimation of the number of clusters. In: Proceedings of ICML (2000)

22. Pellegrini, L., Graffieti, G., Lomonaco, V., Maltoni, D.: Latent replay for real-time continual learning. In: Proceedings of IROS (2020)

23. Riemer, M., et al.: Learning to learn without forgetting by maximizing transfer and minimizing interference. In: Proceedings of ICLR (2019)

24. Rusu, A.A., et al.: Progressive neural networks (2016)

25. Serra, J., Suris, D., Miron, M., Karatzoglou, A.: Overcoming catastrophic forgetting with hard attention to the task. In: Proceedings of ICML (2018)

26. Singh, P., Verma, V.K., Mazumder, P., Carin, L., Rai, P.: Calibrating CNNs for lifelong learning. In: Proceedings of NeurIPS (2020)

27. Wortsman, M., et al.: Supermasks in superposition. In: Proceedings of NeurIPS (2020)

28. Zenke, F., Poole, B., Ganguli, S.: Continual learning through synaptic intelligence. In: Proceedings of ICML (2017)

Dual-State Knowledge Tracing Model with Mutual Information Maximization

Haodong Meng[1], Changzhi Chen[2], Hongyu Yi[2], and Xiaofeng He[1,3](✉)

[1] School of Computer Science and Technology, East China Normal University, Shanghai, China
51205901079@stu.ecnu.edu.cn, hexf@cs.ecnu.edu.cn
[2] Born to Learn Education Technology, Chengdu, China
{ccz,yhy}@sxw.cn
[3] Shanghai Key Laboratory of Trustworthy Computing, Shanghai, China

Abstract. Knowledge tracing aims to trace students' knowledge states and predict their future performance based on their historical learning processes. Most existing methods of characterizing a student's state are not effective enough, using only global representation or knowledge concept level representation. Such representation methods cannot consider the characteristics of knowledge concepts and the relations between concepts at the same time. In this paper, we propose a Dual-State Knowledge Tracing (DSKT) Model with Mutual Information Maximization. DSKT uses dynamic routing to extract knowledge commonalities from original knowledge concepts, updates the knowledge state at the concept and commonality levels, and predicts future performance by fusing two states. In addition, to incorporate the relationship between exercises and knowledge concepts, we use the principle of mutual information maximization to learn their representations. Extensive experimental results show the effectiveness of our model.

Keywords: Knowledge tracing · Dynamic routing · Mutual information maximization · Knowledge commonality

1 Introduction

Online education has gradually become popular in recent years. Tracking the students' knowledge mastery level plays an important role in online education so that education platforms can provide students with more personalized learning schedules. Knowledge tracing (KT) [5] is designed to model the students' learning states and predict their future learning performance by analyzing lots of historical logs produced during the learning process.

Although there has been considerable research on knowledge tracing [9,17], their abilities to characterize students' states are still limited, and these methods can be mainly divided into two categories. The first kind of method is to represent the learner's knowledge state at the latent global level, such as DKT [18] summarizes the overall state of the student with the hidden state in RNN.

J. Gama et al. (Eds.): PAKDD 2022, LNAI 13280, pp. 380–392, 2022.
https://doi.org/10.1007/978-3-031-05933-9_30

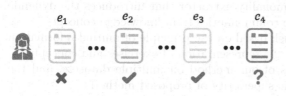

Exercise	Knowledge Concept
e_1	Ordering Integer
e_2	Ordering Fraction; Fraction Division;
e_3	Ordering Real Number; Mean;
e_4	Ordering Integer

Fig. 1. A case of a student's exercising sequence. She initially answered e_1 about *"Ordering Integer"* incorrectly, but in the follow-up learning process, she answered the questions about *"Ordering Fraction"* and *"Ordering Real Number"* correctly, we could infer the correct response on e_4 with a great probability.

Practically, a straightforward idea of judging how a student will perform on a future exercise is to consider how well he has mastered the knowledge concept contained in that exercise. However, the abstract fusion state representation cannot give such an intuitive modeling process. DKVMN [26] extends the single hidden state into several latent memory slots, but still lacks the use of correspondence between exercises and knowledge concepts. Another type of method is to represent the learner's knowledge state at the concept level such as BKT [5] uses a corresponding binary variable to indicate whether the student has mastered each knowledge concept, but modeling knowledge states separately for each knowledge concept tends to ignore the correlation between knowledge concepts. To illustrate this, we give a case in Fig. 1. A certain concept may not be practiced by the student for a relatively long period, the results of the student's interactions at other concepts will affect the state of this concept, the concept-level state is hard to capture this effect. Some works like GKT [13], SKT [23] have tried to solve this problem by modeling on concept graphs, but the effect is not satisfactory due to the lack of explicit knowledge structure annotation.

To address the above problems, in this paper, we propose a novel knowledge tracing method called Dual-State Knowledge Tracing (DSKT) Model with Mutual Information Maximization to model the evolution of the student's knowledge state at different levels simultaneously. To capture the relevance between knowledge concepts and incorporate this relevance when updating the knowledge state, we use dynamic routing [19] to discover commonalities between knowledge concepts and model the state evolution on them. The concept-level state provides a more refined tracing capability and the commonality level state representation can take into account the associations between knowledge skills. Besides, to introduce the many-to-many relationship between exercises and concepts into their representations, we minimize the InfoNCE [14] loss based on the idea of maximizing mutual information between exercises and concepts. In summary, the contributions of this paper are as follows:

1) We propose a new method to depict the student's abilities by simultaneously modeling the student's knowledge state at the specific knowledge concept level and abstract knowledge commonality level.

2) We design a knowledge commonality extractor that introduces the dynamic routing to extract knowledge commonalities from knowledge concepts.
3) We propose a pre-training task based on the principle of mutual information maximization [7] to enhance the representation of exercises and concepts.
4) We validate the effectiveness of our method on multiple datasets, and the experimental results show the superiority of proposed method.

2   Related Work

Early studies on knowledge tracing are based on traditional machine learning methods. BKT [5] is the most representative work among them, which uses the hidden Markov model to track the student's learning state and models the student's mastery as a binary variable. In addition, some factor analysis models based on logistic functions, such as LFA [1] and PFA [16], evaluate the probabilities of students answering questions correctly by considering some factors (e.g. the number of correct responses and incorrect responses) in the learning process.

The methods of deep learning have been introduced into the field of knowledge tracing in recent years. DKT [18] uses a recurrent neural network for knowledge tracing for the first time and represents the knowledge state of the student by the hidden state. Many subsequent works have made extensions and improvements to DKT, such as considering students' forgetting process [12] or incorporating some feature engineering [27]. DKVMN [26] introduces the key-value memory network [11] into the KT task, which maintains a key matrix and a value matrix, and uses read and write operations to store and update the state. These works do not utilize the text information of the exercise. EERNN [22] uses a bidirectional LSTM [8] to learn semantic information in text of the question to enhance the performance of the model. The outstanding performance of Transformer [24] in the field of NLP also attracted the attention of researchers. SAKT [15] introduces the multi-head self-attention mechanism into the KT task for the first time, and SAINT [3] directly applies the encoder and decoder structures of the Transformer to the model. In addition, in order to use the structural information between the knowledge concepts, GKT [13] uses the potential graph structure formed by the knowledge concepts to model the temporal knowledge state, SKT [23] introduces the transfer of knowledge theory [21] on this basis, emphasizing the influence propagate between related knowledge concepts.

Previous methods used to represent the learner's state cannot fully consider the specificity of each concept and the potential relationship between them. Our work tries to solve this problem by fusing the knowledge state at different levels.

3   Problem Formulation

In knowledge tracing, we have an exercise set $\mathcal{E}$ containing M exercises, a knowledge concept set $\mathcal{C}$ containing N concepts, and an exercise-concept correlation matrix $\mathbf{Q} \in \mathbb{R}^{M \times N}$ consisting only of zeros and ones, If exercise i contains knowledge concept j, $\mathbf{Q}_{ij} = 1$; otherwise $\mathbf{Q}_{ij} = 0$. A student's history record of doing exercises is $\mathcal{X} = \{(e_1, r_1), (e_2, r_2), ..., (e_t, r_t)\}$, where $e_t \in \mathcal{E}$ means that the student answered the exercise e_t at time step t, $r_t = 1$ means a correct response and $r_t = 0$ means an incorrect one.

The knowledge tracing task can be formalized as: At time step T, given an exercise e_T, we will predict the student's performance on this exercise based on the past performance up to time $T - 1$, that is, $p(r_T = 1 | e_T, \mathcal{X}_{T-1})$, which means the probability of answering e_T correctly.

4   Method

In this section, we will present the proposed DSKT model in detail. The general framework of the model is shown as Fig. 2.

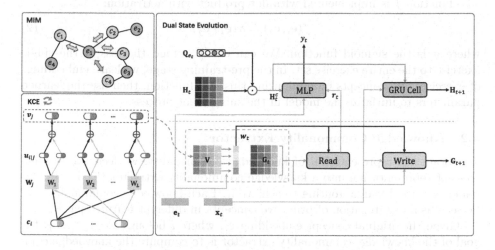

Fig. 2. The architecture of the DSKT model.

In DSKT, the MIM module aims to obtain pre-trained representations of exercises and knowledge concepts based on the principle of mutual information maximization. The KCE module uses dynamic routing to extract knowledge commonalities from the original knowledge concepts. Then, in the dual state evolution module, a GRU [2] unit is used to model the changes in the student's concept-level state (noted as concept state). At the same time, we use the *read* and *write* process to maintain and update the student's knowledge commonality-level state (noted as commonality state).

4.1   Representation Learning with MIM

Exercises and knowledge concepts can be regarded as two views of the content that the student interacts with. For each exercise, the knowledge concepts that it contains are generalizations of its characteristics. Likewise, for a knowledge concept, all the exercises associated with it are its concrete presentation.

Mutual information measures the dependence between two random variables, but it is not easy to calculate directly in practice. Previous work [14] established a connection between mutual information maximization and InfoNCE loss. Given an exercise i, the set of knowledge concepts it contains is $C_i = \{j|Q_{ij} = 1\}$. Now, let $\mathbf{e}_i \in \mathbb{R}^d$ denote the embedding of exercise i and $\mathbf{c}_j \in \mathbb{R}^d$ denote the embedding of knowledge concept j. With the help of InfoNCE loss, we can maximize the mutual information of the exercise and concept representations. That is, we minimize the loss function as follows:

$$L_{MIM}(i) = -\sum_{j \in C_i} [log \frac{f(\mathbf{e}_i, \mathbf{c}_j)}{\sum_{k \in C} f(\mathbf{e}_i, \mathbf{c}_k)}] \tag{1}$$

The function f is implemented with dot product with activation:

$$f(\mathbf{e}_i, \mathbf{c}_j) = \sigma(\mathbf{e}_i \cdot \mathbf{c}_j) \tag{2}$$

where σ is the sigmoid function. We can easily extend the loss of the single exercise to the entire exercise set. In the pre-training stage, we learn embeddings of exercises and concepts by minimizing this loss function, then use the learned parameters to initialize the model in the subsequent process.

4.2 Knowledge Commonality Extractor

To obtain the representation of knowledge commonality, we feed the original concept representation into a knowledge commonality extractor (KCE). In this paper, we use dynamic routing to achieve it. Each knowledge commonality can be seen as an aggregation of primitive concepts in different feature spaces.

Given the original concept embedding $\mathbf{c}_i$, where i belongs to $\{1, ..., N\}$, the goal of the knowledge commonality extractor is to compute the knowledge commonality matrix $\mathbf{V} \in \mathbb{R}^{L \times d}$, where L is the number of knowledge commonalities.

KCE is carried out in an iterative manner. We will first calculate the linear transformation of each knowledge concept:

$$\mathbf{u}_{i|j} = \mathbf{W}_j \mathbf{c}_i \tag{3}$$

The transformation matrix $\mathbf{W}_j \in \mathbb{R}^{d \times d}$ transforms the concept $\mathbf{c}_i$ into the potential contributor of the jth knowledge commonality.

Then, at r-th iteration, the candidate knowledge commonality vector $\tilde{\mathbf{v}}_j^r$ will be computed as a weighted sum of all the transformations of concepts:

$$\tilde{\mathbf{v}}_j^r = \sum_{i=1}^{M} d_{ij}^r \mathbf{u}_{i|j} \tag{4}$$

A normalization operation is applied to $\tilde{\mathbf{v}}_j^r$ to obtain the bounded results of this iteration:

$$\mathbf{v}_j^r = \frac{\tilde{\mathbf{v}}_j^r}{\|\tilde{\mathbf{v}}_j^r\|} \tag{5}$$

d_{ij}^r is the coupling coefficient, calculated as:

$$d_{ij}^r = Softmax(b_{ij}^r)$$

$$= \frac{exp\ b_{ij}^r}{\sum_{i=1}^{M} exp\ b_{ij}^r} \tag{6}$$

b_{ij}^0 is initialized to zero before the first iteration and accumulates the agreement of the candidate knowledge commonality and the transformations of concepts at the end of each iteration:

$$b_{ij}^{r+1} = b_{ij}^r + \mathbf{u}_{i|j} \cdot \mathbf{v}_j^r \tag{7}$$

After multiple iterations of the above process, we will finally get the commonality matrix $\mathbf{V}$ by stacking all commonality vectors $\mathbf{v}_j$ of the last iteration.

4.3   Dual State Evolution

In DSKT, at time step t, we assume that students have two kinds of knowledge state, concept state $\mathbf{H}_t \in \mathbb{R}^{N \times d}$ and commonality state $\mathbf{G}_t \in \mathbb{R}^{L \times d}$. In this section, we will combine the two states to predict student performance and update them dynamically.

Performance Prediction. The embedding of the input exercise $\mathbf{e}_t \in \mathbb{R}^d$ will be dot-producted with the knowledge commonality matrix to get its importance on each knowledge commonality. After a softmax unit, the final set of weights for the exercise is obtained:

$$w_t^i = Softmax(\mathbf{V}^i \cdot \mathbf{e}_{e_t}) \tag{8}$$

where $\mathbf{V}^i$ is the i-th row of the knowledge commonality matrix. Using this set of weights, we can *read* the commonality state related to the exercise from the commonality state matrix $\mathbf{G}_t$:

$$\mathbf{r}_t = \sum_{i=1}^{L} w_t^i \mathbf{G}_t^i \tag{9}$$

Suppose that the set of knowledge concepts of size n corresponding to the exercise e_t is C_{e_t}. The probability of a student answering e_t correctly is

$$y_t = \frac{1}{n} \sum_{c \in C_{e_t}} \sigma(\mathbf{W}_p(\mathbf{s}_t^c \oplus \mathbf{e}_{e_t}) + \mathbf{b}_p) \tag{10}$$

$$\mathbf{s}_t^c = \sigma(\mathbf{W}_s(\mathbf{H}_t^c \oplus \mathbf{r}_t) + \mathbf{b}_s) \tag{11}$$

where $\mathbf{H}_t^c \in \mathbb{R}^d$ indicates the concept state corresponding to concept c (i.e., the c-th row of $\mathbf{H}_t$), $\oplus$ denotes the concatenation operator, $\mathbf{W}_p \in \mathbb{R}^{2d \times 1}$, $\mathbf{W}_s \in \mathbb{R}^{2d \times d}$ are weight matrices, $\mathbf{b}_p \in \mathbb{R}^1$, $\mathbf{b}_s \in \mathbb{R}^d$ are bias terms.

State Update Module. At each time step, the embedding $\mathbf{x}_t$ of the student's interaction (e_t, r_t) is constructed as:

$$\mathbf{x}_t = \begin{cases} \mathbf{e}_{e_t} \oplus \mathbf{1}, & if\ r_t = 1 \\ \mathbf{e}_{e_t} \oplus \mathbf{0}, & if\ r_t = 0 \end{cases} \tag{12}$$

where $\mathbf{1} = (1, 1, ..., 1)$ with the same dimension of $\mathbf{e}_{e_t}$, and $\mathbf{0}$ has a similar form. For the concept state, we use a GRU unit to update it as:

$$\mathbf{H}_{t+1}^c = \begin{cases} GRU(\mathbf{H}_t^c, \mathbf{x}_t), & if\ c \in C_{e_t} \\ GRU(\mathbf{H}_t^c, \mathbf{0}), & otherwise \end{cases} \tag{13}$$

For the commonality state, we update it with *write* process. To forget the historical information in the commonality state and introduce the latest information, an erase vector $\mathbf{z}_t$ and an add vector $\mathbf{a}_t$ are calculated as:

$$\mathbf{z}_t = Sigmoid(\mathbf{W}_z\mathbf{x}_t + \mathbf{b}_z) \tag{14}$$

$$\mathbf{a}_t = Sigmoid(\mathbf{W}_a\mathbf{x}_t + \mathbf{b}_a) \tag{15}$$

the new commonality state matrix $\mathbf{G}_{t+1}$ will be updated as follows:

$$\tilde{\mathbf{G}}_{t+1}^i = \mathbf{G}_t^i[\mathbf{1} - w_t^i\mathbf{z}_t] \tag{16}$$

$$\mathbf{G}_{t+1}^i = \tilde{\mathbf{G}}_{t+1}^i + w_t^i\mathbf{a}_t \tag{17}$$

where $\mathbf{W}_z, \mathbf{W}_a \in \mathbb{R}^{2d \times d}$, $\mathbf{b}_z, \mathbf{b}_a \in \mathbb{R}^d$ are weight matrices and bias items.

4.4   Training

According to the prediction of the probabilities of the student answering the question correctly at each step, all trainable parameters will be trained by minimizing the binary cross-entropy loss:

$$loss = -\sum_t (r_t log\ y_t + (1 - r_t) log\ (1 - y_t)) \tag{18}$$

5   Experiment

In this section, we carry out experiments to verify the effectiveness of DSKT.

5.1   Dataset

Our experiments are conducted on the following four datasets: ASSIST-ments2009[1], ASSISTments2012[2], ASSISTments Challenge[3], Ednet[4], and their detailed statistical information is shown in Table 1.

Table 1. Statistics of datasets.

	ASSIST09	ASSIST12	ASSISTCha	Ednet
Students	4,217	46,674	1,709	784,309
Exercises	26,688	179,999	3,162	13,169
Concepts	123	265	102	188
Responses	346,860	6,123,270	942,816	95,293,926
Avg. skills per exercise	1.197	1.000	1.036	2.260

The ASSISTments2009 and ASSISTments2012 datasets are derived from student interaction logs collected from online educational platforms in the school year 2009–2010 and 2012–2013. The ASSISTments Challenge was used in the 2017 ASSISTments Datamining Competition, which is collected from students' use of the ASSISTments blended learning platform in middle school from 2004–2007. EdNet is a large-scale hierarchical dataset consisting of student interaction logs collected over more than 2 years from *Santa* [4]. Due to the large scale of the EdNet, we randomly selected 5000 students' data to use.

For all the above datasets, we remove the sequences with lengths less than 5 and the exercises without knowledge concept annotation. In addition, instead of dividing a question with multiple knowledge concepts into multiple questions and simply treating the concept as the question like some previous works [18, 26], which will introduce unreasonable noise, we use the original question sequence as input. We split 80% of the dataset as the training set and 20% as the test set, five-fold cross-validation is applied on the training set to select the hyperparameter setting. The area under the receiver operating characteristics curve (AUC) and root mean squared error (RMSE) are used as the evaluation metrics.

5.2   Performance Comparison

We compare the performance of DSKT with the following models:

1) **BKT** [5] models the students' skill mastery levels as binary variables and traces them with hidden Markov model.

---

[1] https://sites.google.com/site/assistmentsdata/home/assistment-2009-2010-data.

[2] https://sites.google.com/site/assistmentsdata/home/2012-13-school-data-with-affect.

[3] https://sites.google.com/view/assistmentsdatamining/dataset.

[4] https://github.com/riiid/ednet.

2) **DKT** [18] utilizes recurrent neural network to model the student's learning process and uses hidden layer vectors to represent the knowledge states.
3) **DKT+** [25] tries to solve the problem of failing to reconstruct the observed input and inconsistent predicted performance across time-steps in DKT.
4) **DKVMN** [26] uses a key-value memory network to store and update students' knowledge levels.
5) **SAKT** [15] applies the multi-head self-attention mechanism to the knowledge tracking task.
6) **AKT** [6] uses a novel monotonic attention mechanism to capture relations between questions that the learner has interacted with in the past.
7) **LPKT** [20] monitors students' knowledge states from the perspective of learning gains, and considers the impact of answer time and interval time.

Table 2. Results of the KT methods on student performance prediction.

Method	ASSIST09		ASSIST12		ASSISTCha		Ednet	
	AUC	RMSE	AUC	RMSE	AUC	RMSE	AUC	RMSE
BKT	0.6308	0.4849	0.5912	0.5403	0.5767	0.4917	0.5409	0.4880
DKT	0.7015	0.4538	0.7316	0.4253	0.7564	0.4443	0.6892	0.4542
DKT+	0.7144	0.4536	0.7201	0.4281	0.7581	0.4404	0.6927	0.4522
DKVMN	0.6917	0.4511	0.6956	0.5216	0.7219	0.4571	0.6933	0.4535
SAKT	0.6385	0.4875	0.7179	0.4281	0.7129	0.4626	0.6640	0.4631
AKT	0.7426	0.4361	0.7417	0.4197	0.7497	0.4470	0.7192	0.4437
LPKT	–	–	0.6894	0.4368	0.6976	0.4652	0.6205	0.4674
DSFKT	**0.7753**	**0.4174**	**0.7734**	**0.4070**	**0.7671**	**0.4376**	**0.7639**	**0.4282**

The experimental results are shown in Table 2. As we can see, our proposed DSKT model performs the best on all four datasets, indicating the effectiveness of our model in terms of student performance prediction ability. Specifically, models based on deep learning such as DKT outperform traditional machine learning methods like BKT, which shows the advantage of deep learning for knowledge tracing. Besides, our model outperforms DKT and DKVMN, indicating the limitation of using only abstract fusion state to represent students' mastery levels. In addition, DSKT performs better than models based on the self-attentive mechanism. The possible reason is that DSKT considers the state at concept level. Even if the two exercises are a long time apart, DSKT can directly establish a connection through their shared concepts without relying on the attention mechanism. Although LPKT additionally models the time effects on learning gain and forgetting, it still performs worse than our model due to the lack of key information of the concepts contained in exercises. It is worth noting that although DSKT is designed to enhance the representation of questions containing multiple knowledge concepts, it still performs well in cases where the question contains only a single knowledge concept as the result shown on ASSISTment2012.

5.3   Ablation Analysis

In order to investigate the effectiveness of each module in DSKT, we construct several variants of DSKT and conduct some ablation experiments.

1) **DSKT/MIM** removes the pre-training process of embeddings of exercises and concepts, uses an embedding layer to learn the representation automatically.
2) **DSKT/CS** removes the knowledge commonality state, i.e., only the student's state at each knowledge concept is considered.
3) **DSKT/KS** removes the knowledge states at knowledge concept level.
4) **DSKT/DR** removes the dynamic routing-based knowledge commonality extractor module and uses randomly initialized knowledge commonality representation.

Table 3. Ablation study performance (AUC) of DSKT.

Method	ASSIST09	ASSIST12	ASSISTCha	Ednet
DSKT/MIM	0.7662	0.7620	0.7605	0.7300
DSKT/KS	0.7644	0.7566	0.7622	0.7289
DSKT/CS	0.7618	0.7590	0.7322	0.7115
DSKT/DR	0.7627	0.7700	0.7614	0.7298
DSKT	**0.7753**	**0.7734**	**0.7671**	**0.7639**

The performance of the above variants of the model is shown in Table 3. From Table 3, we can see that (1) the original DSKT performs best, indicating each module's effectiveness in the model. (2) After removing the pre-training of exercise and knowledge concept representations, the performance of the model deteriorates on the four datasets, which indicates that the MIM module helps to enhance the representations of our model. (3) Removing either the commonality state or the concept state leads to a decline in model performance, which means that it makes sense to consider knowledge states at these two levels. (4) From the comparison of DSKT/DR and DSKT, the knowledge commonality extractor based on dynamic routing can indeed learn better knowledge commonalities.

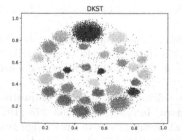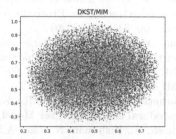

Fig. 3. Visualization of exercise embeddings with and without MIM.

5.4 Embedding Visualization

In order to show the effect of DSKT on representation learning vividly, we use the t-SNE [10] algorithm to visualize the embeddings of the exercises. To facilitate differentiation, we select the top 30 knowledge concepts with the most exercises from ASSISTment2012 and draw their associated exercises in different colors.

As shown in Fig. 3, the distribution of exercise embeddings learned by the model without MIM is relatively chaotic, while the exercise embeddings of DSKT are split into several clusters. The exercises with the same knowledge concept are located in the same cluster and close to each other, and unrelated exercises are well separated.

6 Conclusion

In this paper, we propose a dual-state knowledge tracing model to enrich the student's knowledge state representation. By mining the knowledge commonalities, we establish the knowledge state at the commonality and concept levels to jointly predict the student's performance. Based on the principle of mutual information maximization, we use the pre-training method to integrate the connection between exercises and knowledge concepts into their representations. Extensive experiments on several datasets verify the effectiveness of our proposed model.

In the future, we will consider more diversified relations (such as hierarchical relations, etc.) to enrich the student's knowledge state representation. Further, we will explore the learning process and provide a stronger explanatory for the knowledge tracing process by combining relevant theories of pedagogy.

References

1. Cen, H., Koedinger, K., Junker, B.: Learning factors analysis – a general method for cognitive model evaluation and improvement. In: Ikeda, M., Ashley, K.D., Chan, T.-W. (eds.) ITS 2006. LNCS, vol. 4053, pp. 164–175. Springer, Heidelberg (2006). https://doi.org/10.1007/11774303_17
2. Cho, K., et al.: Learning phrase representations using rnn encoder-decoder for statistical machine translation. arXiv preprint arXiv:1406.1078 (2014)
3. Choi, Y., et al.: Towards an appropriate query, key, and value computation for knowledge tracing. In: Proceedings of the Seventh ACM Conference on Learning@ Scale, pp. 341–344 (2020)
4. Choi, Y., et al.: EdNet: a large-scale hierarchical dataset in education. In: Bittencourt, I.I., Cukurova, M., Muldner, K., Luckin, R., Millán, E. (eds.) AIED 2020. LNCS (LNAI), vol. 12164, pp. 69–73. Springer, Cham (2020). https://doi.org/10.1007/978-3-030-52240-7_13
5. Corbett, A.T., Anderson, J.R.: Knowledge tracing: modeling the acquisition of procedural knowledge. User Model. User-Adapt. Interact. 4(4), 253–278 (1994)
6. Ghosh, A., Heffernan, N., Lan, A.S.: Context-aware attentive knowledge tracing. In: Proceedings of the 26th ACM SIGKDD International Conference on Knowledge Discovery & Data Mining, pp. 2330–2339 (2020)

7. Hjelm, R.D., et al.: Learning deep representations by mutual information estimation and maximization. In: International Conference on Learning Representations (2018)
8. Hochreiter, S., Schmidhuber, J.: Long short-term memory. Neural Comput. **9**(8), 1735–1780 (1997)
9. Liu, Q., Shen, S., Huang, Z., Chen, E., Zheng, Y.: A survey of knowledge tracing. arXiv preprint arXiv:2105.15106 (2021)
10. Van der Maaten, L., Hinton, G.: Visualizing data using t-sne. J. Mach. Learn. Res. **9**(11), 1–27 (2008)
11. Miller, A.H., Fisch, A., Dodge, J., Karimi, A.H., Bordes, A., Weston, J.: Key-value memory networks for directly reading documents. In: EMNLP (2016)
12. Nagatani, K., Zhang, Q., Sato, M., Chen, Y.Y., Chen, F., Ohkuma, T.: Augmenting knowledge tracing by considering forgetting behavior. In: The World Wide Web Conference, pp. 3101–3107 (2019)
13. Nakagawa, H., Iwasawa, Y., Matsuo, Y.: Graph-based knowledge tracing: modeling student proficiency using graph neural network. In: 2019 IEEE/WIC/ACM International Conference on Web Intelligence (WI), pp. 156–163. IEEE (2019)
14. Oord, A.V.d., Li, Y., Vinyals, O.: Representation learning with contrastive predictive coding. arXiv preprint arXiv:1807.03748 (2018)
15. Pandey, S., Karypis, G.: A self-attentive model for knowledge tracing. arXiv preprint arXiv:1907.06837 (2019)
16. Pavlik Jr, P.I., Cen, H., Koedinger, K.R.: Performance factors analysis-a new alternative to knowledge tracing. Online Submission (2009)
17. Pelánek, R.: Bayesian knowledge tracing, logistic models, and beyond: an overview of learner modeling techniques. User Model. User-Adapt. Interact. **27**(3), 313–350 (2017)
18. Piech, C., Bassen, J., Huang, J., Ganguli, S., Sahami, M., Guibas, L.J., Sohl-Dickstein, J.: Deep knowledge tracing. Adv. Neural Inf. Process. Syst. **28**, 505–513 (2015)
19. Sabour, S., Frosst, N., Hinton, G.E.: Dynamic routing between capsules. arXiv preprint arXiv:1710.09829 (2017)
20. Shen, S., et al.: Learning process-consistent knowledge tracing. In: Proceedings of the 27th ACM SIGKDD Conference on Knowledge Discovery & Data Mining, pp. 1452–1460 (2021)
21. Simons, P.R.J.: Transfer of learning: Paradoxes for learners. Int. J. Educ. Res. **31**(7), 577–589 (1999)
22. Su, Y., et al.: Exercise-enhanced sequential modeling for student performance prediction. In: Proceedings of the AAAI Conference on Artificial Intelligence, vol. 32 (2018)
23. l Tong, S., et al.: Structure-based knowledge tracing: an influence propagation view. In: 2020 IEEE International Conference on Data Mining (ICDM), pp. 541–550. IEEE (2020)
24. Vaswani, A., et al.: Attention is all you need. In: Advances in Neural Information Processing Systems, pp. 5998–6008 (2017)
25. Yeung, C.K., Yeung, D.Y.: Addressing two problems in deep knowledge tracing via prediction-consistent regularization. In: Proceedings of the Fifth Annual ACM Conference on Learning at Scale, pp. 1–10 (2018)

26. Zhang, J., Shi, X., King, I., Yeung, D.Y.: Dynamic key-value memory networks for knowledge tracing. In: Proceedings of the 26th International Conference on World Wide Web, pp. 765–774 (2017)
27. Zhang, L., Xiong, X., Zhao, S., Botelho, A., Heffernan, N.T.: Incorporating rich features into deep knowledge tracing. In: Proceedings of the Fourth (2017) ACM Conference on Learning@ Scale, pp. 169–172 (2017)

Multi-granularity Evolution Network for Dynamic Link Prediction

Yi Yang[1,2], Xiaoyan Gu[1], Haihui Fan[1(✉)], Bo Li[1], and Weiping Wang[1]

[1] Institute of Information Engineering, Chinese Academy of Sciences, Beijing, China
{yangyi1,guxiaoyan,fanhaihui,libo,wangweiping}@iie.ac.cn
[2] School of Cyber Security, University of Chinese Academy of Sciences, Beijing, China

Abstract. Dynamic link prediction target to predict future new links in a dynamic network, is widely used in social networks, knowledge graphs, etc. Some existing dynamic methods capture structural characteristics and learn the evolution process from the entire graph, which pays no attention to the association between subgraphs and ignores that graphs under different granularity have different evolve patterns. Although some static methods use multi-granularity subgraphs, they can hardly be applied to dynamic graphs. We propose a novel Temporal K-truss based Recurrent Graph Convolutional Network (TKRGCN) for dynamic link prediction, which learns graph embedding from different granularity subgraphs. Specifically, we employ k-truss decomposition to extract multi-granularity subgraphs which preserve both local and global structure information. Then we design a RNN framework to learn spatio-temporal graph embedding under different granularities. Extensive experiments demonstrate the effectiveness of our proposed TKRGCN and its superiority over some state-of-the-art dynamic link prediction algorithms.

Keywords: Dynamic graph · Link prediction · Network embedding

1 Introduction

Link prediction, as a task of predicting the relationship between entities, plays a vital role in many graph mining applications, such as social networks [21] and biology network [18]. It can be divided into two categories. One is to predict missing links on static graph, and the other predicts new links that may appear in the future on dynamic graph. Since many real-world networks are dynamic, whose nodes and edges appear or disappear over time, dynamic link prediction [4] can keenly capture the variation trend and achieve better prediction effect, therefore attracts wide attention. For instance, in social networks, we predict future interactions between users for friend recommendation; In academic networks, we study the cooperation of scholars to predict their future co-workers.

Dynamic link prediction aims to learn the evolution of the graph from historical information and predict future links. Existing methods mainly extract

J. Gama et al. (Eds.): PAKDD 2022, LNAI 13280, pp. 393–405, 2022.
https://doi.org/10.1007/978-3-031-05933-9_31

features (structure and attribute information) at different time from the entire graph and use those time-stamped features to model graph dynamic, such as GCRN [16], EvolveGCN [14], DynamicTriad [22]. However, all these methods ignore the fact that the entire graph usually contains diverse structures, and the evolution of different structures over time is different. This may lead to sub-optimal performance of link prediction. For example, Fig. 1 shows the evolve of subgraphs with varying structures from time t to time t+1. In blue and yellow dense subgraphs, more links will appear at next time. In green sparse subgraph, it is unlikely to have more future node interactions. Simply learning the evolution of the entire graph without distinguishing structures will affect the accuracy of link prediction. Therefore, it is necessary to use a multi-granularity graph instead. The graphs of different granularities contain different structures, which facilitates better learning graph structural characteristics and dynam-ics. However, existing dynamic link prediction methods can't well divide graph to multi-granularity subgraphs to learn the evolution of graphs under different granularities. Although in static link prediction there are some methods learn graph structural characteristics on multiple granularities such as mlink [1] and PME [2], this kind of method only build multi-granularity graphs on the local subgraph composed of nodes and their neighbors, without dividing global sim-ilar structures to same granularity. Thus, it can hardly be applied to dynamic link prediction to learn the co-evolution pattern of the global similar structures. A recent method CTGCN [11] uses multi-granularity graphs to capture richer hierarchical structure features for dynamic link prediction. However, it does not distinguish the structure evolution under different granularities.

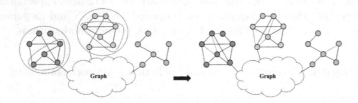

Fig. 1. It shows the changes of the subgraphs from time t to $t+1$. The red line represents the new link, and the circle represents the local multi-granularity subgraph divided by the one-hop and two-hop neighbors of the middle node (best see in color).

Multi-granularity graphs can mine richer structure characteristics. Still, when applied to dynamic link prediction tasks, the inherent difficulty mainly originates from two aspects: 1) How to divide multi-granularity graphs? 2) How to learn structural features and evolution patterns on multi-granularity graphs? To better explain the first problem, the circle in Fig. 1 shows a partition way to get the multi-granularity subgraphs by node multi-order neighbors. Still, this method only focuses on the local multi-level structure, ignoring that, on the entire graph, the structures in the blue and yellow subgraphs are similar and have similar evolution patterns. Thus we attempt to seek a graph partition method that can retain both local and global information. For the second problem, the diversity of the structural characteristic and dynamics of multi-granularity dynamic graphs forces us to design a unified framework to aggregate them.

To materialize our idea, we present a novel Temporal K-truss based Recurrent Graph Convolutional Network (TKRGCN) for dynamic link prediction, which learns structural characteristics and dynamics from different granularity. Specifically, we employ k-truss decomposition to extract multi-granularity subgraphs which preserve both local and global structure information. To better extract features to capture diverse structural information, we modify GCN to alleviate the problem of over-smoothing as the number of layers deepens, enabling GCN to propagate high-order features effectively. Then we design a framework to learn the evolution process of subgraphs of different granularities. Subgraphs of different granularities make different contributions to the evolution of the entire graph. Discriminatively treating different subgraphs helps to model the complete evolution process. We conduct extensive experiments on six real-world datasets and the result shows that our model performs better than current state-of-the-art methods. The main contributions of this paper are as follows:

- We propose TKRGCN for dynamic link prediction, which learns structural characteristics and dynamic evolution from different granularity subgraphs while preserving both local and global similar subgraph features.
- We decouple GCN and deepen the propagation depth of GCN to alleviate the performance degradation so that GCN can effectively extract high-order features from the subgraph.
- The experiment results demonstrate that TKRGCN outperforms the state-of-the-art benchmark in link prediction.

2 Related Work

The dynamic link prediction method needs to capture both structural properties and time evolution patterns. It mainly falls into two broad categories: discrete methods and continuous methods. Discrete methods pay more attention to changes in structural characteristics of dynamic graphs. Many methods use the architecture of combining GNNs [15] and RNNs [13] such as GCRN [16], RgCNN [17] and GGNN [10]. EvolveGCN [14] adapts to GCN in the time dimension by using RNNs to encode the parameters of GCN. DynGEM [6] employs autoencoder to generate highly non-linear node embeddings and makes some improvements on computation. In addition, continuous methods more consider the graph evolving process. DynamicTriad [22] models dynamic network evolution through modeling the triadic closure process. Dyrep [19] defines topological evolution and node interaction to simulate the evolution of dynamic graphs. However, these methods do not consider the multi-granularity dynamic graph evolution. Although CTGCN [11] divides out the multi-granularity graph to better capture the structural information, it does not distinguish the different evolution modes under the multi-granularity.

In static link prediction, there are some methods to learn graph structural characteristics on multiple granularities. For instance, mlink [1] proposes a node aggregation method that can transform the enclosing subgraph into different scales to learn scale-invariant features. PME [2] integrates first-order and second-order proximities and projects feature to different spaces to model nodes and links. But, they only build multi-granularity graphs on the local subgraphs and can hardly be applied to dynamic link prediction to learn the co-evolution pattern of the global similar structures.

3   Preliminaries

Consider a static undirected graph as $G = (V, E)$, where $V = \{v_1, \ldots, v_N\}$ denotes the node set with N nodes and E is the link set. We denote the dynamic graph G as an ordered set of snapshots $\{G^1, G^2, \ldots, G^T\}$ from time step 1 to T. $G_t = (V, E^t)$ is the state of the graph at time step t with a shared node set V and E^t contains the links that appear at time step t. The adjacency matrix $A^t \in \mathbb{R}^{n \times n}$ can be either weighted or unweighted.

Given a series of snapshots represented by $A = \{A^1, A^2, \ldots, A^t\}$ and node features $X = \{X^1, X^2, \ldots, X^t\}$, the goal is to predict A^{t+1} at time $t + 1$. In our method, we learn the mapping series $F = \{f_1, f_2, \ldots, f_t\}$ that f_t encodes each node in G^t into an embedding space with $d(d \ll N)$ dimension. The node embeddings at time step $t + 1$ will be utilized to predict links.

4   The Proposed Method

We propose Temporal K-truss based Recurrent Graph Convolutional Network (TKRGCN) shown in Fig. 2, our method consists the following two parts: 1) Multi-granularity graph partition and feature extraction: To mine richer graph information, we apply the k-truss decomposition algorithm to divide multi-granularity subgraphs. This algorithm retains the local similar structure and reflects the global structural similarity, which is conducive to better learning the evolution of dynamic graphs later. Besides, to capture diverse structural information, we decouple GCN and deepen the depth of feature propagation, thus alleviating the problem of deep GCN performance degradation and enabling GCN to extract high-order features. 2) Spatio-temporal evolution embedding: To learn the structural characteristics and temporal evolution process from multi-granularity subgraphs, we design a novel architecture composed of RNNs and Attention, which learn structural information and temporal evolution of different granularities. Discriminatively treating different subgraphs helps to model the entire graph embedding.

4.1 Multi-granularity Graph Partition and Feature Extraction

Multi-granularity Graph Partition. To mine rich multi-granularity struc-tures, we utilize k-truss decomposition [8] to obtain subgraphs. The definition is defined as follows:

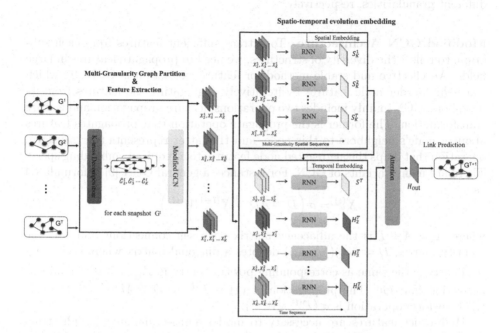

Fig. 2. Schematic illustration of TKRGCN.

Theorem. k-truss decomposition: Given a graph G and $k \in \mathbb{N}$, a k-truss subgraph $\widehat{G}_k$ of G is the largest subgraph such that $\forall e \in E(\widehat{G}_k)$, $sup_{\widehat{G}_k}(e) \geq k - 2$.

$sup_{\widehat{G}_k}(e)$ is the support of edge e, defined as the number of triangles contain-ing e. G is divide to a series of nested hierarchical subgraphs $\{\widehat{G}_2, \widehat{G}_3, \ldots, \widehat{G}_{k_{max}}\}$ by k-truss decomposition, where k_{max} is the max subgraph truss number. According to the definition, $\widehat{G}_1 = \widehat{G}_2$, so the index of subgraphs start from 2.

From the definition, we know that a subgraph with a high value of k has a denser structure and fewer nodes. Thus the probability of links between nodes in $\widehat{G}_k$ is greater. Intuitively, the more friends two people have in common, the stronger their relationship will be. In social networks, many groups are inter-connected with dense structures, such as peer groups. People usually have close connections with members of the same group and have little contact with other groups. Meanwhile, the evolution of structurally similar groups, such as peer groups from different companies, is similar. K-truss decomposition can divide these similar structures into subgraphs under the same granularity, which is convenient for learning the common evolution process of these global similar structures.

In summary, the multi-granularity subgraphs based on k-truss decomposition retain both local and global similar structures, helping to capture rich hierarchical structure information and learn the evolution process of structures under different granularities, respectively.

Modified GCN Architecture. To extract sufficient features from each subgraph to reflect the diversity of structures, we need to propagate features in large fields. An effective and stable method for feature extraction is GCN [9], which learns high-order node features by iteratively aggregating the features from its neighbors. GCN mainly includes two operations: feature propagation and feature transformation. The former is the propagate operation that propagates features about a node's neighbors to this node, and the latter represents the transform operation that maps the aggregated node features to a required embedding space. There are many variants of GCN. For instance, a general GCN [9] is formulated as

$$X^{(l)} = \sigma \left(\widetilde{D}^{-\frac{1}{2}} \widetilde{A} \widetilde{D}^{-\frac{1}{2}} X^{(l-1)} W^{(l)} \right), \tag{1}$$

where $\widetilde{A} = A + I$ is the adjacency matrix with self-connections and I is the identity matrix. $\widetilde{D} = diag(\sum_j \widetilde{A}_{ij})$ denotes a diagonal matrix where each diagonal entry is the same as corresponding position entry in $\widetilde{A}_{ij}$. $\sigma()$ is a nonlinear activation function. The propagation of (1) is $P^{(l)} = \widetilde{D}^{-\frac{1}{2}} \widetilde{A} \widetilde{D}^{-\frac{1}{2}} X^{(l-1)}$ while the transform operation is $\sigma \left(P^{(l-1)} W^{(l)} \right)$.

High-order features are necessary to model multi-granularity graph structures. A one-layer GCN only considers the direct neighbors of nodes, while the multi-layer stacking GCN can learn high-order structural features, but the problem of over-smoothing may occur. Over-Smoothing refers to the fact that as the layer deepens, the features of all nodes in the same connected component tend to be the same. Thus GCN performs worse as it goes deeper. Some works [7,12] point out that the excessive entanglement of transformation and propagation in current GCN is the key factor that affects the performance of the model. Decoupling these two operations can effectively alleviate the over-smoothing problem. Inspired by this, We design a mGCN with disentangling propagation and transformation operation as follows:

$$X^0 = MLP(X_{init}), \tag{2}$$

$$X^j = \widetilde{D}^{-\frac{1}{2}} \widetilde{A}^j \widetilde{D}^{-\frac{1}{2}} X^0, \quad j = 1, 2, \cdots, J, \tag{3}$$

$$att^j = \sigma(MLP(X^j)), \quad j = 0, 1, 2, \cdots, J, \tag{4}$$

$$X_{out} = softmax(sum(att^0 \circ X^0, \cdots, att^J \circ X^J)), \tag{5}$$

where J is the depth of the propagation, $X^0 \in \mathbb{R}^{N \times d}$ maps the initial node feature X_{init}. $att^j \in \mathbb{R}^{N \times 1}$ is trained to adaptively adjust the information that each node should retain in each propagation depth. $\circ$ means the element of att is multiplied by the corresponding d-dimensional node vector. X_{out} is calculated by combining each propagation layer X^j. We apply our mGCN on k-truss subgraphs at each time t and get

$$X_k^t = mGCN(A_k^t, X_{init,k}^t). \tag{6}$$

In our mGCN, we decouple the feature propagation and transformation and deepen the propagation depth separately. By learning parameters att, we can adjust the information that the node should retain in different propagation layers so that mGCN can be applied to larger receptive fields without affecting performance.

4.2   Spatio-temporal Evolution Embedding

To learn spatial and temporal embedding in multi-granularity graphs, we design the following architecture shown in Fig. 2.

Spatial Embedding. In spatial dimension, we intend to capture structure embedding from multi-granularity subgraphs. It can be seen from the previous section that there is a strong connection between the multi-granularity subgraphs $\{X_2^t, X_3^t, \cdots, X_K^t\}$ obtained by k-truss decomposition. RNN is exactly suitable for processing highly correlated sequences. On snapshot G^t, we reverse the sequence as $\{X_K^t, X_{K-1}^t, \cdots, X_2^t\}$ and feed it into RNNs to learn structure information in spatial dimension. We have a mathematical representation as:

$$\begin{aligned}
S_2^t &= RNN(X_K^t \quad, S_1^t), \\
S_3^t &= RNN(X_{K-1}^t, S_2^t), \\
&\cdots \\
S_K^t &= RNN(X_2^t \quad, S_{K-1}^t),
\end{aligned} \tag{7}$$

where $S^1 = 0$ is the initial matrix. S_k^t represents the hidden state and X_k^t denotes the input feature. In the input sequence, the subgraph with a high value of k has a denser structure and fewer nodes. Therefore, the reverse order sequence input into RNN is to learn the structural development pattern from the dense small graph to the large sparse graph. In this process, the structure information under different granularities is merged. The final hidden state output S_K^t contains the structural properties of the entire snapshot. We do this on each snapshot and ultimately obtain spatial node embeddings at each time $\{S_K^1, S_K^2, \cdots, S_K^T\}$. Besides, RNN has many variants, we use the Gated Recurrent Unit (GRU) [3].

Temporal Embedding. The k-truss based multi-granularity subgraph extracts the similar structure on the whole graph into the same granularity subgraph, which retains the local and global similar structure. In temporal dimension, subgraphs of different granularities make different contributions to the evolution of the entire graph. Discriminatively treating different subgraphs helps to model the complete evolution process. We use RNN to learn how the subgraphs evolve over time of different granularities. Under each granularity k, we input $\{X_k^1, X_k^2, \cdots, X_k^T\}$ into RNN as follows:

$$
\begin{aligned}
H_k^1 &= RNN(X_k^1, H_k^0), \\
H_k^2 &= RNN(X_k^2, H_k^1), \\
&\cdots \\
H_k^T &= RNN(X_k^T, H_k^{T-1}),
\end{aligned}
\tag{8}
$$

where T is the length of the time sequence. The temporal module is similar to the spatial module, and the difference is their input. Note that the structure embedding sequence $\{S_K^1, S_K^2, \cdots, S_K^T\}$ is also fed into the temporal module with the same formula as (8) and output S^T.

To obtain the final node representation for link prediction, we use the multi-head attention mechanism to learn the importance of the temporal embedding and spatial embedding and get the final node embedding H_{out}. The formula is defined as follows:

$$
H_{out} = MultiHeadAtten(S^T, H_2^T, \cdots, H_K^T).
\tag{9}
$$

4.3 Optimization

To estimate the parameters of our model, we need to specify an objective function to optimize. We design an unsupervised loss function described as:

$$
\mathcal{L} = \sum_{t=1}^{T} \sum_{u \in V} (\mathcal{L}_{u+}^t - \mathcal{L}_{u-}^t),
\tag{10}
$$

$$
\mathcal{L}_{u+}^t = \sum_{v+ \in \mathcal{N}^+(u)} \sigma(< h_u, h_{v+} >),
\tag{11}
$$

$$
\mathcal{L}_{u-}^t = \sum_{v- \in \mathcal{N}^-(u)} \sigma(< h_u, h_{v-} >).
\tag{12}
$$

The "positive" node set $\mathcal{N}^+(u)$ includes the nodes that sampled in fixed-length random walks where node u has appeared and the "negative" nodes in $\mathcal{N}^-(u)$ are randomly sampled from the entire graph. $<, >$ denotes the Hadamard product. Such a design guarantees the representations of closely related nodes are close while the irrelevant nodes are far away from each other.

5   Experiments

5.1   Datasets and Baselines

We experiment on six dynamic datasets in KONECT[1] and SNAP[2]. Details are summarized in Table 1. All the datasets are split by month.

We compare our model with two static methods and five dynamic methods.

GCN [9]: It can simultaneously perform end-to-end learning of node features and structures.

GAT [20]: As an optimization method of GCN, GAT introduces an attention mechanism to calculate the attention coefficient of the current node and its neighbors to reduce the impact of noise.

Table 1. Datasets.

Datasets	#Nodes	#Edges	K	#Timesteps
UCI	1899	59835	6	7
AS	6828	1947704	11	100
MATH	24740	323357	11	77
FACEBOOK	60730	607487	7	27
ASKU	74924	356822	11	21
ENRON	87036	530284	18	38

GCRN [16]: A direct dynamic embedding method that lets GCN process each snapshot and provides the output of GCN to the time series component RNN to learn the temporal patterns.

DynGEM [6]: It uses a deep autoencoder to get non-linear graph embedding and proposes PropSize to increase the scale of the neural network dynamically.

Dyngraph2vec [5]: A continuation of DynGEM which consider the historical information of the past l snapshots. It has three variants: dyngraph2vecAE, dyngraph2vecRNN, and dyngraph2vecAERNN.

EvolveGCN [14]: EvolveGCN uses RNNs to evolve GCN parameters. It has better results in extreme situations where nodes change frequently.

CTGCN [11]: It uses k-core and GCN to capture the hierarchical nature of graphs and extends it to dynamic graphs.

Settings. We use the previous $l = 5$ snapshots $G^{t-l+1} - G^t$ to predict the link in G^{t+1}. The k-truss decomposition is used to extract 3 subgraphs from each snapshot, namely, 2-truss to 4-truss subgraphs. For fair comparisons, we set the embedding dimension $d = 128$ and uniformly utilize 2 layers in GCN, GAT, GCRN, EvolveGCN.

---

[1] http://konect.uni-koblenz.de/.
[2] http://snap.stanford.edu/.

5.2   Performance Comparison

We conduct several experiments on link prediction. Each link feature vector is calculated by the Hadamard product of the node-pair vectors. We train a logistic regression classifier with L2 regularization to classify the positive and negative links. In addition, the static version of our method KRGCN(remove the temporal module) can also perform static link prediction, which predicts the missing links using only the known information at time t. The area under the curve(AUC) is employed as the evaluation metric. We take the average of AUC as the final result.

Table 2 demonstrates the link prediction results on six datasets in static and dynamic models. The best results are shown in bold. Due to memory limitations, the AUC of DynAE and DynAERNN are not available on some datasets indicated by '-'. Our method TKRGCN outperforms other methods on each dataset, which strongly proves the effectiveness of our approach in using multi-granularity subgraphs to capture structural and temporal information. Moreover, our static method KRGCN surpasses some dynamic methods, showing that our multi-granularity strategy can capture more effective structural properties.

Table 2. Average AUC scores for link prediction.

Methods	UCI	AS	MATH	FACEBOOK	ASKU	ENRON
GCN [9]	0.7082	0.7451	0.7887	0.5928	0.7741	0.8068
GAT [20]	0.7906	0.7027	0.7246	0.5553	0.6793	0.8601
Ours(static)	**0.9266**	**0.9366**	**0.8857**	**0.7336**	**0.8361**	**0.9082**
GCRN [16]	0.8258	0.9309	0.7929	0.6512	0.7818	0.9247
DynGEM [6]	0.9053	0.9413	0.8500	0.6023	0.8032	0.8767
DynAE [5]	0.9231	0.9284	0.9462	0.7401	–	–
DynAERNN [5]	0.9019	0.8972	0.8383	–	–	–
EvloveGCN [14]	0.9126	0.9294	0.8954	0.7435	0.9279	0.9361
CTGCN [11]	0.9368	0.9544	0.9598	0.8158	0.9468	0.9855
Ours	**0.9825**	**0.9608**	**0.9689**	**0.8500**	**0.9838**	**0.9941**

An essential hyper-parameter is K. It determines the granularity level of the graph. To analyze the influence of K on TKRGCN, we design a small-scale experiment, take the last 20 snapshots of four datasets for training and shorten the embedding dimensions to 32. The results are shown in Fig. 3. In the beginning, with the increase of K, link prediction performance has a specific improvement, especially in UCI.

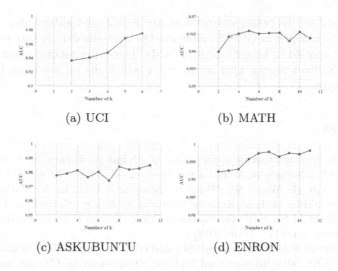

(a) UCI   (b) MATH

(c) ASKUBUNTU   (d) ENRON

Fig. 3. The AUC performance of various k numbers on four datasets.

5.3 Ablation Study

To further investigate the impact of k-truss decomposition and modified GCN module for TKRGCN, we reconstruct the architecture as a)TKRGCN-sGCN: replace the modified GCN with a simple GCN [9] in TKRGCN; b)TKRGCN-single: TKRGCN without multi-granularity subgraphs. As shown in Table 3, TKRGCN performs better than the other two methods on all datasets, illustrating that both the modified GCN and the k-truss decomposition have a positive effect on TKRGCN. Additionally, compared with TKRGCN-sGCN, the performance of TKRGCN-single drops more, which indicates that k-truss based multi-granularity subgraph partition strategy is essential to dynamic link prediction.

Table 3. AUC for ablation study.

Methods	UCI	AS	MATH	FACEBOOK	ASKU	ENRON
TKRGCN	**0.9825**	**0.9608**	**0.9689**	**0.8500**	**0.9838**	**0.9941**
TKRGCN-sGCN	0.9583	0.9487	0.9349	0.8261	0.9505	0.9638
TKRGCN-single	0.9332	0.9418	0.8924	0.7815	0.9394	0.9471

6 Conclusion

In this paper, we propose a novel framework named TKRGCN for dynamic link prediction, which learns structural characteristics and dynamic evolution from multi-granularity subgraphs while preserving both local and global similar subgraph features. To better extract features in different subgraphs, we deepen the propagation depth of GCN to alleviate the over-smoothing problem so that GCN can be applied to larger receptive fields. The experimental results validate our method's effectiveness. Our future work will focus on studying large-scale dynamic graphs.

References

1. Cai, L., Ji, S.: A multi-scale approach for graph link prediction. In: Proceedings of the AAAI Conference on Artificial Intelligence, vol. 34, pp. 3308–3315 (2020)
2. Chen, H., Yin, H., Wang, W., Wang, H., Nguyen, Q.V.H., Li, X.: PME: projected metric embedding on heterogeneous networks for link prediction. In: Proceedings of the 24th ACM SIGKDD International Conference on Knowledge Discovery and Data Mining, pp. 1177–1186 (2018)
3. Dey, R., Salem, F.M.: Gate-variants of gated recurrent unit (GRU) neural networks. In: 2017 IEEE 60th International Midwest Symposium on Circuits and Systems (MWSCAS), pp. 1597–1600. IEEE (2017)
4. Eppstein, D., Galil, Z., Italiano, G.F.: Dynamic graph algorithms. Algorithms Theor. Comput. Handb. **1**, 9-1 (1999)
5. Goyal, P., Chhetri, S.R., Canedo, A.: dyngraph2vec: capturing network dynamics using dynamic graph representation learning. Knowl.-Based Syst. **187**, 104816 (2020)
6. Goyal, P., Kamra, N., He, X., Liu, Y.: Dyngem: deep embedding method for dynamic graphs. arXiv preprint arXiv:1805.11273 (2018)
7. He, X., Deng, K., Wang, X., Li, Y., Zhang, Y., Wang, M.: Lightgcn: Simplifying and powering graph convolution network for recommendation. In: Proceedings of the 43rd International ACM SIGIR Conference on Research and Development in Information Retrieval, pp. 639–648 (2020)
8. Huang, X., Lakshmanan, L.V., Xu, J.: Community search over big graphs. Synthesis Lect. Data Manage. **14**(6), 1–206 (2019)
9. Kipf, T.N., Welling, M.: Semi-supervised classification with graph convolutional networks. arXiv preprint arXiv:1609.02907 (2016)
10. Li, Y., Tarlow, D., Brockschmidt, M., Zemel, R.: Gated graph sequence neural networks. arXiv preprint arXiv:1511.05493 (2015)
11. Liu, J., Xu, C., Yin, C., Wu, W., Song, Y.: K-core based temporal graph convolutional network for dynamic graphs. IEEE Trans. Knowl. Data Eng. (2020)
12. Liu, M., Gao, H., Ji, S.: Towards deeper graph neural networks. In: Proceedings of the 26th ACM SIGKDD International Conference on Knowledge Discovery and Data Mining, pp. 338–348 (2020)
13. Mikolov, T., Karafiát, M., Burget, L., Černockỳ, J., Khudanpur, S.: Recurrent neural network based language model. In: Eleventh Annual Conference of the International Speech Communication Association (2010)
14. Pareja, A., et al.: Evolvegcn: evolving graph convolutional networks for dynamic graphs. In: Proceedings of the AAAI Conference on Artificial Intelligence, vol. 34, pp. 5363–5370 (2020)
15. Scarselli, F., Gori, M., Tsoi, A.C., Hagenbuchner, M., Monfardini, G.: The graph neural network model. IEEE Trans. Neural Networks **20**(1), 61–80 (2008)

16. Seo, Y., Defferrard, M., Vandergheynst, P., Bresson, X.: Structured sequence modeling with graph convolutional recurrent networks. In: Cheng, L., Leung, A.C.S., Ozawa, S. (eds.) ICONIP 2018. LNCS, vol. 11301, pp. 362–373. Springer, Cham (2018). https://doi.org/10.1007/978-3-030-04167-0_33
17. Te, G., Hu, W., Zheng, A., Guo, Z.: RGCNN: regularized graph CNN for point cloud segmentation. In: Proceedings of the 26th ACM International Conference on Multimedia, pp. 746–754 (2018)
18. Theocharidis, A., Van Dongen, S., Enright, A.J., Freeman, T.C.: Network visualization and analysis of gene expression data using Biolayout express 3D. Nat. Protoc. 4(10), 1535–1550 (2009)
19. Trivedi, R., Farajtabar, M., Biswal, P., Zha, H.: Dyrep: learning representations over dynamic graphs. In: International Conference on Learning Representations (2019)
20. Veličković, P., Cucurull, G., Casanova, A., Romero, A., Lio, P., Bengio, Y.: Graph attention networks. arXiv preprint arXiv:1710.10903 (2017)
21. Wasserman, S., Faust, K., et al.: Social network analysis: Methods and applications (1994)
22. Zhou, L., Yang, Y., Ren, X., Wu, F., Zhuang, Y.: Dynamic network embedding by modeling triadic closure process. In: Proceedings of the AAAI Conference on Artificial Intelligence, vol. 32 (2018)

A Two-Tower Spatial-Temporal Graph Neural Network for Traffic Speed Prediction

Yansong Shen[1], Lin Li[1(⊠)], Qing Xie[1], Xin Li[2], and Guandong Xu[3]

[1] Wuhan University of Technology, Wuhan, China
{yaso9527,cathylilin,felixxq}@whut.edu.cn
[2] University of Science and Technology of China, Hefei, China
leexin@ustc.edu.cn
[3] University of Technology Sydney, Sydney, NSW 2008, Australia
Guandong.Xu@uts.edu.au

Abstract. Recently, the remarkable effect of applying Dynamic Graph Neural Networks (DGNNs) to traffic speed prediction has received wide attention. Existing DGNN-based researches usually use a pre-defined or an adaptive matrix to capture the spatial correlations in traffic data. However, these static matrices are not enough to match the dynamic characteristics of spatial correlations. We argue that the global changes and local fluctuations of spatial correlations are dynamic with different frequencies. To this end, in this paper, we propose a Two-Tower DGNN (T^2-GNN) framework which divides the traffic data into a seasonal static component and an acyclic dynamic component, thus enhancing traffic speed prediction. The two components generated by an auto-decomposing block reflect global changes and local fluctuations of spatial correlations, respectively. Moreover, we use two parallel dynamic graph generation layers to construct a seasonal graph and an acyclic graph at each time step. In this way, the high-level representations of these two kinds of dynamic changes are learned through two dynamic graph convolution layers. Besides, the impact of fixed road network structure is modeled on the pre-defined graph and added to the spatial correlations. And we capture temporal correlations in temporal block before modeling spatial correlations. Finally, skip connections are used to converge the spatial-temporal correlations for final prediction. Experimental results on an urban dataset and two highway datasets show our proposed framework achieves the state-of-the-art prediction performances in terms of Mean Average Error (MAE) and Root Mean Squared Error (RMSE).

Keywords: Traffic speed prediction · Dynamic graph neural network · Spatial correlation

This work is supported by the China Postdoctoral Science Foundation (Grant No. 2021M693101).

J. Gama et al. (Eds.): PAKDD 2022, LNAI 13280, pp. 406–418, 2022.
https://doi.org/10.1007/978-3-031-05933-9_32

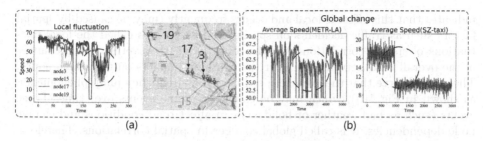

(a)   (b)

Fig. 1. The local fluctuations and global changes of spatial correlations. Figure (a) is the speed of some nodes in METR-LA dataset for a period of time, and the node distribution is on the right. The content inside the circle shows the rapid decline and recovery of speed, about 50 time steps (4 h). Figure (b) is the average speed in two datasets. In METR-LA, the average speed dropped from 65 to 60 and recovered to 65 after about 1500 time steps (125 h). In SZ-taxi, the average speed dropped from 17 to 10 and this change lasted for 2000 time steps (500 h).

1 Introduction

Traffic speed prediction has been studied for decades. Traditional methods mostly require data to satisfy stationary assumption for each time series. Therefore, they are limited by their ability to capture temporal and spatial correlations [8,9]. Later, deep learning models are widely used for modeling temporal correlations, like Convolutional Neural Network (CNN) and Recurrent Neural Network (RNN), and spatial correlations, like CNN [3,17,19].

Most recently, GNNs have proven effective in modeling non-euclidean spatial structure data due to their permutation, local connectivity and compositionality [2,4,5]. Graph Convolution Network (GCN) is widely applied to learn spatial correlations for traffic speed prediction .The effect of spatial correlations depends on the adjacency matrix that describes the relationships between traffic nodes. *(1) Early GNN-based methods* use a pre-defined static matrix as the description of node dependencies [10,12,23]. These matrices are mostly constructed by calculating the similarity between pairs of nodes [10]. *(2) The methods of adding self-adaptive matrix* try to learn the hidden node dependencies to generate an adaptive matrix as a supplement or alternative to the pre-defined matrix [15,16]. However, these static matrices are not enough to reflect the spatial correlations of the traffic network. Node dependencies have not only the inherent impact caused by road network structure but also the dynamic impacts caused by traffic condition changes [11]. *(3) The methods of constructing dynamic graph* are proposed to capture the dynamic changes of spatial correlations [6,15]. What is inadequate is that they attempt to use dynamic graphs to learn spatial correlations without discriminating different types of dynamics.

We observe that different dynamic changes of spatial correlations have different frequencies and impact ranges. For example, Fig. 1(a) shows that a traffic congestion at node 15 (the orange speed drops first) affected nodes 3, 17 and 19 on the same road segment over a period of time and disappeared quickly. It indicates that this effect is local and occurs frequently (may be caused by spatial factors such as traffic accidents on specific nodes), so it is called the local fluctuations of spatial correlations. On the other hand, the average speed changes of the overall road network can infer the common impacts on the traffic nodes. Then, we observe that the average speed has a low-frequency jump phenomenon as shown in Fig. 1(b), which may be caused by some events over time. These events (such as bad weather or holidays) can affect road conditions and change node dependencies. It is called global changes in spatial correlations. Therefore, we assume that the traffic flow can be divided into a seasonal static component and an acyclic dynamic component. Specifically, the **seasonal static component** includes the impact of **fixed road network structure** and the **global changes** of spatial correlations, and **acyclic dynamic component** represents the **local fluctuations** of spatial correlations. A fine modeling of the dynamic spatial correlations can further enhance DGNN's ability to improve the accuracy of traffic speed prediction.

In summary, we propose a Two-Tower Graph Neural Networks (T^2-GNN) for traffic speed prediction. The main contributions of our work are as follows:

- We propose a novel two-tower DGNN framework which consists of stacked spatial-temporal layers containing a temporal block and a two-tower spatial block. The spatial block includes an auto-decomposing block and two following D-GCN (dynamic graph convolution network) blocks. Each D-GCN block contains a dynamic graph generation and a graph convolution layer.
- Our D-GCN block utilizes dynamic graph generation to construct a seasonal graph and an acyclic graph for seasonal static components and acyclic dynamic components at each time step. Following these dynamic graphs, GCN can capture global changes and local fluctuations in spatial correlations.
- Conducted on an urban and two highway datasets, experimental results show that our framework outperforms the state-of-the-art methods in terms of MAE and RMSE.

The rest of this paper is organized as follows. Section 2 discusses the related work. The problem definition is in Sect. 3. Section 4 describes our framework. The evaluation results are present in Sect. 5. Finally, Sect. 6 concludes our work.

2   Related Works

2.1   Traffic Speed Prediction

After years of efforts, the researches on traffic speed forecasting have achieved great accomplishments. Previous studies build a data-driven statistical model for prediction, such as ARIMA [8,9]. However, these methods are generally designed for small datasets, and cannot deal with complex changes in traffic data well. Later, deep learning methods perform better than traditional methods. RNN is used to capture temporal correlations [10,13]. In order to make better long-term predictions, Wu et al. designed CNN-based dilated causal convolution to model temporal dependency, which can achieve better results with lower consumption than standard 1D convolution [16]. Another important work direction is how to learn the spatial correlations in traffic data. Viewing the road network as a grid graph is a way to learn the spatial correlations via CNN [19,20]. However, compared with CNN, GNN is better at extracting node correlations due to its ability to process non-European data.

2.2   Graph Neural Network

From the first application of GNN to traffic speed prediction, it is always a key issue that how to construct a suitable graph to represent the node dependencies. Bruna et al. first applied GCN to capture spatial dependencies based on a symmetric matrix [1]. Considering the directionality of traffic data, Li et al. applied bidirectional random walks on an asymmetric matrix to learn the node dependencies [10]. Following them, pre-defined adjacency matrix is adopted to model the spatial correlations [12,18,22]. However, this static matrix is not enough to describe the dynamic spatial correlations.

In order to overcome the limitation of the pre-defined matrix, self-adaptive matrix is proposed to preserve hidden node dependencies [14,16,21]. However, it is essentially a static matrix due to lack of dynamic adjustment. Wu et al. proposed a graph learning layer that generates a dynamic graph by calculating similarities among learnable node embeddings [15]. Inspired by the Trucker decomposition, Han et al. designed a dynamic graph construction layer to generate dynamic graphs for time-specific node dependencies during a day [6]. We argue that global changes and local fluctuations of spatial dependencies have different characteristics. Therefore, the power of DGNN can be further released by constructing the fine topology of the dynamic graph at each time step.

3   Problem Definition

The traffic speed prediction is usually regarded as a multivariate time series prediction with auxiliary prior knowledge. The traffic network, the prior knowledge, is represented by a weighted directed graph $\mathcal{G} = (\mathcal{V}, \mathcal{E}, A)$. $\mathcal{V}$ is a set of nodes $|\mathcal{V}| = N$, and $\mathcal{E}$ is a set of edges. $A \in \mathbb{R}^{N \times N}$ denotes the adjacency matrix.

At each time step t, the speed of traffic node is $X_t \in \mathbb{R}^{N \times D}$, where D is the features length of each node. Given the graph $\mathcal{G}$ and T steps historical traffic speed $X_{(t-T+1):t}$, our task is to learn a function $f(\cdot)$ which is able to map $X_{(t-T+1):t}$ and $\mathcal{G}$ to next Q steps traffic speed $X_{t+1:(t+Q)}$, represented as follow:

$$\left[X_{(t-T+1):t}, \mathcal{G} \right] \xrightarrow{f(\cdot)} X_{t+1:(t+Q)} \tag{1}$$

4   Methodology

4.1   Our Framework

The overall framework of our T^2-GNN is shown in Fig. 2 with k stacked spatial-temporal layers and an output layer. In each spatial-temporal layer, the temporal block (Sect. 4.2) and subsequent two-tower spatial block (Sect. 4.3) are designed to capture the temporal correlations and spatial correlations of

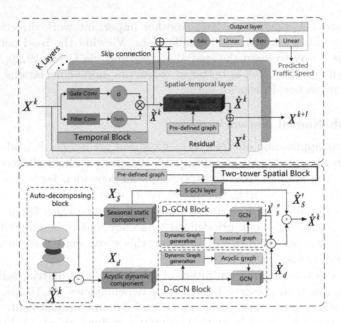

Fig. 2. Our framework.

traffic data, respectively. To avoid the problem of gradient vanishing, the residual connections are added from the input of the temporal block to the output of the two-tower spatial block. The output of current spatial-temporal layer will be used as the input of the next spatial-temporal layer. Finally, skip connections are added after each temporal block to connect features for predicting future traffic speed.

As shown in the bottom part of Fig. 2, our proposed two-tower spatial block contains an auto-decomposing block, a S-GCN layer and two D-GCN blocks with dynamic graph generation and a GCN layer. The auto-decomposing block divides input into a seasonal static component X_s and an acyclic dynamic component X_d by extracting key features of the data. S-GCN layer and D-GCN block are used to model fixed features and dynamic changes of spatial correlations, respectively.

4.2 Temporal Block

The gated dilated causal convolution is used as our temporal convolution layer to learn temporal trend [16]. It allows us to handle longer sequences with less layers. The temporal block is described in the golden of the upper part of Fig. 2. Given the input $X^k \in \mathbb{R}^{T \times N \times D}$, it takes the form

$$\widetilde{X}^k = tanh\left(f_k * X^k\right) \odot \sigma\left(g_k * X^k\right), \tag{2}$$

where $*$ denotes convolution operator, f_k and g_k are the learnable matrices in k-th spatial-temporal layer, $\odot$ is the element-wise Hadamard product. And $tanh\left(\cdot\right)$ is an activation function, $\sigma\left(\cdot\right)$ is the sigmoid function which control the rate of information flow to the next layer.

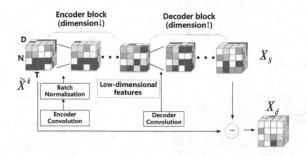

Fig. 3. The process of auto-decomposing block.

4.3   Two-Tower Spatial Block

As mentioned before, spatial correlations are complex, because they not only have the fixed impact from the road network structure but also have dynamic changes determined by traffic characteristics. To capture the dynamic changes of spatial correlations, the following two points need to be considered: (1) The dependencies between traffic nodes can fluctuate rapidly; (2) The entire road network will be affected seasonally due to the traffic conditions changes.

Therefore, we propose a two-tower spatial block, as shown in the bottom part of Fig. 2. We first design an auto-decomposing block to divide traffic data into two components, and then following two D-GCN blocks are used to capture two kinds of changes in spatial correlations. S-GCN layer applies diffusion convolution on the pre-defined graph to model the inherent influence of road network structure.

Auto-Decomposing Block. As shown in Fig. 3, the auto-decomposing block divides the input data into a seasonal static component and an acyclic dynamic component. The input traffic speed data $\widetilde{X}^k \in \mathbb{R}^{T \times N \times D}$ passes the encoder block and decoder block composed of multilayer CNNs to get the seasonal static component $X_s \in \mathbb{R}^{T \times N \times D}$. Besides, batch normalization is added behind every encode convolution to accelerate convergence [7]. Finally, we use the input minus the seasonal static component as the acyclic dynamic component $X_d \in \mathbb{R}^{T \times N \times D}$.

The X_s includes the inherent impact of the road network structure and the global changes of spatial correlations, while the X_d represents the local fluctuations of node dependencies. As a result, we can separately model the different changes of the spatial correlations.

D-GCN Block. Taking into account the different characteristics of global changes and local fluctuations of spatial correlation, we design two D-GCN blocks with the same structure to handle different components. Each D-GCN block includes a dynamic graph generation and a GCN layer. The dynamic graph generation constructs a dynamic graph based on the traffic data at the current time step, and then the GCN layer relies on this dynamic graph to capture the dynamic changes of spatial correlations. Same technology is replicated

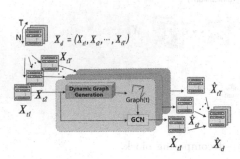

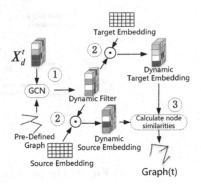

Fig. 4. D-GCN block for X_d, X_s is the same process.

Fig. 5. Dynamic graph generation for X_d^t, the same goes for X_s^t.

to process different input components and generate seasonal graph or acyclic graph, so that we take the processing of acyclic dynamic components X_d as an example to introduce our D-GCN block, as shown in Fig. 4.

(1) Dynamic Graph Generation. The process of generating dynamic graph can be divided into three steps, as shown in Fig. 5.

In The First Step, we apply graph convolution to process incoming traffic data based on the pre-defined graph. This allows each node to perceive the information of neighbor nodes for generating a more suitable real-time graph, as the Eq. 3.

$$DF_t = \mathcal{G} \star f_\theta X_d^t, \tag{3}$$

where X_d^t is the acyclic dynamic component at t time step, and $DF_t \in \mathbb{R}^{N \times D}$ denotes dynamic filter. $\star$ represents the GCN which is defined in later Eq. 6.

In The Second Step, we employ two static node embedding dictionaries $E_{in}, E_{out} \in \mathbb{R}^{N \times D}$ to represent source node embedding and target node embedding, respectively. These two embedding dictionaries are randomly initiated, and whose parameters are learnable. The element-wise Hadamard operation is then computed between dynamic filter DF_t and these two node embeddings:

$$DE_{in}^t = tanh\left(\lambda\left(DF_t \odot E_{in}\right)\right),$$
$$DE_{out}^t = tanh\left(\lambda\left(DF_t \odot E_{out}\right)\right). \tag{4}$$

In The Last Step, we construct a dynamic graph (acyclic graph for X_d, seasonal graph for X_s) $DG_t \in \mathbb{R}^{N \times N}$ by calculating node similarities. It is showed as Eq. 5.

$$DG_t = softmax\left(ReLU\left(tanh\left(\lambda\left(DE_{in}^t DE_{out}^{t}{}^T - DE_{out}^t DE_{in}^{t}{}^T\right)\right)\right)\right), \tag{5}$$

where λ is a hyper-parameter for controlling the saturation rate of the activation function, and $softmax$ function is used to normalize the matrix.

(2) GCN Layer. The diffusion convolution is adopted by as our GCN [10]. The process of diffusion convolution can be defined as Eq. 6, where $\theta \in \mathbb{R}^{K \times 2}$ are filters, k is diffusion step, and $\mathcal{D}_f^k, \mathcal{D}_b^k$ represent the forward and backward transition matrixes, respectively.

$$\hat{X} = \mathcal{G} \star f_\theta X = \sum_{k=0}^{K} \left(\theta_{k,1} \mathcal{D}_f^k + \theta_{k,2} \mathcal{D}_b^k \right) X. \tag{6}$$

Thus, the process of D-GCN block for X_d is shown as Eq. 7, where DG_t is the acyclic graph constructed by dynamic graph generation at t time step.

$$\hat{X}_d^{t,:,:} = DG_t \star g_\theta X_d^t. \tag{7}$$

4.4   Loss Function

We learn temporal and spatial correlations in temporal block and the two-tower spatial block, respectively. And skip connections are added after the temporal block to transmit the spatial-temporal correlations learned in each layer, then are passed to output layer for final prediction. The output layer contains two groups of ReLU and Linear layers to get desired prediction length. Residual is used to speed up convergence [16]. They are shown in the upper part of Fig. 2.

The mean absolute error (MAE) is our loss function, described in Eq. 8.

$$Loss = \frac{1}{TND} \sum_{t=1}^{T} \sum_{i=1}^{N} \sum_{j=1}^{D} \left| X_{ij}^t - \hat{X}_{ij}^t \right|, \tag{8}$$

where X and $\hat{X}$ represent the ground truth and the predicted result, respectively.

5   Experiment

5.1   Datasets and Evaluation Measures

In order to verify our proposed framework, we conducted a series of comparative experiments on two highway datasets[1] and an urban dataset[2] [10,15,16,22].

- **METR-LA**: A public traffic speed dataset collected from loop detectors deployed on highway segments in Los Angeles. It contains 207 sensors and records the traffic speed in 4 months, ranging from Mar 1*st* 2012 to Jun 30*th* 2012. The total number of timestamps is 34,272.
- **PEMS-BAY**: The traffic speed dataset is collected by the California Transportation Agencies (CalTrans). It selects 325 sensors ranging from Jan 1*st* 2017 to May 31*st* 2017. The total number of timestamps is 52,116.

---

[1] https://github.com/liyaguang/DCRNN.
[2] https://github.com/lehaifeng/T-GCN.

- **SZ-taxi**: This is the taxi trajectory of ShenZhen from Jan 1st to Jan 31st, 2015. It contains 156 major roads of Luohu District, the speed on each road is calculated every 15 min. The total number of timestamps is 2,976.

All data input is normalized by the Z-Score method. These datasets are split in chronological order with 70% for training, 10% for validation, and 20% for testing [10]. All experiments use historical traffic information in the past hour to predict the traffic speed in the next 15, 30 and 60 min. Pre-defined static adjacency matrix A is constructed based on pairwise road network distance between sensors with thresholded Gaussian kernel [10]. It is defined as Eq. 9,

$$A_{ij} = \begin{cases} \exp\left(-\frac{d_{v_i,v_j}^2}{\sigma^2}\right) & d_{v_i,v_j} \leq \kappa \\ 0 & d_{v_i,v_j} > \kappa \end{cases} \tag{9}$$

where d_{v_i,v_j} is the distance between sensors. σ is the deviation of distances, and κ is the threshold. MAE and RMSE are used as evaluation metrics [10,15,16].

5.2   Baselines and Experimental Settings

Five baseline schemes are from traditional methods, static matrix-based methods and dynamic graph-based methods. Following their parameters setting, DMST-GCN is reproduced and applied to highway datasets, and some results are reported in their paper [10,15,16]. Besides, all baseline models have been tested on the urban dataset (SZ-taxi) to select their best results.

- **ARIMA** [9]: It uses traditional time series forecasting methods to predict traffic speeds.
- **DCRNN** [10]: It considers RNN to capture temporal correlations and GCN based on a pre-defined graph to model the inherent spatial correlations caused by the road network structure.
- **Graph WaveNet** [16]: It integrates GCN with gated 1D dilated convolutions and adds a self-adaptive matrix to preserve hidden node dependencies.
- **MTGNN** [15]: It designs a graph learning layer to generate a dynamic graph by using external features and employs mix-hop propagation layers and dilated inception layers.
- **DMSTGCN** [6]: It proposes a dynamic graph construction method to learn the time-specific spatial dependencies.

About the parameters setting of our two-tower spatial block, based on our experimental results, the auto-decomposing block has 4 layers, and the out channels are 16, 8, 16, 32. In the GCN layer, diffusion step K is 2 and dropout is 0.5. The hyper-parameter of dynamic graph generation λ is 3. The dimensions of node embeddings E_{in} and E_{out} are 32. The out channel of spatial block is 32. As the parameters setting of temporal block, following in [16], the number of spatial-temporal layers is 8, and the temporal block's diffusion parameters are 1, 2, 1, 2, 1, 2, 1, 2, respectively. Then, the out channel of temporal block is 32.

Our experiments are compiled and tested by Pytorch 1.7.1 on a computer environment (GPU: NVIDIA Titan XP). The optimizer is the Adam algorithm, and the initial learning rate is 0.001 with a decay rate of 0.1 after 70 epochs.

Table 1. Performance comparsion of T^2-GNN with baseline models. The smaller error, the better performance.

	T	Metric	ARIMA	DCRNN	GraphWaveNet	MTGNN	DMSTGCN	T^2-GNN	
METR-LA	15 min	MAE	3.99	2.77	2.69	2.69	<u>2.65</u>	**2.49**	6.0%
		RMSE	8.21	5.38	5.15	5.18	<u>4.94</u>	**4.57**	7.4%
	30 min	MAE	5.15	3.15	3.07	3.05	<u>2.94</u>	**2.73**	7.1%
		RMSE	10.45	6.45	6.22	6.17	<u>5.66</u>	**5.24**	7.4%
	60 min	MAE	6.90	3.60	3.53	3.49	<u>3.32</u>	**3.04**	8.4%
		RMSE	13.23	7.60	7.37	7.23	<u>6.52</u>	**6.06**	7.0%
PEMS-BAY	15 min	MAE	1.62	1.38	1.30	1.32	<u>1.15</u>	**1.09**	5.2%
		RMSE	3.30	2.95	2.74	2.79	<u>2.31</u>	**2.16**	6.4%
	30 min	MAE	2.33	1.74	1.63	1.65	<u>1.40</u>	**1.31**	6.4%
		RMSE	4.76	3.97	3.70	3.74	<u>2.99</u>	**2.79**	6.6%
	60 min	MAE	3.38	2.07	1.95	1.94	<u>1.72</u>	**1.57**	8.7%
		RMSE	6.50	4.74	4.52	4.49	<u>3.78</u>	**3.50**	7.4%
SZ-taxi	15 min	MAE	4.98	3.92	3.32	<u>3.04</u>	3.17	**2.47**	18.7%
		RMSE	7.24	5.39	4.70	<u>4.31</u>	4.49	**3.69**	14.3%
	30 min	MAE	4.67	4.03	3.36	<u>3.05</u>	3.20	**2.48**	18.6%
		RMSE	6.78	5.50	4.47	<u>4.32</u>	4.53	**3.72**	13.8%
	60 min	MAE	4.66	4.29	3.42	<u>3.08</u>	3.24	**2.50**	18.8%
		RMSE	6.77	5.77	4.83	<u>4.35</u>	4.57	**3.74**	14.0%

5.3   Experimental Results

Table 1 shows that our T^2-GNN performs better than all baselines on three datasets. For the 60 min traffic speed prediction on highway datasets, the DMST-GCN is better than other baselines, and our model can further reduce its MAE and RMSE by 8.4%, 7.0% on METR-LA, and 8.7%, 7.4% on PEMS-BAY, respectively. As the urban dataset, MTGNN performs better than other baselines, our model can further reduce its MAE and RMSE by 18.8%, 14.0%.

- **Traditional method**: The traditional ARIMA is limited by its own methods, is unable to deal with complex nonlinear temporal correlations and lacks of modeling spatial correlations. Thus, its prediction errors are the largest.
- **Deep learning methods with static adjacency matrix**: The deep learning models are used in temporal block, therefore, these methods all have higher prediction accuracy than ARIMA. Compared with DCRNN on a pre-defined matrix, Graph WaveNet adds an adaptive matrix to improve the ability of model and thus obtains better prediction results.
- **Deep learning methods with constructing dynamic graphs**: Compared with Graph WaveNet(static matrix), MTGNN and DMSTGCN can better capture the dynamic spatial correlations by constructing dynamic graphs. Our T^2-GNN inherits the method of generating dynamic graphs and further proposes fine modeling of the dynamic changes of spatial correlations, which shows the lowest prediction errors.

Table 2. Ablation study on SZ-taxi.

T	Metric	$\mathbf{T^2}$-**GNN**	w/o seasonal graph	w/o dynamic graphs	w/o pre-defined graph
15 min	MAE	**2.47**	2.55	2.7	2.56
	RMSE	**3.69**	4.07	4.31	4.09
30 min	MAE	**2.48**	2.57	2.74	2.59
	RMSE	**3.72**	4.1	4.35	4.13
60 min	MAE	**2.50**	2.61	2.79	2.64
	RMSE	**3.74**	4.15	4.42	4.14

5.4 Ablation Study

Affected by more external factors, the spatial correlations of urban road is more complex. Compared with the highway datasets, our T^2-GNN has larger improvement effect on the urban dataset, reaching 18.8%. Therefore, we conduct an ablation study on the SZ-taxi dataset. The variants of T^2-GNN are as follows:

- **w/o seasonal graph (variant I)**: We ignore the different characteristics of the two dynamic changes of spatial correlations by removing the generation of seasonal graphs (remove the upper D-GCN block in Fig. 2).
- **w/o dynamic graphs (variant II)**: We do not consider the dynamic changes of spatial correlations (remove all D-GCN blocks in Fig. 2).
- **w/o pre-define graphs (variant III)**: We do not consider the impact of the fixed road network structure by removing the S-GCN layer in Fig. 2.

Compared with the complete T^2-GNN on the 60-mins prediction, in terms of the MAE and RMSE, the variant I increases by 4.4% and 10.9%. The variant II increases by 11.6% and 18.1%, the variant III increases by 5.6% and 10.6%.

For the variant I and variant II, deteriorating experimental results show that constructing dynamic graphs at each time step can help to capture global changes and local fluctuations in spatial correlations. And Compared with the variant I, the prediction effect of the variant II has a more serious decline, about 7.2% and 7.2%. It shows that local fluctuations occur at a higher frequency than global changes. Therefore, generating dynamic graphs and the fine modeling of spatial dependences together improve the prediction effect of our framework.

Variant III only considers two dynamic changes of spatial correlations and its forecast accuracy slightly decreases. It shows that the dynamic changes of spatial correlations are more critical than the road network structure, which is also in line with the complex and dynamic characteristics of traffic data.

6 Conclusions and Future Work

In this paper, we propose a novel T^2-GNN model for traffic speed prediction. To fine modeling node dependencies, D-GCN block is designed to process two components which contain different changes of spatial correlations. Our T^2-GNN outperforms state-of-art baselines on three datasets. In future, we will consider the different effects of historical information to improve traffic speed prediction.

References

1. Bruna, J., Zaremba, W., Szlam, A., et al.: Spectral networks and locally connected networks on graphs. In: ICLR 2014 (2014)
2. Diao, Z., Wang, X., Zhang, D., et al.: Dynamic spatial-temporal graph convolutional neural networks for traffic forecasting. In: AAAI 2019, pp. 890–897 (2019)
3. Ermagun, A., Levinson, D.: Spatiotemporal traffic forecasting: review and proposed directions. Transp. Rev. **38**(6), 786–814 (2018)
4. Fang, S., Zhang, Q., Meng, G., et al.: GstNET: global spatial-temporal network for traffic flow prediction. In: IJCAI, pp. 2286–2293 (2019)
5. Guo, S., Lin, Y., Feng, N., et al.: Attention based spatial-temporal graph convolutional networks for traffic flow forecasting. In: AAAI 2019, pp. 922–929 (2019)
6. Han, L., Du, B., Sun, L., et al.: Dynamic and multi-faceted spatio-temporal deep learning for traffic speed forecasting. In: SIGKDD 2021, pp. 547–555 (2021)
7. Ioffe, S., Szegedy, C.: Batch normalization: accelerating deep network training by reducing internal covariate shift. In: PMLR 2015, pp. 448–456 (2015)
8. Jeong, Y.S., Byon, et al.: Supervised weighting-online learning algorithm for short-term traffic flow prediction. IEEE Trans. Intell. Transp. Syst. **14**(4), 1700–1707 (2013)
9. Kumar, S.V., Vanajakshi, L.: Short-term traffic flow prediction using seasonal ARIMA model with limited input data. Eur. Transp. Res. Rev. **7**(3), 1–9 (2015)
10. Li, Y., Yu, R., Shahabi, et al.: Diffusion convolutional recurrent neural network: data-driven traffic forecasting. In: ICLR 2018 (2018)
11. Li, Z., Li, et al.: A two-stream graph convolutional neural network for dynamic traffic flow forecasting. In: ICTAI 2020, pp. 355–362 (2020)
12. Pan, Z., Liang, et al.: Urban traffic prediction from spatio-temporal data using deep meta learning. In: SIGKDD 2019, pp. 1720–1730 (2019)
13. Sutskever, I., Vinyals, et al.: Sequence to sequence learning with neural networks. In: NIPS 2014, pp. 3104–3112 (2014)
14. Wang, X., Ma, et al.: Traffic flow prediction via spatial temporal graph neural network. In: WWW 2020, pp. 1082–1092 (2014)
15. Wu, Z., Pan, et al.: Connecting the dots: Multivariate time series forecasting with graph neural networks. In: SIGKDD 2020, pp. 753–763 (2020)
16. Wu, Z., Pan, et al.: Graph wavenet for deep spatial-temporal graph modeling. In: IJCAI 2019, pp 1907–1913 (2019)
17. Yao, H., Wu, et al.: Deep multi-view spatial-temporal network for taxi demand prediction. In: AAAI 2018, pp. 2588–2595 (2018)
18. Yu, B., Yin, et al.: Spatio-temporal graph convolutional networks: a deep learning framework for traffic forecasting. In: IJCAI 2018, pp. 3634–3640 (2018)
19. Zhang, J., Zheng, et al.: Deep spatio-temporal residual networks for citywide crowd flows prediction. In: AAAI 2017, pp. 1655–1661 (2017)
20. Zhang, J., Zheng, Y., Qi, D., Li, R., Yi, X., Li, T.: Predicting citywide crowd flows using deep spatio-temporal residual networks. Artif. Intell. **259**, 147–166 (2018)

21. Zhang, Q., Chang, et al.: Spatio-temporal graph structure learning for traffic fore-casting. In: AAAI 2020, pp. 1177–1185 (2020)
22. Zhao, L., Song, et al.: T-GCN: A temporal graph convolutional network for traffic prediction. IEEE Trans. Intell. Transp. Syst. **21**(9), 3848–3858 (2020)
23. Zheng, C., Fan, et al.: GMAN: a graph multi-attention network for traffic prediction. In: AAAI 2020, pp. 1234–1241 (2020)

Memory-Efficient Minimax Distance Measures

Fazeleh Hoseini$^{(\boxtimes)}$ and Morteza Haghir Chehreghani

Chalmers University of Technology, Göteborg, Sweden
{fazeleh,morteza.chehreghani}@chalmers.se

Abstract. Minimax distance measure is a transitive-aware measure that allows us to extract elongated manifolds and structures in the data in an unsupervised manner. Existing methods require a quadratic memory with respect to the number of data points to compute the pairwise Minimax distances. In this paper, we investigate two memory-efficient approaches to reduce the memory requirement and achieve linear space complexity. The first approach proposes a novel hierarchical representation of the data that requires only $\mathcal{O}(N)$ memory and from which the pairwise Minimax distances can be derived in a memory-efficient manner. The second approach is an efficient sampling method that adapts well to the proposed hierarchical representation of the data. This approach accurately recovers the majority of Minimax distances, especially the most important ones. It still works in $\mathcal{O}(N)$ memory, but with a substantially lower computational cost, and yields impressive results on clustering benchmarks, as a downstream task. We evaluate our methods on synthetic and real-world datasets from a variety of domains.

Keywords: Unsupervised learning · Representation learning · Memory efficiency · Minimax distance measure · Sampling

1 Introduction

The complexity of data processing tasks often differs significantly depending on how the data representation is derived. Therefore, deriving an appropriate data representation is crucial. Representation learning is about developing machine learning algorithms to discover useful representations and latent features from given data. Traditional feature learning and deep learning models are two types of representation learning [21]. Traditional feature learning algorithms can be global or local. Global approaches generally aim to extract global information from the data in the learned feature space, while local methods focus on preserving the local similarity between the data while learning the new representations [18]. In the context of Deep Learning, some representation learning methods [9] have been developed that are highly parameterized and require a large amount of labeled data for training which is not available for some applications.

In various applications, the underlying patterns of data are better represented by the transitive relationships between data points than by their direct (dis)similarities. The transitive relation between two data points is derived from the different paths that connect them. In particular, for clustering applications, the Minimax distance between a

© The Author(s), under exclusive license to Springer Nature Switzerland AG 2022
J. Gama et al. (Eds.): PAKDD 2022, LNAI 13280, pp. 419–431, 2022.
https://doi.org/10.1007/978-3-031-05933-9_33

pair of data points, i.e., the minimum largest distance among all possible paths between them, allows for easier separation of elongated manifolds and structures [17]. The Minimax distance is an unsupervised and nonparametric representation learning method that does not require prior knowledge of the shape of the clusters and the structure of the data. However, the major drawback of the Minimax distance technique is that it incurs high computational and storage costs. There are recent studies to improve the computational aspect [10, 14]. Thus, we focus on reducing the storage cost, for which, to the best of our knowledge, there are no previous studies. The contributions of this paper are as follows:

- We propose a hierarchical representation for the data with a linear memory requirement with respect to the number of data points, from which the exact pairwise Minimax distances can be derived in a memory-efficient manner.
- We propose a memory-efficient sampling method that is able to exactly recover the majority of Minimax distances, especially the most important ones, by exploiting the proposed hierarchical representation of the data. The proposed sampling method is adaptable to Minimax distances. Then, we investigate the effectiveness of the sampling method on several synthetic and real-world datasets and compare it with the alternatives.

2   Background

A category of unsupervised representations and distance measures, called *link-based* distance [1, 6, 19], consider all *paths* between data points represented by a graph. These distance measures are often obtained by inverting the Laplacian of a basic distance matrix in the context of a Markov diffusion kernel [5]. However, computing link-based distances for all pairs of objects requires a running time of $\mathcal{O}(N^3)$, where N is the number of objects. A more promising distance measure, called *Minimax distance* (or *path-based* distance [4]), looks for the minimum largest gap among all feasible paths between the objects in the graph [13]. The Minimax distance measure, I) detects the underlying characteristics of the data in an unsupervised manner, II) extracts elongated structures and manifolds by considering transitive relations, and III) appropriately adapts to the shape of different structures. It has demonstrated sophisticated results in several clustering and classification applications [4, 10, 13, 14].

Computing pairwise Minimax distances can be expensive both in terms of computation and memory. The first methods on pairwise Minimax distance were based on the Floyd-Warshall algorithm with a computational complexity of $\mathcal{O}(N^3)$. Later, more efficient methods for computing pairwise Minimax distances were developed [11, 13], including the sparse setting in [12]. Spectral clustering [16], clustering for hyper-spectral data [15], and K-nearest neighbor search are other tasks where Minimax distances have been successfully applied by several computationally efficient methods [10, 14].

Despite the recent advances in the computational efficiency of Minimax distances, the memory efficiency of Minimax distances has not yet been studied in depth. The computation of pairwise Minimax distances requires $\mathcal{O}(N^2)$ memory; however, such memory capacity may not be available in many applications, such as embedded devices and systems with limited memory. Therefore, in this paper, we investigate a memory-efficient solution for computing pairwise Minimax distances by developing two approaches for a memory-efficient Minimax computation with a memory complexity of $\mathcal{O}(N)$ instead of $\mathcal{O}(N^2)$.

3 Definitions and Notations

We are given a dataset $\mathbf{D}$ consisting of N objects with indices $\mathbf{O} = \{1,...,N\}$ and the respective measurements $\mathbf{D} = \{\mathbf{d}_1, \mathbf{d}_2, ..., \mathbf{d}_N\}$ where $\mathbf{d}_i$ is the feature vector of the object $i \in \mathbf{O}$. We assume a function $f(i, j)$ that returns the pairwise dissimilarity of objects i and j, $\forall i, j \in \mathbf{O}$. We represent the dataset by a graph $G(\mathbf{O}, \mathbf{E})$, where the nodes $\mathbf{O}$ represent the indices of the objects and the edge weights $\mathbf{E}$ indicate the pairwise dissimilarities obtained by f. The function f is required to meet semi-metrics conditions.

In this work, we compute the transitive relations between the objects in order to extract the elongated manifolds and structures, i.e., if object i is similar to l, and l is similar to j, then objects i and j are considered as similar objects, even if their direct pairwise similarity is low. To compute such a transitive-aware distance measure, we can look for the smallest largest gap among all possible paths between i and j on graph $\mathbf{G}$. For each particular path, we compute the largest gap (maximal edge weight), and then we choose the minimum of the largest gaps among different paths. Thus, this distance measure, known as Minimax or path-based distance [4,13] can be formulated as

$$\mathbf{M}_{ij} = \min_{p \in P_{i,j}} \max_{(n,m) \in p} \mathbf{E}_{n,m}, \qquad (1)$$

where $\mathbf{M}_{ij}$ yields the Minimax distance between i and j, $\forall i, j \in \mathbf{O}$; and $P_{i,j}$ is the set of possible paths connecting i and j over graph $G(\mathbf{O}, \mathbf{E})$. A path $p \in P$, is characterized by a set of consecutive edges, each denoted by (n,m) where $n,m \in \mathbf{O}$ and $n \neq m$ is a pair of vertices at two ends of edge $\mathbf{E}_{n,m}$.

As shown in Eq. 1, to compute the Minimax distance between two objects, we need to trace all possible paths between them, which is computationally expensive. As studies in [11,13], given graph $G(\mathbf{O}, \mathbf{E})$, a minimum spanning tree (MST) over G provides the necessary and sufficient information to obtain the pairwise Minimax distances for all pairs. Thus, the pairwise Minimax distances on an arbitrary graph are equal to the pairwise Minimax distances computed on any MST over the graph. However, since we assume the available memory is linear with the number of objects, we cannot have a complete graph requiring memory of $\mathcal{O}(N^2)$. Thus, in our approach, we assume that only objects $\mathbf{D}$ are available, and instead of edge weights of a complete graph we assume that we have access to an oracle that gives us the dissimilarity of objects $i, j \in \mathbf{O}$, $f(i,j)$, upon request. We define the dissimilarity function f as squared Euclidean distance between objects i, and j; which can be interpreted as edge weights of the G.

4 Memory-Efficient Minimax Framework

We propose memory-efficient approaches for computing pairwise Minimax distances where the memory complexity is linear to the number of objects, i.e., $\mathcal{O}(N)$. We study

Algorithm 1. Incremental Prim's algorithm

Input: Dataset $\mathbf{D}$ and an oracle returns $f(i,j)$ for $\forall\, i,\ j \in \mathbf{O}$ upon request, where $f(.,.)$.
Output: Vector $\mathbf{E}_T \in \mathbb{R}^{(N-1)\times 1}$ in tree $T(\mathbf{O}, \mathbf{E_T})$ holding edge weights of T, and $\mathbf{P}_T \in \mathbb{N}^{(N-1)\times 2}$ denoting $\mathbf{E}_T$ edges' endpoints in T.

1:	$\mathbf{C} \leftarrow \{v\}$	8:	$\mathbf{P}_T(n,:) \leftarrow \big(\mathbf{S}(j)\,,\, j\big)$
2:	$\mathbf{I} \leftarrow f(v,i)$ for all $i \in \mathbf{O}$	9:	**for** $m = 1$ *to* N **do**
3:	$\mathbf{I}(v)$, $\mathbf{S} \leftarrow +\infty$, v	10:	**if** $\big(\mathbf{I}_m > f(j,m)$ and $\mathbf{I}_m \neq \infty\big)$ **then**
4:	**for** $n = 1$ *to* $N-1$ **do**	11:	$\mathbf{I}(m)$, $\mathbf{S}(m) \leftarrow f(j,m)$, j
5:	$\mathbf{E}_T(n)$, $j \leftarrow \min(\mathbf{I})$, $\arg\min(\mathbf{I})$	12:	**end if**
6:	$\mathbf{C} \leftarrow \mathbf{C} \cup \{j\}$	13:	**end for**
7:	$\mathbf{I}(j) \leftarrow +\infty$	14:	**end for**

the memory efficient Minimax distances in two settings: I) We maintain the exact Minimax distances between all pairs of objects, addressing the memory constraints at the cost of higher computational cost. II) We propose an efficient sampling method based on Minimax distances and preserve Minimax distances only between samples. We argue that our sampling procedure is able to capture the structures of all objects.

An MST, provides all necessarily and sufficient information to compute the pairwise Minimax distances [11, 13]. Thus, both approaches get a tree $T(\mathbf{O}, \mathbf{E_T})$ as an input. We define a tree by set of objects $\mathbf{O}$, a vector $\mathbf{E}_T \in \mathbb{R}^{(N-1)\times 1}$ representing edge weights, and a matrix $\mathbf{P}_T \in \mathbb{N}^{(N-1)\times 2}$ denoting the two corresponding endpoints of the edges in $\mathbf{E}_T$. Different MST algorithms usually assume a graph of a dataset like G is given. However, such a graph can require an $\mathcal{O}(N^2)$ memory, which cannot be afforded in our settings. Therefore, to compute the minimum spanning tree, we employ the Prim's algorithm in an incremental way. Rather than assuming the graph edges are given, we assume instead there is an oracle that gives $f(i,j)$ for $\forall\, i,\ j \in \mathbf{O}$ at the request.

Algorithm 1, describes the incremental Prim's algorithm, which the dataset $\mathbf{D}$ and the mentioned oracle are the only inputs, and the outputs are $\mathbf{E}_T$ in $T(\mathbf{O}, \mathbf{E}_T)$ and $\mathbf{P}_T$. Let list $\mathbf{C}$ shows the current objects in the MST, initialized by a random object $v \in \mathbf{O}$, i.e., $\mathbf{C} = \{v\}$. Also, vector $\mathbf{I} \in \mathbb{R}^{N\times 1}$ and $\mathbf{S} \in \mathbb{N}^{N\times 1}$ indicate the minimum dissimilarities between $\mathbf{C}$ and each of the objects in $\mathbf{O} \setminus \mathbf{C}$, and their corresponding indices (the vertex in $\mathbf{C}$ which has the minimum dissimilarity from a vertex in $\mathbf{O} \setminus \mathbf{C}$), respectively. At first, $\mathbf{I}$ is initialized with the pairwise dissimilarities between v and the objects in $\mathbf{O}$ using the oracle (returning $f(.,.)$). We set a element in $\mathbf{I}$ to $+\infty$ once the respective index is included in $\mathbf{C}$, thus, $\mathbf{I}(v) = +\infty$, and then, set all the elements in $\mathbf{S}$ to v. Then, the algorithm at each step finds the minimum element in $\mathbf{I}$ and adds the element to the tree edge weights $\mathbf{E}_T$. It obtains the index of the minimal element in $\mathbf{I}$, i.e., j, and adds the index element to $\mathbf{C}$, and then sets the respective index in $\mathbf{I}(j)$ to $+\infty$. Next, it adds two endpoints of the newly added edge to $\mathbf{P}_T$. Then, it updates $\mathbf{I}$ which is the minimum dissimilarity between the unselected objects in $\mathbf{O}$ and $\mathbf{C}$, and then updates the indices in $\mathbf{S}$. Note that elements in $\mathbf{I}$ which are already set to $+\infty$, stay unchanged. The algorithm stops after $N-1$ iteration and returns $T(\mathbf{O}, \mathbf{E_T})$ and edges' endpoints $\mathbf{P}_T$.

Algorithm 2. Maintaining exact pairwise Minimax dissimilarity

Input: Tree $T(\mathbf{O}, \mathbf{E}_T)$ described by a vector of edges $\mathbf{E}_T \in \mathbb{R}^{(N-1) \times 1}$, and $\mathbf{P}_T \in \mathbb{N}^{(N-1) \times 2}$ holding the two endpoints of edges in T.

Output: Array $\mathbf{A}_{(N-1) \times 4}$

1: $\mathbf{E}_T$, *sorted_ind* ← Ascending sort the $\mathbf{E}_T$,   9: $\mathbf{C}_A(s) \leftarrow N + s$
    and obtain sorted indices,   10:   **for** $r = s + 1$ *to* $N - 1$ **do**
2: $\mathbf{P}_T \leftarrow \mathbf{P}_T(\text{sorted\_ind})$ 11:   **if** $\mathbf{N}_A(r) = \mathbf{N}_A(s)$ or $\mathbf{N}_A(r) = \mathbf{M}_A(s)$
3: Create $\mathbf{A}_{(N-1) \times 4} = [\mathbf{W}_A, \mathbf{N}_A, \mathbf{M}_A, \mathbf{C}_A]$ **then**
    where the columns hold edge weights,   12: $\mathbf{N}_A(r) = \mathbf{C}_A(s)$
    component IDs of two ends of edges, and   13:   **end if**
    component IDs of edges.   14:   **if** $\mathbf{M}_A(r) = \mathbf{N}_A(s)$ or $\mathbf{M}_A(r) = \mathbf{M}_A(s)$
4: **for** $i = 1$ *to* $N - 1$ **do**   **then**
5:    $\mathbf{W}_A(i)$, $\mathbf{C}_A(i) \leftarrow \mathbf{E}_T(i)$, *0*   15: $\mathbf{M}_A(r) = \mathbf{C}_A(s)$
6:    $\mathbf{N}_A(i), \mathbf{M}_A(i) \leftarrow \mathbf{P}_T(i, 1), \mathbf{P}_T(i, 2)$ 16:   **end if**
7: **end for**   17:   **end for**
8: **for** $s = 1$ *to* $N - 1$ **do**   18: **end for**

4.1 A Memory-Efficient Algorithm for Exact and Complete Minimax Distances

In this setting, we compute exact pairwise Minimax distances for all pairs with a linear memory capacity. Naturally, given linear memory capacity, the pairwise distances cannot be stored directly. Instead, we propose an algorithm that organizes the data hierarchically so that all pairwise Minimax distances are encoded and the Minimax distance between any pair of objects can be retrieved by a separate procedure.

Algorithm 2, works in a bottom-up fashion, and groups objects into a hierarchy of what we will refer to as *components*. Each component is a sub-tree of the entire tree T, and is specified by a *component ID*. Initially, the components are simply the objects themselves. We start from sorting the edges of the tree T, with ascending weights. This can be done via an in-place sorting method without a need for significant extra memory [2]. Starting with the smallest edge, the algorithm traverses all edges, and in each step it merges the two components connected with the current edge. Referring to the two components as "child" components, a new "parent" component is created. The sub-tree of the parent component will then consist of the sub-trees of both child components, as well as the connecting edge between them. We assign a *component ID* to the parent component, and let the connecting (inner) edge carry a reference to it. Furthermore, any outer edges previously connected to the child components, are redirected to the new parent component. To conclude, the algorithm successively combines components into larger components based on their similarities, while assigning a *component ID* to each of them. The algorithm terminates when all components have merged into one, and the final result is a hierarchy of nested components, from which Minimax distances between all objects can then be retrieved. A couple of properties for the final hierarchy of components can be observed: I) If going higher up in the hierarchy, the edge weights are increasing. II) Intra-component objects (i.e., objects that are contained within the same component) are treated similarly in the sense that they all have the same Minimax distance to any other inter-component objects (i.e., objects outside of that component). III) The Minimax distance between any pair of objects, is exactly the edge weight of the lowest level component containing both objects, since that is the maximum edge weight along the path between the objects. Note that the path itself is unique when considering a tree, and that an MST preserves Minimax distances.

Algorithm 2 illustrates our approach in greater detail. The algorithm starts with storing the edge weights $\mathbf{E}_T$ of tree T in ascending order, and then, permutes the endpoints of the edges $\mathbf{P}_T$ accordingly. In practice, the algorithm works with a matrix $\mathbf{A} \in \mathbb{R}^{(N-1) \times 4}$ where each row corresponds to an edge in the tree T. The columns of $\mathbf{A}$ are denoted as $\mathbf{A} = \begin{bmatrix} \mathbf{W}_A, \mathbf{N}_A, \mathbf{M}_A, \mathbf{C}_A \end{bmatrix}$, where for any given row, $\mathbf{W}_A$ holds the corresponding edge weights, $\mathbf{N}_A$ and $\mathbf{M}_A$ hold *component IDs* of the connected components, and $\mathbf{C}_A$ holds the *component ID* of the corresponding "parent" component, which can also be considered as the *component ID* of the edge itself.

As an example, assuming tree T' with $O_{T'} = [1, 2, 3, 4]$ and $\mathbf{P}_{T'} = [< 1, 2 >, < 1, 3 >, < 2, 4 >]$ and $\mathbf{E}_{T'} = [2, 3, 4]$. Table 1 shows the steps of Algorithm 2 on the example tree T'. In the initialization step, $\mathbf{A}$ is filled by the edge weights $\mathbf{E}'_T$, and indices of the two ends of each edge, $\mathbf{N}_A$ and $\mathbf{M}_A$, get values in $\mathbf{P}_{T'}$, and *component ID* are initialized by *zero*. After initialization, the algorithm is performed sequentially on the sorted edge weights $\mathbf{W}_A$, whereat step s we first set the *component ID* of edge $\mathbf{W}_A(s)$ to $\mathbf{C}_A(s) = N + s$, corresponding to the new parent component resulting from the merge. Then, we search through the remaining edges $\{r : s < r < N\}$ (that is the ones with larger weights), and for each edge r connected to any of the child components for s – $\mathbf{N}_A(s)$ or $\mathbf{M}_A(s)$ – we replace the matching *component ID* – $\mathbf{N}_A(r)$ or $\mathbf{M}_A(r)$ – with the parent component for s, $N + s$. Algorithm 2 terminates when it has processed all edges on $\mathbf{W}_A$, and consequently all components have merged into one. The final version of $\mathbf{A}$ is returned, which then encodes the hierarchy of components as well as the exact pairwise Minimax distances between all pairs of objects. The memory requirement of Algorithm 2 is $\mathcal{O}(N)$; however, its computational complexity is $\mathcal{O}(N^2)$.

Once we obtain the matrix $\mathbf{A}$, any query about Minimax pairwise distance of two arbitrary objects i and j can be acquired by Algorithm 3, with the memory requirement of $\mathcal{O}(N)$ and computational cost of $\mathcal{O}(N^2)$. In order to determine the Minimax distance between i and j, we find the lowest level component such that both objects are contained, noting that the edge weight of that component is exactly the Minimax distance we are looking for, as mentioned previously. In more detail, the algorithm finds one list of 2-tuples $< componentID, weight >$ for each of the objects i and j, representing the

Table 1. Values of the matrix $\mathbf{A}$, while running Algorithm 2 on the sample graph.

	$A_{s=1}$			$A_{s=1}$			$A_{s=2}$			$A_{s=3}$		
W_A	2	3	4	2	3	4	2	3	4	2	3	4
N_A	1	1	2	1	5	2	1	5	6	1	5	6
M_A	3	2	4	3	2	4	3	2	4	3	2	4
C_A	0	0	0	5	0	0	5	6	0	5	6	7

Algorithm 3. Query on exact pairwise Minimax distance of a pair(i, j)

Input: Two arbitrary object i and $j \in O$, and a two-dimensional array $\mathbf{A}_{(N-1)\times 4}$. The columns hold $\mathbf{W}_A$, $\mathbf{N}_A$, $\mathbf{M}_A$, and $\mathbf{C}_A$.
Output: Minimax distance of i and j.

1: $i\_list, j\_list = [< i, null >], [< j, null >]$ 7: $j\_list.append(< \mathbf{C}_A(k), \mathbf{W}_A(k) >)$
2: **for** $k = 1$ *to* $N - 1$ **do**   8:   **end if**
3:    **if** $i\_list.lastItem.componentID$ =   9: **end for**
      $\mathbf{N}_A(k)$ or $\mathbf{M}_A(k)$ **then**   10: **for** $k = 1$ *to* $|i\_list|$ **do**
4:      $i\_list.append(< \mathbf{C}_A(k), \mathbf{W}_A(k) >)$ 11:   **if** $i\_list(k)$ *exists in* j_list **then**
5:    **end if**   12:   **return** $i\_list.weight(k)$
6:    **if** $j\_list.lastItem.componentID$ =   13:   **end if**
      $\mathbf{N}_A(k)$ or $\mathbf{M}_A(k)$ **then**   14: **end for**

---

chain from the corresponding lowest level component (the object itself), to the highest level component (the one "root" component containing all others). The lists are initialized as $[< i, null >]$ and $[< j, null >]$, and are gradually extended by the algorithm. The lists are iteratively appended with new tuples, corresponding to components at higher levels. In each iteration, we take the *componentID* of the current last item in the list of tuples, and search through $\mathbf{N}_A$ and $\mathbf{M}_A$ for a match. Assuming there is a match in the k-th iteration, we append $\mathbf{C}_A(k)$ and $\mathbf{W}_A(k)$ as $< componentID, weight >$ to the list. We perform the same steps for both i and j and obtain a list of 2-tuples for each. Finally, go through one of the lists tuple by tuple, and for each of them look for a matching tuple in the other list. Once we find a matched pair of tuples, these will correspond to the lowest level component, containing both i and j. Consequently, as discussed before, the *weight* of that tuple will be the desired Minimax distance between i and j. For the mentioned example tree T' with the obtained matrix $\mathbf{A}$ represented in Table 1, to compute the Minimax distance between $i = 4$ and $j = 3$, first we form the $i\_list = [< 4, null >, < 7, 4 >]$ and $j\_list = [< 3, null >, < 5, 2 >, < 6, 2 >, < 7, 4 >]$, then we find the matching pair between these two list according to the Algorithm 3 which is $< 7, 4 >$. Thus, the Minimax distance between $i = 4$ and $j = 3$ is 4. Also, the algorithm works the same way to obtain the Minimax distance between two adjacent nodes.

4.2 A Computationally and Memory-Efficient Sampling-Based Algorithm

Due to the high computational cost of the previous approach for querying pairwise Minimax distances, in this section we extend our approach to a memory-efficient sampling framework that is also computationally efficient. The approach is hierarchical and focuses on samples of the data. A sample in this context refers to a representative of a subset of the objects. We present the sampling framework for the downstream task of clustering, for which it is well suited, but note that it could be applied similarly to many tasks. To evaluate our method, we apply a clustering task to the samples and then generalize the labels of the samples to the out-of-samples objects. For other tasks, the same procedure can be applied: First, ruining the task on the samples and then generalize the results to all objects. However, the nature of the final step depends on the task at hand.

Algorithm 4 shows an overview of the different steps of our framework given a dataset $\mathbf{D}$ of N objects and an oracle giving the squared Euclidean distance between every two objects in $\mathbf{O}$. the algorithm contains the following steps:

- **Step1:** We obtain the MST $T(\mathbf{O}, \mathbf{E}_T)$ and the matrix $\mathbf{P}_T$ by applying the incremental Prim's algorithm as described in Algorithm 1.
- **Step2:** We collect $\lceil \sqrt{N} \rceil$ samples from $\mathbf{D}$ by Minimax Sampling (MM) and store the sample indices in $\mathbf{S} \in \mathbb{N}^{\lceil \sqrt{N} \rceil \times 1}$. Our sampling divides the dataset into $|\mathbf{S}|$ disjoint components. We discuss this step in greater details in Subsect. 4.2.1.
- **Step3:** We compute the pairwise Minimax distances between the $\lceil \sqrt{N} \rceil$ components (samples) denoted by $\mathbf{M}_s \in \mathbb{R}^{\lceil \sqrt{N} \rceil \times \lceil \sqrt{N} \rceil}$. Since every object in a given component is similar, in the sense that all have equal Minimax distance to any object outside of the component, we can pick any member to represent a component. Therefore, we randomly pick one object from each component to obtain $\lceil \sqrt{N} \rceil$ samples and let $\mathbf{S} \in \mathbb{N}^{\lceil \sqrt{N} \rceil \times 1}$ represent the sample objects indices. Then, using the algorithm proposed in [11, 13] we compute the pairwise Minimax distances between the samples in $\mathbf{S}$ with spaces and run time complexity of $\mathcal{O}(N)$.
- **Step4:** Some applications require vector space representations. To obtain a vector representation, we employ an embedding of the pairwise $\mathbf{M}_s$ with Multidimensional Scaling (MDS) [20] to $\mathbf{E}_d \in \mathbb{R}^{|\mathbf{S}| \times d}$ where $d \leq |\mathbf{S}|$. Therefore, the space complexity after the embedding is $\mathcal{O}(|\mathbf{S}| \times |\mathbf{S}|) = \mathcal{O}(N)$.
- **Step5:** As mentioned, Minimax Sampling (MM) approach (step2), splits the dataset into $|\mathbf{S}|$ disjoint components. Thus, each sample represents a group of objects sharing the same *component ID*, and every object is represented by one and only one such sample. The samples can be utilized by different machine learning tasks. We select clustering; a fundamental task in machine learning, to assess the capacity of our sampling method to capture underlying patterns in the dataset. So, we cluster the samples and obtain label vector $\mathbf{L}_S \in \mathbb{N}^{\lceil \sqrt{N} \rceil \times 1}$, $\mathbf{L}_S = \{l_1, l_2, ..., l_S\}$.
- **Step6:** Since each sample belongs to one and only one component, we can consider a sample as the representative of out-of-sample objects within the same com-

---

Algorithm 4. Clustering over sampling-based Minimax distances.

Input: Dataset $\mathbf{D} = \{\mathbf{d}_1, \mathbf{d}_2, ..., \mathbf{d}_N\}$, and an oracle gives the square Euclidean distance of any two arbitrary objects in $\mathbf{D}$ on request.
Output: Set of cluster labels $\mathbf{L} = \{l_1, l_2, ..., l_N\}$
1: Obtain tree $T(\mathbf{O}, \mathbf{E}_T)$, and vector P_T by described algorithm in Algorithm 1.
2: Collect $\lceil \sqrt{N} \rceil$ samples with indices hold by a vector $\mathbf{S} \in \mathbb{N}^{\lceil \sqrt{N} \rceil \times 1}$ from the entire dataset $\mathbf{D}$.
3: Obtain the pairwise Minimax distances $\mathbf{M}_s \in \mathbb{R}^{|\mathbf{S}| \times |\mathbf{S}|}$ of the objects in $\mathbf{S}$.
4: Embed $\mathbf{M}_s$ with Multidimensional Scaling (MDS) to $\mathbf{E}_d \in \mathbb{R}^{|\mathbf{S}| \times d}$ where $d \leq |\mathbf{S}|$.
5: Apply the clustering algorithm on $\mathbf{E}_d$ and obtain label vector of samples $\mathbf{L}_S = \{l_1, l_2, ..., l_S\}$.
6: Extend the samples' labels in $\mathbf{L}_S$ to all other objects in $\mathbf{D}$ to obtain labels $\mathbf{L} = \{l_1, l_2, ..., l_N\}$.

ponent. Therefore, we can generalize the labels of the samples to the labels of their corresponding out-of-samples objects to obtain a label vector $\mathbf{L} \in \mathbb{N}^{N \times 1}$, $\mathbf{L} = \{l_1, l_2, ..., l_N\}$.

4.2.1 Minimax Sampling (MM)

Algorithm 5 describes different steps of MM Sampling given the tree $T(\mathbf{O}, \mathbf{E}_T)$, and pairs of edge endpoints denoted by $\mathbf{P}_T$. We use the same definition for a component as we defined in Sect. 4.1, and we construct components of objects in a bottom-up fashion. The main difference between Algorithm 2 in Sect. 4.1 and Algorithm 5 is that in Algorithm 2 we traverse all edges in $\mathbf{E}_T$, but here in Algorithm 5, we only traverse the $\lceil \sqrt{N} \rceil$ smallest edges in $\mathbf{E}_T$ and form $\lceil \sqrt{N} \rceil$ components. Then, we assign other edges to one of these components. In the following, we will describe the algorithm in greater detail.

In Algorithm 5, first, we start with ascending sorting edge weights and corresponding two endpoints of edges. In each step, the algorithm merge two child components, can be seen as two sub-trees, and forms a new parent component or parent sub-tree. Therefore, two sub-trees and the connecting edge between them shape a parent subtree. In each step, the algorithm assigns a new *component ID* to two child components, and the connecting edge, after merging them. Vector $\mathbf{C} \in \mathbb{N}^{N \times 1}$ holds the *component IDs*, initialized with the nodes indices. Thus, the algorithm starts with the number of components equal to number of objects N. The algorithm performs iteratively on the sorted edge weights $\mathbf{E}_T$ in T. In the $\mathbf{i} - th$ iteration, the algorithm considers two child components connected by $\mathbf{E}_T(i)$, with the corresponding *Component IDs* represented by $\mathbf{P}_T(i,:)$. For simplicity, lets n and m hold the corresponding child *component IDs* of these two endpoints, then the algorithm finds objects belongs to the child components with *component ID* equal to n, and calls it $\mathbf{S}_1$, and similarity, $\mathbf{S}_2$ shows the objects with the *component ID* equal to m. Then, the algorithm merges $\mathbf{S}_1$ and $\mathbf{S}_2$ and shapes the parent component and assigns a new *component ID* to the objects in the parent component, i.e., $C(j) = N + i$, for $\forall j \in (\mathbf{S}_1 \cup \mathbf{S}_2)$. The algorithm iterates until the number of (nonempty) components becomes equal to the desired number of samples, i.e., $\lceil \sqrt{N} \rceil$. The output of the algorithm is $\mathbf{C}$ which shows the *component IDs* of all objects. Objects within the same component, shared their *component IDs* and there are $\lceil \sqrt{N} \rceil$ non-identical *component IDs* in $\mathbf{C}$.

Algorithm 5. Minimax Sampling (MM)

Input: $T(\mathbf{O}, \mathbf{E}_T)$ characterized by a vector of edge wights $\mathbf{E}_T$ and two endpoints of edges $\mathbf{P}_T$.
Output: $\mathbf{C} \in \mathbb{N}^{N \times 1}$ representing *Component IDs* of objects in $\mathbf{O}$.

1: $\mathbf{E}_T$, *sorted_ind* $\leftarrow$ Ascending sort the $\mathbf{E}_T$, and obtain sorted indices.
2: $\mathbf{P}_T \leftarrow \mathbf{P}_T(sorted\_ind)$
3: $\mathbf{C} \leftarrow \mathbf{O}$
4: **for** $i = 1$ *to* $N - \lceil \sqrt{N} \rceil$ **do**
5: $\quad n, m \leftarrow \mathbf{C}(\mathbf{P}_T(i,1)), \mathbf{C}(\mathbf{P}_T(i,2))$
6: $\quad$ **for** $j = 1$ *to* N **do**
7: $\quad\quad$ **if** $\mathbf{C}(j) = n$ or $\mathbf{C}(j) = m$ **then**
8: $\quad\quad\quad \mathbf{C}(j) = N + i$
9: $\quad\quad$ **end if**
10: $\quad$ **end for**
11: **end for**

As mentioned, an MST sufficiently contains the edges that represent the pairwise Minimax distances. Thus, when we combine two child components, then the weight of the respective edge represents the Minimax distance between the set of objects in the child component S_1 and the set of objects in the child component S_2. Because, I) this edge weight is the largest dissimilarity on the (only) MST path between the objects in S_1 and those in S_2, II) all the other edge weights inside the components are smaller (or equal) as they are visited earlier. This implies that our sampling method computes the components (samples) such that internal Minimax distances (intra-sample Minimax distances) are kept minimal. In other words, we discard the largest $\lceil\sqrt{N}\rceil - 1$ edges of the MST to produce $\lceil\sqrt{N}\rceil$ samples. Thus, this method is adaptive and consistent with the Minimax distances on the entire data.

5   Experimental Results

In this section, we demonstrate the performance of our framework on clustering of several synthetic and real-world datasets. We apply the different sampling methods on three synthetic datasets [7] (Pathbased, Spiral, Aggregation), and five real-world datasets from UCI repository [3] (Banknote Authentication, Cloud, Iris, Perfume, Seeds).

Our generic sampling framework in Algorithm 4 allows us to investigate different sampling strategies for pairwise Minimax distances (step 2 of Algorithm 4). Here, we study two other sampling methods, k-means and Determinantal Point Processes Sampling (DPPS) presented by [8]. Table 2 shows the memory complexity of each sampling methods. Note that for k-means, the space complexity is $\mathcal{O}(N(D+k))$, where the data dimension is D and the number of clusters, k, is $\lceil\sqrt{N}\rceil$. To obtain $\lceil\sqrt{N}\rceil$ samples, we apply k-means with k equal to $\lceil\sqrt{N}\rceil$. Then, we consider the centroids of the clusters as our samples, letting each sample represents the objects belonging to its cluster.

For each dataset, we first sample the dataset with the mentioned approaches. Then, we apply a clustering task on the samples using the most common clustering methods, i.e., GMM, k-means, and spectral clustering. To evaluate the clustering results after generalizing samples' labels to out-of-samples objects, we use ground truths of our datasets and apply three commonly used metrics: Rand score, mutual information, and v-measure denoted by RS, MI, and VM.

For embedding (step 4 of Algorithm 4), first we transform the Minimax matrix into a Mercer kernel and then perform an eigenvalue decomposition. Then, we sort normalized eigenvalues of $\mathbf{M}_s$. The eigenvalues drop in magnitude after a specific eigen index which we consider this point as the proper value for d; so, the selected d is

Table 2. The space complexity of the sampling methods. N is the number of objects.

Sampling methods	Space complexity
Minimax Sampling (MM)	$\mathcal{O}(N)$
k-means Sampling (KMS)	$\mathcal{O}\left(N(D+\sqrt{N})\right)$
DPP Sampling (DPPS)	$\mathcal{O}(N^2)$

Table 3. Quantitative GMM results of different sampling schemes.

Sampling method	Metric	Pathbased	Spiral	Aggregation	Banknote	Cloud	Iris	Perfume	Seeds
MM	RS	**0.61**	**1.0**	0.80	**0.62**	**0.96**	0.56	**0.64**	0.48
	MI	**0.68**	**1.0**	0.79	**0.61**	**0.93**	0.58	0.74	0.45
	VM	**0.70**	**1.0**	0.88	**0.61**	**0.93**	0.71	**0.85**	0.50
KMS	RS	0.43	0.32	**0.92**	0.07	0.73	**0.74**	0.59	**0.75**
	MI	0.47	0.44	**0.91**	0.05	0.68	**0.78**	**0.77**	**0.70**
	VM	0.52	0.63	**0.93**	0.05	0.69	0.70	0.81	0.69
DPPS	RS	0.42	0.04	0.88	0.08	0.15	0.64	0.40	0.65
	MI	0.43	0.10	0.90	0.06	0.22	0.66	0.51	0.67
	VM	0.48	0.13	0.92	0.08	0.26	0.69	0.53	**0.69**

adjusted according to the dynamics of eigenvalues (using the elbow rule), resulting in a $(d \times \sqrt{\lceil N \rceil})$-dimensional vector representation of samples.

Eventually, we apply GMM, k-means, and spectral clustering over the embedded vectors $\mathbf{E}_d$. Then, we extend the sample labels to the out-of-samples objects represented by the corresponding sample. Next, we use the mentioned metrics to evaluate the clustering with different sampling methods as shown in Table 3 for GMM. The results for k-means and spectral clustering yield the consistent results. However, due to space limit, we could not include them in this paper. In Table 3, we observe that for two out of three synthetic datasets, MM sampling outperforms the other methods. However, even for *Aggregation* dataset the results from MM sampling are acceptable. Similarly, on UCI datasets, MM sampling yields often the best or close to best results.

Figure 1 illustrates GMM clustering on *Spiral* dataset which are two-dimensional and thus suitable for visualization. Different colors denote different predicted cluster labels, and dots and crosses indicate if predicted labels are correct or not. Figure 1a shows the data points and their true labels. Figure 1b shows GMM results obtained over samples from MM sampling. In Fig. 1b all objects are correctly clustered. MM sampling adapts well with the elongated structures in the data. Figure 1c, and Fig. 1d illustrate the clustering results with KMS and DPPS methods. Both methods mistakenly assign almost all data to a single cluster.

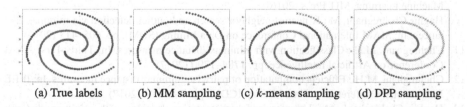

  (a) True labels   (b) MM sampling   (c) *k*-means sampling   (d) DPP sampling

Fig. 1. Illustration of clustering results on *Spiral* datasets by GMM clustering.

In terms of space complexity, MM sampling satisfy a linear space complexity $\mathscr{O}(N)$. k-means sampling requires $\mathscr{O}(N(D+\sqrt{N}))$ memory, and as discussed, DPPS requires computing samples offline with the space complexity of $\mathscr{O}(N^2)$.

6 Conclusion

We developed two approaches for memory-efficient computation of Minimax distances. In the first approach, we compute the exact Minimax distances of all pairs of given dataset in a linear memory with respect to the size of the dataset; however, the respective computational cost is high. In the second approach, we developed a framework for memory-efficient computation of Minimax distances based on effective sampling schemes. Within this framework, we developed an adaptive and memory-efficient sampling method consistent with the pairwise Minimax distances on the entire datasets. We evaluated the framework and the sampling methods on clustering of several datasets with GMM, k-means, and spectral clustering.

References

1. Chebotarev, P.: A class of graph-geodetic distances generalizing the shortest-path and the resistance distances. Discrete Appl. Math. **159**(5), 295–302 (2011)
2. Cormen, T.H., Stein, C., Rivest, R.L., Leiserson, C.E.: Introduction to Algorithms. McGraw-Hill Higher Education (2001)
3. Dua, D., Graff, C.: UCI machine learning repository (2017). http://archive.ics.uci.edu/ml
4. Fischer, B., Buhmann, J.M.: Path-based clustering for grouping of smooth curves and texture segmentation. IEEE Trans. Pattern Anal. Mach. Intell. **25**(4), 513–518 (2003)
5. Fouss, F., Francoisse, K., Yen, L., Pirotte, A., Saerens, M.: An experimental investigation of kernels on graphs for collaborative recommendation and semisupervised classification. Neural Netw. **31**, 5372 (2012)
6. Fouss, F., Pirotte, A., Renders, J.M., Saerens, M.: Random-walk computation of similarities between nodes of a graph with application to collaborative recommendation. IEEE TKDE **19**(3), 355–369 (2007)
7. Fränti, P., Sieranoja, S.: K-means properties on six clustering benchmark datasets. Appl. Intell. **48**(12), 4743–4759 (2018). https://doi.org/10.1007/s10489-018-1238-7
8. Gautier, G., Polito, G., Bardenet, R., Valko, M.: DPPy: DPP sampling with python. JMLR-MLOSS (2019)
9. Goodfellow, I.J., Bengio, Y., Courville, A.C.: Deep Learning, Adaptive Computation and Machine Learning. MIT Press (2016)
10. Haghir Chehreghani, M.: K-nearest neighbor search and outlier detection via minimax distances. In: SIAM International Conference on Data Mining (SDM), pp. 405–413 (2016)
11. Chehreghani, M.H.: Classification with minimax distance measures. In: Thirty-First AAAI Conference on Artificial Intelligence, pp. 1784–1790 (2017)
12. Chehreghani, M.H.: Efficient computation of pairwise minimax distance measures. In: IEEE International Conference on Data Mining, ICDM, pp. 799–804 (2017)
13. Haghir Chehreghani, M.: Unsupervised representation learning with minimax distance measures. Mach. Learn. **109**(11), 2063–2097 (2020). https://doi.org/10.1007/s10994-020-05886-4

14. Kim, K., Choi, S.: Walking on minimax paths for k-NN search. In: Twenty-Seventh AAAI Conference on Artificial Intelligence (2013)
15. Le Moan, S., Cariou, C.: Minimax bridgeness-based clustering for hyperspectral data. Remote Sens. **12**(7), 1162 (2020)
16. Little, A.V., Maggioni, M., Murphy, J.M.: Path-based spectral clustering: guarantees, robustness to outliers, and fast algorithms. J. Mach. Learn. Res. **21**, 6:1–6:66 (2020)
17. Moscovich, A., Jaffe, A., Boaz, N.: Minimax-optimal semi-supervised regression on unknown manifolds. In: Artificial Intelligence and Statistics, pp. 933–942. PMLR (2017)
18. Yan, S., Xu, D., Zhang, B., Zhang, H.J., Yang, Q., Lin, S.: Graph embedding and extensions: a general framework for dimensionality reduction. IEEE TPAMI **29**(1), 40–51 (2006)
19. Yen, L., Saerens, M., Mantrach, A., Shimbo, M.: A family of dissimilarity measures between nodes generalizing both the shortest-path and the commute-time distances. In: Proceedings of the 14th ACM SIGKDD, pp. 785–793 (2008)
20. Young, G., Householder, A.: Discussion of a set of points in terms of their mutual distances. Psychometrika **3**(1), 19–22 (1938). https://doi.org/10.1007/BF02287916
21. Zhong, G., Wang, L.N., Ling, X., Dong, J.: An overview on data representation learning: from traditional feature learning to recent deep learning. JFDS **2**(4), 265–278 (2016)

Joint Feature and Labeling Function Adaptation for Unsupervised Domain Adaptation

Fengli Cui, Yinghao Chen, Yuntao Du, Yikang Cao, and Chongjun Wang(✉)

State Key Laboratory for Novel Software Technology at Nanjing University, Nanjing, China
{mf1933014,mf1933011,duyuntao,mf1933002}@smail.nju.edu.cn
chjwang@nju.edu.cn

Abstract. Unsupervised domain adaptation aims to transfer knowledge from a labeled source domain to an unlabeled target domain. Although having achieved remarkable progress, most existing methods only focus on learning domain-invariant features and achieving a small source error. They ignore the discrepancy between labeling functions which will also cause discrepancy across domains. Inspired by this observation, we propose a novel method to simultaneously perform feature adaptation and labeling function adaptation. Specifically, for the feature adaptation, a domain discriminator is trained to reduce the discrepancy between feature distributions across domains. For the labeling function adaptation, we introduce a target predictor and a predictor discriminator. The target predictor is trained on target samples with pseudo-labels. The predictor discriminator is a novel component and is trained to distinguish whether the prediction output is from the source or the target predictor while the feature extractor and the label predictors try to confuse the predictor discriminator in an adversarial manner. Additionally, the intrinsic characteristics of the target domain are expected to be exploited thanks to the task-specific training. Comprehensive experiments are conducted and results validate the effectiveness of labeling function adaptation and demonstrate that our approach outperforms state-of-the-art methods.

Keywords: Transfer learning · Unsupervised domain adaptation · Labeling function adaptation

1 Introduction

Transfer learning aims to transfer knowledge from a related but different source domain to a target domain, such that we can get better performance in the target domain [24,34]. Unsupervised domain adaptation is a sub-field of transfer learning, where there are labeled data in the source domain and unlabeled data

---

F. Cui and Y. Chen—The first two authors contributed equally.

J. Gama et al. (Eds.): PAKDD 2022, LNAI 13280, pp. 432–446, 2022.
https://doi.org/10.1007/978-3-031-05933-9_34

in the target domain [24,34]. As the source data and target data are drawn from different distributions, by reducing the discrepancy across domains, the source predictor (classifier) is able to generalize well in the target domain, which is the key to unsupervised domain adaptation [2,24,40].

Many theories and algorithms have been proposed for unsupervised domain adaptation [2,39,40]. A generalization error bound is proposed in a classical theory [2]. As the theory revealed, the generalization error in the target domain is bounded by three terms: the empirical error in the source domain, the feature distribution discrepancy across domains, and the ideal joint error λ^*. For the last term, it is assumed that there exists an ideal joint hypothesis that can achieve small classification errors in both domains. In such a case, the ideal joint error can be regarded as a small constant term. Based on the classical theory [2], many methods have focused on learning domain-invariant features to decrease the feature distribution discrepancy across domains and achieving a small classification error in the source domain at the same time [9,18].

However, recent researches show that transforming the feature representations to be domain-invariant may enlarge the error of the ideal joint hypothesis [16]. Such a phenomenon reminds us that the ideal joint error λ^* can not be ignored. Recently, a general and interpretable generalization upper bound without the pessimistic term λ^* for domain adaptation has been proposed in a new theory [40]. Compared with the previous theory [2], the theory proposes a new term named the shift between labeling functions (i.e., optimal predictors) across domains which also cause the discrepancy between domains, instead of simply regarding the ideal joint error λ^* as a constant term.

Inspired by the new theory [40], two major insights are concluded. Firstly, instead of assuming that there exists a predictor which can simultaneously perform well in both domains, we could learn different predictors for the source domain and the target domain such that the intrinsic characteristics of each domain can be exploited. Secondly, not only the feature distribution discrepancy, but also the shift between the labeling functions can cause the discrepancy across domains. Therefore, besides reducing the discrepancy between feature distributions, the discrepancy between labeling functions across domains should also be reduced.

In this paper, we propose a novel method called *Joint Feature and Labeling function adaptation* (JFL) based on the above observations. The goal of JFL is to perform both feature adaptation to learn domain-invariant features and labeling function adaptation to pull the labeling functions across domains close besides minimizing the source error. Notably, the labeling functions discussed before are the "true" labeling functions in the source and the target domain and are intrinsic to the feature space of the samples, so they are difficult to learn and optimize. In this paper, we assume that the empirical optimal predictor learned by our model is close to the ground-truth labeling function in each domain (In the following sections, we refer to this assumption as *Labeling Function Assumption*). Thus, minimizing the distance between labeling functions across domains can be transformed into reducing the discrepancy between the learned predictors, where the latter is more feasible in practice.

To be specific, we train a source predictor to minimize the source error. For feature adaptation, we adopt a widely used domain discriminator to measure and decrease feature distribution discrepancy [9,19]. For labeling function adaptation, we introduce a target predictor and a label predictor discriminator to reduce the discrepancy between the source and target predictors. The target predictor is trained on target samples with pseudo-labels [20] provided by the source predictor. Following the idea of GAN [11], we propose an adversarial learning method to pull the predictors in both domains close. Specifically, the label predictor discriminator takes the prediction outputs from the source and target predictors as input and is trained to distinguish which predictor the prediction output is given. While the feature extractor and predictors try to confuse the predictor discriminator. To the best of our knowledge, this is the first time that a predictor discriminator is proposed to reduce the shift between labeling functions in an entirely adversarial learning manner. Additionally, during the learning procedure, the intrinsic characteristics of the target domain can be explored by utilizing the task-specific training on the target domain. Moreover, we perform discriminative feature learning to obtain discriminative information of both domains for better adaptation and classification.

The results on three public datasets show that the proposed method outperforms state-of-the-art methods. And the performances on some difficult tasks are significantly improved (e.g. mnist → svhn). Furthermore, the experimental results show that the labeling function adaptation could improve the performance of the previous methods by applying this objective to these methods which only focus on feature adaptation and source error minimization. To sum up, there are three contributions in this paper:

1. We propose a new method, which not only performs feature adaptation but also performs labeling function adaptation to reduce the discrepancy across domains more thoroughly, while the latter is ignored by existing methods.
2. To perform labeling function adaptation, we introduce a novel component named *label predictor discriminator* and design an adversarial mechanism to train the model. Moreover, we apply the labeling function adaptation to previous methods and the results show that it can further improve the performance, which empirically validates the effectiveness of this objective.
3. We conduct extensive experiments on three widely used datasets. Results show that the proposed method outperforms baselines, especially on difficult tasks.

2   Related Work

Domain adaptation theory. The theory in [2] is one of the pioneering theoretical works in this field. A new statistics named $\mathcal{H}\Delta\mathcal{H}$-divergence is proposed as a substitution of traditional distribution discrepancies (such as L_1 distance, KL-divergence), and a generalization error bound is presented. The theory shows that the target error is bounded by the source error, the feature distribution discrepancy across domains, and the optimal joint error. It is usually assumed that

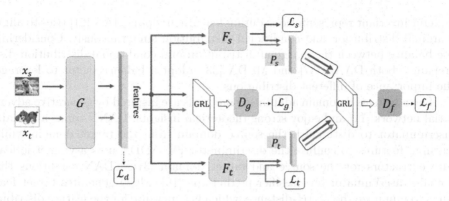

Fig. 1. The architecture of JFL. G represents the feature extractor, F_s and F_t are the source predictor and the target predictor respectively. D_g is the domain discriminator which takes features from feature extractor G as input and aims to distinguish whether the features are from the source domain or the target domain. P_s and P_t are the prediction outputs for the samples of both domains given by F_s and F_t respectively. D_f is the predictor discriminator which takes the prediction outputs from F_s and F_t as input and aims to distinguish whether the input is from the source predictor or the target predictor. $\mathcal{L}_s$ and $\mathcal{L}_t$ are the classification losses in the source and the target domain respectively. $\mathcal{L}_g$ is the domain classification loss, $\mathcal{L}_f$ is the predictor classification loss and $\mathcal{L}_d$ is the discriminative learning loss. GRL is the gradient reversal layer [9].

there exists a classifier that can perform well in both domains. In such a case, the optimal joint error becomes a small constant term. So most domain adaptation methods aim to minimize the source error and the feature distribution discrepancy across domains. A general class of loss functions satisfying symmetry and subadditivity are considered in [23], and a new generalization theory concerning the newly proposed discrepancy distance is developed. A margin-aware generalization bound based on asymmetric margin loss is proposed in [39], and reveals the trade-off between generalization error and the choice of margin. Recently, a theory considering labeling functions is proposed in [40], which shows that the error in the target domain is bounded by three terms: the source error, the discrepancy between feature distributions, and the distance between the labeling functions across domains. The third term also cause the discrepancy across domains, which is ignored by existing methods. JFL is able to optimize all the three terms of the error bound simultaneously.

Domain adaptation algorithm. The mostly well-known domain adaptation approaches include statistics matching [4,18,21,37] and adversarial domain adaptation [1,5,8,17,19,28]. The statistics matching methods assume that there exists a common space where the distributions of two domains are similar and focus on finding a feature transformation that projects features of two domains into another latent shared subspace with less distribution discrepancy. DAN [18] tries to align marginal distribution across domains, which learns domain-invariant representations during feature mapping. JAN [21] tries to align marginal distribution and conditional distribution simultaneously. Considering the balance between the marginal distribution and conditional distribution discrepancy, both DAAN [37] and MEDA [33] adopt a balance factor to leverage the importance of different distributions.

The adversarial domain adaptation methods are inspired by generative adversarial network [11] and enjoy strong theoretical insights. DANN trains a domain discriminator to distinguish the source domain from the target domain while learning features to confuse the discriminator [9]. ADDA uses asymmetric feature extractors for the source and target domain [31]. CDAN conditions the domain discriminator on classifier predictions [19]. MCD generates target features to minimize the $H\Delta H$-distance, which is a measure for the feature distribution discrepancy [28]. MCD also adopts two predictors, but both predictors are trained with source labeled samples. While in our method these two predictors are trained with the source samples and the target samples, respectively.

3   Method

3.1   Overview

In unsupervised domain adaptation, there are a set of labeled samples $\mathcal{D}_s = \{(x_s^i, y_s^i)\}_{i=1}^{n_s}$ drawn from the source domain and a set of unlabeled samples $\mathcal{D}_t = \{x_t^j\}_{j=1}^{n_t}$ drawn from the target domain, where $\mathcal{X}$ and $\mathcal{Y}$ denote the feature space and the label space of both domains respectively, i.e., $x_s^i, x_t^j \in \mathcal{X} = \mathbb{R}^d$, $y_s^i \in \mathcal{Y} = \{1, 2, ..., K\}$. Let $P(\mathcal{X}, \mathcal{Y})$ and $Q(\mathcal{X}, \mathcal{Y})$ be the joint distribution of the source and the target domain, respectively. We assume the joint distributions across domains are different, i.e., $P(\mathcal{X}, \mathcal{Y}) \neq Q(\mathcal{X}, \mathcal{Y})$. The goal of this work is to train a model with both labeled source samples and unlabeled target samples to achieve better classification performance in the target domain.

Most previous methods [9,18,19] only focus on learning domain-invariant features and achieving a small source error. Such a strategy is not sufficient for the success of domain adaptation, which is revealed by the theory proposed in [40]:

$$\varepsilon_t(f) \leq \varepsilon_s(f) + d(\mathcal{D}_s, \mathcal{D}_t) + \min\{\mathbb{E}_{\mathcal{D}_s}[|f_s - f_t|], \mathbb{E}_{\mathcal{D}_t}[|f_s - f_t|]\} \tag{1}$$

where $\varepsilon_t(f) = \mathbb{E}_{x,y \sim Q(\mathcal{X}, \mathcal{Y})}[f(x) \neq y]$ and $\varepsilon_s(f) = \mathbb{E}_{x,y \sim P(\mathcal{X}, \mathcal{Y})}[f(x) \neq y]$ are the expected errors in the target and the source domain, respectively. $d(\mathcal{D}_s, \mathcal{D}_t)$ represents the feature distribution discrepancy across domains. The third term measures the discrepancy between the labeling functions f_s and f_t in the source and the target domain.

Theoretically, most existing methods focus on optimizing the first two terms while ignoring the discrepancy between the labeling functions across domains, which may also lead to a larger upper bound of $\varepsilon_t(f)$. Therefore, we should not only adapt the feature distributions but also adapt labeling functions to further reduce the discrepancy across domains.

Inspired by the above analysis, we propose a method named *Joint Feature and Labeling function adaptation* (JFL), to simultaneously adapt the feature representations and the labeling functions between the source domain and the target domain as well as minimizing the source error. As shown in Fig. 1, the architecture of JFL is composed of five components: the feature extractor G, the source predictor F_s, the target predictor F_t, the domain discriminator D_g, and the label predictor discriminator D_f.

For the source error minimization, we train a source predictor F_s to minimize the source classification loss. For the feature adaptation, we adopt a domain discriminator D_g to minimize the domain classification loss while the feature extractor G is trained to fool domain discriminator D_g, which are widely used in domain adaptation [9,19].

For the labeling function adaptation, we introduce a target predictor F_t and a predictor discriminator D_f. The target predictor F_t is trained on the target samples with the pseudo-labels predicted by the source predictor. Thus, the target predictor can explore the inherent characteristics of the target domain thanks to the task-specific training. Given a sample, the predictor discriminator D_f aims to distinguish whether the prediction output for the sample is from the source predictor or the target predictor while the feature extractor G and predictors F_s and F_t are trained to confuse the predictor discriminator. Moreover, as there are only unlabeled data in the target domain, we perform discriminative feature learning to obtain discriminative information of both domains.

3.2 Source Error Minimization and Feature Adaptation

To minimize the source error, we train the feature extractor G and the source predictor F_s to classify the source samples. The classification loss in the source domain is:

$$\min_{G,F_s} \mathcal{L}_s(X_s, Y_s) = \mathbb{E}_{(x_s,y_s)\sim(X_s,Y_s)} l(F_s(G(x_s)), y_s) \qquad (2)$$

where $l(\cdot,\cdot)$ is the cross-entropy loss. $F_s(G(x_s)) \in \mathbb{R}^K$ represents the label prediction output of F_s for sample x_s and y_s is the ground truth label of x_s.

To extract domain-invariant features, we adapt the feature representations to reduce the feature distribution shift between different domains in an adversarial manner. Similar to previous methods [9,11], the adversarial learning procedure is a two-player game. The first player, domain discriminator D_g, is trained to distinguish whether each example belongs to the source or the target domain while the second player, feature extractor G, tries to confuse the domain discriminator. To be specific, the domain discriminator D_g aims to minimize the domain classification loss, while the feature extractor G is trained to maximize the domain classification loss. As a result, the feature extractor will extract domain-invariant features to achieve feature adaptation. Technically, the objective of feature adaptation is:

$$\min_{D_g} \max_{G} \mathcal{L}_g(X_s, X_t) \qquad (3)$$

$$\mathcal{L}_g(X_s, X_t) = \mathbb{E}_{x \sim X_s}[\log D_g(G(x))] + \mathbb{E}_{x \sim X_t}[\log(1 - D_g(G(x)))] \qquad (4)$$

where $G(x)$ is the deep feature mapping of two domains extracted by feature extractor G and $D_g(G(x)) \in \mathbb{R}$ is the domain classification output of the domain discriminator D_g.

Through the adversarial leaning procedure, the discrepancy between feature distributions across domains will be reduced. However, this is not sufficient to guarantee a satisfying performance in the target domain based on the previous analysis. Hence we will introduce the labeling function adaptation in the next subsection.

3.3   Labeling Function Adaptation

Most existing methods ignore reducing the discrepancy between labeling functions of different domains. To address this issue, besides feature adaptation, our method further performs labeling function adaptation to pull the label predictors across domains close so that the discrepancy across domains can be further reduced.

In order to adapt the labeling functions, a target predictor F_t and a predictor discriminator D_f are introduced and we design an adversarial learning method to train the predictor discriminator D_f to pull the predictors of different domains close. Additionally, the proposed target predictor and predictor discriminator are helpful to exploit the intrinsic characteristics of the target domain itself by utilizing the task-specific training on target samples [28].

The target predictor F_t takes features from the feature extractor G as input and is trained on target samples with the pseudo-labels provided by the source predictor F_s. As learning domain-invariant features can destroy the inherent structure of the target domain, the supervised classification training on target samples which is task-specific could help the feature extractor exploit the characteristics of the target domain and combine the structural information of the source and the target domain better. Formally, in the target domain, we minimize the classification loss as:

$$\min_{G, F_t} \mathcal{L}_t(X_t, \hat{Y}_t) = \mathbb{E}_{x_t \sim X_t} l(F_t(G(x_t)), \hat{y}_t) \qquad (5)$$

where $l(\cdot, \cdot)$ is the cross-entropy loss, and $\hat{y}_t$ is the pseudo-label of sample x_t predicted by source predictor F_s:

$$\hat{y}_t = arg\,max_{y \in \mathcal{Y}} F_s(G(x_t)) \qquad (6)$$

The predictor discriminator D_f takes the prediction outputs of a sample from both predictors F_s and F_t as input, regardless of which domain the sample comes from. The predictor discriminator is trained to distinguish whether the prediction outputs for samples are given by the source or the target predictor while the feature extractor and label predictors try to confuse the predictor discriminator in an adversarial manner. Specifically, the predictor discriminator is essentially a binary classifier. By this min-max game, the predictor discriminator can hardly distinguish which predictor gives the prediction, so that the discrepancy between labeling functions can be reduced.

Technically, the optimization objective for labeling function adaptation is:

$$\min_{D_f} \max_{G,F_s,F_t} \mathcal{L}_f(X_s, X_t)$$

$$\mathcal{L}_f(X_s, X_t) = \mathbb{E}_{x \sim X_s \cup X_t}[\log D_f(P_s(x)) + \log(1 - D_f(P_t(x)))] \tag{7}$$

where $D_f(P_s(x)) \in \mathbb{R}$ and $D_f(P_t(x)) \in \mathbb{R}$ are the predictor classification outputs of the predictor discriminator D_f, and $P_s(x) = F_s(G(x)), P_t(x) = F_t(G(x))$ are the predictions for the samples given by F_s and F_t, respectively.

To further understand the adversarial learning process, we illustrate the min-max process intuitively. For the process of maximizing $\mathcal{L}_f(X_s, X_t)$, the feature extractor and the predictors try to confuse the predictor discriminator to make the predictor discriminator D_f can not distinguish which predictor gives the prediction output. In other words, maximizing $\mathcal{L}_f(X_s, X_t)$ is equivalent to directly minimizing the distance between the labeling functions. For the process of minimizing $\mathcal{L}_f(X_s, X_t)$, the two predictors are expected to obtain the intrinsic characteristics of each domain so that they can firstly achieve a better classification performance before they are completely pulled close, which exactly accords with the *Labeling Function Assumption*. Otherwise, if the process is discarded, the source predictor F_s and the target predictor F_t will become identical immediately in the case of very poor performance which is not conform to the *Labeling Function Assumption*. Therefore, the labeling function adaptation can be better accomplished by the adversarial learning process, which is validated in the experiments.

Remark: Labeling function adaptation is a general objective, it can not only be used in our proposed method, but also be employed to previous methods to improve the performance by more completely reducing the domain discrepancy, which is validated by the results in Sect. 4.4.

3.4 Discriminative Feature Extraction

Besides the previous objectives, we also perform discriminative feature learning to explore the discriminative information in both domains for better classification.

At the sample level, a discriminative clustering loss $\mathcal{L}_{d_c}$ is introduced for each domain: $\mathcal{L}_{d_c}(X_s, X_t) = \mathcal{L}_{d_c}(X_s) + \mathcal{L}_{d_c}(X_t)$. This loss encourages the features from the same class to gather together, and pushes the features of different classes to be far away from each other. Similar to the previous method [22], we employ the following discriminative clustering loss for each domain:

$$\mathcal{L}_{d_c}(X) = \sum_{x_i, x_j \in X}[\delta_{ij}d(G(x_i), G(x_j)) + (1 - \delta_{ij})max(0, \mu - d(G(x_i), G(x_j)))] \tag{8}$$

where δ_{ij} is an indicator function which outputs 1 only if the sample x_i and x_j have the same ground-truth label (source domain) or pseudo-label (target domain), $d(\cdot, \cdot)$ is the distance between features (such as Euclidean distance), and μ is a pre-defined margin.

At the class level, we further introduce a discriminative alignment loss $\mathcal{L}_{d_a}$ to align the class centers of the same categories between domains for better adaptation. Technically, the discriminative alignment loss $\mathcal{L}_{d_a}$ is:

$$\mathcal{L}_{d_a}(X_s, X_t) = \frac{1}{K}\sum_{k=1}^{K}\left|\left|\frac{1}{|X_{s,k}|}\sum_{x_s^i \in X_{s,k}} G(x_s^i) - \frac{1}{|X_{t,k}|}\sum_{x_t^j \in X_{t,k}} G(x_t^j)\right|\right|^2 \quad (9)$$

where $X_{s,k}$ denotes the set of the source samples belonging to the k-th class, and $X_{t,k}$ denotes the set of the target samples whose pseudo-labels are k. To sum up, the overall objective of discriminative learning is:

$$\min_G \mathcal{L}_d(X_s, X_t) = \mathcal{L}_{d_c}(X_s, X_t) + \mathcal{L}_{d_a}(X_s, X_t) \quad (10)$$

3.5 Overall Objective

Combining the above objectives discussed in Subsects. 3.2–3.4 together, we get the overall objective as:

$$\min_{G, F_s, F_t} \lambda_s \mathcal{L}_s(X_s, Y_s) + \lambda_t \mathcal{L}_t(X_t, \hat{Y}_t) + \lambda_d \mathcal{L}_d(X_s, X_t)$$
$$- \lambda_g \mathcal{L}_g(X_s, X_t) - \lambda_f \mathcal{L}_f(X_s, X_t) \quad (11)$$

$$\min_{D_g} \mathcal{L}_g(X_s, X_t) \quad (12)$$

$$\min_{D_f} \mathcal{L}_f(X_s, X_t) \quad (13)$$

Following the previous method [9], the min-max training procedure is accomplished by applying a Gradient Reversal Layer (GRL). GRL behaves as the identity function during the forward propagation and inverts the gradient sign during the backward propagation, hence driving the parameters to maximize the output loss.

4 Experiments

We evaluate JFL on three datasets against state-of-the-art domain adaptation methods. The code is available at https://github.com/yuntaodu/JFL.

4.1 Datasets

Digital dataset contains four datasets of 10 categories: MNIST is composed of 60,000 images, MNIST-M [8] consists of 60,000 images, SVHN is composed of 73,257 images and SynthDigits (DIGITS) [8] consists of 479,400 images. The images in MNIST and MNIST-M are gray while the images in SVHN and DIGITS are in color. **Office-31** [26] is a standard benchmark for visual domain adaptation, containing 4,652 images of 31 categories from three diverse domains: **A**

Table 1. Accuracy (%) on digital dataset.

Source Target	MNIST SVHN	SVHN MNIST	MNIST MNIST-M	DIGITS SVHN
Source-only	45.9	72.1	71.7	85.8
DANN [9]	60.6	68.3	94.6	90.1
DRCN [10]	40.1	82.0	–	–
ATT [27]	52.8	86.2	94.2	92.9
d-SNE [35]	77.6	97.6	94.1	–
AADA [36]	–	98.6	95.7	92.6
VADA [30]	47.5	97.9	97.7	94.8
DIRT-T [30]	54.5	99.4	98.9	96.1
Co-DA [13]	81.7	99.0	99.0	**96.4**
Co-DA + DIRT-T [13]	88.0	99.4	99.1	**96.4**
RCA [3]	89.2	99.3	**99.4**	96.2
JFL	**91.7**	**99.6**	**99.4**	96.2

Table 2. Accuracy (%) on Office-31 for unsupervised domain adaptation (ResNet-50).

Method	A → W	D → W	W → D	A → D	D → A	W → A	Avg
ResNet-50 [12]	68.4	96.7	99.3	68.9	62.5	60.7	76.1
DAN [18]	80.5	97.1	99.6	78.6	63.6	62.8	80.4
ADDA [31]	86.2	96.2	98.4	77.8	69.5	68.9	82.9
JAN [21]	85.4	97.4	99.8	84.7	68.6	70.0	84.3
GTA [29]	89.5	97.9	99.8	87.7	72.8	71.4	86.5
MCD [28]	88.6	98.5	**100.0**	92.2	69.5	69.7	86.5
SHOT [15]	90.1	98.4	99.9	**94.0**	74.7	74.3	88.6
JFL	94.5	98.7	**100.0**	93.4	**75.7**	**74.5**	**89.4**
DANN [9]	82.0	96.9	99.1	79.7	68.2	67.4	82.2
DANN + LFA	86.8	98.5	**100.0**	90.0	70.2	72.6	86.4
CDAN [19]	94.1	98.6	**100.0**	92.9	71.0	69.3	87.7
CDAN + LFA	**95.4**	99.2	**100.0**	93.1	72.3	72.5	88.8

(Amazon), **W** (Webcam) and **D** (Dslr). **Office-Home** [32] is a more complex dataset, which consists of 15,500 images with 65 object classes from four different domains: Real-world (**Rw**), Clipart (**Cl**), Product images (**Pr**), and Artistic images (**Ar**).

4.2 Setup

We follow the widely used protocol for unsupervised domain adaptation [19]. We report the average accuracies of three independent experiments.

For Office-31 and Office-Home dataset, we adopt ResNet-50 [12] pre-trained on ImageNet [25] as well as a newly introduced bottleneck layer as the feature extractor. The source predictor and target predictor are both 2-layer neural networks with width 1024. The domain discriminator and the predictor discriminator are both 3-layer neural networks with width 1024. For optimization, we use the mini-batch Adam with the learning rate of 7.5e-5. The learning rate of the other components except the feature extractor are set 10 times to that of the feature extractor. And the learning rate is adjusted according to [9]. For the digital dataset, we train the model with the learning rate of 1e-3 optimized by mini-batch Adam. In all the experiments, we set $\mu = 30$ to conduct discriminative feature extraction. At the test time, we combine the F_s and F_t to obtain the predicted labels for the target samples by computing largest prediction confidence in the probability prediction outputs of both the source predictor and the target predictor.

Table 3. Accuracy (%) on Office-Home for unsupervised domain adaptation (ResNet-50).

Method	Ar→Cl	Ar→Pr	Ar→Rw	Cl→Ar	Cl→Pr	Cl→Rw	Pr→Ar	Pr→Cl	Pr→Rw	Rw→Ar	Rw→Cl	Rw→Pr	Avg
ResNet-50 [12]	34.9	50.0	58.0	37.4	41.9	46.2	38.5	31.2	60.4	53.9	41.2	59.9	46.1
DAN [18]	43.6	57.0	67.9	45.8	56.5	60.4	44.0	43.6	67.7	63.1	51.5	74.3	56.3
DANN [9]	45.6	59.3	70.1	47.0	58.5	60.9	46.1	43.7	68.5	63.2	51.8	76.8	57.6
JAN [21]	45.9	61.2	68.9	50.4	59.7	61.0	45.8	43.4	70.3	63.9	52.4	76.8	58.3
CDAN [19]	50.7	70.6	76.0	57.6	70.0	70.0	57.4	50.9	77.3	70.9	56.7	81.6	65.8
ETD [14]	51.3	71.9	85.7	57.6	69.2	73.7	57.8	51.2	79.3	70.2	57.5	82.1	67.3
SymNets [38]	47.7	72.9	78.5	64.2	71.3	**74.2**	64.2	48.8	**79.5**	74.5	52.6	82.7	67.6
MDD [39]	**54.9**	**73.7**	77.8	60.0	71.4	71.8	61.2	**53.6**	78.1	72.5	**60.2**	82.3	68.1
JFL	52.1	**73.7**	**78.6**	**65.6**	**72.7**	73.7	**65.2**	52.9	78.3	**74.6**	58.5	**83.1**	**69.1**

4.3 Results

The results on the digital dataset are shown in Table 1. The results of source-only are the accuracies of the model trained only with the labeled source samples. For this dataset, JFL achieves the best or the second best results after RCA [3] and Co-DA [30] in all the experiments. Especially in the most challenging tasks MNIST → SVHN, for which state-of-the-art accuracies are below 90%, JFL outperforms all the previous methods and reaches 91.7%.

The results on the Office-31 dataset are shown in Table 2. JFL achieves state-of-the-art accuracies on four of six transfer tasks. We clearly observe that on W→ A and D → A with relatively large domain shift and imbalanced domain scales, JFL exceeds all feature adaptation methods by a large margin and even performs better than models incorporating complex generative architectures. This testifies that by matching the labeling functions across domains, JFL is able to further decrease the discrepancy across domains and therefore mitigate the imbalance across domains.

The results on the Office-Home dataset are shown in Table 3. Compared with previous methods, JFL makes remarkable performance boost when domain discrepancy is significant. The average accuracy is 1% higher than the second best method. Most baselines only focus on performing feature adaptation and minimizing the source error while JFL further adapts the labeling functions, making a remarkable improvement.

4.4 Insight Analysis

In this subsection, we conduct extensive experiments to analyse the effect of our method.

Table 4. Accuracy (%) on digital dataset of ablation study.

Source	MNIST	SVHN	MNIST	DIGITS
Target	SVHN	MNIST	MNIST-M	SVHN
JFL (SEM)	45.9	72.1	71.7	85.8
JFL (SEM + FA)	68.8	86.2	93.0	91.4
JFL (SEM + FA + DFL)	83.3	97.8	89.9	95.3
JFL (SEM + FA + LFA)	88.9	96.7	99.0	95.4
JFL (ALL)	**91.7**	**99.6**	**99.4**	**96.2**

Ablation Study. It is interesting to investigate the contribution of each part of JFL: Source Error Minimization (SEM), Feature Adaptation (FA), Labeling Function Adaptation (LFA), and Discriminative Feature Learning (DFL). To enable ease of use, we integrate these parts into a coherent loss. Results in Table 4 justify that each part has its indispensable contribution. Moreover, the results indicate that labeling function adaptation and feature adaptation are both beneficial for bridging cross-domain discrepancy, and learning discriminative features can further improve the performance. Finally, a combination of all losses can achieve the best result.

Improvement for Existing Methods. To verify the effectiveness of performing Labeling Function Adaptation (LFA) more comprehensively, we apply LFA to previous methods. In this paper we choose DANN and CDAN as they are representative domain adaptation methods, where only the feature adaptation and the source error minimization are performed without any extra training tricks. We denote these two methods as DANN + LFA and CDAN + LFA. The results are shown in Table 2. As we can see, after applying LFA to DANN and CDAN, the performance is improved by 4.2 % and 1.1%, respectively. And CDAN + LFA achieves the best result in task A → W. The results show that performing labeling function is not only useful in our method but also can improve the performance for the previous methods.

Effectiveness of Adversarial Learning for Labeling Function Adaptation. Intuitively, we could adapt the labeling functions by directly minimizing

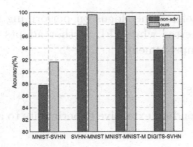

Fig. 2. Comparison with non-adversarial baseline.

the pre-defined distance measurement for the discrepancy between the source and the target predictor in a non-adversarial manner. Technically, the non-adversarial loss for labeling function adaptation computed by the distance measurement proposed in [28] is:

$$\mathcal{L}'_f = \mathbb{E}_{x \sim (X_s \cup X_t)} \sum_{k=1}^{K} |P_s^k(x) - P_t^k(x)| \tag{14}$$

where $P_s^k(x)$ and $P_t^k(x)$ are the prediction outputs of the k-th class given by the source and target predictors respectively. We replace the loss $\mathcal{L}_f$ with $\mathcal{L}'_f$ and denote the baseline as non-adv. We compare the baseline with JFL on the digital dataset and the results are shown in Fig. 2. As we can see, JFL outperforms this baseline in all the tasks, especially in the most challenging task, i.e., MNIST $\rightarrow$ SVHN. Such results also prove that the adversarial based method can perform better than the non-adversarial method for labeling function adaptation.

5   Conclusion

In this paper, we focus on unsupervised domain adaptation. Previous methods only aim to learn domain-invariant features and achieve a small source error while they ignore the shift between the labeling functions. Thus, we propose a method to perform feature adaptation, labeling function adaptation and minimize the source error simultaneously. Especially, for labeling function adaptation, a novel component named label predictor discriminator is introduced and is trained in an adversarial manner. By optimizing these three objectives, the discrepancy across domains can be reduced. The results on three real-world datasets show that the proposed method outperforms state-of-the-art methods. In the future, we would like to explore labeling function adaptation in semi-supervised domain adaptation and multi-source domain adaptation. Besides, we will explore domain adaptation in complex scenes such as source-free domain adaptation [6,15] and activate domain adaptation [7].

Acknowledgement. This paper is supported by the National Key Research and Development Program of China (Grant No. 2018YFB1403400), the National Natural

Science Foundation of China (Grant No. 61876080), the Key Research and Development Program of Jiangsu (Grant No. BE2019105), the Collaborative Innovation Center of Novel Software Technology and Industrialization at Nanjing University.

References

1. Acuna, D., Zhang, G., Law, M.T., Fidler, S.: f-domain-adversarial learning: theory and algorithms. In: ICML (2021)
2. Ben-David, S., Blitzer, J., Crammer, K., Kulesza, A., Pereira, F.C., Vaughan, J.W.: A theory of learning from different domains. Mach. Learn. **79**, 151–175 (2009)
3. Cicek, S., Soatto, S.: Unsupervised domain adaptation via regularized conditional alignment. In: 2019 IEEE/CVF International Conference on Computer Vision (ICCV), pp. 1416–1425 (2019)
4. Du, Y., Chen, Y., Cui, F., Zhang, X., Wang, C.J.: Cross-domain error minimization for unsupervised domain adaptation. In: DASFAA (2021)
5. Du, Y., Tan, Z., Chen, Q., Zhang, X., Yao, Y., Wang, C.J.: Dual adversarial domain adaptation. ArXiv abs/2001.00153 (2020)
6. Du, Y., Yang, H., Chen, M., Jiang, J., Luo, H., Wang, C.J.: Generation, augmentation, and alignment: a pseudo-source domain based method for source-free domain adaptation. ArXiv abs/2109.04015 (2021)
7. Fu, B., Cao, Z., Wang, J., Long, M.: Transferable query selection for active domain adaptation. In: CVPR, pp. 7268–7277 (2021)
8. Ganin, Y., Lempitsky, V.: Unsupervised domain adaptation by backpropagation. In: ICML (2015)
9. Ganin, Y., et al.: Domain-adversarial training of neural networks. J. Mach. Learn. Res. **17**, 59:1–59:35 (2016)
10. Ghifary, M., Kleijn, W., Zhang, M., Balduzzi, D., Li, W.: Deep reconstruction-classification networks for unsupervised domain adaptation. In: ECCV (2016)
11. Goodfellow, I.J., et al.: Generative adversarial nets. In: NIPS (2014)
12. He, K., Zhang, X., Ren, S., Sun, J.: Deep residual learning for image recognition. In: 2016 IEEE Conference on Computer Vision and Pattern Recognition (CVPR), pp. 770–778 (2016)
13. Kumar, A., et al.: Co-regularized alignment for unsupervised domain adaptation. In: NeurIPS (2018)
14. Li, M., Ming Zhai, Y., Luo, Y.W., Ge, P., Ren, C.X.: Enhanced transport distance for unsupervised domain adaptation. In: CVPR (2020)
15. Liang, J., Hu, D., Feng, J.: Do we really need to access the source data? Source hypothesis transfer for unsupervised domain adaptation. In: ICML (2020)
16. Liu, H., Long, M., Wang, J., Jordan, M.I.: Transferable adversarial training: a general approach to adapting deep classifiers. In: ICML (2019)
17. Liu, X., et al.: Adversarial unsupervised domain adaptation with conditional and label shift: infer, align and iterate. In: ICCV (2021)
18. Long, M., Cao, Y., Wang, J., Jordan, M.I.: Learning transferable features with deep adaptation networks. In: ICML (2015)
19. Long, M., Cao, Z., Wang, J., Jordan, M.I.: Conditional adversarial domain adaptation. In: NeurIPS (2018)
20. Long, M., Wang, J., Ding, G., Sun, J., Yu, P.: Transfer feature learning with joint distribution adaptation. In: 2013 IEEE International Conference on Computer Vision, pp. 2200–2207 (2013)

21. Long, M., Zhu, H., Wang, J., Jordan, M.I.: Deep transfer learning with joint adaptation networks. ArXiv abs/1605.06636 (2017)
22. Luo, Y., Zhu, J., Li, M., Ren, Y., Zhang, B.: Smooth neighbors on teacher graphs for semi-supervised learning. In: CVPR, pp. 8896–8905 (2018)
23. Mansour, Y., Mohri, M., Rostamizadeh, A.: Domain adaptation: learning bounds and algorithms. In: COLT (2009)
24. Pan, S., Yang, Q.: A survey on transfer learning. IEEE Trans. Knowl. Data Eng. **22**, 1345–1359 (2010)
25. Russakovsky, O., et al.: Imagenet large scale visual recognition challenge. Int. J. Comput. Vision **115**, 211–252 (2015)
26. Saenko, K., Kulis, B., Fritz, M., Darrell, T.: Adapting visual category models to new domains. In: ECCV (2010)
27. Saito, K., Ushiku, Y., Harada, T.: Asymmetric tri-training for unsupervised domain adaptation. In: ICML (2017)
28. Saito, K., Watanabe, K., Ushiku, Y., Harada, T.: Maximum classifier discrepancy for unsupervised domain adaptation. In: CVPR, pp. 3723–3732 (2018)
29. Sankaranarayanan, S., Balaji, Y., Castillo, C.D., Chellappa, R.: Generate to adapt: aligning domains using generative adversarial networks. In: CVPR, pp. 8503–8512 (2018)
30. Shu, R., Bui, H.H., Narui, H., Ermon, S.: A DIRT-T approach to unsupervised domain adaptation. In: ICLR (2018)
31. Tzeng, E., Hoffman, J., Saenko, K., Darrell, T.: Adversarial discriminative domain adaptation. In: CVPR, pp. 2962–2971 (2017)
32. Venkateswara, H., Eusebio, J., Chakraborty, S., Panchanathan, S.: Deep hashing network for unsupervised domain adaptation. In: CVPR, pp. 5385–5394 (2017)
33. Wang, J., Feng, W., Chen, Y., Yu, H., Huang, M., Yu, P.S.: Visual domain adaptation with manifold embedded distribution alignment. In: MM 2018 (2018)
34. Wang, M., Deng, W.: Deep visual domain adaptation: a survey. Neurocomputing **312**, 135–153 (2018)
35. Xu, X., Zhou, X., Venkatesan, R., Swaminathan, G., Majumder, O.: d-SNE: domain adaptation using stochastic neighborhood embedding. In: CVPR, pp. 2492–2501 (2019)
36. Yang, J., Zou, H., Zhou, Y., Zeng, Z., Xie, L.: Mind the discriminability: asymmetric adversarial domain adaptation. In: ECCV (2020)
37. Yu, C., Wang, J., Chen, Y., Huang, M.: Transfer learning with dynamic adversarial adaptation network. In: 2019 IEEE International Conference on Data Mining (ICDM), pp. 778–786 (2019)
38. Zhang, Y., Tang, H., Jia, K., Tan, M.: Domain-symmetric networks for adversarial domain adaptation. In: CVPR, pp. 5026–5035 (2019)
39. Zhang, Y., Liu, T., Long, M., Jordan, M.I.: Bridging theory and algorithm for domain adaptation. In: ICML (2019)
40. Zhao, H., des Combes, R.T., Zhang, K., Gordon, G.J.: On learning invariant representation for domain adaptation. In: ICML (2019)

Protoformer: Embedding Prototypes for Transformers

Ashkan Farhangi[✉], Ning Sui, Nan Hua, Haiyan Bai, Arthur Huang, and Zhishan Guo

University of Central Florida, Orlando, FL, USA
{ashkan.farhangi,zhishan.guo}@ucf.edu

Abstract. Transformers have been widely applied in text classification. Unfortunately, real-world data contain anomalies and noisy labels that cause challenges for state-of-art Transformers. This paper proposes Protoformer, a novel self-learning framework for Transformers that can leverage problematic samples for text classification. Protoformer features a selection mechanism for embedding samples that allows us to efficiently extract and utilize anomalies prototypes and difficult class prototypes. We demonstrated such capabilities on datasets with diverse textual structures (e.g., Twitter, IMDB, ArXiv). We also applied the framework to several models. The results indicate that Protoformer can improve current Transformers in various empirical settings.

Keywords: Text classification · Twitter analysis · Class prototype

1 Introduction

For real-world textual datasets, anomalies are known as samples that depart from the standard samples. Such anomalies tend to have scattered textual distributions, which can cause performance drops for state-of-art Transformer models [13]. Moreover, models that rely on supervised learning can suffer from incorrect convergence when provided with noisy labeled data gathered from Internet [14]. Hence, there is a need to automatically detect the anomalies and adjust noisy labels to make the model more robust to complex noisy datasets.

As human annotations can be highly time-and-cost inefficient, it is more common that noisy labels are gathered from the Internet. For instance, Twitter has been increasingly adopted to understand human behavior [3]. However, such data tend to complex and often contain noisy labels. This can make the standard supervised learning objective lead to incorrect convergence [4].

One of the applications of this study is to classify college students' academic major choices based on their historical Tweets. When students follow a certain college's official account, it might indicate that the student belongs to that major. However, there are uncertainties about the correctness of the labels. Therefore, the supervised model's results can become untrustworthy.

© The Author(s), under exclusive license to Springer Nature Switzerland AG 2022
J. Gama et al. (Eds.): PAKDD 2022, LNAI 13280, pp. 447–458, 2022.
https://doi.org/10.1007/978-3-031-05933-9_35

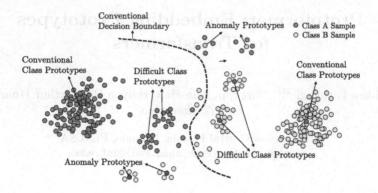

Fig. 1. Distribution of embeddings for real world data samples is often scatterd. Although conventional class prototypes are easier to select, difficult class prototypes and anomaly prototypes require a more careful approach in selection and play a critical role in improving the decision boundary.

There are some prior works on prototype embeddings. CleanNet [7] proposes providing extra supervision for the training. Subsequently, SMP [5] proposes using multiple prototypes to capture embeddings with high density without extra human supervision. However, both approaches do not provide a solution for troublesome embeddings that are scattered and are often minorities, as shown in Fig. 1. To alleviate this issue, we select prototypes through their contextual embeddings in a way to not only cover the difficult-to-classify samples but also represent minority samples of the dataset (i.e., anomalies).

We propose Protoformer framework that selects and leverages multiple embedding prototypes to enable Transformer's specialization ability to classify noisy labeled data populated with anomalies. Specifically, we improve the generalization ability of Transformers for problematic samples of a class through *difficult class prototypes* and their specialization ability for minority samples of a class through *anomaly prototypes.* We show that the representations of both prototypes are necessary to improve the model's performance. Protoformer leverages these prototypes in a self-learning procedure to further improve the robustness of textual classification. To our best knowledge, this is the first study that extracts and leverages anomaly prototypes for Transformers.

In summary, the contributions are threefold:

- We propose a novel framework that learns to leverage harder to classify and anomaly samples. This acts as a solution for classifying datasets with complex samples crawled from the Internet.
- The framework contains a label adjustment procedure and thus is robust to noise. This makes the framework suitable for noisy Internet data and can be used to promote a more robust Transformer model. Leveraging the similarity in the embedding space and a ranking metric, we can identify questionable labels and provide a certain level of adjustment. This mitigates the potential negative impact on the training.

- We evaluate the framework based on multiple datasets with both clean and noisy labels. Results show that our model improves the testing accuracy from 95.7% to 96.8% on the IMDB movie review dataset. For a self-gathered Twitter dataset with noisier labels, the classification accuracy improved with a greater margin (from 56.7% to 81.3%).

2   Problem Formulation

Given a sample text as x_i, $X = \{x_1, x_2, \cdots, x_N\}$ represents all the N samples of the dataset, while $\hat{Y} = \{\hat{y}_1, \hat{y}_2, \cdots, \hat{y}_N\}$ indicates the corresponding noisy labels from the Internet. The noisy label $\hat{y}_i \in \{0, 1\}^{\bar{c}}$ is a binary vector format with only one non-zero element, indicating the class label of x_i, where $\bar{c}$ is the total number of classes. A Transformer model $\mathcal{F}_W$ can be used as a classification model to produce an estimated label $\mathcal{F}_W(x_i) \in [0, 1]^{\bar{c}}$, where W represents the parameters. The optimization strategy is based on the cross-entropy loss function:

$$\mathcal{L}(\mathcal{F}_W(x_i), \tilde{y}_i) = - \sum_{j=1}^{\bar{c}} \tilde{y}_{i,j} \log \left(\mathcal{F}_W(x_i)_j \right), \qquad (1)$$

In addition, labels from the internet are often noisy. Hence, as detailed in Sect. 3.4, the labels can be adjusted according to the similarities of the class prototypes, resulting in adjusted labels $\tilde{y}_i \in [0, 1]^{\bar{c}}$—it is a probability distribution, and thus $\sum_{j=1}^{\bar{c}} \tilde{y}_{i,j} = 1$. Even when we have sufficient confidence in the original labels, we can use it as a complementary supervision.

Specifically, for each batch with m samples, we would pursue the following optimization problem:

$$W^* = \operatorname{argmin}_W \frac{1}{m} \sum_{i=1}^{m} \mathcal{L} \left(\mathcal{F}_W(x_i), \tilde{y}_i \right) \qquad (2)$$

3   Design of Protoformer

This section provides the details of Protoformer. Specifically, we describe a procedure for extracting the difficult class prototypes (Sect. 3.1). Subsequently, we describe a procedure for extracting anomaly prototypes (Sect. 3.2). Both types of prototypes are then used in a multi-objective self-learning training process that optimizes the network parameters for robust text classification (Sect. 3.3). In order to handle noisy labeled data, we adjust the noisy labels through a label adjustment procedure that uses the prototype similarities (Sect. 3.4).

3.1   Difficult Class Prototypes

Difficult class prototypes act as the representatives for the problematic samples of the dataset. For example, Fig. 2 showcases the fine-tuned embeddings of a benchmark dataset gathered from the Internet (i.e., IMDB). Although the

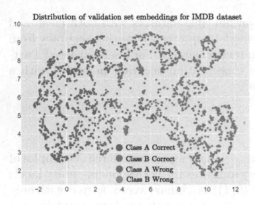

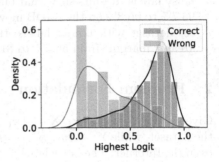

Fig. 2. Left: distribution of the embedding for IMDB dataset. Presence of anomalies and problematic samples cause misclassification. **Right:** Distribution of the highest output logits (scaled 0–1) for the same model. The higher values of the largest logits can represent the confidence of the network's classification.

majority of samples of each class are located closely together, there are anomaly samples that are scattered and often far from the majority. Unfortunately, these harder-to-classify samples are not the target focus of the state-of-art models in text classification. Moreover, traditional clustering methods (e.g., K-means) are not designed to capture or cluster such samples that are scattered and distributed throughout the embedding space.

Intuitively, these problematic samples can cause the greatest error. For instance, Fig. 2 also shows the classification error of the fine-tuned BERT [2] model where the majority of the classification error stems from harder-to-classify samples (over 51%). Such error arises when the highest classification logit values are still low and in between classes, which indicates the indecisiveness of the Transformer. Following [5], we define the similarity of the extracted embeddings through **pairwise similarity score** (i.e., cosine distance) of any two inputs $\mathbf{x}_i$ and $\mathbf{x}_j$ as:

$$s_{ij} = \frac{\mathbf{e}(\mathbf{x}_i)^T \cdot \mathbf{e}(\mathbf{x}_j)}{\|\mathbf{e}(\mathbf{x}_i)\|_2 \|\mathbf{e}(\mathbf{x}_j)\|_2}, \tag{3}$$

where $\mathbf{e}(\mathbf{x})$ is the embedding vector of sample $\mathbf{x}$, extracted from the first layer of the Transformer[1].

To determine the closeness of embeddings, we also define the **proximity** metric p for each embedding as:

$$p_i = \sum_{j=1}^{m} sign(s_{ij} - s_c), \tag{4}$$

---

[1] For large-scale datasets, one can randomly choose a limited number (e.g., q) of samples per class to develop a triangular similarity matrix $S^{q \times q}$ which can enhance the computational efficiency.

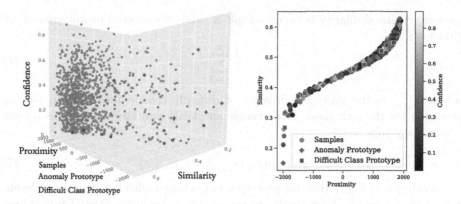

Fig. 3. Selected embedding prototypes of a single class of Twitter dataset. Difficult class prototypes have higher proximity, while anomaly prototypes suffer from low proximity due to their complex nature.

where $sign(x)$ is a sign function[2] and s_c is an arbitrary value from the similarity matrix (default as 20-percentile). Intuitively, a higher proximity indicates that the textual embeddings have more similar embeddings around them and are 'closer' to every other sample in the embedding space.

Follwing [12], problematic samples cause low confidence in output logits of the model. Hence, we define the **confidence** metric c as:

$$c_i = |\overbrace{\max_{\hat{c}^1} \mathcal{F}_W(\mathbf{x}_i)_{\hat{c}^1}}^{\text{largest logit}} - \overbrace{\max_{\hat{c}^2} \mathcal{F}_W(\mathbf{x}_i)_{\hat{c}^2}}^{\text{second largest logit}}| \qquad (5)$$

where logits are scaled (0–1 range) and are taken from the output before the softmax layer after a preliminary training stage. Intuitively, when the confidence is low (near zero), the model indecisivess is the highest.

We can now represent the embeddings in a three-dimensional space as shown in Fig. 3 (similarity-proximity-confidence). The difficult class prototype selection follows three general rules: (i) it should prioritize low confidence samples (i) it should be 'far' enough from existing prototypes (if any), (iii) it should have high 'proximity' when possible. To this end, the first prototype with the lowest confidence, highest proximity, and highest similarity is chosen. Then, the subsequent difficult class prototypes are chosen in a logsparse [8] manner for every round with an exponential selection step of sample size $(\log_2(N))$. Note that the samples are selected based on the low confidence, then high proximity but should have the lowest average similarity with the previously selected prototypes to be distinctive from each other. This strategy ensures us that the difficult class prototype are well represent problematic samples of the dataset.

Next, at a certain round (t), a prototype set $\mathbf{X}_c = \left\{\mathbf{x}_c^{(1)}, \ldots, \mathbf{x}_c^{(t)}\right\}$ is already formed for the c-th class, $c = 1, \ldots, \bar{c}$. Given any text $\mathbf{x}_i$, we can calculate the

---

[2] $sign(x) = 1$ for $x > 0$, $sign(x) = 0$ for $x = 0$, and $sign(x) = -1$ otherwise.

average cosine similarity between sample $\mathbf{x}_i$ and the selected prototype embeddings as:

$$s_{i,(c)}^c = \frac{1}{t} \sum_{j=1}^{t} s_{i,c^{(j)}}, \tag{6}$$

where $s_{i,(c)}^c$ is the average similarity of difficult class embeddings in the j_{th} iteration for the c-th class. This average similarity can then be used as a complementary supervsion:

$$z_i^c = \operatorname{argmax}_c \{s_{i,(c)}^c | c = 1, ..., \bar{c}\}. \tag{7}$$

As shown in Fig. 3, difficult prototypes are chosen with low confidence levels, where they have the least similarity among the previously selected prototypes. During this process, we ensure that the subsequent prototypes stay far enough from existing prototypes so that there are limited redundant representations of the similar samples.

3.2 Anomaly Prototypes

Anomaly prototypes are the selected sample prototypes that represent the scattered and shattered minority samples of a dataset. Such samples are often harder to detect and tend to deviate from normal samples.

Given that the remaining classification error can be caused by such anomalies, it's important to not only capture such anomalies robustly but also leverage them for the optimization objectives of Transformers.

So far, difficult class prototypes can cover the problematic samples as they are detected by having high proximity and similarity. However, a certain portion of prototypes may be located 'far' from the difficult class prototypes and often represent the minority members of a class, as indicated by the red dots in the in Fig. 3. Such prototypes represent the minority of samples as they have a lower density.

The prototype with the least proximity p_{min} is selected in the first round. This ensures us that the elected prototype is representative of the minority samples. We then select the subsequent ones in the same logsparse manner as before, ensuring that the prototypes have the least similarity. The similarity score is calculated in a similar manner to Eq. (6) while including the anomaly prototypes in the summation.

Figure 2 also illustrates the process, where gray dots represent all sample embeddings, and red dots indicate the embeddings of selected anomaly prototypes.

3.3 Multi-objective Self-learning

Transformers used in text classification often rely on a single source of supervision which is the given labels. However, such design choice limits the Transformer's ability to perform well when the datasets are noisy labeled. Moreover,

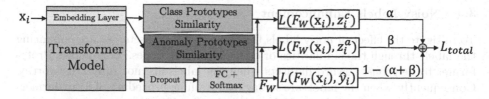

Fig. 4. Protoformer leverages the embedding space to derive the difficult class and anomaly prototypes. The network is trained jointly on Transformer and similarity of embedding prototypes. The total loss is dependent on the α and β values which are estimated in the training phase.

anomaly samples appear less in training compared to samples with high similarity. Note that majority of self-learning objectives for Transformers are to provide the greatest level of classification accuracy for all samples regardless of whether they are in the majority or minority. Intuitively, such a self-learning objective does not guarantee that the model suits well for minority classes due to their lower occurrence. In order to incorporate our prototypes during the training and test stage, we introduce a multi-objective self-learning mechanism to Protoformer.

As shown in Fig. 4, the similarities of embedding prototypes are used as self-supervision to train the Protoformer $\mathcal{F}_\mathcal{W}$ after its fine-tuning state. The self-supervision is provided by the class prototype as below:

$$\mathcal{L}_{\text{proto}} = \frac{1}{m} \sum_{i=1}^{m} (\alpha \cdot \mathcal{L}(\mathcal{F}_W(\mathbf{x}_i), z_i^c) + \beta \cdot \mathcal{L}(\mathcal{F}_W(\mathbf{x}_i), z_i^a)), \tag{8}$$

where the weight factors $\alpha, \beta \in [0,1)$ and $\alpha + \beta < 1$ indicate the concentration of Transformer on the similarities of self-supervision of difficult class prototypes z_i^c and anomaly prototypes z_i^a. Hence, the overall loss is calculated by minimizing the classification loss based on three components:

$$\mathcal{L}_{\text{total}} = (1 - (\alpha + \beta)) \cdot \frac{1}{m} \sum_{i=1}^{m} (\mathcal{L}(\mathcal{F}_W(\mathbf{x}_i), \hat{y}_i) + \mathcal{L}_{\text{proto}}, \tag{9}$$

To this end, when the network's predictions are in between classes, the network can improve its training by the self-supervision provided by the similarity of difficult class prototype z_i^c and anomaly prototype z_i^a. Hence, we continue the training procedure iteratively until convergence: $W^{(t+1)} \leftarrow W^{(t)} - \xi \nabla (\mathcal{L}_{total})$, where the gradient descent vector $\nabla(\mathcal{L}_{total})$ holds the partial derivatives of weights and biases of the total loss function, and ξ is the learning rate. We use a fully connected layer over the final hidden state corresponding to the output token of the Transformer (i.e., [CLS] token). The softmax activation function is then applied to the hidden layer to provide classification. It is important to note that this procedure can also be implemented solely during the test stage, which can make the calculation timing complexity of Protoformer similar to the fine-tuning process.

3.4 Noisy Labels Enhancement

To mitigate the effect of noisy labels throughout the datasets, we are enhancing the labels through the similarities of embedding prototypes. This allows Protoformer to be robust toward datasets when the labels are not fully trustworthy. Consequently, when the labels are wrong, the training procedure of Transformers provides suboptimal weights, which makes the classification results untrustworthy.

Specifically, we can obtain the adjusted label of the a noisy labeled sample through maximum similarity to the difficult class prototype:

$$\tilde{y}_i = \mathrm{argmax}_c\{s_{i,(c)}|c = 1, ..., \bar{c}\}, \tag{10}$$

where $s_{i,(c)}$ is the cosine similarity defined in Eq. (6) and the enhanced labels $\tilde{y}$ can be used as a replacement for the noisy labels. Thus, the overall loss is calculated in a similar manner as Eq. (9), while we are replacing the original noisy labels with the adjusted label.

4 Experiments

In this section, we provide descriptions for the datasets. We also describe the experimental settings and evaluation results. Lastly, we provide an analysis section that further discusses the effectiveness of Protoformer components.

4.1 Benchmark Datasets and Baselines

We have experimented with three challenging real-world datasets[3]. The brief discussion for each dataset is as follows:

Table 1. Summary statistics of the evaluation dataset.

Dataset	Twitter-Uni	IMDb	Arxiv-10
# Examples	25,000	25,000	100,000
# Train	20,000	20,000	80,000
# Validation	2,500	2,500	10,000
# Test	2,500	2,500	5,000
# Classes	8	2	10

Twitter-Uni (See footnote 3). We crawled over 12 million historical Tweets of 25,000 Twitter profiles from 8 U.S. college followers. As an example, the college of engineering holds near 3000 followers, which are labeled as engineering. Note that

---

[3] Self-gathered datasets are accessible at https://github.com/ashfarhangi/Protoformer.

most existing benchmark Twitter datasets fail to hold high-quality labels that are provided by the original Twitter users. To alleviate this issue, we extracted a set of students that stated their major in their Twitter bio. This set can serve as ground truth of the clean labels. We made this challenging new dataset available online, which can be used for future text classification or noisy label correction studies.

ArXiv-10 (See footnote 3). We also crawled the abstracts and titles of 100 thousand ArXiv scientific papers on ten research categories that include subcategories of computer science, physics, and math. The dataset is downsampled to contain exactly 10 thousand samples per category.

IMDB. The third dataset is the benchmark IMDb movie reviews [10]. The dataset is widely used as the sentiment classification task. It contains 25 thousand samples per sentiment (positive or negative). Both IMDb and ArXiv-10 datasets are originally labeled by the authors. It is however good to note that the labels are still susceptible to noisy labels.

The **baseline** methods for comparison include:

- SVM [11], supervised learning with a linear separator to maximize the margin between classes, with the fine-tuned embeddings derived from the Transformers.
- HAN [6], a hierarchical attention network for textual classification with word and sentence-level attention mechanisms.
- DocBERT [1], a document Transformer model with an LSTM architecture rather than a fully connected layer.
- RoBERTa [9], a Transformer with an improved pretraining procedure. Specifically, showing improvement by removing the next sentence prediction pretraining objective.

Table 2. Hyperparameters of the Protoformer used for each dataset.

Parameter	Twitter-Uni	IMDb	Arxiv-10
Batch size	32	64	32
Learning rate	5×10^{-5}	3×10^{-5}	5×10^{-5}
Weight decay	5×10^{-5}	1×10^{-5}	1×10^{-4}
Preliminary training epochs	5	3	2
Fine-tuning epochs	20	10	10
Training time	1:49 h	1:32 h	1:45 h
Transformer	DistilBERT	BERT	RoBERTa

4.2 Experimental Settings

To showcase the generalization ability of our framework, we selected a unique Transformer for each dataset (Table 2). The hyperparameters are based on the

Table 3. Evaluation of the Protoformer and baseline methods.

Model	Twitter			IMDb			ArXiv		
	Ma-F1	Recall	Acc	Ma-F1	Recall	Acc	Ma-F1	Recall	Acc
SVM [11]	0.384	0.361	0.391	0.744	0.733	0.748	0.691	0.654	0.708
HAN [15]	0.412	0.392	0.425	0.894	0.882	0.896	0.732	0.696	0.746
DocBERT [1]	0.521	0.506	0.534	0.932	0.921	0.936	0.752	0.727	0.764
RoBERTa [9]	0.555	0.531	0.567	0.952	0.941	0.957	0.769	0.732	0.779
Protoformer	**0.802**	**0.784**	**0.813**	**0.964**	**0.952**	**0.968**	**0.784**	**0.744**	**0.794**

highest Macro-F1 score obtained on the validation set for all models (following the standard 80-10-10 split). We used a grid search approach to explore the hyperparameters: size of fully connected layer $H^D \in \{256, 512, 768, 1024\}$ and dropout $\delta \in \{0.0, 0.1, \cdots, 0.9\}$. The experiments are conducted using PyTorch on a cloud workstation using Nvidia Tesla A100 GPU.

4.3 Experimental Results

For a less noisy labeled datasets such as IMDB and Arvix, the evaluated methods performed comparatively. Note that the majority of the classification error appears when the network does not show confidence in its classification, as was previously shown for the IMDB dataset in Fig. 5. The Protoformer is also able to provide a competitive accuracy for cleaner datasets such as IMDb and ArXiv-10. Among the baselines, the performance of RoBERTa [9] is favorable compared to others. This is partly due to the different pretraining objectives from DocBERT. As shown in Table 3, Protoformer resulted in the highest margin of accuracy for a noisy dataset, improving the Macro-F1 score from 55.5% to 80.2% for the Twitter-Uni dataset. We observed that this dataset provides the greatest difficulties for baseline methods where the models often misclassify problematic samples. To this end, we report a detailed accuracy breakdown for the Twitter dataset in Fig. 5. The fine-tuning process for Transformers such as DocBERT, RoBERTa results in suboptimal classification. Leveraging the selected prototypes, Protoformer was able to improve its classification accuracy on the harder and more complex samples (e.g., management students that are similar to other classes). To this end, the fine-tuning process alone does not result in adequate accuracy due to the noise of the dataset. The combination of both embedding prototypes allows the Transformer to have a solution for anomalies and problematic samples of the dataset and further improves its generalization ability through difficult class prototypes.

4.4 Analysis

In this section, we provide an extensive analysis of the performance of Protoformer, as well as the role of each type of prototype on the overall performance. Hence, we limited the number of prototypes per class for the Twitter dataset

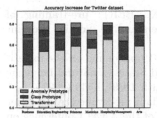
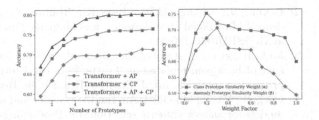

Fig. 5. Left: Accuracy increase from the initial (light blue), class protoype (blue) and class anomalies (red), for Twitter dataset. Right: Influence of anomaly labeling of hurricanes for Collier county gross hotel sales revenue. **Middle:** Number of anomaly prototypes (AP) and difficult class prototypes (CP) per class for Twitter dataset. Higher number of prototypes resulted in marginal improvement while the combination of both category of prototypes gives us the optimal accuracy. **Right:** Testing accuracy with respect to the weight factors (α and β) ranging from 0 to 1. (Color figure online)

and reported the changes. The results in Fig. 5 show that a single prototype is not sufficient to provide competitive accuracy even with the help of a fine-tuned Transformer. However, as the number of prototypes increased, we observed improvements in the accuracy of the Protoformer. The prototype selection procedure previously discussed ensures that there are multiple prototypes for every proximity metric, and the calculation of them is computationally expensive even for the large-scale dataset. Moreover, the weight factors are reported separately to showcase the effect of their self-supervision for the Twitter dataset. The results show that relying on the noisy labels (α and $\beta = 0$) during training would be suboptimal and perform poorly on confirmed test data. Moreover, the accuracy would be optimal when weight factors sum to 0.5 (i.e., $\alpha = 0.2$, $\beta = 0.3$).

5   Conclusion

In this work, we developed a novel Transformer framework, Protoformer, that leverages the embedding prototypes of the dataset to enhance its generalization and specialization abilities. It also includes a procedure for handling noisy labels. Various experiments are conducted to demonstrate the effectiveness of Protoformer over state-of-art topic and sentiment classification methods. For future work, we are interested in applying Protoformer for the image recognition tasks. We also like to explore the use of Protoformer on spherical and hyperbolic embedding space.

Acknowledgement. Our work has been supported by the US National Science Foundation under grants No. 2028481, 1937833, and 1850851.

References

1. Adhikari, A., Ram, A., Tang, R., Lin, J.: DocBERT: BERT for document classification. arXiv preprint arXiv:1904.08398 (2019)

2. Devlin, J., Chang, M.W., Lee, K., Toutanova, K.: BERT: pre-training of deep bidirectional transformers for language understanding. arXiv preprint arXiv:1810.04805 (2018)
3. Fiok, K., et al.: A study of the effects of the COVID-19 pandemic on the experience of back pain reported on Twitter® in the United States: a natural language processing approach. Int. J. Environ. Res. Public Health **18**(9), 4543 (2021)
4. Garg, S., Vu, T., Moschitti, A.: TandA: transfer and adapt pre-trained transformer models for answer sentence selection. In: Proceedings of the AAAI Conference on Artificial Intelligence. vol. 34, pp. 7780–7788 (2020)
5. Han, J., Luo, P., Wang, X.: Deep self-learning from noisy labels. In: Proceedings of the IEEE/CVF International Conference on Computer Vision, pp. 5138–5147 (2019)
6. Krishnan, R., Shalit, U., Sontag, D.: Structured inference networks for nonlinear state space models. In: Proceedings of the AAAI Conference on Artificial Intelligence. vol. 31 (2017)
7. Lee, K.H., He, X., Zhang, L., Yang, L.: CleanNet: transfer learning for scalable image classifier training with label noise. In: Proceedings of the IEEE Conference on Computer Vision and Pattern Recognition, pp. 5447–5456 (2018)
8. Li, S., et al.: Enhancing the locality and breaking the memory bottleneck of transformer on time series forecasting. Adv. Neural Inf. Process. Syst. **32** (2019)
9. Liu, Y., et al.: RoBERTa: A robustly optimized BERT pretraining approach. arXiv preprint arXiv:1907.11692 (2019)
10. Maas, A., Daly, R.E., Pham, P.T., Huang, D., Ng, A.Y., Potts, C.: Learning word vectors for sentiment analysis. In: Proceedings of the 49th Annual Meeting of the Association for Computational Linguistics: Human Language Technologies, pp. 142–150 (2011)
11. Meyer, D., Leisch, F., Hornik, K.: The support vector machine under test. Neurocomputing **55**(1–2), 169–186 (2003)
12. Pleiss, G., Zhang, T., Elenberg, E., Weinberger, K.Q.: Identifying mislabeled data using the area under the margin ranking. Adv. Neural Inf. Process. Syst. **33**, 17044–17056 (2020)
13. Vaswani, A., et al.: Attention is all you need. Adv. Neural Inf. Process. Syst. **30** (2017)
14. Wei, H., Feng, L., Chen, X., An, B.: Combating noisy labels by agreement: a joint training method with co-regularization. In: Proceedings of the IEEE/CVF Conference on Computer Vision and Pattern Recognition, pp. 13726–13735 (2020)
15. Yang, Z., Yang, D., Dyer, C., He, X., Smola, A., Hovy, E.: Hierarchical attention networks for document classification. In: Proceedings of the 2016 Conference of the North American Chapter of the Association for Computational Linguistics: Human Language Technologies, pp. 1480–1489 (2016)

Graph Multi-Head Convolution for Spatio-Temporal Attention in Origin Destination Tensor Prediction

Manish Bhanu[1]([✉]), Rahul Kumar[1], Saswata Roy[1], João Mendes-Moreira[2,3], and Joydeep Chandra[1]

[1] Indian Institute of Technology Patna, Patna, India
{manish.pcs16,rahul_2121cs10,saswata.pcs17,joydeep}@iitp.ac.in
[2] LIAAD-INESC TEC, Porto, Portugal
[3] Faculty of Engineering, University of Porto, Porto, Portugal
jmoreira@fe.up.pt

Abstract. Capturing complex spatio-temporal features of thousands of correlated taxi-demand time-series in the city makes the traffic flow prediction problem a challenging task. Hence, several Deep Neural Network (DNN) models have been developed to mimic the latent spatio-temporal behaviour of taxi-demand time-series in a city to improve the prediction results. Despite, good performance of recent DNN based traffic prediction techniques, such models can only identify either adjacent or connected regions with direct or transitive connection; hence they fail to capture spatio-temporal correlation among regions that exhibit implicit or latent connection. Additionally, the dependency of the recent DNN models on recursive components facilitates error propagation during feature aggregation without any counter strategy for it. In view of these existing glitches, we introduce a novel DNN model, graph Multi-Head Convolution for Spatio-Temporal Aggregation (gMHC-STA) which supports capturing spatio-temporal correlation among regions with explicit and implicit connection both. Moreover, gMHC-STA aggregates both spatial and temporal characteristics using multi-head attention; thus overriding recursive RNN or its variant approach to prevent noise propagation. The experimental results of gMHC-STA on two real-world city taxi-demand datasets report *minimum* of 6.5–10% improvement over the best state-of-the-art on standard benchmark metric in varying experimental conditions.

Keywords: Spatio-temporal · Graph convolution · Multi-step prediction · Time series · Taxi-demand

1 Introduction

Traffic demand prediction is important for transportation planners, traffic engineers, and policymakers. More precisely, they are used by many Departments

J. Gama et al. (Eds.): PAKDD 2022, LNAI 13280, pp. 459–471, 2022.
https://doi.org/10.1007/978-3-031-05933-9_36

of Transportation (DOTs) and highway agencies to help plan, build, and main-tain transportation infrastructure at the country, provincial, and national levels. Taxi-demand prediction being a similar objective is of prime importance to the taxi-companies to reduce cost on resources and meet commuters' maximum sat-isfaction. Hence, it becomes an essential challenge to predict the taxi-demand of city traffic networks when thousands of daily commuters are dependent on it for their professional needs. Many research have been carried in this direction from classical machine learning (ML) [1,10–12,16] algorithms to trendy Artifi-cial Neural Networks (AI) approach [7,8,22]. The initial time series statistical predictive models like ARIMA [11], SARIMA [10], Bayesian Update [16] and many of their variants have been found applicable to these prediction problems having linear or simpler features. Their capability is challenged when the data is multi-dimensional or has complex latent spatio-temporal features. Taxi-demand of city traffic network comprise of multiple correlated time series data with com-plex latent spatio-temporal features. Tensor-based [2,4,14,16] and Deep Neural Networks models [5,19–22] have been very efficient in handling high dimen-sionality as well as complex latent features in data. Tensor-Based models like DTC [16], tucker-cur [4], CP-ALS [14] could efficiently predict short-term traffic but are not able to harness traffic network characteristics which vary vastly for different cities. Tensor models like TeDCaN [2] could efficiently predict long-term prediction as well as capture city specific traffic characteristics. Despite all, the high computational complexity of Tensor models [14] and their limited performance on long-term predictions, DNN models have been now the first choice of predictive models in this domain. Moreover, DNN [5,19–22] models usually provide high predictive accuracy and are efficient for both short and long-term prediction. The capability of DNN models to include any form of additional features as attention make them a suitable predictive model for this domain. For example recent DNN models like STGCN [20], GEML [19], ST-MGCN [5], att-ConvLSTM [22], GMAN [21] have already shown their high pre-dictive performance. These models use Convolutional Neural Networks (CNN) or Graph Convolution Networks (GCN) to capture spatial features in the taxi-demand data. While to capture temporal features, these models use Recurrent Neural Networks (RNN) or its variants LSTM, GRU. Despite these advanced techniques, these models show strategic limitations like the inability to capture implicit connection, preventing high error accumulation due to recursive com-ponents. Another major pitfall of these recent models is their limitation to pre-dict region-pair prediction. Since taxi-demand from a region (s) to a region (d) would differ vastly from that of d to s. Also, these to-and-fro demands depend on explicit as well as implicit connections of many other regions. Except for GEML, other models like att-convLSTM, ST-MGCN, GMAN, STGCN do not support region to region taxi-demand prediction either due to their strategic or structural limitations. att-ConvLSTM uses city-grid structure while STGCN, ST-MGCN, GMAN would need large channel size for the region to region pre-diction. Even then these models could not capture region pair correlation during prediction owing to their structural limitations. With the view of culminating these constraints in taxi-demand predictions that are prevalent in recent models, we propose gMHC-STA, a DNN based model that can override recursive approach to mitigate error accumulation and use modified graph convolution [18] that can help in capturing all explicit as well implicit connections to provide attention with every region under study. Structurally superior to ST-MGCN, GMAN, the proposed model can incorporate region-region interaction during prediction for region pair prediction. Figure 1 demonstrates the implicit connection between two distant regions in the city and their similar demand pattern. In the paper, we present how the proposed model is able to bring these regions closer in the embedding space. Hence, the proposed model is able to retrieve each region pair taxi-demand predictions using embedded features of region pair interaction. We summarize the contributions of our work as follows:

- The proposed model introduces spatial and temporal convolution by means of parallel multi-head convolution which overrides error prone recursive approach and is able to identify spatio-temporal correlations among regions with both implicit and explicit connection.
- The proposed model is capable of making region to region prediction which varies both directions, unlike many recent predictive models which often overlook the problem.
- The experimental findings on two real-world city taxi-demand dataset for varying experimental condition presents the model's distinct advantages over the state-of-the-art in reducing prediction error for every region in the city.

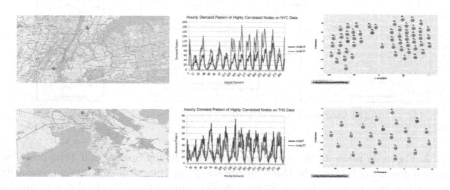

Fig. 1. NYC (above), THS. The leftmost map shows two regions (GRs: grid representatives/nodes) that are not explicitly connected in data and are far away. The graphs indicate the demand pattern of these GRs. Last column, TSNE plots by the proposed model exhibit that those GRs (red) are closer in the plot despite no explicit connection between them. The proposed model is able to capture implicit relation between these GRs which GCN could have missed. (Color figure online)

2   Problem Definition

An Origin-Destination Tensor ($\mathcal{A} \in \mathbb{R}^{t \times n \times n}$) prediction problem can be stated as
a group of mutually correlated time series prediction problem where taxi-demand
data being time series varies in different spatial and temporal snapshot of the
city. Taxi-demand prediction problem detailed for a region pair can be stated as
provided $Historic\,ODT\,\mathcal{A} = \{A_0, A_1, A_2 \ldots A_t\}$, the objective is to find a model
that can well approximate to the real future ODT $\mathcal{A}_\circ = \{A_{t+1}, A_{t+2} \ldots A_{t+r}\}$
using $Pridicted\,ODT\,\hat{\mathcal{A}} = \{\hat{A}_{t+1}, \hat{A}_{t+2} \ldots \hat{A}_{t+r}\}$ with minimum possible error.
The above problem can be mathematically formulated as follows:

$$\underset{\theta}{\operatorname{argmin}} \sum_{i=1}^{r} ||A_{t+i} - \hat{A}_{t+i}||_2, \tag{1}$$

where θ is the model parameter that needs to be learnt. $A \in \mathbb{R}^{n \times n}$, $\hat{A} \in \mathbb{R}^{n \times n}$
are real and predicted ODMs respectively and prediction is made for the r times-
tamps ahead.

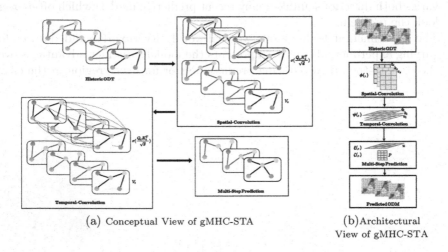

(a) Conceptual View of gMHC-STA   (b) Architectural
                                          View of gMHC-STA

Fig. 2. Figure 2a presents the concept of Attention during Spatial and Temporal Convolution module. All the dashed lines are not shown in Temporal-convolution for clarity.
The dashed line presents the attention (implicit connection) involved during prediction. Figure 2b shows the shared kernels and operations in each block of the model. The
shared kernels help to capture common features across spatial-temporal dimension.

3   Proposed Methodology

In this section, we present a detailed constructional view of gMHC-STA. Initially, we give a summarised view of the proposed model which consists of three
operational units (or modules) which are as follows:

- **Spatial-Convolution:** The first module of gMHC-STA convolves over the *spatial features* of each temporal snapshot of all the regions of the city. Unlike many previous models where attention is based on neighbouring regions or explicit connections [5,19], this module helps to provide attention based on both explicit as well as latent behaviour of the nearby and distant regions.
- **Temporal-Convolution:** The second module convolves over the *temporal features* of each region based on correlation attention among the temporal components of the region; thus overriding the recursive requirement such as utilizing RNN, GRU, LSTM etc. used in many state-of-the-art models [5,19, 20].
- **Multi-Step Prediction:** The final module is used to carry multi-step prediction through a common kernel. It emphasizes on shared spatio-temporal properties of the city regions.

3.1 Spatial-Convolution

Firstly, we attempt to aggregate spatial features of region pairs in a temporal snapshot of the city. We convolve over the demand pattern of each region pair irrespective of whether these regions are connected or not. In the process, the strength of the convolution is decided by the correlation of demand pattern (σ in Eq. 3) of region pair. This module leverages modified graph convolution [17]. A generic Graph Convolution Networks (GCN) has already shown their effectiveness in aggregating graphs features in node domain depending up on the connected neighborhood [5,19]. Such convolution is effective in mapping contextual relations of nodes with its neighbors. Despite, generic graph convolution can possibly miss many transitive or implicit relationships between nodes which do not appear in graph connections. For example, if an important region is connected to city airport through multiple other regions, that important region and city airport forms an implicit connections. Such connections can not be directly visible though they exhibit strong correlation. Hence, we attempt to aggregate spatial features of each node depending on their correlation with any other existing nodes either in explicit or implicit connection; attention by using "complete graph analog of ODM" in Fig. 2. The modified graph convolution operation on an ODM ($A \in \mathbb{R}^{n \times n}$) and node feature matrix ($X \in \mathbb{R}^{n \times f}$) is represented below:

$$H = AXW \tag{2}$$

where $W \in \mathbb{R}^{f \times d}$ is GCN kernel mapping f features to d dimension. $H \in \mathbb{R}^{n \times d}$ is the spatially convolved graph network (demand traffic network in our case). n is the number of nodes (regions) in the network.

The above convolution can only aggregate neighboring features of the node while we need to aggregate features of each node based on all other nodes' correlation values. To achieve this, we apply 3 graph convolution kernels $q_s \in \mathbb{R}^{f \times d}$, $k_s \in \mathbb{R}^{f \times d}$ and $v_s \in \mathbb{R}^{f \times d}$. Now, the aggregation of graph (A) features (X) are carried as follows:

$$Q_s = AXq_s, \quad K_s = AXk_s, \quad V_s = AXv_s, \quad H = \sigma(\frac{Q_s K_s^T}{\sqrt{d}})V_s \tag{3}$$

where σ is softmax activation on last axis which provides attention to V_s and is a measure of correlation strength between each region pair demand patterns. H is generated as a result of *multiple graph convolutions (heads)*. The obtained convolved graph representation $H \in \mathbb{R}^{n \times d}$ has each node whose features is a weighted sum of all other nodes' features based on the degree of their correlation strength with them. Ultimately, the presented graph convolution can successfully capture implicit as well as explicit correlation between each node pair irrespective of their connection in the graph. We represent the above spatial convolution operation over a temporal element (i) of the sequence of input of size t as follows:

$$H_i = \phi(A_i, X_i) \tag{4}$$

The model's first module uses the sliding window strategy (for broadcast operation) to carry spatial convolution on each element of the temporally ordered sequence of ODMs and its features as follows:

$$\{H_1, H_2 \ldots H_t\} = \{\phi(A_1, X_1), \phi(A_2, X_2) \ldots \phi(A_t, X_t)\} \tag{5}$$

$$\mathcal{H} = \{H_1, H_2 \ldots H_t\} \tag{6}$$

The output of the first module of the model is a temporally ordered spatially aggregated graph representative ($\mathcal{H} \in \mathbb{R}^{t \times n \times d}$). The output is fed to the next module to carry temporal convolution, which is explained later.

3.2   Temporal-Convolution

Each region can be represented as a temporal sequence of its state. Using temporal-convolution, we aggregate a region's features in its different temporal states into a single representation. Hence, we represent each region embedding by aggregating that region's features in all of its previous temporal states. Here again, the strength between two temporal state depends on their correlation value (Eq. 7). To present, the operations in this module, we define a temporal slice of the input as $\widetilde{H} \in \mathcal{H}$ where $\widetilde{H} \in \mathbb{R}^{t \times d}$. It should be noted, $\widetilde{H}$ is slice of $\mathcal{H}$ on spatial axis while H_i in Eq. 6 is slice of same $\mathcal{H}$ on temporal axis. The following presents the convolution of each temporal slice of the input by 3 kernels ($q_t \in \mathbb{R}^{d \times g}$, $k_t \in \mathbb{R}^{d \times g}$, $v_t \in \mathbb{R}^{d \times g}$) of this module:

$$Q_t = \widetilde{H}q_t, \quad K_t = \widetilde{H}k_t, \quad V_t = \widetilde{H}v_t, \quad \overline{H} = \sigma(\frac{Q_t K_t^T}{\sqrt{g}})V_t \tag{7}$$

The above temporal convolution can be represented on each spatial element (i) of the input of size n as follows:

$$\overline{H}_i = \psi(\widetilde{H}_i) \tag{8}$$

For each node, the above temporal convolution is carried using the sliding window strategy and is demonstrated as follows:

$$\{\overline{H}_1, \overline{H}_2 \ldots \overline{H}_n\} = \{\psi(\widetilde{H}_1), \psi(\widetilde{H}_2) \ldots \psi(\widetilde{H}_n)\} \tag{9}$$

$$\mathcal{T} = \{\overline{H}_1, \overline{H}_2 \ldots \overline{H}_n\} \tag{10}$$

In the above equation, $\mathcal{T} \in \mathbb{R}^{t \times n \times g}$ is the spatio-temporal convolution of the input ODT which can be used for the prediction in the next phase. g is the dimension of the kernels in this module, bringing transformation of the input from dimension d to g.

3.3 Multi-step Prediction

The output of the previous module $\mathcal{T} \in \mathbb{R}^{t \times n \times g}$ preserves all the spatio-temporal characteristics of the ODT in latent space which might not be explicitly evident in the raw input data. Hence, $\mathcal{T}$ is the sufficient information that can now be used to predict the future ODT. Though, the prediction of r windows needs a proper aggregation of t windows information. Hence, we facilitate this process by a kernel ($u \in \mathbb{R}^{t \times r}$) that transforms the t dimensional information into r dimensional information. Firstly, we represent $\mathcal{T}$ as a sequence of n slices, i.e., $\mathcal{T} = \{T_{g \times t}\}_n$. Then, we apply matrix multiplication with the kernel $u \in \mathbb{R}^{t \times r}$ on each slice, i.e., the broadcast operation (Eqs. 11 and 12). This matrix multiplication is the transformation of the t dimensional space into the r dimensional space. We represent the final result $\widetilde{\mathcal{T}}$ as a 3-D tensor (Eq. 13). The following transformation represents this:

$$\widetilde{\mathcal{T}} = \{T_{g \times t}\}_n u_{t \times r} \tag{11}$$

$$\widetilde{\mathcal{T}} = \{\dot{T}_{g \times r}\}_n \tag{12}$$

$$\widetilde{\mathcal{T}} = \widetilde{\mathcal{T}}_{r \times n \times g} \tag{13}$$

In the above equations, subscript refers to the dimension and $\dot{T}_{g \times r}$ is one of nth slice of $\widetilde{\mathcal{T}} \in \mathbb{R}^{r \times n \times g}$. Equation 12 is the result of a broadcast operation on each slice whose transformation can be seen as:

$$\dot{T} = \xi(T) \tag{14}$$

The result $\widetilde{\mathcal{T}}$ can be seen as sequence of r slices as $\widetilde{\mathcal{T}} = \{\bar{T}_{n \times g}\}_r$. We have a broadcast operation of matrix multiplication using each slice, each slice transpose and the kernel $P \in \mathbb{R}^{g \times g}$ as follows:

$$\hat{A} = \{\bar{T}_{n \times g}\}_r P_{g \times g} \{\bar{T}_{g \times n}^T\}_r \tag{15}$$

$$\hat{A} = \{\hat{A}_{n \times n}\}_r \tag{16}$$

$$\hat{A} = \hat{A}_{r \times n \times n} \tag{17}$$

The kernel in the Eq. 15 undergoes a broadcast operation and is implemented using the sliding window strategy on each prediction-temporal element (i) of input size r as follows:

$$\hat{A}_i = \zeta(\bar{T}_i) \tag{18}$$

The sequence of the prediction-temporal slice are seen to undergo the following convolution:

$$\{\hat{A}_1, \hat{A}_2 \ldots \hat{A}_r\} = \{\zeta(\bar{T}_1), \zeta(\bar{T}_2) \ldots \zeta(\bar{T}_r)\} \tag{19}$$

$$\hat{A} = \{\hat{A}_1, \hat{A}_2 \ldots \hat{A}_r\} \tag{20}$$

where $\hat{A} \in \mathbb{R}^{n \times n}$ is a predicted ODM. Ultimately, we obtain the predicted ODT ($\hat{\mathcal{A}}$) for the real ODT ($\mathcal{A}_o$) by the model using the input ODT ($\mathcal{A}$) provided.

4   Experimentation

4.1   Datasets

We evaluate our proposed approach on two benchmark real-world taxi demand datasets [2,3]: New York City (NYC: January-March, 2014) and Thessaloniki (THS: January-March, 2015). NYC dataset contains GPS information of both the source and destination along with the time of journey, while the THS data has only source GPS information together with the time of journey and taxi-id. To extract the destination GPS information in THS data, we use the source GPS information and timestamp of the next starting trip by the same taxi-id. Both NYC and THS traffic regions are divided into equal size rectangular grids. Central GPS of a grid is termed Grid Representative (GR). In ODT [2,3,13,15] formation, many vacant and less significant grids are discarded. The top 55 and 25 grids are used for the experimental study for New York City and Thessaloniki based on mobility count. The grid size of 5×5 km^2 [2,5,22] is used and time interval of 1 h is chosen; thus each ODT is composed of 2160 ODMs.

4.2   Experimental Setups

We use 80% of data for training, 20% for testing and 10% of training data for validation. Model parameters are tuned using grid-search and the best parameters are reported as follows: dimension size (d) is 10, epochs 600 (NYC) and 200 (THS), batch 32. Learning rate is 10^{-4} with optimizer ADAM and regularization is l2(10^{-3}). Activation functions are leakyReLU, sigmoid, softmax. Historic window (HWND) size is $H = \{12, 24, 168\}$ and prediction window (PWND) size is $r = \{1, 2, 4, 6, 8, 10, 12\}$, both in hours. Loss function is mse on metric $accuracy$. Percentage of missing information under study is $mi = \{5, 10, 20, 40, 50\}\%$.

4.3   Baseline Techniques

We compared the proposed model with the trendy Deep Neural Networks predictive models in this domain. **ST-MGCN** [5] is a temporal attention based DNN model which uses ChebNets as GCN for spatial convolution and RNN for temporal Convolution. **GEML** [19] uses GCN and LSTM with Multi-Task Learning (MTL) strategy. It follows the concept of graph-sage [6] for improved spatial convolution. **ForGAN** [9] is a generative adversarial predictive model which uses the concept of Conditional-GAN for the prediction of traffic time-series data. Recently, **GMAN** [21] uses the notion of multi-head attention [17] mechanism instead of graph convolution for spatio-temporal attention.

5   Result Analysis

We present the experimental results in this section. For evaluation of the test case, we have used Root Mean Squared Error (RMSE) [19], Mean Absolute Percentage Error (MAPE) [21], Symmetric Mean Absolute Percentage Error (SMAPE) [19]. Reported MAPE, SMAPE are scaled between [0, 1].

Table 1. RMSE of multi-step prediction with Historic ODT of 24 on NYC and THS datasets.

Models	Prediction window (NYC)							Prediction window (THS)						
	1	2	4	6	8	10	12	1	2	4	6	8	10	12
ST-MGCN	4.321	4.618	3.824	4.283	3.991	4.102	4.485	4.229	4.579	6.719	5.796	4.636	4.531	4.756
GEML	**3.221**	3.625	3.906	3.824	4.468	3.975	4.119	4.794	4.489	**4.177**	4.638	4.588	5.026	4.836
GMAN	6.054	6.059	6.031	6.050	5.533	5.720	6.053	10.23	10.21	10.22	10.22	10.21	10.21	10.20
ForGAN	6.038	6.031	6.036	6.020	6.006	5.989	5.971	5.489	5.712	5.748	5.854	5.846	5.869	5.871
gMHC-STA	3.623	**3.621**	**3.614**	**3.615**	**3.616**	**3.615**	**3.615**	4.190	4.187	4.186	**4.186**	**4.189**	4.195	4.190

Table 2. MAPE of multi-step prediction with Historic ODT of 24 on NYC and THS datasets.

Models	Prediction window (NYC)							Prediction window (THS)						
	1	2	4	6	8	10	12	1	2	4	6	8	10	12
ST-MGCN	0.116	0.135	0.100	0.133	0.111	0.123	0.148	0.112	0.133	0.193	0.140	0.116	0.129	0.129
GEML	0.115	0.128	0.151	0.152	0.238	0.159	0.158	0.127	0.127	0.127	0.144	0.125	0.129	0.123
GMAN	0.589	0.589	0.564	0.588	0.443	0.473	0.588	0.536	0.519	0.519	0.509	0.502	0.498	0.488
ForGAN	0.378	0.378	0.375	0.372	0.369	0.369	0.369	0.561	0.541	0.526	0.524	0.523	0.523	0.522
gMHC-STA	**0.080**	**0.079**	**0.079**	**0.079**	**0.079**	**0.079**	**0.079**	**0.098**	**0.098**	**0.097**	**0.097**	**0.097**	**0.097**	**0.096**

Table 3. SMAPE of multi-step prediction with Historic ODT of 24 on NYC and THS datasets.

Models	Prediction window (NYC)							Prediction window (THS)						
	1	2	4	6	8	10	12	1	2	4	6	8	10	12
ST-MGCN	0.099	0.104	0.096	0.101	0.096	0.099	0.106	0.135	**0.131**	0.146	0.138	0.136	0.136	0.136
GEML	0.102	0.105	0.108	0.109	0.112	0.109	0.107	0.136	0.135	0.138	0.142	0.136	0.140	0.139
GMAN	0.696	0.696	0.680	0.695	0.647	0.655	0.695	0.964	0.964	0.963	0.962	0.962	0.961	0.961
ForGAN	0.744	0.743	0.737	0.733	0.726	0.727	0.726	0.895	0.854	0.826	0.832	0.730	0.739	0.828
gMHC-STA	**0.087**	**0.094**	**0.094**	**0.094**	**0.094**	**0.094**	**0.094**	**0.112**	0.137	**0.137**	**0.136**	**0.136**	**0.136**	**0.135**

Table 4. Average value of RMSE, MAPE, SMAPE for Historic ODT size-{12,168}.

Models	NYC dataset						THS dataset					
	RMSE		MAPE		SMAPE		RMSE		MAPE		SMAPE	
	12	168	12	168	12	168	12	168	12	168	12	168
ST-MGCN	4.065	5.07	0.112	0.15	0.098	0.107	5.276	6.182	0.107	0.116	0.135	0.141
GEML	4.083	5.245	0.167	0.266	0.111	0.15	4.756	4.72	0.133	0.143	0.139	0.142
GMAN	5.227	5.396	0.380	0.466	0.619	0.602	10.23	9.882	0.483	0.485	0.962	0.964
ForGAN	6.005	5.829	0.374	0.364	0.736	0.715	5.705	5.697	0.532	0.529	0.846	0.829
gMHC-STA	**3.623**	**3.624**	**0.078**	**0.079**	**0.096**	**0.096**	**4.191**	**4.192**	**0.098**	**0.097**	**0.135**	**0.138**

Table 5. RMSE values with different spatial (left) and temporal embedding dimensions at PWND 12 & HWND 24.

Dataset	Spatial-embedding					Temporal-embedding				
	5	10	20	40	50	4	8	16	32	128
NYC	3.672	3.672	3.672	3.6272	3.6280	3.6724	3.672	3.6723	3.6270	3.6281
THS	4.1915	4.1915	4.1908	4.1908	4.1910	4.1917	4.1915	4.1910	4.1906	4.1910

5.1 Performance on Varying Historic-Window Size

Referring to the Tables 1, 2, 3 and 4, we observe that RMSE, MAPE, SMAPE of prediction error for prediction-windows 1–12 hr of gMHC-STA is comparatively lower than the baseline methods. The lower value of SMAPE (Tables 3 and 4) indicates that the proposed model can be well suited for many other similar datasets. Among the baseline models, we find that GEML and ST-MGCN perform better than ForGAN, GMAN, stating that it is crucial to capture region pairwise correlation. One potential reason is that GEML and ST-MGCN utilize GCN+RNN/LSTM modules that can well adopt to region pair wise demand prediction similar to gMHC-STA. Additionally, there is an evident strategic difference in the prediction module of GEML, ForGAN, GMAN models. Another pitfall is that ForGAN being a generative model requires a huge amount of training data while GMAN despite using similar convolution strategy, does not have sophisticated channel free "multi-step prediction module" which can incorporate region-region correlation during prediction (ζ). The proposed model reports *minimum* of 6.5–10%, 49.7–24.63%, 13.08–4.3% improvement in RMSE, MAPE, SMAPE over the best performing baseline model on NYC, THS respectively. All the differences with the best performing baseline model is statistically significant with p value 0.032012, 0.00001 ($p \leq 0.05$) and 0.091367 ($p \leq 0.1$) for RMSE, MAPE, SMAPE respectively. An example of prediction performance of the proposed model is demonstrated in Fig. 4.

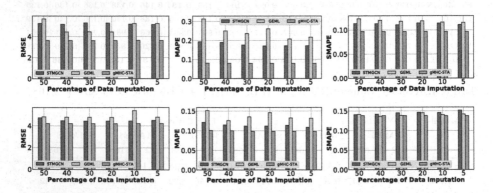

Fig. 3. Comparison of mean RMSE, MAPE and SMAPE score of different approaches with varying % of missing information at HWND 24 on NYC (top) and THS.

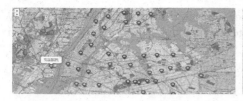

Fig. 4. NYC (left) and THS showing correctly predicted demands GRs (red) over the missed GRs (blue) for a test case. (Color figure online)

5.2   Performance on Varying Percentage of Missing Information

We also compare the proposed model in the scenario when a part of information is missing due to noise or some other unavoidable interruptions. Following the similar approach in [2, 16]. Figure 3 shows the prediction results for the varying % of missing information. gMHC-STA is capable of making better prediction than the comparative models owing to its non-recursive approach. To maintain the clarity of results, we have omitted ForGAN and GMAN for their lower predictive performance than the others.

5.3   Performance on Varying Spatio-Temporal Dimension, Variants and Latent Embedding Representation

We evaluate the effect of varying spatial and temporal embedding dimensions. As shown in Table 5, we observe that the proposed model shows a consistent performance with minute fluctuation in the reported results. Also, the results minutely increase beyond the dimensions mentioned owing to increase noise due to sparsity. We also find that the performance (RMSE) of the proposed model's variant with multiple sets of spatio-temporal convolution kernels at historic ODT $hr = 24$ and predicted ODT $hr = 12$ are 6.09, 10.24 for NYC, THS respectively. TSNE plots in Fig. 1 present that the proposed model is capable to bring GRs with similar demand pattern nearer to each other in the embedding space. TSNE is plotted on the GRs embeddings produced by the proposed model before multi-step prediction to retain both spatio-temporal latent features.

6   Conclusion

In this work, we present a novel deep Graph Multi-Head Convolution for Spatio-Temporal network model for taxi demand prediction. To capture all explicit as well as implicit correlations among different nodes in a graph representation of ODM, we propose a unique spatial-convolution module that utilizes Multi-Head GCN. We further develop a Multi-Head based temporal-convolution unit for capturing the temporal properties on each region; thus reducing the need for error prone recursive models. Moreover, our approach is able to perform region to region prediction and it can handle missing data comparatively better than the competitive models. Extensive experiments show that gMHC-STA consistently outperforms all the baselines with a good margin on two standard city taxi datasets.

References

1. Beiraghi, M., Ranjbar, A.: Discrete Fourier transform based approach to forecast monthly peak load. In: 2011 Asia-Pacific Power and Energy Engineering Conference, pp. 1–5. IEEE (2011)
2. Bhanu, M., Mendes-Moreira, J., Chandra, J.: Embedding traffic network characteristics using tensor for improved traffic prediction. In: IEEE Transactions on Intelligent Transportation Systems, pp. 1–13 (2020)
3. Bhanu, M., Chandra, J., Mendes-Moreira, J.: Enhancing traffic model of big cities: network skeleton & reciprocity. In: 2018 10th International Conference on Communication Systems & Networks (COMSNETS), pp. 121–128. IEEE (2018)
4. Bhanu, M., Priya, S., Dandapat, S.K., Chandra, J., Mendes-Moreira, J.: Forecasting traffic flow in big cities using modified tucker decomposition. In: International Conference on Advanced Data Mining and Applications, pp. 119–128. Springer (2018)
5. Geng, X., et al.: Spatiotemporal multi-graph convolution network for ride-hailing demand forecasting. In: Proceedings of the AAAI Conference on Artificial Intelligence. vol. 33, pp. 3656–3663 (2019)
6. Hamilton, W., Ying, Z., Leskovec, J.: Inductive representation learning on large graphs. In: Advances in Neural Information Processing Systems, pp. 1024–1034 (2017)
7. Hoang, M.X., Zheng, Y., Singh, A.K.: FCCF: forecasting citywide crowd flows based on big data. In: Proceedings of the 24th ACM SIGSPATIAL International Conference on Advances in Geographic Information Systems, pp. 1–10 (2016)
8. Ke, J., Qin, X., Yang, H., Zheng, Z., Zhu, Z., Ye, J.: Predicting origin-destination ride-sourcing demand with a spatio-temporal encoder-decoder residual multi-graph convolutional network. Transp. Res. Part C: Emerg. Technol. **122**, 102858 (2021)
9. Koochali, A., Schichtel, P., Dengel, A., Ahmed, S.: Probabilistic forecasting of sensory data with generative adversarial networks-forGAN. IEEE Access **7**, 63868–63880 (2019)
10. Kumar, S.V., Vanajakshi, L.: Short-term traffic flow prediction using seasonal arima model with limited input data. Eur. Transp. Res. Rev. **7**(3), 21 (2015)
11. Lee, S., Fambro, D.B.: Application of subset autoregressive integrated moving average model for short-term freeway traffic volume forecasting. Transp. Res. Rec. **1678**(1), 179–188 (1999)
12. Lv, Y., Duan, Y., Kang, W., Li, Z., Wang, F.Y., et al.: Traffic flow prediction with big data: a deep learning approach. IEEE Trans. Intell. Transp. Syst. **16**(2), 865–873 (2015)
13. Moreira-Matias, L., Gama, J., Ferreira, M., Mendes-Moreira, J., Damas, L.: Time-evolving OD matrix estimation using high-speed GPS data streams. Expert Syst. Appl. **44**, 275–288 (2016)
14. Ren, J., Xie, Q.: Efficient OD trip matrix prediction based on tensor decomposition. In: 2017 18th IEEE International Conference on Mobile Data Management (MDM), pp. 180–185. IEEE (2017)

15. Sobral, T., Galvão, T., Borges, J.: Knowledge-assisted visualization of multi-level origin-destination flows using ontologies. IEEE Trans. Intell. Transp. Syst. **22**(4), 2168–2177 (2021)
16. Tan, H., Wu, Y., Shen, B., Jin, P.J., Ran, B.: Short-term traffic prediction based on dynamic tensor completion. IEEE Trans. Intell. Transp. Syst. **17**(8), 2123–2133 (2016)
17. Vaswani, A., et al.: Attention is all you need. In: Advances in Neural Information Processing Systems, pp. 5998–6008 (2017)
18. Wang, S., Miao, H., Chen, H., Huang, Z.: Multi-task adversarial spatial-temporal networks for crowd flow prediction. In: Proceedings of the 29th ACM International Conference on Information & Knowledge Management, pp. 1555–1564 (2020)
19. Wang, Y., Yin, H., Chen, H., Wo, T., Xu, J., Zheng, K.: Origin-destination matrix prediction via graph convolution: a new perspective of passenger demand modeling. In: Proceedings of the 25th ACM SIGKDD International Conference on Knowledge Discovery and Data Mining, pp. 1227–1235 (2019)
20. Yu, B., Yin, H., Zhu, Z.: Spatio-temporal graph convolutional networks: a deep learning framework for traffic forecasting. In: IJCAI, pp. 3634–3640 (2018)
21. Zheng, C., Fan, X., Wang, C., Qi, J.: GMAN: a graph multi-attention network for traffic prediction. In: Proceedings of the AAAI Conference on Artificial Intelligence. vol. 34, pp. 1234–1241 (2020)
22. Zhou, X., Shen, Y., Zhu, Y., Huang, L.: Predicting multi-step citywide passenger demands using attention-based neural networks. In: Proceedings of the Eleventh ACM International Conference on Web Search and Data Mining, pp. 736–744 (2018)

Evolution-Based Online Automated Machine Learning

Cedric Kulbach[1]([✉]), Jacob Montiel[2], Maroua Bahri[3], Marco Heyden[1], and Albert Bifet[2]

[1] Research Center for Information Technology (FZI), Haid-Und-Neu-Str. 10-14, 76131 Karlsruhe, Germany
kulbach@fzi.de, marco.heyden@kit.edu
[2] University of Waikato, Private Bag 3105, Hamilton 3240, New Zealand
{jmontiel,abifet}@waikato.ac.nz
[3] Inria Paris, 2 Rue Simone IFF, 75012 Paris, France
maroua.bahri@inria.fr

Abstract. Automated Machine Learning (AutoML) deals with finding well-performing machine learning models and their corresponding configurations without the need of machine learning experts. However, if one assumes an online learning scenario, where an AutoML instance executes on evolving data streams, the question for the best model and its configuration with respect to occurring changes in the data distribution remains open. Algorithms developed for online learning settings rely on few and homogeneous models and do not consider data mining pipelines or the adaption of their configuration. We, therefore, introduce *EvoAutoML*, an evolution-based online learning framework consisting of heterogeneous and connectable models that supports large and diverse configuration spaces and adapts to the online learning scenario. We present experiments with an implementation of *EvoAutoML* on a diverse set of synthetic and real datasets, and show that our proposed approach outperforms state-of-the-art online algorithms as well as strong ensemble baselines in a traditional test-then-train evaluation.

Keywords: Incremental learning · Ensemble learning · Evolutionary algorithm · Data stream

1 Introduction

Automated Machine Learning (AutoML) has shown impressive performance on offline learning tasks in which the whole data are available at once. In contrast to stand-alone offline learning approaches, AutoML automates the data mining pipeline by concatenating different algorithms and applying hyperparameter optimization (HPO) techniques to find the best performing combination and configuration of models. The success of AutoML has lead to the development of multiple well-known frameworks, such as *autosklearn* [15,16,23], *TPOT* [26],

J. Gama et al. (Eds.): PAKDD 2022, LNAI 13280, pp. 472–484, 2022.
https://doi.org/10.1007/978-3-031-05933-9_37

GAMA [18] or H_2O [34]. However, many real-world environments generate data continuously and indefinitely in the form of never-ending data streams [2,17]. Unlike the batch setting, the unbounded nature of these data raises some practical and technical requirements that need to be addressed, where a stream algorithm [6]:

- **R1:** processes a single instance at a time,
- **R2:** processes each instance in a limited amount of time,
- **R3:** uses a limited amount of memory,
- **R4:** is ready to predict at any time,
- **R5:** is able to adapt to changes in the data distribution[1].

When retraining AutoML or other offline learning algorithms, a major part of these requirements is infringed. Data patterns may change in unforeseen ways leading to a decrease in the predictive performance of the machine learning model because the current learned model may be no more representative for the next upcoming data. As a result, offline learning AutoML algorithms might not recommend suitable models for future data without retraining the entire model (**R1**, **R2** infringed). In order to enable adaption to ever-evolving data streams current approaches; we either use (i) change detectors to decide if a model should be retrained [13,24] or (ii) homogeneous ensemble learning techniques [31,33]. Both approaches are not applicable in practice due to two main reasons: retraining AutoML algorithms is often computational expensive [13] (**R1**, **R4** infringed), especially in large search spaces. The second reason is that if large search spaces are acquired, pure ensemble techniques would lead to a large increase in the number of parallel trainings (**R3**, **R4** infringed).

Data streams evolve over time, just like natural environments change, so survival of the fittest, mutations, and offspring allow populations to adapt to such environmental changes. Evolutionary algorithms follow a similar concept, where by creating offspring and allowing for mutations, they mimic natural selection and let the fittest individuals move over to the next generation. In this manner, they enable the system to adapt to changing data patterns which makes them particularly well suited for Online Automated Machine Learning.

Our approach, *EvoAutoML*, takes up this idea and naturally adapts the population of algorithms and configurations if changes occur in the data. As a result, we are able to avoid expensive retraining of an AutoML learner and take advantage of ensemble learning techniques.

The proposed offline AutoML approaches are unable to work with evolving data streams because they allow, among others, several access to data instances and therefore broke the requirements of the streaming framework. Thus, for fair comparison, we evaluate our approach by comparing the relevant performance metrics on established datasets to related state-of-the-art online learning approaches through compliance with the defined requirements. The main contributions of this paper are summarized as follows:

---

[1] Changes in data distributions or patterns are also referred to as *concept drift* [36].

- We provide a formalization and implementation for adapting large algorithm and configuration search spaces to evolving data streams.
- We conduct a broad evaluation of the proposed approach against state-of-the-art algorithms.
- To foster reproducibility, the code and datasets employed in our work are available on GitHub[2].

2   Related Work

In this section, we present an overview of the related work for online AutoML. We first discuss relevant offline AutoML methods and then relevant (ensemble) algorithms for the Online Learning setting.

2.1   Automated Machine Learning

Generally, AutoML aims to automate a Machine Learning (ML) pipeline containing the steps of (i) data cleaning, (ii) feature engineering and (iii) algorithm modelling. It can be defined as the problem of automatically (without human intervention) producing test set predictions for a new dataset within a fixed computational budget [16]. To automate the data analysis pipeline, AutoML addresses the Combined Algorithm Selection and Hyperparameter (CASH) optimization problem [16]. The idea of AutoML was initially developed in [35], which combines the *WEKA* ML framework [21] with Bayesian optimization [12] to search for the best ML instance for a given dataset.

The most established frameworks for offline AutoML are *Auto-Weka 2.0* [25], *autosklearn* [15,16,23], *TPOT* [26], *GAMA* [18], and H_2O [34], that mainly differ in their search space and HPO technique. As HPO technique, *Auto-Weka 2.0* exploits a random forest algorithm, *autosklearn* a Baysian optimization [23] approach, *TPOT* and *GAMA* employs evolutionary algorithms, and H_2O a grid search approach. Our approach, *EvoAutoML*, extends current batch AutoML approaches in order to make them applicable and suitable with evolving data streams.

2.2   Online Learning

Since data streams are potentially infinite and new observations may arrive with a high frequency, stream algorithms must be efficient in terms of resource usage, i.e. time **R3** and memory **R4** consumption. Figure 1 exemplary shows how an online learning framework is able to comply with the stream requirements for a supervised learning task [6]. It processes each instance from an evolving data stream S, updates the underlying model, and is ready to predict at any time.

---

[2] https://github.com/kulbachcedric/EvOAutoML.git.

However, some algorithms have been specifically created and/or adapted to operate on data streams. For instances, *Hoeffding Tree (HT)* [14], *Hoeffding Adaptive Tree (HAT)* [5], *Logistic Regression*, ensemble methods such as *Online Bagging (OB)* [29]. Other algorithms require adaption so that they can be used in

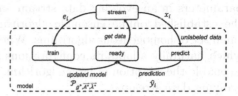

Fig. 1. Online learning, following [28]

an online fashion, such as creating mini-batches or introducing a sliding window [3,30]. If concept drift occurs in the online learning setting, the initially selected model may not longer be the optimal one. Ensemble techniques, such as *OB* with or without an ADWIN change detector [31], *Leveraging Bagging (LB)* [8] or *Adaptive Random Forest (ARF)* [19] have shown to be competent in adapting to temporal changes. *OB* [29,30] propose an approach that updates a set of models by weighting each instance from the stream with a *Poisson*(1) distributed number. Adding an ADWIN [4] change detector to *OB* enables dealing with concept drifts. *LB* [8] improves the *OB* approach by adding more randomization to the input and output of the classifier and therefore leverages the predictive performance. *ARF* [19], an adaption to the random forest algorithm [10], includes an effective resampling method that handles different types of concept drifts. Streaming Random Patches (SRP) [20] is also a ensemble method that combines random subspaces and bagging while using a strategy to detect drifts similar to the one introduced in ARF [19]. To adapt the configuration of heterogeneous algorithms to changing data streams, in [13], authors proposed an AutoML approach that uses different adaption strategies to retrain AutoML instances, such as H_2O, *Autosklearn* and *GAMA* [18], but without taking into account costly retrainings of offline AutoML instances. However, the presented ensemble approaches employ homogeneous algorithms with identical configurations. Assuming algorithm and hyperparameter search spaces, such as in *autosklearn* ($|A| = 110$ possible configurations), training base algorithms in the manner of ensembles (e.g., *OB* [31] and *LB* [8]) becomes increasingly inefficient and does not consider a combination of algorithms within the algorithm search space. Our approach therefore introduces an evolutionary adaption strategy that consider heterogeneous algorithms and configuration spaces to cope with different types of concept drifts.

3   Approach

The question of changing and adapting the configuration of an algorithm as well as the orchestration of models without infringing the requirements [6] for online learning remains open. Our online AutoML framework is inspired by the CASH problem, a Genetic Algorithm (GA) approach and extends *OB* [29,30] to enable online training in a high-dimensional algorithm- and hyperparameter-search space. However, the CASH solution does not consider the adaption of parameters in an evolving data stream environment so far, on the other hand, the established online ensemble algorithms are only capable of processing a small set of homogeneous algorithms. Whence, our proposal uses a GA approach which naturally adapts its configurations within a small ensemble (population) to enable the adaption of large algorithm- and hyperparameter-search spaces to evolving data streams.

3.1   Online CASH

We first define the online CASH problem to adapt to the online learning scenario. Following the definition from [37], a ML pipeline structure $g \in G$ can be modelled as an arbitrary directed acyclic graph (DAG), where each node represents an algorithm $A \in \mathcal{A}$.

Definition 1. *Online CASH, adapted from [16,24]*
Let $\mathcal{A} = \{A^{(1)}, \ldots, A^{(R)}\}$ be a set of step independent algorithms, and let the hyperparameters of each algorithm $A^{(j)}$ have a domain $\Lambda^{(j)}$. Further, let $S = e_1, e_2, \ldots, e_t, \ldots$ be an ordered sequence of examples of possibly infinite length and let t be the current observed example. Further, let $S^- = e_0, \ldots, e_t$ be an ordered sequence of past examples. Each example $e_i = \{x_i, y_i\}$ is a tuple of p predictive attributes $x_i = (x_{i,1}, \ldots, x_{i,p})$ and the corresponding label y_i. Let $\mathcal{L}(\mathcal{P}_{g, \overrightarrow{A}, \overrightarrow{\lambda}}(S^T), S^V)$ denote the loss that algorithm combination $P^{(j)}$ achieves on a subset of validation examples $S^V \subset S^-$ when trained on $S^T \subset S^-$ with hyperparameters $\overrightarrow{\lambda}$. Denote that $S^T \cap S^V = \emptyset$.

Then the Online CASH problem is to find the joint algorithm combination and hyperparmeter setting that minimizes the loss:

$$g^*, \overrightarrow{A}^*, \overrightarrow{\lambda}^* \in \underset{P^{(j)} \in \mathcal{P}, \lambda \in \Lambda^{(j)}, A \in \mathcal{A}, g \in G}{\arg \min} \mathcal{L}(\mathcal{P}_{g, \overrightarrow{A}, \overrightarrow{\lambda}}(S^T), S^V) \tag{1}$$

Existing online (ensemble) algorithms do not fully cover the Online CASH problem. On the one hand, they do not consider a structure $g \in G$ and only cover a small range of hyperparameters $\lambda \in \Lambda$. On the other hand, their hyperparamters are usually set at the start of the stream and are not changed as the stream evolves. The range of covered hyper-parameters is restricted by the number of trained algorithms within the ensemble, whereas OB [29,30], ARF [19], and other ensembles are based on homogeneous algorithms [27]. The graph structure g enables a combination and stacking of algorithms within $\mathcal{A}$ e.g. classifying a set of features x, after they have been scaled. Furthermore, Online CASH considers the configuration space Λ for each algorithm $A \in \mathcal{A}$.

Assuming large search spaces, such as those inherent in the number of existing algorithms and their configurations in the stream setting, a scaleable and adaptable approach becomes necessary. Therefore, by following Fig. 1, the introduced Definition 1, and the requirements defined in [6], we propose in Algorithm 1 an AutoML training algorithm, that adapts to concept drifts in an online learning manner and is capable to handle large search spaces.

3.2 EvoAutoML

The core of our training and adaption procedure is a GA inspired by [32]. Algorithm 1 shows the *EvoAutoML* algorithm. The input consists of a data stream S, a population size P, and a sampling rate f_{SS}. The length of the incoming stream S gives the number of updates $\lfloor t/f_{SS} \rfloor$ and is potentially infinite. Furthermore, our algorithm requires a sampling rate f_{SS}, which controls the rate at which a mutation is applied. Finally, we need a loss function $\mathcal{L}$ which estimates the performance of a pipeline configuration on given examples e_i (interleaved test-then-train evaluation) and a search space, containing all possible graph structures $g \in G$, algorithms $A \in \mathcal{A}$ and their configurations Λ. The algorithm is initialized (lines 7–12) by building a random population of algorithm pipelines $\mathcal{P}_{g,\vec{A},\vec{\lambda}}$. Notice that by initializing p with random online learning pipelines, the algorithm is able to predict at any time (**R4**).

Algorithm 1. EvoAutoML Training

1: **Input:**
2: Data stream S,   population size P,   sampling rate f_{SS},   loss function $\mathcal{L}$, configuration space $\mathcal{A}, \Lambda, G$
3: **Output:**
4: Set of suited algorithms configurations:
5: $p^* = \{\mathcal{P}^{(1)}, \ldots, \mathcal{P}^{(P)}\}$
6:
7: $p \leftarrow \emptyset$ ▷ Initialization
8: **while** $|p| < P$ **do**
9:      $\mathcal{P} \leftarrow \text{Random}(G, \mathcal{A}, \Lambda)$
10:      $p \leftarrow p \cup \mathcal{P}$
11: **end while**
12: $t \leftarrow 0$
13: **if** e_t **then**   ▷ Start Datastream
14:      **if** $t \mod f_{SS} == 0$ **then**
15:          $\mathcal{P}^{best} \leftarrow \min_{\mathcal{P} \in p} \mathcal{L}(\mathcal{P}(S^T), S^V)$
16:          $\mathcal{P}^{weak} \leftarrow \max_{\mathcal{P} \in p} \mathcal{L}(\mathcal{P}(S^T), S^V)$
17:          $\mathcal{P}^{mut} \leftarrow \text{Mutate}(\mathcal{P}^{best})$
18:          $p \leftarrow p \cup \mathcal{P}^{mut}$
19:          $p \leftarrow p \backslash \mathcal{P}^{weak}$
20:      **end if**
21:      $\omega \leftarrow Poisson(6)$
22:      **for** $\mathcal{P} \in p$ **do**   ▷ Update Population
23:          **loop** ω
24:             $\mathcal{P}.\text{fit}(e_t)$
25:          **end loop**
26:      **end for**
27:      $t \leftarrow t + 1$
28: **end if**

In line 13, the data stream starts and a mutation is applied with a rate of f_{SS} (lines 14–20). Within the mutation steps the algorithm selects, in the first step, the best $\mathcal{P}^{best}$ and weakest $\mathcal{P}^{weak}$ pipeline configuration. Based on the best pipeline $\mathcal{P}^{best}$ configuration the mutation is applied in line 17, where similar to [32] a random parameter of $\mathcal{P}^{best}$ is changed within $\mathcal{A}$ and Λ, passed to $\mathcal{P}^{mut}$ and added to p. The weakest pipeline $\mathcal{P}^{weak}$ is removed from p (line 19). After the mutation step, the population is trained on the new instance e_i (lines 21–25) similar to OB [30] with a $\omega \sim Poisson(6)$ distribution. The choice for the distribution results from LB [8].

Our approach respects the requirements of [6] (**R1–R5**) by processing each example e_i at a time (**R1**) and in a limited amount of time (**R2**), e.g. by adjusting the population size P (see also Sect. 4). Since our approach has access to a population of trained pipelines at each point in time of the data stream *EvoAutoML* is also able to predict at any time (**R4**). Algorithm 1 updates the population in an ensemble manner within the training process to search for suited configurations that can also be used for prediction. To predict for an unlabelled instance (see Fig. 1), our approach uses a hard majority voting approach of the algorithm configurations in p to predict the label $\hat{y}_i = mode\{\mathcal{P}.predict(e_i) \in p\}$.

In contrast to existing techniques which only consider the problem of algorithm selection, our approach also takes into account the configuration space Λ from a range of algorithms $\mathcal{A}$ and the pipeline structure g. As a result, *EvoAutoML* addresses the complete Online CASH problem while other approaches can only deliver partial solutions.

4 Experiments

In this section, we describe our evaluation, present the baseline algorithms, introduce the datasets used for evaluation, and discuss the experimental setup. To evaluate our approach, we apply the interleaved test-then-train evaluation, which is a commonly used approach in data stream settings. Here, each incoming instance first serves for testing the current performance of the algorithm and afterwards for training and updating the algorithm. In addition to the fulfillment of the requirements **R1** and **R4** (see Sect. 1), we show that our approach is able to outperform related algorithms by evaluating (i) the final accuracy (**R5**), (ii) the avg. time required (**R2**) to process selected datasets, and (iii) the memory consumption (**R3**). We show that *EvoAutoML* is compatible with recent online algorithms and thus fulfills the requirements of [6] (**R1–R5**). In Table 1, we present the stream datasets as well as the synthetic data stream generators used within the evaluation.

Table 1. Datasets

Name		Variables	#Samples	#Features	#Classes
RBF(a, b)	[7]	a: #centroids	1M	50	5
		b: moving speed			
SEA(a)	[23]	a: changing width	1M	3	2
Agrawal(a)	[1]	a: changing width	1M	9	2
LED()	[11]		1M	24	7
HYP(a, b)	[22]	a: #features	1M	50	2
		b: magnitude change			
SINE()	[17]		1M	2	2
Covertype	[7]		581,012	54	7
Elec	[7]		45,312	6	2

4.1   Search Space

Our approach uses two algorithm types that can be categorised into (i) *prepro-cessors* $A_{(i)}$ and (ii) *predictors* $A_{(ii)}$, and can be variably linked with each other. The preprocessing step can either be a *missing value cleaner, min-max scaler,* or a *standard scaler* ($|A_{(i)}| = 3$). The prediction step contains *Gaussian Naive Bayes (GNB), HT, k-Nearest Neighbors (KNN)*, and *Logistic Regression* classifiers. In total, the classification step contains $|A_{(ii)}| = 4$ and therefore $3 \times 4 = 12$ possible algorithm configurations.

All algorithms $A^{(i)}$ can be parametrized by their domain $\Lambda^{(i)}$. For example, the *KNN* classifier can be parameterized by the number of neighbors, or the *HT* classifier by its maximal depth or the tie threshold as well as by the binary parameters if a binary split strategy should be applied or if poor attributes should be removed. On the whole, our domain space contains 174 possible pipeline configurations. Here, the advantage of our approach (Algorithm 1) comes apparent. While current ensemble and boosting methods are based on homogeneous models (pipelines $\mathcal{P}$), *EvoAutoML* is capable of handling a diverse set of pipelines and pipeline configurations $G, \mathcal{A}$.

4.2   Experimental Setup

We implemented *EvoAutoML* on top of *River* [27], the source code is made publicly available[3]. We evaluated our approach with a population size $P = 10$ and a sampling rate $f_{SS} = 1000$. The population size is chosen equal to the size of the ensemble learners. Furthermore, the choice for the population size and rate is two-folded: First, all algorithms within the population are trained in an ensemble and thus a high population size or sampling rate would lead to computational expensive training updates. Second, during the implementation, the configuration $P = 10$, $f_{SS} = 1000$ was found to be a compromise between predictive performance and the amount of resources required. To compare our approach with established ensemble learners, one can set (i) an equal algorithm space $\mathcal{A}$ to all ensemble learners or (ii) compare our approach to the preset configurations. However, by setting an equal algorithm space $\mathcal{A}$, with or without consideration of further configuration Λ, we pursue the question of the search for the best performing parameterization, that ensemble learners answer by training all algorithms in $\mathcal{A}$ in a parallel manner. In contrast, by using the default configuration of each ensemble-learner, we pursue the question for the best performing approach. Regarding the computational complexity for large search spaces, we evaluated the related algorithms in their proposed configuration to pursue the question for the best performing approach.

To cover a broad range and the most suitable incremental algorithms, we evaluate *EvoAutoML* against *HT* [14], Gaussian NB, and *KNN* classifiers. Since *EvoAutoML* contains a population of pipelines $P_{g,\vec{A},\vec{\lambda}}$, we also evaluate our approach against ensemble learners such as *ARF* [8], *LB* [8] and *OB* [19,31]

---

[3] https://github.com/kulbachcedric/EvOAutoML.git.

using *HT* as base classifier. Each ensemble learner contains 10 base classifiers. Each instance from the stream is transformed using a standard scaler (zero mean and unit variance) before passing it to the ML algorithm. All baseline algorithms are carried out with their default configuration.

To conclude our experimental setup, we evaluated the predictive performance, the total running time, and memory consumption of our approach with the search space proposed in Sect. 4.1 against the proposed baseline algorithms.

5   Results

In this section, we present the results of our approach and show that our approach is able to adapt to temporal changes, outperforms the state-of-the-art algorithms in predictive performance and has comparable computational costs as other online ensemble approaches. To show the adaption to temporal changes by following the requirements of [6] (esp. **R5**), we exemplary depict, in Fig. 2, the learning curves of our approach and the baseline approaches presented in Sect. 4.2 for the Covertype dataset.

The Covertype dataset [7] contains labeled instances of forest cover type (7 classes) from the US Forest Service, where 581,012 samples (measured in 30 × 30 m cells) are characterized by 54 attributes. It has been used in several papers on data stream classification [9] and shows exemplary the adaptanility of *EvoAutoML* against other streaming classifiers (see Sect. 4.2). One can see in Fig. 2 that *EvoAutoML* adapts to changes faster than

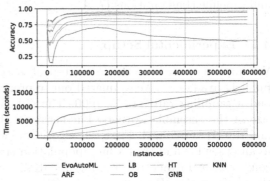

Fig. 2. Accuracy curve and time (in seconds) for *EvoAutoML* and baseline algorithms

LB, *ARF* and *OB*. While the other approaches show decreases in their accuracy at ~250,000 evaluated instances, *EvoAutoML* remains stable. Furthermore, we compare the training and testing time incurred by each model. The graph is in agreement with the statements from [19], and shows that single algorithms (*HT*, *GNB* and *KNN*) have the the lowest running time. For the ensemble learners, *ARF* has the lowest running time, followed by *LB*, *EvoAutoML* and *OB*. The evaluation of *EvoAutoML* takes slightly more time than *LB*, whereby the running time of *OB* increases faster than *LB* and *EvoAutoML*. In addition to other ensemble learners, *EvoAutoML* is able to adapt its parameters during the datastream based on a given loss function $\mathcal{L}$. This allows the adaptation to the data stream in terms of accuracy, but could also incorporate metrics such as latency or memory consumption. However, Table 2 and Table 3 already show that *EvoAutoML* is competitive in terms of the final percentage of correctly classified examples, memory and the time consumption with an underlying accuracy loss. Table 2 compares *EvoAutoML* against the baseline approaches presented in Sect. 4.2. It shows that *EvoAutoML* outperforms the baseline algorithms with an average final avg. accuracy of 93.32%. Comparing *EvoAutoML* against the single best algorithms, *EvoAutoML* performs 10.71% better on average. However, beside the strong results of our approach and the chosen ensemble learners, Table 2 also shows, that in the case of *KNN* classifiers on the RBF dataset, single best algorithms might perform marginally better than the ensemble learners. This slightly better performance on the RBF dataset may be the result of (i) an unsuitable baseline algorithm for the ensemble learners, or the transition gap to a suitable pipeline in the case of *EvoAutoML*. Comparing our approach against the ensemble algorithms, *EvoAutoML* slightly outperforms them with 0.32% on average. Taking the average rank into account *EvoAutoML* performs best with an avg. rank of 2.08. Furthermore, all ensemble algorithms perform better on the avg. accuracy and the avg. rank than the chosen single algorithms. Table 3 records the memory consumption, as well as the used RAM-hours and the avg. time consumption. One RAM-Hour equals to 1 Gb of RAM deployed for 1 h and is accumulated over the generators and datasets. It shows that the significantly better performance of the ensemble learners is accompanied by higher memory and time consumption than with single algorithms. However, comparing the deployed RAM-hours and the time consumption of the ensemble learners, our approach consumes a fraction of the memory in terms of deployed RAM-hours and manages to iterate the quickest over the data stream.

Table 2. Accuracy comparison of *EvoAutoML* against baselines. Accuracy is measured as the final percentage of examples correctly classified. The best individual accuracies are indicated in boldface

Dataset	EvoAutoML	HT	GNB	KNN	ARF	LB	OB
Agrawal(50)	99.02 ±0.01	98.09 ± 0.01	62.31 ± 0.09	55.73 ± 0.02	94.98 ± 0.95	**99.69 ± 0.00**	98.46 ± 0.01
Agrawal(50000)	94.43 ± 0.02	91.84 ± 0.02	62.33 ± 0.09	55.52 ± 0.02	93.03 ± 0.93	**97.52 ± 0.01**	92.89 ± 0.02
HYP(50,0.0001)	**87.51 ± 0.02**	84.38 ± 0.00	91.01 ± 0.01	67.93 ± 0.00	71.19 ± 0.71	84.54 ± 0.01	87.14 ± 0.01
HYP(50,0.001)	83.69 ± 0.01	81.79 ± 0.01	80.83 ± 0.02	68.01 ± 0.00	71.67 ± 0.72	**83.95 ± 0.01**	84.43 ± 0.01
LED()	**76.49 ± 0.01**	75.95 ± 0.01	76.48 ± 0.01	66.6 ± 0.00	76.47 ± 0.76	76.48 ± 0.01	76.42 ± 0.01
RBF(10,0.0001)	99.82 ± 0.00	89.32 ± 0.03	65.86 ± 0.09	**100 ± 0.00**	99.85 ± 0.01	99.64 ± 0.00	98.07 ± 0.00
RBF(10,0.001)	99.63 ± 0.00	77.61 ± 0.02	39.75 ± 0.11	**99.99 ± 0.00**	99.22 ± 0.99	99.01 ± 0.00	93.68 ± 0.01
RBF(50,0.0001)	97.51 ± 0.01	83.05 ± 0.03	35.26 ± 0.13	**99.83 ± 0.00**	98.21 ± 0.98	98.71 ± 0.01	96.17 ± 0.01
RBF(50,0.001)	96.99 ± 0.01	48.15 ± 0.04	25.32 ± 0.07	**99.80 ± 0.00**	94.31 ± 0.94	93.56 ± 0.01	71.87 ± 0.03
SINE()	**99.87 ± 0.00**	99.63 ± 0.01	93.62 ± 0.00	98.75 ± 0.00	99.74 ± 0.01	99.68 ± 0.01	99.77 ± 0.01
SEA(50)	98.99 ± 0.00	97.78 ± 0.01	95.65 ± 0.00	97.23 ± 0.00	99.64 ± 0.01	**99.67 ± 0.01**	98.34 ± 0.01
Elec	**88.09 ± 0.01**	79.61 ± 0.02	72.87 ± 0.03	79.53 ± 0.01	87.79 ± 0.88	87.32 ± 0.01	81.74 ± 0.02
Covertype	**91.09 ± 0.07**	66.67 ± 0.10	63.64 ± 0.11	73.74 ± 0.12	89.7 ± 0.09	90.41 ± 0.08	83.66 ± 0.12
Avg. acc.	**93.32**	82.61	66.58	81.74	90.45	93.09	89.43
Avg. rank	2.08	5.31	5.92	4.77	3.46	2.54	3.85

In summary, we show beside the requirements **R1** and **R4** (see Sect. 3) that *EvoAutoML* meets the requirements **R2** and **R3** of [6] by consuming less time and memory as state-of-the-art ensemble learners. *EvoAutoML* outperforms these ensemble learners in a common test-then-train evaluation, which shows the ability to adapt (**R5**) to changes in the data distribution.

Table 3. Comparison of memory consumption (in MB) and Avg. Time (in s). One RAM-Hour equals to 1 Gb of RAM deployed for 1 h.

Dataset	EvoAutoML	HT	GNB	KNN	ARF	LB	OB
Agrawal(50)	17.609	0.604	0.013	0.455	11.093	12.205	6.008
Agrawal(50000)	56.854	2.223	0.013	0.455	12.920	37.600	21.501
HYP(50,0.0001)	104.576	18.287	0.066	2.020	229.900	528.847	180.870
HYP(50,0.001)	127.877	18.516	0.066	2.020	356.600	395.203	187.146
LED()	35.954	2.104	0.048	0.379	10.133	39.723	18.570
RBF(10,0.0001)	24.527	13.359	0.133	2.020	25.803	22.897	134.988
RBF(10,0.001)	36.107	30.530	0.133	2.020	11.668	4.893	291.346
RBF(50,0.0001)	64.458	24.165	0.166	2.020	27.117	35.643	236.124
RBF(50,0.001)	29.288	9.173	0.166	2.020	25.453	8.023	98.340
SINE()	9.760	0.421	0.004	0.169	14.622	11.128	4.211
SEA(50)	17.833	0.716	0.005	0.205	8.408	14.070	7.454
Elec	12.697	0.205	0.012	0.417	6.850	1.729	1.938
Covertype	12.082	0.125	0.080	2.170	4.750	15.549	19.368
Avg. time	33,638	4,635	1,489	2,119	56,786	58,347	35,243
RAM-hours	7.19	0.32	0	0.01	50.38	44.55	24.35

6 Conclusion

In this paper, we propose an approach for evolution-based online automated machine learning that extends the CASH problem to the stream setting and adapts the hyperparameter search to work with data streams. The adaption of hyperparameters and the possibility of algorithm pipelines, showed that an evolutionary approach is able to outperform state-of-the-art single and ensemble-based methods. We evaluated *EvoAutoML* on performance metrics, as well as the total running time and the used memory as efficiency metrics on several common online learning generators and datasets.

References

1. Agrawal, R., Imielinski, T., Swami, A.N.: Database mining: a performance perspective. IEEE TKDE **5**(6), 914–925 (1993)
2. Alberg, D., Last, M., Kandel, A.: Knowledge discovery in data streams with regression tree methods. Wiley Interdisc. DMKD **2**(1), 69–78 (2012)
3. Bahri, M., Bifet, A., Gama, J., Gomes, H.M., Maniu, S.: Data stream analysis: foundations, major tasks and tools. Wiley Interdisc.: DMKD **11**(3), e1405 (2021)
4. Bifet, A., Gavaldà, R.: Learning from time-changing data with adaptive windowing. In: SIAM ICD, pp. 443–448 (2007)
5. Bifet, A., Gavaldà, R.: Adaptive learning from evolving data streams. In: Adams, N.M., Robardet, C., Siebes, A., Boulicaut, J.-F. (eds.) IDA 2009. LNCS, vol. 5772, pp. 249–260. Springer, Heidelberg (2009). https://doi.org/10.1007/978-3-642-03915-7_22

6. Bifet, A., Gavaldà, R., Holmes, G., Pfahringer, B.: Machine Learning for Data Streams: With Practical Examples in MOA. MIT Press, Cambridge (2018)
7. Bifet, A., Holmes, G., Kirkby, R., Pfahringer, B.: MOA: massive online analysis. JMLR **11**, 1601–1604 (2010)
8. Bifet, A., Holmes, G., Pfahringer, B.: Leveraging bagging for evolving data streams. In: Balcázar, J.L., Bonchi, F., Gionis, A., Sebag, M. (eds.) ECML PKDD 2010. LNCS (LNAI), vol. 6321, pp. 135–150. Springer, Heidelberg (2010). https://doi. org/10.1007/978-3-642-15880-3_15
9. Bifet, A., Read, J., Žliobaitė, I., Pfahringer, B., Holmes, G.: Pitfalls in benchmarking data stream classification and how to avoid them. In: Blockeel, H., Kersting, K., Nijssen, S., Železný, F. (eds.) ECML PKDD 2013. LNCS (LNAI), vol. 8188, pp. 465–479. Springer, Heidelberg (2013). https://doi.org/10.1007/978-3-642-40988-2_30
10. Breiman, L.: Random forests. ML **45**(1), 5–32 (2001)
11. Breiman, L., Friedman, J.H., Olshen, R.A., Stone, C.J.: Classification and Regression Trees. Wadsworth (1984)
12. Brochu, E., Cora, V.M., de Freitas, N.: A tutorial on Bayesian optimization of expensive cost functions, with application to active user modeling and hierarchical reinforcement learning. CoRR (2010)
13. Celik, B., Vanschoren, J.: Adaptation strategies for automated machine learning on evolving data. CoRR abs/2006.06480 (2020)
14. Domingos, P.M., Hulten, G.: Mining high-speed data streams. In: Ramakrishnan, R., Stolfo, S.J., Bayardo, R.J., Parsa, I. (eds.) SIGKDD, pp. 71 80. ACM (2000)
15. Feurer, M., Eggensperger, K., Falkner, S., Lindauer, M., Hutter, F.: Auto-Sklearn 2.0: the next generation. CoRR (2020)
16. Feurer, M., Klein, A., Eggensperger, E.A.: Efficient and robust automated machine learning. In: Cortes, C., Lawrence, N.D., Lee, D.D. (eds.) Advances in Neural Information Processing Systems 28: NIPS, pp. 2962–2970 (2015)
17. Gama, J., Medas, P., Castillo, G., Rodrigues, P.: Learning with drift detection. In: Bazzan, A.L.C., Labidi, S. (eds.) SBIA 2004. LNCS (LNAI), vol. 3171, pp. 286–295. Springer, Heidelberg (2004). https://doi.org/10.1007/978-3-540-28645-5_29
18. Gijsbers, P., Vanschoren, J.: GAMA: genetic automated machine learning assistant. J. Open Sour. Softw. **4**(33), 1132 (2019)
19. Gomes, H.M., et al.: Adaptive random forests for evolving data stream classification. ML **106**(9–10), 1469–1495 (2017)
20. Gomes, H.M., Read, J., Bifet, A.: Streaming random patches for evolving data stream classification. In: ICDM. IEEE (2019)
21. Hall, M.A., Frank, E., Holmes, G., Pfahringer, B., Reutemann, P.: The WEKA data mining software: an update. SIGKDD Explor. **11**(1), 10–18 (2009)
22. Hulten, G., Spencer, L., Domingos, P.: Mining time-changing data streams. In: SIGKDD, pp. 97–106. ACM (2001)
23. Hutter, F., Kotthoff, L., Vanschoren, J. (eds.): Automated Machine Learning - Methods, Systems, Challenges. Springer, Cham (2019). https://doi.org/10.1007/978-3-030-05318-5
24. Imbrea, A.: An empirical comparison of automated machine learning techniques for data streams. B.S. thesis, University of Twente (2020)
25. Kotthoff, L., Thornton, C., Hoos, H.H., Hutter, F., Leyton-Brown, K.: Auto-WEKA 2.0: automatic model selection and hyperparameter optimization in WEKA. JMLR **18**, 25:1–25:5 (2017)

26. Le, T.T., Fu, W., Moore, J.H.: Scaling tree-based automated machine learning to biomedical big data with a feature set selector. Bioinformatics **36**(1), 250–256 (2020)
27. Montiel, J., et al.: River: machine learning for streaming data in Python (2020)
28. Montiel, J., Read, J., Bifet, A., Abdessalem, T.: Scikit-multiflow: a multi-output streaming framework. JMLR **19**, 72:1–72:5 (2018)
29. Oza, N.C.: Online bagging and boosting. In: ICSMC, pp. 2340–2345. IEEE (2005)
30. Oza, N.C., Russell, S.J.: Experimental comparisons of online and batch versions of bagging and boosting. In: Lee, D., Schkolnick, M., Provost, F.J., Srikant, R. (eds.) ACM SIGKDD, pp. 359–364. ACM (2001)
31. Oza, N.C., Russell, S.J.: Online bagging and boosting. In: Richardson, T.S., Jaakkola, T.S. (eds.) Workshop on AISTATS (2001)
32. Real, E., Aggarwal, A., Huang, Y., Le, Q.V.: Regularized evolution for image classifier architecture search. In: AAAI, pp. 4780–4789. AAAI (2019)
33. van Rijn, J.N., Holmes, G., Pfahringer, B., Vanschoren, J.: Having a blast: meta-learning and heterogeneous ensembles for data streams. In: Aggarwal, C.C., Zhou, Z., Tuzhilin, A., Xiong, H., Wu, X. (eds.) ICDM, pp. 1003–1008 (2015)
34. Stetsenko, P.: Machine learning with Python and H2O (2020). http://docs.h2o.ai/h2o/latest-stable/h2o-docs/booklets/PythonBooklet.pdf
35. Thornton, C., Hutter, F., Hoos, H.H., Leyton-Brown, K.: Auto-WEKA: combined selection and hyperparameter optimization of classification algorithms. In: SIGKDD, pp. 847–855. ACM (2013)
36. Widmer, G., Kubat, M.: Learning in the presence of concept drift and hidden contexts. ML **23**(1), 69–101 (1996)
37. Zöller, M., Huber, M.F.: Survey on automated machine learning. CoRR abs/1904.12054 (2019)

Smooth Perturbations for Time Series Adversarial Attacks

Gautier Pialla[1(✉)], Hassan Ismail Fawaz[1], Maxime Devanne[1],
Jonathan Weber[1], Lhassane Idoumghar[1], Pierre-Alain Muller[1],
Christoph Bergmeir[2], Daniel Schmidt[2], Geoffrey Webb[2],
and Germain Forestier[1,2]

[1] Université de Haute-Alsace, Mulhouse, France
{gautier.pialla,hassanismail.fawaz,maxime.devanne,jonathan.weber,
lhassane.idoumghar,pierre-alain.muller,germain.forestier}@uha.fr
[2] Monash University, Melbourne, Australia
{christoph.bergmeir,daniel.schmidt,geoffrey.webb,
germain.forestier}@monash.edu

Abstract. Adversarial attacks represent a threat to every deep neural network. They are particularly effective if they can perturb a given model while remaining undetectable. They have been initially introduced for image classifiers, and are well studied for this task. For time series, few attacks have yet been proposed. Most that have are adaptations of attacks previously proposed for image classifiers. Although these attacks are effective, they generate perturbations containing clearly discernible patterns such as sawtooth and spikes. Adversarial patterns are not perceptible on images, but the attacks proposed to date are readily perceptible in the case of time series. In order to generate stealthier adversarial attacks for time series, we propose a new attack that produces smoother perturbations. We find that smooth perturbations are harder to detect by the naked eye. We also show how adversarial training can improve model robustness against this attack, thus making models less vulnerable.

Keywords: Time series · Adversarial attack · Smooth perturbations · InceptionTime · BIM

1 Introduction

A time series is a set of data points ordered in time. Time series have become a growing field of research in deep learning and more globally in artificial intelligence. Nowadays, thanks to the presence of sensors, they have become abundant and we can find use cases in almost all sectors of industry. For example, time series are used in healthcare [12], for weather forecasting [13] and for predictive maintenance [7].

Time series classification (TSC), refers to the task of classifying time series according to the presence or not of phenomena. Szegedy et al. [6] have found

© The Author(s), under exclusive license to Springer Nature Switzerland AG 2022
J. Gama et al. (Eds.): PAKDD 2022, LNAI 13280, pp. 485–496, 2022.
https://doi.org/10.1007/978-3-031-05933-9_38

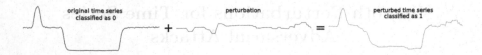

Fig. 1. Scheme of adversarial attack. Time series from the BME dataset, perturbation generated with SGM, not represented at scale.

that adding a small perturbation to an input sample can change a classifier's output. This is known as an *adversarial attack*. It is illustrated on Fig. 1.

As adversarial attacks are a vulnerability present in every neural network, many attacks were proposed but first for image classification tasks. It is necessary to study them in order to assess the robustness of the models, and to prevent them on critical systems. For example, Eykolt et al. [5] showed an application on real-world road sign classification, which is an obvious threat for autonomous vehicles.

Fawaz et al. [9] introduced and adapted some of them for time series classification. The main difference between adversarial attacks on images and time series lies in the visualization and the interpretation of the data. When sightly changing the value of one or few pixels, an image will always look the same and have the same appearance. Theses changes only affect how the neural network will process the data, but not how we, humans, perceive the image. For images, the human classifier is a competitive benchmark, often used as gold standard. For TSC it is not, because time series data are more complex to analyze.

The attacks introduced by Fawaz et al. [9] are effective to perturb time series of the UCR Archive [3]. But when we look at their visual appearance, it is sometimes easy to distinguish the disturbed series from the original ones. Indeed, theses perturbed samples often contain patterns like spikes of a sawtooth. Because the presence of such elements can easily be spotted, they can warn about the presence of an attack.

In this paper, we will introduced a novel adversarial attack based on a gradient method. We will show that it outperforms BIM's performance over most of the UCR archive datasets. But unfortunately this method generates perturbations that also contain spike and sawtooth patterns. We will then explain how we reduced these patterns, by enforcing a smoothness condition. Finally, we will show how adversarial training is a good way to improve a time series classifier's robustness against smoothed perturbations.

Our main contributions are:

- A novel adversarial attack for time series classifiers that outperforms BIM
- An altered version of the first attack, that produces smooth perturbations
- A benchmark of our two methods along with BIM over the UCR archive
- We showed how smoothed perturbations are harder to detected
- We showed that adversarial training is a good counter measure against smooth attacks.

2   Related Work

Given a neural network trained on an image classification task, such as ImageNet, Szegedy et al. [20] showed that it is possible to change the model output by adding low magnitude noise, small enough to be imperceptible to the human eye. It was also shown that this vulnerability is present regardless of the number of layers, activation functions or training data and thus affects all deep neural networks.

Goodfellow et al. [6] proposed a single step attack called Fast Gradient Sign Method (FGSM). Then, Kurakrin et al. [14] presented the Basic Iterative Method (BIM), an iterative version of FGSM. Inspired by them, many similar attacks were proposed, like M-IGSM [4] or vr-IGSM [21].

Other approaches where studied, like adding black and white strips on stop signs [5] or stickers on objects [15]. These real life attacks raised the issue of security threat for sensitive applications like autonomous vehicles. Along with new attacks, multiple defensive strategies have also emerged, including leveraging denoisers [16], randomization [23] and adversarial training [11,22].

Adversarial training trains a model using both normal and perturbed samples. Rathore et al. [18] shows how adversarial training can help a model to become more robust.

Most of the work on adversarial attacks was first done on image classification, as it is a trending topic in deep learning. It is only later that Fawaz et al. [9] introduced adversarial attacks for time series classification.

It is sometimes quite straightforward to adapt adversarial attacks from images to times series. However, some attacks that work well on images can't be used, or are ineffective on time series. For example Su et al. [19] describes attacks where only one pixel of an image is affected. An equivalent perturbation for time series would modify the value of only a single data point. But such modifications would be very noticeable as it takes extreme values to sufficiently perturb a sample based solely on a single data point.

Adversarial attacks can be categorized into black and white-box strategies. Black-box attacks, like presented in [1,17], don't use any knowledge of the architecture, the parameters or the weights of the model. They have also no access of the datasets used for the training. Huan et al. [8] showed that even in these conditions, many current models are still at risk. In contrast, white-box attacks may use any of those elements to perform the attack. Some attacks have both black-box and white-box variants, like the Carlini & Wagner method [2]. In this paper we will focus exclusively on white-box attacks.

3   Background Material

3.1   Mathematical Description

In this paper, we only use univariate time series. We can describe each time series as a vector $\mathbf{x}$ such as $\mathbf{x} \in \mathbb{R}^T$, $\mathbf{x} = [x_1, ..., x_T]$ with T denoting its length.

Given a time series classifier f and a time series $\mathbf{x}$, the aim of an adversarial attack is to perturb the classifier by adding a small variation $\mathbf{r}$ to a time series $\mathbf{x}$. $\mathbf{r}$ will be referred as noise or perturbation. We call the perturbed time series $\mathbf{x}^{adv} = \mathbf{x} + \mathbf{r}$ an *adversarial sample*. The attack is successful if the class predicted for the original time series is different from the class predicted for the adversarial sample, $\arg\max f(\mathbf{x}) \neq \arg\max f(\mathbf{x}^{adv})$. The added noise $\mathbf{r}$ must be imperceptible by design, thus we need that $\mathbf{x}$ and $\mathbf{x}^{adv}$ remain close to each other.

3.2   Basic Iterative Method

In order to improve the success rate of FGSM, Kurakin et al. [14] developed BIM. At each iteration N, the gradient is computed and then added to the input, in the same way as for FGSM. Instead of minimizing the loss function, the aim is to maximize it by taking a step in the direction of the gradient. At each iteration, the values are clipped using an ϵ parameter. This ensures that each value of $\mathbf{x}^{adv}$ will stay close to $\mathbf{x}$ within a ϵ-neighbourhood.

$$\mathbf{x}_0^{adv} = \mathbf{x}$$
$$\mathbf{x}_{N+1}^{adv} = \text{Clip}_{\mathbf{x},\epsilon} \left\{ \mathbf{x}_N^{adv} + \alpha \, \text{sign}(\Delta_{\mathbf{x}} J(\Theta, \mathbf{x}_N^{adv}, y_{true})) \right\} \quad (1)$$

y_{true} denotes the label of the time series $\mathbf{x}$. If we don't know y_{true}, as in a real attack scenario, we replace it by $f(\mathbf{x})$. The noise clipping is done for $\mathbf{r} = \mathbf{x}^{adv} - \mathbf{x}$ as follow:

$$\forall r_i \in \mathbf{r}, r_i = \begin{cases} \epsilon, & \text{if } r_i > \epsilon \\ -\epsilon, & \text{if } r_i < -\epsilon \end{cases}$$

By adding iterations, BIM becomes more effective than FGSM to perturb time series. But BIM requires clipping in order to control the amount of the noise. This method had two main disadvantages. First, clipping the noise in such way often produce sawtooth shapes between $-\epsilon$ and $+\epsilon$ as we can see on Fig. 5. This particular pattern can easily be detected when added to a smooth time series and is therefore to be avoided.

With BIM, in order to obtain a stealthier noise, we need to reduce to value of ϵ. By doing this, the saw-tooth shapes will be harder to be noticeable, but this will result in a lower attack success rate. This trade-off prevents the perturbation that are both hard to detect and have a high attack success rate.

4   Proposed Methods

4.1   Gradient Method (GM)

In order to correct the flaws of BIM, we need to design a method that, given a model, can perturb a time series while optimizing the quantity of noise according to the L2 norm.

Ensuring $f(\mathbf{x}) \neq f(\mathbf{x}')$ can be written as a maximization problem of the KL-divergence between the two probability distributions, as follows:

$$\max D_{KL}(f(\mathbf{x}), f(\mathbf{x}')) \equiv \sum_{}^{c} f(\mathbf{x}) \log \frac{f(\mathbf{x})}{f(\mathbf{x}')}, \qquad (2)$$

with c denoting the classes in the dataset.

Generating an adversarial example can then be written as follows where the primary addition is the term $(-\|\mathbf{x} - \mathbf{x}'\|_2)$ to be maximized:

$$\max \{\mu D_{KL}(f(\mathbf{x}), f(\mathbf{x}')) - \|\mathbf{x} - \mathbf{x}'\|_2\}, \qquad (3)$$

with μ denoting a hyper-parameter to control the penalty of miss-classification.

Let us consider the generated time series $\mathbf{x}' = \mathbf{x} + \mathbf{r}$. Then the maximization problem is equivalent to the following minimization problem:

$$\min \{-\mu D_{KL}(f(\mathbf{x}), f(\mathbf{x} + \mathbf{r})) + \|\mathbf{r}\|_2\} \qquad (4)$$

We can add an hyper-parameter α in order to control the regularization of $\|\mathbf{r}\|$. Finally, we have:

$$\mathbf{x}^{adv} = \min \{-\mu D_{KL}(f(\mathbf{x}), f(\mathbf{x} + \mathbf{r})) + \alpha\|\mathbf{r}\|_2\} \qquad (5)$$

4.2   Smooth Gradient Method (SGM)

The previous method manages to generate adversarial samples while optimizing the L2 norm of $\mathbf{r}$. But it does not prevent the appearance of sawtooth. In order to obtain smoother perturbations, we need to ensure a smoothness condition on $\mathbf{r}$. This can be done by adding a fused lasso term to the minimization. The equation can now be written as:

$$\min\{-\mu D_{KL}(f(\mathbf{x}), f(\mathbf{x} + \mathbf{r})) + \alpha\|\mathbf{r}\|_2 + \lambda \sum_{i=1}^{T-1} \|r_i - r_{i+1}\|_1\} \qquad (6)$$

In Eq. 6, $\|.\|_1$ denotes the L1 norm. λ is a hyper-parameter that controls the penalty for the smoothness condition. To minimize the latter equation, we will use the gradient descent by computing the gradient with respect to $\mathbf{r}$ (which will be initialized randomly).

5   Experimental Setup

In this section, we present the data, models and the parameters we used during our experiments.

5.1   Classifier and Datasets

We used InceptionTime [10] for all our experiments. InceptionTime is a TS classifier, that was the state-of-the-art model on the UCR archive, when published in 2019. All the weights used are the InceptionTime defaults, as used and presented in its paper.

In order to demonstrate our results over several datasets, we used the well know TSC benchmark UCR Archive [3]. The 2018 version of this archive comprises 128 univariate time series datasets.

Each dataset of the UCR archive is split between the training and the test set. When generating adversarial samples, we used the samples of the test set, as the model has only been trained on the training set.

5.2   Reproductibility

The code used and all our results are publicly available in our companion repository[1].

All experiments were done by leveraging the computation power of a remote GPU cluster containing Nvidia GTX 1080 Ti graphic cards. Reproducing the results on a single graphic card takes roughly 7 days of computing time.

5.3   Hyper-parameters

For BIM we set the number of iterations at 1000. For the noise clipping we use the value $\epsilon = 0.1$. We use the same value of ϵ, when applying the noise clipping to the Gradient Method.

In the case of GM and SGM both μ and α parameters are always set to 1. In the case of SGM, when nothing is specified, λ is also equal to 1.

5.4   Comparison Metrics

Average Success Rate. For evaluating the relative success of adversarial attacks, we used the Average Success Rate (ASR). The ASR, corresponds to the rate of reclassified samples. In other words, it is equal to the percentage of cases where the attack was able to alter the output of the network $(f(\mathbf{x}) \neq f(\mathbf{x}^{adv}))$.

L2 Norm. The L_∞ norm is commonly used to quantify the noise for adversarial attacks. This is especially true in the case of attacks on images. The L_∞ norm of a time series is equal to $\|\mathbf{x}\|_\infty = \max_t |x_t|$. As explain earlier, our aim is to design smooth perturbations that are hard to detect by the naked eye. Moreover attacks designed for images are easily detectable when adapted to times series. Thus, we needed to evaluate the overall quantity of noise, not just its maximum value and choose to use the L2 norm over the L_∞ norm.

---

[1] https://github.com/Gpialla/SmoothPerturbationsTSAA.

5.5   Adversarial Training

We will present an example of adversarial training using adversarial samples generated by SGM. For each dataset, we doubled the size of the training set, by adding the corresponding adversarial samples of the original training set. The validation is done with the original test set, without additional adversarial samples. Finally, we will show how adversarial training is effective at reducing a classifier's susceptibility to adversarial attack.

6   Results

In this section, we will first compare SGM with the other methods according to the two metrics we selected: the ASR and the L2 norm. The benchmark between the others methods is available in our companion repository. In a second study, we will vary the SGM's λ parameter and see its influence on the ASR. Finally will perform an adversarial training, in order to propose a counter measure against SGM attacks.

6.1   SGM Benchmark

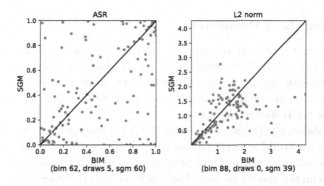

Fig. 2. Win/Draw/Loss diagram. BIM vs SGM. On the left: average success rate, on the right: L2 norm of the perturbation

Figure 2 represents a Win/Draw/Loss diagram comparing BIM and SGM. Each blue dot represent a single dataset. If a dot lies above the median line in the upper left triangle, it means that this dataset has an average value bigger for SGM than for BIM for the given metric.

As we want to maximize ASR, in the corresponding plot, the most successful method is the one with the most dots on its side of the median line. For the L2 norm, however, the reasoning is reversed as we want to minimize the metric.

Given Fig. 2, as the dots are evenly distributed, we conclude that SGM as an overall ASR as good as BIM on the UCR archive. This also means that

SGM manages to perturb datasets that BIM can not and vice-versa. But for an equivalent efficiency, BIM introduces an higher quantity of noise than SGM, in a majority of datasets.

Figure 3 compares GM with SGM. We can see that for almost all datasets, GM has a better ASR than SGM. This shows that the sawtooth and spikes which can only be produced by GM are decisive elements in order to perturb a TSC.

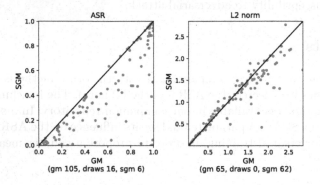

Fig. 3. Win/Draw/Loss diagram. GM vs SGM. On the left: average success rate, on the right: L2 norm of the perturbation

6.2   Varying the λ Parameter

According to our previous results, the best case scenario would be an attack with GM's ASR and SGM's smoothness. As the only difference between the two methods is the adding of the smoothness condition, it is interesting to vary the λ parameter. If λ is equal to zero, the attack is GM and if it's equal to 1, we have SGM as we tested it previously.

Figure 4 shows the impact of varying the λ parameter over two datasets, Beef and Car. As we could expect, the more we enforce the smoothness condition, the fewer the samples the method manages to perturb successfully.

This parameter should be tuned for each dataset in order to get the optimal trade-off between smoothness and ASR.

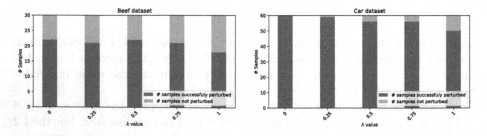

Fig. 4. Varying SGM's λ parameter. For each value of λ is displayed the number of samples successfully perturbed (blue) or not (orange). (Color figure online)

6.3   Visual Comparison

In order to remain undetectable by the naked eye, an attack performed on a time series must be as smooth as possible. As we did not find any suitable metric to assess the smoothness of a time series, we propose a visual comparison between the four methods presented, on the same test sample of the Beef dataset. To be fair, we picked a time series which is successfully perturbed by all the attacks.

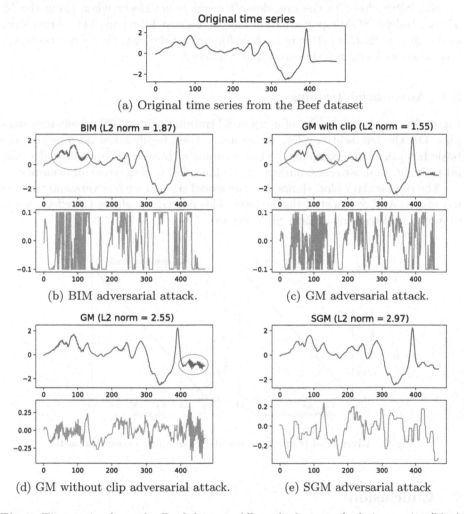

(a) Original time series from the Beef dataset

(b) BIM adversarial attack.

(c) GM adversarial attack.

(d) GM without clip adversarial attack.

(e) SGM adversarial attack

Fig. 5. Time series from the Beef dataset. All methods perturbed time series (blue) and generated noise (red). The purple circles show the presence of sawtooth on the perturbed time series. (Color figure online)

In this example, shown on Fig. 5, we plotted in green the original time series, and for each method, in blue the perturbed time series and in red the perturbation.

We plotted a second version of GM with a clipped perturbation in the same way as BIM. As expected, for BIM and the GM methods, the perturbations are clearly visible, in particular the parts containing sawtooth patterns that are circled in purple. The example of GM with clipping shows that clipping the noise reduce indeed the amount of noise and the visual impact, but not sufficiently enough. SGM is the only attack that produced an adversarial sample with a perturbation that is not noticeable when judging with the naked the eye.

But being closer to the eye, doesn't mean being closer when using the L2 metric. Indeed, SGM's perturbation has the biggest L2 norm. This shows that, although a method is better in average for a given dataset, this is not necessary true when we look at each sample independently.

6.4   Adversarial Training

Figure 6 presents the results of adversarial training using SGM adversarial samples. On the left scatter plot, we compare the classification accuracy of the basic InceptionTime compared to the accuracy of InceptionTime with adversarial training. In most cases, adversarial training led to a decrease of accuracy.

The right scatter plot, shows that the model trained with adversarial training led to zero ASR for most of the datasets. This huge drop, shows the effectiveness of adversarial training against SGM attacks.

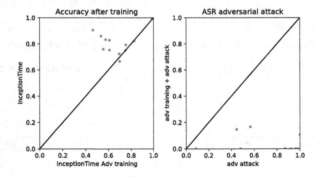

Fig. 6. Adversarial training results of 13 randomly chosen datasets.

7   Conclusion

In this paper, we explained that adapting adversarial attacks from image classifiers to time series classifier is not trivial. The attacks are more likely to be detected on time series, and thus need smoother perturbations.

We introduced two novel adversarial attacks for time series classification. The Gradient Method (GM) and a smooth version, called Smooth Gradient Method (SGM). We used the Basic Iterative Method (BIM), a well known adversarial attack, as a baseline to have a benchmark over the entire UCR archive. We showed that GM, has the higher success rate on perturbing an InceptionTime classifier, followed by BIM and SGM.

Through examples, we illustrated that GM, like BIM produces perturbations which have recognizable patterns like spikes and sawtooth. On one hand, these patterns can help the attack to fool the network, but on the other hand, they can be easily detected, even by the naked eye.

Our second method SGM, is based on GM but has an added fuzed lasso regularization. It has the effect of smoothing the generated perturbations. Smoothing the noise makes it harder to differentiate perturbed and original time series by the naked eye. But smoothed adversarial samples are less effective for attacking the neural network. This highlights the current trade off between having a stealth attack and an effective one.

Finally, we showed that adversarial training is an effective way of countering SGM attacks.

For the future works, we would like to find a metric that can measure the smoothness of a time series. This would help in order to find a new smooth adversarial attack that is better than SGM at fooling time series classifiers.

Acknowledgement. This work was funded by ArtIC project "Artificial Intelligence for Care" (grant ANR-20-THIA-0006-01) and co-funded by Région Grand Est, Inria Nancy - Grand Est, IHU of Strasbourg, University of Strasbourg and University of Haute-Alsace. The authors would like to thank the providers of the UCR archive as well as the Mésocentre of Strasbourg for providing access to the GPU cluster.

References

1. Bhambri, S., Muku, S., Tulasi, A., Buduru, A.B.: A survey of black-box adversarial attacks on computer vision models (2020)
2. Carlini, N., Wagner, D.: Towards evaluating the robustness of neural networks. In: 2017 IEEE Symposium on Security and Privacy (SP), pp. 39–57. IEEE (2017)
3. Dau, H.A., et al.: The UCR time series archive (2019)
4. Dong, Y., et al.: Boosting adversarial attacks with momentum. In: Proceedings of the IEEE Conference on Computer Vision and Pattern Recognition, pp. 9185–9193 (2018)
5. Eykholt, K., et al.: Robust physical-world attacks on deep learning visual classification. In: Proceedings of the IEEE Conference on Computer Vision and Pattern Recognition, pp. 1625–1634 (2018)
6. Goodfellow, I.J., Shlens, J., Szegedy, C.: Explaining and harnessing adversarial examples. arXiv preprint arXiv:1412.6572 (2014)
7. Guillaume, A., Vrain, C., Wael, E.: Time series classification for predictive maintenance on event logs. arXiv preprint arXiv:2011.10996 (2020)
8. Huan, Z., Wang, Y., Zhang, X., Shang, L., Fu, C., Zhou, J.: Data-free adversarial perturbations for practical black-box attack. In: Lauw, H.W., Wong, R.C.-W., Ntoulas, A., Lim, E.-P., Ng, S.-K., Pan, S.J. (eds.) PAKDD 2020. LNCS (LNAI), vol. 12085, pp. 127–138. Springer, Cham (2020). https://doi.org/10.1007/978-3-030-47436-2_10

9. Ismail Fawaz, H., Forestier, G., Weber, J., Idoumghar, L., Muller, P.A.: Adversarial attacks on deep neural networks for time series classification. In: 2019 International Joint Conference on Neural Networks (IJCNN), July 2019

10. Ismail Fawaz, H.: InceptionTime: finding AlexNet for time series classification. Data Min. Knowl. Disc. **34**(6), 1936–1962 (2020)

11. Jiang, Y., Ma, X., Erfani, S.M., Bailey, J.: Dual head adversarial training (2021)

12. Kaushik, S., et al.: Ai in healthcare: time-series forecasting using statistical, neural, and ensemble architectures. Front. Big Data **3**, 4 (2020)

13. Kumar, N., Jha, G.K.: A time series ANN approach for weather forecasting. Int. J. Control Theory Comput. Model (IJCTCM) **3**(1), 19–25 (2013)

14. Kurakin, A., Goodfellow, I.J., Bengio, S.: Adversarial examples in the physical world. In: Artificial Intelligence Safety and Security, pp. 99–112. Chapman and Hall/CRC (2018)

15. Li, J., Schmidt, F., Kolter, Z.: Adversarial camera stickers: a physical camera-based attack on deep learning systems. In: International Conference on Machine Learning, pp. 3896–3904. PMLR (2019)

16. Liao, F., Liang, M., Dong, Y., Pang, T., Hu, X., Zhu, J.: Defense against adversarial attacks using high-level representation guided denoiser. In: Proceedings of the IEEE Conference on Computer Vision and Pattern Recognition, pp. 1778–1787 (2018)

17. Papernot, N., McDaniel, P., Goodfellow, I., Jha, S., Celik, Z.B., Swami, A.: Practical black-box attacks against machine learning (2017)

18. Rathore, P., Basak, A., Nistala, S.H., Runkana, V.: Untargeted, targeted and universal adversarial attacks and defenses on time series. In: 2020 International Joint Conference on Neural Networks (IJCNN), July 2020

19. Su, J., Vargas, D.V., Sakurai, K.: One pixel attack for fooling deep neural networks. IEEE Trans. Evol. Comput. **23**(5), 828–841 (2019)

20. Szegedy, C., et al.: Intriguing properties of neural networks. arXiv preprint arXiv:1312.6199 (2013)

21. Wu, L., Zhu, Z., Tai, C., et al.: Understanding and enhancing the transferability of adversarial examples (2018)

22. Xie, C., Tan, M., Gong, B., Yuille, A., Le, Q.V.: Smooth adversarial training (2021)

23. Xie, C., Wang, J., Zhang, Z., Ren, Z., Yuille, A.: Mitigating adversarial effects through randomization. arXiv preprint arXiv:1711.01991 (2017)

Misleading Inference Generation via Proximal Policy Optimization

Hsien-Yung Peng[1,2], Ho-Lam Chung[1,2], Ying-Hong Chan[1,2],
and Yao-Chung Fan[1,2(✉)]

[1] Department of Computer Science and Engineering,
National Chung Hsing University, Taichung, Taiwan
yfan@nchu.edu.tw
[2] Smart Sustainable New Agriculture Research, Center (SMARTer),
Taichung, Taiwan

Abstract. In this paper, we propose to investigate *Misleading Inference Generation*, a new natural language generation task. The goal is to generate a counterfactual sentence for a context and a factual sentence. This paper proposes a framework based on BART and reinforcement learning for the misleading inference generation task. The experiment results show our model significantly outperforms the compared models, making our solution a necessary and strong baseline for future research toward misleading inference generation.

Keywords: Text generation · Educational application

1 Introduction

In this paper, we propose to investigate Misleading Inference Generation (MIG) task, which is formulated as follows. Given a context (a factual sentence and two contextual sentences before and after the factual sentence), the MIG task is to generate a misleading sentence with respect to the factual sentence. As a concrete example, given the context illustrated in Table 1, our task is to generate a misleading inference (i.e., "Ben got over his fear") with respect to the contextual sentences and the factual sentence.

The MIG task is motivated by automatic preparation of educational reading comprehension assessment. A direct MIG application is to generate a distractor (wrong option) for Cloze questions in English exams. Since the cost of counterfactual rewriting is quite expensive (e.g., roughly 4 or 5 min per counterfactual [6]), automatic misleading sentence generation is therefore desired.

A naive baseline for the MIG task is to apply counterfactual (CF) generation techniques [13,17]. In the CF generation, the goal is to transform a given factual sentence y to a CF sentence $\hat{y}$. However, in misleading inference, we target at the mapping $\{x, y, z\} \rightarrow \hat{y}$. Note that the misleading inference generation is needed to consider the factual sentence y and contextual sentences x, z. In comparison, the CF generation considers only factual sentence y. The MIG result

J. Gama et al. (Eds.): PAKDD 2022, LNAI 13280, pp. 497–509, 2022.
https://doi.org/10.1007/978-3-031-05933-9_39

considers the context for misleading demands, which are more challenging than CF generation.

Table 1. Examples of misleading inference generation

First contextual sentence	Ben was afraid of the dark
Factual sentence	Ben bought himself a nightlight
Second contextual sentence	The light helped ben sleep much easier
Misleading inference	Ben got over his fear

In this paper, we propose a model called MIG-PPO (**M**isleading **I**nference **G**eneration via **P**roximal **P**olicy **O**ptimization). Our MIG-PPO is featured by the following designs. First, we propose two novel training strategies (*negative training* (See Subsect. 3.2) and *incoherent training* (See Subsect. 3.3)) for boosting the performance. Second, we propose to enhance the MIG-PPO performance by reinforcement learning (See Subsect. 3.4).

The contributions of this paper are summarized:

1) We address Misleading Inference Generation (MIG) task.
2) We establish baseline performances using BART with negative training and incoherent training on the MIG task.
3) We demonstrate that proximal policy optimization improves the performance on the misleading ability and maintains sentence fluency at the same time.

The rest of this paper is organized as follows. Section 2 reviews the existing literature and discusses the difference between the existing works and our study. In Sect. 3 we introduce our MIG-PPO model and discuss its design intuition. In Sect. 4, we conduct a series of experiments to evaluate the performance of the compared models and present case studies for qualitative comparison. We conclude this research in Sect. 5.

2   Related Work

In this section, we discuss the existing works related to our MIG task. The related works can be categorized into two categories: Counterfactual Generation and Abductive Inference Generation.

Table 2. Related works comparison

	Task	Study	Input	Target
Counterfactual generation	Counterfactual generation	[13,17]	y	$\hat{y}$
	Controllable counterfactual generation	[16]	control code, y	$\hat{y}$
	Distractor generation	[2,9]	passage, question, y	$\hat{y_1}, \hat{y_2}, \hat{y_3}, \dots$
	Misleading inference generation	this work	x, y, z	$\hat{y}$
Abductive inference generation	Story rewriting	[12]	$x, y, z, \hat{y}$	$\hat{z}$
	Abductive reasoning	[1]	x, z	y

Counterfactual Generation. Counterfactual is useful for many applications, such as strengthening the language model's ability by closing the systematic gaps between training and testing data [4], empowering natural language understanding ability on generated counterfactual inferences [3], or mitigating gender stereotypes by counterfactual data augmentation [20].

In recent years, counterfactual generation based on neural models are reported [13,17]. The main idea behind the works is to use language models and various sampling strategies to generate counterfactuals. Following our notation used in the introduction section, the goal of the counterfactual generation is to generate a counterfactual $\hat{y}$ with respect to a factual y. As mentioned, our goal is to generate counterfactuals by further considering contextual information. Thus, the counterfactual techniques can not fit for our MIG task. As will be seen in the experiment section, directly using the counterfactual techniques leads to poor performance.

Further, the authors [16] present a controllable framework called *Polyjuice* for generating counterfactual of various types. Polyjuice enables the control of generation type with respect to the various type defined. Essentially, Polyjuice is still a variant of counterfactual generation by further allowing the generation type control, which is still not a good fit for our MIG task.

Another types of counterfactual generation is distractor generation [2,9]. The distractor generation is motivated by the need for reading comprehension assessment preparation. Given a passage, question, and answer, the goal is to generate distractors (plausible wrong options) with respect to the given context. The work [9] was a pioneer for neural-based distractor generation investigation. Following the work, [2] proposes to focus on *multiple* distractor generation by proposing a negative training strategy to enhance the performance of distractor generation. The setting and application of distractor generation are different from the MIG task setting. The different applications will have different design considerations. The technique for distractor generation cannot be directly employed for our task.

Abductive Inference Generation. The second category of our related works is abductive inference generation whose goal is to generate plausible sentences with respect to a given context. One representative work for abductive inference generation is the story rewriting task proposed by [12]. The goal of the task is to rewrite/generate a new story with respect to a given original story and an intervening counterfactual event/sentence; the generated story will have different endings with respect to the original story. Following our notation used in the introduction section, the goal of story rewriting is to take a story x, y, z and a counterfactual sentence $\hat{y}$ to generate a new story ending $\hat{z}$. In comparison, as mentioned, our MIG goal is to take x, y, z to generate $\hat{y}$.

Yet another interesting work most related to our work is [1] for abductive commonsense reasoning. Abductive reasoning is inference to the most plausible explanation for incomplete observations. Again, following our notation, the goal of [1] is take x, z to generate y. The goal is to generate a plausible sentence y with respect to the contextual sentence. This work can be viewed as a factual generation, while our work is counterfactual generation.

In Table 2, we summarize the related tasks mentioned above. Counterfactual Generation generates the counterfactual $\hat{y}$ from factual y. Controllable Counterfactual Generation additionally adds control code for different types of counterfactual. Distractor Generation generates multiple distractors from a passage, question, and answer. Counterfactual Story Rewriting generates the consecutive ending $\hat{z}$ from the original story x, y, z and a counterfactual $\hat{y}$. Abductive Reasoning generates the most plausible explanation y from the before and after contextual sentences x, z.

3   Methodology

3.1   MIG-Base

A baseline (called MIG-Base) for the MIG task is to fine-tune BART [7] by taking x, y, z as input and training/predicting $\hat{y}$ as output. However, our experiment results indicate this baseline does not work as intended; the MIG results in many cases are exactly the same as the given factual sentence. To better see this problem, Table 3 shows the token scores (BLEU-1, BLEU-4, and ROUGE-L) between y and $\hat{y}$. The high scores raise the concern of low misleading capability. For ease of discussion, we call this problem as *copying problem* in the following discussion (Fig. 1).

Table 3. The token score between factual sentence y and misleading inference $\hat{y}$.

	BLEU-1	BLEU-4	ROUGE-L
MIG-Base	85.81	68.44	83.56

3.2   Negative Training

We investigate the idea of *negative training* to address the copying problem. The negative training is inspired by the unlikelihood loss strategy proposed by [8] to regularize the text generation. The loss function of the negative training is as follows.

$$\mathcal{L}_{Neg.Training} = \mathcal{L}_{x,y,z\rightarrow\hat{y}}^{Base} + \alpha\mathcal{L}_{x,y,z\rightarrow y}^{Neg},$$

where $\mathcal{L}_{Neg.Training}$ is the sum of Base loss $\mathcal{L}_{x,y,z\rightarrow\hat{y}}^{Base}$ and Negative loss $\mathcal{L}_{x,y,z\rightarrow y}^{Neg}$, and $\alpha \in \mathbb{R}$ is a hyper-parameter for weighting the two losses.

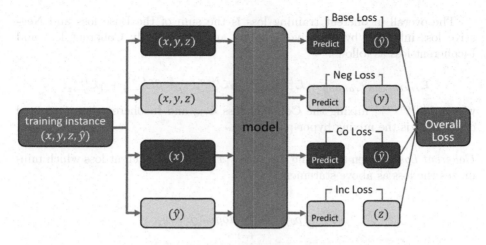

Fig. 1. The overall loss of incoherent training

Base Loss Given a training instance $D = (\{x, y, z, \} \hat{y})$, the Base loss is computed as follows:

$$\mathcal{L}^{Base}_{x,y,z \to \hat{y}}(p_\theta) = - \sum_{i=1}^{|\hat{y}|} log p_\theta(\hat{y}_i | x, y, z, \hat{y}),$$

where x, y, z is the input and $\hat{y}$ is a gold label for likelihood training, and $\hat{y}_i$ is the i-th token of $\hat{y}$.

Negative Loss We employ the unlikelihood loss [8] to regulate the copying problem. The idea of the Neg loss is to penalize the results similar to the factual sentence y.

$$\mathcal{L}^{Neg}_{x,y,z \to y}(p_\theta) = - \sum_{i=1}^{|y|} \sum_{y_i \in y} log(1 - p_\theta(y_i | x, y, z)),$$

3.3 Incoherent Training

Furthermore, we propose *incoherent training* by extending the negative training idea with a data augmentation mechanism as follows.

The incoherent training proceeds as follows. Given a training instance $(\{x, y, z\}, \hat{y})$, we expand it into four instances:

- $(\{x, y, z\}, \hat{y})$
- $(\{x, y, z\}, y)$
- $(\{x\}, \hat{y})$
- $(\{\hat{y}\}, z)$

The main idea is to maximize two likelihoods: (1) generating $\hat{y}$ given $\{x, y, z\}$ and (2) generating $\hat{y}$ given x while minimize the unlikelihoods (3) generating y given $\{x, y, z\}$ and (4) generating z given $\hat{y}$

The overall incoherent training loss is the sum of the Base loss and Negative loss inherited by Negative Training and additionally Coherent loss and Incoherent loss as follows:

$$\mathcal{L}_{IncoherentTraining} = \mathcal{L}_{x,y,z\to\hat{y}}^{Base} + \alpha\mathcal{L}_{x,y,z\to y}^{Neg} + \beta\mathcal{L}_{x\to\hat{y}}^{Co} + \gamma\mathcal{L}_{\hat{y}\to z}^{Inc},$$

where $\mathcal{L}^{Incoherent}$ mixing the Coherent loss $\mathcal{L}_{Co}$ and Incoherent loss $\mathcal{L}_{Inc}$ and $\alpha, \beta, \gamma \in \mathbb{R}$ is the mixing hyper-parameter.

Coherent Loss. Given a training instance $(\{x\}, \hat{y})$, the Coherent loss which minimizes the loss as above statement:

$$\mathcal{L}_{x\to\hat{y}}^{Co}(p_\theta) = -\sum_{i=1}^{|\hat{y}|} logp_\theta(\hat{y}_i|x, \hat{y}),$$

where x is the input and $\hat{y}$ is a gold label for our training, and $\hat{y}_i$ is the i-th token of $\hat{y}$.

Incoherent Loss. Follow the Negative loss mentioned on Sect. 3.2, we construct Incoherent loss to penalize the incoherent sentence in our settings. The form of the Incoherent loss which the misleading inference $\hat{y}$ is coherent with the first contextual sentence x, but incoherent with the second contextual sentence z shows as follow:

$$\mathcal{L}_{\hat{y}\to z}^{Inc}(p_\theta) = -\sum_{i=1}^{|z|}\sum_{z_i\in z} log(1 - p_\theta(z_i|\hat{y}, z)),$$

where $\mathcal{L}^{Inc}$ is Incoherent loss, z is the second contextual sentence, and z_i is the i-th token in z.

3.4   Proximal Policy Optimization

In our previous method, we established a MIG model on different training strategies with a strong performance on the token-based metric. While the model can generate a fluent and sensible result, we would like to further improve the misleading ability of our MIG model. We propose to employ Proximal Policy Optimization.

Proximal Policy Optimization (PPO) [14] is a popular deep policy gradient algorithm in reinforcement learning which was proposed by OpenAI in 2017. PPO alternate between sampling data through interaction with the environment, and optimizing a surrogate objective function using stochastic gradient ascent.

We use PPO in the decoding layer which aims to replace the undecidable distribution with a learnable random variable, predicted by an RL agent that takes the last hidden state, the previously decoded token, and the context as input. In other words, PPO learns an RL actor to manipulate the model's hidden states for a suitable outcome.

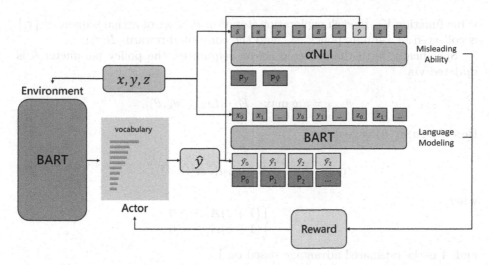

Fig. 2. The RL Architecture for MIG task

The goal of PPO fits the reinforcement learning objective, which is to maximize expected reward (misleading ability) over trajectories (action sequences) induced by policy (actor). We formulate the trainable decoding algorithm with PPO as a sequential decision-making process.

At each time step i, let the state s_i be the last hidden representation given from our $M(C, \hat{y}_{[1:i-1]})$, which M stands for our model, C stands for input context and $\hat{y}_{[1:i-1]}$ is the current generated tokens. In RL formulation, we seek to find a stochastic policy π, which maps s_i to a probability distribution $\pi_\theta(a_i|s_i)$ over the action of decoding vocabulary $\mathcal{V}$, where $a_i \in \mathcal{V}$.

Thus, the generation process can be viewed as an action sequence $\tau = (a_1, ..., a_{|\hat{y}|})$. The probability of an action sequence is:

$$P(\tau|\pi) = \prod_{t=1}^{|\hat{y}|} P(s_{t+1}|s_t, a_t)\pi(a_t|s_t)$$

Further, let $R(\tau)$ the reward of the action sequence τ. The expected return $J(\tau)$ can then be expressed by:

$$J(\tau) = \mathop{E}_{\tau \sim \pi}[R(\tau)] = \int_\tau P(\tau|\pi)R(\tau)$$

The RL objective is to maximize the expected reward by finding the optimal policy π^*:

$$\pi^* = \arg\max_\pi J(\pi)$$

We employ proximal policy optimization for our RL formulation. In PPO, there are two set of parameters to learn: the parameters θ of policy π_θ and ϕ of value function V_ϕ. In each exploration iteration k, a set of action sequences $\{\tau_i\}$ is collected by running policy π_{θ_k}, and the computed rewards $R(\tau_i)$.

Specifically, with the collected action sequences, the policy parameter θ is updated via

$$\theta_{k+1} = \arg\max_\theta \mathop{E}_{s,a\sim\pi_{\theta_k}} [L(s, a, \theta_k, \theta)],$$

Here $L(s, a, \theta_k, \theta)$ is given by

$$\min(\frac{\pi_\theta(a|s)}{\pi_{\theta_k}(a|s)} \hat{A}(s, a), g(\varepsilon, \hat{A}(s, a))),$$

where

$$g(\varepsilon, A) = \begin{cases} (1 + \varepsilon)A, A \geq 0 \\ (1 - \varepsilon)A, A < 0 \end{cases}$$

and $\hat{A}$ is the estimated advantage based on V_ϕ.

Further, the value function parameters ϕ is updated via

$$\phi_{k+1} = \arg\min_\phi \sum_{\forall\tau_i} \sum_{t=1}^{|\tau_i|} (V_\phi(s_t) - R(\tau_i))^2$$

Figure 2 shows the RL architecture for the MIG task. We set the environment as our fine-tuned BART model, actor as the linear layer to decode the token from the last hidden state in BART, and reward function for adjusting actor. The RL architecture aims to strengthen the Misleading Ability and maintain the basic MIG task-oriented ability.

Reward Function. With the RL formulation, we propose to use *Misleading Ability* and *Language Modeling* as reward for RL. The goal is to enhance the misleading ability but maintain the original model generating ability. In the following discussion, we denote $\hat{y}_\tau$ be the misleading inference generated according to the action sequence τ given by the RL policy and $\hat{y}$ as the non-PPO misleading inference.

1) *Misleading Ability*: To examine the misleading ability of the misleading inference, we propose to use classification confidence of the original classification task αNLI in the ART dataset. We fine-tuned an αNLI classifier with DeBERTa pre-trained model [5], which was the state-of-the-art open-source model on αNLI leaderboard. We evaluate misleading ability by the confidence of misleading inference generated by our MIG model. αNLI's confidence is formulated as

$$p(y_i|x, z) = softmax(\alpha\text{NLI}(y_i, \theta)),$$

where y_i represents the classified sentence in αNLI. θ is the parameters of the αNLI, fixed during fine-tuning. The reward is

$$R_{MA} = p(\hat{y}|x, z) - p(y|x, z)$$

where $\hat{y}$ represents the misleading inference and y is the factual sentence in MIG. If the probability of $\hat{y}$ is higher than the probability of y that implies that our misleading inference has a better misleading ability, then the actor will receive a positive reward. In contrast, if the probability of $\hat{y}$ is lower than the probability of y then the actor will receive a negative reward i.e. actor encourages the current action.

2) *Language Modeling*: We propose language modeling as a reward for our RL formulation. Language modeling stands for the probability of a given sequence of words (including input and generated sentence) occurring in a sentence. We compute the language modeling score of misleading inference $\hat{y}$ in our model to prevent model collapse. The reward of language modeling is

$$R_{LM}(q) = \exp(\frac{1}{|q|} \sum_{i=1}^{|q|} log(p(t_{q,i}|t_{q,<i})))$$

where q is the sentence generated in RL, and $t_{q,i}$ is the $i-th$ token of misleading inference token.

We conclude our overall reward given by a weighted sum of the individual rewards.

$$R = \delta_{MA} \cdot R_{MA} + \delta_{LM} \cdot R_{LM}$$

with hyperparameters δ_{MA} and δ_{LM}.

4 Experiment

4.1 Datasets

In this study, we adopt the dataset (referred to as ART, Abductive Reasoning in narrative Text) released by [1] for our MIG task. ART is a dataset with two sub task usages, Abductive Natural Language Inference (αNLI) and Abductive Natural Language Generation (αNLG). The instances in ART are as follows:

- x: The observation at time t_1.
- z: The observation at time $t_2 > t_1$
- y: A factual sentence that explains the observations x and z.
- $\hat{y}$: An implausible hypothesis for observations x and z we called it misleading inference in the following discussion.

We use the original ART dataset setting. There are 169654 training instances and 1532 testing instances for performance.

4.2 Automatic Metric

The most common metrics used to evaluate text generation's performance is overlap metrics, based on the textual overlap between candidate-generated sentence and reference sentence. We chose BLEU [11] and ROUGE-L [10] to measure

Table 4. Performance comparison on token scores

	BLEU-1(↑)	BLEU-2(↑)	BLEU-3(↑)	BLEU-4(↑)	ROUGE-L(↑)	BertScore(↑)	Self-BLEU (↓)	Self-ROUGE(↓)
CF-Base	16.62	6.57	3.42	2.00	16.09	88.75	12.61	22.56
MIG-Base	21.21	12.32	8.77	6.81	20.25	89.54	68.43	83.56
MIG-Neg	19.92	10.85	7.54	5.80	18.72	89.25	61.19	76.14
MIG-Inc	20.14	11.11	7.66	5.81	19.71	89.38	60.25	79.87
MIG-PPO	19.53	10.81	7.35	5.48	19.48	89.32	50.49	69.36

the number of overlapping n-grams and the length of the longest common subsequence. We also use BERTScore [18] to evaluate our model ability. BERTScore computes each token score between the candidate sentence and the inference sentence, instead of exact matches, it will compute token similarity using contextual embeddings.

For evaluating diversity between factual sentences and misleading inference, we adopt self-BLEU and self-ROUGE to calculate the similarity between factual sentence y and misleading inference $\hat{y}$ to specify the diversity of them. Self-BLEU score based on the BLEU-4 score calculating which was proposed in [19]. Since the generated sentences' length may be restricted to training data, we add self-ROUGE as our diversity metrics which are based on ROUGE-L calculating.

4.3 Implementation Details

All models are implemented based on huggingface Transformer [15] architecture. The transformer is an open-source library consisting of state-of-the-art model implementation under a unified API. The BART model we employed is one of the model implementations in huggingface projects. To compare the quality of the MIG model, we implement several methods fixed on the same parameters as follows.

We train our models on two GeForce RTX^{TM} 3090 s with a memory of 48 GB. For all MIG models, we use the BART model released by huggingface and limit the maximum input length to 256 tokens. The AdamW optimizer is applied with the learning rate of 5e-5. All models are set to run 5 epochs with batch size = 16.

4.4 MIG Models Comparison

Before we evaluate the following metrics, we introduce the MIG models we use in following comparison.

- **CF-Base**: A simple counter-factual generation model trained directly by $y \rightarrow \hat{y}$.
- **MIG-Base**: We implement MIG-Base with only base loss in Subsect. 3.2.
- **MIG-Neg**: The MIG-Negative model is an implementation with the negative training loss in Subsect. 3.2.
- **MIG-Inc**: The MIG implementation with the incoherent training loss introduced in Sect. 3.3.

Table 5. Case study

Example 1	
First contextual sentence(x)	Daisy was at her middle school graduation
Factual sentence(y)	She lost her hat
Second contextual sentence(z)	She decided to forget about it, and went home
MIG-Inc	She found her hat
MIG-PPO	Daisy found her hat in the trashcan
Example 2	
First Contextual Sentence(x)	Sandy lived in New York
Factual Sentence(y)	It stormed in New York
Second Contextual Sentence(z)	Sandy was prepared
MIG-Inc	It was sunny in New York
MIG-PPO	Sandy was glad she had gone to New York

- **MIG-PPO**: The implementation with the Proximal Policy Optimization introduced in Subsect. 3.4.

4.5   Evaluation Results

We summarize the automatic evaluation result in Table 4. We have the following observations to note about the experiment results. First, by looking only at BLEU scores, it seems that MIG-Base is a winner. However, we would like to note that the BLEU scores cannot reflect the true performance of the models. As mentioned, by observing the real generated sentences, we find the MIG results are similar to the original factual sentences. We can also see this fact from the high Self-BLEU scores of MIG-Base shown in Table 4.

Second, by comparing MIG-Base and MIG-Neg, we can see that the negative training indeed brings performance improvement (a Self-BLEU improvement from 68.43 to 61.19). This validates our negative training design for mitigating the copying problem.

While the improvement on Self-BLEU was observed, we see a side effect for MIG-Neg that the BLEU-1 score is degraded from 21.21 to 19.92 when MIG-Neg is used. In comparison, our MIG-Inc design shows improvement both on

BLEU and Self-BLUE, indicating MIG-Inc is a better choice than MIG-Neg. Last, we see our MIG-PPO design significantly improve the Self-BLEU scores with a slight performance degradation on BLEU scores.

4.6   Case Studies

In this subsection, we present two case studies for the qualitative comparison in Table 5. We also show the results with the PPO employment. In Example 1, we see that the factual sentence is modified by replacing the verb (*lost* to *found*). We can also see that with the PPO employment, the more diverse result is generated (i.e., *Daisy found her hat in the trashcan*). We see a similar observation in Example 2. We see that MIG-Inc again changes the verb to generate a misleading result. Furthermore, we see MIG-PPO generates a nice misleading result, which is different from the factual sentence both on a semantic and syntactical structure.

5   Conclusion

This paper proposes to investigate the Misleading Inference Generation task, which is a challenging task for the current state-of-the-art natural language generation model with a conditional generation setting. We established baseline performances of state-of-the-art language models on MIG-Inc using the ART dataset with different strategies. Furthermore, we propose to use PPO to strengthen the misleading ability. A strong baseline for the targeted MIG task is recorded in this paper.

Acknowledgement. This work is supported by MOST 110-2634-F-005-006 - project Smart Sustainable New Agriculture Research Center (SMARTer) and MOST Project under grant No. 109-2221-E-005-058-MY3.

References

1. Bhagavatula, C., et al.: Abductive commonsense reasoning. arXiv preprint arXiv:1908.05739 (2019)
2. Chung, H.L., Chan, Y.H., Fan, Y.C.: A bert-based distractor generation scheme with multi-tasking and negative answer training strategies. arXiv preprint arXiv:2010.05384 (2020)
3. Feng, F., Zhang, J., He, X., Zhang, H., Chua, T.S.: Empowering language understanding with counterfactual reasoning. arXiv preprint arXiv:2106.03046 (2021)
4. Gardner, M., et al.: Evaluating models' local decision boundaries via contrast sets. arXiv preprint arXiv:2004.02709 (2020)
5. He, P., Liu, X., Gao, J., Chen, W.: Deberta: decoding-enhanced bert with disentangled attention. arXiv preprint arXiv:2006.03654 (2020)
6. Kaushik, D., Hovy, E., Lipton, Z.C.: Learning the difference that makes a difference with counterfactually-augmented data. arXiv preprint arXiv:1909.12434 (2019)

7. Lewis, M., et al.: Bart: Denoising sequence-to-sequence pre-training for natural language generation, translation, and comprehension. arXiv preprint arXiv:1910.13461 (2019)
8. Li, M., et al.: Don't say that! making inconsistent dialogue unlikely with unlikelihood training. arXiv preprint arXiv:1911.03860 (2019)
9. Liang, C., Yang, X., Dave, N., Wham, D., Pursel, B., Giles, C.L.: Distractor generation for multiple choice questions using learning to rank. In: Proceedings of the Thirteenth Workshop on Innovative Use of NLP for Building Educational Applications, pp. 284–290 (2018)
10. Lin, C.Y.: Rouge: A Package for Automatic Evaluation of Summaries. In: Text summarization branches out, pp. 74–81 (2004)
11. Papineni, K., Roukos, S., Ward, T., Zhu, W.J.: Bleu: a method for automatic evaluation of machine translation. In: Proceedings of the 40th Annual Meeting of the Association for Computational Linguistics, pp. 311–318 (2002)
12. Qin, L., Bosselut, A., Holtzman, A., Bhagavatula, C., Clark, E., Choi, Y.: Counterfactual story reasoning and generation. arXiv preprint arXiv:1909.04076 (2019)
13. Ren, S., Deng, Y., He, K., Che, W.: Generating natural language adversarial examples through probability weighted word saliency. In: Proceedings of the 57th Annual Meeting of the Association for Computational Linguistics, pp. 1085–1097 (2019)
14. Schulman, J., Wolski, F., Dhariwal, P., Radford, A., Klimov, O.: Proximal policy optimization algorithms. arXiv preprint arXiv:1707.06347 (2017)
15. Wolf, T., et al.: Huggingface's transformers: state-of-the-art natural language processing. arXiv preprint arXiv:1910.03771 (2019)
16. Wu, T., Ribeiro, M.T., Heer, J., Weld, D.S.: Polyjuice: automated, general-purpose counterfactual generation (2021)
17. Zhang, H., Zhou, H., Miao, N., Li, L.: Generating fluent adversarial examples for natural languages. arXiv preprint arXiv:2007.06174 (2020)
18. Zhang, T., Kishore, V., Wu, F., Weinberger, K.Q., Artzi, Y.: Bertscore: evaluating text generation with bert. arXiv preprint arXiv:1904.09675 (2019)
19. Zhu, Y., et al.: Texygen: a benchmarking platform for text generation models. In: The 41st International ACM SIGIR Conference on Research & Development in Information Retrieval, pp. 1097–1100 (2018)
20. Zmigrod, R., Mielke, S.J., Wallach, H., Cotterell, R.: Counterfactual data augmentation for mitigating gender stereotypes in languages with rich morphology. arXiv preprint arXiv:1906.04571 (2019)

Instance-Guided Multi-modal Fake News Detection with Dynamic Intra- and Inter-modality Fusion

Jie Wang[1,2], Yan Yang[1,2(✉)], Keyu Liu[1,2], Peng Xie[1,2], and Xiaorong Liu[1]

[1] School of Computing and Artificial Intelligence, Southwest Jiaotong University,
Chengdu 611756, China
{jackwang,keyu_liu,pengxie,xr_liu}@my.swjtu.edu.cn, yyang@swjtu.edu.cn
[2] Manufacturing Industry Chains Collaboration and Information Support Technology
Key Laboratory of Sichuan Province, Southwest Jiaotong University,
Chengdu 611756, China

Abstract. Multi-modal Fake News Detection (MFND), which aims to identify fake news by integrating texts and attached images, has attracted considerable attention in recent years. Existing works on MNFD have made a great progress by enhancing text-only fake news detection with visual information. However, most prior efforts focus on conducting multi-modal fusion yet largely ignore the significance of multi-modal representation, which is insufficient to explore various semantic interactions between images and texts. In this paper, we propose an instance-guided multi-modal graph fusion method by jointly modeling the intra- and inter-modality relationships between image and text. Specifically, considering that the content of multi-media news is always narrated around instances, we extract instance-level features of images to represent visual contents. After that, we construct a unified graph to enhance the multi-modal representation for improving fake news detection. In addition, we utilize multiple fusion layers to learn the graph embeddings, which is able to capture the intra-modality relationship within each modality and the inter-modality relationship between textual and visual instances simultaneously. Finally, we devise a fake news detector with hierarchical multi-modal representations to identify the fake news. Experimentation on two benchmark datasets demonstrates the superiority of our model.

Keywords: Fake news detection · Multi-modal representation · Intra- and inter-modality

1 Introduction

Social media posts become increasingly multi-modal recently, which engage more readers and provides them with a better reading experience [5]. However, multi-modal contents always foster various forms of fake news, which brings harmful social impacts. Accordingly, it is urgent and crucial to detect fake news involved the image on social media [19].

J. Gama et al. (Eds.): PAKDD 2022, LNAI 13280, pp. 510–521, 2022.
https://doi.org/10.1007/978-3-031-05933-9_40

Fig. 1. An example for multi-modal fake news detection with (a) the region-guided visual cues, (b) global-guided visual cues and (c) instance-guided visual cues.

As with other Vision-and-Language task [16], existing prominent MFND models mainly follow an extract-then-fuse paradigm: extracting visual and textual features from image and text respectively, and then fusing them by attention mechanism to obtain the multi-modal information. From the perspective of visual feature extraction, we classify existing approaches into the two main categories: (i) Region-guided methods, which segment the whole image into multiple regions and make them interact with the text sequence [4,15]. (ii) Global-guided methods, which encode the whole image into a global feature vector to perform cross-modal interaction with sentence-level features [7,14,18]. Despite their compelling success, these methods involve two limitations.

The first limitation is that they ignore the mapping relations between visual instances and textual tokens. Actually, the contents of multi-modal news are mostly instance-centric description [9,12], such as "Ane Lee" in Fig. 1. Accordingly, capturing this instance-level alignment is beneficial for fake news detection. However, both of these methods fail to capture such instance-level semantic interactions between the two modalities. Taking Fig. 1 (a) as an example, region-guided methods consider fine-grained visual features but fail to maintain the integrity of visual instance, such as "trophy" in the image. In the global-guided methods shown in Fig. 1 (b), all visual objects and relations are fused in a single vector, making it hard to produce fine-grained semantic mapping relations between vision and language. Thus, it is necessary to utilize the instance-level features for detecting fake news rather than region- and global-level features.

The other limitation is that the above works focus on conducting various multi-modal fusion mechanisms but largely neglect the importance of multi-modal representation. Indeed, a unified multi-modal representation can exploit the complementarity of different modalities and lay a solid foundation for the following fusion stage [1]. Nevertheless, current MFND models only consider intra-modal and inter-modal semantic interactions during fusion, which lacks such informative interactions during the multi-modal representation learning stage. Therefore, we believe that constructing a unified multi-modal representation is beneficial for fake news detection since this way can model intra- and inter- semantic correspondences simultaneously.

In this work, to tackle above challenges, we propose an instance-guided multi-modal fake news detection framework, which fuses multi-modal clues to detect fake news. Specifically, as illustrated in Fig. 1 (c), considering the characteristics of multi-modal fake news, we introduce the visual instances to capture the fine-grained semantics of images. In addition, to build a comprehensive multi-modal representation, we construct a unified graph to connect the visual and textual instances. After that, we leverage multiple graph-based fusion layers to conduct graph encoding, which can capture semantic relationships within the same modality (intra-modality) and semantics interactions between different modalities (inter-modality). Finally, we employ these representations with a fake news detector to recognize fake posts. Overall, the major contributions of our work are listed as follows:

- We propose a novel instance-guided multi-modal graph fusion model which enhances the multi-modal representation for improving fake news detection. To the best of our knowledge, our work is the first to construct a instance-level based multi-modal graph in MFND.
- We design a dynamic multi-modal fusion module that is able to jointly capture the semantic interactions of intra- and inter-modality for fake news detection.
- Without the need of incorporating extra knowledge as other works, experimental results on two public datasets demonstrate the superiority of our model, and further analysis illustrates the effectiveness of our proposed modules.

2   Related Work

With the quick development of social media on the Internet, perceiving and recognizing fake news with texts and images become progressively prominent [19]. To date, several previous works have attempted to perform fake news detection using multi-modal contents and obtain superior performance. Jin et al. [4] extracted the multi-modal information including textual, visual and social context features, and then fused them by local attention mechanism. Wang et al. [14] developed an adversarial network to learn event-invariant features and then obtained the multi-modal features of each post. Khattar et al. [7] utilized a multi-modal variational autoencoder to learn the shared representation of the text and the attached image. Zhou et al. [18] explored the textual and visual information along with their relationship are jointly learned and used to detect fake news. Wu et al. [15] took a further step to introduce frequency-domain features of image and stack multiple co-attention layers to fuse spatial, frequency and text feature for MFND.

Different from above approaches, we focus on visual instances that appear in the image and represent the multi-media post as a unified graph, where various semantic connections between multi-modal elements can be effectively captured for MFND.

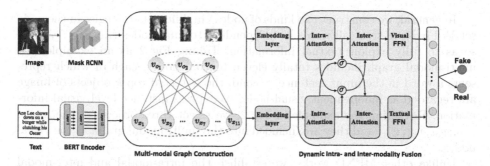

Fig. 2. The overall architecture of our model. In the multi-modal graph, the green and yellow solid circles denote textual nodes and visual nodes respectively. An intra-modal edge (dotted line) connects two nodes in the same modality, and an inter-modal edge (solid line) links two nodes in different modalities.

3   Methodology

Given a multi-modal news that includes a textual sentence X and image I associated with the text, our model aims to learn a comprehensive multi-modal representation by jointly considering semantic interactions of intra- and inter-modality for MFND. As shown in Fig. 2, it principally consists of *instance representation, multi-modal graph construction, dynamic intra- and inter-modality fusion*. As what follows, this section is dedicated to introducing them elaborately.

3.1   Instance Representation

Text Encoder. Due to the capability of giving different representations for the same word in different contexts, we employ the recent contextualized representations from BERT [6] as our text encoder. Given a sentence $X = \{x_1, x_2, \ldots, x_n\}$, where n denotes the number of words in the text, we obtain the transformed feature as $T = \{t_1, t_2, \ldots, t_n\}$, where $t_i \in \mathbb{R}^{d_w}$ and d_w is the dimension of word embedding.

Image Encoder. Previous researches show that visual contents can provide complementary clues in MFND [4,7,14,15,18]. Different from these methods, we extract the visual instances to model the high-level semantics of images. More concretely, given an image I, we apply the object detection model Mask R-CNN [2] which is pre-trained on the COCO dataset [10] to identify the objects in images. In most cases, only the salient objects are related to the event mentioned in a sentence. As a result, we consider the top k objects with the highest classification probabilities, denoting as $O = \{o_1, o_2, \ldots, o_k\}$, where $o_i \in \mathbb{R}^{d_v}$ and d_v is the dimension of the image embedding.

3.2   Multi-modal Graph Construction

In this section, we describe how we formulate the multi-modal news into a graph $\mathcal{G} = (\mathcal{V}, \mathcal{E})$, where $\mathcal{V}$ denotes the nodes and $\mathcal{E}$ is a set of undirected edges.

In general, we construct two kinds of nodes via different strategies in the node set $\mathcal{V}$. First, we treat all words as independent textual nodes (i.e., $v_{t_1}, \ldots, v_{t_{11}}$) so as to fully exploit textual information. Taking Fig. 2 as an example, the multi-modal graph includes totally eleven textual nodes, each of which represents a word in the input sentence. Second, the detected top k objects of image are regarded as independent visual nodes to express instance-level visual information. Revisiting the example in Fig. 2, we select three objects of the image (i.e., $v_{o_1}, v_{o_2}, v_{o_3}$) and they are included into the multi-modal graph as the visual nodes.

Unlike other MFND works, we establish the intra-modal and inter-modal semantic connections at the multi-modal representation stage. Specifically, we also consider two kinds of edges in the edge set E. First, any two nodes within the same modality are connected by an **intra-modal** edge. In Fig. 2, all visual nodes are connected to each other, as well as all textual nodes. Second, as the description of news is narrated an event consisting of multiple noun phrases [19], we consider these phrases as textual instances and employ the Stanford parser toolkit to identify all noun phrases in the input sentence, then these nodes and visual nodes are connected with an **inter-modal** edge[1]. Back to Fig. 2, only textual nodes $v_{t_1}, v_{t_2}, v_{t_7}$ and $v_{t_{11}}$ are connected with visual nodes by inter-modal edges. It follows that we directly build the cross-modal semantic connections between noun phrases and visual objects which are essential for identifying the truthfulness of the sample, and explicitly alleviate the alignment among many function words such as a, *while* with any visual object.

3.3 Dynamic Inter-modality and Intra-modality Fusion

As illustrated in Fig. 2, before inputting the graph into the dynamic fusion layer, we introduce an embedding layer to obtain the initial node states. Specifically, we employ two MLP with ReLu active function to project different representations from two modalities onto the same dimension d. As a result, for each textual node v_{t_i} and visual node v_{o_j}, we define their initial state as $S_{t_i}^{(0)} \in \mathbb{R}^d$ and $S_{o_j}^{(0)} \in \mathbb{R}^d$ respectively.

After embedding layer, we stack L fusion layers to encode the above-mentioned multi-modal graph. Inspired by the success of Transformer [13], we conduct dynamic intra- and inter-modality fusion based on multi-head attention, to model the intra- and inter-relationships of different nodes. Particularly, considering that visual nodes and textual nodes are two types of semantic units, we apply similar operations but with different parameters to model their state update process respectively inspired by studies [3,16]. Concretely, in the l-th fusion layer, both updates of textual node states $\boldsymbol{S}_t^{(l)} = \{S_{t_i}^{(l)}\}$ and visual node states $\boldsymbol{S}_o^{(l)} = \{S_{o_j}^{(l)}\}$ mainly involve the following steps:

---

[1] If no noun is detected for a sentence, all textual and visual nodes are connected with fully connection.

Intra-modality Fusion. We employ Multi-Head Attention [13] (MHA) to generate the contextual representation of each node by collecting the message from its neighbors of the same modality. Specifically, the contextual representations $H_t^{(l)}$ of all textual nodes are calculated as follows:

$$H_t^{(l)} = \text{MHA}(S_t^{(l-1)}, S_t^{(l-1)}, S_t^{(l-1)}) \tag{1}$$

Similarly, we generate the contextual representations $H_o^{(l)}$ of all visual nodes as:

$$H_o^{(l)} = \text{MHA}(S_o^{(l-1)}, S_o^{(l-1)}, S_o^{(l-1)}) \tag{2}$$

where we omit the descriptions of layer normalization and residual connection for simplicity.

Inter-modality Fusion. We apply a gating mechanism to gather the semantic interactions of the cross-modal neighbors of node, which not only filters irrelevant information adaptively but also captures the high-level interactions between visual and textual information better. Concretely, we generate the representation $H_{t_i \to o_j}^{(l)}$ of a text node v_{t_i} in the following way:

$$H_{t_i \to o_j}^{(l)} = H_{t_i}^{(l)} + \sum_{i \in \mathcal{N}(v_{t_i})} G_{i \to j} \odot H_{o_j}^{(l)} \tag{3}$$

$$G_{i \to j} = \sigma\left(W_1^{(l)} H_{t_i}^{(l)} + W_2^{(l)} H_{o_j}^{(l)}\right) \tag{4}$$

where $\mathcal{N}(v_{t_i})$ is the set of neighboring visual nodes of v_{t_i}, $\odot$ is an element-wise operation, and σ is sigmoid function. The $W_1^{(l)}$ and $W_2^{(l)}$ are parameter matrices.

Similarly, we produce the representation $H_{o_j \to t_i}^{(l)}$ of a visual node v_{o_j} as follows:

$$H_{o_j \to t_i}^{(l)} = H_{o_j}^{(l)} + \sum_{i \in \mathcal{N}(v_{o_j})} G_{j \to i} \odot H_{t_i}^{(l)} \tag{5}$$

$$G_{j \to i} = \sigma\left(W_3^{(l)} H_{o_j}^{(l)} + W_4^{(l)} H_{t_i}^{(l)}\right) \tag{6}$$

where $\mathcal{N}(v_{o_j})$ is the set of neighboring textual nodes of v_{o_j}, and $W_3^{(l)}$ and $W_4^{(l)}$ are parameter matrices.

After that, we adopt position-wise feed forward networks FFN($\cdot$) [13] to generate the textual node states $S_t^{(l)}$ and visual node states $S_o^{(l)}$:

$$S_t^{(l)} = \text{FFN}(H_{t \to o}^{(l)}) \tag{7}$$

$$S_o^{(l)} = \text{FFN}(H_{o \to t}^{(l)}) \tag{8}$$

where $H_{t \to o}^{(l)} = \{H_{t_i \to o_j}^{(l)}\}$ and $H_{o \to t}^{(l)} = \{H_{o_j \to t_i}^{(l)}\}$ denote the above updated representations of all textual nodes and visual nodes respectively.

Finally, we concatenate the output of the two representations to obtain the final multi-modal feature representation:

$$\mathcal{M}_f = \text{concat}(S_t^{(l)}, S_o^{(l)}) \tag{9}$$

3.4   Fake News Detector

We take the multi-modal representation $\mathcal{M}_f$ as input, and obtain the probability of a tweet being fake based on the proposed model:

$$\hat{y}_i = \text{softmax}(W_f\mathcal{M}_f + b) \tag{10}$$

where W_f and b are parameters of the fully connected layer. Then the cross-entropy is leveraged to measure the classification loss as follows:

$$\mathcal{L} = -\sum_{i=1}^{N} [y_i * \log(\hat{y}_i) + (1 - y_i) * \log(1 - \hat{y}_i)] \tag{11}$$

where N is the number of posts, y_i is the ground-truth label of the ith post.

4   Experiment

4.1   Experimental Setup

Dataset. We conduct experiments on two public real-world datasets, Twitter [11] and Weibo [4], respectively. The Twitter dataset consists of tweets contain texts, attached images and social context information. In this work, we focus on text and image information. Therefore, following previous work [15], we reserve the tweets both contain texts and images and remove the others. After that, 512 images are shared by the remaining data in Twitter dataset. The Weibo dataset is collected from the Xinhua News Agency and we follow the preprocessing steps in [4]. We keep the same data split scheme as the benchmark on these two datasets.

Implementation Details. The max length of the text is 25 on Twitter and 160 on Weibo. We set $d_w = 768$, $d_v = 2048$ and $d = 512$. The head number H in MHA is set as 4 and the layer number of multi-modal graph fusion L is 2. The parameters of Mask R-CNN and BERT are frozen when training on Twitter dataset due to overfitting, but not on Weibo dataset. The BERT model used on Twitter dataset is multilingual cased BERT-based model and the one used on Weibo dataset is Chinese BERT-based model. Our model is trained for 100 epochs with a learning rate of 1e−4, drop out rate of 0.5 and the mini-batch size is set to 128. Our algorithms are implemented on Pytorch deep learning framework and are trained with Adam [8] optimizer. For a fair comparison, we adopt official Accuracy scores as the main evaluation metrics and report the fine-grained Precision, Recall, and F1 scores yielded on Twitter and Weibo datasets.

Baselines. We compare our model with two types of baseline models: single-modal and multi-modal models, including (1) **Textual**: we use a pre-trained

Table 1. Performance comparison between the proposed model and the baselines on test sets of Twitter and Weibo (the best values are highlighted in bold).

Dataset	Method	Accuracy	Fake news			Real news		
			Precision	Recall	F1	Precision	Recall	F1
Twitter	Visual	0.621	0.783	0.536	0.636	0.430	0.631	0.511
	Textual	0.633	0.656	0.762	0.705	0.587	0.459	0.515
	EANN	0.648	0.810	0.498	0.617	0.584	0.759	0.660
	att-RNN	0.664	0.749	0.615	0.676	0.589	0.728	0.651
	MCAN	0.741	0.728	0.733	0.730	0.705	0.726	0.715
	MVAE	0.745	0.801	0.719	0.758	0.689	0.777	0.730
	SAFE	0.766	0.777	0.795	0.786	0.752	0.731	0.742
	DIIF (ours)	**0.783**	**0.810**	**0.803**	**0.806**	**0.758**	**0.786**	**0.772**
Weibo	Visual	0.716	0.702	0.718	0.710	0.713	0.726	0.719
	SAFE	0.763	0.833	0.659	0.736	0.717	0.868	0.785
	att-RNN	0.772	0.854	0.656	0.742	0.720	**0.889**	0.795
	EANN	0.782	0.827	0.697	0.756	0.752	0.863	0.804
	MVAE	0.824	0.854	0.769	0.809	0.802	0.875	0.837
	MCAN	0.872	0.883	0.872	0.877	0.862	0.878	0.870
	Textual	0.876	0.885	0.871	0.878	0.865	0.878	0.871
	DIIF (ours)	**0.890**	**0.912**	**0.894**	**0.903**	**0.905**	0.879	**0.892**

BERT [6] to obtain the representation of the given piece of news and a fully connected layer to make classifications. (2) **Visual**: Following [17], we employs a convolutional neural network to learn the feature representation. (3) **att-RNN** [4]: att-RNN uses an attention mechanism to combine textual, visual and social content information. To make a fair comparison, we remove the component processing social context information. (4) **EANN** [14]: EANN is composed of the multi-modal feature extractor, the fake news detection, and the event discriminator. For fairness of the comparison, we conduct experiments with a simplified version of EANN that removes the event discriminator. (5) **MVAE** [7]: MVAE learns multi-modal representations of text and image using a variational autoencoder along with a binary classifier. (6) **SAFE** [18]: SAFE is a similarity-aware multi-modal method for fake news detection, which investigates their relationships to obtain the multi-modal representation. (7) **MCAN** [15]: MCAN develops multiple co-attention layers to fuse text, spatial domain, and frequency domain for MFND. To make a fair comparison, we remove the frequency domain information in our experiments.

4.2 Result and Discussion

Comparison with Existing Models. Our proposed model is referred to as DIIF (<u>D</u>ynamic <u>I</u>ntra- and <u>I</u>nter-modality <u>F</u>usion) for convenience. As shown in Table 1, it is impressive to see that the proposed DIIF nearly outperforms all the baselines over all metrics across two datasets. Besides, there are some interesting observations.

On one hand, there are many similar trends on the two datasets. Apparently, single-modal approaches mostly perform much worse than multi-modal joint methods, which indicates that fusing visual and linguistic features improves model performance, but it is not always relevant. As illustrated in Table 1, the performance of Textual method performs better than other multi-modal approaches on Weibo dataset. Although their effectiveness, it is worth noting that our proposed DIIF beats all baselines in accuracy, which demonstrates that our strategy of multi-modal representation and fusion is indeed better than others.

On the other hand, there are also some differences on the two datasets. Fine-tuning BERT on Weibo datasets performs much better than on Twitter dataset so as to Textual approach gets better performance on Weibo dataset, as exhibited in Table 1. The reason is that the average length of each sentence on Weibo dataset is about 10 times longer than that on Twitter dataset, which probably makes BERT perform better on Weibo dataset. In addition, more than 70% of tweets on Twitter dataset are related to a single event, probably leading to the overfitting on Twitter dataset. But Weibo dataset has no such imbalanced issue. This is also the reason why we fine-tuned BERT on Weibo dataset but not on Twitter dataset.

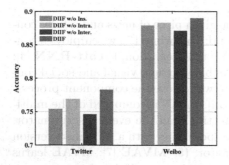

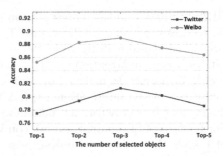

Fig. 3. Ablation analysis of DIIF.

Fig. 4. Results of our proposed model on different numbers of objects.

Ablation Analysis. To investigate the importance of each component in our DIIF, we perform comparison between the full DIIF and its ablated approaches:

- **DIIF w/o Ins.**, a variant of our approach, which replaces the visual instances guidance with region-aware methods as the same as **MCAN**.
- **DIIF w/o Intra.**, a variant of our approach, which removes the intra-modality semantic correlations of visual nodes.
- **DIIF w/o Inter.**, a variant of our approach, which removes the inter-modality relations between vision and language.

As shown in Fig. 3, we can see that all the components in **DIIF** make important contributions to the final results on both two datasets. Specifically, compared with **DIIF w/o Obj.**, **DIIF** adopts visual objects as nodes rather than the averagely segmented regions, and obtains better performance on both datasets, which shows the importance of objects-aware visual guidance. In addition, **DIIF** outperforms **DIIF w/o Intra.** which lacks the intra-modal interactions of visual contents, indicating that the effectiveness of intra-modal correlations among visual semantic units. Furthermore, **DIIF w/o Inter.** brings in a significant performance degradation, which demonstrates that it is vital to exploit the semantic interactions between different modalities. Besides, on Weibo dataset, the performance of **DIIF** does not drop significantly on Twitter dataset when removing any of the components. This also benefits from the balanced data distribution on Weibo dataset and the stability of fine-tuned BERT.

Parameter Sensitivity. As mentioned before, we choose the top k detected objects as visual instances according to the detection possibility of each object in an image. Figure 4 shows the results of our model with different number of selected objects. As can be noted, the two datasets show a similar trend that the accuracy firstly increases and then decreases with the changes of the selected objects, and we achieve the best performance when setting the number of objects to 3. Performance degradation is the greatest when we select only the most prominent object. The reason is that including both salient object and contextual regions can favor the learning of better visual feature representations. For example, in Fig. 2, the "trophy" object in image is small, however, it is useful to recognize the "Oscar" as an award name so as to connect each other. Meanwhile, too many visual objects may bring noise into the prediction of entities. However, even if given 5 visual objects, the performance is only decreased little. The main reason is that our graph-based fusion module can filter out irrelevant visual information to a certain extent.

Case Study. To further verify the effectiveness of our model, we compare and analyze the prediction results of **Textual**, **Visual**, **MCAN**, and **DIIF** methods for some fake news. As shown in Fig. 5, all these cases are fake news, and only our **DIIF** model predict all of them correctly. For the first sample, it is incorrectly identified as true news since the **Visual** method is only based on the visual signal. Thus, the **Visual** method recognize it wrongly and others are right. For the second sample, as only considering textual modality, the **Textual** method give a wrong prediction. However, it is obvious that the associated image was fake. For the last sample, all methods predict wrongly except **DIIF**. Neither the image or the text along can prove it is fake news, but we infer that it is probably a fake news from the mismatched image and textual contents. Compared with **MCAN**, our **DIIF** model can capture fine-grained semantic relationships between different modalities, which is beneficial for us to detect multi-modal fake news.

Multi-modal Samples	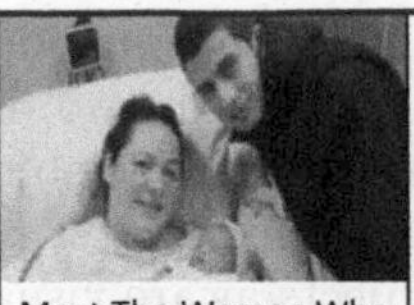 Meet The Woman Who Has Given Birth To 14 Children From 14 Different Fathers!	 A group of dolphins brought a dog that fell into a canal to safe area.	 American Scientists test the F1 rocket before the launch of Apollo 11, 1969.
Textual	False	True	True
Visual	True	False	True
MCAN	False	False	True
DIIF	False	False	False
Ground Truth	**False**	**False**	**False**

Fig. 5. Prediction results for some cases obtained from various methods, where the wrong prediction are marked red. (Color figure online)

5 Conclusion

In this work, we proposed a novel instance-guided multi-modal graph fusion method for MFND, which exploits intra- and inter-modality relationships simultaneously between multimodal semantic units. Without introducing any extra information, experiment results on two benchmark datasets demonstrated the superiority of our model. In the future, we plan to incorporate more external knowledge such as social context and common sense, to enrich the multi-modal graphs for MFND.

Acknowledgement. The authors would like to thank the anonymous reviewers for their constructive comments. This research was supported by the National Natural Science Foundation of China (No. 61976247).

References

1. Baltrusaitis, T., Ahuja, C., Morency, L.: Multimodal machine learning: a survey and taxonomy. IEEE Trans. Pattern Anal. Mach. Intell. **41**(2), 423–443 (2019)
2. He, K., Gkioxari, G., Dollár, P., Girshick, R.B.: Mask R-CNN. In: ICCV 2017, pp. 2980–2988 (2017)
3. Huang, Z., et al.: Audio-oriented multimodal machine comprehension via dynamic inter-and intra-modality attention. In: AAAI 2021, pp. 13098–13106 (2021)
4. Jin, Z., Cao, J., Guo, H., Zhang, Y., Luo, J.: Multimodal fusion with recurrent neural networks for rumor detection on microblogs. In: MM 2017, pp. 795–816 (2017)

5. Jin, Z., Cao, J., Zhang, Y., Zhou, J., Tian, Q.: Novel visual and statistical image features for microblogs news verification. IEEE Trans. Multimedia **19**(3), 598–608 (2017)
6. Kenton, J.D.M.W.C., Toutanova, L.K.: BERT: pre-training of deep bidirectional transformers for language understanding. In: NAACL-HLT 2019, pp. 4171–4186 (2019)
7. Khattar, D., Goud, J.S., Gupta, M., Varma, V.: MVAE: multimodal variational autoencoder for fake news detection. In: WWW 2019, pp. 2915–2921 (2019)
8. Kingma, D.P., Ba, J.: Adam: a method for stochastic optimization. In: ICLR 2015 (2015)
9. Li, P., Sun, X., Yu, H., Tian, Y., Yao, F., Xu, G.: Entity-oriented multi-modal alignment and fusion network for fake news detection. IEEE Trans. Multimedia (2021)
10. Lin, T., et al.: Microsoft COCO: common objects in context. CoRR abs/1405.0312 (2014)
11. Maigrot, C., Claveau, V., Kijak, E., Sicre, R.: Mediaeval 2016: A multimodal system for the verifying multimedia use task. In: MediaEval 2016: Verfiying Multimedia Use task (2016)
12. Qi, P., et al.: Improving fake news detection by using an entity-enhanced framework to fuse diverse multimodal clues. In: MM 2021, pp. 1212–1220 (2021)
13. Vaswani, A., et al.: Attention is all you need. In: NeurlPIS 2017, pp. 6000–6010 (2017)
14. Wang, Y., et al.: Eann: event adversarial neural networks for multi-modal fake news detection. In: SIGKDD 2018, pp. 849–857 (2018)
15. Wu, Y., Zhan, P., Zhang, Y., Wang, L., Xu, Z.: Multimodal fusion with co-attention networks for fake news detection. In: ACL-IJCNLP 2021, pp. 2560–2569 (2021)
16. Yin, Y., et al.: A novel graph-based multi-modal fusion encoder for neural machine translation. In: ACL 2020, pp. 3025–3035 (2020)
17. Yu, F., Liu, Q., Wu, S., Wang, L., Tan, T.: A convolutional approach for misinformation identification. In: IJCAI 2017, pp. 3901–3907 (2017)
18. Zhou, X., Wu, J., Zafarani, R.: SAFE: similarity-aware multi-modal fake news detection. In: PAKDD 2020, pp. 354–367 (2020)
19. Zhou, X., Zafarani, R.: A survey of fake news: fundamental theories, detection methods, and opportunities. ACM Comput. Surv. **53**(5), 1–40 (2020)

Simulate Human Thinking: Cognitive Knowledge Graph Reasoning for Complex Question Answering

Hong Zhao[1,2], Yao Fu[1(✉)], Weihao Jiang[1], Shiliang Pu[1], and Xiaoyu Cai[1]

[1] Hikvision Research Institute, Hikvision, Hangzhou, China
{zhaohong7,fuyao,jiangweihao5,pushiliang.hri,
caixiaoyu6}@hikvision.com
[2] College of Computer Science and Technology, Zhejiang University, Hangzhou, China

Abstract. Question answering over knowledge graph has attracted increasing attention. Though the previous algorithms have achieved competitive performance, they fail to solve problems like humans resulting in the bottleneck of reasoning. However, it is difficult for machines to simulate the question answering process of humans. In order to address this challenge, we propose a novel Cognitive Knowledge Graph Reasoning (CKGR) model based on the cognitive architecture for complex question answering. The CKGR processes information hierarchically with a three-level framework. To fully analyze the question, the first level is proposed to transform the question into features according to different aspects. Then, the relative knowledge graph (KG) regions are activated to simulate the human unconscious thinking process by a memory mapping module. Finally, the CKGR goes deeper to infer the correct answer over KG considering the both semantic and logical parsing of the questions. The CKGR successfully narrows the gap between humans and machines. Extensive experiments on three real-world datasets demonstrate that the proposed method achieves better performance compared with the state-of-the-art methods and provides the reasoning score to find the reasonable path for the answer.

Keywords: Question answering · Cognitive knowledge graph · Knowledge reasoning

1 Introduction

With the improvement in building large-scale knowledge graphs (KGs), question answering over knowledge graphs (KGQA) has emerged as a hot topic in artificial intelligence over the last few years. The early researches of KGQA mainly focus on simple questions where only a single triple in KG is involved. The performance of answering simple questions is competitive [1, 2], and thus increasing researchers have begun to pay attention to complex questions. Complex questions usually have multiple topic entities, relations, and some constraints, so that they consist of many different situations. In order to solve complex questions, there are two kinds of methods: semantic parsing (SP) methods [3,

J. Gama et al. (Eds.): PAKDD 2022, LNAI 13280, pp. 522–534, 2022.
https://doi.org/10.1007/978-3-031-05933-9_41

4] and information retrieval (IR) methods [5–7]. SP methods can give interpretable reasoning results due to the symbolic logic form. However, the logic forms are not easy to design, which is the choke point of improving the method performance. IR methods learn the representations of questions and entities by using end-to-end models which are easier to train, but the predicted results are less interpretable.

The above methods face obvious bottlenecks because they fail to think like humans to solve problems. In this paper, we simulate human cognitive processes to enable machines to solve QA task in a natural way. Cognition refers to the mental processes involved in acquiring knowledge and processing information [8]. The cognitive processes works together to integrate the new knowledge using existing knowledge and create an interpretation of the world. QA task is a proper situation that introduces human cognition to improve the decision-making level of machines. It involves the ability to understand through natural language questions and make decisions by retrieving information from knowledge stored in memory [9, 10]. The questions are transformed into several features that the brain can understand. They are subsequently input to the human brain establishing relationships between features and knowledge to activate relative brain regions. Finally, the brain generates behavior by making use of declarative knowledge and working memory to perform reasoning and synthesis to make decisions [11]. Overall, the cognitive process for QA happens naturally to synthesize question information, integrate it with prior knowledge and reason over the memory to get accurate answers. Implementation of human intelligence for QA task helps to improve the accuracy and interpretability of the model. Therefore, the promising methods should simulate the human cognitive processes to answer questions.

In this paper, we propose a novel cognitive knowledge graph reasoning (CKGR) method for complex question answering, which is a hierarchical information processing mechanism to simulate human thinking. The mechanism is equipped with a three-level framework as shown in Fig. 1. For answering a complex question, people will first try to understand the questions by extracting useful question features. This process is simulated by the first level that transforms a given question into a series of features from different aspects instead of using a simple pre-trained language model in previous work [12–14]. The second level is designed for memory mapping to simulate the activation process of the brain region. The extracted features and knowledge stored in KG are in different spaces so that a semantic aware module is proposed to integrate the language space and graph space into the same space. To find answers in KG, people would locate some regions where likely exists answers. This process is even ignored by the latest methods such as teacher-student method of NSM [7] and link prediction of EmbedKGQA [13]. Most question answers may not be far from the topic entities in KG, so different hop neighbors of the topic entities are activated for different scores. And the entities whose relations are relevant to the question relations are also activated. Finally, the third level conducts a deep reasoning process. A more expressive representation of question improves the reasoning results. So the questions are analyzed from both logical parsing and semantic parsing aspects, and are used to find the correct answers over KG. The main contributions are summarized as follows:

- The CKGR is proposed to integrate the human cognitive process in QA area, which simulates human thinking with a hierarchical information processing mechanism.

- The CKGR consists of a three-level framework including question feature extraction, memory mapping, and answer reasoning. The extracted question features are mapped to the KG for activating relevant regions. The final reasoning process uses more expressive question representations from both semantic and logical parsing.
- We evaluate CKGR through extensive experiments on multiple datasets. The results show that CKGR achieves better performance than the state-of-the-art baselines.

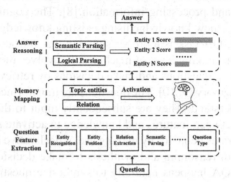

Fig. 1. The three-level framework of the CKGR for QA task based on human cognitive process.

2 Related Work

Complex Question Answering over KG. In recent years, many works have been devoted to complex question answering based on the knowledge graph. They can be divided into two groups: information retrieval (IR) methods and semantic parsing (SP) methods. IR methods capture the information contained in questions to learn representations and rank entities in KG to get final answers. Key-Value Memory Network [5] allows encoding prior knowledge in the key and value memories using knowledge bases. VRN [6] adopts an end-to-end variational learning method to jointly recognize the topic entity and learn reasoning graph embedding. EmbedKGQA [13] uses pre-trained entity embeddings to perform multi-hop KGQA over sparse KGs, which regards QA task as link prediction. NSM [7] is proposed to adopt a novel teacher-student approach to alleviate the issue of spurious reasoning. MULTIQUE [3] is an SP method, which uses simple queries to map a complex question to a complex query pattern. However, there are few models to simulate human cognitive processes to solve problems, which helps to reduce the bottlenecks of the previous methods.

Human Cognitive Processes. Cognition [9, 10] is a term referring to the mental action of acquiring knowledge and comprehension. There are many cognitive processes relevant to QA task including learning, memory, language, reasoning, decision making, etc. When answering questions, people would integrate the above cognitive processes to synthesize question information, incorporate it into prior knowledge, and reason over a certain memory to get accurate answers. In order to have artificial computational

system processes that act like humans, researchers have proposed many cognitive architectures, such as ACT-R [15] and Soar [16]. They all try to construct the structures of
the human mind by combining cognitive science and artificial intelligence (AI). Implementing aspects of human intelligence in computers is one of the practical goals of
AI. Cognition has been introduced into visual reasoning to narrow the gap between
humans and machines [17]. However, limited researchers attempt to introduce cognitive
processes into QA task, which may help improve the performance of QA and provide
interpretability justifying the answers. In this work, we propose a three-level framework
simulating human thinking with a hierarchical information processing mechanism for
QA task.

3 Methodology

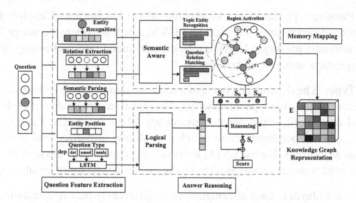

Fig. 2. The overall architecture of CKGR model.

To simulate the human cognitive process for the QA task, we propose a hierarchical
information processing mechanism with a three-level framework as shown in Fig. 2.
The first level aims to transform the question into several features. The second level
attempts to activate relative KG regions. The recognized entities and extracted relations
in questions are denoted as *topic entities* and *question relations*, and the entities and
relations in KG are denoted as *graph entities* and *graph relations* in the following steps.
The third level generates reasoning over KG to find correct answers.

3.1 Question Feature Extraction

When people answer questions, they will first try to understand the questions by extracting several features from the questions. This process is crucial since useful features are
helpful to find the answers in the brain while useless features hamper the answering
process. Thus, we design the first level of CKGR to transform a given natural language
question into a series of features which we believe are important for QA tasks. Certainly, the proposed CKGR is a general framework that allows the addition of more
useful features for QA tasks in future work.

Entity Recognition: The entity recognition module presented in this paper uses the spaCy library [18]. Spacy not only achieves Entity Recognition but also provides integrated word vectors **e**, which may not be realized both by other NLP tools.

Relation Extraction: The question relations may not be determined by topic entities. For example, the questions "what is the nationality of Obama" and "what is the nationality of Newton" both concern about the relation "nationality". The entities "Obama" and "Newton" provide useless information for relation extraction. Hence, we mask the entities recognized by the above step in the questions, such as "what is the nationality of <NE>". The masked question is then input to spaCy to get the predicted question relation representation **r**.

Entity Position: The entity position in the question is another useful feature that is adopted for logical parsing of the question. Each position in question is assigned a position embedding which will be learned by the end-to-end network.

Semantic Parsing: The semantic information of text has been well explored by some natural language processing, like Bert and GloVe. In this paper, we adopt RoBERTa [19] as the semantic parsing, which is a robustly optimized Bert pretraining approach to obtain the question semantic embedding q_s.

Question Type: The dependency parse of a sentence denotes its grammatical structure, which reflects the question type information. The questions are analyzed by dependency parsing and each word would get its dependency labels by spaCy. Each dependency label is assigned a dependency vector which is randomly initialized and will be learned by the whole algorithms. We use LSTM to encode the dependency parse of the question, and the last hidden state is considered to be the question type representation q_t.

The above features extract information from different aspect of questions, and thus they will be useful for memory mapping and question answering.

3.2 Memory Mapping

After acquiring question features, people will map them to the relative knowledge. Then people will activate the surrounding brain regions of relative knowledge unconsciously, since these areas likely include correct answers. This process is generalized as the second level Memory Mapping. It consists of two modules: semantic aware module that establishes relationships between questions and KG, and region activation module that activates relative graph entities.

Semantic Aware Module: The representations learned from the language model and graph model are in different spaces. Thus, it is improper to directly calculate the similarity between the topic entity and all graph entities. The question information is also crucial for entity linking, since it provides semantic meaning to find the corresponding graph entity more accurately. So we propose a semantic aware module to get the topic entities and most relevant graph relations to question relation. The graph entity embedding **E** and relation embedding **R** are first mapped to the language space:

$$\tilde{\mathbf{E}} = \sigma(\mathbf{E}\mathbf{W}_1), \quad \tilde{\mathbf{R}} = \sigma(\mathbf{R}\mathbf{W}_2), \tag{1}$$

where $\tilde{\mathbf{E}}$ and $\tilde{\mathbf{R}}$ are the mapped graph entity and relation embeddings in language space. $\mathbf{W_1}$ and $\mathbf{W_2}$ are parameter matrices, σ denotes ReLU function. Such operators transform the embeddings from graph space to language space. Then the question semantic information is considered to calculate the topic entity distribution p_t and question relation distribution p_r by the following equations:

$$p_t = \text{softmax}\left((\mathbf{q_s} \odot \mathbf{e})\tilde{\mathbf{E}}^{\mathrm{T}}\right), p_r = \text{softmax}\left((\mathbf{q_s} \odot \mathbf{r})\tilde{\mathbf{R}}^{\mathrm{T}}\right), \tag{2}$$

where $\mathbf{e}$ and $\mathbf{r}$ are the topic entity vector and question relation vector. The recognized results would be more accurate since the distributions are incorporated with the question semantic information.

Region Activation Module: This module simulates the process of the brain solving the problem to activate relative memory regions. Graph entities are activated in consideration of two aspects with the activation scores. First, most answers may not be far from the topic entities in KG so that the neighbors of the topic entities should be activated with basic scores. However, the fact that graph entities in which hop of topic entities are essential to question are not determined, and thus it needs adjustable parameters to learn. For a given question, the basic scores s_b are defined as follows:

$$s_b = \sum_{j=1}^{3} \lambda_j p_t \mathbf{A_j}, \tag{3}$$

where $\mathbf{A_j}$ denotes the j-*hop* adjacency matrix, λ_j is a learnable parameter. Here, we only activate the neighbors of topic entity within 3-hop neighbors.

Second, those graph entities whose graph relations are similar to the question relations should also be activated with mapping scores, since these entities are likely to be the answers. The mapping scores s_m are defined as follows:

$$s_m = p_r \mathbf{H}^{\mathrm{T}}, \tag{4}$$

where $\mathbf{H} \in \mathbb{R}^{N_E \times N_R}$ is an incidence matrix, N_E and N_R are the numbers of graph entities and relations respectively. H_{ij} is set to be 1 if the i-th graph entity is connected with j-th graph relation. Overall, the activation scores s_a are obtained based on basic score and mapping score, which help to perform reasoning to find the final answers.

$$s_a = s_b + s_m. \tag{5}$$

3.3 Answer Reasoning

The more accurate reasoning process is based on the more expressive representations of questions and graph entities. Thus, we fully analyze the questions from both logical parsing and semantic parsing aspects. And the graph entity embedding can change dynamically in terms of different questions, since the aggregation process incorporates the question information. The answer is finally obtained via a reasoning module.

Logical Parsing Module: Question contains two aspects of information: semantic information and logical information. The semantic question embedding is obtained by the previous step. Here, we analyze the question from logical aspects. The entity position and question type are typical logical features. The absolute entity position is used to model how the entity position influences other words at different positions, thus the integrated information combined by question type is defined as follows:

$$
\mathbf{q}_l = \frac{1}{|\mathcal{N}_p|} \sigma \left(\sum_{p \in \mathcal{N}_p} \sum_{i=1}^{k} (([\mathbf{x_i}; \mathbf{x_p}; \mathbf{q_t}]) \mathbf{W}_3) \right) \quad i \notin \mathcal{N}_p, \tag{6}
$$

where $\mathcal{N}_p$ is the set of the entity position numbers, $\mathbf{q_t}$ is the question type representation, $\mathbf{x_p}$ is the position vector of entity, $\mathbf{x_i}$ is the vector of position i, k is the maximum length of the question, $\mathbf{W}_3$ is a parameter matrix. Finally, we get the more expressive question representation based on logical parsing and semantic parsing:

$$
\mathbf{q} = \mathrm{FFN}([\mathbf{q}_s; \mathbf{q}_l]), \tag{7}
$$

where $\mathrm{FFN}(\cdot)$ is a feed forward neural networks.

Reasoning Module: Once the question representations are obtained, we can use them to find the answers in KG. The graph entity embedding at k-th reasoning step is learned from not only its graph entity neighbors and relation neighbors but also the question information:

$$
\mathbf{h}^{(k)} = \sum_{(h,r,t) \in \mathcal{N}_h} p_t^{(k-1)} \cdot \left(\mathbf{q} \odot \left(\mathbf{r}^{(k-1)} \mathbf{W_r} + \mathbf{t}^{(k-1)} \mathbf{W_t} \right) \right), \tag{8}
$$

where $\mathcal{N}_h$ is the set of triples whose head entity is h, $p_t^{(k-1)}$ is the answer probability of tail entity t at $(k-1)$-th reasoning step, $\mathbf{r}^{(k-1)}$ and $\mathbf{t}^{(k-1)}$ are the embeddings of r and t at $(k-1)$-th step, $\mathbf{W_r}$ and $\mathbf{W_t}$ are parameter matrices. The learned embedding is able to capture relation and entity information in KG, and can be different for different questions, which is more powerful for answering the specific questions. The reasoning score is calculated as follows:

$$
s_r = \mathbf{w} E^{(k)^{\mathrm{T}}}, \tag{9}
$$

where $E^{(k)}$ is the updated graph entity embedding matrix whose row vector is the entity embedding learned by Eq. (8) at k-th step, $\mathbf{w}$ is a parameter matrix. The answer probability of graph entities at k-th reasoning step is obtained as follows:

$$
p^{(k)} = \mathrm{softmax}(s_r + \beta s_a), \tag{10}
$$

where $\beta \in (0,1)$ is a hyper-parameter, $p^{(0)}$ is set to be p_t according to Eq. (2). The answer probability involves a preliminary reasoning process to find a relative regions and a deeper reasoning process to find the final answers, that is similar to the human cognitive process for question answer.

The loss function is to minimize the following binary cross entropy loss:

$$\mathcal{L} = -\frac{1}{N_E} \sum_i \left(y_i \cdot \log\left(p_i^{(n)}\right) + (1 - y_i) \cdot \log\left(1 - p_i^{(n)}\right)\right),\tag{11}$$

where $p^{(n)}$ denotes the final answer probability of graph entities, y_i denotes the label for the i-th graph entity.

4 Experiments

4.1 Datasets

MetaQA [6] is a largescale movie domain dataset, which contains more than 400k questions generated by several templates. The KG of MetaQA includes 43k entities, 9 relations, and 135k triples.

WebQSP [20] has a smaller scale of questions containing 4737 questions that can be answered by the Freebase knowledge graph. The KG dataset includes 1.8 million entities, 572 relations, and 5.7 million triples.

CWQ [21] is an extended dataset of WebQSP. Some constraints are added to the questions that require up to 4-hops of reasoning over KG.

The detailed statistics of all used datasets are shown in Table 1.

Table 1. Statistics of the three datasets

Dataset	Train	Valid	Test
MetaQA 1-hop	96,106	9,992	9,947
MetaQA 2-hop	118,980	14,872	14,872
MetaQA 3-hop	114,196	14,274	14,274
WebQSP	2,848	250	1,639
CWQ	27,639	3,519	3,531

4.2 Baselines

We compare the CKGR with the following baselines to demonstrate the effectiveness of the proposed method.

KV-MemNN [5] stores the knowledge in the key and value memories to retrieve the answers.

VRN [6] is an end-to-end variational learning method to handle multi-hop reasoning.

GraftNet [12] adopts a graph convolution network to find answers on the subgraph.

PullNet [14] adopts an iterative process to construct a subgraph by a graph CNN. After that, a similar graph CNN is used to conduct reasoning over the subgraph.

EmbedKGQA [13] regards the QA task as a link prediction task.

MULTIQUE [3] utilizes simple queries each targeting a specific KB to construct query patterns for complex questions.

NSM [7] uses a teacher-student approach to alleviate the issue of spurious reasoning.

4.3   Experimental Setup

We adopt the widely used metric Hits@1 and F1 to evaluate the proposed method. To achieve optimal performance of CKGR, the parameters are determined by a grid search during the training. The parameter ranges are manually specified as follows: learning rate among (0.0005, 0.001, 0.005), embedding size among (50, 80, 100), batch size among (20, 80, 120), β among (0.001, 0.01, 0.1, 1). We use early stopping according to Hits@1 for different datasets in the validation set to determine the optimal parameters to avoid overfitting. The learning rate and β are set to 0.0005 and 0.001 for all datasets. Embedding size and batch size are set to 80, 80 for MetaQA, and 50, 20 for other datasets. The reasoning step is set to 4 for CWQ and 3 for other datasets. All results are the average of 10 runs.

4.4   Results

Table 2. Experimental results of Hits@1 on three datasets

Models	MetaQA			WebQSP	CWQ
	1-hop	2-hop	3-hop		
KV-MemNN	96.20	82.70	48.90	46.70	21.10
VRN	97.50	89.90	62.50	–	–
GraftNet	97.00	94.80	77.70	66.40	32.80
PullNet	97.00	99.90	91.40	68.10	45.90
EmbedKGQA	97.50	98.80	94.80	66.60	–
MULTIQUE	–	–	–	69.80	41.20
NSM	97.30	99.90	98.90	74.30	48.80
CKGR	**98.02**	**99.98**	**99.16**	**76.28**	**50.50**

Experimental results of Hits@1 for different methods are shown in Table 2. Here we use Hits@1 to compare with the baselines, since many baselines only use Hits@1 in their papers [5, 6, 13, 14]. The results of baselines are implemented according to the illustration reported in their papers. For the MetaQA dataset, our method achieves better performance than other baselines. Though the performances of 1-hop, 2-hop, and 3-hop datasets already reach 97.50%, 99.90%, and 98.90%, CKGR still improves the performance to 98.02%, 99.98%, and 99.16%, respectively. Compared with the MetaQA dataset, WebQSP and CWQ have less QA training set, where the questions have more constraints, and KG has more relations and triples, which makes the task more difficult. CKGR also outperforms the above baselines on WebQSP and CWQ. The performance improvement strongly indicates that the three-level framework of our method is beneficial to the QA task.

Region Activation Evaluation: To evaluate whether the graph entities activated by CKGR are in the correct hop of topic entities, we take the MetaQA dataset as an example. Figure 3 shows the learnable parameters λ of different hops of topic entities in Eq. (3) for MetaQA dataset. For different hops datasets, the weights of the corresponding hop neighbors of topic entity are the largest. Thus, the CKGR automatically learns which hop of topic entities are essential to questions, which endows the model human intelligence due to the simulation of human cognitive processes.

Ablation Study: In order to get a better understanding of the contributions of each model component to the proposed model, we further compare CKGR with its variants: (1) CKGR-s_b is CKGR without activating the neighbors of the topic entities with basic scores; (2) CKGR-s_m is CKGR without considering the similarity between the graph relations of entities and the question relations; (3) CKGR-q_l is CKGR without using logical parsing module; (4) CKGR-s is CKGR without using semantic aware module, but directly calculates the similarity between the topic entity and graph entities, the question relation and graph relations. The results of Hits@1 and F1 are reported in Table 3. We can conclude that 1) the performance of CKGR-s is the worst of the four variants, which indicates that integrating the language space and graph space into the same space plays a more important role in QA task. 2) CKGR-s_b underperforms CKGR, which indicates basic scores help to improve reasoning accuracy since they activate the neighbors of the topic entities to give the preliminary knowledge. Mapping scores give a greater performance improvement by comparing CKGR to CKGR-s_m, because they utilize the question information to calculate similarity. It proves that taking advantages of more question information improves the reasoning results. 3) CKGR-q_l performs better than two memory mapping variants for MetaQA while worse for WebQSP and CWQ. Because the questions of WebQSP and CW have more constraints, they cannot be well analyzed only based on semantic parsing. The logical parsing provides different aspect information which benefits more for complex questions. Above all, each model component is useful for QA task.

4.5 Case Study

In this section, we provide a case study in MetaQA 3-hop as an example to illustrate the effectiveness of the CKGR. Given the question "who are the directors that the actors in

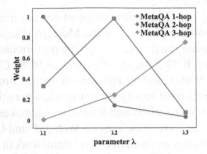

Fig. 3. The learnable parameters λ of each hop on different MetaQA datasets.

Table 3. Ablation study of CKGR (Hits@1/F1)

Models	MetaQA			WebQSP	CWQ
	1-hop	2-hop	3-hop		
CKGR	**98.02/99.79**	**99.98/99.81**	**99.16/89.06**	**76.28/68.90**	**50.50/45.34**
CKGR-s_b	97.24/99.05	99.41/99.08	98.65/88.38	75.75/68.12	49.63/44.70
CKGR-s_m	97.11/98.90	99.32/98.95	98.52/88.09	75.52/67.77	49.42/44.41
CKGR-q_l	97.32/99.33	99.55/99.29	98.73/88.56	75.34/67.55	49.28/44.21
CKGR-s	96.96/98.54	98.94/98.51	98.37/87.86	74.77/67.34	48.73/44.02

their movies also appear in the movie Finding Amanda", the activating and reasoning process is shown in Fig. 4. The blue and red circles denote the topic entity and final answer. The deeper color of the orange circle, the higher the activation level obtained by the region activation module. For this question, the model prefers to give the instruction signals to activate the names of people and movies, and thus relevant people and movies get higher activation levels, which is consistent with the human cognitive process of finding the answer. The purple scores are the similarities between the graph relations and the question relation. It can be seen that the graph relation "*directed_by*" and "*starred_actors*" have higher scores since this question involved the information of "director" and "actors". The green scores are the answer probabilities. The red lines denote the reasoning paths for the final answer. The CKGR finds the correct answer with the reasonable path (*Finding Amanda, starred_actors, Brittany Snow*), (*Prom Night, starred_actors, Brittany Snow*) and (*Prom Night, directed_by, Paul Lynch*). The example gives the reasonable path in KG to the correct answer, which benefits the QA task.

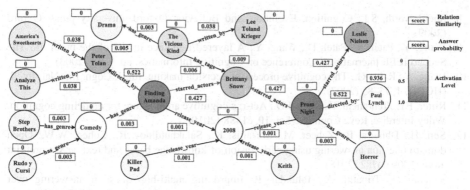

Fig. 4. A case example of activating and reasoning process for a given question "who are the directors that the actors in their movies also appear in the movie Finding Amanda".

5 Conclusion

In this paper, we propose a novel CKGR model for complex question answering, which introduces the human cognition concept into KGQA. The CKGR adopts a three-level framework to process information hierarchically, including question feature extraction, memory mapping, and answer reasoning. The extracted question features are utilized to gradually activate relevant KG regions. The questions are analyzed from both semantic parsing and logical parsing. Finally, the topic entities and encoded questions are integrated to find the correct answers over KG. The results show that CKGR achieves better performance than previous methods on real-world datasets. In addition, the ablation study verifies the contributions of each model component. The case study proves that CKGR can provide the activating and reasoning processes to find the path for the answer.

References

1. Yani, M., Krisnadhi, A.A.: Challenges, techniques, and trends of simple knowledge graph question answering: a survey. Information **12**, 271 (2021)
2. Wang, Z., Ng, P., Nallapati, R., Xiang, B.: Retrieval, re-ranking and multi-task learning for knowledge-base question answering. In: EACL, pp. 347–357 (2021)
3. Bhutani, N., Zheng, X., Qian, K., Li, Y., Jagadish, H.: Answering complex questions by combining information from curated and extracted knowledge bases. In: Proceedings of the First Workshop on Natural Language Interfaces, pp. 1–10 (2020)
4. Chen, Y., Li, H., Hua, Y., Qi, G.: Formal query building with query structure prediction for complex question answering over knowledge base. In: IJCAI, pp. 3751–3758 (2020)
5. Miller, A., Fisch, A., Dodge, J., Karimi, A.-H., Bordes, A., Weston, J.: Key-value memory networks for directly reading documents arXiv preprint arXiv:1606.03126 (2016)
6. Zhang, Y., Dai, H., Kozareva, Z., Smola, A.J., Song, L.: Variational reasoning for question answering with knowledge graph. In: AAAI (2018)
7. He, G., Lan, Y., Jiang, J., Zhao, W.X., Wen, J.-R.: Improving multi-hop knowledge base question answering by learning intermediate supervision signals. In: WSDM, pp. 553–561 (2021)

8. Shettleworth, S.J.: Cognition, Evolution, and Behavior. Oxford university press, Oxford (2009)
9. Wang, Y., Patel, S., Patel, D., Wang, Y.: A layered reference model of the brain. In: The Second IEEE International Conference on Cognitive Informatics, pp. 7–17 (2003)
10. Wang, Y., Ruhe, G.: The cognitive process of decision making. Int. J. Cogn. Inf. Nat. Intell. (IJCINI) **1**, 73–85 (2007)
11. Ritter, F.E., Tehranchi, F., Oury, J.D.: Act-r: a cognitive architecture for modeling cognition. Wiley Interdisc. Rev.: Cognitive Sci. **10**, e1488 (2019)
12. Sun, H., Dhingra, B., Zaheer, M., Mazaitis, K., Salakhutdinov, R., Cohen, W.W.: Open domain question answering using early fusion of knowledge bases and text. arXiv preprint arXiv:1809.00782 (2018)
13. Saxena, A., Tripathi, A., Talukdar, P.: Improving multi-hop question answering over knowledge graphs using knowledge base embeddings. In: ACL, pp. 4498–4507 (2020)
14. Sun, H., Bedrax-Weiss, T., Cohen, W.W.: Pullnet: open domain question answering with iterative retrieval on knowledge bases and text arXiv preprint arXiv:1904.09537 (2019)
15. Anderson, J.R., Matessa, M., Lebiere, C.: Act-r: a theory of higher level cognition and its relation to visual attention. Hum.-Comput. Interact. **12**, 439–462 (1997)
16. Laird, J.E., Newell, A., Rosenbloom, P.S.: Soar: an architecture for general intelligence. Artif. Intell. **33**, 1–64 (1987)
17. Zellers, R., Bisk, Y., Farhadi, A., Choi, Y.: From recognition to cognition: visual commonsense reasoning. In: CVPR, pp. 6720–6731 (2019)
18. Honnibal, M., Montani, I.: Spacy 2: natural language understanding with bloom embeddings, convolutional neural networks and incremental parsing. To appear **7**, 411–420 (2017)
19. Liu, Y., et al.: A robustly optimized bert pretraining approach arXiv preprint arXiv:1907.11692 (2019)
20. Yih, S.W.-T., Chang, M.-W., He, X., Gao, J.: Semantic parsing via staged query graph generation: Question answering with knowledge base (2015)
21. Talmor, A., Berant, J.: The web as a knowledge-base for answering complex questions arXiv preprint arXiv:1803.06643 (2018)

Separate then Constrain: A Hierarchical Network for End-to-End Triples Extraction

Huizhao Wang, Yao Fu$^{(\boxtimes)}$, Linghui Hu, Weihao Jiang, and Shiliang Pu

Hikvision Research Institute, Hikvision, HangZhou, China
{wanghuizhao,fuyao,hulinghu,jiangweihao5,pushiliang.hri}@hikvision.com

Abstract. In recent years, end-to-end triples extraction based on multi-task learning has achieved promising performance. The existing methods typically use the same sentence representation generated by pretrained language models to address different subtasks. They are either hard to capture the subtask-specific features, or hard to make deep associations among different subtasks. In this paper, we propose a Separate then Constrain Network (SCN) that contains two main layers, i.e., separation layer and constraint layer. Specifically, separation layer first transfers the sentence representation into three different subtask spaces, respectively. Then, constraint layer further refines all sentence representations by simulating the inherent dependencies among three parts of a triple. In addition, to alleviate the negative impact of the error entity prediction on relation classification, we design a simple but effective way, called *Entity-Derive Checker*. On three public datasets, SCN shows significant improvement over existing methods.

Keywords: Triples extraction · Joint model · Hierarchical network

1 Introduction

Extracting triples from the unstructured text is an essential issue for automatic construction of knowledge graph, where the facts contained in the text will be reorganized in the form of ⟨*head, relation, tail*⟩.

Some works [23,25,26] take a pipeline approach, training a named entity recognition (NER) model to extract entities and a relation extraction (RE) model to classify relations between each pair of those entities. Such an approach makes the task easy to deal with, and each model may be more flexible. However, due to the complete separation of NER and RE, it tends to suffer from the error propagation problem, resulting in poor performance. Recently, to mitigate the above issue, some researchers try to bring the two subtasks closer together, and propose the joint methods [12,22,24]. Different from pipelined methods, these methods use a single model to extract entities and relations simultaneously.

Multi-task learning is one popular idea for end-to-end triples extraction [7,13,15]. This family of models essentially builds multiple related subtasks

© The Author(s), under exclusive license to Springer Nature Switzerland AG 2022
J. Gama et al. (Eds.): PAKDD 2022, LNAI 13280, pp. 535–547, 2022.
https://doi.org/10.1007/978-3-031-05933-9_42

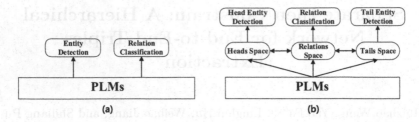

Fig. 1. A Comparison of existing methods and our method. (a): The existing methods. (b): Our method.

and optimizes them together through parameter sharing. In recent years, the paradigm of pretrained language models (PLMs) is thriving. There have also been some works trying to introduce PLMs into this task [7,13]. As shown in Fig. 1(a), these methods simply treat PLMs as the shared module, and different subtasks extract their own required features from the same sentence representation generated PLMs. Thanks to a large amount of linguistic knowledge encoded by PLMs from pretrained corpus, such approaches have achieved the promising results. Unfortunately, they still generally suffer from two major drawbacks.

First, it is hard to capture task-specific features. These methods directly use the same sentence representation to address all different subtasks. However, the representation generated by PLMs is mixed and messy, it contains all information required for each subtask. During the training process, in order to reduce the overall loss, the model tends to capture the common features shared by all subtasks from the sentence embedding. At the extreme, features learned for one subtask may conflict with those for another subtask, which will confuse the training process.

Second, it is hard to make deep associations among different subtasks. It is widely recognized that a well-designed multi-tasks combination architecture is beneficial to integrate the information of different subtasks, and thus one task can provide more complementary information for another task [5]. However, PLMs aim to learn the more universal representation of language by jointly conditioning on both left and right context of each word, which are not specifically designed for triples extraction. In other words, the architecture of PLMs (transformer) is more inclined to capture the co-occurrence pattern between different words in the sentence, rather than the correlation among different subtasks of triples extraction.

To address the problem mentioned above, we propose a Separate then Constraint Network (SCN) for joint entities-relations extraction. As shown in Fig. 1(b), compared with the existing methods, we decompose the whole task into three subtasks. Besides, our model has two special parts, i.e., separation layer and constraint layer. Specifically, the sentence embedding encoded by PLMs is transferred into three feature spaces by separation layer, which ensures each subtask can be addressed in different spaces. Then, considering the inherent dependencies among head entity, relation, and tail entity in a triple, in constraint layer, different subtask spaces interact and help each other.

In addition, we argue that if a span is an entity, then most spans that share several consecutive words are not entities. Thus, we design a simple but effective way, called *Entity-Derivate Checker*, to further improve the quality of extracted triples. Specifically, the credibility of several consecutive words as an entity is re-verified, and the candidate triples containing low-confidence entities will be screened more rigorously. Therefore, the negative impact of the error entity prediction on relation classification can be alleviated.

In summary, SCN can offer several distinct advantages:

- SCN decomposes entities-relations extraction task into three subtasks and encourages to address the different subtasks in different feature spaces by separation layer, which helps to capture the subtask-specific features.
- SCN cleverly utilizes constraint layer to model the complex dependency pattern among different parts of a triple, which helps to make the deep association among different subtasks.
- SCN introduces a simple but effective way, called *Entity-Derivate Checker*, which helps to further improve the quality of extracted triples.
- We evaluate SCN on three benchmark datasets. The experiment results show that SCN can achieve better performance over the state-of-the-art methods.

2 Related Work

These existing joint methods can be grouped into two categories, i.e., structured prediction and multi-task learning. Structured prediction approaches [12,17] cast NER and RE into one unified framework, which can be formulated in various ways. For example, NayakN et al. [17] propose a representation scheme, where all gold triples in a text are reorganized into a target sequence. In addition, multiple specific markers are inserted into the sequence to separate different components in one triple and distinguish different triples. Therefore, the decoder generates one word at a time like machine translation models, thus all triples can be found. Multi-task learning [1,7,15] essentially builds multiple separate models and optimizes them together through parameter sharing. For example, Miva et al. [15] use a sequence tagging model for NER and a tree-based LSTM model for RE. The two models share one LSTM layer for contextualized word representations.

There are two challenges in entities and relations extraction task: (1) Overlapping entities that have nested structures are common. Think of the sentence "Ford's Chicago plant employs 4,000 workers", where both "Chicago" and "Chicago plant" are entities. (2) Overlapping relations which share the same entities in the sentence are common. Consider this sentence: "Jackie was born in Washington, the capital city of America.", it contains two triples, i.e., $\langle Jackie, Birth\_in, Washington \rangle$ and $\langle Washington, Capital\_of, America \rangle$. These two triples are overlapping, because they share the same entity "*Washington*". Meeting the above difficult situations, researchers have made some efforts. For example, Li et al. [12] formulate entities and relations as a multi-turn question answering problem and generates questions by relation-specific templates. It is worth emphasizing that our model proposed in this work allow for the above two challenges.

3   Model

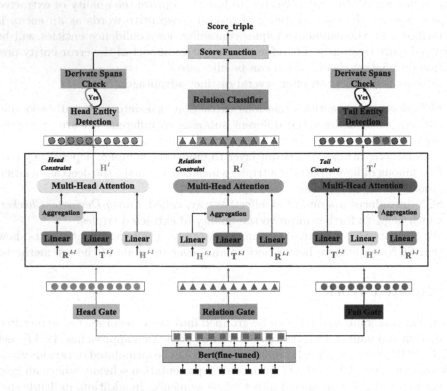

Fig. 2. An illustration of the proposed model SCN.

3.1   Sentence Embedding

Same as the existing methods, given a sentence X with n words $[x_1, x_2, ..., x_n]$, a pre-trained BERT [6] model first is used to map each word x_i into a low-dimensional contextual vector m_i. Thus, the sentence X can be encoded as a 2-D matrix $M \in \mathbb{R}^{n \times d_e}$.

$$M = (m_1, m_2, \cdots m_n), \tag{1}$$

3.2   Separation Layer

As shown in Fig. 2, different from the existing methods, we decompose the entities-relations extraction into three subtasks, i.e., head entity detection, tail entity detection, and relation classification for head-tail entity pairs. Considering that using a single sentence representation for three different tasks may cause feature confusion, where information extracted for one subtask may coincide with those for other subtasks, thus confusing the learning model. We adopt the gate mechanism to map the sentence representation into three different information spaces, belong to heads, relations, and tails, respectively.

Taking the head information space as an example. The gate mechanism consists of two steps. In first step, a trainable vector $\boldsymbol{f}_h \in \mathbb{R}^{d_e}$ is initialized, and then it is concatenated with word embedding $\boldsymbol{m}_i$ together as the input of a full connected layer, finally a vector of weights $\boldsymbol{s}_i \in \mathbb{R}^{d_e}$ can be obtained:

$$\boldsymbol{s}_i = \sigma(\boldsymbol{W}[\boldsymbol{m}_i; \boldsymbol{f}_h] + \boldsymbol{b}), \tag{2}$$

where $\boldsymbol{W} \in \mathbb{R}^{d_e \times 2d_e}$ and $\boldsymbol{b} \in \mathbb{R}^{d_e}$ are the parameters of the fully connected layer. σ denotes the sigmoid function, which makes each dimension value of vector $\boldsymbol{s}_i$ between 0 and 1. In second step, word embedding $\boldsymbol{m}_i$ and $\boldsymbol{f}_h$ are fused together through the corresponding weight vector $\boldsymbol{s}_i$, thus the sentence representation matrix $\boldsymbol{M}$ can be transformed into another matrix within head information space, denoted as $\boldsymbol{H} \in \mathbb{R}^{n \times d_e}$:

$$\boldsymbol{h}_i = \boldsymbol{s}_i \circ \boldsymbol{m}_i + (1 - \boldsymbol{s}_i) \circ \boldsymbol{f}_h, \tag{3}$$

$$\boldsymbol{H} = (\boldsymbol{h}_1, \boldsymbol{h}_2, \cdots \boldsymbol{h}_n), \tag{4}$$

where "∘" refers to the element-wise product between two vectors with the same shape. Suppose $[*]_k$ indexes the kth dimension value of vector $*$, then the larger $[\boldsymbol{s}_i]_k$ means that the kth abstract feature of $\boldsymbol{m}_i$ is more informative in head information space. Vividly, the vector $\boldsymbol{f}_h$ act as a beacon, which can guide the model to pick out the important features from the sentence representation.

Similarly, with the help of the other two trainable vectors, we can obtain the unique representations of one sentence in relation and tail information spaces, denoted as $\boldsymbol{R} \in \mathbb{R}^{n \times d_e}$ and $\boldsymbol{T} \in \mathbb{R}^{n \times d_e}$, respectively. Thus separation layer can be understood as a high level representation of one query "**which features are informative or crucial for three different subtasks, respectively?**".

3.3 Constraint Layer

There is the complex constraint pattern among head, relation, and tail information spaces. For example, given the head "Paris" and the relation "Located_in", the type of tail will be restricted to geographic location. As shown in Fig. 2, the constraint pattern can be grouped into three categories, including head constraint, relation constrain, and tail constraint. In other words, any two information spaces will have a natural restriction on the third information space.

Taking the head constraint as an example, as shown in Fig. 2, matrices $\boldsymbol{R}$ and $\boldsymbol{T}$ are regraded together as queries, matrix $\boldsymbol{H}$ is treated as both keys and values, then the representation of the sentence in head information space is updated, denoted as $\widehat{\boldsymbol{H}}$.

$$\boldsymbol{Q} = \text{mean\_pooling}(\boldsymbol{R}, \boldsymbol{T}), \quad \boldsymbol{K} = \boldsymbol{V} = \boldsymbol{H}, \tag{5}$$

$$\widehat{H} = \text{multihead\_att}(Q, K, V), \tag{6}$$

where $Q \in \mathbb{R}^{n \times d_e}$, $\widehat{H} \in \mathbb{R}^{n \times d_e}$, the function multihead_att() means the computing process of the multi-head attention [21]. Similarly, through relation constraint and tail constraint, we can obtain the new representations $\widehat{R}$ and $\widehat{T}$, respectively.

In attention mechanism, for each query, its result is a weighted sum of the values, where the weight assigned to each value is determined by the relevance of the query with all the keys. Vividly, the above attention acts as the glue, it allows different information spaces to make associations, and thus forming a stronger attraction among head, relation, and tail in a gold triple.

To promote a deeper interaction among different information spaces, we imitate Transformer [21] to stack the constraint network N layers. The output matrices of the previous layer are aggregated to generate those of the next layer. In addition, the position-wise feedforward neural network (FFNN), the residual connection, and layer normalization are also used to compute the output of the l-th layer.

$$H^l = \text{LayerNorm}(H^{l-1} + \text{FFNN}(\widehat{H^l})), \tag{7}$$

where H^{l-1} is the output of the $(l-1)^{th}$ constraint layer, $\widehat{H^l}$ is got by Eq. 5 and Eq. 6. Therefore, we can obtain the unique representations of the sentence in three information spaces, denoted as H^l, R^l and $T^l \in \mathbb{R}^{n \times d_e}$, respectively. Thus constraint layer can be understood as a high level representation of another query "**How do different subtasks help and influence each other?**".

3.4   Subtask Modules

Head Entity Detection. Let $\mathcal{E}_h$ denote a pre-defined head entity types set. Given a sentence $X = [x_1, x_2, ..., x_n]$, we first collect all spans length up to L, namely $S = \{[x_i, ..., x_j] : 1 \leq i \leq j \leq n\}$. Then, for each span $s \in S$, its representation is fed into a softmax classifier to predict the probability distribution of head entity types $e_h \in \mathcal{E}_h \cap \{none\} : P_h(e_h|s)$, where *none* represents the corresponding span is not a head entity. Finally, we can get a head entity set, denoted $E_{head} = \{(s, e_h) : s \in S, e_h \in \mathcal{E}_h\}$. Especially, our approach follows the popular way to construct the span representation, more details refer to SpERT [7]. In addition, it is worth emphasizing that our approach uses the matrix H^l obtained by constraint layer as the input to construct the span representation, while other methods use the matrix M generated by PLMs as the input.

Tail Entity Detection. Let $\mathcal{E}_t$ denote a pre-defined tail entity types set. Like head entity detection, taking the matrix T^l as the input to construct span representation, a tail entity set can be got, denoted $E_{tail} = \{(s, e_t) : s \in S, e_t \in \mathcal{E}_t\}$.

Relation Classification. Let $\mathcal{R}$ be a set of pre-defined relation types. Following the existing work [7], for each entity pair (h_e, t_e), drawn from $E_{head} \times E_{tail}$, a vector representation can be constructed from the matrix R^l, then a sigmoid classifier takes it as input to predict the probability distribution of the relation type $r \in \mathcal{R} : P_r (r|h_e, t_e)$. The high scores in r indicates that the corresponding relations may be holden between h_e and t_e. Given a threshold α, any triple with a score $\geq \alpha$ is considered correct. As a result, a triple set can be got, denoted $T = \{(h, r, t) : h \in E_{head}, t \in E_{tail}, r \in \mathcal{R}\}$.

3.5   Training and Inference

Our model is trained jointly based on multi-task learning strategy, where the losses of different subtasks are added together. Specifically, the head and tail entity detection adopt the cross-entropy loss, and the relation classification uses the binary cross-entropy loss.

$$\mathcal{L}_h = - \sum_{s_i \in S} logP_h (e_h^*|s_i), \quad \mathcal{L}_t = - \sum_{s_j \in S} logP_t (e_t^*|s_j), \tag{8}$$

$$\mathcal{L}_r = - \sum_{s_i, s_j \in S, s_i \neq s_j} logP_r (r^*|s_i, s_j), \quad \mathcal{L} = \lambda_h \mathcal{L}_h + \lambda_t \mathcal{L}_t + \lambda_r \mathcal{L}_r, \tag{9}$$

where e_h^* and e_t^* represent the gold head and tail entity type of s_i and s_j, respectively. r^* represents the gold relation type of span pair (s_i, s_j). λ_h, λ_t and λ_r are three learnable weights, more details refer to [9].

Entity-Derivate Checker. In inference, each span is processed independently, and some spans that are not entities may be mixed in set E_{head} (or E_{tail}), which will have a negative impact on the quality of the extracted triples. Thus, we introduce the concept derived samples and design a simple but effective rule, called *Entity-Derivate Checker*, to verify the credibility of each extracted entity. And then, those entities with low confidence will be marked and the involved candidate triples will be screened more rigorously.

Table 1. An example of derived samples

The United States President *Joseph Robinette Biden will visit the Apple Inc founded by Steven Paul Jobs.*		
Derived Samples	*Joseph Robinette*	truncation
	President Joseph Robinette	offset
	President Joseph Robinette Biden will	expansion
	...	

Derived samples are slightly changed from the original sample, and they share several consecutive words. As shown in Table 1, for entity "*Joseph Robinette Biden*", its derived samples can be generated through offset, truncation or expansion. Obviously, a derived sample is essentially a span, and it has been identified whether it is a head (or tail) entity. Obviously, some potential connections exist between a gold entity and its derived samples. Simply put, if a span is an entity, then most of its derived samples cannot be entities. Therefore, we can compute a confidence score for each extracted entity by evaluating the prediction results of the corresponding derived samples.

Suppose span s_i has been recognized as an entity, and D^{s_i} is the corresponding derived sample set, in which t spans are also identified as entities. Thus, the confidence score that the span s_i is an entity can be computed by $\frac{|D^{s_i}|-t}{|D^{s_i}|}$. The higher the score, the higher the credibility that span s_i can be detected as an entity. Given a threshold η, the entity with a score $\leq\eta$ will be marked as the dispute entity and added to set $\widehat{E_{head}}$ (or $\widehat{E_{tail}}$).

In order to further ensure the quality of the extracted triples, the entity pair containing the dispute entity will be more rigorously screened. Specifically, if $s_i \in \widehat{E_{head}}$ or $s_j \in \widehat{E_{tail}}$, then a larger threshold $\beta(>\alpha)$ is adopted to determine the relations between the two entities.

4   Experiments

4.1   Datasets and Implementation

Our approach is evaluated on three public datasets CoNLL04 [19], ADE [8], and SciERC [13]. These three datasets not only list the triples included in each sentence, but also provide the entity type. Table 2 shows the data statistics of each dataset. Specifically, CoNLL04 and SciERC datasets are collected from news articles and AI paper abstracts, respectively. The ADE dataset is constructed from medical reports that describe the adverse effects arising from drug use. We follow previous work and use the same preprocessing procedure and splits for all datasets, more details refer to [7].

Table 2. The statistics of the datasets.

Data	#Ent_Types	#Rel_Types	#Sentences	
			Train	Test
CoNLL04	4	5	1153	288
SciERC	6	7	2136	551
ADE	2	1	4272(10-fold)	

F1 measure is used to evaluate the performance of SCN, and the average score of 5 runs is reported. For NER, the detection results of head and tail entities are merged together, and a predicted entity is considered correct if its span boundaries and entity type all match the ground truth. For RE, there are two evaluation metrics [25]: (1) *Rel*, a predicted relation is considered as correct if the span boundaries of head and tail entities are correct, and the relation type

Table 3. Performance comparison on the three datasets. SCN outperforms the state-of-the-art in both entities and relations extraction. (*micro* : †, *macro* : ‡, *not stated* : ∗).

Data	Model	Ent	Rel	Rel+
CoNLL04	Global Optimization [23]†	85.60	-	67.80
	Multi-turn QA [12]†	87.80	-	68.90
	Table-filing [16]∗	80.70	-	61.00
	Hierarchical Attention [4]∗	86.51	-	62.32
	Multi-head + AT [1]‡	83.61	-	61.95
	Multi-head [2]‡	83.09	-	62.04
	Relation-Metric [20]‡	84.57	-	62.68
	Biaffine Attention [18]‡	86.20	-	64.40
	SpERT [7]†	88.94	-	71.47
	SpERT [7]‡	86.25	-	72.87
	SCN†	**89.66**	**73.81**	**73.71**
	SCN‡	**87.18**	**74.72**	**74.63**
ADE	GNN + Global features [11]‡	79.50	-	63.40
	BiLSTM + SDP [10]‡	84.60	-	71.40
	Multi-head [2]‡	86.40	-	74.58
	Multi-head + AT [1]‡	86.73	-	75.52
	Relation-Metric [20]‡	87.11	-	77.29
	SpERT [7]‡	89.28	-	78.84
	SCN‡	**89.87**	**79.80**	**79.80**
SciERC	SciIE [13]†	64.20	39.30	-
	DyGIE [14]†	65.20	41.60	-
	DyGIE++ [22]†	67.50	48.40	-
	PRUE [25]†	68.90	50.10	36.80
	SCN†	**69.91**	**50.72**	**39.21**

is correct; (2) *Rel+*: in addition to what is required in *Rel*, the types of head and tail also must be correct.

We compare SCN with several state-of-the-art methods, among which **PRUE** [25] and **SpERT** [7] are the most promising methods, and they all rely on PLMs as the core. In addition, to evaluate the impact of each part of our model, we also compare SCN with its several variants. Specifically, SCN_ner_re follows the conventional way that decomposing triples extraction into both NER and RE, then in constraint layer, two different subtask spaces interact with each other. SCN_separate and SCN_constraint remove separation layer and constraint layer from SCN, respectively.

We specifically tune our model on CoNLL04, and the same hyperparameters are used for the other two datasets. The pre-trained model *bert-base-cased* [6] is used as the default sentence encoder. For a fair comparison, *scibert-scivocab-cased* [3] (fine-tuned on a large corpus of scientific papers) is adopted on SciERC. In addition, we only consider spans up to $L = 10$ words. Compared with the boundaries of the original sample, the left and right sides of the derived samples are shifted by up to 3 words. We stack the constraint network six layers, i.e., $N = 6$. According to the gird search in $\{0.4, 0.5, 0.6, 0.7\}$, $\{0.6, 0.65, 0.7, 0.75, 0.8\}$, and $\{0.85, 0.9, 0.95\}$, the special thresholds α, β and η are set to 0.5, 0.75, and 0.9, respectively.

4.2 Main Results

Table 3 reports the results of our model against other baseline methods on all datasets. It can be seen that SCN consistently outperforms all the baselines in terms of F1-score. Especially, our NER performance is increased by 0.93, 0.59 and 1.01 absolute F1 points over the previous best methods in three datasets respectively. Besides, we also observe significant performance increases in the RE task (Rel+), which is 1.76, 0.96, and 2.41 absolute F1 points, respectively. This clearly demonstrates the superiority of our method.

4.3 Ablation Study

We design several additional experiments to understand the effectiveness of several settings in our approach. Table 4 lists the experiment results.

Table 4. The ablation experiment results under different settings

	CoNLL04			ADE			SciERC		
	Ent	Rel	Rel+	Ent	Rel	Rel+	Ent	Rel	Rel+
SCN_ner_re	86.27	73.42	73.38	89.30	78.86	78.86	68.92	50.01	38.23
SCN_separate	86.01	73.29	73.21	89.05	78.00	78.00	67.91	49.11	37.50
SCN_constraint	86.18	73.35	73.30	89.19	78.50	78.50	68.49	49.86	37.99
SCN	**87.18**	**74.72**	**74.63**	**89.87**	**79.80**	**79.80**	**69.91**	**50.72**	**39.21**

Instead of the conventional way that decomposing triples extraction into both NER and RE, our approach further disassembles NER into two more detailed subtasks, i.e., head entity detection and tail entity detection. To investigate the effect of the above modification, we compare SCN with the variant model SCN_ner_re. From the Table , we can see that SCN can achieve better performance. This indicates that our strategy that designing a corresponding subtask for each part of a triple might be more suitable for triples extraction.

Separation layer and Constraint layer are two main parts of our network. In order to investigate their effect, we compare SCN with its variants, i.e.,

SCN_separate and SCN_constraint. We can observe that neither of the two variants can achieve prediction results comparable to SCN. In some metrics, they even perform worse than baselines. This indicates that Separation layer and Constraint layer are indispensable, and only when the two work together, our model can achieve the best performance. From the perspective of multi-task learning, the two parts help to obtain a balance of commonality (features shared by all subtasks) and individuality (task-specific features).

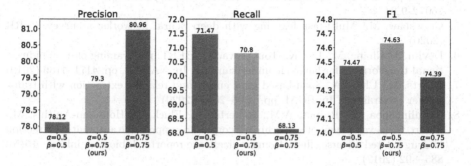

Fig. 3. Performance comparison on CoNLL04 under different thresholds

In addition, we also investigate the effect of *Entity-Derivate Checker*. Figure 3 shows the performance of SCN under different relation thresholds, i.e., α and β. Obviously, our strategy (red bar) can get a higher Precision, but it does not leads to a drastic reduction in Recall, finally F1 is the highest. The above result demonstrates that *Entity-Derivate Checker* can alleviate the negative impact of the error entity prediction on relation classification, and then helps our approach achieve better overall performance.

5   Conclusion

In this paper, we decompose triples extraction into three subtasks, i.e., head entity detection, tail entity detection and relation classification. Correspondingly, a novel hierarchical network (SCN) is proposed, where different subtasks are addressed in respective feature spaces. In addition, we introduce the constraint network to aggregate the information of different subtasks, which is consistent with capturing the inherent constraint pattern among heads, relations, and tails. Finally, we design a simple but effective way to further ensure the quality of triples. The experimental results on three public datasets demonstrate the efficacy of our approach.

References

1. Bekoulis, G., Deleu, J., Demeester, T., Develder, C.: Adversarial training for multi-context joint entity and relation extraction. In: EMNLP, pp. 2830–2836 (2018)

2. Bekoulis, G., Deleu, J., Demeester, T., Develder, C.: Joint entity recognition and relation extraction as a multi-head selection problem. Expert Syst. Appl. **114**, 34–45 (2018)
3. Beltagy, I., Lo, K., Cohan, A.: Scibert: a pretrained language model for scientific text. In: EMNLP-IJCNLP, pp. 3613–3618 (2019)
4. Chi, R., Wu, B., Hu, L., Zhang, Y.: Enhancing joint entity and relation extraction with language modeling and hierarchical attention. In: Shao, J., Yiu, M.L., Toyoda, M., Zhang, D., Wang, W., Cui, B. (eds.) APWeb-WAIM 2019. LNCS, vol. 11641, pp. 314–328. Springer, Cham (2019). https://doi.org/10.1007/978-3-030-26072-9_24
5. Crawshaw, M.: Multi-task learning with deep neural networks: a survey. CoRR (2020)
6. Devlin, J., Chang, M., Lee, K., Toutanova, K.: BERT: pre-training of deep bidirectional transformers for language understanding. In: NAACL, pp. 4171–4186 (2019)
7. Eberts, M., Ulges, A.: Span-based joint entity and relation extraction with transformer pre-training. In: ECAI, pp. 2006–2013 (2020)
8. Gurulingappa, H., Rajput, A.M., Roberts, A., Fluck, J., Hofmann-Apitius, M., Toldo, L.: Development of a benchmark corpus to support the automatic extraction of drug-related adverse effects from medical case reports. J. biomed. inform. **45**(5), 885–892 (2012)
9. Kendall, A., Gal, Y., Cipolla, R.: Multi-task learning using uncertainty to weigh losses for scene geometry and semantics. In: CVPR, pp. 7482–7491. Computer Vision Foundation/IEEE Computer Society (2018)
10. Li, F., Zhang, M., Fu, G., Ji, D.: A neural joint model for entity and relation extraction from biomedical text. BMC **18**, 198:1–198:11 (2017)
11. Li, F., Zhang, Y., Zhang, M., Ji, D.: Joint models for extracting adverse drug events from biomedical text. In: IJCAI, pp. 2838–2844 (2016)
12. Li, X., et al.: Entity-relation extraction as multi-turn question answering. In: ACL, pp. 1340–1350 (2019)
13. Luan, Y., He, L., Ostendorf, M., Hajishirzi, H.: Multi-task identification of entities, relations, and coreference for scientific knowledge graph construction. In: EMNLP, pp. 3219–3232 (2018)
14. Luan, Y., Wadden, D., He, L., Shah, A., Ostendorf, M., Hajishirzi, H.: A general framework for information extraction using dynamic span graphs. In: NAACL-HLT, pp. 3036–3046 (2019)
15. Miwa, M., Bansal, M.: End-to-end relation extraction using lstms on sequences and tree structures. In: ACL (2016)
16. Miwa, M., Sasaki, Y.: Modeling joint entity and relation extraction with table representation. In: EMNLP, pp. 1858–1869 (2014)
17. Nayak, T., Ng, H.T.: Effective modeling of encoder-decoder architecture for joint entity and relation extraction. In: AAAI, pp. 8528–8535 (2020)
18. Nguyen, D.Q., Verspoor, K.: End-to-end neural relation extraction using deep biaffine attention. In: Azzopardi, L., Stein, B., Fuhr, N., Mayr, P., Hauff, C., Hiemstra, D. (eds.) ECIR 2019. LNCS, vol. 11437, pp. 729–738. Springer, Cham (2019). https://doi.org/10.1007/978-3-030-15712-8_47
19. Roth, D., Yih, W.: A linear programming formulation for global inference in natural language tasks. In: CoNLL 2004, pp. 1–8 (2004)
20. Tran, T., Kavuluru, R.: Neural metric learning for fast end-to-end relation extraction. CoRR abs/1905.07458 (2019)
21. Vaswani, A., et al.: Attention is all you need. In: NIPS, pp. 5998–6008 (2017)

22. Wadden, D., Wennberg, U., Luan, Y., Hajishirzi, H.: Entity, relation, and event extraction with contextualized span representations. In: EMNLP-IJCNLP, pp. 5783–5788 (2019)
23. Zhang, M., Zhang, Y., Fu, G.: End-to-end neural relation extraction with global optimization. In: EMNLP. pp. 1730–1740 (2017)
24. Zheng, S., Wang, F., Bao, H., Hao, Y., Zhou, P., Xu, B.: Joint extraction of entities and relations based on a novel tagging scheme. In: ACL, pp. 1227–1236 (2017)
25. Zhong, Z., Chen, D.: A frustratingly easy approach for entity and relation extraction. In: NAACL-HLT, pp. 50–61 (2021)
26. Zhou, G., Su, J., Zhang, J., Zhang, M.: Exploring various knowledge in relation extraction. In: ACL, pp. 427–434 (2005)

Sparse Imbalanced Drug-Target Interaction Prediction via Heterogeneous Data Augmentation and Node Similarity

Runze Wang[1], Zehua Zhang[1(✉)], Yueqin Zhang[1], Zhongyuan Jiang[2], Shilin Sun[1], and Chenwei Zhang[3]

[1] Taiyuan University of Technology, Taiyuan 030024, China
wangrunze0317@link.tyut.edu.cn, {zhangzehua,zhangyueqin}@tyut.edu.cn
[2] Xidian University, Xian 710068, China
zyjiang@xidian.edu.cn
[3] Amazon, Seattle, WA 98109, USA
cwzhang@amazon.com

Abstract. Drug-Target Interaction (DTI) prediction usually devotes to accurately identify the potential binding targets on proteins so as to guide the drug development. However, the sparse imbalance of known drug-target pairs remains a challenge for high-quality representation learning of drugs and targets, interfering with accurate prediction. The labeled drug-target pairs are far less than the missed since the obtained DTIs are recorded with pathogenic proteins and sophisticated bio-experiments. Therefore, we propose a deep learning paradigm via **H**eterogeneous graph data **A**ugmentation and node **S**imilarity (**HAS**) to solve the sparse imbalanced problem on drug-target interaction prediction. Heterogeneous graph data augmentation is devised to generate multi-view augmented graphs through a heterogeneous neighbors sampling strategy. Then the consistency across different graph structures is captured using graph contrastive optimization. Node similarity is calculated on the heterogeneous entity association matrices, aiming to integrate similarity information and heterogeneous attribute gain for drug-target interaction prediction. Extensive experiments show that HAS offers superior performance in sparse imbalanced scenarios compared state-of-the-art methods. Ablation studies prove the effectiveness of heterogeneous graph data augmentation and node similarity.

Keywords: Sparse imbalanced DTI prediction · Heterogeneous graph data augmentation · Graph contrastive optimization · Node similarity

1 Introduction

Drug-Target Interaction prediction plays an essential role in the drug discovery process [1]. And it often leads to the next stages of pharmacological in vitro experiments [2]. The growing clinical demands pose the challenges to drug

J. Gama et al. (Eds.): PAKDD 2022, LNAI 13280, pp. 548–561, 2022.
https://doi.org/10.1007/978-3-031-05933-9_43

screening based on traditional experiments. The emergence of machine learning has brought a new boom in computer-aided drug design which reduces the time-consuming and expensive bio-experiments [3]. Some computational approaches for DTI prediction were proposed in supervised learning view, such as applying deep learning techniques to extract chemical features from known structure data [4–6] or analyze the potential correlation among labeled drug-target pairs [7,8]. Several studies attempted to perform the semi-supervised tasks with known and unknown drug-target pairs, including modeling the tripartite relations of drug-target-disease [9], constructing the heterogeneous information networks and leveraging the diverse biological entity properties to alleviate the negative impact of missed DTI labels [10,11]. Although the efforts have been made on respective tasks, the supervised learning methods rely on the both chemical structure data and labels, and the semi-supervised learning methods are based on the hypothesis of balanced positive and negative samples (i.e., the known drug-target pairs are treated as positive samples, while the unknown interacting pairs are regarded as negative samples), neglecting the realistic issue that positive samples are far less than negative samples. Drug discovery usually builds on the pathogenic proteins [12]. Pharmaceutical researchers screen the candidate drugs that change the proteins bioactivity. Only if the changes of proteins bioactivity meet the clinical and research needs, can DTIs be recorded. Large number of DTIs are missed and the obtainable DTI labels are limited by the amount of discovered pathogenic proteins. Furthermore, the real-world drug-target interactions far exceed the recorded, causing the observed DTIs are extra sparse compared with the whole drug-target pairs space. The sparse imbalanced interacting drug-target pairs are insufficient to learn high-quality feature representations for drugs and targets which leads to inaccurate prediction.

To sum up, we propose a deep learning paradigm by integrating **H**eterogeneous Data **A**ugmentation and Node **S**imilarity for sparse imbalanced DTI prediction, named as **HAS**. Especially, we present a heterogeneous graph data augmentation module to generate multi-view augmented graphs on constructed heterogeneous graph involving the node types of drug and target. Differentiate from the recent studies of contrastive learning on homogeneous graph [13,14], a heterogeneous contrastive learning strategy is designed to capture the agreement between different graph structures. Based on the general assumption that drugs with higher similarity are more likely to have common linked target [3], node similarity is calculated on the heterogeneous entity associated matrices. So far, the intrinsic topological structure information and node similarity information from different attribute spaces are acquired to supplement the sparse supervised signal. The main works are summarized as follows:

- We formalize the sparse imbalance problem on drug-target interaction prediction, and present a novel deep learning method via heterogeneous graph data augmentation and node similarity to solve.
- Heterogeneous graph data augmentation is designed to capture intrinsic and universal structure patterns between multi-view augmented graphs. The

similarity information and heterogeneous attribute information are incorporated to strengthen the features of drugs and targets.
- Empirical studies on the real-world datasets demonstrate that HAS has significant improvement in sparse imbalanced DTIs scenario compared with the state-of-the-art methods.

2 Related Work

DTI prediction has attracted much attention in recent years. Numerous studies are dedicated to reducing the search space of drug candidates and facilitating drug discovery process [15]. And the related methods can be mainly divided into three aspects: bio-feature extraction, pairwise similarity discovery and bioinformatics network mining.

Bio-feature extraction takes chemical structure data as the input of deep learning framework to extract the main features of drugs and targets respectively, and finally fusing the features of both to predict DTI. For example, the works DeepDTA [4] designs deep learning models to predict the binding affinity (one type of DTI) using sequential data of drugs and proteins. DeepConv-DTI [5] ensembles local residue patterns of proteins. Graph neural network (GNN) is reported as a powerful tool in graph embedding tasks [16], a computational approaches named GraphDTA [6] is proposed to capture molecular topological features of drugs with GNN to improve the prediction performance. Such methods inevitably rely on known drug-target pairs and structure data.

Pairwise similarity discovery mainly measures the similarity between multiple drug-target pairs, which is used as the interaction information. MATT_DTI [17] introduce multi-head attention mechanism to obtain the similarity information of different drug-target pairs. Chen et al. [8] present to utilize a transformer-based decoder that extract interaction features substructure pairs of drugs and proteins. These methods have great effort by incorporating the similarity information into interaction prediction, but the complex network relations are unconsidered, e.g. drug-drug.

Bioinformatics network mining aims at using graph representation learning methods to predict drug-target interactions on the heterogeneous network. The work NeoDTI [10] constructs heterogeneous network with drug, target, disease, etc. and predict drug-target interactions in graph reconstruction way. Multi-DTI [11] maps all the heterogeneous biological entities to common feature space, so the space distances between nodes are regarded as prediction scores of DTI. EEG_DTI [18] applies graph neural network to learn embedding vectors of drugs and targets for DTI prediction. However, these studies make the number of positive and negative samples approximate the balanced, ignoring the realistic problem that the known drug-target interactions are sparse in the whole drug-target pairs space. HAS focuses on the sparse imbalanced DTI prediction that belongs to an urgent real-world issue. Heterogeneous graph data augmentation and node similarity are proposed from topology-level and node-level to alleviate the negative impact brought by sparse known drug-target pairs.

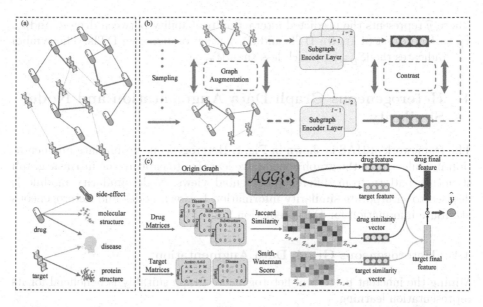

Fig. 1. Illustration of the proposed HAS. (a) The upper half is heterogeneous information network including drugs and targets. The bolded red edges represent the drug-target interactions, the blue and yellow edges show drug-drug interactions and target-target interactions, respectively. The lower half depicts heterogeneous relations about drug and target. (b) Heterogeneous graph data augmentation module first generates multi-view augmented graphs through heterogeneous neighbors sampling, then encodes subgraph structure and learns the nodes features, finally maximizes the agreement of same node from different views via contrastive learning. (c) Node similarity information is calculated on the heterogeneous associated matrices. The learned features of drugs and targets on the origin graph are learned using the aggregate function. Next two types of features are fused as final feature representation to predict the DTI probability. (color figure online)

3   Sparse Imbalanced DTI Prediction

The final goal is to predict interactions between drugs and targets, so heterogeneous information network with only drug and target nodes is defined as $\mathcal{HG} = \{\mathcal{D}, \mathcal{T}, \mathcal{E}, \mathcal{R}\}$, where $\mathcal{D}$ and $\mathcal{T}$ denote sets of drugs and targets. $\mathcal{E}$ and $\mathcal{R}$ are sets of edges and edge types, which are associated with relational matrices, drug-drug matrix $\mathcal{M}_{\mathcal{D}\_\mathcal{D}}$, drug-target matrix $\mathcal{M}_{\mathcal{D}\_\mathcal{T}}$, target-target matrix $\mathcal{M}_{\mathcal{T}\_\mathcal{T}}$. For matrix element $m_{(i,j)} \in \{0,1\}$, if $m_{(i,j)} = 1$, existing $e_{(i,j)} \in \mathcal{E}$.

Given the heterogeneous graph $\mathcal{HG}$, the known edges between drugs and targets are far less than the unknown drug-target interaction edges since only DTIs meeting the clinical needs will be recorded. The final DTI prediction can be cast as an edge classification task via learning the prediction function $\mathcal{F}\{\cdot\}$ under the sparse imbalanced condition, which is formulated as follows:

$$\hat{y} = \mathcal{F}\{(d_i, t_j), \mathcal{HG} | d_i \in \mathcal{D}, t_j \in \mathcal{T}, \mathcal{E}_{\mathcal{D}\_\mathcal{T}}^{+} << \mathcal{E}_{\mathcal{D}\_\mathcal{T}}^{-}\} \tag{1}$$

where $\hat{y}$ represents the predicted interaction probability between drug d_i and t_j, $\mathcal{E}_{\mathcal{D\_T}}^+$ denotes set of known DTIs and $\mathcal{E}_{\mathcal{D\_T}}^-$ is set of unknown DTIs. The symbol '$\ll$' indicates the $|\mathcal{E}_{\mathcal{D\_T}}^+|$ is much less than $|\mathcal{E}_{\mathcal{D\_T}}^-|$.

4   Heterogeneous Graph Data Augmentation and Node Similarity

The framework of HAS is shown in Fig. 1. Heterogeneous graph data augmentation module adopts the graph contrastive learning to capture the intrinsic graph structure pattern from different augmented views. Node similarity module is devised to incorporate similarity information between nodes and heterogeneous attribute information for DTI prediction.

4.1   Heterogeneous Graph Data Augmentation

Mining the inherent pattern of heterogeneous graph suitably is beneficial for its representation learning.

Multi-view Graph Augmentation. Different from the recent works of graph contrastive learning that build generators on homogeneous graph, HAS focuses on generating augmented graphs on the heterogeneous graph including various node types. Besides, the imbalanced distribution of multi-typed edges causes the number of neighboring nodes varies from each node. Overall, the multi-view generator is designed through a heterogeneous neighbors sampling strategy, the sampled drugs and targets are derived by random walk with restart. This way of augmented graph generation can avoid the imbalanced problem that edges with heterogeneous types and establishing message propagation with high-order nodes as far as possible. The implementation process takes drug nodes as example:

1. Taking current drug node d_0 as starting point of random walk with restart, the iterative walk is performed to its neighboring node which is either drug or target, and the next step could be itself with probability π. The walk will stop until set Γ_{d_0} about d_0 successfully collects nodes with fixed number.
2. According to the node set Γ_{d_0}, walking path and their related edges on the original graph, a random heterogeneous subgraph $\mathcal{G}_{d_0}$ is generated. $\mathcal{G}_{d_0}$ is regarded as an augmented version in a view with the core node d_0. Repeat the above process twice to obtain two augmented graphs $\mathcal{G}_{d_0}^{(1)}$, $\mathcal{G}_{d_0}^{(2)}$.

Similarly, if the target node t_0 is used as the 'hub' node, the generated augmented graphs are denotes as $\mathcal{G}_{t_0}^{(1)}$, $\mathcal{G}_{t_0}^{(2)}$.

Heterogeneous Subgraph Encoding. The researches of heterogeneous graph learning [19,20] analyze the inherent heterogeneity that the features of different types of nodes may fall in different feature space. In this paper, we consider that drugs and targets are heterogeneous on the sampled subgraphs, a heterogeneous graph neural network is adopted to aggregate the neighboring attribute with different types. Since different types of nodes contribute differently to its embedding, and so do the different nodes with the same type, we employ attention mechanism in GNN layers to weight the aggregated neighbors messages for each node. First the embeddings of nodes are initialized. Then for each drug node d_i, the attention coefficients are calculated with its neighboring drug nodes:

$$\alpha_{(d_i,d_j)}^{(l)} = \frac{\exp\{\text{LeakyReLU}(\mathbf{a}_{\mathcal{D}}^{(l)^T}[\mathbf{h}_i^{(l)} \oplus \mathbf{h}_j^{(l)}])\}}{\sum_{k \in \mathcal{N}_{\mathcal{D}}(i)} \exp\{\text{LeakyReLU}(\mathbf{a}_{\mathcal{D}}^{(l)^T}[\mathbf{h}_i^{(l)} \oplus \mathbf{h}_k^{(l)}])\}} \tag{2}$$

where $\alpha_{(d_i,d_j)}^{(l)}$ is the attention coefficient between drug d_i and it neighboring drug d_j, l denotes the current layer of heterogeneous graph neural network, LeakyReLU$(\cdot)$ is the nonlinear activation function, $\mathbf{h}_i^{(l)}$ and $\mathbf{h}_j^{(l)}$ represent the hidden feature vectors of d_i and d_j at $l-$ th layer. $\mathbf{a}_{\mathcal{D}}^T$ is transposed attention vector between drug nodes, $\oplus$ defines the concatenation of two vectors and $\mathcal{N}_{\mathcal{D}}(i)$ defines the set for d_i with drug type neighbors. If the neighbors are target nodes, the heterogeneous attention scores are computed as follows:

$$\beta_{(d_i,t_j)}^{(l)} = \frac{\exp\{\text{LeakyReLU}(c_{\mathcal{D}}^{(l)^T}[\mathbf{h}_i^{(l)} \oplus \mathcal{W}_{\mathcal{D}}^{(l)} \mathbf{p}_j^{(l)}])\}}{\sum_{k \in \mathcal{N}_{\mathcal{T}}(i)} \exp\{\text{LeakyReLU}(c_{\mathcal{D}}^{(l)^T}[\mathbf{h}_i^{(l)} \oplus \mathcal{W}_{\mathcal{D}}^{(l)} \mathbf{p}_k^{(l)}])\}} \tag{3}$$

where $\beta_{(d_i,t_j)}^{(l)}$ defines the computed heterogeneous attention score between drug d_i and t_j at $l-$ th layer. $c_{\mathcal{D}}^T$ and $\mathcal{N}_{\mathcal{T}}(i)$ denote the transposed attention vector and the set for d_i with target type neighbors, respectively. $\mathbf{p}_j^{(l)}$ and $\mathcal{W}_{\mathcal{D}}^{(l)}$ are the learned hidden feature of target and the feature mapping matrix from target space to drug space. Finally, the feature aggregation in a grouping way is performed to update the 'hub' drug node feature according calculated homogeneous and heterogeneous attention coefficients:

$$\mathbf{h}_i^{(l+1)} = \text{ReLU}((\sum_{d_j \in \mathcal{N}_{\mathcal{D}}(i)} \alpha_{(d_i,d_j)}^{(l)} \mathbf{h}_j^{(l)} + \sum_{t_j \in \mathcal{N}_{\mathcal{T}}(i)} \beta_{(d_i,t_j)}^{(l)} \mathcal{W}_{\mathcal{D}}^{(l)} \mathbf{p}_j^{(l)}) \mathcal{W}^{(l)} + \mathbf{b}^{(l)}) \tag{4}$$

where ReLU$(\cdot)$ is a nonlinear activation function, $\mathcal{W}^{(l)}$ and $\mathbf{b}^{(l)}$ define the learnable feature transformation matrix and bias vector. After L-layer graph neural network, the drug feature representation $\mathbf{h}_i^{(1)}$ of d_i is obtained in subgraph $\mathcal{G}_{d_0}^{(1)}$, as well as $\mathbf{h}_i^{(2)}$ in $\mathcal{G}_{d_0}^{(2)}$. Analogously, the learned representation of any target node t_i is calculated in two views of data augmentation as $\mathbf{p}_i^{(1)}$, $\mathbf{p}_i^{(2)}$.

Graph Contrastive Optimization. By this, the nodes features of drugs and targets are learned containing the multi-view subgraph structure information. The recent studies use contrastive learning [21] to optimize the self-supervised learning task that maximize the agreement between positive samples. In order to discover the universal graph topological feature between two augmented graphs, we devise the optimizer in a graph contrastive learning manner. The features under different views of the same node are defined as the positive pairs and the features under different views of different nodes are defined as the negative pairs. For example, $(\mathbf{h}_i^{(1)}, \mathbf{h}_i^{(2)})$ of drug d_i is regarded as positive pair, $(\mathbf{h}_i^{(1)}, \mathbf{h}_j^{(2)})$ of drug d_i and d_j is regarded as negative pair. Then, the contrastive loss $\mathcal{L}_\mathcal{D}$ related to drug is calculated as follows:

$$\mathcal{L}_\mathcal{D} = \sum_{d_i \in \mathcal{D}} - \log \frac{\exp(\mathrm{sim}(\mathbf{h}_i^{(1)}, \mathbf{h}_i^{(2)})/\tau)}{\sum_{d_j \in \mathcal{D}, i \neq j} \exp(\mathrm{sim}(\mathbf{h}_i^{(1)}, \mathbf{h}_j^{(2)})/\tau)} \tag{5}$$

where $\mathrm{sim}(\cdot)$ is the cosine similarity function and τ defines the temperature parameter. Similarly, the contrastive loss about target can be obtained as $\mathcal{L}_\mathcal{T}$.

4.2 Node Similarity

Based on the general assumption that drugs with high similarity may share common interactions with the same target, we incorporate drug-drug, target-target similarity information to enrich the feature of drugs and targets. The direct associated biological entities can be viewed as heterogeneous attributes, so we calculate the node similarity on the associated matrices. The chemical structures of drugs are comprised of SMILES strings, a cheminformatics tool named RDKit is used to convert SMILES strings to morgan fingerprints that are expressed as binary vectors. Each entry demonstrates the presence or absence of certain chemical substructure. Then the substructure feature matrix $\mathcal{M}_{D\_sub}$ of all drugs can be acquired. Given the biological association matrices $\mathcal{M}_{D\_sid}$ for drug-side effect, $\mathcal{M}_{D\_dis}$ for drug-disease, $\mathcal{M}_{T\_dis}$ for target-disease, and drug substructure matrix $\mathcal{M}_{D\_sub}$, the principal components analysis algorithm is employed to tackle the negligible vector sparsity and high-dimensional issues. Next the similarities of drug-drug or target-target are calculated by the Jaccard similarity measure. After that, similarity matrices from different heterogeneous attribute spaces can be obtained: $\mathcal{Z}_{D\_sid}$ in side effect space, $\mathcal{Z}_{D\_dis}$ in disease space, $\mathcal{Z}_{D\_sub}$ in substructure space and a target similarity matrix $\mathcal{Z}_{T\_dis}$ in disease space. The protein structure consists of amino acids sequence. Considering the co-occurrence of local functional fragments in different protein, we choose the Smith-Waterman score measure as the similarity calculation means between proteins. The protein substructure similarity matrix is denoted as $\mathcal{Z}_{T\_sub}$. Finally, the respective similarity matrices are fused:

$$\mathcal{Z}_{sim}^D = \mathcal{Z}_{D\_sid} \oplus \mathcal{Z}_{D\_dis} \oplus \mathcal{Z}_{D\_sub}, \quad \mathcal{Z}_{sim}^T = \mathcal{Z}_{T\_dis} \oplus \mathcal{Z}_{T\_sub} \tag{6}$$

$\mathcal{Z}_{sim}^D$ and $\mathcal{Z}_{sim}^T$ are the fused similarity matrices of drug and target and the row vector contains the similarity and heterogeneous attribute information.

4.3 DTI Prediction Task and Optimization

Here we aim to perform the DTI prediction on the original graph $\mathcal{HG}$. Duo to the existing heterogeneity on $\mathcal{HG}$, a node feature aggregation function with attention

---

Algorithm 1. Sparse Imbalanced DTI prediction based on HAS.

Input: Graph $\mathcal{HG} = \{\mathcal{D}, \mathcal{T}, \mathcal{E}, \mathcal{R}\}$, Matrices $\mathcal{M}_{\mathcal{D}\_sid}, \mathcal{M}_{\mathcal{D}\_dis}, \mathcal{M}_{\mathcal{D}\_sub}, \mathcal{M}_{\mathcal{T}\_dis}, \mathcal{M}_{\mathcal{T}\_sub}$

Output: Predicted drug-target interaction probability $\widehat{y}$

1: Generate multi-view augmented graphs $\mathcal{G}_{d_i}^{(1)}$, $\mathcal{G}_{d_i}^{(2)}$, $\mathcal{G}_{t_j}^{(1)}$, $\mathcal{G}_{t_j}^{(2)}$

2: Heterogeneous attention subgraph encoding using Equation (2)(3)(4)

3: Maximize the agreement of positive pairs from different views using Equation (5)

4: Get node similarity information and heterogeneous attribute information via similarity computing on the matrices $\mathcal{M}_{\mathcal{D}\_sid}$, $\mathcal{M}_{\mathcal{D}\_dis}$, $\mathcal{M}_{\mathcal{D}\_sub}$, $\mathcal{M}_{\mathcal{T}\_dis}$, $\mathcal{M}_{\mathcal{T}\_sub}$

5: Apply weighted aggregation function (7) and feature fusing function (8) to learn node embeddings on $\mathcal{HG}$

6: Predict the interaction probability using Equation (9)

---

weights is applied to feature learning on augmented graphs similarly:

$$\mathbf{h}_i^{\mathcal{HG}} = \mathcal{AGG}_{j \in \mathcal{N}_\mathcal{D}(i), k \in \mathcal{N}_\mathcal{T}(j)}^{\mathcal{HG}} \{\mathbf{h}_j^{\mathcal{HG}}, \mathbf{p}_k^{\mathcal{HG}}, \alpha_{(d_i, d_j)}^{\mathcal{HG}}, \beta_{(d_i, t_k)}^{\mathcal{HG}}\} \tag{7}$$

where $\mathcal{AGG}\{\cdot\}$ denotes the weighted node aggregation function, $\alpha_{(d_i, d_j)}^{\mathcal{HG}}$ is the attention score between drug d_i and drug d_j on the original graph, $\beta_{(d_i, t_k)}^{\mathcal{HG}}$ is the attention score between drug d_i and target t_k. Analogously, the target nodes features on the original graph can be acquired using function $\mathcal{AGG}\{\cdot\}$. For the purpose of taking full advantage of known drug-target interaction information and similarity information, we utilize a multi-layer fusion function to fuse the learned nodes features on $\mathcal{HG}$ and the computed similarity features:

$$\mathbf{h}_i^{final} = \mathcal{FC}_\Theta(\mathbf{h}_i^{\mathcal{HG}} \oplus \mathbf{z}_i^\mathcal{D}), \quad \mathbf{p}_j^{final} = \mathcal{FC}_\Theta(\mathbf{p}_j^{\mathcal{HG}} \oplus \mathbf{z}_j^\mathcal{T}) \tag{8}$$

where $\mathbf{h}_i^{final}$ and $\mathbf{p}_j^{final}$ denote the final features of drug d_i and target t_j. $\mathcal{FC}_\Theta(\cdot)$ is the multi-layer fusion function and Θ is set of trainable parameters. $\mathbf{z}_i^\mathcal{D}$ and $\mathbf{z}_j^\mathcal{T}$ represent the similarity vectors of drug d_i and target t_j. The final layer predicts the probability via calculating the inner product of vectors:

$$\widehat{y}_{(d_i, t_j)} = \text{Sigmoid}(\mathbf{h}_i^{final} \odot \mathbf{p}_j^{final}) \tag{9}$$

where $\widehat{y}_{(d_i, t_j)}$ denotes the predicted probability between drug d_i and target t_j, $\odot$ and $\text{Sigmoid}(\cdot)$ represent the dot product measure and sigmoid nonlinear function. As the final DTI prediction task is treated as edge classification, we adopt the cross-entropy loss to fit prediction score and the label value:

$$\mathcal{L}_{pre} = -\sum\nolimits_{(d_i, t_j) \in \mathcal{E}_{\mathcal{D}\_\mathcal{T}}^+} \log(\widehat{y}_{(d_i, t_j)}) - \sum\nolimits_{(d_i, t_k) \in \mathcal{E}_{\mathcal{D}\_\mathcal{T}}^-} \log(1 - \widehat{y}_{(d_i, t_k)}) \tag{10}$$

To complete the whole optimization task that the DTI prediction under sparse imbalance condition, we combine the loss of both data augmentation and DTI prediction together, which is defined as follows:

$$\mathcal{L} = \mathcal{L}_{pre} + \xi_1(\mathcal{L}_\mathcal{D} + \mathcal{L}_\mathcal{T}) + \xi_2 ||\Theta||_2^2 \tag{11}$$

$||\Theta||_2^2$ is the L_2-norm term that prevents training overfitting. ξ_1 and ξ_2 are hyperparameters that control the loss of data augmentation and the L_2-norm term. Algorithm 1 shows the DTI prediction procedure of our proposed framework.

5   Experiments

To evaluate the effectiveness of the proposed method and discuss the reasons, we conduct extensive experiments with different sparsity settings.

5.1   Datasets and Experiment Setup

Experiments are conducted on the constructed drug-target network and associated matrices following Luo *et al.* [22], where drug-drug interactions and drug-target interactions are extracted from DrugBank (Version 3), protein-protein interactions are extracted from HPRD database Release 9. Others are that associated disease data from Comparative Toxicogenomics Database, related side effect data from SIDER database Version 2. The SMILES strings for drugs and amino acid sequences for proteins are obtained following Zhou [13]. The details of heterogeneous entities are summarized in Table 1.

Table 1. The statistics of datasets

Entity type	Numbers	Relation type	Numbers	Sparse ratio
Drug	708	Drug-target	1923	0.00179
Target	1512	Drug-drug	10036	
Disease	5603	Drug-disease	199214	
Side effect	4192	Drug-side effect	80164	
		Target-target	7363	
		Target-disease	1596745	

The compared baselines cover the recent state-of-art methods and traditional deep learning-based models, which all perform drug-target interaction prediction on heterogeneous biological networks. DTINet (2017) combines the unsupervised feature learning from heterogeneous biological network and matrix completion for DTI prediction. NeoDTI (2019) tends to train the model by reconstructing the edge on the heterogeneous graph. MultiDTI (2021) maps the biological entities into vector space aiming to minimize the distance between the entities features. In addition, we consider Graph Attention Network (GAT) and Deep Neural Network (DNN) algorithms as contrast group.

Experiments were conducted on Inspur heterogeneous cluster GPU:12 *32 G Tesla V100 s, memory 640 G DDR2. We deploy the HAS framework with Pytorch and DGL. About the training process of model, we use Adam optimizer with the learning rate of 0.005. The dimension of the initialized features is set as 128, the restart probability π and the temperature parameter τ are set as 0.8 and 0.07, respectively. The final task is denoted as edge classification, we evaluate the DTI prediction performance using Area Under the Receiver Operating Characteristic Curve (AUROC) and Area Under the Precision-Recall Curve (AUPRC).

5.2 Results Discussion

Comparison with Baselines in Different Sparse Ratios. We examine the DTI prediction performance of HAS under sparse imbalanced condition. To further explore its robustness, the positive-negative ratio is adjusted to simulate different sparse DTIs scenarios. The 10-fold cross-validation is implemented on all positive samples and randomly negative samples that are selected according to sparse ratio. We split 90% positive and negative samples in each fold dataset for training, 10% for test purposes.

In addition to simulate the realistic issue that the known DTIs are far less than unknown DTIs, we also consider the experimental setup of baselines and design the experiment with balanced positive-negative samples. Table 2 shows the result comparison with baselines. 1:10 is that the negative samples are 10 times to positive samples, 1:all represents all negative samples are used. Particularly, we have the following observations:

Table 2. Performance comparison with baselines in different sparsity setting

Method	AUPRC						AUROC					
	1:1	1:10	1:30	1:50	1:100	1:all	1:1	1:10	1:30	1:50	1:100	1:all
DNN	0.765	0.691	0.645	0.582	0.441	0.326	0.776	0.755	0.712	0.646	0.597	0.535
GAT	0.873	0.800	0.724	0.612	0.533	0.405	0.825	0.801	0.761	0.723	0.662	0.496
DTINet	0.932	0.865	0.816	0.757	0.671	0.507	0.914	0.873	0.845	0.789	0.692	0.522
NeoDTI	NA	0.874	0.835	0.784	0.726	0.602	NA	**0.943**	0.890	0.839	0.790	0.662
MultiDTI	**0.947**	0.921	0.878	0.837	0.782	0.656	**0.961**	0.891	0.866	0.818	0.730	0.633
HAS	0.938	**0.931**	**0.906**	**0.865**	**0.817**	**0.706**	0.945	0.926	**0.911**	**0.874**	**0.832**	**0.715**
Improv.	NA	1.10%	3.19%	3.35%	4.48%	7.62%	NA	NA	2.36%	4.17%	5.32%	8.01%

(1) The sparse imbalanced interactions between drugs and targets limit efficient prediction performance. We can see that all the models perform well on the balanced DTI prediction. However, with the negative sample increases, the results show a significant decreasing trend. When the negative pairs are sampled to 100 times, model performance drops more than 15%. Until all negative pairs are joined, the metrics drop dramatically again by nearly 15% compared with 1:100 sparse scenario. It confirms the aforementioned statement that a large number of missed drug-target pairs have negative impact on learning high-quality features representation for drugs and targets. Because the rare drug-target interactions cause the weak supervised signals on heterogeneous graph, message propagation between nodes is less to represent graph structure.
(2) HAS expresses the superior improvement. We find that the improvements of HAS mainly come from the sparse imbalanced DTIs scenarios. For example, a 3.35% gain (AUPRC) and 4.17% gain (AUROC) over MultiDTI when the negative pairs are sampled up to 50 times. Furthermore, HAS significantly outperforms alternative approaches by 7.62% (AUPRC) and 8.01% (AUROC).

We conclude that HAS is less affected by negative effect of sparse imbalanced drug-target interactions than the compared baselines. It may be that HAS could capture the intrinsic and universal graph structure feature from topology level as well as similarity information between nodes from node level. All above are used to enhance feature learning when a large amounts of DTIs are missed. MultiDTI achieve the best on the balanced DTI prediction as it adopts a oversampling strategy that oversamples the positive samples by 10 times and under-samples the negative samples. NeoDTI tends to perform DTI prediction under sparse condition, we use 'NA' to label it in Table 2.

Benefits of Heterogeneous Graph Data Augmentation. Heterogeneous graph data augmentation module is proposed to learn the intrinsic graph patterns. We examine the effectiveness of the module in sparse imbalanced DTIs scenarios. The results are shown in Fig. 2(a) and Fig. 2(b), where 'Non-augmentation' means without using heterogeneous graph data augmentation.

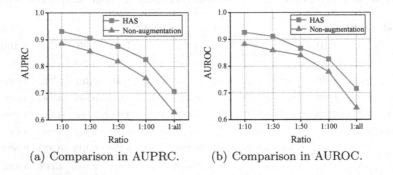

(a) Comparison in AUPRC.   (b) Comparison in AUROC.

Fig. 2. Effectiveness of heterogeneous graph data augmentation.

As expected, the prediction performance of the model without data augmentation drops significantly compared to the overall HAS. Specifically, a clear trend is emerging that the sparser the DTIs data is, the better the data augmentation performs. We observe quite significant drop in AUPRC and AUROC if all the negative drug-target pairs are used. It illustrates that the heterogeneous graph data augmentation contributes more to performance improvements under sparse imbalanced DTIs condition. The augmented graphs encompass the multi-view graph structure information and the contrastive learning optimizes the association between augmented graphs to capture the universal graph structure, which can be used to supplement the missed interactions information.

Benefits of Node Similarity. To test the effectiveness of node similarity information with sparse known drug-target pairs, we first simulate three different sparse imbalanced scenarios as shown in Fig. 3(a) (1:10), Fig. 3(b) (1:50) and Fig. 3(c) (1:100). The experiments between node similarity (**HAS**) and non-similarity (**non-s**) are conducted. Besides, only using drug similarity (**non-ds**) and only using target similarity (**non-ts**) are set so as to explore the importance of drug similarity information and target similarity information for DTI prediction. The results indicate average 4% drop (AUPRC) and 3% drop (AUROC) without using node similarity information. This verifies that the joined similarity information provide positive impact with sparse known drug-target pairs. And diverse attribute information from heterogeneous entities can better characterize the latent properties of drugs and targets. In addition, we can find that both drug nodes similarity and target nodes similarity can make a contribution, which can be used to enrich the learned features.

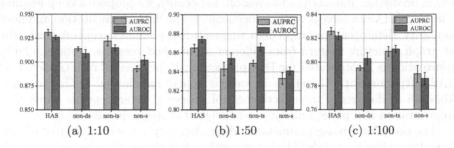

(a) 1:10   (b) 1:50   (c) 1:100

Fig. 3. Effectiveness of node similarity.

Performance Comparison with Different Layers. Considering that GNN with various layers have differences in node feature learning, we perform multi-combination of heterogeneous layers in order to seek the most beneficial setting for the DTI prediction. The experimental results can be seen in Table 3. The performance of HAS achieve the best if setting 2 layers for augmented graphs and origin graph. The setting of 2 layers on the augmented graphs outperforms the setting of 3 layers. It may be smaller size of node data on subgraphs, the aggregation of 2-hop neighboring nodes covers enough drugs and targets. The stacking of aggregated layers will no longer perform significantly better.

Table 3. HAS performance with different layers.

Augmented graph	Origin graph	AUPRC	AUROC
Layer = 2	Layer = 1	0.880	0.898
	Layer = 2	0.931	0.926
	Layer = 3	0.922	0.919
Layer = 3	Layer = 1	0.911	0.923
	Layer = 2	0.920	0.923
	Layer = 3	0.908	0.917

6 Conclusion

In this work, we formulate the sparse imbalance problem on drug-target inter-action prediction and analyze the reason. Especially, we propose a deep learning framework HAS to solve it via heterogeneous graph data augmentation and node similarity. Heterogeneous graph data augmentation pursues to capture the intrin-sic graph structure pattern from different augmented versions. Node similarity information is incorporated for DTI prediction. Experimental results show that HAS outperforms the baselines in various sparse imbalanced DTIs scenarios. Ablation studies verify the effectiveness of proposed heterogeneous graph data augmentation and node similarity to alleviate the sparse imbalance issue.

The complicate bio-experiments in drug discovery cause that the real labeled drug data is less accessible. The future work will explore and construct the other heterogeneous biological networks to strengthen generalization of model. And the impact of augmented graphs scale will be further investigated.

Acknowledgements. This work was supported by the National Natural Science Foundation of China (61503273, 61702356), Industry-University Cooperation Educa-tion Program of the Ministry of Education, and Shanxi Scholarship Council of China.

References

1. Sun, M., Zhao, S., Gilvary, C.: Graph convolutional networks for computational drug development and discovery. Briefings Bioinform. **21**(3), 919–935 (2020)
2. Vamathevan, J., Clark, D., Czodrowski, P.: Applications of machine learning in drug discovery and development. Nat. Rev. Drug Discov. **18**(6), 463–477 (2019)
3. Bagherian, M., Sabeti, E., Wang, K.: Machine learning approaches and databases for prediction of drug-target interaction: a survey paper. Briefings Bioinf. **22**(1), 247–269 (2021)
4. Hakime, Ö.: DeepDTA: deep drug-target binding affinity prediction. Bioinformatics **34**(17), 821–829 (2018)
5. Lee, I., Keum, J., Nam, H.: DeepConv-DTI: prediction of drug-target interac-tions via deep learning with convolution on protein sequences. PLoS Comput. Biol. **15**(6), e1007129 (2019)

6. Nguyen, T., Le, H., Quinn, T.P.: GraphDTA: predicting drug-target binding affinity with graph neural networks. Bioinformatics **37**(8), 1140–1147 (2021)
7. Huang, K., Xiao, C., Glass, L.M.: MolTrans: molecular interaction transformer for drug-target interaction prediction. Bioinformatics **37**(6), 830–836 (2021)
8. Chen, L., Tan, X., Wang, D.: TransformerCPI: improving compound-protein interaction prediction by sequence-based deep learning with self-attention mechanism and label reversal experiments. Bioinformatics **36**(16), 4406–4414 (2020)
9. Chen, H., Li, J.: Modeling relational drug-target-disease interactions via tensor factorization with multiple web sources. In: WWW (2019)
10. Wan, F., Hong, L., Xiao, A.: NeoDTI: neural integration of neighbor information from a heterogeneous network for discovering new drug-target interactions. Bioinformatics **35**(1), 104–111 (2019)
11. Zhou, D., Xu, Z., Li, W.T.: MultiDTI: drug-target interaction prediction based on multi-modal representation learning to bridge the gap between new chemical entities and known heterogeneous network. Bioinformatics **37**(23), 4485–4492 (2021)
12. Xia, X.: Bioinformatics and drug discovery. Curr. Top. Med. Chem. **17**(15), 1709–1726 (2017)
13. Qiu, J., Chen, Q., Dong, Y.: Gcc: graph contrastive coding for graph neural network pre-training. In: KDD, pp. 1150–1160 (2020)
14. You, Y., Chen, T., Sui, Y.: Graph contrastive learning with augmentations. In: NeurIPS, pp. 5812–5823 (2020)
15. Jung, L.S., Cho, Y-R.: Survey of network-based approaches of drug-target interaction prediction. In: BIBM, pp. 1793–1796 (2020)
16. Wu, Z., Pan, S., Chen, F.: A comprehensive survey on graph neural networks. IEEE Trans. Neural Netw. Learn. Syst. **32**(1), 4–24 (2020)
17. Zeng, Y., Chen, X., Luo, Y.: Deep drug-target binding affinity prediction with multiple attention blocks. Briefings Bioinform. **22**(5), bbab117 (2021)
18. Peng, J., Wang, Y., Guan, J.: An end-to-end heterogeneous graph representation learning-based framework for drug-target interaction prediction. Briefings Bioinform. **22**(5), bbaa430 (2021)
19. Zhang, C., Song, D., Huang, C.: Heterogeneous graph neural network. In: KDD, pp. 793–803 (2019)
20. Wang, X., Ji, H., Shi, C.: Heterogeneous graph attention network. In: WWW, pp. 2022–2032 (2019)
21. Wu, J., Wang, X., Feng, F.: Self-supervised graph learning for recommendation. In: SIGIR, pp. 726–735 (2021)
22. Luo, Y., Zhao, X., Zhou, J.: A network integration approach for drug-target interaction prediction and computational drug repositioning from heterogeneous information. Nat. Commun. **8**(1), 1–13 (2017)

Structure-Aware Reasoning
for Knowledge Base Question Answering

Lu Ma[1,2], Peng Zhang[1(✉)], Xi Zhu[1,2], Dan Luo[1,2], and Bin Wang[3]

[1] Institute of Information Engineering, Chinese Academy of Sciences, Beijing, China
{malu,pengzhang,zhuxi,luodan}@iie.ac.cn
[2] School of Cyber Security, University of Chinese Academy Sciences, Beijing, China
[3] Xiaomi AI Lab, Beijing, China
wangbin11@xiaomi.com

Abstract. Answering question according to knowledge base (i.e. KBQA) has attracted extensive attention recently. Information retrieval is one of the mainstream methods for the KBQA task that first finds the topic entity in the question via entity linking systems, and then selects the most related entities as answers from the subgraph (nodes in it are called candidate answers) of topic entity on the knowledge base (KB). However, existing methods generally separately perform reasoning over every candidate answer by considering the semantic relationships between question and the features extracted from KB, breaking away from the graphical structure of the KB and suffering from long-term dependency problem of entities. To address that, we propose a structure-aware reasoning method, which enables to exploit the graphical structure of entities on KB via Graph Convolutional Network and capture deep semantic relationships between question and candidate answers. Our method reasons about the correct answer by jointly considering information of all candidate answers, and focusing on important components in the question and on KB . We conduct experiments on the WebQuestions dataset, and the results demonstrate the effectiveness of our proposed method.

Keywords: Question answering · Knowledge base · Graph convolutional network

1 Introduction

Knowledge base question answering (KBQA) task aims to answer natural language question based on a knowledge base (KB) automatically. For example, given *"Where is Mali located?"* as the input question and Freebase [4] as the KB, entities *Africa* and *West Africa* from Freebase comprise the answers. Generally, the solutions of KBQA can be divided into semantic parsing based (SP-based) and information retrieval based (IR-based) methods. This work focus on IR-based methods, which directly retrieve and rank answers from KB according to the semantic information in the question [19]. The general process is to select

© The Author(s), under exclusive license to Springer Nature Switzerland AG 2022
J. Gama et al. (Eds.): PAKDD 2022, LNAI 13280, pp. 562–573, 2022.
https://doi.org/10.1007/978-3-031-05933-9_44

a set of entities from KB as candidate answers and then use a reasoning module to rank them. The crucial step is how to rank these candidate answers.

Historically, researches perform reasoning by representing question and candidate answer independently [5,6,12], whereas the relatedness between the question and candidate answer is neglected. Recently, researchers start putting more emphasis on the mutual attention between the question and candidate answers [8,13,28], which helps learn adequate question representation and adjust the question-answer weight for better reasoning. They generally perform reasoning separately over every candidate answer considering the extracted aspects information including answer path, answer context and answer type from KB, which contain entity information and neighborhood information at most. We call this strategy local reasoning, and observe that the semantic relationship between question and candidate answer is inadequate, as models omit the structural information on KB, and suffer from long-term dependency problem of entities.

To address the above issue, we propose a Structure-aware Reasoning method based on Graph Convolutional Network (SRGCN), which performs a local reasoning and a global reasoning over the subgraph of the topic entity on KB. The subgraph, called *reason graph*, is constructed with all candidate answers as nodes and relations between them as edges, which retains the structural information on KB. Our global reasoning module performs reasoning about a candidate answer by taking into account information from multiple entities in the reason graph via Graph Convolutional Network (GCN) [18]. The GCN can jointly assess the suitability of all candidate answers, capturing deep semantic information from KB. Moreover, we introduce the personalized PageRank method [14] to extract more related entities to the question as candidate answers, aiming to alleviate the noise problem of the k-hop subgraph and decrease the computational cost.

The contributions of our work are as follows:

(1) We propose a Structure-aware Reasoning method (SRGCN) for KBQA task based on the global reasoning module and local reasoning module to handle the inadequacy problem of semantic relationship between the question and KB.
(2) We present a global reasoning module to exploit the graphical structure of the KB and share the information between candidate answers, which can jointly assess the suitability of all candidate answers.
(3) We demonstrate our proposed method on the WebQuestions dataset [3], outperforming the state-of-the-art methods.

2 Related Works

There are two popular categories for researches on KBQA. For SP-based method, recent works parse questions into logical forms by utilizing predefined rules or templates [1,15] or focusing more on neural networks [17,20,21,31]. The former limits the scalability and coverage for wise-open domains, and the latter suffers from poor efficiency due to the heavily querying on KB. Some works

[11,32] model semantic parsing as an question to logical form machine translation problem by employing neural Sequence-to-Sequence models, which have a weak-coupling with KB and hardly exploit structure and semantic constraints.

The IR-based methods first extract candidate answers from KB according to the given question, and then map the question and candidate answers into a semantic space. The similarity between question and a candidate answer is used to measure the probability that the candidate answer is the correct answer. Bordes et al. [5] encode a subgraph for every candidate answer to predict correct answers. Dong et al. [12] calculate the similarity for the question-candidate answer pair by using multi-column convolutional neural networks. Xu et al. [30] introduce a multi-channel convolutional neural network to compute the semantic similarity for the question-candidate relation pair so that the answers can be obtained with the optimal relation. Recent approaches [8,13] analysis the semantic relationship between question and candidate answers via attention mechanism. Such approaches extract the features of a candidate answer out of the KB, and conduct reasoning over candidate answers by calculating the similarity of each question-candidate answer pair respectively, as a result that the structural information from KB is lost and information between candidate answers can not interact with each other.

Therefore, we propose a global reasoning strategy to alleviate the above problem. It learns representations for every node in the graph with all candidate answers as nodes, which enables to jointly consider information of surrounding entities by Graph Convolutional Networks (GCN). By stacking multiple layers, the model can gather information from nodes further away. Kipf and Welling [18] introduced GCN to arbitrarily connected undirected graphs. Marcheggiani et al. [23] employed GCN for natural language processing for the first time. Recent studies based on GCN have established that GCN is applied successfully for question answering task [10,25,29] and reading comprehension task [7].

3   Methods

3.1   Task Setup

In the KBQA task, we define the knowledge base as a multi-relational graph $\mathcal{K} = (\mathcal{V}, \mathcal{E})$. Here $\mathcal{V}$ is the set of nodes that denotes entities or properties from KB; $\mathcal{E}$ is the set of edges that connect nodes in $\mathcal{V}$, corresponding to relation types in the KB. Each entity or property or relation type has a textual description (a sequence of words) in KB, which constitutes the label of a node or an edge. Given a natural language question q, the goal is to find a set of nodes $\mathcal{Y} \subseteq \mathcal{V}$ that can answer the question.

We apply a topic entity predictor [8] to find the best topic entity of q that links the question to KB. For instance, the topic entity of the question "*Where is Mali located*" is *Mali*. All nodes related to the topic entity within k-hop are regarded as candidate answers $\mathcal{Y}$. Then, the model predicts a score $S(q, y), y \in \mathcal{Y}$, determining whether y is a correct answer or not. The architecture of our method is shown in Fig. 1.

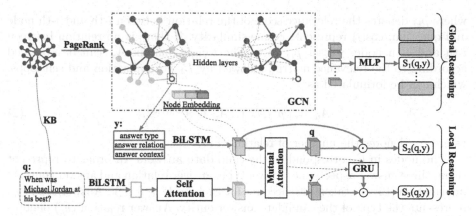

Fig. 1. Overall architecture of our proposed Structure-aware Reasoning method (SRGCN) for KB-QA task. q: input question, y: an candidate answer for q. In the reason graph, the red node is the topic entity; blue nodes represent candidate answers for q; gray nodes are irrelevant candidate answers filtered out by PageRank. $\odot$ denotes inner product. (color figure online)

3.2 Reason Graph Generation

Many candidate answers in $\mathcal{Y}$ can be irrelevant to the question, resulting in introducing unnecessary reasoning and overfitting. In response, we further extract a subset from $\mathcal{Y}$ to select more related entities to the question, which can prune the candidate answer set. We apply the Personalized PageRank (PPR) method [14] around the topic entity to compute the probability that a candidate answer is the answer to the question. The edge-weights around each candidate answer entity are weighed based on the label of the corresponding edge, as a result that edges more relevant to the question receive higher weights.

Specifically, we use a pre-trained word embedding matrix $E_w \in \mathbb{R}^{|V_w| \times d_v}$ to map each relation $r = \{o_1, o_2, ..., o_{|r|}\}$ into word embeddings $\{\mathbf{o}_1, \mathbf{o}_2, ..., \mathbf{o}_{|r|}\}$, and then apply average operation to compute a relation vector v^r. Here o_i denotes the i-th word in r; $|V_w|$ denotes the number of words in vocabulary; d_v denotes the embedding dimension. Similarly, we get the question vector v^q of $q = \{w_i | i \in [1, |q|]\}$, where w_i is the i-th word in the question. The edge weight is calculated via cosine similarity between the question and relation. Each candidate answer has a PPR score after running PPR, and the top-$|\bar{\mathcal{Y}}|$ candidate answers are extracted to generate a novel candidate answer set $\bar{\mathcal{Y}}$, $\bar{\mathcal{Y}} \subseteq \mathcal{Y}$, where $|\bar{\mathcal{Y}}|$ denotes the number of extracted candidate answers.

A *reason graph* $\mathcal{G} = (\bar{\mathcal{Y}}, \bar{\mathcal{R}})$ is defined for a question, consisting of candidate answers with high probability as nodes and relations between them as edges if exist in KB. The edge-weight is denoted as a weight matrix A_w, which shows the relevance between the question and relation. We compute the relevance matrix ρ of question representation v^q and relation representations v^r as:

$$\rho_{ij} = \cos(v^q, v^r_{ij}) \tag{1}$$

where ρ_{ij} denotes the relevance score of the relation between i-th and j-th node and question. $\cos(\cdot)$ represents cosine similarity. If there is no relation between i-th and j-th node, we set ρ_{ij} to 0. The weight matrix A_w can be obtained according to the relevance matrix, followed by L_1 normalization and self loops, which can be formulated as:

$$A_w = \| \rho^2 / N_r \|_1, A_w \in \mathbb{R}^{|\bar{\mathcal{Y}}| \times |\bar{\mathcal{Y}}|}. \tag{2}$$

where N_r denotes the number of relations in $\mathcal{G}$.

The nodes in $\mathcal{G}$ correspond to the candidate answers. In order to represent them, three answer aspects (i.e. answer type, answer relation and answer context) information [12] are taken into account for each node y, $y \in \bar{\mathcal{Y}}$. Answer type (y_t) represents the type of the candidate answer entity. Answer relation (y_r) denotes a relation path from topic entity to y, and y'_r is a set of relations from y_r. Answer context is a set of surrounding entities of y from $\mathcal{G}$. We look up the word embedding matrix E_w to map each aspect into word embeddings (e_t, e_r, e_c, e'_r), where e'_r denotes the average of relation embeddings in y'_r. Then the first three are fed into three bidirectional long short-term memory (BiLSTM, [2]) networks to produce hidden state vectors (h_t, h_r, h_c). Following [8], we apply a key-value memory network [24] to store answer aspects information, and obtain key representations of nodes as follows:

$$H^{k_t} = f_t(h_t); H^{k_r} = f_r([h_r; e'_r]); H^{k_c} = f_c(h_c)$$
$$H^k = [H^{k_t}; H^{k_r}; H^{k_c}] \tag{3}$$

where f denotes linear projection. Similarly, we get value representations $H^v = [H^{v_t}; H^{v_r}; H^{v_c}]$ for a node, $H^k, H^v \in \mathbb{R}^{3 \times |\bar{\mathcal{Y}}| \times d}$. d denotes the dimensionality of BiLSTM hidden state. $[;]$ denotes column-wise concatenation.

3.3   Answer Prediction

Given a reason graph $\mathcal{G} = (\bar{\mathcal{Y}}, \bar{\mathcal{R}})$ contains the candidate answer set, our goal is to reason about the answer, producing a score for each candidate answer via global and local reasoning respectively.

Global Reasoning. In order to capture deep semantic relationships between the question and KB, we propose a global reasoning mechanism to jointly assess the suitability of all candidate answers via Graph Convolutional Network (GCN) [18]. Given the answer aspect representation of nodes in $\mathcal{G}$ and an weight adjacency matrix A_w describing the graph structure $\bar{\mathcal{R}}$ as the input to the GCN, the goal is to learn how to combine these representations and output the feature representation.

In our case, each node $y \in \bar{\mathcal{Y}}$ is represented by the concatenation of the corresponding question, answer aspects and entity type representations, i.e. $u_y = [H^q; H^k_y; h'_{t_y}]$, where h'_{t_y} is the entity type embedding of node. Such a combination ensures that each candidate answer depends on question. To represent the question q, we fed the word embeddings v^q into a BiLSTM net, which enables to integrate forward and backward contextual information into text embeddings, and obtain $h^q \in \mathbb{R}^{|q| \times d}$. Furthermore, a self-attention is applied over all words of question to get the question representation:

$$H^q = \text{BiLSTM}([h^q(\text{softmax}((h^q)^T h^q)), h^q]), H^q \in \mathbb{R}^d \tag{4}$$

where the softmax function is applied over the last dimension of the input tensor.

We initialize the hidden states for the i-th node via: $U_i^{(0)} = u_i$. At the l-th layer, the i-th node representation is updated by:

$$U_i^{(l)} = \text{RELU}\left(\sum_{j=1}^{E} A_w^{ij} W_l U_j^{(l-1)} + b^l\right), U_i^{(l)} \in \mathbb{R}^{d_{\mathcal{G}}} \tag{5}$$

where $d_{\mathcal{G}}$ is the dimension of hidden state in GCN, W_l is a weight matrix, and b^l is a bias term. We stack such networks for L layers and obtain representations of all nodes $\hat{U} = \{\hat{U}_1, \hat{U}_2, ..., \hat{U}_{|\bar{\mathcal{Y}}|}\}$. After obtaining the output feature $\hat{U}$ from the GCN, we obtain the normalized representation by linear transformation and a residual connection as:

$$O = \text{LayerNorm}(\hat{U} + f(u)), O \in \mathbb{R}^{|\bar{\mathcal{Y}}| \times d_{\mathcal{G}}} \tag{6}$$

where f is a full-connected layer. u denotes the initial representation of nodes. We pass the normalized representation of nodes through a multi-layer perceptron (MLP) to predict the probability $S_1(q, y)$ that node y is the answer to q.

Local Reasoning. We use a mutual attention module to embed the answer information into the question representation, and focus on important aspects by the guidance of question. As a result, model can put emphasis on important components of question and answer aspects. The process is formulated as:

$$
\begin{aligned}
&\mu = \text{softmax}(\max_{|\bar{\mathcal{Y}}|}(h^{qT}\{Att_{add}(H^q, H^{k_x}) \cdot H^{v_x}\}_{x \in \{t,r,c\}})) \\
&\hat{H}^q = \mu h^q, \hat{H}^q \in \mathbb{R}^d \\
&\omega = \text{softmax}(\max_{|q|}(H^k h^q)^T)^T \\
&\hat{H}^k = \omega H^k, \hat{H}^k \in \mathbb{R}^{|\bar{\mathcal{Y}}| \times d}
\end{aligned} \tag{7}
$$

where μ, ω denote the aspect-aware attention and question-aware attention, respectively. $Att_{add}(p_1, p_2) = \text{softmax}(\tanh([p_1^T, p_2]W_1)W_2)$, with $W_1 \in \mathbb{R}^{2d \times d}$ and $W_2 \in \mathbb{R}^{d \times 1}$ being trainable weights. $\{\cdot\}$ indicates concatenation. $\max_{|\bar{\mathcal{Y}}|}(\cdot)$ and $\max_{|q|}(\cdot)$ denote *max*-pooling over candidate answer dimension and question word dimension, respectively. Finally, a score function is used to predict the probability that each node $y \in \mathcal{Y}$ is an answer to q: $S_2(q, y) = \hat{H}^{qT} \cdot H^k$.

Moreover, we apply a Gated Recurrent Unit (GRU) [9] and batch normalization (BatchNorm) [16] to enhance the question vector as follows:

$$
\begin{aligned}
&\bar{H}^q = \text{GRU}(\hat{H}^q, Att_{add}(\hat{H}^q, \hat{H}^k) \cdot H^v) \\
&H_n^q = \text{BatchNorm}(\hat{H}^q, \bar{H}^q)
\end{aligned} \tag{8}
$$

where $H_n^q \in \mathbb{R}^d$. The probability that a node is the answer to q is calculated via: $S_3(q, y) = H_n^{qT} \cdot \hat{H}^k$, where $\hat{H}^k$ is derived by Eq. (7).

Table 1. Average F_1 scores of models on WebQ test set. The bests are in bold.

Method		Average F_1 (%)
SP-based	Sempre [3]	35.7
	STAGG [31]	52.5
	QUINT [1]	51.0
	NFF [15]	49.6
	SeMat Model [22]	52.0
IR-based	Subgraph + C$_2$ [5]	39.2
	MCCNN [30]	47.1
	C-ATT [13]	42.9
	BAMnet [8]	51.8
	LMKB-QA [28]	52.7
Ours	SRGCN	**53.0**
	SRGCN w/gold topic entity	**56.6**

3.4   Learning

The answer predictor's parameters are comprised of weights from the question and candidate answer embeddings, GCN, MLP, mutual attention module and GRU. These are trained end-to-end. We adopt the sampling strategy following [8] and use a triplet-based loss function as follows:

$$
\mathcal{L} = \sum_{(\bar{y})}[1 + S_1(q, y') - S_1(q, y)]_+ +
$$
$$
\sum_{(\bar{y})}[1 + S_2(q, y') - S_2(q, y)]_+ + \sum_{(\bar{y})}[1 + S_3(q, y') - S_3(q, y)]_+ \tag{9}
$$

where y and y' denote positive and negative answer respectively. $[\cdot]_+$ is equivalent to $\max[\cdot, 0]$. $S_1(\cdot)$ is the predicted score of a candidate answer computed by global reasoning, and $S_2(\cdot)$, $S_3(\cdot)$ are computed by local reasoning.

At testing time, we rank the scores of the candidate answers $S_3(q, y)$ to select the optimal answer to the question, and a hyper-parameter θ is used to handle questions with multiple answers. The candidate answers whose scores are close to the highest score within θ constitute the answer set.

4   Experiments

4.1   Experimental Setup

We evaluate our proposed method on the WebQuestions (WebQ) [3] dataset, which is split into 3,023 training examples, 755 validation examples, and 2,032 testing examples. Freebase [4] is used as the knowledge base. We adopt

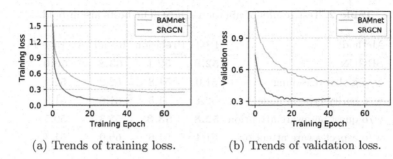

(a) Trends of training loss.   (b) Trends of validation loss.

Fig. 2. Trends of loss during training.

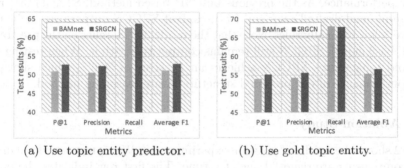

(a) Use topic entity predictor.   (b) Use gold topic entity.

Fig. 3. Comparison with BAMnet on WebQ test set.

precision@1 (P@1), precision, recall and average F_1 score as the evaluation metrics [3,5].

We implement our method in PyTorch [26]. We employ Adam optimizer with the initial learning rate as 0.001 to train our model on a TITAN RTX 24G GPU. GloVe [27] is used as word embedding matrix E_w with dimensions set to $d_v = 300$. The dimensions of hidden state vectors in BiLSTM, GCN, MLP are set as 128, 150 and 128, respectively. The number of GCN layers is set to $L = 2$. The dropout rates on the word embedding layer, question encoder side, candidate answer encoder side and GCN side are 0.3, 0.3, 0.2 and 0.1, respectively. The threshold θ is set to 0.62. The number of candidate answers is set to $|\bar{\mathcal{Y}}| = 200$.

4.2 Comparison with State-of-the-Arts

Table 1 shows the performance of our method (SRGCN) and previous state-of-the-art methods. We can observe that SRGCN consistently outperforms all SP-based methods and IR-based methods. Specifically, compared with BAMnet [8] and C-ATT [13], which are the most close works to SRGCN, SRGCN outperforms them by a margin of 1.2% and 10.1% in terms of average F_1 score. To compare the performance of BAMnet and SRGCN, we re-implement BAMnet in our environment and report the training process in Fig. 2 and detail results in Fig. 3, showing that our model has quicker convergence rapidity and achieves

Table 2. Test results of ablation study. The bests are in bold.

Method	P@1	Precision	Recall	Average F_1
SRGCN	**52.8**	52.4	**63.8**	**53.0**
w/o global reasoning	51.0	51.8	62.2	52.2
w/o entity type in node	52.1	51.5	61.6	51.9
w/o question-aware attention	**52.8**	**53.3**	55.3	50.5
w/o aspect-aware attention	51.4	51.6	60.0	51.3

better performance. As the previous best SP-based method, STAGG [31] relies on careful hand-drafted rules and features. But SRGCN outperforms it by 0.5% with very few manually rules. Note that the model in [28] relies on the pre-trained BERT base model, SRGCN still remains competitive with it. When testing with the gold topic entity, SRGCN achieves an F_1 score of 56.6%, which shows the pure performance of our proposed structure-aware reasoning method.

4.3   Ablation Study

Table 2 shows the results of ablation experiments on WebQ test set where important components are removed one at a time. The first row indicates the performance of our proposed SRGCN method. The results in the 'w/o global reasoning' row only use local reasoning module, as a result that the average F_1 score drops by 0.8 points. This indicates that information of surrounding entities for a candidate answer contributes to reason about the answers. The third row ('w/o entity type in node') indicates that global reasoning module removes entity type embedding from initial node embedding. We found that the average F_1 score of it is lower than when removing the whole global reasoning module. This means global reasoning module without entity type as a feature is limited in its ability to capture the semantic relationship between question and node (i.e. candidate answer). The last two rows ('w/o question-aware attention' and 'w/o aspect-aware attention') represent the results without using question-aware attention and aspect-aware attention mechanism in the mutual attention module, respectively. Results show that these modules are crucial as well.

We apply a PageRank method to prune the original candidate answer set Sect. (3.2). We use the overall recall of answers among the candidate answers as the metric to compare the results under different conditions on WebQ test set. When the nodes from the subgraph of topic entity within 2-hop on KB are selected as the candidate answer set, it achieves a overall recall of 86.9%. After using PageRank to prune the subgraph, we select the top-200 related nodes as candidate answers. The overall recall is improved by 0.7% with less nodes, which enables to decrease the computational cost. If we directly truncate the subgraph and take the top-200 candidate answers, the overall recall decreases by 14%. We also investigate the effect of the layer number L of GCNs. The F_1 scores of $L = 1$, $L = 2$, $L = 3$ are 51.6%, 53.0% and 51.8%, respectively.

Table 3. Case study. Four examples from the test set of WebQ. We use ellipsis due to space constraints.

Question	BAMnet	SRGCN	Ground truth
Who is the president of Peru now?	Ollanta Humala, César Villanueva, Juan Jiménez Mayor, Simón Bolívar	Ollanta Humala	Ollanta Humala
What team will Michael Vick play for in 2011?	United States Penitentiary, Leavenworth	Philadelphia Eagles	Philadelphia Eagles
When was Michael Jordan at his best?	1995−96 NBA season, 1997−98 NBA season, 1984−85 NBA season,	1995−96 NBA season	1995−96 NBA season
Who is the current leader of France 2010?	Nicolas Sarkozy, Serge Dassault, Lionel Jospin,	Nicolas Sarkozy	Nicolas Sarkozy

4.4 Case Study

To demonstrate our model's capability, we show several typical examples from WebQ test set in Table 3. In the first case, SRGCN accurately answer the question with a time constraint *now*, while BAMnet predicts other wrong answers. For instance, *Simón Bolívar* was the president of Peru from 1824 to 1827. As for the second case, there are entity type and time constraints (i.e. *team, in 2011*), but BAMnet predicts a prison *United States Penitentiary, Leavenworth*, which can not match the entity type constraint. Our method can successfully handle it. As for the third case, the *argmax* constraint (*at his best*) is expressed by a phrase, which makes it difficult to capture the deep semantic information. Seasons with lower points than the correct answer are also selected as answers by BAMnet. However, the predicted answer of SRGCN is the same as ground truth. In the last case, models are expected to understand the semantic of *learder* and handle a time constraint. Except for the ground truth, BAMnet also predicts many wrong answers. For example, *Serge Dassault* was a Senator, and *Serge Dassault* was the Prime Minister of France from 1997 to 2002. Results of these cases demonstrate that our method performs well on capture deep semantic information between question and candidate answers, especially for complex questions.

5 Conclusions

We presented SRGCN, a structure-aware reasoning method for the KBQA task. SRGCN enables to jointly assess the suitability of all candidate answers by a global reasoning mechanism based on graph convolutional network, and capture deep semantic relationship between the question and KB using a local reasoning mechanism based on mutual attention. Extensive experiments on WebQuestions dataset showed SRGCN's improvements over existing approaches on KBQA task.

Acknowledgements. This work is supported by the National Natural Science Foundation of China (grant No. 61876223), Youth Innovation Promotion Association, Chinese Academy of Sciences (No. 2020163), National Natural Science Foundation of China (grant No. 61832004), Projects of International Cooperation and Exchanges NSFC (grant No. 62061136006).

References

1. Abujabal, A., Yahya, M., Riedewald, M., Weikum, G.: Automated template generation for question answering over knowledge graphs. In: WWW, pp. 1191–1200 (2017)
2. Bahdanau, D., Cho, K., Bengio, Y.: Neural machine translation by jointly learning to align and translate. In: ICLR (2015)
3. Berant, J., Chou, A., Frostig, R., Liang, P.: Semantic parsing on Freebase from question-answer pairs. In: EMNLP, pp. 1533–1544 (2013)
4. Bollacker, K.D., Evans, C., Paritosh, P., Sturge, T., Taylor, J.: Freebase: a collaboratively created graph database for structuring human knowledge. In: ACM SIGMOD, pp. 1247–1250 (2008)
5. Bordes, A., Chopra, S., Weston, J.: Question answering with subgraph embeddings. In: EMNLP, pp. 615–620 (2014)
6. Bordes, A., Weston, J., Usunier, N.: Open question answering with weakly supervised embedding models. In: Calders, T., Esposito, F., Hüllermeier, E., Meo, R. (eds.) ECML PKDD 2014. LNCS (LNAI), vol. 8724, pp. 165–180. Springer, Heidelberg (2014). https://doi.org/10.1007/978-3-662-44848-9_11
7. Cao, Y., Fang, M., Tao, D.: BAG: bi-directional attention entity graph convolutional network for multi-hop reasoning question answering. In: NAACL-HLT, pp. 357–362 (2019)
8. Chen, Y., Wu, L., Zaki, M.J.: Bidirectional attentive memory networks for question answering over knowledge bases. In: NAACL-HLT, pp. 2913–2923 (2019)
9. Cho, K., et al.: Learning phrase representations using RNN encoder-decoder for statistical machine translation. In: EMNLP, pp. 1724–1734 (2014)
10. De Cao, N., Aziz, W., Titov, I.: Question answering by reasoning across documents with graph convolutional networks. In: NAACL-HLT, pp. 2306–2317 (2019)
11. Dong, L., Lapata, M.: Language to logical form with neural attention. In: ACL, pp. 33–43 (2016)
12. Dong, L., Wei, F., Zhou, M., Xu, K.: Question answering over freebase with multi-column convolutional neural networks. In: ACL-IJCNLP, pp. 260–269 (2015)
13. Hao, Y., et al.: An end-to-end model for question answering over knowledge base with cross-attention combining global knowledge. In: ACL, pp. 221–231 (2017)
14. Haveliwala, T.H.: Topic-sensitive pagerank: a context-sensitive ranking algorithm for web search. IEEE Trans. Knowl. Data Eng. $15(4)$, 784–796 (2003)
15. Hu, S., Zou, L., Yu, J.X., Wang, H., Zhao, D.: Answering natural language questions by subgraph matching over knowledge graphs. IEEE Trans. Knowl. Data Eng. $30(5)$, 824–837 (2017)
16. Ioffe, S., Szegedy, C.: Batch normalization: accelerating deep network training by reducing internal covariate shift. In: ICML, pp. 448–456 (2015)
17. Jia, R., Liang, P.: Data recombination for neural semantic parsing. In: ACL, pp. 12–22 (2016)
18. Kipf, T.N., Welling, M.: Semi-supervised classification with graph convolutional networks. In: ICLR (2017)

19. Lan, Y., He, G., Jiang, J., Jiang, J., Zhao, W.X., Wen, J.R.: A survey on complex knowledge base question answering: methods, challenges and solutions. In: IJCAI, pp. 4483–4491 (2021)
20. Lan, Y., Jiang, J.: Query graph generation for answering multi-hop complex questions from knowledge bases. In: ACL, pp. 969–974 (2020)
21. Liang, C., Berant, J., Le, Q., Forbus, K.D., Lao, N.: Neural symbolic machines: learning semantic parsers on Freebase with weak supervision. In: ACL, pp. 23–33 (2017)
22. Luo, K., Lin, F., Luo, X., Zhu, K.: Knowledge base question answering via encoding of complex query graphs. In: EMNLP, pp. 2185–2194 (2018)
23. Marcheggiani, D., Titov, I.: Encoding sentences with graph convolutional networks for semantic role labeling. In: EMNLP, pp. 1506–1515 (2017)
24. Miller, A., Fisch, A., Dodge, J., Karimi, A.H., Bordes, A., Weston, J.: Key-value memory networks for directly reading documents. In: EMNLP, pp. 1400–1409 (2016)
25. Narasimhan, M., Lazebnik, S., Schwing, A.G.: Out of the box: reasoning with graph convolution nets for factual visual question answering. In: NeuraIPS, pp. 2659–2670 (2018)
26. Paszke, A., et al.: Automatic differentiation in pytorch. In: NeuraIPS (2017)
27. Pennington, J., Socher, R., Manning, C.: Glove: global vectors for word representation. In: EMNLP, pp. 1532–1543 (2014)
28. Sharath, J.S., Banafsheh, R.: Question answering over knowledge base using language model embeddings. In: IJCNN, pp. 1–8 (2020)
29. Sun, H., Dhingra, B., Zaheer, M., Mazaitis, K., Salakhutdinov, R., Cohen, W.: Open domain question answering using early fusion of knowledge bases and text. In: EMNLP, pp. 4231–4242 (2018)
30. Xu, K., Reddy, S., Feng, Y., Huang, S., Zhao, D.: Question answering on freebase via relation extraction and textual evidence. In: ACL, pp. 2326–2336 (2016)
31. Yih, W.t., Chang, M.W., He, X., Gao, J.: Semantic parsing via staged query graph generation: question answering with knowledge base. In: ACL-IJCNLP, pp. 1321–1331 (2015)
32. Zhang, H., Cai, J., Xu, J., Wang, J.: Complex question decomposition for semantic parsing. In: ACL, pp. 4477–4486 (2019)

Detecting Anchors' Opinion in Hinglish News Delivery

Siddharth Sadhwani[✉], Nishant Grover, Md Shad Akhtar,
and Tanmoy Chakraborty

Department of CSE, IIIT, Delhi, India
{siddharth18313,nishant18399,shad.akhtar,tanmoy}@iiitd.ac.in

Abstract. Humans like to express their opinions and crave the opinions of others. Mining and detecting opinions from various sources are beneficial to individuals, organisations, and even governments. One such organisation is news media, where a general norm is not to showcase opinions from their side. Anchors are the face of the digital media, and it is required for them not to be opinionated. However, at times, they diverge from the accepted norm and insert their opinions into otherwise straightforward news reports, either purposefully or unintentionally. This is primarily seen in debates as it requires the anchors to be spontaneous, thus making them vulnerable to add their opinions. The consequence of such mishappening might lead to biased news or even supporting a certain agenda at the worst. To this end, we propose a novel task of anchors' opinion detection in debates. We curate code-mixed news debates and develop the ODIN dataset. A total of 2054 anchors' utterances in the dataset are marked as opinionated or non-opinionated. Lastly, we propose DetONADe – an interactive attention-based framework for classifying anchors' utterances and obtain the best weighted-F1 score of 0.703. A thorough analysis and evaluation show many interesting patterns in the dataset and predictions.

Keywords: Anchors' opinion · Opinion detection · Code-mixed conversations

1 Introduction

News bulletins play a significant role in educating, informing, spreading awareness, and influencing the masses about important current affairs. Recent estimates show that the Indian news channels are broadcast over 161 million TV households, and more than 200 million internet users [5]. This puts a lot of responsibility on the news channels as they are the primary source of the masses' knowledge about current affairs.

Common citizens expect their news to be free of opinions and only based on facts. However, in recent years, we have seen countless instances where

---

S. Sadhwani and N. Grover: Equal Contribution

J. Gama et al. (Eds.): PAKDD 2022, LNAI 13280, pp. 574–586, 2022.
https://doi.org/10.1007/978-3-031-05933-9_45

Table 1. An annotated snippet of a dialog (debate) in ODIN. For brevity, we do not show the full conversation. Anchor's opinion spans are highlighted in blue. [A]: Anchor's utterance, whereas, [S$_j$]: jth speaker's utterance.

Utterance	Opinion
[S$_1$] $\cdots$	
[A] *"Nahi aap aap party ke prvakta ke taur par baithe hai ya political analyst vkhyatigat roop se baithe hai?"* (Are you representing yourself as a party representative or a political analyst?)	No
[S$_2$] *"maine aapse kya kaha sabko is laxman rekha ka samman karna chahiye yeh nirdesh sab par prabhavi roop se lagu hoga"* (What did I say to you? We all have to respect the rule, this rule will be implemented effectively)	–
[A] *"aap thodi na teh karenge jo aapko apmaanjanak lag jaye kuch maine puch liya meri nazar main nahi hai <name> <name> <name> <name> <name> <name> aap abhi lage huye hai brashtachaar ke bhism pitamah doordanth apradhiyo ke sanghrakshan karta <name> avedh sarkar ke kamjoor mukhiya tanashah aap log abhi bhi lage huye hai aap log sudhar hi nahi rahe hai <name> itna samjha rahe aapko"* (You won't decide even if you feel that something is disrespectful, if it is not in my eyes <name> sir <name> sir <name> sir <name> sir <name> sir <name> you are also involved in corruption. You are still not understanding, <name> is trying so hard to make you understand)	Yes
[S$_3$] $\cdots$	–
[A] *"to ab jail bhejiyega na aapko bhi fayda milega unko jail bhejiyega ab agar woh aisa kuch kahe jail bhejiyega unko"* (Send him to jail you will also profit from it. Send him to jail if he says something like this again.)	Yes
[S$_3$] *"pehle sun lijiye to"* (Listen to me first)	–
[A] *"sab ek doosre ko jail bhejiyega sab mile huye hai aapne koi kasar thodi na chhodi hai aap logo ne kya kuch kaha hai <name> ji yeh to ab aapko bhi fayda milega unko jail bhejiyega"* (Send each other to jail. You all are in a cohoots. You have'nt left any stone unturned, you people have said too much, <name> sir, now you will also get the benefit, send them to jail)	Yes
[S$_1$] $\cdots$	

reporters/news anchors either purposefully or unintentionally insert their opinions into otherwise straightforward news articles, thus earning the tags of biased news or biased reporting. They do so by blurring the line between fact-based news reporting and edited-agenda based news reporting. As a consequence, readers and viewers often get confused or get exposed to the targeted viewpoints of others. This practice is even more prevalent in live news reporting or during news debates wherein the content on display is unstructured and somewhat spontaneous compared to written news articles that get processed by many people, including editors, content managers, etc., making the reporter vulnerable to voice out their opinions. To this end, we propose **a novel task of anchor's opinion detection in code-mixed conversations**. The objective of the task is to identify the anchor's utterance when they posit their own opinion in between. Table 1 shows one such instance where the anchor appends their opinion in a news debate. The highlighted blue text signifies the opinionated spans.

Such utterances should not be presented by an anchor, as the reporter is the face of the news media, and news media is supposed to be unbiased and unopinionated. To help and cater to the news anchors and media in promoting unbiased and unopinionated news, our work aims to help organisations detect opinionated utterances. To this end, we first curate `ODIN` (**O**pinion **D**etection **I**n **N**ews) – a dataset developed by transcribing different code-mixed Hinglish news debates from several mainstream national news channels and annotating the utterances of the debate as opinionated/unopinionated. Then, we propose `DetONADe` (**Det**ecting **O**pinion in **N**ews **A**nchor **De**livery) to detect utterances as opinionated and benchmark the task. We also present a detailed analysis of the dataset and necessary evaluation of the obtained results.

In summary, we make the following contributions:

1. We explore opinion detection in a code-mixed (Hinglish) dialogue environment, a **novel task** which, to the best of our knowledge, has never been attempted before.
2. We curate a **new dataset**, `ODIN`, by transcribing various Hinglish news debates from three national news channels and annotating these captioned utterances as opinionated/unopinionated.
3. We perform **extensive analysis** of `ODIN` and provide interesting insights.
4. We benchmark `ODIN` using `DetONADe` and report the necessary results and error analyses.

The source codes and datasets are available at https://github.com/LCS2-IIITD/ODIN-PAKDD.

2   Related Work

Opinion expression is an integral part of opinion mining, and it was first defined as either Direct Subjective Expression (DSEs) or Expressive Subjective Expressions (ESEs) [16]. Following this definition, a fine-grained opinion Mining corpus, namely Multi-Perspective Question Answering (MPQA) was curated for annotating expressions as opinion. Apart from already present datasets, researchers also explored social media, blogs and news articles as opinion mining from heterogeneous information sources can be of great use for individuals, organisations or governments. Ku et al. [8] dealt with the task of opinion extraction, summarisation and tracking on news and blogs corpora, and a lexicon-based feature modelling technique was proposed to extract opinions from documents. Support Vector Machines (SVM) and Decision Trees (C5) were used to predict the results. Breck et al. [2] proposed a Conditional Random Fields (CRF) based model where identifying opinion expression was assumed as a sequence labelling task and achieved expression-level performance within 5% of human inter-annotator agreement. Raina [13] proposed an opinion mining model that leveraged common-sense knowledge from ConceptNet and SenticNet to perform sentiment analysis in news articles, achieving an F1-score of 59% and 66% for positive and negative sentences, respectively. Recently, researchers explored deep learning for opinion detection [6,17].

The scope of this task has always been limited to English language and monolingual settings. There has not been any significant research work on opinion detection in code-mix or Indic languages. But code-mixing is an increasingly common occurrence in today's multilingual society and poses a considerable challenge in various NLP based downstream tasks. Accordingly, there have been some helpful developments in the field of code-mix in the form of sentiment analysis task, humour, sarcasm and hate-speech detection. A dual encoder based model for Sentiment Analysis on code-mixed data was proposed wherein the network consisted of two parallel BiLSTMs, namely the collective and the specific encoder [10]. This model particularly generated sub-word level embeddings with the help of Convolutional Neural Networks (CNNs) to capture the grammar of code-mixed words. Recently, pretrained monolingual and cross-lingual deep learning models were also leveraged for detection of hate-speech and sarcasm on code-mixed data [12] wherein they used fine-tuned RoBERTa and ULMFit for English and Hindi data streams, respectively. For cross-lingual setting, XLM-RoBERTa was fine-tuned on transliterated Hindi to Devanagri text.

The works mentioned above do leverage code-mix text for common downstream tasks. However, no research has been done on opinion detection on code-mix text in sequential data streams. Most opinion detection and sentiment analysis studies have focused on news articles, blogs and movie reviews. In online news articles, every piece is reviewed by multiple people, and thus the scope of opinions is limited compared to news coverage on live media sources. Moreover, biased and opinionated live news anchoring can significantly impact our society and go against the essence of free and fair news reporting. Therefore, we aim to detect opinions amongst news anchors. We focus on code-mix news anchoring mainly in live video debates through national news channels that stick to a mix of Hindi and English language (Hinglish) for news distribution. Moreover, our deep learning model not only leverages context but does so in a sequential manner, thus focusing on the text and the utterances before a statement to classify the text as accurately and robustly as possible.

3   Dataset

In this section, we lay out the details of the dataset development process. First, we extract debate videos from three popular Indian Hindi news Youtube's channels – ABP News[1], Aaj Tak[2], and Zee News[3]. Subsequently, the collected videos were processed to extract the romanized Hinglish code-mixed utterances. Each utterance is uttered either by the anchor or by the invited speakers. To ensure sanity, we do not identify the utterances with the speaker's name; instead, we assign ids (A for the anchor, and $\{S_1, S_2, \cdots, S_n\}$ for the invited speakers) to each utterance. Next, we annotate the anchor's utterances as opinionated[4] or non-opinionated depending upon the dialog conversation. A high-level dataset development process is outlined in Fig. 1.

---

[1]   https://www.youtube.com/c/abpnews.
[2]   https://www.youtube.com/c/aajtak.
[3]   https://www.youtube.com/c/zeenews.
[4]   A personal sentiment, which describes the anchor's feeling on the topic [11].

Table 2. Dataset statistics of `ODIN`.

Features	Value
Number of dialog (debate) videos	46
Average length of the videos	33 mins
Number of utterances	4490
Number of anchor utterances	2054
Number of opinionated anchor utterances	**597**
Number of tokens	261811
Number of unique tokens (vocabulary)	20023
Average number of utterances per dialog	97.6
Maximum number of utterances in a dialog	233
Average number of words per utterance	58
Maximum number of words in an utterance	1192

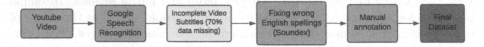

Fig. 1. Dataset annotation pipeline.

Prepossessing. We collect 46 debate[5] videos for two broad topics as religious and political. Initially, we obtain transcriptions of these videos using the Google Speech Recognition[6] tool. The obtained output had many missing words, possibly due to the background noise or due to the code-mixed nature of the conversation; therefore, we manually add the missing words to complete the utterances. Furthermore, we observe many spelling mistakes in English words – '*laiv*' for '*live*', '*ophis*' for '*office*', '*daunalod*' for '*download*', etc. To fix these spelling mistakes, we try mapping words in the English dictionary to the words in question based on word similarity. We use phonological similarity to achieve this. We employ Libindic's Soundex library[7] to obtain the correct mapping.

Annotation. Each debate has a series of utterances – some of them were uttered by the anchor and others by the invited speakers. We employ two annotators to annotate the anchor's utterances as opinionated or non-opinionated. Since the objective of the current work is to identify opinions of anchors, we do not annotate speaker's utterances. Both annotators read the utterances of the debate and annotate the whole data. To check the inter-rater agreement, we compute Cohen's κ value of 0.88. Subsequently, we perform a consolidated step to include only those annotations where both annotators agree on the opinionated label –

---

[5] Henceforth, we will use debate, dialogue, and conversation interchangeably to signify a sequence of utterances.
[6] https://pypi.org/project/SpeechRecognition/.
[7] https://libindic.org/Soundex.

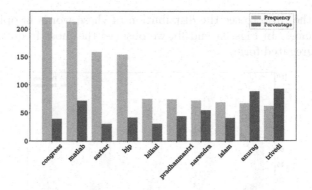

Fig. 2. Top 10 most occurring words in opinionated anchor utterances.

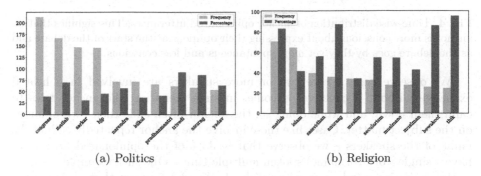

(a) Politics   (b) Religion

Fig. 3. Topic-wise top 10 most occurring words in opinionated anchor utterances.

we treat disagreement as non-opinionated. Table 1 shows an annotated dialog. For brevity, we show only a snippet of the dialog. There are three speakers and one anchor debating over a topic. Out of all utterances in the dialog, we show the annotated anchor's utterances as opinionated and non-opinionated.

Statistics. A detailed statistic of ODIN is listed in Table 2. There are total 46 debate videos with an average length of ~33 mins. In total, there are 4490 utterances – 2054 anchor utterances and 2436 other speakers utterances. Out of 2054 anchor utterances, 597 of them are opinionated, accounting for approximately 30% of the utterances.

Dataset Insights. We analyze the dataset to gain insight of the inherent pattern in opinionated utterances. Apart from various topic-related terms such as *BJP*[8], *Congress*(see footnote 8), *Islam*, etc., corresponding to the political and religious topics, we observe opinionated words like '*bilkul*' (certainly), '*matlab*' (means), '*theek*' (ok), etc., have a significant presence in opinionated utterances. We depict a bar graph of the top-10 most frequent words in opinionated utterances in Fig. 2. Moreover, to comprehend whether the frequent words are opinion specific or not,

---

[8] BJP and Congress are two major political parties in India.

we also plot the ratio to see the distribution of these words in opinionated v/s all the utterances. In Figs. 3a and 3b, we observe the most frequent words in a topic-wise segregated form.

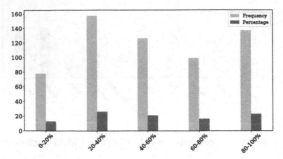

Fig. 4. Time-wise distribution of anchor opinionated utterances. This signifies that an anchor is more conscious about expressing their opinions at the start of the debate and as the debate goes by they get more spontaneous and less conscious.

We observe cases where two or more speakers are involved in a heated exchange without a concrete outcome. In such scenarios, the anchor tries to calm them down, and while doing so, the anchor often slide their own opinions on the subject matter. Cases like these involve the anchor repeatedly calling the name of the speakers – we observe that ~ 33% of the opinionated utterances have a single word (or name) spoken multiple times. One such example is shown in the anchor's second utterance in Table 1. We also find out that anchors tend to ask more questions in an opinionated utterance – on average, 1.45 question words are present in an opinionated utterance, whereas, only 0.9 question words are there in a non-opinionated utterance.

On careful analysis, we observe that an anchor is relatively more likely to express personal opinion at the later stage of the debate rather than at the beginning of the debate. We plot the distribution of the opinionated utterances on the time scale in Fig. 4. As we can see that only 78 utterances are opinionated during the first 20% of the debate duration, which increases to 157 during 20–40%, 126 during 40–60%, 99 during 60–80%, and 137 during 80–100% of the debate duration. This signifies that an anchor is more conscious about expressing their opinions at the start of the debate and as the debate goes by they get more spontaneous and less conscious.

4   Proposed Benchmark Model

In this section, we describe our proposed benchmark model, DetONADe that we adopt for the anchor opinion detection task. Since the number of the opinionated anchor utterances are significantly few compared to the total number of utterances in the dataset, we adopt an instance-based modeling for the detection. For each anchor's utterance u_t, we create an instance that contains all previous utterances $(u_1, u_2, \cdots, u_{t-1})$ of the dialog as context and the target utterance

u_t as the last utterance of an instance. For each instance, we aim to classify the target utterance as opinionated or non-opinionated. We hypothesize that the fixed context will provide appropriate clue about the debate and, at the same time, restrict the model not to overwhelm itself in comprehending the desired context rather than focusing on the opinion discovery. A high-level architecture diagram for the anchor's opinion detection task is depicted in Fig. 5.

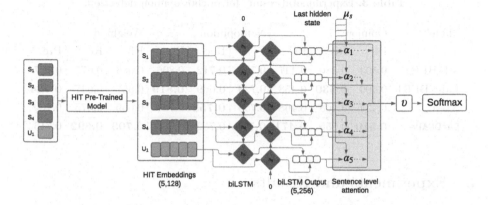

Fig. 5. The proposed `DetONADe` architecture for the anchor opinion detection.

We feed each instance one-by-one to `DetONADe` as input. Since code-mixed texts are susceptible to the spelling variations and various other kinds of noise, representations learned at the sub-word level often counter such variation quite efficiently. Recent literature shows that a wide range of character and sub-word level code-mixed representation models outperform word-level representation models for numerous tasks. Some of them are HIT [14], CS-ELMO [1], CNN_LSTM [7], etc. We employ HIT (Hierarchically attentive Transformer), the most recent and robust representation learning method for code-mixed text among them, to capture the semantic and syntactical features of the debate. It encodes the code-mixed utterance in the embedding space where the semantic difference among various spelling variations of the same word is minimal.

We obtain representation for each utterance in an instance and feed them through a biLSTM layer for sequence learning. The biLSTM layer captures the cross-sentence relationships across the utterances by exploiting the conversation dynamics of the dialog and subsequently learns latent representations $\overrightarrow{h_i}$ for each utterance u_i. Next, we apply the multi-headed self-attention mechanism [15] to identify the importance of contextual utterances considering the target utterance u_t. To this end, we treat the target utterance h_t as the instance-level context vector μ_s and compute the interactions between the context vector and every utterance in the dialog through an interactive attention mechanism. The intuition is to obtain an abstract view of the instance that should help in exploiting the dialog dynamics corresponding to the target utterance in a better way. Subsequently, we accumulate the attention weights through a weighted summation and obtain the final vector as v.

$$\hat{h}_i = tanh(h_i); \quad \alpha_i = \frac{\exp(\hat{h}_i^T \mu_s)}{\sum_j \exp(\hat{h}_j^T \mu_s)}; \quad v = \sum_i \alpha_i \hat{h}_i$$

Finally, we feed the vector v to the softmax classifier for classifying the target utterance as opinionated or non-opinionated.

Table 3. Experimental results for anchor opinion detection.

Model	Opinion			Non-opinion			Weighted		
	F1	Rec	Pre	F1	Rec	Pre	F1	Rec	Pre
ML-BERT	0.503	0.580	0.447	0.756	0.712	0.806	0.686	0.675	0.707
Indic-BERT	0.500	**0.630**	0.458	0.723	0.669	**0.819**	0.657	0.647	**0.724**
XLM	0.424	0.468	0.468	0.759	**0.758**	0.787	0.669	0.675	0.700
DetONADe	**0.510**	0.555	**0.471**	**0.778**	0.752	0.806	**0.703**	**0.692**	0.715

5   Experiments and Results

In this section, we report our experimental results and error analysis.

Baselines. Since opinion mining in code-mixed conversations is relatively unexplored arena, we include various code-mixed representation learning-based system as our baselines. In particular, we employ multi-lingual BERT (mBERT) [4], XLM-RoBERTa [3], and IndicBERT [9] based embedding models to extract the utterance representation. Subsequently, we fine-tune each of these systems through a biLSTM layer followed by a linear layer with softmax classification.

Experiment Setup. Since ODIN is skewed towards the non-opinionated anchor's utterance, we perform oversampling at the instance-level for the opinionated utterances and obtain the equal number of opinionated and non-opinionated instance. For experiments, we perform 3-fold cross-validation and report the average for each case. Note that the oversampling is performed only for the training set in each fold. All experiments are performed on a 12GB K80 Tesla GPU server.

For creating an instance, we vary the size of context from 1 to 7 and observe the best performance with context 5, i.e., $\langle (u_{t-5}, u_{t-4}, u_{t-3}, u_{t-2}, u_{t-1}), u_t \rangle$ as an instance. Furthermore, during experiments, we face a subtle challenge in obtaining the utterance representation for lengthy utterance (number of tokens > 512), since most of the pre-trained language models (PLM) do not comprehend sentences more than 512 tokens. In such cases, a typical solution is to clip the utterance at index 512. However, in this work, we exploit an alternative without omitting the content. We split the lengthier utterances into k chunks of 512 tokens. Subsequently, we obtain representations for each of these k chunks and consolidate them by taking an average of the k representations.

Table 4. Token-level confusion matrix. We show performance w.r.t. a few critical words in anchors' utterances.

Words	TP	FN	TN	FP
Congress	33	19	85	34
Bjp	18	11	46	15
Modi	24	16	77	27
Gandhi	33	18	63	37
Bilkul	14	15	45	20
Hindu	14	10	50	20
Muslim	6	4	14	6
Question-based (kyu, kya, kab, kaha, kaun, kitne, kaise)	70	47	235	102

Results and Comparative Study. We report the results of DetONADe along with other baselines in Table 3. For each case, we compute the weighted-F1 scores. Moreover, we report the class-wise precision, recall, and F1 for the opinionated and non-opinionated cases as well. We observe that DetONADe records the best weighted F1-score of 0.703 in comparison with 0.686 weighted F1-score of the best baseline, ML-BERT. Among all baselines, Indic-BERT has the least score at 0.657 weighted F1-score.

We further observe the class-wise performance of all systems. For the opinion class, DetONADe yields 0.510 F1-score, whereas, it obtains F1-score of 0.778 for the non-opinionated class. Similar to the weighted case, we obtain inferior results for all baselines in both classes. Another observation is that the performance for the non-opinionated class, irrespective of the model, is better than the opinionated class. We relate this to the complex nature of the anchor opinion detection task, where it is extremely challenging to comprehend the intended opinion especially in a conversational setup.

Error Analysis. In this section, we both quantitatively and qualitatively analyse of the results obtained from DetONADe. As we observe in Table 3, a relatively lower F1-score for the opinionated class suggests a significant number of false positives and false negatives. Moreover, we also observe a relatively higher false positives than the false negatives, thus reporting an inferior precision score. This could be due to presence of a few words which are highly inclined towards one class of utterance, as depicted in Fig. 2.

Therefore, in Table 4, we investigate the words that were prevalent in the dataset, and their distribution in the results we obtained. We observe that utterances with reference to the two major Indian political parties (*viz.* 'congress' and 'bjp') caused more false positives than false negatives. On the other hand, question-based utterances (ones that usually start with words like 'kyu' (why), 'kya' (what), 'kab' (when), 'kaha' (where), 'kitne' (how many), 'kaise' (how)) have a very high false negative rate as compared to the false positive. For other words like 'modi' and 'gandhi' (who are political figures) proportionally have similar false positive and negative values. We also observe similar trends for the words representing two major religions in India, e.g., 'hindu', 'muslim', etc.

In Table 5, we report two mis-classified instances – one for the false positive, while another for the false negative. We speculate that, in the first case, DetONADe focuses too much on the contextual utterances and due to the presence of allegations in the second utterance of the instance, it classifies the instance as opinionated rather than non-opinionated. On the other hand, in the second example, a small portion of the target utterance (i.e., 'tabhi yeh haal hai') suggests an opinion, which the model could not comprehend as the opinionated instance. Moreover, we observe a few mis-classified examples that are at the opposite end of the spectrum – one requires context to get a sense of opinion whereas, for others, context is creating noise in the model. Such observation also signifies the subtleness of the proposed task.

Table 5. Examples of test instances which were wrongly predicted by DetONADe.

		Debate	Gold	Predicted
Instance#	1	Context	[A] *"Aapko kya lagta hai <name1> nahi toh <name2>?"* (What do you think, if it is not sonia then it is rahul?)		
			[S₅] *"jeet haar ek vishe hai aaj jeet ke bhi log haar ja rahe hai kyu aap main wahi bata raha hun aapki baat ka jawaab de raha hun jeet haar ek vishe hai aaj log jeet ke bhi haar jaate hai bjp mla kharid leti hai"* (Winning and losing is one topic, today, even after winning people are losing. I am answering to your question only. Winning and losing is one topic, today, even after winning people are losing, bjp buys mlas (Member of the Legislative Assembly).)		
			[A] *"nahi nahi aapko kya lagta hai <name1> ya <name2> aapko kya lagta hai <name1> <name1> ke baad ab <name2>"* (No no what do you think? <name1> or <name2>, what do you think? Will it be <name2> after <name1>?)		
			[S₅] *"isliye toh loktantr ka loktantr ka ki hatya kar di hai lekin ab sawaal yeh paida hota hai ki woh chahe rahul ho chahe sonia ho koi varshit neta ya jo bhi log congress ke bhavishay mein us kursi ko sambhalenge woh vishe abhi baad mein aata hai"* (This is the reason for the death of democracy. The question is be it rahul or sonia or some other politician or anyone who will lead and take the responsiblity of that position in congress in the future, this topic is of future.)		
		Target	[A] *"<name1> ke <name1> ke baad ab laut ke party <name2> pe aayegi"* (Will the party fallback to <name2> after <name1> leaves?)	No	Yes
	2	Context	[A] *"<name> thik thik <name> <name> <name>"* (<name> fine. fine. <name>, <name>, <name>)		
			[S₁] *"Ji"* (Yes.)		
			[A] *"Ji Ji toh main supreme court ko argue kar raha hun aapko supreme court pe bharosa nahi hai aapko air chief marshal par bharosa nahi hai thik hai <name> thik"* (Yes, Yes, I am arguing the supreme court. Don't you trust the supreme court? Don't you trust the air chief marshal? Ok <name>)		
			[S₆] *"apna jo opposition hai woh weak hai the congress was weak"* (Our opposition is weak. Congress was weak.)		
		Target	[A] *"sir yeh weak hai tabhi yeh haal hai chaliye thik hai <name> main aapke paas aa raha hun <name> kya kya rafal ko thik kya rafal ko bofors banane ki koshish ho rahi hai kya yeh rafal ko bofors banane ki koshish hai kya <name>"* (Sir, it is weak that's why the conditions are like this. Anyway, <name> I am coming over to you. Is this an attempt to make Rafael deal like Bofors scam?)	Yes	No

6 Conclusion and Future Work

In this work, we proposed a novel task of anchor's opinion detection in code-mixed conversations. To this end, we curated ODIN, a first of its kind dataset by transcribing various debate videos from mainstream Indian news channels. We performed extensive analyses on ODIN, and reported interesting findings. Furthermore, we benchmark the ODIN dataset using DetONADe – an interactive attention-based framework on top to several pretrained code-mixed representation models. Moreover, we conducted error analysis on the outputs of DetONADe. In future work, we plan to extend the dataset with more opinionated samples as well as other varieties of debates. We also wish to explore the multi-modality for the opinion detection.

Acknowledgement. The authors would like to acknowledge the support of the Ramanujan Fellowship (SERB, India), Infosys Centre for AI (CAI) at IIIT-Delhi, and ihub-Anubhuti-iiitd Foundation set up under the NM-ICPS scheme of the Department of Science and Technology, India.

References

1. Aguilar, G., Solorio, T.: From English to code-switching: transfer learning with strong morphological clues. In: Proceedings of the ACL, pp. 8033–8044 (2020)
2. Breck, E., Choi, Y., Cardie, C.: Identifying expressions of opinion in context. In: Proceedings of the IJCAI, pp. 2683–2688 (2007)
3. Conneau, A., et al.: Unsupervised cross-lingual representation learning at scale. CoRR abs/1911.02116 (2019)
4. Devlin, J., Chang, M., Lee, K., Toutanova, K.: BERT: pre-training of deep bidirectional transformers for language understanding. CoRR abs/1810.04805 (2018)
5. FICCI-KPMG: #shootingforthestars: India media and entertainment industry report 2015 (2015). https://assets.kpmg/content/dam/kpmg/pdf/2015/03/FICCI-KPMG_2015.pdf
6. İrsoy, O., Cardie, C.: Opinion mining with deep recurrent neural networks. In: Proceedings of the EMNLP, pp. 720–728. Doha, Qatar (2014)
7. Joshi, A., Prabhu, A., Shrivastava, M., Varma, V.: Towards sub-word level compositions for sentiment analysis of Hindi-English code mixed text. In: Proceedings of the COLING, pp. 2482–2491 (2016)
8. Ku, L.W., Liang, Y.T., Chen, H.H.: Opinion extraction, summarization and tracking in news and blog corpora (2006)
9. Kunchukuttan, A., Kakwani, D., Golla, S., Bhattacharyya, A., Khapra, M.M., Kumar, P.: AI4Bharat-IndicNLP corpus: monolingual corpora and word embeddings for Indic languages. arXiv:2005.00085 [cs] (2020)
10. Lal, Y.K., Kumar, V., Dhar, M., Shrivastava, M., Koehn, P.: De-mixing sentiment from code-mixed text. In: Proceedings of the ACL: Student Research Workshop, pp. 371–377. Florence, Italy (2019)
11. Munezero, M., Suero Montero, C., Sutinen, E., Pajunen, J.: Are they different? affect, feeling, emotion, sentiment, and opinion detection in text. IEEE Trans. Affect. Comput. **5**, 101–111 (2014)

12. Pant, K., Dadu, T.: Towards code-switched classification exploiting constituent language resources. In: Proceedings of the AACL-IJCNLP: Student Research Workshop, pp. 37–43. Suzhou, China (2020)
13. Raina, P.: Sentiment analysis in news articles using sentic computing. In: Proceedings of the ICDM, pp. 959–962 (2013)
14. Sengupta, A., Bhattacharjee, S.K., Chakraborty, T., Akhtar, M.S.: HIT - a hierarchically fused deep attention network for robust code-mixed language representation. In: Findings of the ACL-IJCNLP, pp. 4625–4639 (2021)
15. Vaswani, A., et al.: Attention is all you need (2017)
16. Wiebe, J., Wilson, T., Cardie, C.: Annotating expressions of opinions and emotions in language. Lang. Resour. Eval. **39**, 165–210 (2005)
17. Xie, X.: Opinion expression detection via deep bidirectional C-GRUS. In: Proceedings of the 28th International Workshop on Database and Expert Systems Applications (DEXA), pp. 118–122. Los Alamitos, CA, USA (2017)

A Hybrid Semantic-Topic Co-encoding Network for Social Emotion Classification

Lu Dai[1], Bang Wang[1], Wei Xiang[1], Minghua Xu[2], and Han Xu[2(✉)]

[1] School of Electronic Information and Communications,
Huazhong University of Science and Technology (HUST), Wuhan, China
{dailu18,wangbang,xiangwei}@hust.edu.cn
[2] School of Journalism and Information Communication,
Huazhong University of Science and Technology (HUST), Wuhan, China
{xuminghua,xuh}@hust.edu.cn

Abstract. Social emotion classification is to predict the distribution of readers' emotions evoked by a document (e.g., news article). Previous work has shown that both semantic and topical information can help improve classification performance. However, many existing topic-based neural models represent the topical feature of document with only topic probabilities, ignoring the fine-grained semantic feature of terms in each topic. Moreover, traditional RNN-based semantic networks often face the disadvantage of slow training. In this paper, we propose a hybrid semantic-topic co-encoding network. It contains a semantics-driven topic encoder to compose topic embeddings, and also utilizes a forward self-attention network to exploit document semantics. Finally, the semantic and topical features of the document are adaptively integrated through a gate layer, which generates the document representation for social emotion classification. Experimental results on three public datasets show that the proposed model outperforms the state-of-the-art approaches in terms of higher accuracy and average Pearson correlation coefficient. Moreover, the proposed model runs fast and with better explainability.

Keywords: Sentiment analysis · Social emotion classification · Topic model · Self-attention

1 Introduction

Social emotion classification is to predict the emotion distribution evoked by an article among numerous readers. Nowadays, many online news websites and social media allow people to express their emotion reactions after reading an article. With the popularity of the Internet, the production and spread of news and reader responses have become much faster than before. Social emotion classification has thus become very valuable in many applications like the analysis of public opinions [8], recommendation system [20], etc.

Supported by National Natural Science Foundation of China (Grant No: 62172167).

The methods for social emotion classification can mainly be divided into three categories, i.e., word-based, document topic-based and document semantics-based models. Word-based models utilize the word-level features (e.g., emotion dictionary) and focus on modeling the direct relations between words and social emotions [1,6]. For example, Katz et al. [6] have proposed the SWAT system to calculate the emotion valence of a given headline based on the manually created emotion dictionary. Because of treating each word separately, these models cannot distinguish between different emotions expressed by a same word in various context. To solve this problem, document topic-based methods try to explore topic-level features [13,14] such as the co-occurrence of topics and emotions. The topic is defined as the probability distribution of a set of words, which is the same as in the *Latent Dirichlet Allocation* (LDA) topic model [2]. These methods usually build a latent topic model with emotional layer to jointly capture the generative process of words, topics and emotions. For example, Rao et al. [14] have presented the affective topic model to design an emotional layer into the LDA. However, both word-based and document topic-based models follow the assumption of bag-of-words, so they have not considered the word-ordering information and the text semantics. Afterwards, due to the powerful ability of neural networks to learn semantic features, many document semantics-based models have been proposed for social emotion classification [10,22]. *Convolutional neural networks* (CNN) or *recurrent neural networks* (RNN) are often adopted as the semantic encoder to generate the vector representation of a document. For example, Zhao et al. [22] have used both the bidirectional long short-term memory (BiLSTM) network and the CNN to learn the word-ordering feature and the local n-gram feature.

However, many document semantics-based models ignore the document topic feature that has been shown to be effective for social emotion classification in previous work. Thus, to simultaneously utilize the topical and semantic information, Li et al. [9] have developed a hybrid neural network that can learn the topic-level feature from the pre-trained topic model (i.e., BTM [3]). Wang et al. [19] have proposed the Gated DR-G-T model that includes a recursive neural network for learning semantics. It also leverages the topical information extracted by the LDA topic model. Nevertheless, there are still two problems to solve: (1) Many existing neural network models only use the probabilities of topics to represent the document topical information, ignoring the fine-grained semantic feature of terms in each topic; (2) although recursive or recurrent neural networks are helpful for extracting semantics, they are usually slow to train.

To address the above problems, we attempt to utilize the semantic feature of terms in each topic to compose the topic embedding and then generate the document topic vector based on the topic probabilities. We also design a light forward self-attention network with fewer parameters and high computational efficiency for learning document semantics.

Specifically, in this paper, we propose a hybrid semantic-topic co-encoding network (STN) for social emotion classification. In the STN, we present a semantics-driven topic encoder that can generate the semantics-aware topic

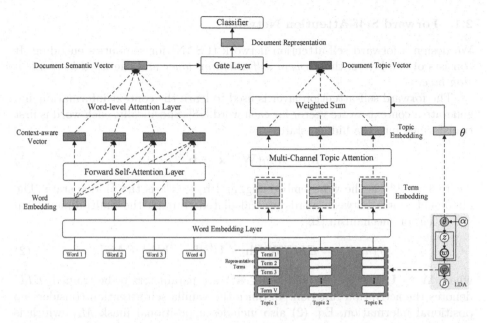

Fig. 1. The overall framework of the proposed STN. The left part is the forward self-attention network (FSAN). The right part is the semantics-driven topic encoder (SDTE). K denotes the number of topics. V is the length of the vocabulary. The top terms with highest probabilities in each topic are regarded as the representative terms of the topic.

embedding by using the feature of terms in each topic. For semantic encoding, we design a forward self-attention network with high computational efficiency, to learn context information and compose the document semantic vector. Finally, the document semantic vector and the document topic vector are integrated through a gate layer, which outputs the document representation used for emotion prediction. Experimental results on three public datasets reveal that the proposed STN outperforms the state-of-the-art methods in terms of higher accuracy and higher average Pearson correlation. It also runs fast and with better explainability.

2 The Proposed Model

The overall framework of our proposed STN is presented in Fig. 1. It contains three components: a semantic encoder based on forward self-attention network, a semantics-driven topic encoder and a gate layer for the fusion of the semantic vector and the topic vector.

2.1   Forward Self-Attention Network

We design a forward self-attention network (FSAN) for semantics encoding. It consists of two layers: the *forward self-attention layer* and the *word-level attention layer*.

The forward self-attention layer is used to learn the context information and generate a context-aware vector for each word [18]. Specifically, each word is first transformed into the hidden state:

$$\mathbf{h}_i = tanh(\mathbf{W}^{(h)}\mathbf{x}_i + \mathbf{b}^{(h)}) \tag{1}$$

where $\mathbf{x}_i \in \mathbb{R}^{d_w}$ is the word embedding, and $\mathbf{h}_i \in \mathbb{R}^{d_h}$ is the hidden state. The alignment scores between words are calculated through the multi-dimensional self-attention mechanism [16]:

$$f(h_i, h_j) = ELU(\mathbf{W}^{(s)}\mathbf{h}_i + \mathbf{U}^{(s)}\mathbf{h}_j + \mathbf{b}^{(s)}) + M_{ij} \tag{2}$$

where $\mathbf{W}^{(s)}, \mathbf{U}^{(s)} \in \mathbb{R}^{d_h \times d_h}$ and $\mathbf{b}^{(s)} \in \mathbb{R}^{d_h}$ are parameters to be trained. ELU denotes the activation function [4]. Since the vanilla self-attention considers no positional information, Eq. (2) also includes a positional mask M_{ij}, which is defined as follows:

$$M_{ij} = \begin{cases} 0, & i \geq j \\ -\infty, & i < j \end{cases} \tag{3}$$

There exists an attention score between the j-th word w_j and the i-th word w_i only if w_j appears before w_i. As a result, the forward positional information is encoded into the alignment scores $f(h_i, h_j)$.

Based on alignment scores, the context-aware vector of w_i is composed as follows:

$$a_{ij} = \frac{exp(f(h_i, h_j))}{\sum_{j'=1}^{N} exp(f(h_i, h_{j'}))}, \quad \mathbf{z}_i = \sum_{j=1}^{N} a_{ij} \odot \mathbf{h}_j \tag{4}$$

where $\odot$ denotes the point-wise product and N represents the length of document. $\mathbf{z}_i \in \mathbb{R}^{d_h}$ is the context-aware vector that captures the context-based semantic feature of w_i.

As not all words are equally important to document semantic representation, the word-level attention layer contains an attention mechanism to compress context-aware vectors of all words into the document vector. Similar to [21], the alignment function is:

$$f(z_i) = \mathbf{q}^T tanh(\mathbf{W}^{(w)}\mathbf{z}_i + \mathbf{b}^{(w)}) + b \tag{5}$$

where $\mathbf{q} \in \mathbb{R}^{d_h \times 1}$ is randomly initialized and updated with network training. Thus, the document semantic vector $\mathbf{v}_d \in \mathbb{R}^{d_h}$ is composed as follows:

$$a_i = \frac{exp(f(z_i))}{\sum_{j'=1}^{N} exp(f(z_{j'}))}, \quad \mathbf{v}_d = \sum_{i=1}^{N} a_i \odot \mathbf{z}_i \tag{6}$$

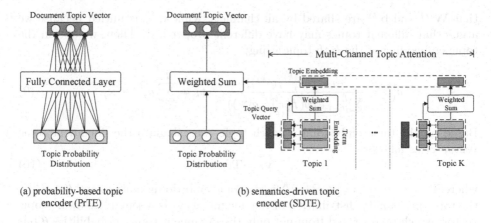

Fig. 2. The structure comparison of the traditional PrTE (a) and our SDTE (b).

2.2 Semantics-Driven Topic Encoder

We propose a novel semantics-driven topic encoder (SDTE) for topic encoding. As presented in the right part of Fig. 1, a LDA model is pre-trained to infer topics. $\theta \in \mathbb{R}^K$ represents the topic probability distribution of the document, and $\varphi = [\varphi_1, \varphi_2, \ldots, \varphi_K]$ represents the topic-term distributions. K is the number of topics.

In previous work, the document topical information is then encoded through a probability-based topic encoder (PrTE) [19], where the document topic vector is transformed from only the topic probabilities θ with a fully connected layer. The fine-grained feature of terms in each topic is not considered. As Fig. 2 shows, our SDTE is very different from the PrTE. It composes the topic embedding for each topic by using the word embeddings of representative terms in the topic. The word embedding captures the semantic feature of the term [12]. Then the topic vector of the document is generated based on both the topic embeddings and the topic probabilities θ.

Specifically, we first build a word embedding matrix for each topic using the top L words in φ_k, $k \in [1, K]$, i.e.,

$$\mathbf{X}^{(k)} = [\mathbf{x}_1^{(k)}, \mathbf{x}_2^{(k)}, \ldots, \mathbf{x}_L^{(k)}] \tag{7}$$

The top L words are regarded as the representative terms in each topic. Afterwards, we develop a multi-channel topic attention layer to compose the topic embedding from $\mathbf{X}^{(k)}$, $k \in [1, K]$. One channel corresponds to one topic. For each channel, the alignment function is defined as follows:

$$f(x_i^{(k)}, q^{(k)}) = \mathbf{q}^{(k)} ELU(\mathbf{W}^{(t)} \mathbf{x}_i^{(k)} + \mathbf{b}^{(t)}) \tag{8}$$

where $\mathbf{q}^{(k)} \in \mathbb{R}^{d_w}$ denotes the query vector of the k-th topic. It is randomly initialized and updated with network training. The alignment function assigns larger weight scores to the words that are more representative in the topic. Notice that $\mathbf{W}^{(t)}$ and $\mathbf{b}^{(t)}$ are shared by all channels, while $\mathbf{q}^{(k)}$ is untied because we argue that different topics may have different properties[1]. Then, we obtain the topic embedding of the k-th topic using:

$$a_i^{(k)} = \frac{exp(f(x_i^{(k)}, q^{(k)}))}{\sum_{i'=1}^{L} exp(f(x_{i'}^{(k)}, q^{(k)}))}, \quad \mathbf{t}_k = \sum_{i=1}^{L} a_i^{(k)} \odot \mathbf{x}_i^{(k)} \tag{9}$$

Based on the topic embedding of each topic, we calculate the document topic vector $\mathbf{v}_t$ as follows:

$$\mathbf{v}_t = \mathbf{T}\theta \tag{10}$$

where $\mathbf{T} = [\mathbf{t}_1, \mathbf{t}_2, ..., \mathbf{t}_K] \in \mathbb{R}^{d_w \times K}$ is the topic embedding matrix, and $\theta \in \mathbb{R}^K$ is the topic probability distribution of the document. $\mathbf{v}_t$ is a semantics-aware topic vector, which is generated from not only the document topic probabilities θ but also the fine-grained semantic feature of the representative terms in each topic.

2.3   Gate Layer

We apply a gate layer to compose the final document representation by integrating the document semantic vector and the document topic vector. The transition functions are as follows:

$$\mathbf{g} = \sigma(\mathbf{W}^{(f)}\mathbf{v}_d + \mathbf{U}^{(f)}\mathbf{v}_t + \mathbf{b}^{(f)}) \tag{11}$$

$$\mathbf{v}_f = \mathbf{g} \odot \mathbf{v}_d + (1 - \mathbf{g}) \odot \mathbf{v}_t \tag{12}$$

where σ denotes the *sigmoid* activation function. $\mathbf{W}^{(f)}$, $\mathbf{U}^{(f)}$ and $\mathbf{b}^{(f)}$ are trainable parameters. $\mathbf{g} \in \mathbb{R}^{d_h}$ is the gate vector to control the importance weights of semantic information and topical information.

3   Experiments

3.1   Datasets

We evaluate the effectiveness of our model on three public datasets.

SinaNews. It contains 5,258 Chinese news articles collected from the Sina News website [9]. Each sample includes the reader votes over 6 emotion labels (i.e., anger, touch, sadness, amusement, curiosity and surprise). 3,109 articles are used as training set, and 2,149 articles are used as testing set.

SemEval. It is provided in SemEval-2007 task 14, and contains 1,250 English news headlines [17]. Each headline is scored from 0 to 100 over 6 emotions (i.e., anger, disgust, fear, joy, sadness and surprise). 1,000 headlines are used as training set, and 246 headlines are used as testing set. Notice that 4 samples without scores are removed.

---

[1] For example, the representative terms may be related to "sadness" in some topics, and may be related to "happiness" in some other topics.

ISEAR. This is a single label English dataset [15]. It has 7,666 instances, each of which is a textual description of the situation that can evoke one of the seven emotions (i.e., joy, fear, anger, sadness, disgust, shame, and guilt). 60% of the dataset is used for training, and the rest is used for testing.

The splitting method for each of the three datasets is the same as the previous work.

3.2 Experimental Settings

Tokenization. We use jieba[2] on the Chinese dataset (SinaNews) and NLTK[3] on the English dataset (SemEval, ISEAR).

Word Embedding. For SinaNews, we pre-train a Word2vec model ($d_w = 100$) over the large-scale Chinese Wikipedia corpus[4]. For SemEval and ISEAR, we use the public English Word2vec model provided by Google[5] ($d_w = 300$). $d_h = d_w$ in our experiments.

Topic Encoder. For each dataset, the vocabulary is built based on the training set with removal of stop words and words occurring no more than three times. L is set to 50, 50, 100 on SinaNews, ISEAR and SemEval, respectively.

Training. We use Adam optimizer [7] with a learning rate of 0.003, and the batch size is 20. All weight matrices are initialized by Glorot Initialization [5], and the biases are initialized to zero. All models are implemented with PyTorch[6] and run on single Nvidia GTX 1080Ti graphic card.

3.3 Evaluation Metrics

Following previous work, we adopt Acc@1 and Pearson's r as our metrics. Acc@1 represents the accuracy of top-ranked emotion. Pearson's r denotes the Pearson correlation between the predicted emotion distribution and the gold distribution. The Acc@1 is computed as follows:

$$Acc@1 = \frac{\sum_{d \in \mathcal{D}_{test}} \mathbb{I}_d}{|\mathcal{D}_{test}|}, \quad \mathbb{I}_d = \begin{cases} 1, & \text{if } \hat{y}_{top} = y_{top} \\ 0, & \text{otherwise} \end{cases}$$

The Pearson correlation is computed as follows:

$$r = \frac{\sum_{\mathcal{D}_{test}} r'(\hat{\mathbf{y}}, \mathbf{y})}{|\mathcal{D}_{test}|}, \quad r'(\hat{\mathbf{y}}, \mathbf{y}) = \frac{cov(\hat{\mathbf{y}}, \mathbf{y})}{\sqrt{var(\hat{\mathbf{y}})var(\mathbf{y})}},$$

where $\hat{y}_{top}$ is the predicted top-ranked emotion, y_{top} is the gold top-ranked emotion, and $|\mathcal{D}_{test}|$ means the size of the testing set. cov denotes the covariance.

---

[2] https://github.com/fxsjy/jieba.
[3] https://www.nltk.org/.
[4] https://dumps.wikimedia.org/.
[5] https://code.google.com/archive/p/word2vec/.
[6] https://pytorch.org/.

Table 1. Classification performance on the three datasets. The first group is baseline methods. The second group includes our proposed STN and its variants. We report mean results of our experiments over five runs ($p < 0.05$, paired t-test).

Models	SinaNews		ISEAR		SemEval	
	Acc@1	Pearson's r	Acc@1	Pearson's r	Acc@1	Pearson's r
SWAT [6]	0.3897	0.40	0.2629	0.21	0.3699	–
CSTM [13]	0.4074	0.43	0.2823	0.19	–	–
ETM [1]	0.5419	0.49	0.4879	0.35	0.3544	–
WMCM [11]	–	–	–	–	0.4171	–
1-HNN-BTM [9]	–	–	0.5121	0.40	–	–
AttBiLSTM [23]	0.6295	0.6814	0.5965	0.5595	0.4783	0.4515
Gated DR-G-T [19]	0.6520	0.7123	0.6044	0.5726	0.5032	0.4866
FSAN	0.6537	0.7144	0.6142	0.5835	0.5016	0.4871
STN(PrTE)	0.6565	0.7237	0.6167	0.5872	0.5236	0.4938
STN	**0.6624**	**0.7273**	**0.6209**	**0.5902**	**0.5285**	**0.5031**

3.4   Comparison Models

SWAT [6] is the best-performed system in SemEval-2007 task 14. It predicts the emotion scores of a document based on a word-emotion mapping dictionary.

ETM [1] is a document topic-based method that adds an emotional layer into the LDA and jointly models emotions and topics.

CSTM [13] tries to explicitly distinguish context-independent topics from both a background theme and a contextual theme.

WMCM [11] introduces the concept of "emotional concentration" to compute the weight of the document for each emotion. It also uses a topic model and infers the social emotions with the Bayesian theory.

1-HNN-BTM [9] builds a fee dforward neural network that exploits the topic feature from BTM [3].

AttBiLSTM [23] is a bidirectional long short-term memory network with attention, which has been widely used as a document semantic encoder.

Gated DR-G-T [19] learns the document semantics with a recursive neural network. A multilayer perceptron is used to encode topical information.

We also compare our STN with two variants to further validate the effectiveness of the semantics-driven topic encoder (SDTE).

- **FSAN:** only use the output of the forward self-attention network for emotion prediction. Topical information is not considered in FSAN.
- **STN(PrET):** replace the SDTE in the STN with a probability-based topic encoder (shown in Fig. 2(a)).

3.5   Results and Analysis

Classification Performance: Table 1 shows the comparison on classification results of different methods. It can be seen that our STN performs the best on all

Table 2. The comparison of time efficiency for different neural models on SinaNews. Params represents the number of parameters. Time (s)/epoch means average training time (second) per epoch.

Models	Params	Time (s)/epoch	Acc@1	Pearson's r
AttBiLSTM	0.16 m	97	0.6295	0.6814
Gated DR-G-T	0.20 m	2969	0.6520	0.7123
FSAN	0.04 m	66	0.6537	0.7144
STN (proposed)	0.07 m	67	0.6624	0.7273

the three datasets. Compared with the strongest baseline model Gated DR-G-T, the STN improves the system performance about 1.04% (Acc@1), 1.5% (Pearson's r) on SinaNews, 1.65% (Acc@1), 1.76% (Pearson's r) on ISEAR and 2.53% (Acc@1), 1.65% (Pearson's r) on SemEval. The reason may be that our STN successfully learns more effective semantic feature and the fine-grained topical feature through the forward self-attention network (FSAN) and the semantics-driven topic encoder (SDTE). The detailed analysis is as follows.

Comparing the topic-emotion model, i.e., CSTM, ETM and WMCM, with the word-emotion model SWAT, we can find the great superiority of those topic-emotion models, which indicates the topic-level features are helpful to improve the performance of social emotion classification. In the comparison of AttBiL-STM and the other baseline models, we observe that AttBiLSTM shows competitive performance, although it does not consider topical information. This means AttBiLSTM has strong ability to learn document semantics and the semantic information is also important for social emotion classification. Furthermore, when comparing FSAN with AttBiLSTM, we can observe that FSAN performs much better. This indicates that the self-attention can capture more useful semantic features because of its flexibility in learning long-term dependencies in between words.

It is also necessary to compare our STN with its two variants, i.e., FSAN and STN(PrTE), for verifying the effectiveness of SDTE. The better results of STN than FSAN show that the topical information learned by SDTE can help improve the classification performance. This trend can also be seen in the comparison of STN(PrTE) and FSAN. Moreover, STN(PrTE) uses a traditional probability-based topic encoder to extract topic features. It performs worse than STN, which reveals the superiority of SDTE than PrTE. The SDTE successfully generates a semantics-aware topic vector for the document, which considers the fine-grained semantic feature of the representative terms in each topic.

Computation Time Comparison: Taking SinaNews as an example, Table 2 presents the time consumption of our STN and several strong neural models. We can see that our STN runs about 44 times faster than the best baseline model Gated DR-G-T and with very few parameters. It is also more time efficient and more than 2 times lighter than the widely-used AttBiLSTM.

Table 3. The comparison of top 10 terms in topic 2 and 12 learned by LDA and the multi-channel topic attention in our STN. The terms that are more likely to be representative in other topics are marked in blue.

#2	LDA	missing youtube taliban 's new ipods ring porn study napping
	STN	Microsoft australia toyota samsung nascar roma ps3 youtube thailand venezuela
#12	LDA	's bid lead wait u.s. gets resume iran iraqi india
	STN	Obama spanish india iran african smith japanese itunes iraqi vista

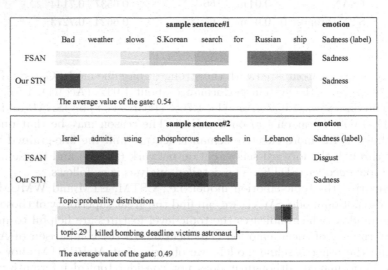

Fig. 3. Case study with two test sample from SemEval. The hot map visualizes the attention weight of each word in the word-level attention layer.

3.6 Visualization and Case Study

In this section, we demonstrate the explainability of our STN from two aspects: what the multi-channel topic attention in STN learns, and how the document topic feature helps the STN to achieve better classification performance.

For the first aspect, we visualize the top-10 terms with the largest attention weights in multi-channel topic attention. Table 3 shows the top-10 terms of two sample topics from SemEval. For comparison, it also presents the top-10 terms learned by LDA. We find that most of the terms learned by STN are about company or country name like "microsoft", "toyota", while LDA learns many words, such as "missing", "new", that are more likely to be representative in other topics. It indicates that the multi-channel topic attention in STN is helpful for extracting representative terms in the topic. As a result, the topic embedding can contain effective topical information.

For the second aspect, we conduct case study with two test samples from SemEval for our STN and the FSAN. Figure 3 visualizes the word-level attention, the gate weight (the average value of the vector **g**) in Eq. (11) and the document topic distribution θ. The emotion of Sample#1 is easy to predict based on semantic information, because there are many words expressing emotions. The FSAN gives large attention weights to *"ship"* and *"search"*. Our STN gives large attention weights to *"bad"*, *"slows"* and *"search"*. These words are all important emotional words, and the two models both predict the correct emotion. The gate layer in STN also learns a large weight (i.e., 0.54) for semantic vector. As for Sample#2, it is hard to predict the correct emotion based on only semantics, because there is no word expressing any emotion. In this case, the FSAN outputs a wrong label disgust, but our STN still reaches the right result by exploiting topical information. As Fig. 3 shows, the topic with the largest probability in Sample#2 is topic 29, where the top 5 words (such as *"killed"*, *"bombing"*) are much related to sadness. Meanwhile, the gate layer assigns a small weight (i.e., 0.49) to semantic information, which allows the final document representation contains more topical information and helps the model output the correct emotion sadness.

4 Conclusion

In this paper, we have proposed a novel semantic-topic co-encoding network model (STN) that can effectively learn the document representation for social emotion classification. The STN constructs a forward self-attention network for semantic encoding and a semantics-driven topic encoder for representing topics from the semantic feature of terms in each topic. The document semantic vector and the document topic vector are fused through a gate layer, which composes the final document representation. Experiment results on three public datasets have validated that our STN can improve the state-of-the-art classification performance. It is also time efficient and with better explainability. In the future, we would like to study how to exploit more features such as common sense knowledge graph for social emotion classification.

References

1. Bao, S., et al.: Mining social emotions from affective text. IEEE Trans. Knowl. Data Eng. **24**(9), 1658–1670 (2012)
2. Blei, D.M., Ng, A.Y., Jordan, M.I.: Latent dirichlet allocation. J. Mach. Learn. Res. **3**(Jan), 993–1022 (2003)
3. Cheng, X., Yan, X., Lan, Y., Guo, J.: BTM: topic modeling over short texts. IEEE Trans. Knowl. Data Eng. **26**(12), 2928–2941 (2014)
4. Clevert, D.A., Unterthiner, T., Hochreiter, S.: Fast and accurate deep network learning by exponential linear units (elus). arXiv preprint arXiv:1511.07289 (2015)
5. Glorot, X., Bengio, Y.: Understanding the difficulty of training deep feedforward neural networks. In: Proceedings of the Thirteenth International Conference on Artificial Intelligence and Statistics, pp. 249–256 (2010)
6. Katz, P., Singleton, M., Wicentowski, R.: SWAT-MP: the semeval-2007 systems for task 5 and task 14. In: Proceedings of the 4th International Workshop On Semantic Evaluations, pp. 308–313. Association for Computational Linguistics (2007)

7. Kingma, D.P., Ba, J.: Adam: a method for stochastic optimization. arXiv preprint arXiv:1412.6980 (2014)
8. Lai, Y., Zhang, L., Han, D., Zhou, R., Wang, G.: Fine-grained emotion classification of Chinese microblogs based on graph convolution networks. World Wide Web 23(5), 2771–2787 (2020). https://doi.org/10.1007/s11280-020-00803-0
9. Li, X., Rao, Y., Xie, H., Lau, R.Y.K., Yin, J., Wang, F.L.: Bootstrapping social emotion classification with semantically rich hybrid neural networks. IEEE Trans. Affect. Comput. 8(4), 428–442 (2017)
10. Li, X., Rao, Y., Xie, H., Liu, X., Wong, T.L., Wang, F.L.: Social emotion classification based on noise-aware training. Data Knowl. Eng. 123, 101605 (2017)
11. Li, X., et al.: Weighted multi-label classification model for sentiment analysis of online news. In: 2016 International Conference on Big Data and Smart Computing (BigComp), pp. 215–222. IEEE (2016)
12. Mikolov, T., Sutskever, I., Chen, K., Corrado, G.S., Dean, J.: Distributed representations of words and phrases and their compositionality. In: Advances in Neural Information Processing Systems 26, pp. 3111–3119 (2013)
13. Rao, Y.: Contextual sentiment topic model for adaptive social emotion classification. IEEE Intell. Syst. 1, 41–47 (2016)
14. Rao, Y., Li, Q., Wenyin, L., Wu, Q., Quan, X.: Affective topic model for social emotion detection. Neural Netw. 58, 29–37 (2014)
15. Scherer, K.R., Wallbott, H.G.: Evidence for universality and cultural variation of differential emotion response patterning. J. Pers. Soc. Psychol. 66(2), 310 (1994)
16. Shen, T., Zhou, T., Long, G., Jiang, J., Pan, S., Zhang, C.: Disan: directional self-attention network for RNN/CNN-free language understanding. In: Thirty-Second AAAI Conference on Artificial Intelligence (2018)
17. Strapparava, C., Mihalcea, R.: Semeval-2007 task 14: affective text. In: Proceedings of the 4th International Workshop on Semantic Evaluations (SemEval-2007), pp. 70–74 (2007)
18. Vaswani, A., et al.: Attention is all you need. In: Advances in Neural Information Processing Systems 30, pp. 5998–6008 (2017)
19. Wang, C., Wang, B., Xiang, W., Xu, M.: Encoding syntactic dependency and topical information for social emotion classification. In: Proceedings of the 42nd International ACM SIGIR Conference on Research and Development in Information Retrieval, pp. 881–884. ACM (2019)
20. Yang, X., Wang, B.: Local matrix approximation based on graph random walk. In: Proceedings of the 42nd International ACM SIGIR Conference on Research and Development in Information Retrieval, pp. 1037–1040 (2019)
21. Yang, Z., Yang, D., Dyer, C., He, X., Smola, A., Hovy, E.: Hierarchical attention networks for document classification. In: Proceedings of the 2016 Conference of the North American Chapter of the Association for Computational Linguistics: Human Language Technologies, pp. 1480–1489 (2016)
22. Zhao, X., Wang, C., Yang, Z., Zhang, Y., Yuan, X.: Online news emotion prediction with bidirectional LSTM. In: Cui, B., Zhang, N., Xu, J., Lian, X., Liu, D. (eds.) WAIM 2016. LNCS, vol. 9659, pp. 238–250. Springer, Cham (2016). https://doi.org/10.1007/978-3-319-39958-4_19
23. Zhou, P., et al.: Attention-based bidirectional long short-term memory networks for relation classification. In: Proceedings of the 54th Annual Meeting of the Association for Computational Linguistics (Volume 2: Short Papers), pp. 207–212 (2016)

Big Data Technologies

Knowledge Lock: Overcoming Catastrophic Forgetting in Federated Learning

Guoyizhe Wei and Xiu Li[✉]

Shenzhen International Graduate School, Tsinghua University, Shenzhen,
People's Republic of China
wgyz19@mails.tsinghua.edu.cn, li.xiu@sz.tsinghua.edu.cn

Abstract. Federated Learning (FL) aims to train machine learning models by decentralized data without direct data sharing. Nevertheless, the heterogeneity of data across FL participants has significantly prevented federated models from competitive performance. In this paper, we consider this issue as the consequence of knowledge forgetting, since the local update process in FL may result in catastrophic forgetting of the knowledge learned from other participants. Motivated by the recent advance in incremental learning techniques, we address this issue by overcoming the sever knowledge forgetting caused by data isolation. We propose a novel method called FedKL (**Fed**erated Learning with **K**nowledge **L**ock), in which knowledge distillation techniques are employed to maintain the previously learned knowledge. Our extensive experiment results demonstrate that FedKL achieves superior performance than prior methods, with over 3.4% and 3.5% accuracy improvements on CIFAR-10 and CIFAR-100 respectively, compared with the popular FL algorithm FedAvg. Furthermore, we also explore the benefits of introducing shared exemplars (a fraction of local data) to FedKL. In the experiments, we select and share 10 samples per class for FedKL and the baseline methods. As a result, FedKL obtains 2.56% accuracy increase on CIFAR-10, instead of the marginal improvements on prior methods (less than 1.5%

Keywords: Federated learning · Incremental learning · Catastrophic forgetting

1 Introduction

Federated learning (FL) techniques [31] offer a safe and efficient solution for multi-party collaborations, by which participants are able to collaboratively train a global model without exposing private data [20]. The underlying idea is to aggregate the local updates of model parameters, instead of a direct data

Supported by the Science and Technology Innovation 2030-Key Project under Grant 2021ZD0201404.

J. Gama et al. (Eds.): PAKDD 2022, LNAI 13280, pp. 601–612, 2022.
https://doi.org/10.1007/978-3-031-05933-9_47

sharing [10,20]. Federated learning breaks the barriers of data cooperation. It works well under the ideal assumption that the training data are independent identically distributed (iid) across the participants.

However, in most application scenarios, participants have widely varying data volumes and class distributions, which is usually called non-iid or statistical heterogeneity [30,34]. It turns out that the heterogeneously distributed data often negatively affect the convergence of federated training and the accuracy of federated models will inevitably decrease [34]. We consider this issue from the perspective that due to the data heterogeneity, it is very easy for local training to forget the knowledge learned from other participants. In this work, we believe that the bottleneck in solving the non-iid problem lies in overcoming the forgetfulness of the global knowledge.

We are inspired by incremental learning approaches [12,16,22], whose key objective is to maintain the previously learned knowledge while learning from newly accessed data. We identify the inner similarity between incremental learning and federated learning that the catastrophic knowledge forgetting is due to the unavailability of the entire dataset. Actually, the primary difference just lies in that this unavailability is due to "temporal causes" in incremental learning since in this scenario the old data cannot be reused once new data comes, while it is due to "spatial causes" in federated learning since the training data is spatially decentralized.

Thus, this similarity motivates us to explore the effectiveness of transferring incremental learning strategies into federated learning. We improve the knowledge maintaining approaches in incremental learning and propose a novel scheme, named Knowledge Lock, in which we decouple the training objective into a classification and a knowledge maintaining term. In the classification term, the local training is supervised by the original label of the data with a softmax-cross-entropy loss applied. In the knowledge maintaining term, the locally unavailable classes are supervised by the distillation [8] of the global model with logic regression loss. Further, we also explore the effect of partially data sharing on our method and the baseline algorithms. Our proposed exemplar selection algorithm (see Algorithm 2) has been empirically proved to be able to yield significant accuracy improvements on Knowledge Lock.

We conduct a series of experiments to evaluate our method FedKL (**Fed**rated Learning with **K**nowledge **L**ock) and compare it with the prior algorithms. As reported in Sect. 4, FedKL outperforms the existing algorithms, including FedAvg [20], FedProx [15], and SCAFFOLD [11] on both VGG [27] and ResNet [7] architectures. Notably, compared with FedAvg [20], FedKL achieves 3.46% and 3.51% higher accuracy with VGG-16 [27] network on CIFAR-10 and CIFAR-100 [14], respectively. Moreover, FedKL shows more powerful adaptation of partially data sharing, with 2.56% and 2.11% accuracy improvements on CIFAR-10 and CIFAR-100 [14] respectively.

2   Related Work

Federated learning's potential in reliable data cooperation has led to a rapidly growing interest on its performance with heterogeneously distributed data. We categorize the existing solutions to the data heterogeneity as parameter-oriented and output-oriented methods. For the former, people try to cope with the heterogeneity by constraining the model parameters. For example, FedProx [15] adds a regularization term of Euclidean distance between the parameters of the local and global model to constrain the local gradient decent. SCAFFOLD [11] also modifies the loss of local back propagation, introducing additional parameters to track the local updates. These methods have obtained considerable improvements on the accuracy, convergence, or communication efficiency of federated models. We re-implement FedProx [15] and SCAFFOLD [11] as competitive baselines for comparison in Sect. 4.

For the output-oriented methods, the constraint is directly applied to model outputs, i.e., they encourage the local models to yield similar outputs as the global model or other stronger teacher models. There have been several attempts of employing knowledge distillation techniques [8] in federated learning tasks [18,19]. However, these methods fail to achieve as competitive performance as the parameter-oriented approaches when data is massively distributed and extremely heterogeneous [20]. This is possibly because these methods ignore the locally unknown classes and apply the same distillation loss to each class. Meta learning strategies are also reported in federated learning methods to enhance the robustness [3]. However, it is also difficult to deal with more challenging federated learning scenarios due to the lack of consideration in the difference of classes.

Notably, in relatively low privacy-sensitivity cases, partially data sharing serves as another effective approach to overcome the data heterogeneity [35]. For example, some of them attempt to modify the data distributions by sharing partial local data [29,33], which can also be improved by data augmentation [2,25]. However, as they randomly choose the samples to be shared, it is neither sufficiently privacy-efficient nor communication-efficient. Instead, we propose an efficient data sharing strategy (see Algorithm 2), in which by sharing only a very small fraction of data we can obtain satisfactory performance improvements.

Moreover, there are also federated learning research works turning to system design [4,5,13,23] or personalized regularization [6,9], with a focus on the trade-off between local and global models. Some methods [1,17] design a flexible network for the participants, part of which is trained collaboratively while the others are trained locally. Also, Multi-task Learning [24,28] and Transfer Learning [21] approaches are employed in federated algorithms in order to enhance the adaption of local training.

Recently, incremental learning, also known as lifelong learning, has been applied to overcome the effects of data heterogeneity. Shoham et al. [26] presents FedCurv algorithm based on Elastic Weight Consolidation. In FedCurv, participants additionally upload the diagonal of the Fisher information matrix, which indicates critical local knowledge. A penalty term is added to the loss function in order to force the local models to converge to a common optimum. However, even with bandwidth optimization tricks, the communication budget in FedCurv is three times higher than that in FedAvg. Yoon et al. [32] develop the Federated Weighted Inter-client Transfer (FedWeIT) method, in which nodes exploit the task-adaptive parameters from other nodes through an attention mask.

3   Federated Learning with Knowledge Lock

3.1   Overview

One of the toughest issues in training federated classifiers is that the complete training data is not available due to its decentralized distribution over different devices. In particular, it has been widely studied that the heterogeneity of training data over devices often incurs considerable decrease in predictive performance and difficulty in optimization. In classification tasks, such heterogeneity is often reflected in imbalanced class distribution [15,28]. Typically, it is a more awkward scenario in which the training samples on each device only belong to partial classes of the entire dataset, i.e., given a complete dataset with m classes, each of the k device only has samples in m_k classes, where $m_k < m$. In other words, there are "unknown" classes in each device during federated training. In this paper, we focus on this scenario and propose our method, Fed-KL, which is inspired by class-incremental learning approaches, to address the issues of training data's heterogeneity.

Given a dataset $\mathcal{D} = \{(x_i, y_i)\}_{i=1}^n$ which distributes across k devices, the goal of federated learning is to train a shared model $f : \mathbb{R}^d \to \mathbb{R}^m$, where $x_i \in \mathbb{R}^d$ is the training data and $y_i \in \mathbb{R}$ denotes its label. Specifically, in a classification task, the model $f(\cdot)$ can be denoted by a representation encoder $g(\cdot)$ followed by a fully-connected layer with m units, where m is the number of total classes.

To fully exploit the data on each device without directly sharing them, the most commonly used federated learning methods such as FedAvg [20] train models in a manner of multiple rounds update-aggregation. In each round, the participants (devices) train the shared global model on their local data for a certain number of epochs in the update step, and the trained local models are then aggregated to a new global model in the aggregation step. Typically, in FedAvg [20], the networks are aggregated by straightforward parameter average, i.e., given the local networks $f_1, f_2, \ldots, f_k$ with parameters $\Theta_1, \Theta_2, \ldots, \Theta_k$ respectively, the parameters of the global network is calculated by

$$\Theta \leftarrow \frac{1}{k} \sum_{j=1}^k \Theta_j \tag{1}$$

Unfortunately, despite their ability to fuse the knowledge learned from different participants, the existing FL methods still suffer from the statistical heterogeneity of the training date across the participants, which prevents the federated models from comparable performance as centralized training. We consider

---

Algorithm 1: Local Update Process of FedKL

Input: Training samples $\{(x_i, y_i)\}_{i=1}^{n_k}$
Input: Global model f with parameters Θ
Input: Unknown classes M', weight of distillation loss λ

1 **for** $i \leftarrow 1$ **to** n_k **do**
2   **for** $m \in M'$ **do**
3 | $\delta_i^m \leftarrow \text{Sigmoid}(f_\Theta^m(x_i))$;
4   **end**
5 **end**
6 $\mathcal{L}_{clf} \leftarrow \frac{1}{n_k} \sum_{i=1}^{n_k} \text{SoftmaxCrossEntropy}(f_\Theta(x_i), y_i)$;
7 $\mathcal{L}_{dis} \leftarrow -\frac{1}{n_k|M'|} \sum_{i=1}^{n_k} \sum_{m \in M'}$
    $[\delta_i^m \log(\text{Sigmoid}(f_\Theta^m(x_i))) + (1 - \delta_i^m)\log(1 - \text{Sigmoid}(f_\Theta^m(x_i)))]$;
8 $\Theta \leftarrow \text{argmin}_\Theta(\mathcal{L}_{clf} + \lambda\mathcal{L}_{dis})$;

---

the issue of data heterogeneity from the aspect of knowledge forgetting, which resembles the theme of incremental learning, online learning, or lifelong learning [14,22]. Specifically, we believe that the considerable gap of performance between centralized training and federated training is due to, or partially due to, the participants' forgetting of knowledge learned from others while training on their local data. Therefore, in this paper, the knowledge maintaining is regarded as an essential theme and serves as the primary strategy of our proposed method, FedKL.

3.2 FedKL with Distillation

A very simple but effective way to overcome catastrophic knowledge forgetting is to apply knowledge distillation [8], with the assumption that the output probabilities of a neural network are able to represent the knowledge it has learnt [14,22]. We follow this key discovery and further propose a novel algorithm, **FedKL-Dist** (**Fed**erated Learning with **K**nowledge **L**ock - **Dist**illation), which adds a distillation loss term beside of the classification loss to "lock" the knowledge learned from other participants.

Formally, in general, a federated classifier is locally trained by the loss

$$\mathcal{L}_{clf} = \frac{1}{n_k} \sum_{i=1}^{n_k} \text{SoftmaxCrossEntropy}(f_\Theta(x_i), y_i), \tag{2}$$

where n_k denotes the number of training samples on the k-th device (participant) and $f(\cdot)$ denotes the network with parameters Θ. This fundamental classification loss forces the network to predict x_i to be as close to its target y_i as possible. However, it may incur significant knowledge forgetting if only adopting such a classification loss. For example, as we supposed in Sect. 3.1, there are always "unknown classes" (the classes of which the samples do not appear on the local device). Thus, the network tends to predict all the training samples to the probability of zero on these classes, which will consequently result in poor predictive performance on such "unknown classes".

Therefore, to constrain the forgetting, we further apply a distillation loss on the locally unknown classes. Supposing that there are M' of the total M classes are unknown on a device, we firstly calculate the Sigmoid[1] probabilities on the M' classes of all the locally available samples. Formally, we have

$$\delta_i^m = \text{Sigmoid}(f_\Theta^m(x_i)), \tag{3}$$

where δ_i^m denotes the i-th sample's probability of the m-th unknown class predicted by the global model $f_\Theta(\cdot)$. Then, the distillation loss is defined as

$$\mathcal{L}_{dis} = -\frac{1}{n_k|M'|} \sum_{i=1}^{n_k} \sum_{m \in M'} [\delta_i^m \log(\text{S}(f_\Theta^m(x_i))) + (1 - \delta_i^m)\log(1 - \text{S}(f_\Theta^m(x_i)))], \tag{4}$$

where $\text{S}(\cdot)$ denotes the Sigmoid function.

Intuitively, $\mathcal{L}_{dis}$ encourages the local optimization to maintain the probabilities on the unknown classes, instead of forcing them to zero, whilst a similar strategy also appears in several incremental learning algorithms and has been empirically proved to effectively prevent knowledge forgetting [14].

The local update process has been summarized in Algorithm 1. Moreover, considering the trade-off between the classification loss $\mathcal{L}_{clf}$ and the distillation loss $\mathcal{L}_{dis}$, we introduce a new hyper-parameter λ and the total loss of local updates can be formulated as

$$\mathcal{L} = \mathcal{L}_{clf} + \lambda\mathcal{L}_{dis}. \tag{5}$$

We have searched the parameter λ and find that setting it to around M'/M leads to better performance.

3.3 Improve FedKL by Class Exemplars

Algorithm 2: Class Exemplar Selection

Input: Training samples from one class $\{x_i\}_{i=1}^{n_c}$
Input: Number of exemplars required L
Input: Representation model $g(\cdot)$
1 $\gamma = \frac{1}{n_c} \sum_{i=1}^{n_c} g(x_i)/\|g(x_i)\|$;
2 **for** $l \leftarrow 1$ **to** L **do**
3 $\quad$ $p_l \leftarrow \text{argmax}_{x \in \mathcal{X}} < \gamma, g(x_i)/\|g(x_i)\| >$;
4 $\quad$ $\mathcal{X} \leftarrow \{x_i | x_i \in \mathcal{X}, x_i \neq p_l\}$;
5 **end**
6 $P \leftarrow \{p_1, \ldots, p_L\}$;

Another popular strategy of overcoming knowledge forgetting to re-train the model with representative exemplars [22]. Specifically, in incremental learning

---
[1] $\text{Sigmoid}(x) = 1/(1 + e^{-x})$.

methods [14, 22], in order to maintain the previously learned knowledge, people try to store some representative samples as exemplars and then train the models with a combination of new data and these exemplars. Similarly, it is also very popular in federated learning to partially share the local data, which improves the performance at the cost of a slice of privacy. Nevertheless, the sharing of small amounts of data is not able to bring about sufficient improvement on the performance, while sharing a large amount of data may violate the basic rules of federated learning.

Therefore, for the efficiency and the least privacy cost, it is crucial to select the most representative samples as exemplars. In this paper, we choose the examples by their representation. Formally, given a global model $f(\cdot)$ with a encoder $g(\cdot)$, the normalized representation of each sample can be calculated by

$$\gamma_i = \frac{g(x_i)}{\|g(x_i)\|_2}. \tag{6}$$

We suppose that the most representative sample is the one being nearest to the averaged representation, which can be formulated as

$$p = \operatorname{argmax}_{x \in \mathcal{X}} < \gamma, g(x_i)/\|g(x_i)\|_2 >, \tag{7}$$

where $\gamma = 1/n_k \sum_{i=1}^{n_k} \gamma_i$ denotes the average of the n_k representations and $\mathcal{X}$ is the dataset of the n_k training samples. As we describe in Algorithm 2, we can select the most L representative samples as exemplars for each class, where L depends on the privacy constraint. Note that for a specific class, we randomly choose participants to share exemplars if L is less than the number of participants who have this class of samples.

Warming up Epochs. Notably, for a higher accuracy, the encoder $g(\cdot)$ should be a well-trained model. For example, in practical training, we can at first train the model with Algorithm 1 for some epochs and then use the trained model as $g(\cdot)$, after which we continue training the model with both local data and the shared exemplars.

4 Experiments

4.1 Experiment Settings

Datasets. We evaluate our method an the baselines on image classification datasets, including:

- MNIST handwritten digit database[2], containing 60k images for training and 10k for evaluation, drawn from ten classes.
- CIFAR-10 and CIFAR-100 [14], the colored images drawn from 10 and 100 classes, respectively. Both CIFAR-10 and CIFAR-100 have 50k training images and 10k evaluation images.

---

[2] http://yann.lecun.com/exdb/mnist/.

We split the complete dataset to 100 parts in order to analogue the 100 participants in federated learning. As we introduce in Sect. 3.1, each participant only accesses partial classes. Specifically, on each device (participant), there are at least two and at most M_{max} classes, where intuitively a higher M_{max} means a lower heterogeneity. Since we do not study the unbalancedness in this paper, all the participants have a similar number of local samples in our experiments.

Neural Network Architectures. As the prior works did [11,15,20], we use convolutional neural networks (CNNs) as base models to evaluate our method and the baselines. Specifically, the 16-layer VGG network [27] and the 18-layer ResNet [7] are employed.

Baselines. In our experiments, we mainly compare our method with the following federated learning algorithms:

- Federated Averaging [20]. Federated Averaging (FedAvg) algorithm is the most commonly used prototype in federated learning, with simply the classification loss in local update steps.
- FedProx [15]. FedProx is one of the most competitive federated learning method, with an additional term of loss function $\mu/2\|\Theta - \Theta^t\|^2$ that constrains the optimization on heterogeneous data[3].
- SCAFFOLD [11]. SCAFFOLD is another competitive federated learning method which tackles the heterogeneity issues with the concentration on "client-drift", where the classification loss is modified to track the optimization path.
- Centralized training. Centralized training serves as the upper bound of federated learning performance, where the model is directly trained on the entire dataset.

Note that the strategies adopted in both FedProx and SCAFFOLD can also be regarded as methods to get over knowledge forgetting. However, the primary difference between our method and these two algorithms lies in that FedKL directly constrains the model outputs, while FedProx and SCAFFOLD focus on constraining the parameter updates. Therefore, it is valuable to compare the output-based method (FedKL) with such parameter-based approaches (FedProx and SCAFFOLD) on their capacity in maintaining knowledge.

4.2   Predictive Accuracy

In this section, we explore the test accuracy of our method FedKL and the baselines on the image classification datasets introduced in Sect. 4.1. For a comprehensive comparison, we explore their performance under and 100-split data partitioning, which is summarized in Table 1. We set $M_{max} = 2$ for all the

---

[3] Θ denotes the parameters of the model and Θ^t denotes the parameters at the t-th round.

Table 1. Test accuracy (%) under 100-split data partitioning with $M_{max} = 2$. We report the average results and standard deviations of five times repeated experiments with different random seeds. The best results for each network architecture are **bolded**.

Network	Dataset	MNIST	CIFAR-10	CIFAR-100
Vgg-16	Centralized (upper bound)	99.13 ± 0.08	91.09 ± 0.47	72.79 ± 0.32
	FedAvg	94.38 ± 1.71	57.21 ± 0.31	24.11 ± 0.44
	SCAFFOLD	94.31 ± 0.54	58.31 ± 1.02	23.12 ± 0.11
	FedProx - $\mu = 0.01$	94.37 ± 1.12	59.25 ± 3.08	25.38 ± 0.72
	FedProx - $\mu = 0.001$	92.12 ± 0.75	58.69 ± 0.73	25.93 ± 0.34
	FedKL - $\lambda = 0.01$ (ours)	$\mathbf{96.12 \pm 0.31}$	$\mathbf{60.67 \pm 0.75}$	27.23 ± 0.47
	FedKL - $\lambda = 0.02$ (ours)	95.57 ± 0.61	59.96 ± 0.23	$\mathbf{27.62 \pm 0.51}$
Res-18	Centralized (upper bound)	99.57 ± 0.14	94.35 ± 0.27	74.28 ± 0.45
	FedAvg	94.81 ± 0.40	59.27 ± 0.76	25.32 ± 0.72
	SCAFFOLD	95.17 ± 0.38	60.34 ± 0.85	24.97 ± 0.49
	FedProx - $\mu = 0.01$	95.08 ± 0.19	60.45 ± 0.42	26.10 ± 0.61
	FedProx - $\mu = 0.001$	93.11 ± 0.70	61.16 ± 0.22	26.14 ± 0.24
	FedKL - $\lambda = 0.01$ (ours)	$\mathbf{96.97 \pm 0.17}$	$\mathbf{62.77 \pm 0.54}$	28.17 ± 0.41
	FedKL - $\lambda = 0.02$ (ours)	96.29 ± 0.72	62.50 ± 0.32	$\mathbf{28.77 \pm 0.1}$

three datasets to simulate a highly heterogeneous data distribution. Moreover, as there are important hyper-parameters that may affect the performance (μ in FedProx and λ in FedKL), we test their accuracy under two parameter settings respectively.

As reported in Table 1, our method outperforms the baselines with both Vgg-16 [27] and ResNet-18 [7] network architectures. Notably, the accuracy gap between FedKL and centralized training is decreased to 2.60% on MNIST with ResNet-18 adopted. On CIFAR-10 and CIFAR-100, FedKL achieves the accuracy of over 62% and 28%, respectively, which significantly outperforms the prior methods.

4.3 Effect of Class Exemplars

Further, we explore the effect of the partially data sharing algorithm we proposed (see Algorithm 2). For direct comparison, we use the same data partitioning settings as Sect. 4.2, i.e., $k = 100$, $M_{max} = 2$, and add 10 (one for CIFAR-100) globally shared samples per class selected by Algorithm 2.

As summarized in Table 2, a total number of 100 shared samples lead to about 1% and 3% accuracy increase on MNIST and CIFAR-10, respectively (10 samples from each of the 10 classes). Moreover, it also indicates that the exemplar sharing algorithm performs better on FedKL than other baseline methods. For example, on CIFAR-100 with Vgg-16 architecture, sample sharing brings about 2.11% accuracy improvement on FedKL, while it is only 1.23% on FedAvg.

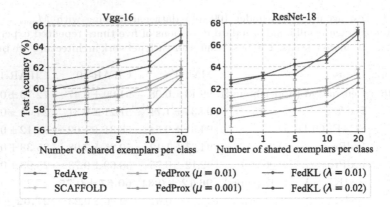

Fig. 1. Performance under different number of shared exemplars in each class.

Furthermore, to fully examine the effect of data sharing, we evaluate its performance with different number of exemplars. As displayed in Fig. 1, the partially data sharing significantly benefits our method and the baseline algorithms, which demonstrates the potential of exemplar sharing in the cases of low privacy-sensitivity.

Table 2. Test accuracy (%) with class exemplars adopted. The datasets are divided into 100 parts with $M_{max} = 2$ and the number of shared exemplars in each class is ten for MNIST, CIFAR-10 and one for CIFAR-100. We report the average results and standard deviations of five times repeated experiments with different random seeds. The best results for each network architecture are **bolded**.

Network	Dataset	MNIST	CIFAR-10	CIFAR-100
Vgg-16	Centralized (upper bound)	99.13 ± 0.08	91.09 ± 0.47	72.79 ± 0.32
	FedAvg	94.89 ± 0.32	58.14 ± 0.50	25.34 ± 0.72
	SCAFFOLD	94.75 ± 0.48	59.85 ± 0.85	24.61 ± 0.49
	FedProx - $\mu = 0.01$	94.83 ± 0.93	60.59 ± 0.84	26.29 ± 0.57
	FedProx - $\mu = 0.001$	92.87 ± 0.36	60.26 ± 0.42	27.42 ± 0.82
	FedKL - $\lambda = 0.01$ **(ours)**	$\mathbf{96.84 \pm 0.45}$	$\mathbf{63.23 \pm 0.66}$	29.35 ± 0.35
	FedKL - $\lambda = 0.02$ **(ours)**	96.21 ± 0.36	62.10 ± 0.49	$\mathbf{29.73 \pm 0.24}$
Res-18	Centralized (upper bound)	99.57 ± 0.14	94.15 ± 0.27	74.98 ± 0.45
	FedAvg	95.34 ± 0.19	60.64 ± 0.12	26.35 ± 0.72
	SCAFFOLD	95.83 ± 0.20	61.80 ± 0.12	26.13 ± 0.56
	FedProx - $\mu = 0.01$	95.76 ± 0.26	61.88 ± 0.34	27.19 ± 0.61
	FedProx - $\mu = 0.001$	93.89 ± 0.61	62.12 ± 0.75	27.83 ± 0.47
	FedKL - $\lambda = 0.01$ **(ours)**	$\mathbf{98.30 \pm 0.13}$	$\mathbf{65.17 \pm 0.21}$	30.63 ± 0.14
	FedKL - $\lambda = 0.02$ **(ours)**	97.46 ± 0.34	64.63 ± 0.32	$\mathbf{30.94 \pm 0.30}$

5   Conclusion

A major challenge in federal learning is overcoming the impact of heterogeneous data distributions on model accuracy and convergence. To tackle this problem, our proposed FedKL maintain the global knowledge by introducing knowledge distillation techniques. FedKL can be applied to any CNN techniques to increase its decentralized learning performance. Furthermore, we explore the benefits of introducing shared exemplars (a fraction of local data) to FedKL. Comprehensive experiments demonstrate the effectiveness of our proposal compared to the existing algorithms.

References

1. Arivazhagan, M.G., Aggarwal, V., Singh, A.K., Choudhary, S.: Federated learning with personalization layers. arXiv preprint arXiv:1912.00818 (2019)
2. Duan, M., et al.: Astraea: self-balancing federated learning for improving classification accuracy of mobile deep learning applications. In: 2019 IEEE 37th International Conference on Computer Design (ICCD), pp. 246–254. IEEE (2019)
3. Fallah, A., Mokhtari, A., Ozdaglar, A.: Personalized federated learning with theoretical guarantees: a model-agnostic meta-learning approach. Adv. Neural Inf. Process. Syst. **33**, 3557–3568 (2020)
4. Ghosh, A., Chung, J., Yin, D., Ramchandran, K.: An efficient framework for clustered federated learning. Adv. Neural Inf. Process. Syst. **33**, 19586–19597 (2020)
5. Ghosh, A., Hong, J., Yin, D., Ramchandran, K.: Robust federated learning in a heterogeneous environment. arXiv preprint arXiv:1906.06629 (2019)
6. Hanzely, F., Richtárik, P.: Federated learning of a mixture of global and local models. arXiv preprint arXiv:2002.05516 (2020)
7. He, K., Zhang, X., Ren, S., Sun, J.: Deep residual learning for image recognition. In: Proceedings of the IEEE Conference on Computer Vision and Pattern Recognition, pp. 770–778 (2016)
8. Hinton, G., Vinyals, O., Dean, J.: Distilling the knowledge in a neural network. arXiv preprint arXiv:1503.02531 (2015)
9. Huang, Y., et al.: Personalized cross-silo federated learning on Non-IID data (2021)
10. Ji, S., Pan, S., Long, G., Li, X., Jiang, J., Huang, Z.: Learning private neural language modeling with attentive aggregation. In: 2019 International Joint Conference on Neural Networks (IJCNN), pp. 1–8. IEEE (2019)
11. Karimireddy, S.P., Kale, S., Mohri, M., Reddi, S., Stich, S., Suresh, A.T.: Scaffold: stochastic controlled averaging for federated learning. In: International Conference on Machine Learning, pp. 5132–5143. PMLR (2020)
12. Kirkpatrick, J., et al.: Overcoming catastrophic forgetting in neural networks. Proc. Natl. Acad. Sci. **114**(13), 3521–3526 (2017)
13. Kopparapu, K., Lin, E.: FedFMC: sequential efficient federated learning on non-iid data. arXiv preprint arXiv:2006.10937 (2020)
14. Krizhevsky, A., Hinton, G., et al.: Learning multiple layers of features from tiny images (2009)
15. Li, T., Sahu, A.K., Zaheer, M., Sanjabi, M., Talwalkar, A., Smith, V.: Federated optimization in heterogeneous networks. arXiv preprint arXiv:1812.06127 (2018)
16. Li, Z., Hoiem, D.: Learning without forgetting. IEEE Trans. Pattern Anal. Mach. Intell. **40**(12), 2935–2947 (2017)

17. Liang, P.P., et al.: Think locally, act globally: Federated learning with local and global representations. arXiv preprint arXiv:2001.01523 (2020)
18. Lin, T., Kong, L., Stich, S.U., Jaggi, M.: Ensemble distillation for robust model fusion in federated learning. arXiv preprint arXiv:2006.07242 (2020)
19. Liu, Y., Kang, Y., Xing, C., Chen, T., Yang, Q.: A secure federated transfer learning framework. IEEE Intell. Syst. **35**(4), 70–82 (2020)
20. McMahan, B., Moore, E., Ramage, D., Hampson, S., y Arcas, B.A.: Communication-efficient learning of deep networks from decentralized data. In: Artificial Intelligence and Statistics, pp. 1273–1282. PMLR (2017)
21. Peng, X., Huang, Z., Zhu, Y., Saenko, K.: Federated adversarial domain adaptation. In: International Conference on Learning Representations (2019)
22. Rebuffi, S.A., Kolesnikov, A., Sperl, G., Lampert, C.H.: ICaRL: incremental classifier and representation learning. In: Proceedings of the IEEE conference on Computer Vision and Pattern Recognition, pp. 2001–2010 (2017)
23. Reddi, S.J., et al.: Adaptive federated optimization. In: International Conference on Learning Representations (2020)
24. Sattler, F., Müller, K.R., Samek, W.: Clustered federated learning: model-agnostic distributed multitask optimization under privacy constraints. IEEE Trans. Neural Networks Learn. Syst. **39**, 3710–3722 (2020)
25. Shin, M., Hwang, C., Kim, J., Park, J., Bennis, M., Kim, S.L.: XOR mixup: privacy-preserving data augmentation for one-shot federated learning. arXiv preprint arXiv:2006.05148 (2020)
26. Shoham, N., et al.: Overcoming forgetting in federated learning on Non-IID data. arXiv preprint arXiv:1910.07796 (2019)
27. Simonyan, K., Zisserman, A.: Very deep convolutional networks for large-scale image recognition. arXiv preprint arXiv:1409.1556 (2014)
28. Smith, V., Chiang, C.K., Sanjabi, M., Talwalkar, A.: Federated multi-task learning. In: Proceedings of the 31st International Conference on Neural Information Processing Systems, pp. 4427–4437 (2017)
29. Tuor, T., Wang, S., Ko, B.J., Liu, C., Leung, K.K.: Overcoming noisy and irrelevant data in federated learning. In: 2020 25th International Conference on Pattern Recognition (ICPR), pp. 5020–5027. IEEE (2021)
30. Wang, S., et al.: Adaptive federated learning in resource constrained edge computing systems. IEEE J. Sel. Areas Commun. **37**(6), 1205–1221 (2019)
31. Yang, Q., Liu, Y., Cheng, Y., Kang, Y., Chen, T., Yu, H.: Federated learning. Synth. Lect. Artif. Intell. Mach. Learn. **13**(3), 1–207 (2019)
32. Yoon, J., Jeong, W., Lee, G., Yang, E., Hwang, S.J.: Federated continual learning with weighted inter-client transfer. In: International Conference on Machine Learning, pp. 12073–12086. PMLR (2021)
33. Yoshida, N., Nishio, T., Morikura, M., Yamamoto, K., Yonetani, R.: Hybrid-FL: cooperative learning mechanism using Non-IID data in wireless networks. arXiv preprint arXiv:1905.07210 (2019)
34. Zhao, Y., Li, M., Lai, L., Suda, N., Civin, D., Chandra, V.: Federated learning with Non-IID data. arXiv preprint arXiv:1806.00582 (2018)
35. Zhu, H., Xu, J., Liu, S., Jin, Y.: Federated learning on Non-IID data: a survey. arXiv preprint arXiv:2106.06843 (2021)

Overcoming Forgetting in Local Adaptation of Federated Learning Model

Shunjian Liu, Xinxin Feng$^{(\boxtimes)}$, and Haifeng Zheng

Fujian Key Laboratory for Intelligent Processing and Wireless Transmission of Media Information, College of Physics and Information Engineering, Fuzhou University, Fuzhou, China
{201127047,fxx1116,zhenghf}@fzu.edu.cn

Abstract. Federated learning allows multiple clients to train a global model without data exchanging. But in real world, the global model is not suitable for all clients because they may hold heterogenous data and have personalized and individual demands, which will directly weaken the motivation of them to participate in federated learning. To make each client benefits from federated learning, researchers propose to train personalized models from global model using local data. However, the lack of raw data in the model retraining process will lead to the challenge of forgetting, which can deprive the personalized model of the benefits gained from federated learning. In extreme cases (e.g., the client lacks certain classes of data), the ability to recognize the lacked data may even be completely forgotten. To this end, we propose a local adaptation method to overcome forgetting, which add the generator synthetic data to local adaptation to realize model updating incrementally. We test our method on real-world datasets, and the results show that when adopting the proposed method on local adaptation, the clients can get flexible adaption ability to new data as well as keep the original recognition capability of the global model even in extreme cases.

Keywords: Federated learning · Local adaptation · Incremental learning · Knowledge distillation

1 Introduction

Data is an important resource to drive machine learning. With the extensive use of intelligent devices (such as mobile phones, intelligent vehicles, and wearables), there are a lot of data stored on these edge devices. Traditional machine learning needs to aggregate data to the central server to complete the model training, which is impossible or undesirable from a data privacy or limited transmission bandwidth point of view. Federated learning, as a decentralized approach, allows clients to complete global model training without exchanging data [1]. In federated learning, clients only need to use local data to train the current global model, and the central server aggregates the update results (such as model

J. Gama et al. (Eds.): PAKDD 2022, LNAI 13280, pp. 613–625, 2022.
https://doi.org/10.1007/978-3-031-05933-9_48

parameters or gradients) of the clients to complete a round of global model update.

As an attractive approach to fully use the large amount of data to improve the intelligence of applications, federated learning also faces challenges brought by heterogeneities of both data and clients. First, clients often hold heterogenous data due to they collect their data independently. For clients with insufficient data to train the model, they will harvest the global model in the federated learning process. However, for clients with sufficient data, it is controversial whether they can benefit from federated learning [2]. Yu et al. [3] point out that some clients can not benefit from federated learning due to the global model is not as accurate as the local model trained only with their own data, which will eventually decrease their motivation to participate in federated learning. Second, clients usually have personalized and individual demands, which are hard to be met by a single global model. For example, since the global model is always not trained across all the clients, a client who owned specified data may not be selected at any round at all [4], to whom the global model is not suitable. In addition, some clients may continuously receive new or even new classes of data (for example, a roadside unit in an intelligent transportation system may receive new traffic data showing an accident caused by a new type of event), which the global model cannot recognize.

To make each client benefit from federated learning, researchers propose to use personalization techniques, so that each client can get a separate personalized model eventually. The work in this field can be divided into two categories. One is to modify the federated learning process for personalization, mainly including federated multi-task learning [5] and federated meta-learning [6]. The other is local adaptation, in which, the local data is adopted to retraining the global model by the methods of transfer learning, domain adaptation, and fine-tuning [7].

The advantage of local adaptation is that it can be integrated into the training process of federated learning naturally. However, the existing local adaptation methods are usually the extension of transfer learning in non-federated learning scenarios. Proposing a local adaptation method according to the characteristic of federated learning is an area received increasing attention [8]. We note that the global model has excellent generalization performance under the training of all client data. However, some abilities of it may lose in local adaptation if the client does not hold all classes of data, which is common for clients due to limited memory.

How to preserve the performance of the global model during local adaptation is a key problem, which is similar to the problem faced in incremental learning. The purpose of incremental learning is to overcoming catastrophic forgetting [9] in machine learning, which requires a model to continuously learn new knowledge from new samples without forgetting the old knowledge. The process of model adaptation can be regarded as incremental learning to realize learning local data without forgetting the knowledge of the global model. In order to apply incremental learning to local adaptation in federated learning, we have to deal with the following challenges: (1) For privacy reasons, the raw data for training the global model is unavailable; (2) This process needs to be integrated into the federated learning process; (3) Considering the heterogeneity of client computing power, we should reduce the computing cost as much as possible. Due to these challenges, most incremental learning methods are difficult to extend to local adaptation.

To this end, we propose a new local adaptation method, which does not need to change the process of federated learning, so that can be integrated into existing federated learning methods naturally. The key idea of our method is allowing clients to use the synthetic data (generated by the generator trained in the central server) for local adaptation to overcome the catastrophic forgetting problem. In addition, we add variable weights to deal with the challenge of imbalance in the number of different classes of local data. To the best of our knowledge, it is the first work to focus on the catastrophic forgetting challenge of local adaptation in federated learning.

The contributions of this paper are summarized as follows.

- We propose a local adaptation method to overcome forgetting, which trains a generator in the central server by using the global model and adds the synthetic data in the process of model adaptation to maintain the benefits gained from federated learning.
- We test our method on real-world datasets. And the results show that our method can bring the personalized model both flexible adaption ability to new data and original recognition capability of the global model even in extreme cases.

2 Related Work

Personalization Techniques for Federated Learning. Chen et al. [10] proposed FedHealth, which implements data aggregation through federated learning, and then trains personalized model through transfer learning. Arivazhagan et al. [11] proposed FedPer, in FedPer, the model is divided into the base layers and the personalization layers. They proposed to train the base layers through federated learning, and the client retraining its own personalization layer. Fallah et al. [6] proposed federated meta-learning, which corresponds the global and personalization model training process to meta-learning and meta-testing.

Incremental Learning Using Synthetic Data. Using synthetic data is an important way to realize incremental learning. Shin et al. [12] proposed train a generator while training the model and use the synthetic data to replay the knowledge of old class data. Yin et al. [13] proposed get synthetic data by model deepInversion for incremental learning. Although their work avoid training a generator in the process of model training, their method is hard to deal with the challenges of federated learning, because the client needs powerful computing power if the synthesis data process is completed on the client, and need to take on huge communication traffic if it is completed on the central server.

Training Generator Without Data. Goodfellow et al. [14] proposed to use adversarial neural network to train the generator, but their method needs data to train the discriminator at the same time. Some work focus on training a generator directly from the classification model without raw data [15]. The main idea of these work is to treat the classification model as a fixed discriminator and let the generator generate images that can be judged as a certain class of images with high confidence.

Our method belongs to local adaptation. Different from the related work, our work focus on overcoming the catastrophic forgetting problem of local adaptation in federated learning. This is similar to the idea of incremental learning. Considering the lack of raw data for clients, inspired by [13] and [15], we propose to train generator in central server. And the clients use the generator to synthesize data to replay the knowledge of global model.

3   Overcoming Forgetting in Local Adaptation of Federated Learning Model

We consider a federated network consisting of N clients, and each client k has their private data x_k. We use federated learning to train the global model w without exchanging data. But a single model can not meet the needs of all clients. Therefore, some clients need to use the global model to further adapt locally. We consider such clients, who have personalized requirements but do not have all classes of data recognized by the global model (this setting is common in some work of federated learning [1]). Their personalized model may lose the ability to identify missed classes of data due to catastrophic forgetting. Moreover, this problem will be difficult to solve because it cannot obtain the information of other clients' data. Our goal is to propose a local adaptation method that can overcome forgetting.

3.1   Overview of Our Method Framework

As shown in Fig. 1, which consists of a cloud server and multiple clients, clients hold heterogeneous data and have personalized and individual demands.

The whole process can be divided into three steps, the first step is federated learning. Our local adaptation method does not need to change the federated learning process, so it can be integrated into various federated learning methods. Here we use the FedAvg [1], in which a small number of clients are randomly selected to participate in each round of training, the selected client will receive the current global model w and update it with local data, then upload the gradient, and the central server will aggregate the update to complete a round of training. The global model is not suitable for all clients due to non-iid data distribution and personalized demands. Therefore, the following local adaptation steps are needed. The second step is completed on the central server. After federated learning, the global model is used as a fixed discriminator to train a generator G. The generator G and global model w are then sent to all clients.

The third step is done in the clients on demand, they use local data and synthetic data generated by generator to train personalized models. Next we will introduce the details of generator training and local adaptation.

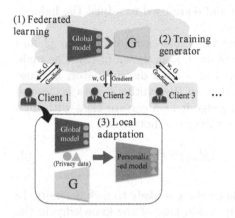

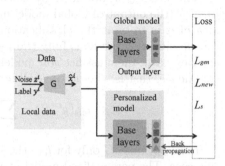

Fig. 1. The process from federated learning to local adaptation. Different shapes represent different class of data.

Fig. 2. Local adaptation.

3.2   Generator Training

If it is impossible to get raw data, training a generator and add the synthetic data to the training process is an important way to realize incremental learning. It can be extended to federated learning by adding generator synthetic data to local adaptation. But the traditional generator training needs data to train the discriminator, which is difficult to accomplish on a central server because of privacy. We notice that some work focus on training the generator in distributed data by federated learning [16]. However, these techniques would bring a large additional cost for clients.

Some work proved it is possible to get data from a trained model and using them training a new model without raw data [13–15]. Yoo et al. [15] take the trained model as a fixed discriminator to train the generator. Inspire by this work, we try to use the trained global model as a fixed discriminator to train the generator.

Our purpose is to train a generator G that outputs $\hat{x}^i$ with input (y^i, z^i), where z^i is noise and y^i is the label of $\hat{x}^i$ that is represented as a one-hot vector, i means index. We hope that $\hat{x}^i$ can be used to replay the knowledge of the global model. To this end, we need to treat the global model as a fixed discriminator and train an decoder D at the same time. The decoder is designed to make the synthetic data more diverse and it will not be used in local adaptation. The objective function is as

$$l(\mathcal{B}) = \sum_{\hat{x}^i \in \mathcal{B}} \left(l_{\text{cls}}(y^i, \hat{x}^i) + a l_{\text{dec}}(z^i, \hat{x}^i) \right) + b l_{\text{div}}(\mathcal{B}), \tag{1}$$

where $\mathcal{B}$ is a batch of sampled variables, a and b are used to adjust the balance between loss items.

The loss l_{cls} is obtained by calculating the cross-entropy between y^i and $y_T(\hat{x}^i)$ (the output of global model under input $\hat{x}^i$), as shown in (2). The main idea of it is to treat the classification model as a fixed discriminator. Because the classification model is fully trained by data, if the synthetic data is close to the raw data, the classification model will judge it as a certain class with high confidence, that is, the output will be close to the one-hot vector.

$$l_{\text{cls}}(y^i, \hat{x}^i) = \sum_i H_{\text{cross}} \left(y^i, y_T(\hat{x}^i) \right) \tag{2}$$

If we optimize G only for l_{cls}, the synthetic data is likely to lack diversity. In this case, The personalized model may only retain part of the knowledge in the global model. To avoid it, we added l_{dec} and l_{div}. l_{dec} is calculated as

$$l_{\text{dec}}(z^i, \hat{x}^i) = \|z^i - D(\hat{x}^i)\|_2^2, \tag{3}$$

where $D(\hat{x}^i)$ is the decoder's output of synthetic data $\hat{x}^i$, $\|\cdot\|_2^2$ is the Euclidean distance. Optimizing D for l_{dec}, the decoder will restore the synthetic data to z^i. The generator will embed the information of z^i in the synthetic data $\hat{x}^i$ so that the decoder can recover successfully.

l_{div} is calculated as

$$l_{\text{div}}(\mathcal{B}) = \exp \left(- \sum_{(y^1, z^1) \in \mathcal{B}} \sum_{(y^2, z^2) \in \mathcal{B}} \| z^1 - z^2 \|_2^2 \cdot \| (\hat{x}^1, \hat{x}^2) \|_1 \right), \tag{4}$$

where $\|\cdot\|_1$ is Manhattan distance. Equation (4) calculates the distance between $G(y^1, z^1)$ and $G(y^2, z^2)$ and is multiplied with $\|z^1 - z^2\|_2^2$. The exponential function makes l_{div} to produce a positive value. l_{div} increases the distance between synthetic data to bring diversity.

3.3 Local Adaptation

Our local adaptation approach is shown in Fig. 2. First, we initialize the output layer of the global model as the initial parameters of personalized model, and then use the local data and the synthetic data generated by generator to fine-tuning.

Soft labels contain more information than hard labels, so we use knowledge distillation [17] to get soft labels and realize knowledge transfer. The synthetic data will input to the global model and the personalized model, and the output of the global model will be used as soft labels to train the personalized model.

If the personalized model has more output nodes than the global model (need to learn new classes of data), a certain number of zeros need to be added to the soft labels to match the new output vector. The soft labels will make the personalized model tend to have the same output as the global model in training, thus transferring the knowledge of the global model to the personalized model. The loss as (5), where t is the category of synthetic data, n_t is the amount of synthetic data in class t, L_t is the loss of synthetic data of class t, $\hat{x}$ is synthetic data, $y_S(\hat{x}_t^i)$ is the personalized model's output of input $\hat{x}_t^i$, $y_T(\hat{x}_t^i)$ is the global model's output of $\hat{x}_t^i$.

$$L_t = \frac{1}{n_t} \sum_i H_{\text{cross}} \left(y_S \left(\hat{x}_t^i \right), y_T \left(\hat{x}_t^i \right) \right) \tag{5}$$

Considering that the client may have an unbalanced amount of data for each category, we design a variable weight to weight the synthetic data according to the category. The calculation definition of the weight is as

$$L_{\text{gen}} = \frac{1}{C} \sum_{t=1}^{C} \frac{1}{e^{\lambda m_t}} L_t, \tag{6}$$

where m_t is the amount of local data in class t, C is the number of different classes, λ is the weight coefficient. For a certain class of t, the weight will decrease with the increase of m_t. When λ increases, the weight will decrease under the same m_t, the effect of local data on model training will increase. When λ set to 0, all synthetic data have the same weight. We can flexibly adjust λ according to the demand of client.

Inspired by [18], in which using new classes of data for knowledge distillation to overcome forgetting, we add local data into the process of knowledge distillation to realize knowledge transfer. The local data will input to the global model and the personalized model. Different from synthetic data, the local data may be divided into a certain class with high confidence through the global model. And the output will be close to the one-hot vector. To get soft labels with more information, we use (7) to replace the activation function of the global model output layer. When $E = 1$, (7) is equivalent to softmax activation function. The larger E is, the farther away the soft label will be from the one-hot vector. a, j are the node of the output layer, z_a is the input of the activation function of the output layer node a, and y_a is the output of the output layer node a. We use (7) to calculate the softened global model output $y_T(x_k)$ of local data x_k, and use (8) to calculate L_s, where m is the amount of local data, $y_S(x_k)$ is the personalized model output of local data.

$$y_a = \frac{e^{\frac{z_a}{E}}}{\sum_{j=1}^{C} e^{\frac{z_j}{E}}} \tag{7}$$

$$L_s = \frac{1}{m} \sum_i H_{cross}(y_S(x_k^i), y_T(x_k^i)) \tag{8}$$

The objective function of local adaptation is as

$$L_{LA} = \alpha L_{gen} + (1 - \alpha)L_{new}(y_S(x_k), y_k) + \beta L_s, \qquad (9)$$

where L_{new} is the cross-entropy between the personalized model output $y_S(x_k)$ and the labels y_k of the local data x_k. α and β is used to adjust the weight of synthetic data and local data. When α set to 0, the synthetic data does not participate in training.

4   Experiment

Here we introduce some details of our experiment.

Datasets: We use the following datasets in experiment.

- MNIST: MNIST is a commonly used dataset, which consists of 60000 training images and 10000 testing images. The images are handwritten digits with 28×28 pixel divided into 10 classes (from numbers 0 to 9).
- SVHN [19]: SVHN consists of 73257 training images and 26032 testing images. These images come from the number in the street view photo with the size of $3 \times 32 \times 32$ and are divided into 10 classes (from numbers 0 to 9).

Model: In the following experiments, we use Lenet-5 for MNIST and ResNet-14 [20] for SVHN.

Default Parameter Settings: We set α to 0.8, β to 0.5, E to 2, λ to 0.002 in all experiments if there is no explanation (we will test the influence of different parameters on the accuracy later).

Other Settings: In the federated learning phase, we simulate 10 clients and train a global model that can identify 8 classes of data (from numbers 0 to 7). We take 2000 images as the testing set.

Comparison Algorithms: We investigate advanced incremental learning methods and extend some methods that can be extended to federated learning local adaptation for comparison.

- LwF: LwF is the incremental learning method proposed by Li et al. [18]. We input the local data into the global model, and the output is added to the training as a soft label to realize LwF (it is equivalent to $\alpha = 0$).
- inFT$_{siw}$: inFT$_{siw}$ is the incremental learning method proposed by Belouadah et al. [21]. We implement inFT$_{siw}$ in federated learning local adaptation through the following steps. Firstly, the output layer weight of the global model is standardized and stored, then the global model is initialized and fine-tuning. Finally, the output layer weight of the model is standardized and the node weight corresponding to the missed classes is replaced with the stored one to obtain the personalized model.
- Global model: the model trained by federated learning.
- FT: retraining the output layer of the global model with local data.

Some terms: Here we defining some terms that appear below.

- Missed classes: the classes of data that can be identified by the global model and lacking in clients.
- New classes: the classes of data that can not be identified by the global model in clients.
- Old classes: all classes except for the new classes.

Clients: We set 5 clients that need local adaptation. Table 1 shows the missed classes and new classes of each client. For MNIST, we build their local dataset by divide 2000 images and remove the classes that are not in Table 1. For SVHN, we use the same method to build local dataset but divide 6000 images.

Table 1. Data category of each client.

Client	1	2	3	4	5
Missed classes	0,1,2	3,4,5	4,5,6,7	0,1,2	0,1,2
New classes	8,9	8,9	8,9	8	9

4.1 Accuracy Comparison

To study the accuracy of our method, we use the local data of 5 clients to train the personalized model. The test set here consists of new classes and the class can be recognized by the global model. Table 2 shows the accuracy of the personalized model trained by our method and competitive algorithms. The addition of synthetic data allows our method to achieve the highest accuracy on each dataset and each client.

Table 2. Accuracy (%) of the personalized model trained by different methods.

	Client	1	2	3	4	5
MNIST	Our Method	**94.4**	**93.58**	**92.15**	**95.22**	**96.09**
	Global Model	78.73	78.73	78.73	78.73	78.73
	LwF	89.21	83.70	82.43	87.53	82.12
	inFT$_{siw}$	89.05	81.00	82.64	85.44	88.26
	FT	66.13	64.40	57.97	64.59	64.23
SVHN	Our Method	**84.93**	**83.96**	**82.43**	**87.57**	**87.38**
	Global Model	75.01	75.01	75.01	75.01	75.01
	LwF	74.49	70.77	75.18	82.38	77.63
	inFT$_{siw}$	79.56	77.03	74.46	82.77	83.59
	FT	66.60	67.43	54.59	63.31	66.85

4.2   The Performance of Overcome Forgetting

We further investigate the performance of our method to overcome forgetting.

Table 3 shows the recognition accuracy of personalized models trained by our method on different clients for different classes of data. It can be seen that although some classes of data are missing from the local dataset of the client, our method can make the personalized model maintain the recognition ability of these data in the global model (take client 3 for an example, although client 3 lacks classes 4, 5, 6, and 7 data, our method can retain the ability to identify these data in the process of local adaptation).

Table 3. Average accuracy (%) of each class of the personalized model trained by our method on 5 clients.

Dataset	MNIST					SVHN				
Client	1	2	3	4	5	1	2	3	4	5
Missed classes	**96.6**	**92.71**	**93.09**	**97.59**	**96.34**	**80.37**	**85.33**	**84.15**	**92.20**	**77.29**
Old classes	97.27	98.36	99.5	96.79	97.83	92.44	92.55	93.66	93.32	92.96
New classes	83.9	82.93	86.98	80.21	86.60	73.00	60.42	52.99	50.72	82.68

To further investigate the performance of our method, we test the recognition accuracy of the personalized model trained by our method in client 1 for various classes of data and compare it with the global model and the local model trained only with local data. The local model we train for the SVHN is a simple convolutional neural network, including two convolution layers and three full connection layers, because the quantity of samples of local dataset is not enough to train the complex Resnet-14 model.

As shown in Table 4, it can be seen that the global model cannot meet the personalized needs of the client because it cannot recognize new classes of data (numbers 8 and 9), while the local model cannot achieve high recognition accuracy because of the small amount of data, and cannot recognize missed classes of data due to the lack of these classes of data (numbers 0, 1 and 2). The personalized model trained by our method can retain a certain identification ability of the global model for missed classes and identify locally owned new classes of data to meet the personalized needs of the client. Moreover, Comparing LwF and inFT$_{siw}$, our method can achieve higher average accuracy.

4.3   Influence of Different Parameters on the Accuracy

We further investigate the impact of different parameters on the accuracy. Table 5 shows the accuracy under different parameters.

α is used to adjust the balance between real data and synthetic data. When α is set to 0, synthetic data does not participate in training, and when set to 1, local data does not participate in training. Because the soft label does not

Table 4. The accuracy of each class of the personalized model trained by our method in client 1.

Class		0	1	2	3	4	5	6	7	8	9
MNIST	Our Method	**98.85**	**98.71**	**92.23**	97.58	97.68	97.76	97.19	96.09	**78.12**	**89.69**
	Global Model	99.42	99.57	99.08	98.06	98.61	98.32	96.62	97.56	0	0
	Local Model	0	0	0	93.71	95.85	96.08	95.50	90.73	89.58	95.36
	LwF	88.57	74.36	75.80	98.55	98.61	95.53	98.31	96.59	78.13	87.63
	inFT$_{siw}$	97.14	62.39	98.17	91.79	95.85	93.85	94.94	76.10	86.78	93.30
SVHN	Our Method	**69.56**	**84.12**	**87.41**	93.22	92.46	94.73	86.18	95.59	**68.84**	**77.16**
	Global Model	93.47	94.70	95.69	94.06	93.96	92.98	92.10	93.08	0	0
	Local Model	0	0	0	76.69	86.43	80.11	77.63	86.79	72.46	70.86
	LwF	3.62	67.99	68.21	88.98	93.67	92.40	86.84	98.11	64.49	80.31
	inFT$_{siw}$	92.03	53.70	72.19	88.98	77.39	91.81	84.87	86.79	65.94	81.89

contain the information of new class data, it can be seen that the recognition accuracy of the model for new class data will decrease with the increase of α or β. λ set to 0 means that the weight of all synthetic data are 1. Under this setting, synthetic data has the same weight under all classes no matter they are or not missed, which leads to lower accuracies of both the missed and the new classes. λ set to ∞ means that the weight of synthetic data corresponding to the data class owned by the client is 0. It can be seen that when the local data owned by client perform uneven distribution across classes, adding variable weights to the synthesis data can improve the average accuracy to a certain extent (take λ equaling 0.002 for an example).

Table 5. Influence of different parameters.

	α			β			λ		
	0.5	0.8	0.9	0.1	0.5	0.9	0	0.002	∞
Missed classes	87.38	96.60	95.69	91.81	96.60	96.49	94.86	96.60	92.27
New classes	84.73	83.90	75.12	82.1	83.90	66.31	70.43	83.90	80.05
All classes	91.99	94.40	92.6	92.46	94.40	91.11	91.51	94.40	92.62

4.4 Synthetic Data

Figure 3 shows the synthetic data and raw data of MNIST and SVHN. Synthetic data can be divided into a certain class with high confidence by the global model, which can be used to replay the knowledge of the global model. It is observed that although the generator synthesizes recognizable images for MNIST data (mainly because they have a simple background), there is a big gap between the synthetic data and raw data of SVHN, which are with complex backgrounds (e.g., the guideboard or the number plate). That is to say, these synthetic data do not contain the private information of the client (such as the writing habits of a certain client), which can avoid privacy problems.

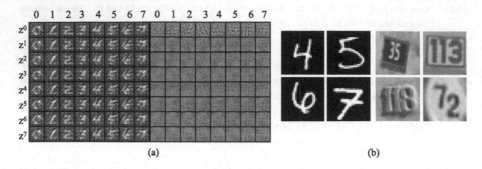

(a)   (b)

Fig. 3. (a) Synthetic data of MNIST (left) and SVHN (right). (b) Raw data of MNIST (left) and SVHN (right).

5   Conclusion

Our work focus on the catastrophic forgetting challenge of local adaptation in federated learning and propose a local adaptation method to solve this problem. Experimental results show that the proposed method can bring the personalized model both flexible adaption ability to new data and original recognition capability of the global model even in extreme cases.

Acknowledgements. This work is supported by NSF China (Nos. 61971139, 61601126), Foundation of Fujian Province (No. 2021J01576), and Research Fund of Fuzhou University (No. GXRC-21012).

References

1. Mcmahan, H.B., Moore, E., Ramage, D., Hampson, S., Arcas, B.: Communication-efficient learning of deep networks from decentralized data. In: PMLR, pp. 1273–1282 (2017)
2. Kulkarni, V., Kulkarni, M., Pant, A.: Survey of personalization techniques for federated learning. In: World S4, pp. 794–797 (2020)
3. Yu, T., Bagdasaryan, E., Shmatikov, V.: Salvaging federated learning by local adaptation. arXiv preprint arXiv:2002.04758 (2020)
4. Zhang, W., Wang, X., Zhou, P., Wu, W., Zhang, X.: Client selection for federated learning with non-IID data in mobile edge computing. IEEE Access **99**, 1 (2021)
5. Smith, V., Chiang, C-K., Sanjabi, M., Talwalkar, A.: Federated multi-task learning. In: NIPS, pp. 4427–4437 (2017)
6. Fallah, A., Mokhtari, A., Ozdaglar, A.: Personalized federated learning: a meta-learning approach. arXiv preprint arXiv:2002.07948 (2020)

7. Cheng, G., Chadha, K., Duchi, J.: Fine-tuning is fine in federated learning. arXiv preprint arXiv:2108.07313 (2021)
8. Kairouz, P., Mcmahan, B., Avent, B., Bellet, A., Bennis, M., et al.: Advances and open problems in federated learning. arXiv e-prints arXiv:1912.04977 (2019)
9. McCloskey, M., Cohen, N.J.: Catastrophic interference in connectionist networks: the sequential learning problem. Psychol. Learn. Motiv. **24**, 109–165 (1989)
10. Chen, Y., Qin, X., Wang, J., Yu, C., Gao, W.: Fedhealth: a Federated transfer learning framework for wearable healthcare. IEEE Intell. Syst. **35**(04), 83–93 (2020)
11. Arivazhagan, M.G., Aggarwal, V., Singh, A.K., Choudhary, S.: Federated learning with personalization layers. arXiv preprint arXiv:1912.00818 (2019)
12. Shin, H., Lee, J.K., Kim, J., Kim, J.: Continual learning with deep generative replay. In: NIPS, pp. 2994–3003 (2017)
13. Yin, H., Molchanov, P., Alvarez, J.M., Li, Z., Kautz, J.: Dreaming to distill: data-free knowledge transfer via deepinversion. In: CVPR, pp. 8712–8721 (2020)
14. Goodfellow, I.J., Pouget-Abadie, J., Mirza, M., Bing, X., Bengio, Y.: Generative adversarial nets. In: NIPS, pp. 2672–2680 (2014)
15. Yoo, J., Cho, M., Kim, T., Kang, U.: Knowledge extraction with no observable data. In: NIPS, pp. 2705–2714 (2019)
16. Fan, C., Liu, P.: Federated generative adversarial learning. In: Peng, Y. (ed.) PRCV 2020. LNCS, vol. 12307, pp. 3–15. Springer, Cham (2020). https://doi.org/10.1007/978-3-030-60636-7_1
17. Hinton, G., Vinyals, O., Dean, J.: Distilling the knowledge in a neural network. arXiv preprint arXiv:1503.02531 (2015)
18. Li, Z., Hoiem, D.: Learning without forgetting. IEEE Trans. Pattern Anal. Mach. Intell. **40**(12), 2935–2947 (2018)
19. Netzer, Y., Wang, T., Coates, A., Bissacco, A., Wu, B., Ng, A.Y.: Reading digits in natural images with unsupervised feature learning. In: NIPS Workshop on Deep Learning and Unsupervised Feature Learning, Granada, 12–17 December 2011, 5 (2011)
20. He, K., Zhang, X., Ren, S., Sun, J.: Deep residual learning for image recognition. In: CVPR, pp. 770–778 (2016)
21. Belouadah, E,. Popescu, A,. Kanellos, I.: Initial classifier weights replay for memoryless class incremental learning. arXiv preprint arXiv:2008.13710 (2020)

A New Skeleton-Neural DAG Learning Approach

Yiwen Cao[1], Kui Yu[1(✉)], Xiaoling Huang[1,2], and Yujie Wang[1]

[1] Hefei University of Technology, Hefei 230601, China
{cyw,yujiewang}@mail.hfut.edu.cn, hxl@chzu.edu.cn, yukui@hfut.edu.cn
[2] Chuzhou University, Chuzhou 239000, Anhui, China

Abstract. Learning a Directed Acyclic Graph (DAG) structure from observational data plays an essential role in causal inference and machine learning. A recent advance in the area is that the DAG learning problem was formulated as a continuous optimization problem and this provides a new research line for leveraging powerful neural networks for DAG learning. Although several DAG learning algorithms using neural networks (called neural DAG learning) have been proposed, they still face some serious challenges, leading to unsatisfactory performance. To tackle this issue, in this paper, we combine ideas from local DAG learning and neural DAG learning techniques and propose a new Skeleton-Neural DAG (SN-DAG) structure learning algorithm. The SN-DAG algorithm first learns the structure skeleton of each variable using a local learning algorithm to initialize the weighted adjacency matrix, then employ multilayer perceptrons to optimize the weighted adjacency matrix for learning a DAG structure, and finally it corrects wrong edges in the learnt DAG using the learnt skeleton and v-structure identifying techniques. By conducting experiments on both synthetic and real datasets, experimental results validate the SN-DAG algorithm compared to traditional DAG learning algorithms and neural DAG learning methods.

Keywords: Directed acyclic graph · Neural DAG structure learning · Local structure learning

1 Introduction

DAG structure learning plays an important role in causal inference [21,26] and machine learning [27,30]. It remains a critical challenge in both areas to develop new methods for learning DAG structures. During the last decades, many approaches to learning DAG structures have been proposed which are mainly categorized into two classes: constraint-based and score-based methods [4,6,7]. Constraint-based methods (e.g. Peter-Clark (PC) algorithm [25]) learn a DAG structure using conditional independence tests, while score-based methods (e.g. Greedy Equivalence Search (GES) algorithm [4]) employ a score function to learn a best DAG structure. Both approaches formulate the DAG learning problem as a combinatorial optimization problem and require various local heuristics

© The Author(s), under exclusive license to Springer Nature Switzerland AG 2022
J. Gama et al. (Eds.): PAKDD 2022, LNAI 13280, pp. 626–638, 2022.
https://doi.org/10.1007/978-3-031-05933-9_49

to search over the combinatorial space of all possible DAGs with the acyclicity constraint. However, this search space is always intractable when the number of nodes in a DAG becomes large. Recently, Zheng et al. proposed NOTEARS [31], a score-based method that formulates the DAG learning problem as a continuous optimization problem. Instead of using traditionally greedy search methods, this new approach can leverage continuous optimization methods to learn a DAG through optimizing a weighted adjacency matrix. Importantly, inspired by it, subsequent work links neural networks with DAG learning. Although several neural DAG learning methods have been proposed [28,32], those existing algorithms still face serious challenges. For instance, those algorithms always learn an adjacency matrix with entries with non zero values, whereas it is difficult to find a proper threshold to identify edges from the estimated entries in the matrix. In addition, for each variable, they set all of the remaining variables to their potential parents, resulting in the neural DAG learning algorithms introducing many extra edges before learning starts. Due to those issues, existing neural DAG learning methods always learn an inaccurate DAG structure with many extra edges and reversing edges, leading to an unsatisfactory performance.

In addition to learning a global DAG structure, many local DAG learning algorithms also have been proposed to learn a local structure around a given target variable. A local structure often refers to the parents and children of a variable in a DAG. Existing local learning approaches mainly employ conditional independence tests and can guarantee their correctness under certain assumptions. Moreover, they can be scaled to high-dimensional data.

Thus a question naturally arises: can we use the idea of local DAG learning to improve neural DAG learning algorithms? In this paper, we present a new algorithm for DAG structure learning, called Skeleton-Neural DAG learning (SN-DAG). The algorithm combines ideas from local DAG learning and neural DAG learning techniques in a principled and effective way. First, our SN-DAG algorithm learns the local structure skeleton of each variable in a dataset respectively and constructs a global DAG skeleton using the learnt local skeletons. Second, it uses the global DAG skeleton to initialize the weighted adjacency matrix, then employs neural networks, i.e. multilayer perceptrons (MLP), to optimize the weighted adjacency matrix for estimating a DAG structure. Third, SN-DAG employs the global DAG skeleton to correct extra edges in the DAG learnt at the second step and further corrects reverse edges in the DAG updated using v-structure identifying techniques.

The remaining paper is organized as follows. Section 2 briefly reviews existing work for DAG learning. Section 3 presents details of our proposed method. Section 4 gives experiment results, and Sect. 5 concludes our work.

2 Related Work

In this section, we briefly review works related to DAG learning. In the past decades, many DAG learning methods have been proposed, and they have formulated the DAG learning problem as a combinatorial optimization problem [4] and continuous optimization problem [31].

In the combinatorial optimization problem, DAG learning methods are subdivided into two types, score-based and constraint-based approaches [8]. Score-based algorithms such as GES [4], GIES [9], bnlearn [15,23], generally use a score to measure the goodness of fit of different graphs over data, and then use a search procedure to find the best graph [3]. In contrast, constraint-based methods, such as PC [25] and FCI [26], adopt conditional independence (CI) tests to first assess whether there is an edge between two variables and then orient edge directions [6]. As the search space of DAGs is combinatorial and exponential with the number of variables, existing global DAG learning methods face scalability issues [5]. To improve the efficiency of DAG learning, local-to-global DAG learning methods were proposed which contain two steps: skeleton learning and edge orientation. In the skeleton learning step, the local-to-global approach first learns the local DAG skeleton of each variable in a dataset independently, then constructs the global DAG skeleton (i.e. the undirected graph) using the learnt local DAG skeletons. Here, the local DAG skeleton usually refers to the parent-child (PC) set or the Markov Blanket (MB) set [21] of a variable in a DAG. The representative local DAG skeleton learning algorithms include PC-simple [13], HITON-PC [1], GSMB [16], and BAMB [14]. In the edge orientation step, edges are oriented in the global DAG skeleton using CI tests or score functions. Based on these MB or PC learning algorithms, several local-to-global DAG learning methods were proposed, such as GSBN [16] and GGSL [7].

To avoid the combinatorial constraint, NOTEARS [31] transfers DAG learning problem to a continuous optimization problem. The authors formulate the acyclic constraint as a smooth term and solve the problem using gradient-based numerical methods. In that NOTEARS is specially designed for linear cases, much subsequent work extends it with different types of neural networks to adapt to nonlinear cases. DAG-GNN [28] reconstructs data using variational auto-encoder and takes Evidence Lower Bound (ELBO) loss as its loss function. GAE [19] abandons the variational part in DAG-GNN, takes graph auto-encoder as its generative model and adopts least square loss. SAM [11] constructs its generative model and optimizes an f-gan loss [20] using generative adversarial network. Leveraging dependency among variable reconstruction residuals, DARING [10] adds residual independence constraint in an adversarial way to reduce reverse edges. Different from previous methods, aiming at leveraging all the parameters of the neural network in representing the weighted adjacency matrix, GraN-DAG [12] uses path products of the weights of its MLP generative model to represent the matrix coefficients. MaskedNN [17] follows GraN-DAG but applies the Gumbel-softmax approach to deal with the threshold problem. To get rid of the problem that GraN-DAG is dependent on the depth of the neural network, NOTEARS-MLP [32] indicates its weighted adjacency matrix by exploiting the weights of the first layer of MLP and proves the validity.

Existing neural DAG learning approaches have tried different types of neural network models, loss functions and representations of adjacency matrix to improve their performance, but still face many problems. Our method focuses on addressing extra edges and reverse edges that exist in the DAG learnt by neural DAG learning approaches and their poor performance in linear cases.

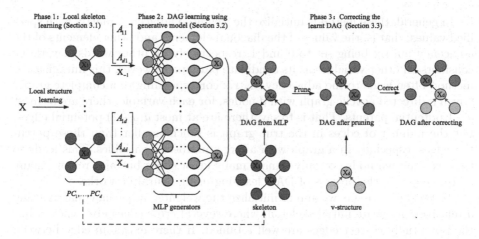

Fig. 1. Overview of the SN-DAG algorithm.

3   Proposed Method

In this section, we give details of the proposed SN-DAG algorithm. As shown in Fig. 1, SN-DAG consists of three learning phases as follows. At Phase 1, SN-DAG first learns the local skeleton (i.e. parents and children (PC)) of each variable and store corresponding separation sets of each pair of variables using an existing local learning algorithm, then obtains an initial skeleton S using the learnt local skeletons. A local skeleton of a variable is an undirected graph structure of the variable without distinguishing parents from children, while a separation set is a set of variables that makes two given variables conditionally independent and is key to identifying v-structures in a DAG skeleton. At Phase 2, SN-DAG first initializes the weighted adjacency matrix and generative model using the local skeletons learnt in the previous phase, then employs a neural network, i.e. multilayer perceptrons, to optimize the generative model with an acyclic constraint to learn the DAG structure. At Phase 3, SN-DAG first leverages the skeleton S achieved at Phase 1 to prune extra edges in the learnt DAG, then corrects reversed edges with the v-structure identifying techniques. The SN-DAG algorithm is described in Algorithm 1.

3.1   Identifying Skeleton Using Local Learning (Steps 1 to 8)

Existing DAG learning approaches with neural networks, such as DAG-GNN [28] and NOTEARS-MLP [32], are built on the assumption that each variable is generated by its parents. For each variable, these methods first construct a generative model based on the functional causal model, and then leverage neural networks to optimize generative models to make the learnt DAG fit the observed data well. These methods encode a DAG using a weighted adjacency matrix A learnt by these generative models, where each column represents the coefficients in the functional causal model.

In general, these methods initialize the weighted adjacency matrix with specified values, that is, the values of the diagonal elements and other elements of the adjacency matrix being set to 0 and 1, respectively. In other words, a variable considers all other variables as its candidate parents. However, this initialization method often leads to the adjacency matrix corresponding to a complete graph. Given a directed acyclic graph with d nodes, for each variable, there are at most $d-1$ potential parents. That is to say, there are at most $d(d-1)$ potential edges, but the number of edges in the true graph is much smaller than those potential edges, especially in a graph with a large number of variables. This leads to that existing methods not only remain many error edges but also remove many correct edges in the process of DAG learning using neural networks.

To tackle this issue, we aim to initialize the weighted adjacency matrix using a relatively accurate initial skeleton, where several error edges are removed but the potentially correct edges are well retained. If there exists an edge between variables X_i and X_j in the learnt local skeleton, then the values of A_{ij} and A_{ji} are set to 1 in the weighted adjacency matrix. In this way, the generated model can focus on the potential correct edges for reducing error edges.

To achieve this goal, a relatively accurate skeleton of each variable is needed to initialize the weighted adjacency matrix. At Phase 1, from Steps 1 to 5, by employing a well-established PC (parent and child) learning algorithm, called HITON-PC [1] (any state-of-the-art PC learning algorithm can be used here), SN-DAG learns the PC set (local skeleton) of each variable in a dataset, respectively store the sepset sets for each pair of variables. Then from Steps 6 to 8, it constructs the initial skeleton of a DAG S by the learnt local skeleton of each variable and set $S_{ij} = 1$ if X_i is in the PC set of X_j.

3.2   Learning a DAG Using Generative Model (Steps 9 to 15)

Given a data matrix $X \in R^{n \times d}$ with n samples and d variables, the relationships between variables in a DAG G and its corresponding weighted adjacency matrix $A \in R^{d \times d}$ are as follows. In A, if the absolute value of the element A_{ij} ($i, j = 1, ..., d$) is bigger than a given threshold value, then there is a directed edge from variable X_i to variable X_j in G. We can obtain the DAG by learning the weighted adjacency matrix A. Assuming that the values of a variable are generated based on its parents as in [31], the data generating procedure can be described as a functional causal model (FCM) as follows:

$$X_j = f_j(X_{Pa(j;g)}) + \varepsilon_j, \quad j = 1, 2, ..., d \tag{1}$$

where f_j is a mapping function and ε_j is an additive noise.

In this paper, in Step 9, MLP generative models [32] are employed to estimate true generative processes $f = (f_1, f_2, ..., f_d)$ and used to generate reconstructed data as follows:

$$\hat{X}_j = MLP_j(X, \theta_j) = \sigma(W_j^{(h)} \sigma(\cdots W_j^{(2)} \sigma(W_j^{(1)} X))), \quad j = 1, 2, ..., d \tag{2}$$

Algorithm 1. SN-DAG
___
Input: Data matrix $X \in R^{n \times d}$; Threshold $T > 0$.
Output: Binary adjacency matrix $A \in R^{d \times d}$.
/*Phase 1: Get initial skeleton */
1: Set all the variables adjacent in $\mathcal{G}$ and initialize A and S to all zeros.
2: **for** i from 1 to d **do**
3:     Find PC of X_i, note it as PC_i, set $S_{ij} = 1$
4:     Get sepset Sep between X_i and X_j, record $sepset(i,j) = sepset(j,i) = Sep$.
5: **end for**
6: **for** each pair (S_{ij}, S_{ji}) in S that $S_{ij} \neq S_{ji}$ **do**
       /*Set X_i and X_j not adjacent in the graph. */
7:     Set $S_{ij} = S_{ji} = 0$
8: **end for**
    /*Phase 2: Optimize generative model */
9: Create an MLP generative model as (2).
10: **for** each $S_{ij} = 0$ in S **do**
11:     Set the weight of the i-th column of the first layer of MLP_j to zero.
12: **end for**
13: Generate $\widehat{X} = MLP(X, \theta)$
14: Compute $F(X; \theta) = L(X; \theta) + \lambda h(A(\theta)) + \frac{\rho}{2}|h(A(\theta))|^2 + c\sum_{j=1}^{d} ||A(\theta)_j^{(1)}||_{1,1}$.
15: Minimize F by updating θ, λ and ρ using augmented lagrangian method.
    /*Phase 3: Prune and fix */
16: Set $A_{ij} = 1$ if $[A(\theta)]_{ij} = ||i\text{th} - \text{column}(A_j^{(1)})||_2 > T$.
17: Find intersection of A and S.
    /*Find v-structures */
18: **for** each triplet(i,j,k) in $\mathcal{G}$ that $A_{ij} + A_{ji} = 1$ and $A_{jk} + A_{kj} = 1$ **do**
19:     **if** $A_{ik} = A_{ki} = 0$ and $X_j \notin sepset(i,k)$ **then**
20:         Set $A_{ij} = A_{kj} = 1$ and $A_{ji} = A_{jk} = 0(*)$
21:     **end if**
22:     **if** $h(A) = e^{A \odot A} - d \neq 0$ **then**
23:         Undo (*)
24:     **end if**
25: **end for**
26: **return** A
___

where $W_j^{(k)}$ is the weight of the k-th layer in MLP_j and $\theta_j = (W_j^{(1)}, \ldots, W_j^{(h)})$. The weighted adjacency matrix $A(\theta)$ implied by the generative model can be constructed as [32]:

$$[A(\theta)]_{ij} = ||i\text{th} - \text{column}(W_j^{(1)})||_2, \quad i,j = 1, ..., d \qquad (3)$$

where $\theta = \theta_1, \ldots, \theta_d$. $W_j^{(1)}$ is the weight matrix between input X and the first hidden layer of MLP_j and its i-th column represents weights between X_i and MLP_j. Thus, $\hat{X}_j$ is independent of X_i if the i-th column of $W_j^{(1)}$ are all zeros.

As we discussed in Sect. 3.1, the initialization values of the weighted adjacency matrix A will affect the quality of the DAG. Therefore, to obtain a high-quality DAG, we initialize A using the skeleton learnt at Phase 1. If the edge between X_i and X_j does not exist in the learnt initial skeleton, we set $[A(\theta)]_{ij} = 0$. To achieve this, from Steps 10 to 12, we set weights in the i-th column of $W_j^{(1)}$ to zeros according to Eq. (3).

We use these generative models to generate constructed data $\hat{X} = (\hat{X}_1, \ldots, \hat{X}_d)$ and minimize the least square loss between original samples X and reconstructed ones $\hat{X}$ to optimize these generative models as follows:

$$L(X; \theta) = \frac{1}{n} ||X - \hat{X}||_F^2 \tag{4}$$

It means the least square loss between original data and reconstructed data. To ensure acyclicity, we adopt the acyclic constraint proposed in NOTEARS [31]:

$$h(A) = e^{A \odot A} - d = 0 \tag{5}$$

where $\odot$ is Hadamard product. It is proved in NOTEARS that $h(A) = 0$ if and only if the graph G is acyclic. The optimization problem is then converted to:

$$\min_\theta \frac{1}{n} ||X - \hat{X}||_F^2 + c \sum_{j=1}^d ||W_j^{(1)}||_{1,1} \tag{6}$$
$$s.t. \ h(A(\theta)) = 0$$

From Steps 13 to 15, we solve (6) with augmented lagrangian method [31].

3.3 Correcting the Learnt DAG (Steps 16 to 25)

Once the weighted adjacency matrix A is learnt, we get the learnt DAG. However, compared to the initial skeleton S learnt at Phase 1, we observe that although we initialize A using the local skeleton to reduce extra edges at Phase 2, the learnt DAG G still often contains extra edges. This further validates that it is hard to select a suitable threshold for determining which variables have an edge between them at Phase 2. To reduce extra edges, from Steps 16 to 17, we propose a simple but effective strategy that only the edges in the learnt DAG are kept if they are the intersection of the learnt DAG and the skeleton S. With the strategy, the number of extra edges in the learnt DAG is reduced to a low level.

Consider a triplet (X, Y, Z), X and Y , Y and Z are adjacent while X and Z are not, v-structure, which is of the form $X \rightarrow Y \leftarrow Z$, plays a key role in the orientation phase of constraint-based methods. For constraint-based methods, it is the only one that can be oriented among the potential structures. If X and Z are dependent but conditionally independent given Y, $X \leftarrow Y \leftarrow Z$, $X \rightarrow Y \rightarrow Z$ and $X \leftarrow Y \rightarrow Z$ all fit the situation but if X and Z are independent but conditionally dependent given Y, only v-structure fits the situation. Thus, v-structure is the only one can be directly identified among the structures.

To correct reverse edges, from Steps 18 to 25, we employ v-structure identifying techniques and the stored sepsets learnt at Phase 1. Given the updated DAG, if a triplet (X_i, X_j, X_k) exists in the DAG satisfying that X_i and X_j, X_j and X_k are adjacent while X_i and X_k are not, SN-DAG employs v-structure identifying techniques and the stored sepsets learnt at Phase 1 to determine whether the triplet (X_i, X_j, X_k) is a v-structure. If the triplet is a v-structure, but the edge directions in the DAG are different from the v-structure, we correct the directions in the triplet and keep the whole DAG acyclicity at the same time.

Meanwhile, in the following, we analyze why many reverse edges appear in the DAG learnt by neural networks. In practice, local uncertainties cannot be eliminated since a joint distribution allows its corresponding FCM to represent either $X_i \to X_j$ or $X_j \to X_i$ [8], leading to that many reverse edges often exist in the DAG learnt at Phase 2. Furthermore, the outputs of neural network based approaches are largely affected by the initialization of the weights. For example, given a DAG with two variables X_i and X_j, the ground truth is $X_i \to X_j$ and their relationship is described as $X_j = mX_i + \varepsilon$, $\varepsilon \sim N(0,1)$ and X_i obeys a uniform distribution from 0 to n. After applying min-max normalization, we compute least square losses for doing linear regression with $X_j = aX_i + b$ ($X_i \to X_j$) and $X_i = cX_j + d$ ($X_j \to X_i$). The losses are referred to as loss-forward (LF) and loss-backward (LB), respectively. According to augmented lagrangian method, the continuous optimization problem is expressed as:

$$\min F = (LB + LF) + \lambda|h| + \frac{\rho}{2}|h|^2 + cL_{sparse} \tag{7}$$

where $\lambda|h| + \frac{\rho}{2}|h|^2$ is the acyclic constraint term and cL_{sparse} is the sparse term. To minimize the acyclic constraint term and the sparse term, only either $X_i \to X_j$ or $X_j \to X_i$ can be kept. To minimize $LB+LF$, the optimizer tends to delete $X_j \to X_i$ because LB is usually larger than LF while mistakes may occur if LB is not much larger than LF. If sample size is 10000, for $m = n = 0.1$ or $m = n = 1$, LB is around 120% larger than LF and the direction is easy to determine. When $m = n = 10$, LB is only around 4% larger than LF and the gap between LB and LF is narrowing with the increase of m and n. Furthermore, if we use neural network as generators, LF may be larger than LB before optimizing because existing methods initialize the weights between hidden units randomly, which means that the optimizer may delete the true edge $X_i \to X_j$ to minimize F.

4 Experiments

Datasets. We adopt datasets generated from five benchmark Bayesian networks (BNs) and a real dataset. Each BN is used to generate 5 datasets with 5000 samples to get average results. The number of variables is chosen from 27 to 441 for benchmark BNs.

Comparison Methods. We compare our method with 2 combinatorial optimization methods, Peter-Clark (PC) [25] and GES [4], 4 state-of-the-art continuous optimization algorithms, NOTEARS [31], NOTEARS-MLP [32], DAG-GNN [28], and ABIC [2]. On the real dataset, we add GOLEM [18], NOCURL [29], SAM [11], MaskedNN [17], LINGAM [24] and bnlearn [15,23] for comparison.

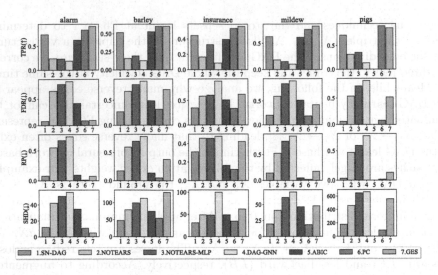

Fig. 2. Results on benchmark BNs.

Evaluation Metrics. We choose the following metrics to evaluate the methods:

(1) Structural Hamming Distance (SHD): SHD is the number of error edges including reverse edges, extra edges and missing edges.
(2) True Positive Rate (TPR): TPR is the proportion of correct edges in the learnt graph to total edges in the true graph.
(3) False Discovery Rate (FDR): FDR is the proportion of false edges in the learnt graph to total edges in the true graph.
(4) Reverse Proportion (RP): RP is the proportion of reverse edges to the sum of true edges and reverse edges in the learnt graph.

In the following figures and tables, (↑) means the higher the better and (↓) means the lower the better.

4.1 Experiment Results of Benchmark BNs

We select five benchmark BNs to generate data, i.e. alarm, barley, insurance, mildew, and pigs. The generative mechanism is as follows: If the predefined DAG is $A \in R^{d \times d}$, then for each X_j, $X_j \sim N(\sum_{i=1}^{d} A_{ij} X_i, 1)$. If X_i is not a parent node of X_j, the coefficient of A_{ij} is zero, else it is sampled from $(-1, -3/4)$ or $(1, 5/4)$ (at a 50 percent rate) and $Z \sim N(0, 1)$. The average results of TPR, FDR, SHD and RP are shown in Fig. 2. Note that ABIC failed to get results on the pigs network and this is why its result is empty there.

From Fig. 2 we can see that:

(1) SN-DAG performs better than other continuous optimization approaches on all metrics and is competitive to the PC algorithm. Our results on TPR, FDR and SHD prove that our skeleton is helpful to focus on true edges and cut extra edges. From RP we can see that the v-structure identifying technique greatly improves the orientation accuracy of the algorithm.

(2) DAG-GNN, NOTEARS-MLP perform poorly partly because their generative models are specially designed for nonlinear cases, ABIC performs better than them but does not adapt to high-dimensional data, such as "pigs'.

(3) PC outperforms the other algorithms on most tasks as conditional independence tests are quite suitable for linear cases while GES is unstable as the number of nodes increases.

4.2   Experiment Results of Real Data

In Table 1, we compare the algorithms on a real dataset [22] corresponding to a protein network problem. The dataset consists of $n = 7466$ observational samples and its ground truth graph with 11 nodes and 17 edges is provided in [22].

We can conclude from Table 1 that:

(1) SN-DAG outperforms its rivals especially on TPR and RP and performs well on FDR and SHD. This illustrates that SN-DAG also performs better than its rivals on real data.

(2) The performance of continuous optimization approaches is comparable to that of traditional methods in the real setting here, which proves that they have their inner advantages.

Table 1. Results on real dataset.

Algorithms	TPR($\uparrow$)	FDR($\downarrow$)	RP($\downarrow$)	SHD($\downarrow$)
SN-DAG	**0.588**	0.583	**0.167**	19
NOTEARS	0.412	**0.500**	0.300	**14**
GOLEM	0.353	0.727	0.400	23
NOCURL	0.294	0.667	0.375	19
NOTEARS-MLP	0.412	0.720	0.462	22
DAG-GNN	0.294	0.800	0.500	27
SAM	0.235	0.818	0.600	26
MaskedNN	0.294	0.800	0.500	27
ABIC	0.471	0.810	0.500	35
PC	0.412	0.720	0.417	23
GES	0.353	0.824	0.600	30
LINGAM	0.412	0.588	0.417	15
bnlearn	0.235	0.846	0.667	27

4.3   The Effect of v-structure Identifying Technique

In Table 2, we compare the results of SN-DAG before and after applying v-structure identifying technique and note them as "BeforeV' and "AfterV' respectively. The comparison in the table shows that v-structure identifying technique can effectively reduce reverse edges.

Table 2. The effect of v-structure identifying technique

	Correct	Reverse	Correct	Reverse	Correct	Reverse
Dataset	Alarm		Barley		Insurance	
BeforeV	21.4	15.6	27.8	24.2	17.4	16.4
AfterV	34.2	2.8	43.4	9	23.2	10.4
Dataset	Mildew		Pigs		Cyto	
BeforeV	16.8	16.8	231.2	206.8	9	3
AfterV	28.8	4.8	423.2	14.8	10	2

5   Conclusion

In this paper, we propose a new SN-DAG method to deal with the problems lying in existing continuous optimization approaches for DAG learning. The experimental results show that our SN-DAG algorithm is robust to adapt to linear and real settings and it is effective to reduce extra and reverse edges for learning a more accurate DAG than existing continuous optimization based DAG learning approaches.

Acknowledgment. This work is supported by the National Key Research and Development Program of China (under grant 2020AAA0106100), National Natural Science Foundation of China (under Grant 61876206), and the Key Project of the Natural Science Foundation of Educational Commission of Anhui Province (under Grant KJ2021A1065).

References

1. Aliferis, C.F., Tsamardinos, I., Statnikov, A.: Hiton: a novel markov blanket algorithm for optimal variable selection. In: AMIA annual symposium proceedings. vol. 2003, p. 21. American Medical Informatics Association (2003)
2. Bhattacharya, R., Nagarajan, T., Malinsky, D., Shpitser, I.: Differentiable causal discovery under unmeasured confounding. In: International Conference on Artificial Intelligence and Statistics. pp. 2314–2322. PMLR (2021)
3. de Campos, C.P., Scanagatta, M., Corani, G., Zaffalon, M.: Entropy-based pruning for learning bayesian networks using bic. Artificial Intelligence **260**, 42–50 (2018)
4. Chickering, D.M.: Optimal structure identification with greedy search. Journal of machine learning research 3, 507–554 (2002)
5. Chickering, D.M., Meek, C., Heckerman, D.: Large-sample learning of bayesian networks is np-hard. arXiv preprint arXiv:1212.2468 (2012)
6. Colombo, D., Maathuis, M.H., et al.: Order-independent constraint-based causal structure learning. J. Mach. Learn. Res. **15**(1), 3741–3782 (2014)
7. Gao, T., Fadnis, K., Campbell, M.: Local-to-global bayesian network structure learning. In: International Conference on Machine Learning. pp. 1193–1202. PMLR (2017)
8. Glymour, C., Zhang, K., Spirtes, P.: Review of causal discovery methods based on graphical models. Frontiers in genetics **10**, 524 (2019)

9. Hauser, A., Bühlmann, P.: Characterization and greedy learning of interventional markov equivalence classes of directed acyclic graphs. The Journal of Machine Learning Research **13**(1), 2409–2464 (2012)
10. He, Y., Cui, P., Shen, Z., Xu, R., Liu, F., Jiang, Y.: Daring: Differentiable causal discovery with residual independence (2021)
11. Kalainathan, D., Goudet, O., Guyon, I., Lopez-Paz, D., Sebag, M.: Structural agnostic modeling: Adversarial learning of causal graphs. arXiv preprint arXiv:1803.04929 (2018)
12. Lachapelle, S., Brouillard, P., Deleu, T., Lacoste-Julien, S.: Gradient-based neural dag learning. arXiv preprint arXiv:1906.02226 (2019)
13. Li, J., Liu, L., Le, T.D.: Practical Approaches to Causal Relationship Exploration. Springer, Cham (2015). https://doi.org/10.1007/978-3-319-14433-7
14. Ling, Z., Yu, K., Wang, H., Liu, L., Ding, W., Wu, X.: Bamb: A balanced markov blanket discovery approach to feature selection. ACM Transactions on Intelligent Systems and Technology (TIST) **10**(5), 1–25 (2019)
15. Margaritis, D.: Learning bayesian network model structure from data. Carnegie-Mellon Univ Pittsburgh Pa School of Computer Science, Tech. rep. (2003)
16. Margaritis, D., Thrun, S.: Bayesian network induction via local neighborhoods. CARNEGIE-MELLON UNIV PITTSBURGH PA DEPT OF COMPUTER SCIENCE, Tech. rep. (1999)
17. Ng, I., Fang, Z., Zhu, S., Chen, Z., Wang, J.: Masked gradient-based causal structure learning. arXiv preprint arXiv:1910.08527 (2019)
18. Ng, I., Ghassami, A., Zhang, K.: On the role of sparsity and dag constraints for learning linear dags. arXiv preprint arXiv:2006.10201 (2020)
19. Ng, I., Zhu, S., Chen, Z., Fang, Z.: A graph autoencoder approach to causal structure learning. arXiv preprint arXiv:1911.07420 (2019)
20. Nowozin, S., Cseke, B., Tomioka, R.: f-gan: Training generative neural samplers using variational divergence minimization. In: Proceedings of the 30th International Conference on Neural Information Processing Systems. pp. 271–279 (2016)
21. Pearl, J.: Probabilistic reasoning in intelligent systems: networks of plausible inference. Morgan kaufmann (1988)
22. Sachs, K., Perez, O., Pe'er, D., Lauffenburger, D.A., Nolan, G.P.: Causal protein-signaling networks derived from multiparameter single-cell data. Science **308**(5721), 523–529 (2005)
23. Scutari, M.: Learning bayesian networks with the bnlearn r package. arXiv preprint arXiv:0908.3817 (2009)
24. Shimizu, S.: Lingam: Non-gaussian methods for estimating causal structures. Behaviormetrika **41**(1), 65–98 (2014)
25. Spirtes, P., Glymour, C.: An algorithm for fast recovery of sparse causal graphs. Social science computer review **9**(1), 62–72 (1991)
26. Spirtes, P., Glymour, C.N., Scheines, R., Heckerman, D.: Causation, prediction, and search. MIT press (2000)
27. Yu, K., et al.: Causality-based feature selection: methods and evaluations. ACM Computing Surveys (CSUR) **53**(5), 1–36 (2020)
28. Yu, Y., Chen, J., Gao, T., Yu, M.: Dag-gnn: dag structure learning with graph neural networks. In: International Conference on Machine Learning. pp. 7154–7163. PMLR (2019)
29. Yu, Y., Gao, T.: Dags with no curl: Efficient dag structure learning. In: Causal Discovery & Causality-Inspired Machine learning Workshop at 34th Conference on Neural InformationProcessing Systems (NeurIPS 2020) (2020)

30. Zhang, K., Gong, M., Stojanov, P., Huang, B., Liu, Q., Glymour, C.: Domain adaptation as a problem of inference on graphical models. arXiv preprint arXiv:2002.03278 (2020)
31. Zheng, X., Aragam, B., Ravikumar, P., Xing, E.P.: Dags with no tears: Continuous optimization for structure learning. arXiv preprint arXiv:1803.01422 (2018)
32. Zheng, X., Dan, C., Aragam, B., Ravikumar, P., Xing, E.: Learning sparse nonparametric dags. In: International Conference on Artificial Intelligence and Statistics. pp. 3414–3425. PMLR (2020)

Rule-Based Collaborative Learning with Heterogeneous Local Learning Models

Ying Pang[1], Haibo Zhang[1(✉)], Jeremiah D. Deng[1], Lizhi Peng[2], and Fei Teng[3]

[1] University of Otago, Dunedin, New Zealand
ying.pang@postgrad.otago.ac.nz, {haibo.zhang,jeremiah.deng}@otago.ac.nz
[2] University of Jinan, Jinan, China
plz@ujn.edu.cn
[3] Southwest Jiaotong University, Chengdu, China
fteng@swjtu.edu.cn

Abstract. Collaborative learning such as federated learning enables to train a global prediction model in a distributed way without the need to share the training data. However, most existing schemes adopt deep learning models and require all local models to have the same architecture as the global model, making them unsuitable for applications using resource- and bandwidth-hungry devices. In this paper, we present CloREF, a novel rule-based collaborative learning framework, that allows participating devices to use different local learning models. A rule extraction method is firstly proposed to bridge the heterogeneity of local learning models by approximating their decision boundaries. Then a novel rule fusion and selection mechanism is designed based on evolutionary optimization to integrate the knowledge learned by all local models. Experimental results on a number of synthesized and real-world datasets demonstrate that the rules generated by our rule extraction method can mimic the behaviors of various learning models with high fidelity (>0.95 in most tests), and CloREF gives comparable and sometimes even better AUC compared with the best-performing model trained centrally.

Keywords: Collaborative learning · Heterogeneous participants · Rule extraction · Federated learning · Knowledge fusion

1 Introduction

Although machine learning has made significant breakthroughs in many domains, existing learning models, especially deep neural networks, rely on the availability of large amounts of quality training data to achieve satisfactory performance [12]. However, in various application fields such as healthcare and finance, regulations and consumer's concerns hinder the organizations to share their data with each other, therefore limiting the opportunity to build powerful predictive models for real-world problem solving.

Federated Learning (FL) has been proposed to address the above challenge [13]. Unlike traditional centralized learning techniques, FL enables

J. Gama et al. (Eds.): PAKDD 2022, LNAI 13280, pp. 639–651, 2022.
https://doi.org/10.1007/978-3-031-05933-9_50

multiple parties to build a high-performance model under the orchestration of a central server that aggregates the parameters of locally trained models, thereby eliminating the need to share local training data. However, a widely accepted assumption for FL is that local models employ the same architecture as the global model so as to produce a single global inference model [11]. This brings several limitations. Firstly, most implementations of federated learning use the deep neural network model, whose number of parameters to be learned grows significantly with the increase of network size [9]. This requires all participating devices to have both high computation capability and large communication bandwidth. However, in reality both capabilities may be limited and can differ significantly among participants. Hence, the global model complexity will be constrained by the most indigent participant. Secondly, the training data held by different participants can vary in both size and distribution, making it reasonable for different participants to apply different types of machine learning models. If all participants have to train the same model, the finally trained model may not achieve the best performance. Moreover, many real-world applications (e.g. IoT and finance) generate tabular data with low or moderate dimensions, making it unnecessary and even unsuitable to use complicated and less-interpretable learning models.

In this paper, we propose a novel "Collaborative learning with optimized Rule Extraction and Fusion" (CloREF) framework, that enables multiple heterogeneous participants to build a powerful global learning model without sharing their training data. Unlike FL, our learning framework allows each participant to train its own best-performing learning model, without needing to conform to the same model type or structure. The **key idea** is to use rules to bridge multiple heterogeneous learning models. We propose a new rule extraction method that uses multiple linear models to approximate the decision boundary of each local model. We further propose a rule fusion method that employs an evolutionary optimization scheme to elect the best rule set by merging the validation outcomes from multiple participants. Experimental results on both synthesized and real-world datasets demonstrate that: 1) Our rule extraction method can approximate complex decision boundaries and mimic the behaviours of a variety of learning models with fidelity >0.95 in most tests. 2) CloREF achieves competitive performance comparable and sometimes even better than the best-performing model trained centrally with the whole training dataset. 3) Compared with FedAvg, CloREF can significantly reduces the communication cost.

2   Related Work

2.1   Distributed Learning and Federated Learning

Traditional distributed learning models typically focus on distributing the load of training one model to multiple processing nodes and are not much concerned about privacy issues [16]. Recently, federated learning was introduced in [13] as a distributed learning model where a loose federation of multiple participating devices are coordinated by a central server to collaboratively train a global model. It embodies the principles of focused collection and data minimization, and mitigates the systemic privacy risks and costs in traditional centralized machine learning. The proposed "FedAvg" algorithm has become the most widely adopted FL baseline where a global model is trained by aggregating the parameters learned by the participants based on stochastic gradient descent [13].

There are few studies on FL with heterogeneous settings. FedMD [10] was proposed for a scenario with different private datasets and a shared public dataset. Each participant incorporates transfer learning to customize its local model and gains the knowledge of others by leveraging the knowledge distillation technique. HeteroFL [4] directly selects subsets of the global model parameters to update the local models. However, this model is restricted to using local neural network models but with different complexity level. In personalized FL such as Per-FedAvg [2] and FML [5], an initial global model is firstly trained and then adapted at each participant using its local data. Our scheme differs from these works on that it allows participants to build truly heterogeneous learning models, and knowledge acquired by participants is integrated at the central server through rule extraction and fusion.

2.2 Rule Extraction

Existing rule extraction techniques can be categorized into whitebox methods and blackbox methods. Whitebox methods extract rules by directly interpreting the strengths of connection weights, model architecture, and other parameters [14]. The main drawback is that they cannot preserve privacy, since both the internal model architecture and the training data are exposed to others. In addition, whitebox methods are usually less generic. Blackbox methods, such as LMENA [8], do not require internal information of the machine learning models. They extract rules by observing the effect of various inputs on the model outputs. However, they can only interpret some local areas of a complex learning model but cannot mimic its overall behaviors. Unlike existing rule extraction schemes, we aim to design a method that can approximate the whole decision boundary generated by a learning model using local linear rules.

3   Overview of CloREF

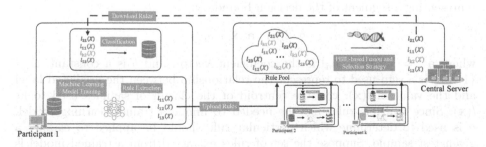

Fig. 1. The CloREF framework for privacy-preserving collaborative machine learning.

We assume that multiple local participants and one central server collaborate to train a global reference model. Each local participant has a private dataset to train a local learning model such as Support Vector Machine (SVM), naive Bayesian classifier (NB), and Multi-Layer Perceptron (MLP). Different local participants can use different local learning models. We assume all local learning models can output prediction probabilities either directly or via probability calibration, e.g. using Platt scaling [15] for SVMs. All participants agree in advance on a common set of data characteristics and a common learning objective. We focus on binary classification since multi-class classification can be decomposed into multiple binary classification problems using transformation techniques.

As illustrated in Fig. 1, each local participant first trains its local model using its private dataset and extracts a set of rules that can mimic the behaviors of the trained local model with high fidelity. Then the extracted rules are uploaded to the central server at which the rules generated by different local models are fused by selecting the best set of rules to form a global decision boundary. Finally, the central server disseminates the rules for the global model to each local participant. In our learning framework, the local models are mainly used for rule extraction. Each local participant performs classification directly based on the set of rules received since the global model is expected to have better performance than each local model after fusing the knowledge learned from different participants. If the training data is updated and new rules are generated at local participants, the above process can be repeated to update the global model.

4   Local Training and Rule Extraction

The **key motivation** of our rule extraction method is: we can mimic the behaviors of a trained learning model if we can model its decision boundary. In this section, we present a rule extraction method that can extract rules to mimic the behaviors of a trained learning model with high fidelity.

4.1   Rule Representation

We propose to use multiple linear models to fit the hypothesis function $H(\mathbf{x})$ of a trained learning model. Denote the n-dim feature vector by $\mathbf{x} = [x_1, x_2, ..., x_n]$. A rule is defined by a triple $L(\mathbf{x})$: $<l(\mathbf{x}), \omega, \mathbf{c}>$. Here, $l(\mathbf{x})$ is a linear function representing a segment of the decision boundary:

$$l(\mathbf{x}) : \mathbf{a}^T \mathbf{x} + b, \tag{1}$$

where $\mathbf{a}^T = [a_1, a_2, ..., a_n]$ is the coefficient vector, and b is a constant. $\omega \in \{1, -1\}$ is a sign used to represent the relationship between the predicted label and the value of $l(\mathbf{x})$. $\mathbf{c}$ is the centroid of the cluster of samples used to fit $l(\mathbf{x})$. Since multiple rules may be needed to mimic a trained learning model, $\mathbf{c}$ is used to determine which particular rule should be applied to classify a given test sample. Suppose the set of rules extracted from a trained model is $\mathbb{L}(\mathbf{x}) = \{L_i(\mathbf{x})\}, i = 1, \cdots, m$, where m is the total number of rules. Given a test sample $\mathbf{x}_j$, the rule used to classify $\mathbf{x}_j$, denoted by $L_k(\mathbf{x}_j)$, is the one that has the minimum Euclidean distance between $\mathbf{x}_j$ and the centriod of a rule $L(\mathbf{x})$:

$$L_k(\mathbf{x}_j) = \arg \min_{L_i(\mathbf{x}) \in \mathbb{L}(\mathbf{x})} \|\mathbf{x}_j - \mathbf{c}_i\|^2. \tag{2}$$

To measure how well $\mathbb{L}(\mathbf{x})$ mimics the behaviors of the trained learning model, we define its fidelity as [7]:

$$\mathrm{Fid}(\mathbb{L}(\mathbf{X})) = \frac{1}{|\mathbf{X}|} \sum_{\mathbf{x}_j \in \mathbf{X}} (1 - |u(H(\mathbf{x}_j) - \theta) - u(\omega l_k(\mathbf{x}_j))|), \tag{3}$$

where $\mathbf{X}$ is the set of training samples, $H(\mathbf{x_j})$ is the probability of $\mathbf{x_j}$ belonging to the positive class, θ is the threshold on prediction probability where a sample $\mathbf{x}$ with $H(\mathbf{x}) = \theta$ is considered as it on the decision boundary, and $u(\cdot)$ is the step function. According to this definition, the higher the fidelity is, the better the extracted rules mimic the trained learning model.

4.2 Rule Extraction

Figure 2 shows a four-step routine employed for rule extraction.

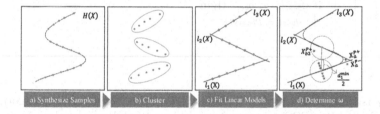

Fig. 2. The flowchart of rule extraction.

Synthesizing Boundary Samples: Since it is infeasible to model $H(\mathbf{x})$ directly with a mathematical function, we propose to detect the sketch of the decision boundary of a trained model using samples. As the training dataset may not contain enough boundary samples, we design a Particle Swarm Optimization (PSO) based algorithm to generate synthetic boundary samples.

PSO is a population-based optimisation technique, in which each individual called a particle learns from both its best historical solution in all previous generations (p_{best}) and the best solution from all personal bests (g_{best}) to iteratively converge to an optimal solution. In our algorithm, each particle represents a candidate boundary sample and its value is randomly initialized in the search space. Hence, we aim to find particles that satisfy

$$\min_{\mathbf{x} \in \chi} |H(\mathbf{x}) - \theta|, \chi \subset \mathbb{R}^n, \tag{4}$$

Algorithm 1: Synthesizing boundary samples

Input: (N_S, N_P, N_G);
ϵ – predefined acceptable threshold; α – inertia weight; c_1, c_2 – acceleration coefficients;
Output: S – Set of synthesized boundary samples

1 **while** $|S| < N_S$ **do**
2 $\quad$ randomly initialize population $P = \{P_1, \cdots, P_{N_P}\}$;
3 $\quad$ **for** $i \leftarrow 1$ **to** N_P **do**
4 $\quad\quad$ $p_{best}(i) = |H(P_i) - \theta|$;
5 $\quad$ $g_{best} = \min_{i=1}^{N_P} p_{best}(i)$;
6 $\quad$ **for** $t \leftarrow 1$ **to** N_G **do**
7 $\quad\quad$ **if** $g_{best} < \epsilon$ **then**
8 $\quad\quad\quad$ $P_k = \arg\min g_{best}$;
9 $\quad\quad\quad$ $S \leftarrow S \cup P_k$;
10 $\quad\quad\quad$ break;
11 $\quad\quad$ **for** $i \leftarrow 1$ **to** N_P **do**
12 $\quad\quad\quad$ $r_1, r_2 \leftarrow \text{rand}(0,1)$;
13 $\quad\quad\quad$ $V_i = \alpha V_i + c_1 r_1 (p_{best}(i) - P_i) + c_2 r_2 (g_{best} - P_i)$; /* update velocity */
14 $\quad\quad\quad$ $P_i = P_i + V_i$;
15 $\quad\quad\quad$ **if** $|H(P_i) - \theta| < p_{best}(i)$ **then**
16 $\quad\quad\quad\quad$ $p_{best}(i) = |H(P_i) - \theta|$;
17 $\quad\quad\quad\quad$ **if** $p_{best}(i) < g_{best}$ **then**
18 $\quad\quad\quad\quad\quad$ $g_{best} = p_{best}(i)$;

---

where χ is the search space of boundary samples. We design our algorithm to repeatedly execute PSO to generate synthetic boundary samples. As shown in Algorithm 1, in each execution of PSO, at most one synthetic boundary sample can be generated. The reason for this design is two-fold: firstly, PSO in each execution can quickly converge due to both the infinite number of boundary samples and the only optimization constraint (fitness value); secondly, the generated synthetic samples will spread over the entire decision boundary instead of huddling together due to the random initialization of the swarm population in each execution of PSO (line 2) and the three random parameters (α, r_1 and r_2) for velocity update. The time complexity of Algorithm 1 is $O(N_S N_G N_P)$, where N_S is the required number of synthetic boundary samples and N_P and N_G are the number of particles and the maximum number of generations in each execution of PSO, respectively.

Theoretically, the more boundary samples are synthesized, the higher fidelity we may obtain. Meanwhile, synthesizing more boundary samples will take more time. The number of generated boundary samples is initially set to nm where n is the feature dimension and m is the expected number of rules. If the fidelity of the generated rules is unsatisfactory, more boundary samples will be synthesized.

Clustering and Linear Fitting: In this step, the synthetic boundary samples are firstly divided into groups using k-means, and then the boundary samples in each group are used to fit a linear model based on the least square method, as illustrated in Fig. 2 where ellipses represent clusters and solid lines represent the fitted linear models.

Since it is difficult to choose an appropriate k to ensure the imitative ability of the generated rules, we use two measures, R^2 score [3] and fidelity (Eq. 3), to control the quality of the fitted linear models. R^2 score measures how well the data aligns with the fitted model.

$$R^2 = 1 - \frac{SS_{res}}{SS_{tot}} = 1 - \frac{\sum_{j=1}^{t}(H(\mathbf{x}_j^S) - l(\mathbf{x}_j^S))^2}{\sum_{j=1}^{t}(H(\mathbf{x}_j^S) - \overline{H(\mathbf{x}^S)})^2}, \tag{5}$$

where SS_{res} represents the sum of squares of residuals with respect to the fitted values, and SS_{tot} represents the sum of squares with respect to the average value.

We employ two thresholds ($T_{split} = 0.75$ and $T_{merge} = 0.95$) to automatically split or merge clusters based on R^2 scores. If R^2 for a cluster is lower than T_{split}, we use k-means to split it into two clusters, and calculate R^2 to check if the two new clusters need to be further split. For merging, a cluster can be merged with a neighboring cluster if the merged cluster has a R^2 no less than T_{merge}.

Determining the Sign ω: Since each $l(\mathbf{x})$ is fitted without using the label information, a sign ω needs be associated with each $l(\mathbf{x})$ to indicate which side is positive. To determine ω for a given $l_i(\mathbf{x})$, a set of synthetic samples called probing samples is generated subject to the following two constraints: (1) the probing samples are distributed on the normal line of $l_i(\mathbf{x})$ that crosses the centroid of the corresponding cluster used to fit $l_i(\mathbf{x})$. This is because linear models may not well fit the corners of the decision boundary. For example, $\mathbf{x}_a^{P+}$ and $\mathbf{x}_a^{P-}$ in Fig. 2(d) are located at a corner of the decision boundary, and using these two samples as probing samples may get incorrect ω. (2) Among all centroids, the probing sample has the shortest distance to the centroid of $l_i(\mathbf{x})$ to void interference from other parts of the decision boundary. For example, $\mathbf{x}_{b2}^{P+}$ shouldn't be used as a probing sample for $l_i(\mathbf{x})$. Let d_i^{min} be the minimum Euclidean distance between the centroid $\mathbf{c}_i$ of $l_i(\mathbf{x})$ and other centroids. The probing samples for $l_i(\mathbf{x})$ are then generated in a pairwise way by repeating the following two steps: (1) randomly choose a value β from the range $(0, d_i^{min}/2)$ as the distance between the probing sample and $\mathbf{c}_i$; (2) generate a pair of probing samples $<\mathbf{x}^{P+}, \mathbf{x}^{P-}>$ on the normal line of $l_i(\mathbf{x})$, where the two samples are located at different sides of $\mathbf{c}_i$. Then the sign ω determined by $<\mathbf{x}^{P+}, \mathbf{x}^{P-}>$ is calculated as follows:

$$\omega = \begin{cases} 1, & \text{if } (H(\mathbf{x}^{P+}) - \theta)l(\mathbf{x}^{P+}) > 0 \text{ and } (H(\mathbf{x}^{P-}) - \theta)l(\mathbf{x}^{P-}) > 0 \\ -1, & \text{if } (H(\mathbf{x}^{P+}) - \theta)l(\mathbf{x}^{P+}) < 0 \text{ and } (H(\mathbf{x}^{P-}) - \theta)l(\mathbf{x}^{P-}) < 0 \end{cases} \tag{6}$$

If the values of ω determined by more than 80% of the probing samples are consistent, this result is accepted; otherwise, the current set of probing samples is discarded and a new set of probing samples is formed. If ω cannot be determined after generating the set of probing samples more than ten times, the current $l(\mathbf{x})$ is considered seriously deviated from the decision boundary and is discarded.

5   Rule Fusion and Selection

Since rules generated by different participants can be redundant or even contradictory, simply concatenating them into a rule set may not work. To address this issue, we propose a PBIL-based optimization scheme to select a set of rules that achieves the best performance. Rules from local participants are pooled and selected at the server, but the evaluation of the selected rules is done at participants in a distributed manner since only participants have training data.

PBIL is a stochastic optimisation technique that combines the mechanisms of a genetic algorithm with simple competitive learning. It is simpler than standard genetic algorithms, but in many cases has better performance. In our scheme, we use the Area Under the ROC Curve (AUC) as the fitness value of the binary classifier as the fitness value of PBIL. Initially, the server merges the rules received from all participants, shuffles them, and then sends the whole set of rules to all participants. During the optimization process, only the gene strings instead of the selected rules needs to be sent to local participants, which can significantly reduce the amount of communication.

Coding: We encode rule selection using a binary "gene" code, $\mathbf{v} = [v_1, v_2, \cdots , v_{N_R}]$, N_R being the total number of rules. For each bit, $v_i = 1$ stands for the i-th rule is tentatively chosen; $v_i = 0$ means it is removed.

Population Generation and Evaluation: PBIL keeps a probability vector $\boldsymbol{\pi} = [\pi_1, \pi_2, \cdots , \pi_{N_R}]$. To generate the genes, v_i is assigned "1" with probability π_i and "0" with probability $1 - \pi_i$. Initially, each π_i is set to 0.5. Each time a new population is generated, the corresponding gene codes are sent to all participants. Each participant firstly decodes the genes to generate the rule set to be evaluated. The rule set is then used to classify the local training data and calculate the AUC values. Finally the calculated AUCs are sent back to the central server. For each gene in the current generation, the sum of AUCs received from all participants is used as the fitness value.

Update: The best solution till the current generation $\mathbf{v}_b$ is used to update the probability vector as follows:

$$\boldsymbol{\pi}^{t+1} = \boldsymbol{\pi}^t(1 - \delta) + \delta\mathbf{v}_b, \tag{7}$$

where $\delta \in (0, 1)$ is the learning rate. To avoid losing the optimum, current optimum is reserved into the next generation.

The generation, evaluation and update procedures are repeated until (1) a given number of generations is reached or, (2) the difference (ϵ_P) of fitness values of the best solutions between two generations is less than a predefined threshold.

6   Evaluation

6.1   Experiment Setup

Datasets and Models: We evaluate CloREF on 8 public two-class datasets provided in KEEL [1] and UCI. The details of these datasets are summarized in Table 1. Five learning models, i.e., SVM, NB, MLP, logistic regression classifier (LR), logistic regression classifier fitted via SGD (SGD), are used as local models. Some models are tested with different configurations, e.g., kernel in SVM, hidden layers and neurons in MLP. All models are implemented using scikit-learn.

Table 1. Detailed specifications of selected data sets.

Dataset	Abbr	#neg	#pos	#attr	Dataset	Abbr	#neg	#pos	#attr
wisconsin	wisc	444	239	9	page-blocks0	pa0	4913	559	10
glass0123_vs_456	glas	163	51	9	vehicle1	vehi1	629	217	18
segment0	se0	1979	329	19	pimaImb	pima	500	265	8
yeast1	yea1	1314	514	8	KDDCup99	kdd	396743	97278	41

Methodology and Metrics: For each experiment, 5-fold cross validation is used. The training dataset is split into sub-datasets with each used as the private dataset for a participant. To ensure each participant has enough samples to train its learning model, the number of participants varies according to the size of the original dataset. We compare CloREF with the following baseline schemes:

- *FedAvg*[13]: Used as a baseline for FL. Specifically, we implement it using the *homo_nn* component of FATE 1.6.0 [6].
- *FedMD* [10]: The state-of-the-art FL algorithm with heterogeneous settings. For a fair comparison, we use fully connected networks to replace CNN based on the source-code shared by the authors. Each participant randomly selects a neural network model from a pre-defined set (given in Table 2) to train.
- *Centralized_Model:* The best-performing learning model centrally trained with all training data.
- *All_Rules:* CloREF with the rule fusion and selection procedure removed.

We use three metrics to evaluate the implemented learning schemes: fidelity, accuracy, and AUC. All experiments run on a server with 40 CPU cores and 64 GB RAM. The CPU model is Intel(R) Xeon(R) E5-2630 v4 @ 2.20 GHz.

Parameter Setting: Table 2 gives the setting of key parameters. θ in Eq. (3) is set to 0.5 as our experimental results show that this generic setting already leads to satisfactory performance. All other parameters are set with the default values given by scikit-learn or FATE 1.6.0.

Table 2. The setting for the key parameters.

Approach	Parameter	Values
Algorithm 1	θ, N_S, N_P, N_G, ϵ, N_G	0.5, 40×n, 20, 50,0.000001, 500
PBIL	population size, δ, ϵ_P	20, 0.02, 0.0001
FedAvg	max_iter, layers, activation function	3000, (20, 20, 20, 1), (relu, sigmoid)
FedMD	max_iter, pre-defined neural networks, activation function	1000, [(20, 20, 1), (10, 10, 10, 1), (10, 10, 1), (5, 5, 5, 1), (5, 5, 1)], (relu, sigmoid)

6.2   Results of Rule Extraction

We use fidelity and accuracy to evaluate how well the extracted rules can mimic the behaviors of the trained learning model. Figure 3 shows the results for three datasets. The y-axis shows the fidelity. One key observation is the achieved high fidelity (>0.95 for most of the tests), which means the extracted rules are able to well mimic the behaviors of different machine learning models.

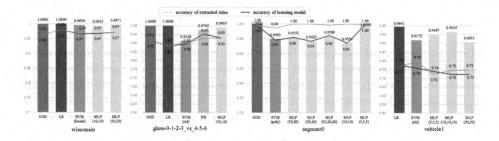

Fig. 3. Fidelity of extracted rules.

6.3   Results on Convergence

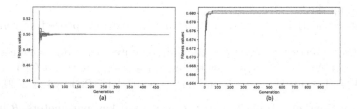

Fig. 4. Convergence on yeast1: (a) Algorithm 1, (b) PBIL-based rule fusion.

The two iterative computational components, *boundary sample generation* (Algorithm 1) and *rule fusion & selection*, are designed based on PSO and PBIL, respectively. Both methods have proven convergence properties. Figure 4 shows the convergence curves on the *yeast1* dataset. For each optimization, we performed it multiple times. Each curve represents one run of the algorithm. It can be seen that both optimization components can quickly converge due to the simple optimization objectives.

6.4 Results of Fusion and Selection

Table 3 compares CloREF with three baseline schemes on eight real-world datasets Due to space limit, we only show the average results of participants (Participants Average) and the number of participants for each dataset is given in the round brackets after the dataset name. Values in the round brackets following average AUCs are the corresponding standard deviation. We can observe that:

- Except for the *wisc* dataset, the performance of CloREF is better than the average of individual participants, which means our fusion and selection strategy can effectively fuse the knowledge learned by different participants, allowing each participant to benefit from the knowledge learned by others.
- CloREF outperforms *All_Rules* on most of the datasets, which demonstrates that our PBIL-based rule fusion and selection strategy can effectively remove the contradictory rules.
- CloREF gains comparable and sometimes (on *yea1* and *vhe1*) even better AUC than that of *Centralized_Model*. This indicates that its has competitive performance despite using distributed learning with heterogeneous models.
- All the results of CloREF are better than those of FedMD, which demonstrates that, in comparison with FedMD, CloREF can achieve competitive performance without relying on a public shared dataset.

Although CloREF gives better AUC values than FedAvg on all the datasets, our exploration of network hyperparameters for FedAvg is not exhaustive. The focus of this comparison study is to show that CloREF can produce competitive performance with heterogeneous models and low complexity.

6.5 Case Study for i.i.d and Non-i.i.d Datasets

We generate a spiral dataset and split it into four subsets in two ways: (1) randomly splitting so that the four subsets have similar distributions; and (2) cross-splitting so that each subset contains a quarter of the whole spiral dataset. Four different models (MLP (5, 5), SVM (RBF), NB, SVM (POLY)) are built on these subsets. The decision boundaries generated by local rules and final fused rule set are shown in Fig. 5, where the accuracy of the corresponding set of rules on the same test dataset is shown in the lower right corner. The results demonstrate that our rule fusion and selection strategy can generate a satisfactory global decision boundary by selecting useful rules and removing redundant and contradictory rules. These results indicate that, regardless of i.i.d or non-i.i.d datasets, CloREF can generate competitive global models to achieve effective collaborative learning.

Table 3. The average results on eight real-world datasets in full feature space.

	Data	Participants Average	All_Rules	CloREF	Centralized Model	FedAvg homo_nn	FedMD fully_con
#Rules	yea1 (5)			4.0 (±3.08)	SGD		
AUC		0.673 (±0.01)	0.642 (±0.06)	**0.684** (±0.03)	0.677 (±0.05)	0.658 (±0.03)	0.643 (±0.02)
#Rules	glas (4)			9.2 (±0.40)	SVM (rbf)		
AUC		0.882 (±0.01)	0.891 (±0.07)	0.907 (±0.05)	**0.923** (±0.09)	0.839 (±0.08)	0.812 (±0.06)
#Rules	se0 (14)			13.2 (±2.23)	MLP (10,10,10)		
AUC		0.923 (±0.05)	0.933 (±0.03)	0.967 (±0.02)	**0.992** (±0.00)	0.942 (±0.02)	0.957 (±0.03)
#Rules	pima (5)			17.2 (±4.35)	SVM (poly)		
AUC		0.684 (±0.02)	0.680 (±0.06)	0.705 (±0.04)	**0.706** (±0.04)	0.654 (±0.02)	0.634(±0.02)
#Rules	veh1 (5)			13.6 (±1.96)	MLP(10,10,10)		
AUC		0.626 (±0.03)	0.640 (±0.04)	**0.673** (±0.05)	0.669 (±0.12)	0.613 (±0.07)	0.576 (±0.04)
#Rules	wisc (5)			8.6(±4.18)	MLP (10,10)		
AUC		0.968 (±0.01)	0.960 (±0.01)	0.964 (±0.01)	**0.970** (±0.01)	0.815 (±0.10)	0.952 (±0.01)
#Rules	kdd (14)			12.4 (±5.57)	SVM (poly)		
AUC		0.939 (±0.15)	0.724 (±0.31)	0.984 (±0.01)	**0.998** (±0.00)	0.919 (±0.14)	0.955 (±0.03)
#Rules	pa0 (19)			25.8 (±3.87)	MLP (10,10)		
AUC		0.745 (±0.03)	0.705 (±0.07)	0.780 (±0.06)	**0.908** (±0.02)	0.778 (±0.04)	0.749 (±0.03)

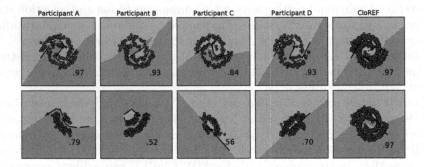

Fig. 5. Classification boundaries on different sub-datasets: (a)–(d) at participants; (e) after rule fusion. First row for random splitting and second row for cross-splitting.

6.6 Analysis on Communication Cost

Table 4 compares the communication cost of CloREF and FedAvg measured in megabytes. The communication cost of CloREF is signficantly lower than that of FedAvg. This is because: (1) CloREF can converge in a few generations due to the limited solution space, whereas FedAvg (implemented by FATE) needs a large number of iterations to converge. (2) one rule only occupies one bit in the gene code whereas a weight for neuron connection needs 4 bytes.

Table 4. Communication cost of CloREF and FedAvg homo_nn (measured in MB).

	glas (4)	yea1 (5)	vehi1 (5)	pima (5)	wis (5)	se0 (14)	kdd (14)	pa0 (19)
CloREF	0.024	0.012	0.030	0.029	0.013	0.020	0.098	0.030
FedAvg	23.368	23.369	27.489	22.911	23.369	27.946	38.017	23.827

7 Conclusion

We proposed a rule-based collaborative learning framework that enables multiple heterogeneous participants to build a global model without sharing their local data. Being lightweight and interpretable, it gives competitive performance on a range of benchmarks. As a new attempt in collaborative learning, it has several limitations. The fidelity of the extracted rules at the participants can be unstable, and we may need to run the program more than once to extract satisfactory rules. We intend to investigate using other distance metrics (e.g. the Mahalanobis distance) to improve the quality of boundary points clustering, which may lead to local linear models that are stable and perform better.

References

1. Alcalá-Fdez, J., et al.: Keel data-mining software tool: data set repository, integration of algorithms and experimental analysis framework. J. Multiple-Valued Log. Soft Comput. **17**(2–3), 255–187 (2011). Citeseer
2. Arivazhagan, M.G., et al.: Federated learning with personalization layers. arXiv (2019)
3. Colin Cameron, A., et al.: An r-squared measure of goodness of fit for some common nonlinear regression models. J. Econom. **77**(2), 329–342 (1997)
4. Diao, E., et al.: HeteroFL: Computation and communication efficient federated learning for heterogeneous clients. arXiv (2021)
5. Fallah, A., et al.: Personalized federated learning: a meta-learning approach. CoRR abs/2002.07948 (2020)
6. FedAI: FATE: an industrial grade federated learning framework. https://fate.fedai. org (2021)
7. Guidotti, R., et al.: A survey of methods for explaining black box models. ACM Comput. Surv. **51**(5), 1–42 (2018)
8. Guo, W., et al.: LEMNA: explaining deep learning based security applications. In: Proceedings of the 2018 ACM SIGSAC Conference on Computer and Communications Security, pp. 364–379 (2018)
9. Konečný, J., et al.: Federated learning: strategies for improving communication efficiency. In: NIPS Workshop on Private Multi-Party Machine Learning (2016)
10. Li, D., Wang, J.: FedMD: heterogenous federated learning via model distillation. In: NeurIPS 2019 Workshop on Federated Learning for Data Privacy and Confidentiality (2019)
11. Li, T., et al.: Federated learning: challenges, methods, and future directions. IEEE Signal Process. Mag. **37**(3), 50–60 (2020)
12. Liu, Y., et al.: A secure federated transfer learning framework. IEEE Intell. Syst. **35**(4), 70–82 (2020)
13. McMahan, B., et al.: Communication-efficient learning of deep networks from decentralized data. In: International Conference on Artificial Intelligence and Statistics. Proceedings of Machine Learning Research, vol. 54, pp. 1273–1282. PMLR (2017)
14. Narendra, T., et al.: Explaining deep learning models using causal inference (2018)
15. Platt, J., et al.: Probabilistic outputs for support vector machines and comparisons to regularized likelihood methods. Adv. Large Margin Classif. **10**(3), 61–74 (1999)
16. Verbraeken, J., Wolting, M.: A survey on distributed machine learning. ACM Comput. Surv. **53**(2), 1–33 (2020)

Author Index